Fundamentals of Powder Diffraction and Structural Characterization of Materials

Peter Zavalij • Vitalij Pecharsky

Fundamentals of Powder Diffraction and Structural Characterization of Materials

Third Edition

Springer

Peter Zavalij
091 Chemistry
University of Maryland, College Park
COLLEGE PARK, MD, USA

Vitalij Pecharsky
AMES, IA, USA

ISBN 978-3-031-91503-1 ISBN 978-3-031-91504-8 (eBook)
https://doi.org/10.1007/978-3-031-91504-8

This Springer imprint is published by the registered company Springer Nature Switzerland AG
The registered company address is: Gewerbestrasse 11, 6330 Cham, Switzerland

To those whose footsteps we have followed,
and to those who will follow ours.

Preface

It has been 15 years since the release of the second edition, making this new edition long overdue—particularly given the advancements in instrumentation (especially X-ray sources and detectors), developments in crystallographic algorithms and software, and significant progress in computing. The latter primarily refers to the substantially increased speed and memory of modern computers compared to 15 years ago. Additionally, artificial intelligence and machine learning[1] are now deployed for solving crystal structures, including but not limited to those derived from powder data.

Work on this edition began more than 5 years ago. Unfortunately, progress was interrupted and significantly delayed by several events: the disastrous pandemic; unprovoked aggression against Vitalij's and my homeland; and, most profoundly, the sudden passing of the first co-author and my dear friend, Vitalij Pecharsky, whom I first met in high school in the early 1970s.

Vitalij and I studied together at L'viv National University in Ukraine, both as undergraduates and graduate students. At one point, he even served as my graduate advisor. We became close friends while working in the Department of Inorganic Chemistry at the same university. After moving to the United States, despite living several states apart, Vitalij and I upheld not only our friendship but also our close scientific collaboration, as evidenced by this book. The most recent manuscript resulting from this collaboration was published in the *Journal of Alloys and Compounds* as part of a special issue[2] dedicated to the memory of Vitalij Pecharsky.[3]

[1] Machine learning in crystallography and structural science, Ed. S.J.L. Billinge and T. Proffen, Acta Cryst. **A80**, 139–145 (2024), https://doi.org/10.1107/S2053273324000172; Special issue: Artificial neural networks: applications in X-ray photon science and crystallography, Ed. T. Ekeberg. J. Appl. Cryst. **56** (2024), https://journals.iucr.org/special_issues/2024/ANNs/

[2] https://www.sciencedirect.com/special-issue/10D56484KX1

[3] Ihor Zavaliy, Ihor Koval'chuck; Roman Denys; Radovan Černý; Vitalij Pecharsky; Peter Zavalij, H-Induced Local Disorder in the Structure of Ti_2Ni-type $Hf(Zr)_4Fe_2O_x$ compounds, J. Alloys Comp. **980**, 173531 (2024), https://doi.org/10.1016/j.jallcom.2024.173531

Notably, the leading co-author of this work is Prof. Ihor Zavaliy—Vitalij's former graduate student and later collaborator—who has since become an academician of the National Academy of Sciences of Ukraine.

This edition includes several additions and changes: a new Sect. 5.6 on magnetic symmetry and its effects on relevant properties; descriptions of novel X-ray sources and detectors in Sects. 6.2 and 6.4; updated and expanded software references and web links, especially those previously sourced from the CCP14 crystallographic software repository, which was extensively cited in earlier editions but is no longer available; added DOI links to all article references; and more descriptive headings for Chaps. 16–25, which feature practical examples. The book concludes with the newly added Chap. 26, which compiles 20 supplementary examples and exercises drawn from the author's published works. These examples are organized by difficulty and are accompanied by experimental data and final structural results, further enhancing the book's practical value. Furthermore, the online materials have been updated with figure captions, selected tables, and equations providing even more comprehensive teaching support. Lastly, the text's readability has been significantly improved by clarifying and refining the content, correcting errors and typos, and updating all figures to color versions.

This book comprises 26 main chapters and 301 sections, spanning approximately 700 pages. It includes 909 figures, tables, and equations; 608 web links, 360 of which are DOI references to articles; and 221 citations for additional readings. For teaching purposes, the book presents 100 problems and 37 examples, 17 of which offer detailed, step-by-step instructions and comprehensive explanations.

The supporting materials include data files for examples and problems, an X-ray safety section, and slides containing the books' illustrative materials—figures, selected tables, and equations. While these files are freely available for download, their content is best understood when used in conjunction with the book. Problem solutions, however, are exclusively available to instructors, who must register with the publisher to gain access. All supporting materials and additional information are hosted on Springer's website.[4]

Since the second edition of this book, several new works on powder diffraction have been published. These include the third edition of the well-known student textbook *Elements of X-Ray Diffraction* by Cullity and Stock (2013), *Powder Diffraction: Theory and Practice* by Dinnebier and Billinge (2008), and *Rietveld Refinement: Practical Powder Diffraction Pattern Analysis using TOPAS* by Dinnebier, Leineweber, and Evans (2019). While these books overlap with this edition to varying degrees and differ in the depth of their coverage, the abundance of examples in this book—featuring detailed, step-by-step analysis of all stages of crystal structure determination from powder data—makes it invaluable for both teaching and self-study.

At times, I doubted my ability to complete this edition on my own, especially after all the distressing events. Nevertheless, it became possible, thanks to the

[4] www.springer.com/FPDSCM3

encouragement of the Springer editors, Vitalij's spouse Olena Pecharsky, and, of course, my family—including the newest addition, my granddaughter Julianna. I extend my special thanks to Lubov Zavalij for reviewing the text and providing valuable suggestions. Equally important was the unwavering support I received from the Department of Chemistry and Biochemistry at the University of Maryland, College Park.

Finally, I would like to express my heartfelt thanks to everyone who supported me and helped me find the time to work on this edition.

Note to Readers
Supplementary files include powder diffraction data for examples and problems, a section on X-ray safety, and slides featuring figures with captions, selected equations, and tables. These files are available at www.springer.com/FPDSCM3.

Note to Instructors
A solutions manual is available to instructors at www.springer.com/FPDSCM3. Instructors must register to access these materials.

College Park, MD, USA — Peter Zavalij
Ames, IA, USA — Vitalij Pecharsky

Preface to the Second Edition

A little over 5 years have passed since the first edition of this book appeared in print. Seems like an instant but also eternity, especially considering numerous developments in the hardware and software that have made it from the laboratory test beds into the real world of powder diffraction. This prompted a revision, which had to be beyond cosmetic limits. The book was and remains focused on standard laboratory powder diffractometry. It is still meant to be used as a text for teaching students about the capabilities and limitations of the powder diffraction method. We also hope that it goes beyond a simple text and, therefore, is useful as a reference for practitioners of the technique.

The original book had seven long chapters that may have made its use as a text inconvenient. So the second edition is broken down into 25 shorter chapters. The first 15 are concerned with the fundamentals of powder diffraction, which makes it much more logical considering a typical 16-week long semester. The last ten chapters are concerned with practical examples of structure solution and refinement, which were preserved from the first edition and expanded by another example—solving the crystal structure of Tylenol®.

Major revisions include an expanded discussion of nonconventional crystallographic symmetry in Chap. 5, a short description of two new types of detectors that are becoming common in laboratory powder diffractometry—real-time multiple strip and multiwire detectors in Chap. 6, a brief introduction to the total scattering analysis in Chap. 10, a short section in Chap. 11 describing nonambient powder diffractometry, an expanded discussion of quantitative phase analysis, including the basics of how to quantify amorphous component in Chap. 13, an update about the recent advancements in the ab initio indexing, together with an example of a difficult pseudo-symmetric case represented by $Li[B(C_2O_4)_2]$, and a major update of Chap. 15 dedicated to the fundamentals of Rietveld analysis, including a brief introduction of the mechanism of restraints, constraints, and rigid bodies. The collection of problems that may be used by instructors to assess students' progress and as self-exercises has also expanded. All problems related to the solution and refinement of crystal structures from powder diffraction data are assembled at the end of Chap. 25.

Considering all these additions, something had to go. A major deletion from the earlier paper version is the section on X-ray safety, which has been moved to the electronic part of the book. Readers familiar with the first edition know that the book included a CD with electronic figures, experimental data, and solutions of all problems. Over the years, both the publisher and we have had numerous inquiries from people who accidentally used the CD as a coaster, clay pigeon, or simply sat on it before making a backup copy. While each and every request about sending a copy of the CD was fulfilled, we thought that it makes more sense to have the electronic files available online. The files are hosted by Springer (https://www.springer.com/gp/book/9780387095783) and they will be made available to everyone who has the book. The files include color figures, powder diffraction data, examples, web links, and solutions to all the problems found throughout the book. Files with the solutions of the problems are only available to instructors, who must register with the publisher.

Finally, we would like to thank everyone who provided critique and feedback. Most important, we thank the readers who opted to buy our book with their hard-earned money, thus providing enough votes for the publisher to consider this second, revised edition. It is our hope that this edition is met with even better acceptance by our readers of students, practitioners, and instructors of the truly basic materials characterization technique, which is the powder diffraction method.

Ames, IA, USA Vitalij Pecharsky
College Park, MD, USA Peter Zavalij

Preface to the First Edition

Without a doubt, crystals such as diamonds, emeralds, and rubies, whose beauty has been exposed by jewelry-makers for centuries, are enjoyed by everybody for their perfect shapes and astonishing range of colors. Far fewer people take pleasure in the internal harmony—atomic structure—which defines shapes and other properties of crystals but remains invisible to the naked eye. Ordered atomic structures are present in a variety of common materials, for example, metals, sand, rocks, or ice, in addition to the easily recognizable precious stones. The former usually consist of many tiny crystals and, therefore, are called polycrystals, for example, metals, ice, or powders, such as sand and snow. Besides external shapes and internal structures, the beauty of crystals can be appreciated from an infinite number of distinct diffraction patterns they form upon interaction with certain types of waves, for example, X-rays. Similarly, the beauty of the sea is largely defined by a continuously changing but distinctive patterns formed by waves on the water's surface.

Diffraction patterns from powders are recorded as numerical functions of a single independent variable, the Bragg angle, and they are striking in their fundamental simplicity. Yet, a well-executed experiment incorporates an extraordinarily rich variety of structural information, which is encoded in a material- and instrument-specific distribution of the intensity of coherently scattered monochromatic waves whose wavelengths are commensurate with lattice spacing. The utility of the powder diffraction method—one of the most essential tools in the structural characterization of materials—has been tested for over 90 years of successful use in both academia and industry. A broad range of general-purpose and specialized powder diffractometers are commonly available today, and just about every research project that involves polycrystalline solids inevitably begins with collecting a powder diffraction pattern. The pattern is then examined to establish or verify phase composition, purity, and the structure of the newly prepared material. In fact, at least a basic identification by employing powder diffraction data as a fingerprint of a substance, coupled with search-and-match among hundreds of thousands of known powder diffraction patterns stored in various databases, is an unwritten mandate for every serious work that involves crystalline matter.

Throughout the long history of the technique, its emphasis underwent several evolutionary and revolutionary transformations. Remarkably, the new developments have neither taken away nor diminished the value of earlier applications of the powder diffraction method; on the contrary, they enhanced and made them more precise and dependable. A noteworthy example is phase identification from powder diffraction data, which dates back to the late 1930s (Hanawalt, Rinn, and Frevel). Over the years, this application evolved into the Powder Diffraction File™ containing reliable patterns of some 300,000 crystalline materials in a readily searchable database format (Powder Diffraction File is maintained and distributed by the International Centre for Diffraction Data, https://www.icdd.com).

As it often happens in science and engineering, certain innovations may go unnoticed for some time but when a critical mass is reached or exceeded, they stimulate unprecedented growth and expansion, never thought possible in the past. Both the significance and applications of the powder diffraction method have been drastically affected by several directly related as well as seemingly unrelated developments that have occurred in the recent past. First was the widespread transition from analogue (X-ray film) to digital (point, line, and area detectors) recording of scattered intensity, which resulted in the improved precision and resolution of the data. Second was the groundbreaking work by Rietveld, Young, and many others, who showed that full-profile powder diffraction data may be directly employed in structure refinement and solution. Third was the availability of personal computers, which not only function as instrument controllers but also provide the much needed and readily available computing power. Computers thus enable the processing of large arrays of data collected in an average powder diffraction experiment. Fourth was the invention and rapid evolution of the Internet, which puts a variety of excellent, thoroughly tested computer codes at everyone's fingertips, thanks to the visionary efforts of many bright and dedicated crystallographers.

Collectively, these major developments resulted in the revolutionary changes and opened new horizons for the powder diffraction technique. Not so long ago, if you wanted to establish the crystal structure of a material at the atomic resolution, virtually the only reliable choice was to grow an appropriate quality single crystal. Only then could one proceed with the collection of diffraction data from the crystal followed by a suitable data processing to solve the structure and refine relevant structural parameters. A common misconception among the majority of crystallographers was that powder diffraction has a well-defined niche, which is limited to phase identification and precise determination of unit cell dimensions. Over the last 10–20 years, the playing field has changed dramatically, and the ab initio structure determination from powder diffraction data is now a reality. This raises the bar and offers no excuse for those who sidestep the opportunity to establish details of the distribution of atoms in the crystal lattice of every polycrystalline material, whose properties are under examination. Indeed, accurate structural knowledge obtained from polycrystals is now within reach. We believe that it will eventually lead to a much better understanding of structure–property relationships, which are critical for future advancements in materials science, chemistry, physics, natural sciences, and engineering.

Before a brief summary recounting the subject of this book, we are obliged to mention that our work was not conducted in a vacuum. Excellent texts describing the powder diffraction method have been written, published, and used by generations of professors teaching the subject and by generations of students learning the trade in the past. Traditional applications of the technique have been exceptionally well-covered by Klug and Alexander (1954), Azaroff and Buerger (1958), Lipson and Steeple (1970), Cullity (1956 and 1978), Jenkins and Snyder (1996), and Cullity and Stock (2001). There has never been a lack of reports describing the modern capabilities of powder diffraction, and they remain abundant in technical literature (*Journal of Applied Crystallography*, *Acta Crystallographica*, *Powder Diffraction*, *Rigaku Journal*, and others). A collective monograph, dedicated entirely to the Rietveld method, was edited by Young and published in 1993. A second collection of reviews, describing the state of the art in structure determination from powder diffraction data, appeared in 2002, and it was edited by David, Shankland, McCusker, and Baerlocher. These two outstanding and highly professional monographs are a part of the multiple-volume series sponsored by the International Union of Crystallography and are solid indicators that the powder diffraction method has been indeed transformed into a powerful and precise, yet readily accessible, structure determination tool. We highly recommend all the books mentioned in this paragraph as additional reading to everyone, although the older editions are out of print.

Our primary motivation for this work was the absence of a suitable text that can be used by both the undergraduate and graduate students interested in pursuing in-depth knowledge and gaining practical experience in the application of the powder diffraction method to structure solution and refinement. Here, we place emphasis on powder diffraction data collected using conventional X-ray sources and general-purpose powder diffractometers, which remain primary tools for thousands of researchers and students in their daily experimental work. Brilliant synchrotron and powerful neutron sources, which are currently operational or in the process of becoming so around the world, are only briefly mentioned. Both may and often do provide unique experimental data, which are out of reach for conventional powder diffraction, especially when high pressure, high and low temperature, and other extreme environments are of concern. The truth, however, is that the beam time is precious, and both synchrotron and neutron sources are unlikely to become available to everyone on a daily basis. Moreover, diffraction fundamentals remain the same, regardless of the nature of the employed radiation and the brilliance of the source.

This book has spawned from our affection and lasting involvement with the technique, which began long ago in a different country, when both of us were working our way through the undergraduate and then graduate programs in Inorganic Chemistry at L'viv State University, one of the oldest and finest institutions of higher education in Ukraine. As we moved along, powder diffraction has always remained on top of our research and teaching engagements. The major emphasis of our research is to obtain a better understanding of the structure–property relationships of crystalline materials, and both of us teach graduate-level powder diffraction courses at our respective departments—Materials Science and Engineering at Iowa State University and Chemistry at the State University of New York (SUNY) at Binghamton. Even before we started talking about this book, we were unanimous in our goals: the syllabi of two different courses were independently designed to be useful for any background, including materials science, solid-state chemistry, physics, mineralogy, and literally any other area of science and engineering, where structural information at the atomic resolution is in demand. This philosophy, we hope, resulted in a text that requires no prior knowledge of the subject. Readers are expected to have a general scientific and mathematical background of the order of the first 2 years of a typical liberal arts and sciences or engineering college.

The book is divided into seven chapters: The first chapter deals with essential concepts of crystallographic symmetry, which are intended to facilitate both the understanding and appreciation of crystal structures. This chapter will also prepare the reader for the realization of the capabilities and limitations of the powder diffraction method. It begins with the well-established notions of the three-dimensional periodicity of crystal lattices and conventional crystallographic symmetry. It ends with a brief introduction to the relatively young subject—the symmetry of aperiodic crystals. Properties and interactions of symmetry elements, including examination of both point and space groups, the concept of reciprocal space, which is employed to represent diffraction from crystalline solids, and the formal algebraic treatment of crystallographic symmetry are introduced and discussed to the extent needed in the context of the book.

The second chapter is dedicated to properties and sources of radiation suitable for powder diffraction analysis and gives an overview of the kinematical theory of diffraction along with its consequences in structure determination. Here, readers learn that the diffraction pattern of a crystal is a transformation of an ordered atomic structure into a reciprocal space rather than a direct image of the former. Diffraction from crystalline matter, specifically from polycrystalline materials is described as a function of crystal symmetry, atomic structure, and conditions of the experiment. The chapter ends with a general introduction to numerical techniques, enabling the restoration of the three-dimensional distribution of atoms in a lattice by the transformation of the diffraction pattern back into direct space.

The third chapter begins with a brief historical overview describing the powder diffraction method and explains the principles, similarities, and differences among the variety of powder diffractometers available today. Since ionizing radiation and highly penetrating and energetic particles are employed in powder diffraction, safety is always a primary concern. Basic safety issues are concisely spelled out using policies and procedures established at the US DOE's Ames Laboratory as a practical example. Sample preparation and proper selection of experimental conditions are exceedingly important in the successful implementation of the technique. Therefore, the remainder of this chapter is dedicated to a variety of issues associated with specimen preparation, data collection, and analysis of most common systematic errors that have an impact on every powder diffraction experiment.

Beginning from Chap. 4, key issues that arise during the interpretation of powder diffraction data, eventually leading to structure determination, are considered in detail and illustrated by a variety of practical examples. This chapter describes preliminary processing of experimental data, which is critical in both qualitative and quantitative phase analyses. In addition to a brief overview of phase identification techniques and quantitative analysis, readers will learn how to determine both the integrated intensities and angles of the observed Bragg peaks with the highest achievable precision.

Chapter 5 deals with the first major hurdle, which is encountered in powder diffraction analysis: unavoidably, the determination of any crystal structure starts from finding the shape, symmetry, and dimensions of the unit cell of the crystal lattice. In powder diffraction, finding the true unit cell from first principles may present considerable difficulty because experimental data are a one-dimensional projection of the three-dimensional reciprocal lattice. This chapter, therefore, introduces the reader to a variety of numerical techniques that result in the determination of precise unit cell dimensions. The theoretical background is followed by multiple practical examples with varying complexity.

Chapter 6 is dedicated to the solution of materials' structures, that is, here we learn how to find the distribution of atoms in the unit cell and create a complete or partial model of the crystal structure. The problem is generally far from trivial, and many structure solution cases in powder diffraction remain unique. Although structure determination from powder data is not a wide-open and straight highway, knowing where to enter, how to proceed, and where and when to exit is equally vital. Hence, in this chapter, both direct and reciprocal space approaches and some practical applications of the theory of kinematical diffraction to solving crystal structures from powder data are explained and broadly illustrated. Practical examples start from simple, nearly transparent cases and end with quite complex inorganic structures.

The solution of a crystal structure is considered complete only when multiple profile variables and crystallographic parameters of a model have been fully refined against the observed powder diffraction data. Thus, the last, the seventh chapter of this book describes the refinement technique, most commonly employed today, which is based on the idea suggested in the mid-1960s by Rietveld. Successful practical use of the Rietveld method, though directly related to the quality of powder diffraction data (the higher the quality, the more reliable the outcome), largely depends on the experience and the ability of the user to properly select a sequence in which various groups of parameters are refined. In this chapter, we introduce the basic theory of Rietveld's approach, followed by a series of hands-on examples that demonstrate the refinement of crystal structures with various degrees of completeness and complexity, models of which were partially or completely built-in Chap. 6.

The book is supplemented by an electronic volume—compact disk—containing powder diffraction data collected from a variety of materials that are used as examples and in the problems offered at the end of every chapter. In addition, electronic versions of some 330 illustrations found throughout the book are also on the CD. Electronic illustrations, which we hope are useful to both instructors and students because electronic figures are in color, are located in a separate folder / Figures on the CD. Three additional folders named /Problems, /Examples, and /

Solutions contain experimental data, which are required for solving problems, as self-exercises, and our solutions to the problems, respectively. The disk is organized as a web page, which makes it easy to navigate. All web links found in the book are included on the CD and can be followed by simply clicking on them. Every link is current as of January 2003. The compact disk is accessible using both Mac's and PC's, and potential incompatibility problems have been avoided by using portable document, HTML, and ASCII formats.

Many people have contributed in a variety of ways in the making of this book. Our appreciation and respect go to all authors of books, monographs, research articles, websites, and computer programs cited and used as examples throughout this text. We are indebted to our colleagues, Professor Karl Gschneidner, Jr. from Iowa State University, Professor Scott R. J. Oliver from SUNY at Binghamton, Professor Alexander Tishin from Moscow State University, Dr. Aaron Holm from Iowa State University, and Dr. Alexandra (Sasha) Pecharsky from Iowa State University, who read the entire manuscript and whose helpful advice and friendly criticism made this book better. It also underwent a common-sense test, thanks to Lubov Zavalij and Vitalij Pecharsky, Jr. Some of the experimental data and samples used as the examples have been provided by Dr. Lev Akselrud from L'viv State University, Dr. Oksana Zaharko from Paul Scherrer Institute, and Dr. Iver Anderson, Dr. Matthew Kramer, and Dr. John Snyder (all from Ames Laboratory, Iowa State University), and we are grateful to all of them for their willingness to share the results of their unpublished work. Special thanks are in order to Professor Karl Gschneidner, Jr. (Iowa State University) and Professor M. Stanley Whittingham (SUNY at Binghamton), whose perpetual attention and encouragement during our work on this book have been invaluable. Finally, yet significantly, we extend our gratitude to our spouses, Alexandra (Sasha) Pecharsky and Lubov Zavalij, and to our children, Vitalij Jr., Nadya, Christina, Solomia, and Michelle, who handled our virtual absence for countless evenings and weekends with exceptional patience and understanding.

Ames, IA, USA
Binghamton, NY, USA

Vitalij Pecharsky
Peter Zavalij

Contents

Chapter 1
Fundamentals of Crystalline State and Crystal Lattice

The concepts of crystalline state and symmetry are just about synonymous today, although the general sense of symmetry is much older than the idea of symmetrical arrangement of atoms in the structures of crystalline solids. Following dictionaries, symmetry can be defined as the "beauty of form arising from balanced proportions" and to be symmetrical is to have the "correspondence in size, shape, and relative position or parts on opposite sides of a dividing line or median plane or about a center or axis."[1]

Humans constantly deal with symmetry, often without even noticing its significance in daily life. For instance, our exposure to symmetry begins every morning with a glimpse in a mirror, and it ends every night when we fall asleep in a bed with balanced proportions. Although intuitive perception of symmetry is familiar to everyone, it has multiple applications in science. A much more comprehensive and formal description of symmetry, when compared to that found in dictionaries, is therefore necessary.

In the first five chapters of this book, we consider basic concepts of crystallographic symmetry, which are essential to the understanding of how atoms and molecules are arranged in space and how they form crystalline solids. Further, the detailed knowledge of crystallographic symmetry is important to appreciate both the capabilities and limitations of powder diffraction techniques when they are applied to the characterization of the crystal structure of solids.

We begin with the well-established notions of the three-dimensional periodicity of crystal lattices and conventional crystallographic symmetry and consider the properties and interactions of both finite and infinite symmetry elements, including an examination of both point and space groups.[2] The formal, algebraic treatment of

[1] Webster's Seventh New Collegiate Dictionary, G. & C. Merriam Company Publ., Springfield, MA, USA (1963).

[2] Finite symmetry elements and point groups are employed to describe relationships among parts of finite objects, such as geometrical figures or shapes of natural and synthetic crystals. Finite and infinite symmetry elements combined and space groups establish symmetrical relationships among

P. Zavalij, V. Pecharsky, *Fundamentals of Powder Diffraction and Structural Characterization of Materials*, https://doi.org/10.1007/978-3-031-91504-8_1

crystallographic symmetry, which is usually omitted in most texts, is introduced and briefly discussed since both the modern crystallography and powder diffraction are for the most part computerized. Furthermore, the algebraic description of crystallographic symmetry makes the subject complete. The topic ends with an introduction to a nonconventional crystallographic symmetry, which has been a poignant subject in crystallography since the discovery of perfectly ordered but clearly aperiodic crystals.[3]

Without a doubt, it is impossible to include all details about crystallographic symmetry in these five chapters that are a part of the book about powder diffraction. We hope, however, that after the main concepts introduced here are understood, the reader is ready to take on a much more comprehensive description of crystallographic symmetry, for example, that found in the International Tables for Crystallography.[4]

1.1 Crystalline State

Matter usually exists in one of the three basic states: gaseous, liquid, or solid. At fixed temperature and pressure, only one of the states is typically stable for any given substance. However, under certain combinations of these thermodynamic variables, two or all three states may coexist in equilibrium. By decreasing temperature and/or increasing pressure, a gas may condense into a liquid and subsequently into a solid. In some cases, however, a gas transitions directly into a solid without forming a liquid phase. The most fundamental differences between gases, liquids, and solids are summarized in Table 1.1.

Gases are formed by weakly interacting, nearly isolated particles—atoms or molecules. Interatomic or intermolecular distances continuously change, and as a

---

components of infinite objects, e.g., two-dimensional wall patterns or three-dimensional arrangements of atoms or molecules in crystals. Although the division of symmetry elements on finite and infinite is not in common use, we employ this terminology both for convenience and to emphasize the nature of the objects that they describe, i.e., finite and infinite objects, respectively. Finite symmetry elements are also known as nontranslational and the infinite ones as translational.

[3]D. Shechtman, I. Blech, D. Gratias and J.W. Cahn, Metallic phase with long-range orientational order and no translational symmetry, Phys. Rev. Lett. **53**, 1951 (1984) https://doi.org/10.1103/PhysRevLett.53.1951. Authors' note: The discovery of aperiodicity dates back to 1982, but as Dan Shechtman recollects, it took nearly 2 years to convince referees that the observation of a fivefold symmetry axis is not an experimental artifact.

[4]International Tables for Crystallography, Vol. A, Space-group symmetry, M. I. Aroyo, Ed., Second online edition, (2016), https://it.iucr.org/Ac/; International Tables for Crystallography, Vol. A1, Symmetry relations between space groups, H. Wondratschek and U. Müller, Eds., Second online edition, (2011), https://it.iucr.org/A1/; International Tables for Crystallography, Vol. B, Reciprocal space, U. Shmueli, Ed., Second online edition, (2010), https://it.iucr.org/B/. More information about the entire nine-volume series of the International Tables for Crystallography is found at https://it.iucr.org/

Table 1.1 Basic characteristics of the three states of matter

State of matter	Fixed volume	Fixed shape	Order	Properties
Gas	No	No	None	Isotropic[a]
Liquid	Yes	No	Short-range[b]	Isotropic
Solid (amorphous)	Yes	Yes	Short-range[b]	Isotropic
Solid (crystalline)	Yes	Yes	Long-range[b]	Anisotropic[c]

[a] A system has same properties in all directions
[b] Short-range order extends over a few atoms, while long-range order spans from approximately 10^3–10^{20} atoms
[c] A system has different properties in different directions

result, gases have no fixed shape or volume, and gaseous matter occupies all available space. As far as the macroscopic properties of a gas are concerned, they remain identical in any direction because its structure, more precisely, the absence of long- or short-range order, is isotropic.

When attraction among atoms or molecules is strong enough to keep them in the immediate vicinity of each other, a gas condenses into a liquid. Since chemical bonding between particles in a liquid remains relatively weak, thermal energy is sufficient to continuously move molecules around and away from their nearest neighbors. When a molecule in a liquid is removed from the assembly of nearest neighbors, another molecule immediately occupies its place, thus preserving only short-range order. Hence, particles in a liquid are not linked together permanently, and liquids have specific volume but no fixed shape. Structures of liquids and, therefore, their properties remain isotropic on a macroscopic scale.

When attractive forces are so strong that the particles cannot easily move away from one another, matter becomes solid. Solids have both shape and volume. Although particles in a solid can be distributed randomly in space, an ordered and repetitive pattern is more likely, as it corresponds to a lower energy state when compared with a random spatial distribution of strongly interacting atoms or molecules. The appearance of long-range order brings about structural anisotropy, and macroscopic properties of crystalline solids become directionally dependent, that is to say, anisotropic.[5]

It is important to recognize that not all solids are crystalline or ordered. For example, glasses have both shape and volume but they also have a high degree of disorder, and therefore, are classified as amorphous solids. Lack of long-range order generally makes macroscopic properties of amorphous solids isotropic. In addition to the glassy state, which is only characterized by short-range order, some solids may have loose (or approximate) long-range order and can be described by a lattice that is distorted to a greater or lesser degree. Similar to crystalline state, nearly ordered solids may be described by a lattice (see Sect. 1.2) that is distorted to a greater or

[5]This statement is true for single crystals and for some polycrystalline materials that exhibit preferred orientation (in other words, are strongly textured), where atomic-scale anisotropy is preserved on a macroscopic scale. Properties of polycrystalline materials, i.e. solids that consist of a large number of randomly oriented single-crystalline grains, generally remain isotropic.

lesser degree. Hence, the boundary between amorphous and crystalline states is generally diffuse, and these intermediate cases are known as semicrystalline solids.

One of the most distinct properties of the crystalline state is, therefore, the presence of long-range order, or in other words, a regular and in the simplest case periodic repetition of atoms or molecules in space. In theory, periodic crystals are infinite, but in practice, their periodicity extends over a distance from ~10^3 to ~10^{20} atomic or molecular dimensions, which occurs because any crystal necessarily has a number of defects and may contain impurities without losing its crystallinity. Further, a crystal is always finite regardless of its size.

Since our surroundings are three-dimensional, we tend to assume that crystals are formed by periodic arrangements of atoms or molecules in three dimensions. However, many crystals are periodic only in two, or even in one dimension, and some do not have three-dimensional periodic structure at all, as, for example, solids with incommensurately modulated and composite structures, certain polymers, and quasicrystals. Materials may assume states that are intermediate between those of a crystalline solid and a liquid, and they are called liquid crystals. Hence, in real crystals, periodicity and/or order extends over a shorter or longer range, which is a function of the nature of the material and conditions under which it was crystallized. Structures of real crystals, for example, imperfections, distortions, defects, dislocations, and impurities, are subjects of separate disciplines, and symmetry concepts considered in Chaps. 1–4 assume an ideal crystal with perfect periodicity.

1.2 Crystal Lattice and Unit Cell

Periodic structure of an ideal crystal is most easily described by a lattice. In a lattice, all elementary parallelepipeds, that is, unit cells, are equal in their shape and content. Most importantly, if the distribution of atoms in one unit cell is known, the structure of the whole crystal, regardless of its physical size, can be reconstructed by simply propagating (translating or shifting) this unit cell along one, two, or three directions independently. Without the lattice, location of every atom in the crystal must be described.[6]

---

[6]Imagine a microscopic crystal of iron in the form of a cube with a side of 1 μm. It has a volume of 10^{-18} m^3. One cubic meter of iron weighs 7.874 metric tons, containing ~141,000 mol = $8.49 * 10^{28}$ iron atoms. The tiny, micrometer size crystal, therefore, consists of nearly 85 billion (84,900,000,000) of iron atoms. Three numerical values (coordinate triplets, see Sect. 2.1) are required to fully define the location of every atom in space. Hence, without the notion of a lattice, one needs over 250,000,000,000 numbers to fully describe the structure of a crystal that is invisible to the naked eye. How about a larger crystal with a volume of only 1 mm^3? The crystal is still very small, but the number of required coordinate triplets increases by a factor of a billion! Clearly, handling such a tremendous amount of numerical data is not only inconvenient, but is absolutely impractical.

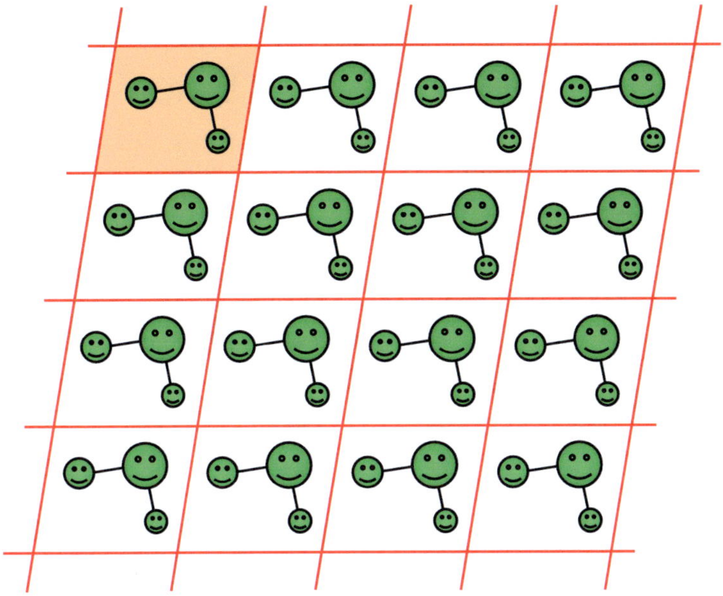

Fig. 1.1 Illustration of a two-dimensional lattice with one unit cell shaded in orange

An example of a simple two-dimensional lattice is shown in Fig. 1.1. Each unit cell, one of which is shaded in orange, is the parallelogram delineated by solid lines, and every unit cell contains one hypothetical molecule that consists of three atoms, shown as happy faces: one large, one medium, and one small. The structure of this molecule, including bond lengths and bond angle, remains identical throughout the whole lattice.

Generally, the origin of the lattice and the origin of the unit cell can be chosen arbitrarily. In Fig. 1.2, an alternative lattice with the origin in the middle of the medium size atom is shown using dash-dotted lines. It is worth noting that both the shape and content of the new unit cell remain the same as in Fig. 1.1.

The lattice itself, including the shape of the unit cell, may be chosen in an infinite number of ways. As an example, a second alternative lattice with a different unit cell is shown in Fig. 1.3. Both the origin of the lattice and the shape of the unit cell have been changed when compared to Fig. 1.1, but the content of the unit cell has not—it encloses the same molecule.

1.3 Shape of the Unit Cell

To fully describe a three-dimensional lattice or its building block—the unit cell—a total of three noncoplanar vectors are required. These vectors (**a**, **b**, **c**) coincide with the three independent edges of the elementary parallelepiped as shown schematically in Fig. 1.4.

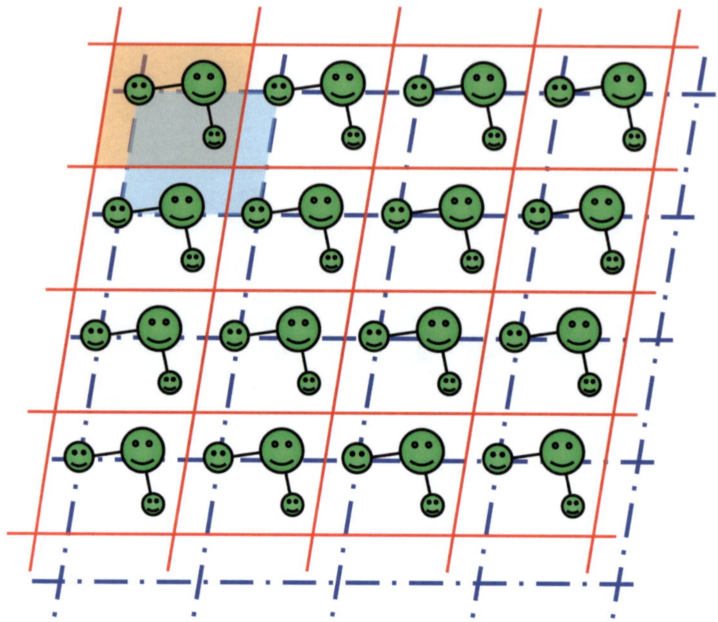

Fig. 1.2 Illustration of an arbitrary origin of a lattice. The original and the alternative lattices are shown using *solid* and *dash-dotted lines*, respectively. The unit cells (shaded in *orange* and *blue*) have identical shapes

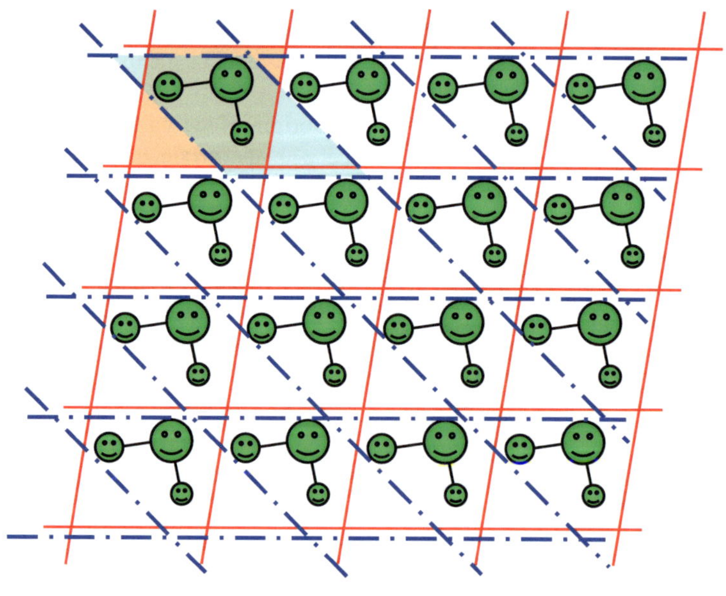

Fig. 1.3 Illustration of an arbitrary unit cell of a lattice. The original and the alternative lattices are shown using *solid* and *dash-dotted lines*, respectively. The unit cells (shaded in *orange* and *blue*) have different shapes but their areas (or volumes in three dimensions) and contents remain identical

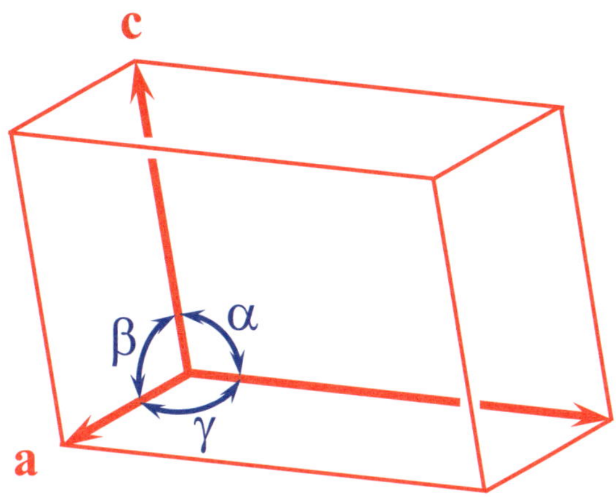

Fig. 1.4 Unit cell in three dimensions

Therefore, any point in a three-dimensional lattice can be described by a vector, **q**, defined in Eq. (1.1), where u, v, and w are integer numbers

$$\mathbf{q} = u\mathbf{a} + v\mathbf{b} + w\mathbf{c} \tag{1.1}$$

The three basis vectors (**a**, **b**, **c**) and all derived vectors (**q**) represent translations in the lattice. They translate the unit cell, including every atom and/or molecule located inside the unit cell, in three dimensions, thus filling the entire space of a crystal. The point with $u = v = w = 0$ is taken as the origin of coordinates; positive and negative u, v, and w define positive and negative directions, respectively. Since the lattice is infinite, any point in the lattice can be chosen as the origin of coordinates.

Instead of three noncoplanar vectors, the unit cell can be completely described by specifying a total of six scalar quantities, which are called the unit cell dimensions or lattice parameters. These are (see also Fig. 1.4):

$$a, b, c, \alpha, \beta, \gamma$$

The first three parameters (a, b, and c) represent the lengths of the unit cell edges, and the last three (α, β, and γ) represent the angles between them. By convention, α is the angle between **b** and **c**, β is the angle between **a** and **c**, and γ is the angle between **a** and **b**.

Unit cell parameters are usually quoted in angströms (Å, where 1 Å = 10^{-10} m = 10^{-8} cm), nanometers (nm, 1 nm = 10^{-9} m), or picometers (pm, 1 pm = 10^{-12} m) for the lengths of the unit cell edges, and in degrees (°) for the angles between basis vectors. To differentiate between basis vectors (**a**, **b**, **c**), which appear in bold, the lengths of the unit cell edges (a, b, c) always appear in italic.

1.4 Crystallographic Planes, Directions, and Indices

The crystallographic plane is a geometrical concept introduced to illustrate the phenomenon of diffraction from ideal crystal lattices since algebraic equations that govern diffraction process are difficult to visualize. It is important to realize and remember that no "real" crystallographic planes exist in real crystals. Moreover, regardless of whether the crystallographic plane is referred to in singular or in plural, the reference is always made to a series, which consists of an infinite number of parallel equally spaced planes.

1.4.1 Crystallographic Planes

A family of crystallographic planes is defined as a set of planes that intersect all lattice points. All planes in the same family are necessarily: (1) parallel to each other and (2) equally spaced. The distance between the neighboring planes is called the interplanar distance or *d*-spacing. The family of crystallographic planes is fully described using three integer indices *h*, *k*, and *l*, which are called crystallographic or Miller indices.[7] When referring to a plane, a triplet of Miller indices is always enclosed in parentheses: (*hkl*). Miller indices indicate that the planes that belong to the family (*hkl*) divide lattice vectors (unit cell edges) **a**, **b**, and **c** into *h*, *k*, and *l* equal parts, respectively. When the planes are parallel to a crystallographic axis, the corresponding Miller index is set to 0.

The meaning of the Miller indices can be better understood after considering Figs. 1.5, 1.6, and 1.7. In Fig. 1.5, both sets of planes are parallel to **b** and **c**. Hence, in both cases $k = l = 0$. The set of planes shown on the left divides **a** into one part, while the planes shown on the right divide **a** into two equal parts. This results in the following Miller indices: (100) for the drawing on the left and (200) for that on the right of Fig. 1.5. It is obvious that the interplanar distance $d_{(200)}$ is ½ that of $d_{(100)}$, and the (200) family of crystallographic planes may be considered as the second order of the (100) family. Following the same rules, the planes shown in Fig. 1.6, left are parallel to **c** and divide both **a** and **b** in one part, which results in the (110) Miller indices for this family of crystallographic planes. Similarly, Fig. 1.6, right illustrates the $(\bar{1}\bar{1}1)$ and Fig. 1.7 illustrates the (213) crystallographic planes.

Assuming that $\alpha = \beta = \gamma = 90°$, the inverse squares of the interplanar distances for these examples of crystallographic planes are shown at the bottom of each figure. In this case, a general formula shown in Eq. (1.2) illustrates dependence of the *d*-

[7] Miller indices were introduced by the British mineralogist William Hallowes Miller (1801–1880). When $\gamma = 120°$, i.e., in the hexagonal and trigonal crystal systems (crystal systems are discussed in Sect. 2.7), a total of four Miller indices may be used to designate a plane: (*hkil*), where $i = -(h + k)$. See Fig. 1.8 for details.

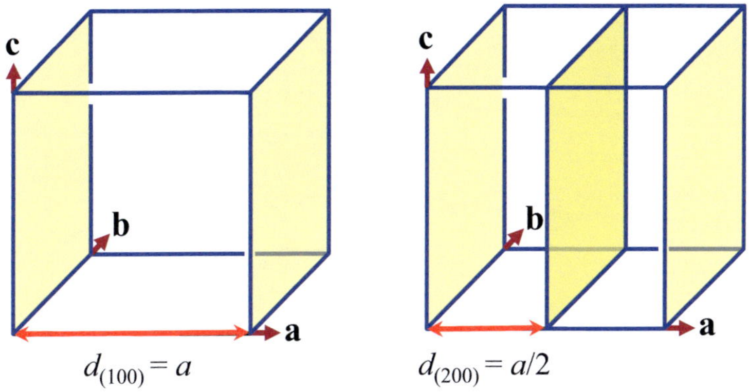

Fig. 1.5 Families of (100) and (200) crystallographic planes

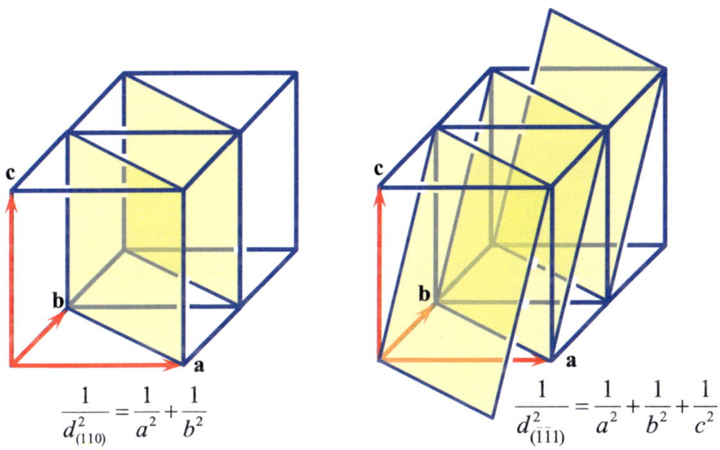

Fig. 1.6 Families of (110) and $(\bar{1}\bar{1}1)$ crystallographic planes

spacing on the Miller indices of the family and the lengths of the three unit cell edges.

$$\frac{1}{d_{hkl}} = \sqrt{\frac{h^2}{a^2} + \frac{k^2}{b^2} + \frac{l^2}{c^2}} \tag{1.2}$$

Equation 1.2 becomes more complicated when interaxial angles (α, β, and/or γ) are different from 90° and should be included in the calculations. The complete description of the corresponding mathematical relationships is given later in the book (Eqs. (8.2)–(8.7) in Sect. 8.4.1).

When $\gamma = 120°$, the fourth index is usually introduced to address the possibility of three similar choices in selecting the crystallographic basis as illustrated in

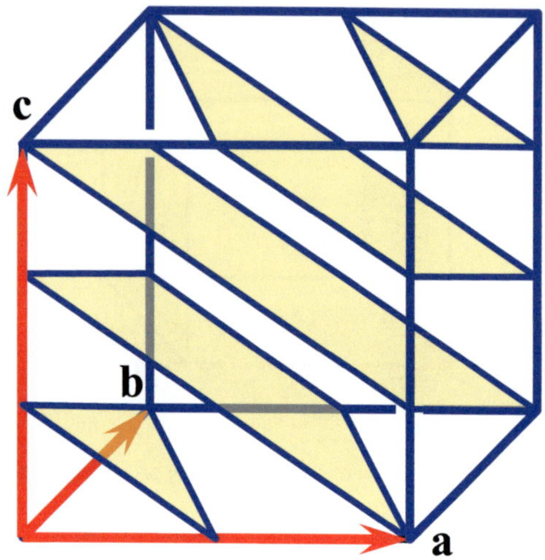

$$\frac{1}{d^2_{(213)}} = \frac{4}{a^2} + \frac{1}{b^2} + \frac{9}{c^2}$$

Fig. 1.7 Family of (213) crystallographic planes

Fig. 1.8. In addition to the unit cell based on the vectors **a**, **b** and **c**, two other unit cells, based on the vectors **a**, −(**a** + **b**), and **c**, and −(**a** + **b**), **b**, and **c** are possible due to the threefold (see Sect. 2.4.3) or the sixfold (see Sect. 2.4.5) rotational symmetry parallel to **c**. Thus, if only three indices are employed to designate related planes (e.g., the orange-, yellow- and brown-colored planes in Fig. 1.8), these are (110), $(1\bar{2}0)$, and $(\bar{2}10)$ planes, respectively, defined in the lattice based on the unit vectors **a**, **b**, and **c**. No apparent relationships are seen between the three planes designated using three indices. When the fourth index, $i = -(h + k)$, is introduced, the three planes $(hkil)$ are only different by a cyclic permutation of the first three indices (see Fig. 1.8), thus emphasizing symmetrical relationships existing between these families of crystallographic planes.

1.4.2 Crystallographic Directions

Directions in the crystal lattice are described using lines that pass through the origin of the lattice and are parallel to the direction of interest. Since the lattice is infinite, a line drawn in any direction from its origin will necessarily pass through the infinite number of lattice points. For example, the line traversing the origin and parallel to the body diagonal of the unit cell passes through the points …, $\overline{333}$, $\overline{222}$, $\overline{111}$, 000, 111, 222, 333, … . The crystallographic direction is therefore indicated by referring to the coordinates (u, v, and w, see Eq. (1.1)) of the first point other than the

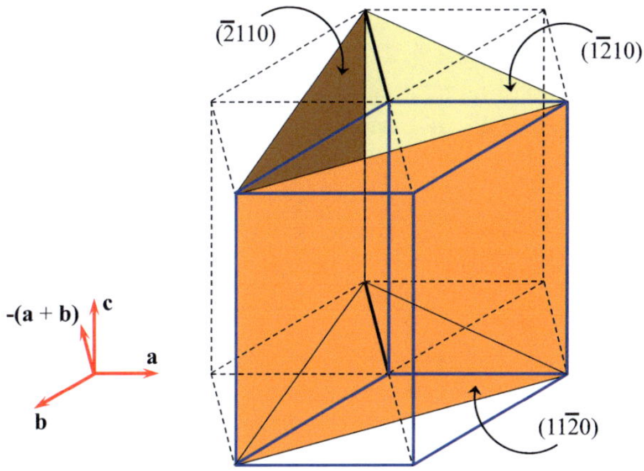

Fig. 1.8 Three possibilities to select the crystallographic basis in hexagonal and trigonal crystal systems and the family of (11$\bar{2}$0) crystallographic planes when $\gamma = 120°$. The indices are shown for the unit cell based on the vectors **a**, **b**, and **c**. Three additional symmetrically related families of planes have indices ($\bar{1}\bar{1}$20), ($\bar{1}$2$\bar{1}$0), and (2$\bar{1}\bar{1}$0) in the same basis and we leave their identification to the reader

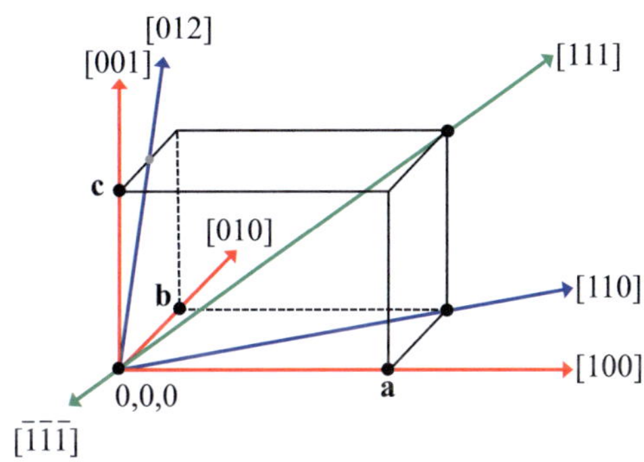

Fig. 1.9 Selected crystallographic directions in the lattice with $\alpha = \beta = \gamma = 90°$

origin, which the line intersects on its way from the origin. To differentiate between indices of the crystallographic planes, indices of the crystallographic directions are enclosed in square brackets [*uvw*], as shown in Fig. 1.9.

1.5 Reciprocal Lattice

The concept of a reciprocal lattice[8] was first introduced by Ewald,[9] and it quickly became an important tool in the illustrating and understanding of both the diffraction geometry and relevant mathematical relationships. Let **a**, **b**, and **c** be the elementary

[8]For additional information, see IUCr teaching pamphlets: A. Authier, The reciprocal lattice, https://www.iucr.org/education/pamphlets/4 and the International Tables for Crystallography, vol. A and vol. B.

[9]Peter Paul Ewald (1888–1985). German physicist, whose work [P.P. Ewald, Das reziproke Gitter in der Strukturtheorie, Z. Kristallogr. **56**, 129 (1921), https://doi.org/10.1524/zkri.1921.56.1.129] is

translations in a three-dimensional lattice (called here a direct lattice), as shown, for example, in Fig. 1.4.

A second lattice, reciprocal to the direct lattice, is defined by three elementary translations $\mathbf{a}^*$, $\mathbf{b}^*$, and $\mathbf{c}^*$,[10] which simultaneously satisfy the following two conditions:

$$\mathbf{a}^* \cdot \mathbf{b} = \mathbf{a}^* \cdot \mathbf{c} = \mathbf{b}^* \cdot \mathbf{a} = \mathbf{b}^* \cdot \mathbf{c} = \mathbf{c}^* \cdot \mathbf{a} = \mathbf{c}^* \cdot \mathbf{b} = 0 \tag{1.3}$$

$$\mathbf{a}^* \cdot \mathbf{a} = \mathbf{b}^* \cdot \mathbf{b} = \mathbf{c}^* \cdot \mathbf{c} = 1 \tag{1.4}$$

All products in Eqs. (1.3) and (1.4) are scalar (or dot) products. As a reminder, the dot product of the two vectors, $\mathbf{v_1}$ and $\mathbf{v_2}$, is a scalar quantity, which is equal to the product of the absolute values of the two vectors and the cosine of the angle α between them:

$$\mathbf{v}_1 \cdot \mathbf{v}_2 = v_1 v_2 \cos \alpha \tag{1.5}$$

Conversely, the vector (or cross) product of the same two vectors ($\mathbf{v_1} \times \mathbf{v_2}$) is a vector, $\mathbf{v_3}$, in the direction perpendicular to the plane of $\mathbf{v_1}$ and $\mathbf{v_2}$, whose magnitude is equal to the product of the absolute values of the two vectors and the sine of the angle α between them, or

$$|\mathbf{v}_1 \times \mathbf{v}_2| = v_3 = v_1 v_2 \sin \alpha \tag{1.6}$$

In other words, the length of the vector $\mathbf{v_3}$ is equal to the area of the parallelogram formed by the vectors $\mathbf{v_1}$ and $\mathbf{v_2}$ (shaded in Fig. 1.10), and its direction is perpendicular to the plane of the parallelogram.

Considering Eqs. (1.3)–(1.6), it is possible to show that the elementary translations in the reciprocal lattice are defined as

$$\mathbf{a}* = \frac{\mathbf{b} \times \mathbf{c}}{V}, \quad \mathbf{b}* = \frac{\mathbf{c} \times \mathbf{a}}{V}, \quad \mathbf{c}* = \frac{\mathbf{a} \times \mathbf{b}}{V} \tag{1.7}$$

and that the inverse relationships are also true, in other words,

$$a = \frac{\mathbf{b}* \times \mathbf{c}*}{V^*}, \quad \mathbf{b} = \frac{\mathbf{c}* \times \mathbf{a}*}{V^*}, \quad \mathbf{c} = \frac{\mathbf{a}* \times \mathbf{b}*}{V^*} \tag{1.8}$$

In Eqs. (1.7) and (1.8), the two scalar quantities V and V^* are the volumes of the unit cells in the direct and reciprocal lattices, respectively. Hence, **a*** is perpendicular to both **b** and **c**; **b*** is perpendicular to both **a** and **c**; and **c*** is perpendicular to

considered a landmark in using reciprocal lattice in X-ray diffraction. See Wikipedia (https://en.wikipedia.org/wiki/Paul_Peter_Ewald) for a brief biography.

[10]Symbol with an asterisk always refers to a parameter of reciprocal lattice.

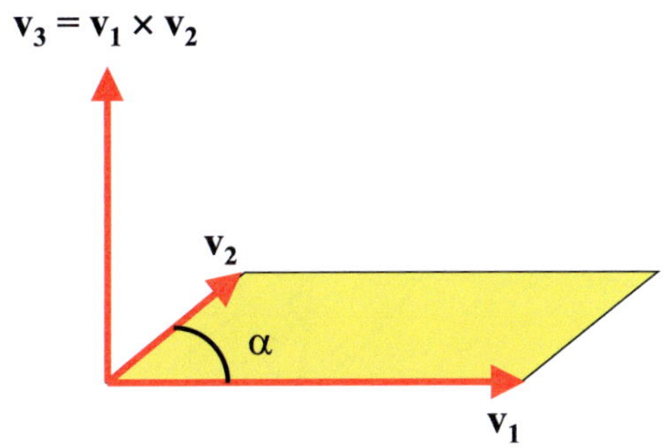

Fig. 1.10 Vector (cross) product of two vectors. The direction of $\mathbf{v}_3$ is determined using the right-hand rule (thumb of the right hand is aligned with $\mathbf{v}_1$, index finger with $\mathbf{v}_2$, then $\mathbf{v}_3$ is aligned with the middle finger; tails of all vectors face the middle of the palm) or the right-hand screw rule (the direction in which a screw advances when turned from $\mathbf{v}_1$ to $\mathbf{v}_2$)

both **a** and **b**. In terms of the interplanar distances, **d*** is perpendicular to the corresponding crystallographic planes, and its length is inversely proportional to d, that is,

$$d^*_{hkl} = \frac{1}{d_{hkl}} \tag{1.9}$$

An important consequence of Eq. (1.9) is that a set, which consists of an infinite number of crystallographic planes in the direct lattice, is represented by a single vector or by a point at the end of the vector in the reciprocal lattice.[11] When interaxial angles are orthogonal, i.e., when $\alpha = \beta = \gamma = 90°$, the relationships between the unit cell dimensions of reciprocal and real lattices are simplified to

$$a^* = 1/a, \;\; b^* = 1/b, \;\; c^* = 1/c \tag{1.10}$$

The two-dimensional example illustrating the relationships between the direct and reciprocal lattices, which are used to represent crystal structures (see Sect. 2.1) and diffraction patterns (see Sects. 7.2.3 and 8.1), respectively, is shown in Fig. 1.11. An important property of the reciprocal lattice is that its symmetry is the same as the symmetry of the direct lattice. However, in the direct-space atoms can be located anywhere in the unit cell, whereas diffraction peaks are represented only by the points of the reciprocal lattice, and the unit cells themselves are "empty" in the reciprocal space. Further, the content of every unit cell in the direct space is the same, but the intensities of diffraction peaks, which are fittingly represented by points in the reciprocal space, vary.

---

[11] Hence, any vector in a three-dimensional reciprocal lattice is determined as $\mathbf{d}^*_{hkl} = h\mathbf{a}^* + k\mathbf{b}^* + l\mathbf{c}^*$. Also see Eq. (8.8) and Fig. 8.10.

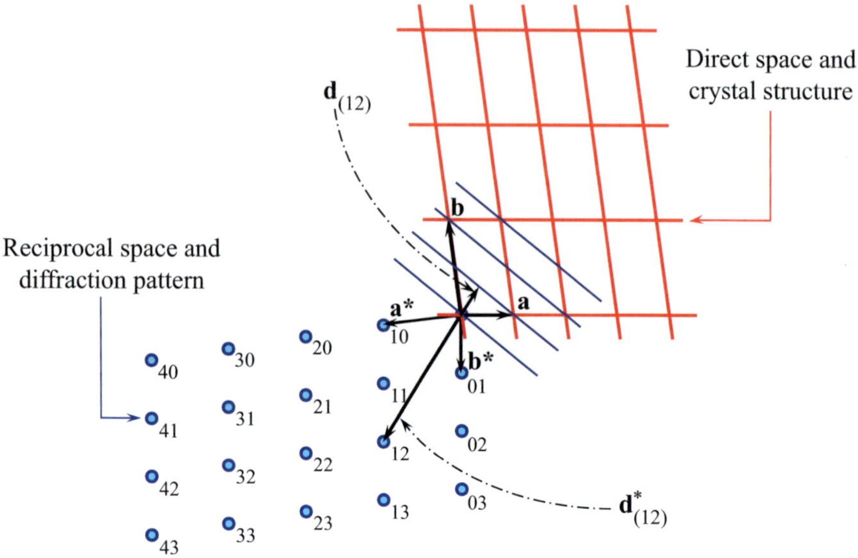

Fig. 1.11 Example of converting crystallographic planes in the direct lattice into points in the reciprocal lattice. The corresponding Miller indices are shown near the points in the reciprocal lattice

We note that reciprocal lattice and unit cell, and reciprocal space itself, are nothing more than mathematical concepts introduced to help with visualizing and describing periodic diffraction patterns. Similarly, direct-space lattice and unit cell, but not direct space itself, are used to describe periodic structures of crystals.

1.6 Additional Reading

1. C. Giacovazzo, H.L. Monaco, G. Artioli, D. Viterbo, M. Milaneso, G. Gilli, P. Gilli, G. Zanotti, G. Ferraris, and M. Catti, Fundamentals of crystallography, 3rd edition, C. Giacovazzo, Ed., IUCr/Oxford University Press (2011), https://doi.org/10.1093/acprof:oso/9780199573653.001.0001.
2. D. Schwarzenbach, Crystallography, Wiley, New York (1996), https://www.wiley.com/en-us/Crystallography-p-9780471955986.
3. C. Hammond, The basics of crystallography and diffraction, 4th edition. IUCr/Oxford University Press (2015), https://doi.org/10.1093/acprof:oso/9780198738671.001.0001.
4. International Tables for Crystallography. 6th edition, Teaching edition, Crystallographic symmetry, M. I. Aroyo, Ed., Wiley (2021), https://www.wiley.com/en-us/-p-9780470974223.
5. International Tables for Crystallography, Vol. A, Space-group symmetry, M. I. Aroyo, Ed., Second online edition, (2016), https://it.iucr.org/Ac/.
6. International Tables for Crystallography, Vol. B, Reciprocal space, U. Shmueli, Ed., Second online edition, (2010), https://it.iucr.org/B/.

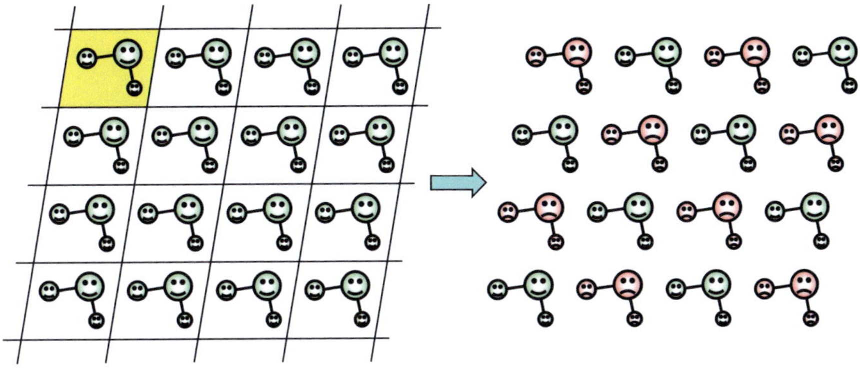

Fig. 1.12 The original lattice containing indistinguishable molecules in which a proper unit cell is shaded in yellow (*left*) and a new lattice derived by switching half of the atoms from happy to sad faces in a regular fashion

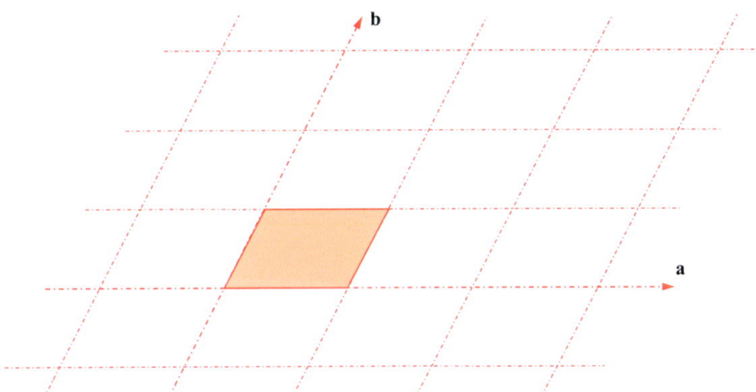

Fig. 1.13 The two-dimensional nonorthogonal lattice with the unit cell shaded in orange

1.7 Problems

1. Consider a two-dimensional lattice shown in Fig. 1.12, left, which was discussed earlier in Sect. 1.2. One-half of the molecules in this lattice have been modified in a regular way so that their atoms now have sad faces, as shown in Fig. 1.12, right. This may be a schematic illustrating the formation of a magnetically ordered structure, where happy and sad faces represent opposite directions of magnetic moments. Suggest the most probable unit cell in this new lattice assuming that the correct unit cell in the original lattice is shown by a yellow-shaded parallelogram.
2. Consider a two-dimensional nonorthogonal lattice shown in Fig. 1.13. Sketch the following sets of crystallographic planes: (12), $(\bar{1}2)$, $(3\bar{2})$.

3. A monoclinic lattice has the following unit cell dimensions: a = 5.00 Å, b = 10.0 Å, c = 8.00 Å, and β = 110°. Calculate the unit cell dimensions of the corresponding reciprocal lattice.

Chapter 2
Finite Symmetry Elements and Crystallographic Point Groups

In addition to simple translations, which are essential for understanding the concept of a lattice, other types of symmetry often exist in most real crystal structures. Here, we begin with considering a single unit cell, as it serves as the fundamental building block of a three-dimensional periodically repeating infinite lattice—underpinning the vast array of crystalline materials.

2.1 Content of the Unit Cell

To completely describe the crystal structure, it is not enough to characterize only the geometry of the unit cell. One also needs to establish the distribution of atoms in the unit cell and, consequently, in the entire lattice. The latter is done by simply translating each point inside the unit cell using Eq. (1.1). Hence, the three noncoplanar vectors **a**, **b**, and **c** form a basis of the coordinate system with three noncoplanar axes *X*, *Y*, and *Z*, which is called the crystallographic coordinate system or the crystallographic basis. The coordinates of a point inside the unit cell, i.e., the coordinate triplets *x*, *y*, *z*, are expressed in fractions of the unit cell edge lengths, and therefore, they vary from 0 to 1 along the corresponding vectors (**a**, **b**, or **c**).[1] Thus, the coordinates of the origin of the unit cell are always 0,0,0 ($x = 0$, $y = 0$, and $z = 0$), and for the ends of **a**-, **b**-, and **c**-vectors, they are 1,0,0; 0,1,0; and 0,0,1, respectively. Again, using capital italic *X*, *Y*, and *Z*, we will always refer to crystallographic axes coinciding with **a**, **b**, and **c** directions, respectively, while small italic *x*, *y*, and *z* are used to specify the corresponding fractional coordinates along the *X*, *Y*, and *Z* axes.

An example of the unit cell in three dimensions and its content given in terms of coordinates of all atoms is shown in Fig. 2.1. Here, the centers of gravity of three

[1] In order to emphasize that the coordinate triplets list fractional coordinates of atoms, in crystallographic literature these are often denoted as *x*/*a*, *y*/*b*, and *z*/*c*.

P. Zavalij, V. Pecharsky, *Fundamentals of Powder Diffraction and Structural Characterization of Materials*, https://doi.org/10.1007/978-3-031-91504-8_2

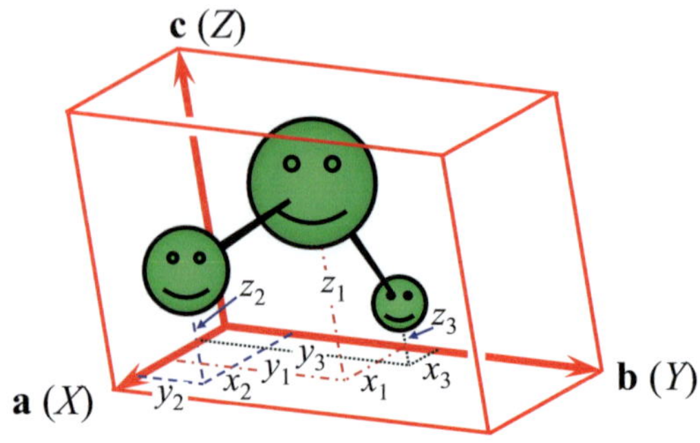

Fig. 2.1 Illustration of the content of the unit cell. The coordinates of the center of gravity of each atom are given as triplets, i.e., x_1, y_1, z_1; x_2, y_2, z_2; and x_3, y_3, z_3

atoms ("large," "medium," and "small" happy faces) have coordinates x_1, y_1, z_1; x_2, y_2, z_2; and x_3, y_3, z_3, respectively. Strictly speaking, the content of the unit cell should be described by specifying other relevant atomic parameters in addition to the position of each atom in the unit cell. These include types of atoms (i.e., their chemical symbols or sequential numbers in a periodic table instead of "large," "medium," and "small"), site occupancy, and individual displacement parameters. All of these quantities are defined and explained later in the book, see Chap. 9.

2.2 Asymmetric Part of the Unit Cell

It is important to realize that the case shown in Fig. 2.1 is rarely observed in reality. Usually, unit cell contains more than one molecule or a group of atoms that are converted into each other by simple geometrical transformations, which are called *symmetry operations*. Overall, there may be as many as 192 transformations in some highly symmetric unit cells. A simple example is shown in Fig. 2.2, where each unit cell contains two molecules that are converted into one another by 180° *rotation* around imaginary lines, which are perpendicular to the plane of the figure. The location of one of these lines (*rotation axes*) is indicated using small filled ellipse. The original molecule, chosen arbitrarily, is green, while the derived, symmetrically related molecule is blue.

The independent part of the unit cell (e.g., the upper half of the unit cell shaded in green in Fig. 2.2) is called the asymmetric unit. It is the only part of the unit cell, for which the specification of atomic positions and other atomic parameters is required. The entire content of the unit cell can be established from its asymmetric unit using the combination of symmetry operations present in the unit cell. Here, this operation is a rotation by 180° around the line perpendicular to the plane of the projection at the center of the unit cell. It is worth noting that the rotation axis shown in the upper left corner of Fig. 2.2 is not the only axis present in this crystal lattice—identical axes are

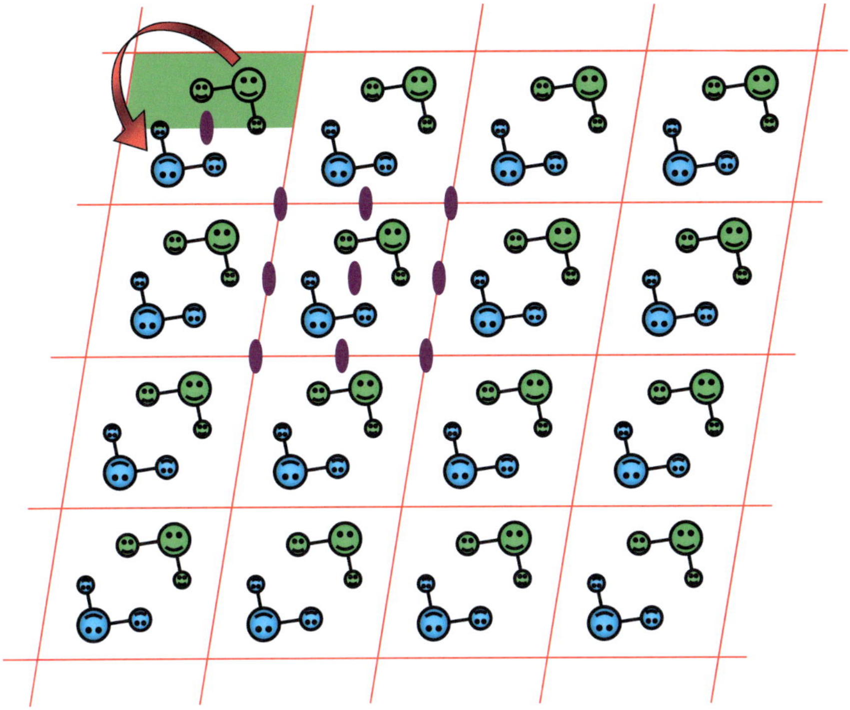

Fig. 2.2 Asymmetric unit (green parallelogram) contains an independent molecule, which is green as well. Blue molecules are related to green molecules in each unit cell via rotation by 180° around the lines perpendicular to the plane of the projection at the center of each unit cell. The difference in color is used only to highlight symmetrical relationships since all molecules are indeed identical. All rotation axes intersecting every unit cell are shown in a neighboring cell

located at the start and middle of every unit cell edge, as shown in one of the neighboring cells.[2]

Symmetry operations, therefore, can be visualized by means of certain symmetry elements represented by various graphical objects. There are four so-called simple symmetry elements: a point to visualize inversion through the point, a line for rotation about the line, a plane for reflection in the plane, and the already mentioned translation is also a simple symmetry element, which can be visualized as a vector. Simple symmetry elements may be combined with one another, producing complex symmetry elements that include roto-inversion axes, screw axes, and glide planes.

[2]The appearance of additional rotation axes in each unit cell is the result of the simultaneous presence of both rotational and translational symmetry, which interact with one another (see Sects. 2.5 and 3.3).

2.3 Symmetry Operations and Symmetry Elements

From the beginning, it is important to acknowledge that a symmetry operation is not the same as a symmetry element. The difference between the two can be defined as follows: a symmetry operation performs a certain symmetrical transformation and yields only one additional object, for example, an atom or a molecule, which is symmetrically equivalent to the original. Conversely, a symmetry element is a graphical or a geometrical representation of one or more symmetry operations, such as a mirror reflection in a plane, a rotation about an axis, or an inversion through a point. A much more comprehensive description of the term "symmetry element" exceeds the scope of this book.[3]

Without the presence of translations, a single crystallographic symmetry element may yield a total from one to six objects symmetrically equivalent to one another. For example, a rotation by 60° around an axis is a symmetry operation, whereas the sixfold rotation axis is a symmetry element which contains six rotational symmetry operations: by 60°, 120°, 180°, 240°, 300°, and 360° about the same axis. The latter is the same as rotation by 0° or any multiple of 360°. As a result, the sixfold rotation axis produces a total of six symmetrically equivalent objects counting the original. Note that the 360° rotation yields an object identical to the original and literally converts the object into itself. Hence, symmetry elements are used in visual description of symmetry operations, while symmetry operations are invaluable in the algebraic or mathematical representation of crystallographic symmetry, for example, in computing.

Four simple symmetry operations—rotation, inversion, reflection and translation—are illustrated in Fig. 2.3. Their association with the corresponding geometrical objects and symmetry elements is summarized in Table 2.1. Complex symmetry elements are shown in Table 2.2. There are three new complex symmetry elements, which are listed in *italics* in this table:

– Roto-inversion axis (usually called inversion axis), which includes simultaneous rotation and inversion through a point located on the axis.[4]

[3]It may be found in: P.M. de Wolff, Y. Billiet, J.D.H. Donnay, W. Fischer, R.B. Galiulin, A.M. Glazer, Marjorie Senechal, D.P. Schoemaker, H. Wondratchek, Th. Hahn, A.J.C. Wilson, and S.C. Abrahams, Definition of symmetry elements in space groups and point groups. Report of the International Union of Crystallography ad hoc committee on the nomenclature in symmetry, Acta Cryst. **A45**, 494 (1989) https://doi.org/10.1107/S0108767389002230; P.M. de Wolff, Y. Billiet, J.D.H. Donnay, W. Fischer, R.B. Galiulin, A.M. Glazer, Th. Hahn, M. Senechal, D.P. Schoemaker, H. Wondratchek, A.J.C. Wilson, and S.C. Abrahams, Symbols for symmetry elements and symmetry operations. Final report of the International Union of Crystallography ad-hoc committee on the nomenclature in symmetry, Acta Cryst. **A48**, 727 (1992) https://doi.org/10.1107/S0108767392003428; H.D. Flack, H. Wondratchek, Th. Hahn, and S.C. Abrahams, Symmetry elements in space groups and point groups. Addenda to two IUCr reports on the nomenclature in symmetry, Acta Cryst. **A56**, 96 (2000) https://doi.org/10.1107/S0108767399012283

[4]Alternatively, roto-reflection axes combining simultaneous rotation and reflection may be used; however, each of them is identical in its action to one of the roto-inversion axes.

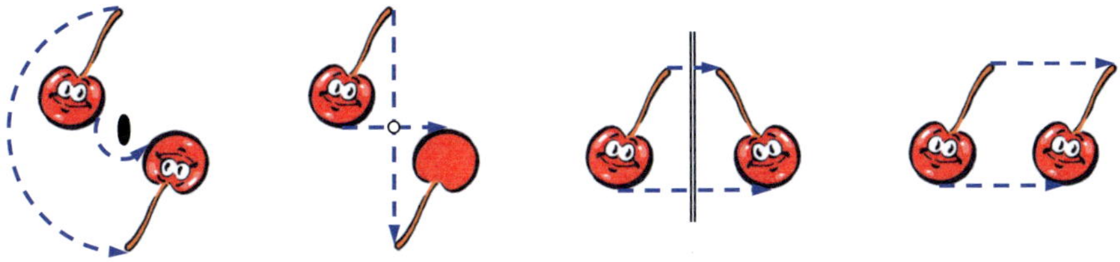

Fig. 2.3 Simple symmetry operations. From *left* to *right*: rotation, inversion, reflection, and translation

Table 2.1 Simple symmetry operations and conforming symmetry elements

Symmetry operation	Geometrical representation	Symmetry element
Rotation	Line (axis)	Rotation axis
Inversion	Point (center)	Center of inversion
Reflection	Plane	Mirror plane
Translation	Vector	Translation vector

Table 2.2 Derivation of complex symmetry elements

Symmetry operation	Rotation	Inversion	Reflection	Translation
Rotation	–	*Roto-inversion axis*[a]	No[b]	*Screw axis*
Inversion	–	–	No[b]	No[b]
Reflection	–	–	–	*Glide plane*
Translation	–	–	–	–

[a] The prefix "roto" is nearly always omitted and these axes are called "inversion axes"
[b] No new complex symmetry element is formed as a result of this combination

– Screw axis, which includes simultaneous rotation and translation along the axis.
– Glide plane, which combines reflection and translation parallel to the plane.

Symmetry operations and elements are sometimes classified by the way they transform an object as proper and improper. An improper symmetry operation inverts an object in a way that may be imaged by comparing the right and left hands: the right hand is an inverted image of the left hand, and if you have ever tried to put a right-handed leather glove on your left hand, you know that it is quite difficult, unless the glove has been turned inside out, or in other words, inverted. The inverted object is said to be enantiomorphous to the direct object and vice versa. Thus, symmetry operations and elements that involve inversion or reflection, including when they are present in complex symmetry elements, are improper. They are: center of inversion, inversion axes, mirror plane, and glide planes. On the contrary, proper symmetry elements include only operations that do not invert an object, such as rotation and translation. They are rotation axes, screw axes, and translation vectors. As is seen in Fig. 2.3, both the rotation and translation, which are proper symmetry operations, change the position of the object without inversion, whereas both the inversion and reflection, that is, improper symmetry operations, invert the object in addition to changing its location.

Another classification is based on the presence or absence of translation in a symmetry element or operation. Symmetry elements containing a translational component, such as a simple translation, screw axis, or glide plane, produce infinite numbers of symmetrically equivalent objects, and therefore, these may be called infinite symmetry elements. For example, the lattice is infinite because of the presence of translations. All other symmetry elements that do not contain translations always produce a finite number of objects, and they may be called finite symmetry elements. Center of inversion, mirror plane, rotation, and roto-inversion axes are all finite symmetry elements. Finite symmetry elements and operations are used to describe the symmetry of finite objects, for example, molecules, clusters, polyhedra, crystal forms, unit cell shape, and any noncrystallographic finite objects, for example, the human body. Both finite and infinite symmetry elements are necessary to describe the symmetry of infinite or continuous structures, such as a crystal structure, two-dimensional wall patterns, and others. We begin the analysis of crystallographic symmetry from simpler finite symmetry elements, followed by the consideration of more complex infinite symmetry elements.

2.4 Finite Symmetry Elements

Symbols of finite crystallographic symmetry elements and their graphical representations are listed in Table 2.3. The full name of a symmetry element is formed by adding "N-fold" to the words "rotation axis" or "inversion axis." The numeral N generally corresponds to the total number of objects generated by the element,[5] and

Table 2.3 Symbols of finite crystallographic symmetry elements

	Rotation axes		Roto-inversion axes	
Rotation angle, φ	International symbol	Graphical symbol[a]	International symbol	Graphical symbol[a]
360°	1	None	$\bar{1}$[b]	⬮
180°	2	○	$\bar{2} = m$[c]	▲
120°	3	═	$\bar{3} = 3 + \bar{1}$	■
90°	4	▲	$\bar{4}$	⬢
60°	6	◪	$\bar{6} = 3 + m\perp 3$	⬡

[a] When the symmetry element is perpendicular to the plane of the projection
[b] Identical to the center of inversion
[c] Identical to the mirror plane

[5]Except for the center of inversion, which results in two objects, and the threefold inversion axis, which produces six symmetrically equivalent objects. See Eqs. (4.27) and (4.28) in Sect. 4.2.4 for an algebraic definition of the order of a symmetry element.

Fig. 2.4 From *left* to *right*: horizontal twofold rotation axis (*top*) and its alternative symbol (*bottom*), diagonal threefold inversion axis inclined to the plane of the projection, horizontal fourfold rotation axis, and horizontal and diagonal mirror planes. *Horizontal* or *vertical lines* are commonly used to indicate axes located in the plane of the projection, and *diagonal lines* are used to indicate axes, which form an angle other than the right angle or zero with the plane of the projection

it is also known as the order or the multiplicity of the symmetry element. Orders of axes are listed in columns two and four in Table 2.3, such as a threefold rotation axis or a fourfold inversion axis.

Note that the onefold inversion axis and the twofold inversion axis are identical in their action to the center of inversion and the mirror plane, respectively. Both the center of inversion and mirror plane are commonly used in crystallography, mostly because they are described by simple geometrical elements: point or plane, respectively. The center of inversion is also often called the "center of symmetry."

Further, as we will see in Sects. 2.4.3 and 2.4.5, transformations performed by the threefold inversion and the sixfold inversion axes can be represented by two independent simple symmetry elements. In the case of the threefold inversion axis, $\bar{3}$, these are the threefold rotation axis and the center of inversion acting independently, and in the case of the sixfold inversion axis, $\bar{6}$, the two independent symmetry elements are the mirror plane and the threefold rotation axis perpendicular to the plane, as denoted in Table 2.3. The remaining fourfold inversion axis, $\bar{4}$, is a unique symmetry element (Sect. 2.4.4), which cannot be represented by any pair of independently acting symmetry elements.

Numerals in the international symbols of the center of inversion and all inversion axes are conventionally marked with the bar on top and not with the dash or the minus sign in front of the numeral (see Table 2.3). The dash preceding the numeral (or the letter "b" following the numeral—shorthand for "bar"), however, is more convenient to use in computing for the input of symmetry data, for example, -1 (or 1b), -3 (3b), -4 (4b), and -6 (6b) rather than $\bar{1}, \bar{3}, \bar{4}$, and $\bar{6}$, respectively.

The columns labeled "Graphical symbol" in Table 2.3 correspond to graphical representations of symmetry elements when they are perpendicular to the plane of the projection. Other orientations of rotation and inversion axes are conventionally indicated using the same symbols to designate the order of the axis with properly oriented lines, as shown in Fig. 2.4. Horizontal and diagonal mirror planes are normally labeled using bold lines as shown in Fig. 2.4 or using double lines in stereographic projections (see Table 2.3 and Sect. 2.8).

When we began our discussion of crystallographic symmetry, we used a happy face and a cherry to illustrate simple concepts of symmetry. These objects are inconvenient to use with complex symmetry elements. Conversely, the commonly

Fig. 2.5 Trigonal pyramid with its apex up (*left*) and down (*right*) relative to the plane of the paper. Different shades of blue are used to emphasize enantiomorphous objects

used empty circles with or without a comma inside to indicate enantiomorphous objects, for example, as in the International Tables for Crystallography,[6] are not intuitive. For example, both inversion and reflection look quite similar. Therefore, we will use a trigonal pyramid, shown in Fig. 2.5. This figure illustrates two pyramids, one with its apex facing upward, where lines connect the visible apex with the base corners, and another with its apex facing downward, which has no visible lines. In addition, the pyramids are colored differently to accentuate the enantiomorphism of the two pyramids.

To review symmetry elements in detail, we must find out more about rotational symmetry, since both the center of inversion and mirror plane can be represented as rotation plus inversion (see Table 2.3). The important properties of rotational symmetry are the direction of the axis and the rotation angle. It is almost intuitive that the rotation angle (φ) can only be an integer fraction (1/N) of a full turn (360°); otherwise, it can be substituted by a different rotation angle that is an integer fraction of the full turn, or it will result in the noncrystallographic rotational symmetry. Hence,

$$\varphi = \frac{360^\circ}{\mathrm{N}} \tag{2.1}$$

By comparing Eq. (2.1) with Table 2.3, it is easy to see that N, which is the order of the axis, is also the number of elementary rotations required to accomplish a full turn around the axis. In principle, N can be any integer number, for example, 1, 2, 3, 4, 5, 6, 7, 8.... However, in periodic crystals only a few specific values are allowed for N due to the presence of translational symmetry. Only axes with N = 1, 2, 3, 4, or 6 are compatible with the periodic crystal lattice, that is, with translational symmetry in three dimensions. Other orders, such as 5, 7, 8, and higher inevitably result in the loss of the conventional periodicity of the lattice, which is defined by Eq. (1.1). The not so distant discovery of fivefold and tenfold rotational symmetry continues to intrigue scientists even today, since it is quite clear that it is impossible to build a periodic crystalline lattice in two dimensions exclusively from pentagons, as depicted in Fig. 2.6, heptagons, octagons, etc. The situation shown in this figure

[6]International Tables for Crystallography, Vol. A, Space-group symmetry, M. I. Aroyo, Ed., Second online edition, (2016). https://it.iucr.org/A/

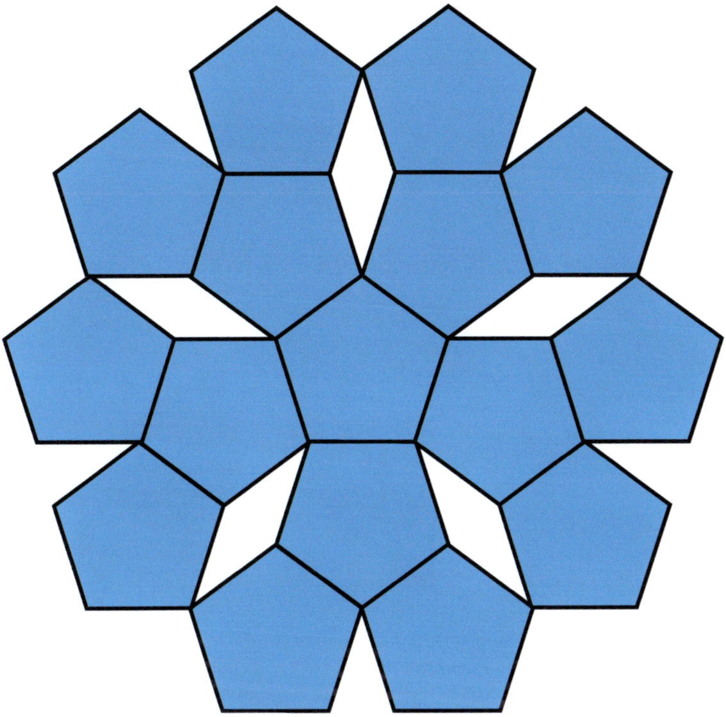

Fig. 2.6 Filling the area with *blue pentagons*. *White parallelograms* represent voids in the two-dimensional pattern of pentagons

may be rephrased as follows: "It is impossible to completely fill the area in two dimensions with pentagons without creating gaps."

It is worth noting that the structure in Fig. 2.6 not only looks ordered, but it is indeed perfectly ordered. Moreover, in recent decades, many crystals with fivefold symmetry have been found and their approximant structures have been determined with various degrees of accuracy. These crystals, however, do not have translational symmetry in three directions, which means that they do not have a finite unit cell and, therefore, they are called quasicrystals: quasi, because there is no translational symmetry; crystals, because they produce discrete, crystal-like diffraction patterns.

2.4.1 Onefold Rotation Axis and Center of Inversion

The onefold rotation axis, shown in Fig. 2.7 on the left, rotates an object by 360°, or in other words converts any object into itself, which is the same as if no symmetrical transformation had been performed. This is the only symmetry element, which does not generate additional objects except the original.

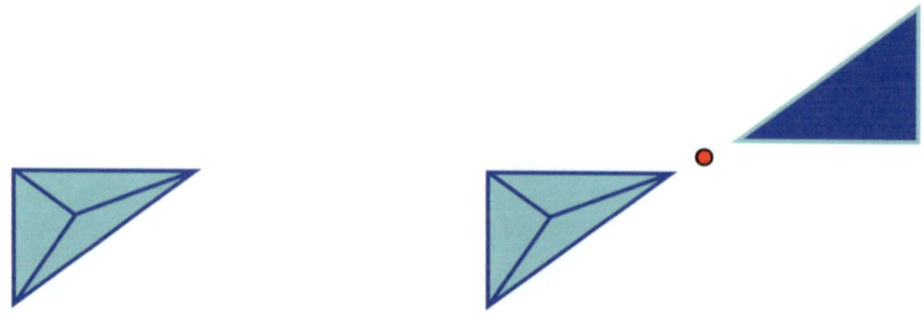

Fig. 2.7 Onefold rotation axis (*left*, unmarked since it can be located anywhere) and center of inversion (*right*)

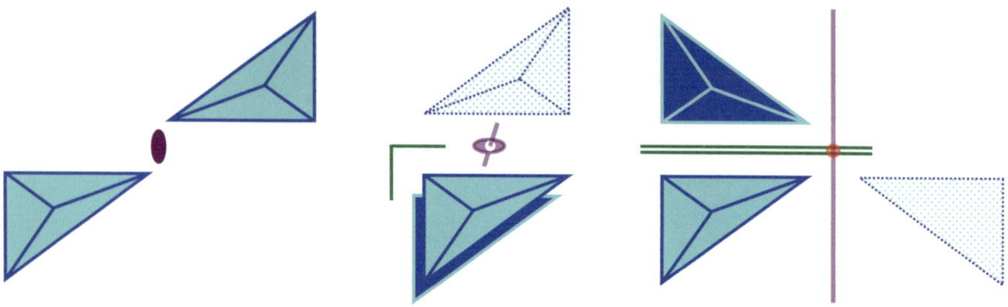

Fig. 2.8 Twofold rotation axis perpendicular to the plane of the projection (*left*) and mirror planes (*middle* and *right*). In the middle—the mirror plane nearly coincides with the plane of the projection (the equivalent twofold inversion axis is tilted by a few degrees away from the vertical) for clarity. On the right—the mirror plane is perpendicular to the plane of the projection. Also shown in the middle and on the right is how the twofold inversion axis, which is perpendicular to the mirror plane, yields the same result as the mirror plane

The center of inversion (onefold inversion axis) inverts an object through a point as shown in Fig. 2.7, right. Thus, the light-blue pyramid with its apex up, which is the original object, is inverted through a point producing its symmetrical equivalent—the dark-blue (enantiomorphous) pyramid with its apex down. The latter is converted back into the original pyramid after the inversion through the same point. The center of inversion, therefore, generates one additional object, giving a total of two related objects.

2.4.2 Twofold Rotation Axis and Mirror Plane

The twofold rotation axis (Fig. 2.8, left) simply rotates an object around the axis by 180°, and this symmetry element results in two symmetrically equivalent objects: original plus transformed. Note that the 180° rotation of the new pyramid around the same axis converts it to the original pyramid. Hence, it is correct to state that the twofold rotation axis rotates the object by 0 (360°) and 180°.

The mirror plane (twofold inversion axis) reflects a light-blue pyramid in a plane to yield the dark-blue pyramid as shown in Fig. 2.8 in the middle and on the right.

Similar to the inversion center and the twofold rotation axis, the same mirror plane reflects the resulting (dark-blue) pyramid yielding the original (light-blue) pyramid. The equivalent symmetry element, that is, the twofold inversion axis first rotates an object (light-blue pyramid) by 180° around the axis as shown by the lighter colored image of a pyramid with its apex up in the middle or apex down on the right of Fig. 2.8. The pyramid does not remain in this position because the rotation is combined with the center of inversion, and the pyramid is immediately (or simultaneously) inverted through the center of inversion located on the axis. The final locations are shown by the dark-blue pyramids in Fig. 2.8. The mirror plane is used to describe this combined operation rather than the twofold inversion axis because of its simplicity and a better graphical representation of the reflection operation versus the roto-inversion. Similar to the twofold rotation axis, the mirror plane results in two symmetrically equivalent objects.

2.4.3 Threefold Rotation Axis and Threefold Inversion Axis

The threefold rotation axis (Fig. 2.9, left) results in three symmetrically equivalent objects by rotating the original object around the axis by 0 (360°), 120°, and 240°.

The threefold inversion axis (Fig. 2.9, right) produces six symmetrically equivalent objects. The original object, for example, any of the three light-blue pyramids with apexes up, is transformed as follows: it is rotated by 120° counterclockwise and then immediately inverted from this intermediate position through the center of inversion located on the axis as shown by the dashed arrows in Fig. 2.9. This

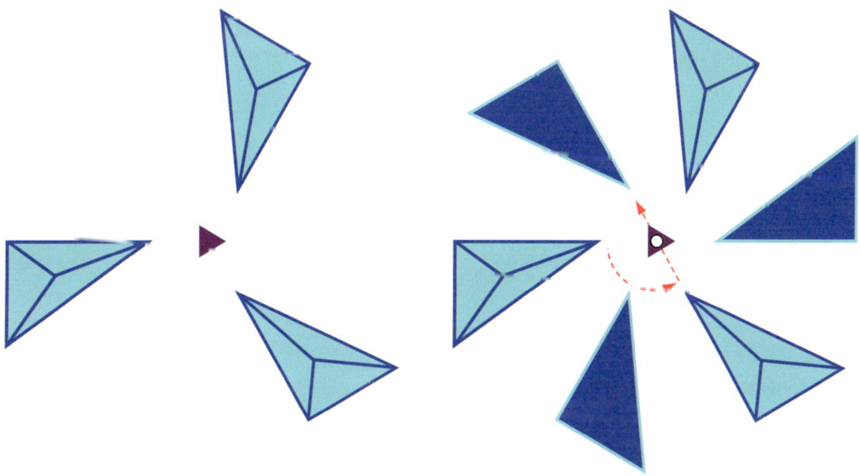

Fig. 2.9 Threefold rotation (*left*) and threefold inversion (*right*) axes perpendicular to the plane of the projection. The *dashed arrows* on the right schematically show the counterclockwise rotation by 120° and a simultaneous inversion through the center of symmetry located on the axis

operation results in a dark-blue pyramid with its apex down positioned 60° clockwise from the original pyramid. By applying the same transformation to this dark-blue pyramid, the third symmetrically equivalent object would be a light-blue pyramid next to the first dark-blue pyramid, rotated by 60° clockwise. These transformations are carried out until the next obtained object repeats the original pyramid.

It is easy to see that the six symmetrically equivalent objects are related to one another by an independent threefold rotation axis that connects the three light-blue pyramids, as well as the three dark-blue pyramids, and by an independent center of inversion, which relates the pairs of opposite pyramids. Hence, the threefold inversion axis is not only the result of two simultaneous operations (3 then $\overline{1}$), but the same symmetrical relationships can be established as a result of two symmetry elements present independently. In other words, $\overline{3}$ is identical to 3 and $\overline{1}$.

2.4.4 Fourfold Rotation Axis and Fourfold Inversion Axis

The fourfold rotation axis (Fig. 2.10, left) results in four symmetrically equivalent objects by rotating the original object around the axis by 0 (360°), 90°, 180°, and 270°.

The fourfold inversion axis (Fig. 2.10, right) also produces four symmetrically equivalent objects. The original object, for example, any of the two light-blue pyramids with apexes up, is rotated by 90° counterclockwise and then it is immediately inverted from this intermediate position through the center of inversion located on the axis. This transformation results in a dark-blue pyramid with its apex down in the position next to the original pyramid clockwise. By applying the same transformation to this dark-blue pyramid, the third symmetrically equivalent object would be a light-blue pyramid next to the dark-blue pyramid clockwise. The fourth object is obtained in the same fashion. Unlike in the case of the threefold

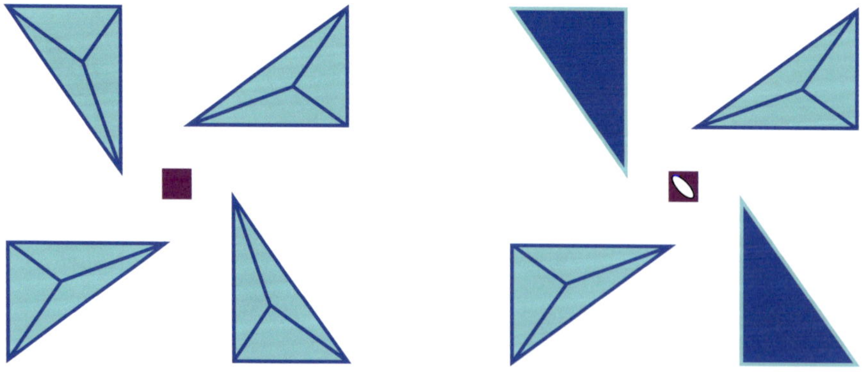

Fig. 2.10 Fourfold rotation (*left*) and fourfold inversion (*right*) axes perpendicular to the plane of the projection

inversion axis (see Sect. 2.4.3), this combination of four objects cannot be produced by applying the fourfold rotation axis and the center of inversion separately, and therefore, this is a unique symmetry element. In fact, the combination of four pyramids shown in Fig. 2.10 (right) does not have an independent fourfold symmetry axis, nor does it have the center of inversion! As can be seen from Fig. 2.10, both fourfold axes contain a twofold rotation axis (180° rotations) as a subelement.

2.4.5 *Sixfold Rotation Axis and Sixfold Inversion Axis*

The sixfold rotation axis (Fig. 2.11, left) results in six symmetrically equivalent objects by rotating the original object around the axis by 0 (360°), 60°, 120°, 180°, 240°, and 300°.

The sixfold inversion axis (Fig. 2.11, right) also produces six symmetrically equivalent objects. Similar to the threefold inversion axis, this symmetry element can be represented by two independent simple symmetry elements: the first one is the threefold rotation axis, which connects pyramids 1-3-5 and 2-4-6, and the second one is the mirror plane perpendicular to the threefold rotation axis, which connects pyramids 1-4, 2-5, and 3-6. As an exercise, try to obtain all six symmetrically equivalent pyramids starting from the pyramid 1 as the original object by applying 60° rotations followed by immediate inversions. Keep in mind that objects are not retained in the intermediate positions because the sixfold rotation and inversion act simultaneously.

The sixfold rotation axis also includes one threefold and one twofold rotation axis, while the sixfold inversion axis includes a threefold rotation and a twofold

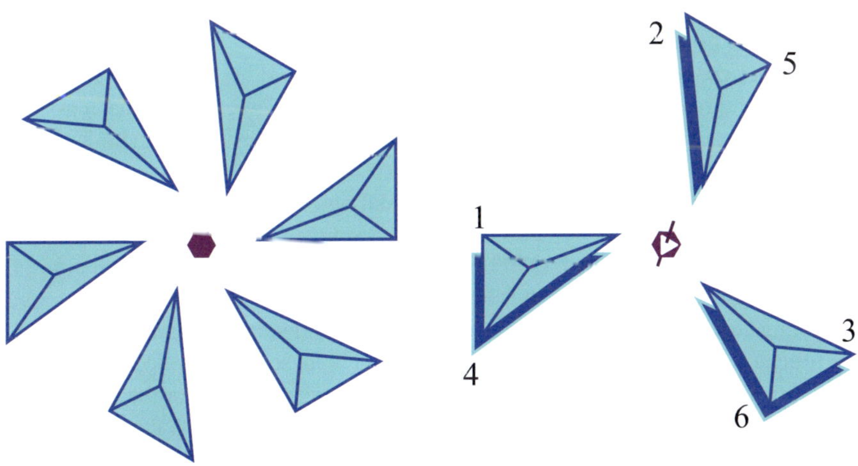

Fig. 2.11 Sixfold rotation (*left*) and sixfold inversion (*right*) axes. The sixfold inversion axis is tilted by a few degrees away from the vertical to visualize all six symmetrically equivalent pyramids. The numbers next to the pyramids represent the original object (1), the first generated object (2), and so on. The odd numbers are for the pyramids with their apexes up

inversion (mirror plane) axis as subelements. Thus, any N-fold symmetry axis with $N > 1$ always includes either rotation or inversion axes of lower orders, which are integer divisors of N.

2.5 Interaction of Symmetry Elements

So far we have considered a total of ten different crystallographic symmetry elements, some of which were combinations of two simple symmetry elements, acting either simultaneously or consecutively. The majority of crystalline objects, for example, crystals and molecules, have more than one nonidentity symmetry element.

Symmetry elements and operations interact with one another producing new symmetry elements and symmetry operations, respectively. When applied to symmetry, an interaction means consecutive (and not simultaneous, as in the case of complex symmetry elements) application of symmetry elements. The appearance of new, that is, derivative, symmetry operations can be understood from a simple deduction, using the fact that a single symmetry operation produces only one new object:

- Assume that symmetry operation No. 1 converts object X into object X_1.
- Assume that another symmetry operation, No. 2, converts object X_1 into object X_2.
- Since object X_1 is symmetrically equivalent to object X, and object X_2 is symmetrically equivalent to object X_1, then objects X and X_2 should also be related to one another.

The question is: what converts object X into object X_2? The only logical answer is: there should be an additional symmetry operation, No. 3, that converts object X into object X_2.

Consider the schematic shown in Fig. 2.12 (left), and assume that initially we have only the twofold rotation axis, 2, and the center of inversion, $\bar{1}$. Also assume that the center of inversion is located on the axis (if not, translational symmetry will result, see Sects. 3.1 and 3.2).

Beginning with the pyramid A as the original object, and after rotating it around the axis by 180° we obtain pyramid B, which is symmetrically equivalent to pyramid A. Since we also have the center of inversion, it converts pyramid A into pyramid D and pyramid B into pyramid C. It is easy to see from Fig. 2.12 (right) that pyramid C is nothing else but the reflected image of pyramid A and vice versa, and pyramid D is the reflected copy of pyramid B. Remembering that these mirror reflection relationships between A and C, and B and D were not present from the beginning, we conclude that a new symmetry element—a mirror plane, m—has emerged as the result of the sequential application of two symmetry elements, that is, 2 and $\bar{1}$, to the original object.

The mirror plane is, therefore, a derivative of the twofold rotation axis and the center of inversion located on the axis. The derivative mirror plane is perpendicular

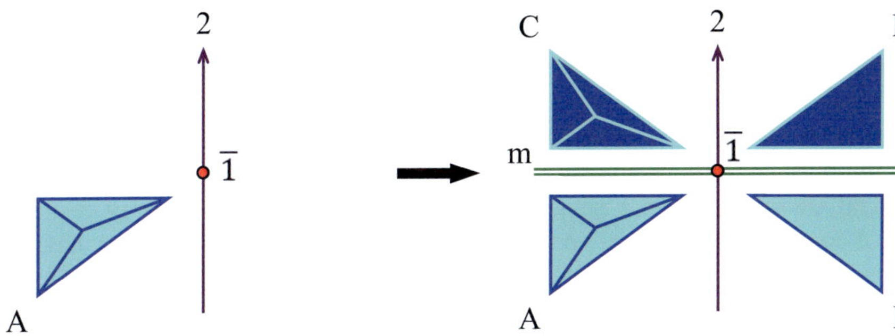

Fig. 2.12 Schematic illustrating the interaction of symmetry elements. A twofold rotation axis (2) and a center of inversion ($\bar{1}$) located on the axis (*left*) result in a mirror plane perpendicular to the axis intersecting it at the center of inversion (*right*). The important difference from Fig. 2.8 (*middle* and *right*), where neither the twofold axis nor the center of inversion is present independently, is that the combination of four pyramids (A, B, C, and D) here includes both of these symmetry elements

to the axis and intersects the axis in a way that the center of inversion also belongs to the plane. If we start from the same pyramid A and apply the center of inversion first (this results in pyramid D) and the twofold axis second (i.e., A → B and D → C), the resulting combination of four symmetrically equivalent objects and the derivative mirror plane remain the same.

This example not only explains how the two symmetry elements interact, but it also serves as an illustration to a broader conclusion deduced at the beginning of this section: any two symmetry operations applied in sequence to the same object create a third symmetry operation, which applies to all symmetrically equivalent objects. Note that if the second operation is the inverse of the first, then the resulting third operation is identity (the onefold rotation axis, 1). For example, when a mirror plane, a center of inversion, or a twofold rotation axis are applied twice, all result in a onefold rotation axis.

The example considered in Fig. 2.12 can be also written in a form of an equation using the international notations of the corresponding symmetry elements (see Table 2.3):

$$2 \times \bar{1}(\text{on } 2) = \bar{1}(\text{on } 2) \times 2 = \text{m}\left(\perp 2 \text{ through } \bar{1}\right) \tag{2.2}$$

where "×" designates the interaction between (successive application of) symmetry elements. The same example (Fig. 2.12) can be considered starting from any two of the three symmetry elements. As a result, the following equations are also valid:

$$2 \times \text{m}\ (\perp 2) = \text{m}\ (\perp 2) \times 2 = \bar{1}\ (\text{at m} \perp 2) \tag{2.3}$$

$$\text{m} \times \bar{1}\ (\text{on m}) = \bar{1}\ (\text{on m}) \times \text{m} = 2\ \left(\perp \text{m through } \bar{1}\right) \tag{2.4}$$

2.5.1 Generalization of Interactions Between Finite Symmetry Elements

In the earlier examples (Fig. 2.12 and Table 2.5), the twofold rotation axis and the mirror plane are perpendicular to one another. However, symmetry elements may in general intersect at various angles (ϕ). When crystallographic symmetry elements are of concern, and since only one-, two-, three-, four- and sixfold rotation axes are allowed, only a few specific angles ϕ are possible. In most cases, they are: 0° (e.g., when an axis belongs to a plane), 30°, 45°, 60°, and 90°. The latter means that symmetry elements are mutually perpendicular. Furthermore, all symmetry elements should intersect along the same line or at the same point; otherwise, a translation occurs, resulting in infinite symmetry.

An example showing that multiple symmetry elements appear when a twofold rotation axis intersects with a mirror plane at a 45° angle is seen in Fig. 2.13. All eight pyramids can be obtained starting from a single pyramid by applying the two symmetry elements (i.e., the mirror plane and the twofold rotation axis), first to the original pyramid and second to the pyramids that appear as a result of symmetrical transformations. As an exercise, try to obtain all eight pyramids beginning from a selected pyramid using only the mirror plane and the twofold axis that are shown in Fig. 2.13 (left). Hints: original pyramid (1), rotate it (2), reflect both (4), rotate all (6), and reflect all (8). Numbers in parenthesis indicate the total number of different pyramids that should be present in the figure after each symmetrical transformation.

So far we have enough evidence that when two symmetry elements interact, they result in additional symmetry element(s). Moreover, when three symmetry elements interact, they will also produce derivative symmetry elements. For example, three mutually perpendicular mirror planes yield a center of inversion in a point, which is common for all three planes, plus three twofold rotation axes along the lines where

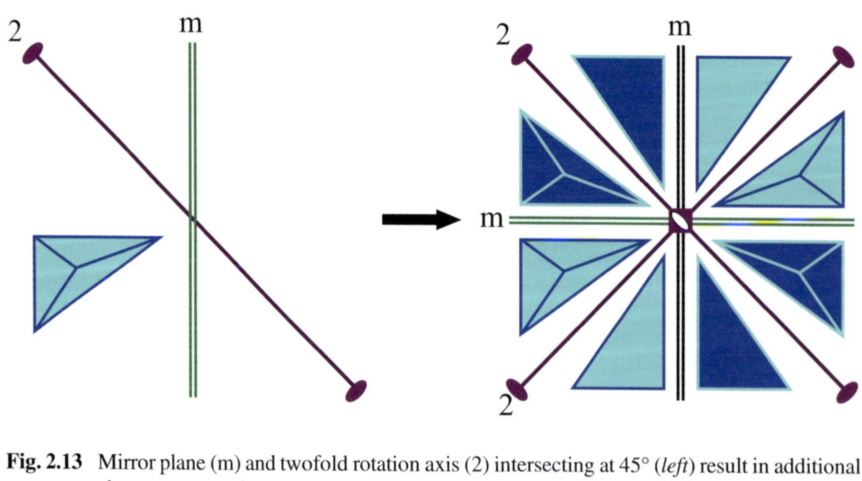

Fig. 2.13 Mirror plane (m) and twofold rotation axis (2) intersecting at 45° (*left*) result in additional symmetry elements: two mirror planes, twofold rotation axis, and fourfold inversion axis (*right*)

Table 2.4 Typical interactions between finite symmetry elements

First element	Second element	Derived element (major)	Comments, examples
$\bar{1}$	N-fold axis	m for even N	$\bar{2}$ = m
		N-fold inversion axis for odd N	$\bar{3}$
2	2 at $\phi = 30$, 45, 60, or 90°	N-fold rotation axis, N = 180/ϕ	6, 4, 3, or 2 perpendicular to first and second axes
m	m at $\phi = 30$, 45, 60, or 90°	Same as above	6, 4, 3, or 2 along the common line
m	2 at 90°	Center of inversion	$\bar{1}$ where m and 2 intersect
m	2 at $\phi = 30$, 45, 60°	N-fold inversion axis, N = 180/(90-ϕ)	$\bar{3}$, $\bar{4}$, or $\bar{6}$ in m and perpendicular to 2
3 or $\bar{3}$	2, 4, or $\bar{4}$ at 54.74°; 3 or 3 at 70.53°	Four intersecting 3 or $\bar{3}$ plus other symmetry elements	Symmetry of a cube or tetrahedron

Table 2.5 Symmetry elements resulting from all possible combinations of 1, $\bar{1}$, 2, and m, when 2 is perpendicular to m, and $\bar{1}$ is located at the intersection of 2 and m

Symmetry operation	**1**	**$\bar{1}$**	**2**	**m**
1	1	$\bar{1}$	2	m
$\bar{1}$	1	1	m	2
2	2	m	1	$\bar{1}$
m	m	2	$\bar{1}$	1

any two of the planes intersect. However, all cases when more than two elements interact with one another can be reduced to the interactions of pairs. The most typical interactions of the pairs of symmetry elements and their results are shown in Table 2.4.

2.5.2 *Symmetry Groups*

As established earlier, the interaction between a pair of symmetry elements (or symmetry operations) results in another symmetry element (or operation). The latter may be new or it may already be present in a given combination of symmetrically equivalent objects. If no new symmetry element(s) appear, and when interactions between all pairs of the existing ones are examined, the generation of all symmetry elements is completed. The complete set of symmetry elements is called a symmetry group.

Table 2.5 illustrates the generation of a simple symmetry group using symmetry elements from Fig. 2.12. The only difference is that in Table 2.5, a onefold rotation axis has been added to the earlier considered twofold rotation axis, center of inversion, and mirror plane for completeness. It is easy to see that no new symmetry elements appear when interactions between all four symmetry elements have been accounted for.

Considering only finite symmetry elements and all valid combinations among them, a total of 32 crystallographic symmetry groups can be constructed. The 32 symmetry groups can be derived in a number of ways, one of which has been illustrated in Table 2.5, but this subject falls beyond the scope of this book. Nevertheless, the family of finite crystallographic symmetry groups, which are also known as the 32 point groups, is briefly discussed in Sect. 2.9.

2.6 Fundamentals of Group Theory

Since the interaction of two crystallographic symmetry elements results in a third crystallographic symmetry element, and the total number of them is finite, valid combinations of symmetry elements can be assembled into finite groups. As a result, mathematical theory of groups is fully applicable to crystallographic symmetry groups.

The definition of a group is quite simple: a group is a set of elements G_1, G_2, …, G_N, …, for which a binary combination law is defined, and which together satisfy the four fundamental properties: closure, associativity, the identity property, and the inverse property. Binary combination law (a few examples are shown at the end of this section) describes how any two elements of a group combine (interact) with one other. When a group contains a finite number of elements (N), it represents a finite group, and when the number of elements in a group is infinite, then the group is infinite. All crystallographic groups composed from finite symmetry elements are finite, that is, they contain a limited number of symmetry elements.

The four properties of a group are: closure, associability, identity, and inversion. They can be defined as follows:

- *Closure* requires that the combination of any two elements, which belong to a group, is also an element of the same group:

$$G_i \times G_j = G_k$$

Note that here and below "×" designates a generic binary combination law and not multiplication. For example, applied to symmetry groups the combination law (×) is the interaction of symmetry elements; in other words, it is their sequential application, as has been described in Sect. 2.5. For groups containing numerical elements, the combination law can be defined as, for example, addition or multiplication. Every group must always be closed, even a group which contains an infinite number of elements.

- *Associativity* requires that the associative law is valid, that is,

$$\left(G_i \times G_j\right) \times G_k = G_i \times \left(G_j \times G_k\right)$$

As established earlier, the associative law holds for symmetry groups. Returning to the example in Fig. 2.12, which includes the mirror plane, the twofold rotation axis, the center of inversion, and onefold rotation axis (the latter symmetry element is not shown in the figure and we did not discuss its presence explicitly, but it is always there), the resulting combination of symmetrically equivalent objects is the same, regardless of the order in which these four symmetry elements are applied. Another example to consider is a group formed by numerical elements with addition as the combination law. For this group, the associative law always holds because the result of adding three numbers is always identical, regardless of the order in which the sum was calculated.

– *Identity* requires that there is one and only one element, E (identity or unity), in a group, such that

$$\mathrm{E} \times \mathrm{G_i} = \mathrm{G_i} \times \mathrm{E} = \mathrm{G_i}$$

for every element of the group. Crystallographic symmetry groups have the identity element, which is the onefold rotation axis—it always converts an object into itself and its interaction with any symmetry element produces the same symmetry element (e.g., see Table 2.5). Further, this is the only symmetry element which can be considered as identity. In a group formed by numerical elements with addition as the combination law, the identity element is 0, and if multiplication is chosen as the combination law, the identity element is 1.

– *Inversion* requires that each element in a group has one and only one inverse element such that

$$G_{\mathrm{i}}^{-1} \times \mathrm{G_i} = \mathrm{G_i} \times G_{\mathrm{i}}^{-1} = \mathrm{E}$$

As far as symmetry groups are concerned, the inversion rule also holds since the inverse of any symmetry element is the same symmetry element applied twice, for example, as in the case of the center of inversion, mirror plane, and twofold rotation axis, or the same rotation applied in the opposite direction, as in the case of any rotation axis of the third order or higher. In a numerical group with addition as the combination law, the inverse element would be the element which has the sign opposite to the selected element, that is, $\mathrm{M} + (-\mathrm{M}) = (-\mathrm{M}) + \mathrm{M} = 0$ (identity), while when the combination law is multiplication, the inverse element is the inverse of the selected element, or $\mathrm{MM}^{-1} = \mathrm{M}^{-1}\mathrm{M} = 1$ (identity).

It may be useful to illustrate how the rules defined here can be used to establish whether a certain combination of elements forms a group or not. The first two examples are noncrystallographic, while the third represents a simple crystallographic group.

1. Consider an integer number 1 and multiplication as the combination law. Since there are no limitations on the number of elements in a group, then a group may consist of a single element. Is this group closed? Yes, $1 \times 1 = 1$. Is the associative rule applicable? Yes, since $1 \times 1 = 1$ no matter in which order you multiply the two ones. Is there one and only one identity element? Yes, it is 1, since $1 \times 1 = 1$.

Is there one and only one inverse element for each element of the group? Yes, because $1 \times 1 = 1$. Hence, this is a group. It is a finite group.

2. Consider all integer numbers (…−3, −2, −1, 0, 1, 2, 3…) with addition as the combination law. Is this group closed? Yes, since a sum of any two integers is also an integer. How about associativity? Yes, since the result of adding three integers is always identical, regardless of the order in which they were added to one another. Is there a single identity element? Yes, this group has one, and only one identity element, 0, since adding 0 to any integer results in the same integer. Is there one and only one inverse element for any of the elements in the group? Yes, for any positive M, the inverse is –M; for any negative M, the inverse is +M, since $M + (-M) = (-M) + M = 0$ (identity). Hence, this is a group. Since the number of elements in the group is infinite, this group is infinite.
3. Consider the combination of symmetry elements shown in Fig. 2.12. The combination law here has been defined as interaction of symmetry elements (or their consecutive application to the object). The group contains the following symmetry elements: 1, $\bar{1}$, 2, and m. Associativity, identity, and inversion properties have been established earlier, when we were considering group rules. Is this group closed? Yes, it is closed as shown in Table 2.5. Therefore, these four symmetry elements form a group as well. This group is finite.

2.7 Crystal Systems

As described earlier, the number of finite crystallographic symmetry elements is limited to a total of ten. These symmetry elements can intersect with one another only at certain angles, and the number of these angles is also limited (e.g., see Table 2.4). The limited number of symmetry elements and the ways in which they may interact with each other leads to a limited number of the completed (i.e., closed) sets of symmetry elements—symmetry groups. When only finite crystallographic symmetry elements are considered, the symmetry groups are called point groups. The word “point” is used because symmetry elements in these groups have at least one common point and, as a result, they leave at least one point of an object invariant to any and all symmetry transformations.

The combination of crystallographic symmetry elements and their orientations with respect to one another in a group defines the crystallographic axes, that is, establishes the coordinate system used in crystallography. Although in general a crystallographic coordinate system can be chosen arbitrarily (e.g., see Fig. 1.3), to keep things simple and standard the axes are chosen with respect to the orientation of specific symmetry elements present in a group. Usually, the crystallographic axes are chosen to be parallel to rotation axes or perpendicular to mirror planes. This choice simplifies both the mathematical and geometrical descriptions of symmetry elements and, therefore, the symmetry of a crystal in general.

As a result, all possible three-dimensional crystallographic point groups have been divided into a total of seven crystal systems, based on the presence of specific

Table 2.6 Seven crystal systems and the corresponding characteristic symmetry elements

Crystal system	Characteristic symmetry element or combination of symmetry elements
Triclinic	No axes other than onefold rotation or onefold inversion
Monoclinic	Unique twofold axis and/or single mirror plane
Orthorhombic	Three mutually perpendicular twofold axes, either rotation or inversion
Trigonal	Unique threefold axis, either rotation or inversion
Tetragonal	Unique fourfold axis, either rotation or inversion
Hexagonal	Unique sixfold axis, either rotation or inversion
Cubic	Four threefold axes, either rotation or inversion, along four body diagonals of a cube

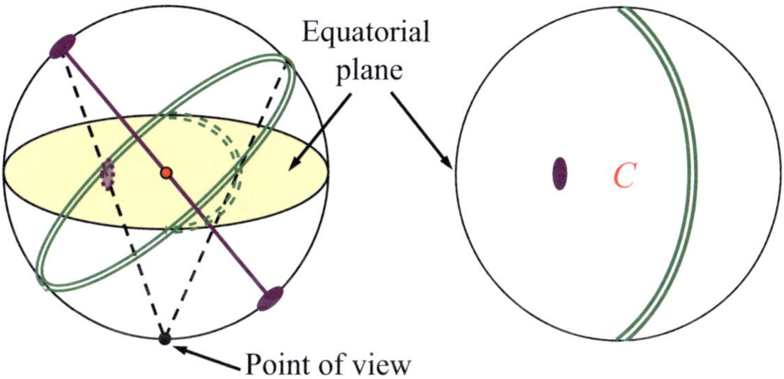

Fig. 2.14 The schematic of how to construct a stereographic projection. The location of the center of inversion is indicated using letter *C* in the middle of the stereographic projection

symmetry element or particular combination of symmetry elements present in the point group. The seven crystal systems are listed in Table 2.6.

2.8 Stereographic Projection

All symmetry elements that belong to any of the three-dimensional point groups can be easily depicted in two dimensions by using the so-called stereographic projections. The visualization is achieved similar to projections of northern or southern hemispheres of the globe in geography. Stereographic projections are constructed as follows:

- A sphere with a center that coincides with the point (if any) where all symmetry elements intersect (Fig. 2.14, left) is created. If there is no such common point, then the selection of the center of the sphere is random as long as it is located on one of the characteristic symmetry elements (see Table 2.6).
- This sphere is split by the equatorial plane into the upper and lower hemispheres.

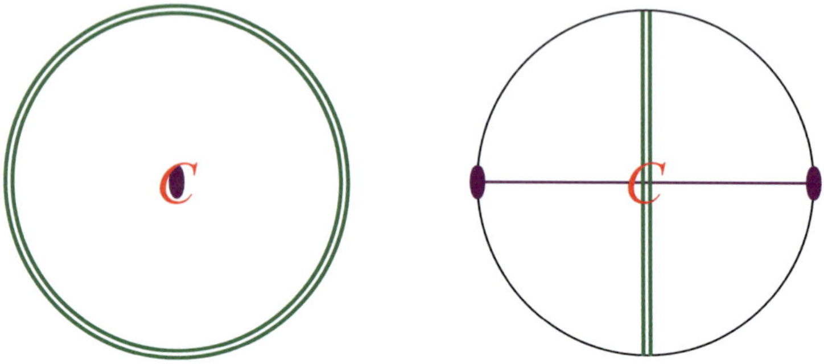

Fig. 2.15 The two conventional stereographic projections of the point-group symmetry containing a twofold axis, a mirror plane, and a center of inversion. The onefold rotation is not shown

- The lines corresponding to the intersections of mirror planes and the points corresponding to the intersections of rotation axes with the upper ("northern") hemisphere are projected on the equatorial plane using the lower ("southern") pole as the point of view.
- The projected lines and points are labeled using appropriate symbols (see Table 2.3 and Fig. 2.4).
- The presence of the center of inversion, if any, is shown by adding letter *C* to the center of the projection.

Figure 2.14 (right) shows an arbitrary stereographic projection of the point-group symmetry formed by the following symmetry elements: twofold rotation axis, mirror plane, and center of inversion (compare it with Fig. 2.12, which shows the same symmetry elements without the stereographic projection). The presence of onefold rotation axis is never indicated on the stereographic projection.

Arbitrary orientations are inconvenient because the same point-group symmetry results in an infinite number of possible stereographic projections. Thus, Fig. 2.15 shows two different stereographic projections of the same point-group symmetry with the horizontal (left) and vertical (right) orientations of the plane, both of which are standard.

Figure 2.16 (left) is an example of the stereographic projection of a tetragonal point-group symmetry containing symmetry elements discussed earlier (see Fig. 2.13). Figure 2.16 (right) shows the most complex cubic point-group symmetry containing three mutually perpendicular fourfold rotation axes, four threefold rotation axes located along the body diagonals of a cube, six twofold rotation axes, nine mirror planes, and a center of inversion. More information about the stereographic projection can be found in the International Union of Crystallography (IUCr) teaching pamphlets[7] and in the International Tables for Crystallography, Vol. A.

[7] E. J. W. Whittaker, The stereographic projection https://www.iucr.org/education/pamphlets/11

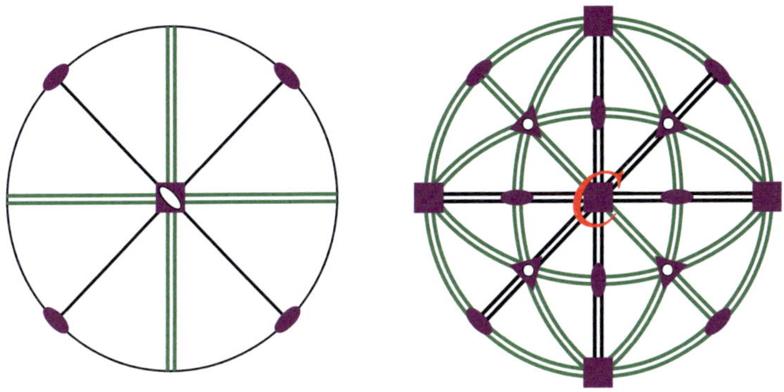

Fig. 2.16 Examples of the stereographic projections with tetragonal (*left*) and cubic (*right*) symmetry

2.9 Crystallographic Point Groups

The total number of symmetry elements that form a crystallographic point group varies from one to as many as 24. However, since symmetry elements interact with one another, there is no need to use each and every symmetry element that belongs to a group in order to uniquely define and completely describe any of the crystallographic groups. The symbol of the point-group symmetry is constructed using the list of basic symmetry elements that is adequate to generate all derivative symmetry elements by applying the first property of the group (closure).

The orientation of each symmetry element with respect to the three major crystallographic axes is defined by its position in the sequence that forms the symbol of the point-group symmetry. The complete list of all 32 point groups is provided in Table 2.7.

The columns labeled "First position," "Second position," and "Third position" describe both the symmetry elements found in the appropriate position of the symbol and their orientation with respect to the crystallographic axes. When the corresponding symmetry element is a rotation axis, it is parallel to the specified crystallographic direction but mirror planes are always perpendicular to the corresponding direction.[8] When the crystal system has a unique axis, for example, 2 in the monoclinic crystal system, 4 in the tetragonal crystal system, and so on, and when there is a mirror plane that is perpendicular to the axis, this combination is always present in the point-group symbol. The axis is listed first and the plane is listed second with the two symbols separated by a slash (/). According to Table 2.7, the crystallographic point group shown in Fig. 2.15 is 2/m, and those in Fig. 2.16 are $\bar{4}$m2 (left) and m$\bar{3}$m (right), respectively.

[8] In fact, since a mirror plane can be represented by a twofold inversion axis, this is the same as the latter being parallel to the corresponding direction, see Fig. 2.8 (right).

Table 2.7 Symbols of crystallographic point groups

Crystal system	First position		Second position		Third position		Point group
	Element	Direction	Element	Direction	Element	Direction	
Triclinic	1 or $\bar{1}$	N/A	None		None		1, $\bar{1}$
Monoclinic	2, m, or 2/m	*Y*	None		None		2, m, 2/m
Orthorhombic	2 or m	*X*	2 or m	*Y*	2 or m	*X*	222, mm2, mmm
Tetragonal	4, $\bar{4}$, or 4/m	*Z*	None or 2 or m	*X*	None or 2 or m	Base diagonal	4, $\bar{4}$, 4/m, 422, 4 mm, $\bar{4}$2m, 4/mmm
Trigonal	3 or $\bar{3}$	*Z*	None or 2 or m	*X*	None		3, $\bar{3}$, 32, 3 m, $\bar{3}$m
Hexagonal	6, $\bar{6}$, or 6/m	*Z*	None or 2 or m	*X*	None or 2 or m	Base diagonal	6, $\bar{6}$, 6/m, 622, 6 mm, $\bar{6}$2m, 6/mmm
Cubic	2, m, 4, or $\bar{4}$	*X*	3 or 3	Body diagonal	None or 2 or m	Face diagonal	23, m3, 432, $\bar{4}$3m, m$\bar{3}$m

Table 2.8 Crystallographic point groups arranged according to their merohedry

Crystal system	N[a]	$\bar{N}$[a]	N⊥m[b]	N⊥2[b]	N ∥ m	$\bar{N}$ ∥ m	N⊥m∥m
Triclinic	*1*	**$\bar{1}$**					
Monoclinic	*2*	m	**2/m**				
Orthorhombic				*222*	mm2		**mmm**
Tetragonal	*4*	$\bar{4}$	**4/m**	*422*	4 mm	$\bar{4}$m2	**4/mmm**
Trigonal	*3*	**$\bar{3}$**		*32*	3 m	**$\bar{3}$m**	
Hexagonal	*6*	$\bar{6}$	**6/m**	*622*	6 mm	$\bar{6}$m2	**6/mmm**
Cubic	*23*		**m$\bar{3}$**	*432*		$\bar{4}$3m	**m$\bar{3}$m**

[a] N and $\bar{N}$ are major N-fold rotation and inversion axes, respectively
[b] m and 2 are mirror plane and twofold rotation axis, respectively, which are parallel (||) or perpendicular (⊥) to the major axis

The list of crystallographic point groups appears not very logical, even when arranged according to the crystal systems as has been done in Table 2.7. Therefore, in Table 2.8 the 32 point groups are arranged according to their merohedry, or in other words, according to the presence of symmetry elements other than the major (or unique) axis.

Another classification of point groups is based on their action. Thus, centrosymmetric point groups or groups containing a center of inversion are shown in Table 2.8 in **bold**, while the groups containing only rotational operation(s) and, therefore, not changing the enantiomorphism (all hands remain either left of right) are in *italic*. Point groups shown in normal font do not have the inversion center; however, they change the enantiomorphism. An empty cell in the table means that the generated point group is already present in a different place in Table 2.8, sometimes in a different crystal system.

2.10 Laue[9] Classes

Radiation and particles, that is, X-rays, neutrons, and electrons, interact with a crystal in a way that the resulting diffraction pattern is always centrosymmetric, regardless of whether an inversion center is present in the crystal or not. This leads to another classification of crystallographic point groups, called Laue classes. The Laue class defines the symmetry of the diffraction pattern produced by a single crystal and can be easily inferred from a point group by adding the center of inversion (see Table 2.9).

[9]Max von Laue (1879–1960). German physicist who was the first to observe and explain the phenomenon of X-ray diffraction in 1912. Laue was awarded Nobel Prize in Physics in 1914 "for his discovery of the diffraction of X-rays by crystals." For more information about Max von Laue, see https://www.nobelprize.org/prizes/physics/1914/laue/facts/

Table 2.9 The 11 Laue classes and six "powder" Laue classes

Crystal system	Laue class	"Powder" Laue class	Point groups
Triclinic	$\bar{1}$	$\bar{1}$	1, $\bar{1}$
Monoclinic	2/m	2/m	2, m, 2/m
Orthorhombic	mmm	mmm	222, mm2, mmm
Tetragonal	4/m	4/mmm	4, $\bar{4}$, 4/m
	4/mmm	4/mmm	422, 4 mm, $\bar{4}$m2, 4/mmm
Trigonal	$\bar{3}$	6/mmm	3, $\bar{3}$
	$\bar{3}$m	6/mmm	32, 3 m, $\bar{3}$m
Hexagonal	6/m	6/mmm	6, $\bar{6}$, 6/m
	6/mmm	6/mmm	622, 6 mm, $\bar{6}$m2, 6/mmm
Cubic	m$\bar{3}$	m$\bar{3}$m	23, m$\bar{3}$
	m$\bar{3}$m	m$\bar{3}$m	432, $\bar{4}$3m, m$\bar{3}$m

For example, all three monoclinic point groups, that is, 2, m, and 2/m will result in 2/m symmetry after adding the center of inversion. In other words, the 2, m, and 2/m point groups belong to the Laue class 2/m, and any diffraction pattern obtained from any monoclinic structure will always have 2/m symmetry. The importance of this classification is easily appreciated from the fact that Laue classes, but not crystallographic point groups, are distinguishable from diffraction data, which is caused by the presence of the center of inversion. All Laue classes (a total of 11) listed in Table 2.9 can be recognized from three-dimensional diffraction data when examining single crystals. However, conventional powder diffraction is fundamentally one-dimensional because the diffracted intensity is measured as a function of one variable (Bragg[10] angle), which results in six identifiable "powder" Laue classes.

As seen in Table 2.9, there is one "powder" Laue class per crystal system, except for the trigonal and hexagonal crystal systems, which share the same "powder" Laue class, 6/mmm. In other words, not every Laue class can be distinguished from a simple visual analysis of powder diffraction data. This occurs because certain diffraction peaks with potentially different intensities (the property which enables us to differentiate between Laue classes 4/m and 4/mmm; $\bar{3}$, $\bar{3}$m, 6/m, and 6/mmm; m$\bar{3}$ and m$\bar{3}$m) completely overlap since they are observed at identical Bragg angles. Hence, only Laue classes that differ from one another in the shape of the unit cell (see Table 2.10) are ab initio discernible from powder diffraction data without a complete structural determination.

[10] Sir William Henry Bragg (1862–1942). British physicist and mathematician who together with his son William Lawrence Bragg (1890–1971) founded X-ray diffraction science in 1913–1914. Both were awarded Nobel Prize in Physics in 1915 "for their services in the analysis of crystal structure by means of X-rays." See https://www.nobelprize.org/prizes/physics/1915/wh-bragg/facts/ and https://www.nobelprize.org/prizes/physics/1915/wl-bragg/facts/ for more details.

Table 2.10 Lattice symmetry and unit cell shapes

Crystal family	Unit cell symmetry	Unit cell shape/parameters
Triclinic	$\bar{1}$	$a \neq b \neq c$; $\alpha \neq \beta \neq \gamma \neq 90°$
Monoclinic	2/m	$a \neq b \neq c$; $\alpha = \gamma = 90°$, $\beta \neq 90°$
Orthorhombic	mmm	$a \neq b \neq c$; $\alpha = \beta = \gamma = 90°$
Tetragonal	4/mmm	$a = b \neq c$; $\alpha = \beta = \gamma = 90°$
Hexagonal and trigonal	6/mmm	$a = b \neq c$; $\alpha = \beta = 90°$, $\gamma = 120°$
Cubic	m$\bar{3}$m	$a = b = c$; $\alpha = \beta = \gamma = 90°$

2.11 Selection of a Unit Cell and Bravais[11] Lattices

The symmetry group of a lattice always has the highest symmetry in the conforming crystal system. Taking into account that trigonal and hexagonal crystal systems are usually described in the same type of the lattice, seven crystal systems can be grouped into six crystal families, which are identical to the six "powder" Laue classes. Different types of lattices, or in general crystal systems, are identified by the presence of specific symmetry elements and their relative orientation. Furthermore, lattice symmetry is always the same as the symmetry of the unit cell shape (except that the lattice has translational symmetry but the unit cell does not), which establishes unique relationships between the unit cell dimensions (a, b, c, α, β, and γ) in each crystal family as shown in Table 2.10. Thus, the fundamental *rule number one* for the proper selection of the unit cell can be formulated as follows: symmetry of the unit cell should be identical to the symmetry of the lattice, excluding translations.

We have already briefly mentioned that in general, the choice of the unit cell is not unique (e.g., see Fig. 1.3). The uncertainty in the selection of the unit cell is further illustrated in Fig. 2.17, where the unit cell in the same two-dimensional lattice has been chosen in four different ways.

The four unit cells shown in Fig. 2.17 have the same symmetry (a twofold rotation axis, which is perpendicular to the plane of the projection and passes through the center of each unit cell), but they have different shapes and areas (volumes in three dimensions). Further, the two unit cells located at the top of Fig. 2.17 do not contain lattice points inside the unit cell, while each of the remaining two has an additional lattice point in the middle. We note that all unit cells depicted in Fig. 2.17 satisfy the rule for the monoclinic crystal system established in Table 2.10. It is quite obvious that more unit cells can be selected in Fig. 2.17, and an infinite number of choices are possible in the infinite lattice, all in agreement with Table 2.10.

Without adopting certain conventions, different unit cell dimensions might and most definitely would be assigned to the same material based on preferences of

[11] Auguste Bravais (1811–1863). French crystallographer, who was the first to derive the 14 different lattices in 1848. A brief biography is found on Wikipedia at https://en.wikipedia.org/wiki/Auguste_Bravais

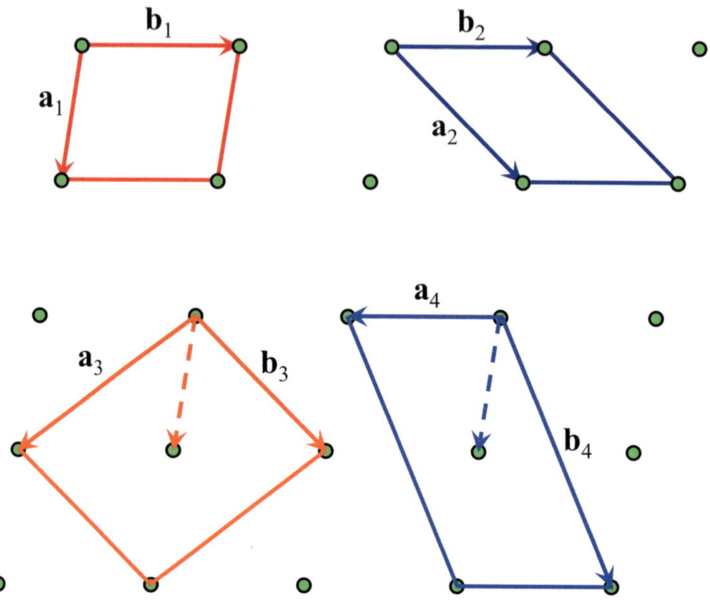

Fig. 2.17 Illustration of different ways to select a unit cell in the same two-dimensional lattice

different researchers. Therefore, long ago the following rules (Table 2.11) were established to designate a standard choice of the unit cell, dependent on the crystal system. This set of rules explains both the unit cell shape and relationships between the unit cell parameters listed in Table 2.10 (i.e., rule number one) and can be considered as *rule number two* in the proper selection of the unit cell.

Applying the rules established in Table 2.11 to two of the four unit cells shown at the top of Fig. 2.17, the cell based on vectors $\mathbf{a}_1$ and $\mathbf{b}_1$ is the standard choice. The unit cell based on vectors $\mathbf{a}_2$ and $\mathbf{b}_2$ has the angle between the vectors much farther from 90° than the first one. The remaining two cells contain additional lattice points in the middle. This type of the unit cell is called *centered*, while the unit cell without a point in the middle is *primitive*. In general, a primitive unit cell is preferred over a centered one; otherwise, it is possible to select a unit cell with any number of points inside, and ultimately it can be made as large as the entire crystal. However, because rule number one requires that the unit cell has the same symmetry as the entire lattice except translational symmetry, it is not always possible to select a primitive unit cell, and so centered unit cells are used.

The *third rule* used to select a standard unit cell is the requirement of the minimum volume (or the minimum number of lattice points inside the unit cell). All things considered, the following unit cells are customarily used in crystallography.

- Primitive, that is, noncentered unit cell. A primitive unit cell is shown schematically in Fig. 2.18 (left). It always contains a single lattice point per unit cell

Table 2.11 Rules for selecting the unit cell in different crystal systems

Crystal family	Standard unit cell choice	Alternative unit cell choice
Triclinic	Angles between crystallographic axes should be as close to 90° as possible but greater than or equal to 90°	Angle(s) less than or equal to 90° are allowed
Monoclinic	*Y*-axis is chosen parallel to the unique twofold rotation axis (or perpendicular to the mirror plane) and angle β should be greater than but as close to 90° as possible	Same as the standard choice, but *Z*-axis in place of *Y* and angle γ in place of β are allowed
Orthorhombic	Crystallographic axes are chosen parallel to the three mutually perpendicular twofold rotation axes (or perpendicular to mirror planes)	None
Tetragonal	*Z*-axis is always parallel to the unique fourfold rotation (inversion) axis. *X*- and *Y*-axes form a 90° angle with the *Z*-axis and with each other	None
Hexagonal and trigonal	*Z*-axis is always parallel to three- or sixfold rotation (inversion) axis. *X*- and *Y*-axes form a 90° angle with the *Z*-axis and a 120° angle with each other	In a trigonal symmetry,[a] threefold axis is chosen along the body diagonal of the primitive unit cell, then $a = b = c$ and $\alpha = \beta = \gamma \neq 90°$
Cubic	Crystallographic axes are always parallel to the three mutually perpendicular two- or fourfold rotation axes, while the four threefold rotation (inversion) axes are parallel to four body diagonals of a cube	None

[a] Instead of a rhombohedrally centered trigonal unit cell shown in Fig. 2.20

(lattice points are located in eight corners of the parallelepiped, but each corner is shared by eight neighboring unit cells in three dimensions).

- Base-centered unit cell (Fig. 2.18, right) contains additional lattice points in the middle of the two opposite faces (as indicated by the vector pointing toward the middle of the base and by the dotted diagonals on both faces). This unit cell contains two lattice points, since each face is shared by two neighboring unit cells in three dimensions
- Body-centered unit cell (Fig. 2.19, left) contains one additional lattice point in the middle of the body of the unit cell. Similar to a base-centered unit cell, the body-centered unit cell contains a total of two lattice points.
- Face-centered unit cell (Fig. 2.19, right) contains three additional lattice points located in the middle of each face, which results in a total of four lattice points in a single face-centered unit cell.
- Rhombohedral unit cell (Fig. 2.20) is a special unit cell that is allowed only in a trigonal crystal system. It contains two additional lattice points located at 1/3,2/3,2/3 and 2/3,1/3,1/3 as shown by the ends of the two vectors inside the unit cell, which results in a total of three lattice points per unit cell.

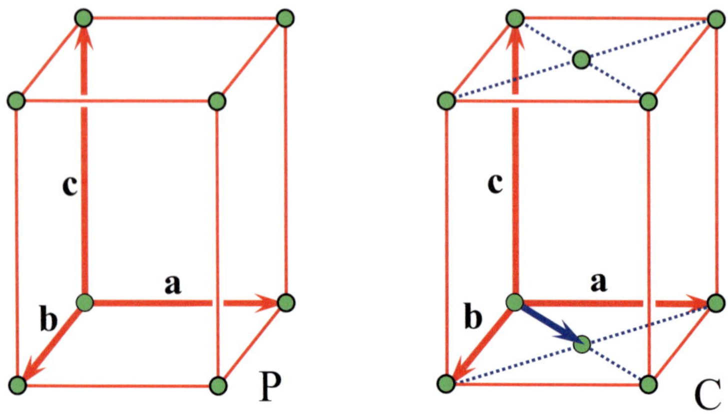

Fig. 2.18 Primitive unit cell (*left*) and base-centered unit cell (*right*)

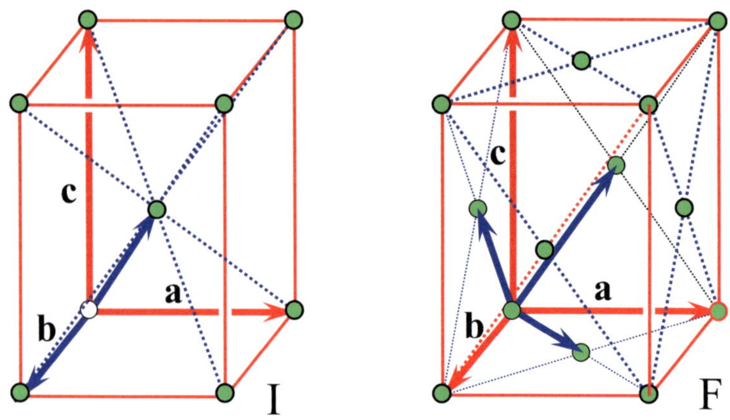

Fig. 2.19 Body-centered unit cell (*left*) and face-centered unit cell (*right*)

Since every unit cell in the crystal lattice is identical to all others, it is said that the lattice can be primitive or centered. We already mentioned (Eq. 1.1) that a crystallographic lattice is based on three noncoplanar translations (vectors), thus the presence of lattice centering introduces additional translations that are different from the three basis translations. Properties of various lattices are summarized in Table 2.12 along with the international symbols adopted to differentiate between different lattice types. In a base-centered lattice, there are three different possibilities to select a pair of opposite faces if the coordinate system is fixed, which is also reflected in Table 2.12.

The introduction of lattice centering makes the treatment of crystallographic symmetry much more elegant when compared to that where only primitive lattices are allowed. Considering six crystal families (Table 2.11) and five types of lattices (Table 2.12)—where three base-centered lattices, which differ only in the orientation

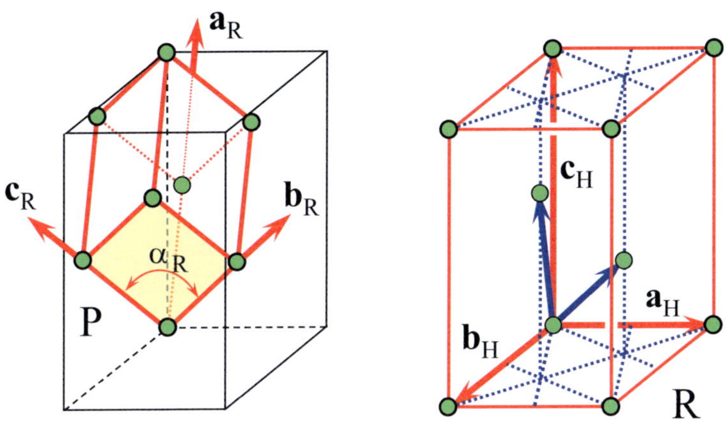

Fig. 2.20 Primitive (*left*) and hexagonal rhombohedral (*right*) unit cells

Table 2.12 Possible lattice centering

Centering of the lattice	Lattice points per unit cell	International symbol	Lattice translation(s) due to centering
Primitive	1	P	None
Base-centered	2	A	½(**b** + **c**)
Base-centered	2	B	½(**a** + **c**)
Base-centered	2	C	½(**a** + **b**)
Body-centered	2	I	½(**a** + **b** + **c**)
Face-centered	4	F	½(**b** + **c**); ½(**a** + **c**); ½(**a** + **b**)
Rhombohedral	3	R	1/3**a** + 2/3**b** + 2/3**c**; 2/3**a** + 1/3**b** + 1/3**c**

of the centered faces relative to a fixed set of basis vectors, are counted as one—it can be shown that only 14 different types of unit cells are needed to describe all lattices using conventional crystallographic symmetry. These are listed in Table 2.13, and they are known as Bravais lattices.

Empty positions in Table 2.13 exist because the corresponding lattices can be reduced to a lattice with different centering and a smaller unit cell (rule number three), or they do not satisfy rules number one or two. For example:

- In the triclinic crystal system, any of the centered lattices can be reduced to a primitive lattice with the smaller volume of the unit cell (rule number three).
- In the monoclinic crystal system, the body-centered lattice can be converted into a base-centered lattice (C), which is standard. The face-centered lattice is reduced to a base-centered lattice with half the volume of the unit cell (rule number three). Even though the base-centered lattice may be reduced to a primitive cell and further minimize the volume of the unit cell, this reduction is incompatible with rule number one since more complicated relationships between the unit cell parameters would result instead of the standard $\alpha = \gamma = 90°$ and $\beta \neq 90°$.

Table 2.13 The 14 Bravais lattices

Crystal system	P	C	I	F	R
Triclinic	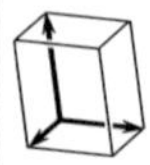				
Monoclinic	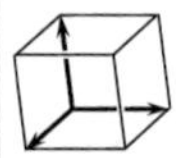	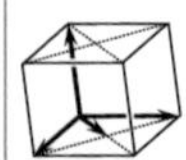			
Orthorhombic	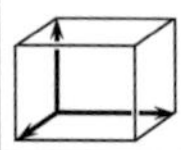	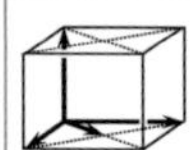	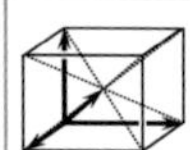	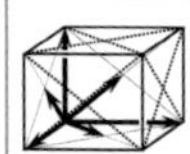	
Tetragonal	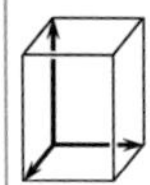		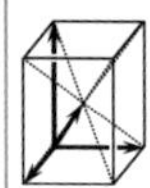		
Hexagonal Trigonal	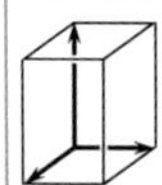				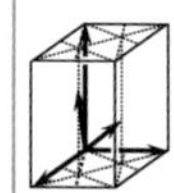
Cubic	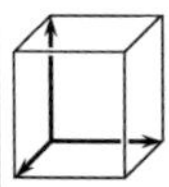		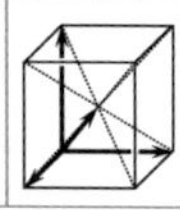	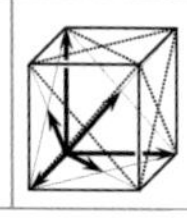	

– In the tetragonal crystal system, the base-centered lattice (C) is reduced to a primitive (P) one, whereas the face-centered lattice (F) is reduced to a body-centered (I) cell; both reductions result in half the volume of the corresponding unit cell (rule number three).

The latter example is illustrated in Fig. 2.21, where a tetragonal face-centered lattice is reduced to a tetragonal body-centered lattice, which has the same symmetry, but half the volume of the unit cell. The reduction is carried out using the transformations of basis vectors as shown in Eqs. (2.5–2.7).

$$\mathbf{a}_{\mathrm{I}} = \tfrac{1}{2}(\mathbf{a}_{\mathrm{F}} - \mathbf{b}_{\mathrm{F}}) \tag{2.5}$$

$$\mathbf{b}_{\mathrm{I}} = \tfrac{1}{2}(\mathbf{a}_{\mathrm{F}} + \mathbf{b}_{\mathrm{F}}) \tag{2.6}$$

$$\mathbf{c}_{\mathrm{I}} = \mathbf{c}_{\mathrm{F}} \tag{2.7}$$

The relationships between the unit cell dimensions and unit cell volumes of the original face-centered (V_{F}) and the reduced body-centered (V_{I}) lattices are:

$$a_{\mathrm{I}} = b_{\mathrm{I}} = \frac{a_{\mathrm{F}}}{\sqrt{2}} = \frac{b_{\mathrm{F}}}{\sqrt{2}}; \qquad c_{\mathrm{I}} = c_{\mathrm{F}} \tag{2.8}$$

$$V_{\mathrm{I}} = V_{\mathrm{F}}/2 \tag{2.9}$$

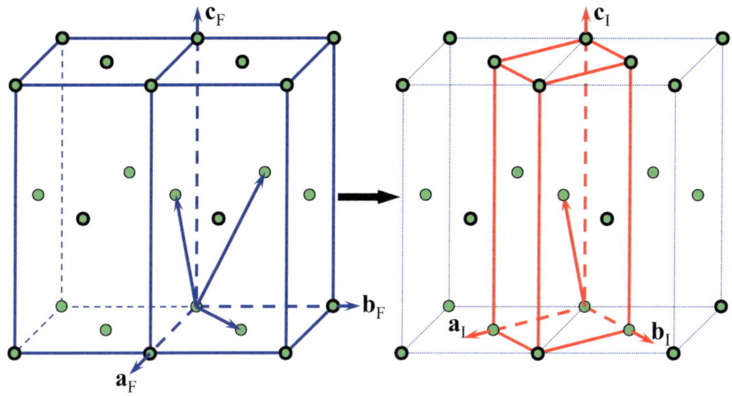

Fig. 2.21 The reduction of the tetragonal face-centered lattice (*left*) to the tetragonal body-centered lattice with half the volume of the unit cell (*right*). Small green circles indicate lattice points

2.12 Additional Reading

1. C. Giacovazzo, H.L. Monaco, G. Artioli, D. Viterbo, M. Milaneso, G. Gilli, P. Gilli, G. Zanotti, G. Ferraris, and M. Catti, Fundamentals of crystallography, 3rd edition, C. Giacovazzo, Ed., IUCr/Oxford University Press (2011), https://doi.org/10.1093/acprof:oso/9780199573653.001.0001.
2. D. Schwarzenbach, Crystallography, Wiley, New York (1997), https://www.wiley.com/en-us/Crystallography-p-9780471955986.
3. C. Hammond, The basics of crystallography and diffraction, 4th edition, IUCr/Oxford University Press (2015), https://doi.org/10.1093/acprof:oso/9780198738671.001.0001.
4. D. Farmer, Groups and symmetry: A Guide to Discovering Mathematics (Mathematical World, Vol. 5) 1st Edition, ISBN: 0821804502, Amer. Math. Soc. (1995), https://bookstore.ams.org/mawrld-5.
5. International Tables for Crystallography, Vol. A, Space-group symmetry, M. I. Aroyo, Ed., Second online edition, (2016), https://it.iucr.org/Ac/.
6. International Tables for Crystallography. 6th edition, Teaching edition, Crystallographic symmetry, M. I. Aroyo, Ed., Wiley (2021), https://www.wiley.com/en-us/-p-9780470974223.
7. IUCr Commission on Crystallographic teaching. Teaching pamphlets, https://www.iucr.org/education/pamphlets.

2.13 Problems

1. Consider two mirror planes that intersect at $\phi = 90°$. Using geometrical representation of two planes, establish which symmetry element(s) appear as the result of this combination of mirror planes. What is(are) the location(s) of new symmetry element(s)? Name point-group symmetry formed by this combination of symmetry elements.

2. Consider two mirror planes that intersect at $\phi = 45°$. Using geometrical representation of two planes, establish which symmetry element(s) appear as the result of this combination of mirror planes. What is(are) the location(s) of new symmetry element(s)? Name point-group symmetry formed by this combination of symmetry elements.
3. Consider the following sequence of numbers: $1, 1/2, 1/3, 1/4, \ldots, 1/N, \ldots$ Is this a group assuming that the combination law is multiplication, division, addition, or subtraction? If yes, identify the combination law in this group and establish whether this group is finite or infinite.
4. Consider the group created by three noncoplanar translations (vectors) using the combination law defined by Eq. (1.1). Which geometrical arrangement can be chosen to illustrate this group? Is the group finite?
5. Determine both the crystal system and point-group symmetry of a parallelepiped (a brick), which is shown schematically in Fig. 2.22 and in which $a \neq b \neq c$ and $\alpha = \beta = \gamma = 90°$?
6. Determine both the crystal system and point-group symmetry of benzene molecule, C_6H_6, which is shown in Fig. 2.23. Treat atoms as spheres, not as dimensionless points.
7. Determine both the crystal system and point-group symmetry of the ethylene molecule, C_2H_4, shown schematically in Fig. 2.24. Using the projection on the left, show all symmetry elements that you were able to identify in this molecule, include both the in-plane and out-of-plane symmetry elements. Treat atoms as spheres, not as dimensionless points.

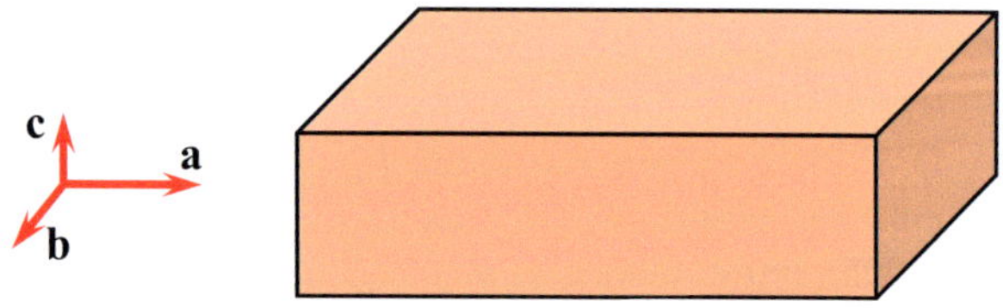

Fig. 2.22 Illustration of a parallelepiped (a *brick*) in which three independent edges have different lengths

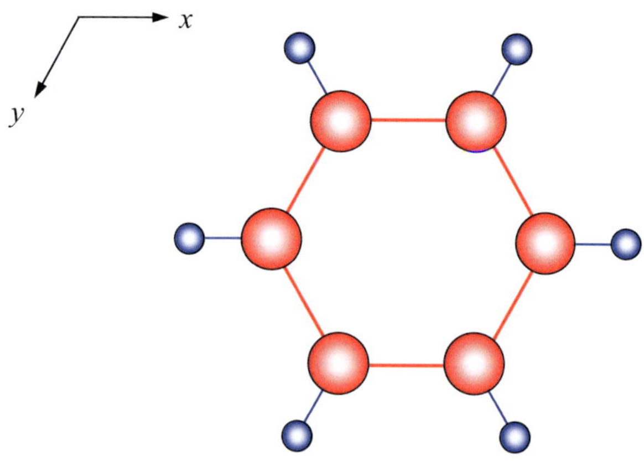

Fig. 2.23 The schematic of benzene molecule. Carbon atoms are *pink* and hydrogen atoms are *blue*

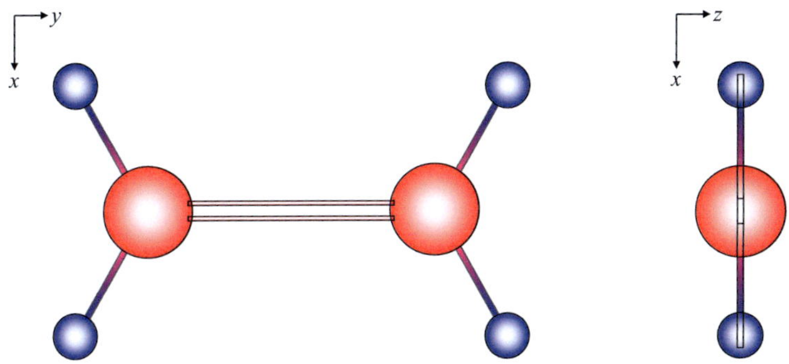

Fig. 2.24 The schematic of ethylene molecule. Carbon atoms are *pink* and hydrogen atoms are *blue*

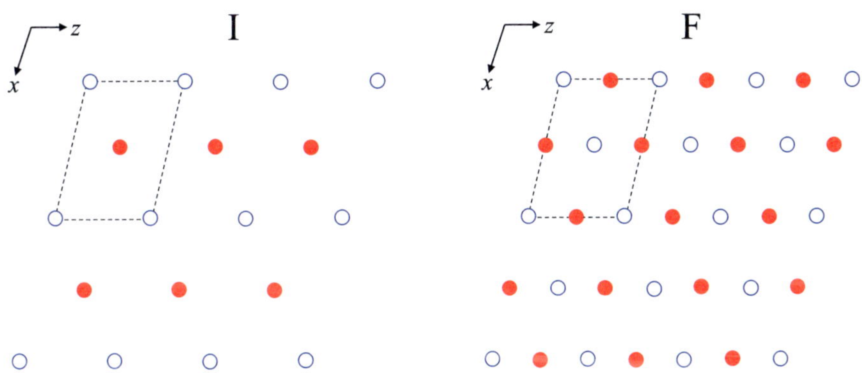

Notes: y is perpendicular to the paper
○ - point at $y = 0, \pm 1, \ldots$
● - point at $y = \pm\frac{1}{2}, \pm\frac{3}{2}, \ldots$

Fig. 2.25 Body-centered (*left*) and face-centered (*right*) monoclinic lattices projected along the *Y*-axis with the corresponding unit cells shown using the *dashed lines*

8. Determine the point-group symmetry of the octahedron. How many and which symmetry elements are present in this point-group symmetry?
9. The following relationships between lattice parameters: $a \neq b \neq c$, $\alpha \neq \beta \neq 90$ or $120°$, and $\gamma = 90°$ potentially define a "diclinic" crystal system (two angles $\neq 90°$). Is this an eighth crystal system? Explain your answer.
10. The relationships $a = b \neq c$, $\alpha = \beta = 90°$, and $\gamma \neq 90°$ point to a monoclinic crystal system, except that $a = b$. What is the reduced (standard) Bravais lattice in this case? Provide equations that reduce this lattice to one of the 14 standard Bravais types.
11. Imagine that there is an "edge-centered" lattice (for example, unit cell edges along *Z* contain lattice points at ½**c**). If this were true, the following lattice translation is present: (0, 0, ½). Convert this lattice to one of the standard lattices.
12. Monoclinic crystal system has primitive and base-centered Bravais lattices (see Table 2.13). Using two-dimensional projections depicted in Fig. 2.25, show how

a body-centered lattice and a face-centered monoclinic lattice (their unit cells are indicated with the dashed lines) can be reduced to a base-centered lattice. Write the corresponding vectorial relationships between the unit cell vectors of the original body-centered and face-centered lattices and the transformed base-centered lattices. What are the relationships between the unit cell volumes of the original body- and face-centered lattices and the resulting base-centered lattices?

Chapter 3
Infinite Symmetry Elements and Crystallographic Space Groups

So far, our discussion of symmetry of the lattice was limited to lattice points and symmetry of the unit cell. The next step is to think about symmetry of the lattice, including the content of the unit cell. This immediately brings translational symmetry into consideration to reflect the periodic nature of crystal lattices, which are infinite objects. As a result, we need to introduce the so-called infinite or translational symmetry elements in addition to the already familiar finite or nontranslational symmetry elements, which can be present in a lattice as well. Translation or shift is a simple infinite symmetry element (see Fig. 2.3). When acting simultaneously, translation and rotation result in screw axes; translation and reflection in a mirror plane produce glide planes (Table 2.2). Screw axes and glide planes are, therefore, complex infinite symmetry elements.

3.1 Glide Planes

The combination of a mirror reflection plane with the corresponding translations that are always parallel to the plane results in a total of five possible crystallographic glide planes.[1] The allowed translations are one-half or one-fourth of the length of the basis vector, parallel to which the shift (i.e., gliding) occurs. All possible glide planes are listed in Table 3.1 together with their graphical symbols. Since each of the glide planes produces an infinite number of symmetrically equivalent objects from the original, the order of the plane indicates the number of symmetrically equivalent objects within the boundaries of one unit cell and is also listed in Table 3.1. Thus, there are three types of glide planes:

[1] A sixth glide plane, e, has been introduced by de Wolff et al. (see the Footnote 3 to Chap. 2 on page 20) to resolve ambiguities occurring in some space groups (see Sect. 3.4). For example, space group Cmca, where the translation after the reflection in the a-plane occurs along **a** and **b**, is identical to Cmcb, but the group becomes unique, i.e., Cmce, using the e plane.

P. Zavalij, V. Pecharsky, *Fundamentals of Powder Diffraction and Structural Characterization of Materials*, https://doi.org/10.1007/978-3-031-91504-8_3

Table 3.1 Crystallographic glide planes

Plane symbol	Order	Graphical symbol[a]	Translation vector
a	2	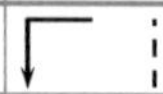	½**a**
b	2	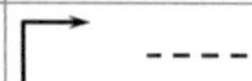	½**b**
c	2		½**c**
n	2	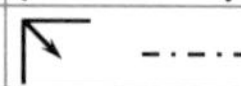	½**d**[b]
d	4	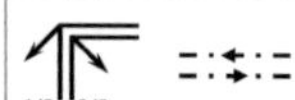	¼**d**[b]

[a] Shown in the following coordinate system: fx; numbers near the symbols indicate displacements of the planes from origin along Z (elevations above the plane of the projection; 0 and ½ are not shown)
[b] **d** is the diagonal vector, e.g., **d** = **a** + **b**, **d** = **a** − **b**, **d** = **a** + **b** + **c**, etc., which depends on the orientation of the plane with respect to crystallographic basis vectors

- Glide planes a, b, and c, which after reflecting in the plane, translate an object by half of the length of **a**, **b**, and **c** basis vectors, respectively. Because of this, for example, glide plane, a, can be perpendicular to either **b** or **c**, but it cannot be perpendicular to **a**. Similarly, glide plane, b, cannot be perpendicular to **b**, and glide plane, c, cannot be perpendicular to **c**. Since the translation is always by half of the corresponding basis vector, these planes produce two symmetrically equivalent objects within one full length of the corresponding basis vector (and within one unit cell), that is, their order is 2.
- Glide plane n, which after reflecting in the plane, translates an object by half of the length of the diagonal between the two basis vectors located in the plane parallel to n. For example, glide plane, n, perpendicular to **c** will translate an object by ½(**a** + **b**). Glide plane n results in two symmetrically equivalent objects within the full length of the diagonal vector (and within one unit cell), and its order is 2.
- Glide planes d, which after reflecting in the plane translate an object by one-fourth of the length of the diagonal between the two basis vectors located in the plane parallel to d. These planes, also known as "diamond" planes since they are present in the diamond crystal structure, are always present in pairs parallel to one another and translate along different diagonals. The length of the translation is one-fourth, which results in a total of four symmetrically equivalent objects per unit cell.

The illustration of how glide planes b, c, and n generate an infinite number of symmetrically equivalent objects is shown in Fig. 3.1.

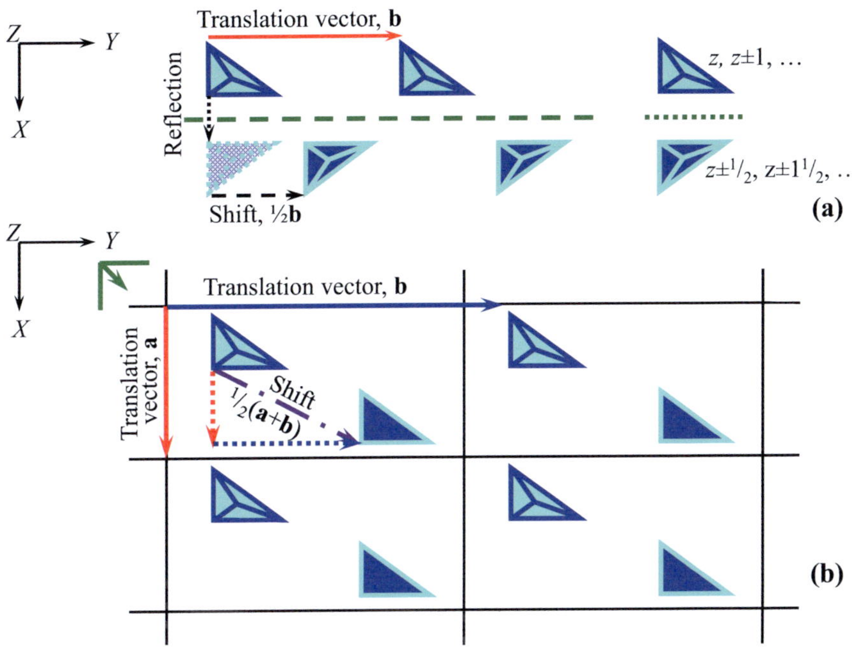

Fig. 3.1 (**a**) Vertical glide plane b, perpendicular to **a** with a horizontal translation by ½ along **b** (*left*); vertical glide plane c, perpendicular to **a** with a vertical translation by ½ along **c** (*right*). (**b**) Horizontal glide plane n, perpendicular to **c** with a diagonal translation by ½(**a** + **b**) parallel to the **ab** plane. The dotted pyramid in (**a**) indicates the intermediate position to which the first pyramid is reflected before it is translated along **b** and along the plane

3.2 Screw Axes

Screw axes perform a rotation simultaneously with a translation along the rotation axis. In other words, the rotation occurs around the axis, while the translation occurs parallel to the axis. Crystallographic screw axes include only two-, three-, four-, and sixfold rotations due to the three-dimensional periodicity of the crystal lattice, which prohibits five-, seven-, and higher-order rotations. Hence, the allowed rotation angles are the same as for both rotation and inversion axes (see Eq. (2.1) on page 24).

Translations, t, along the axis are also limited to a few fixed values, which depend on the order of the axis, and are defined as $t = k/\mathrm{N}$, where N is axis order, and k is an integer number between one and N–1. For instance, for the threefold screw axis, $k = 1$ and 2, and the two possible translations are one-third and two-third of the length of the basis vector parallel to this axis, whereas for the twofold screw axis, $k = 1$, and only half translation is allowed.

The symbol of the screw axis is constructed as N_k to identify both the order of the axis (N) and the length of the translation (k). Thus, the two threefold screw axes have symbols 3_1 and 3_2, whereas the only possible twofold screw axis is 2_1. The international symbols, both text and graphical, as well as the allowed translations

Table 3.2 Crystallographic screw axes

Axis order	Text symbol[a]	Graphical symbol[a]	Shift along the axis[b]
2	2_1	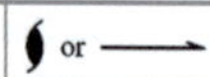	$^1/_2$
3	3_1, 3_2		$^1/_3$, $^2/_3$
4	4_1, 4_2, 4_3		$^1/_4$, $^2/_4$, $^3/_4$
6	6_1, 6_2, 6_3, 6_4, 6_5	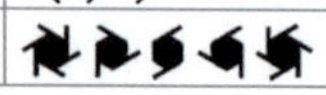	$^1/_6$, $^2/_6$, $^3/_6$, $^4/_6$, $^5/_6$

[a] Pairs of screw axes, in which the sums of the subscripts equal to the order of the axis are called enantiomorphous pairs, since one is the mirror image of another. The latter is reflected in the graphical symbols of the corresponding pairs of the enantiomorphous axes. These are: 3_1 and 3_2; 4_1 and 4_3; 6_1 and 6_5; 6_2 and 6_4. Two enantiomorphous axes differ only by the direction of rotation or, which is the same, by the direction of translation

[b] Given as a fraction of a full translation in a positive direction assuming counterclockwise rotation along the same axis

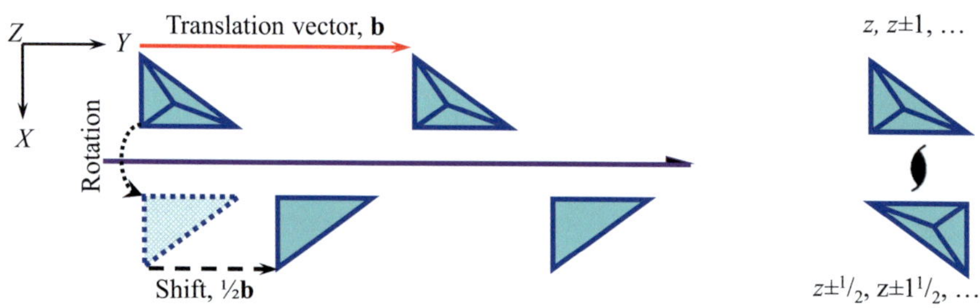

Fig. 3.2 Horizontal (*left*) and vertical (*right*) twofold screw axis, 2_1. The dotted pyramid indicates the intermediate position to which the first pyramid is rotated before it is translated along the axis

for all crystallographic screw axes are listed in Table 3.2. Figure 3.2 illustrates how the twofold screw axis generates an infinite number of symmetrically equivalent objects via rotations by 180° around the axis with the simultaneous translations along the axis by half of the length of the basis vector to which the axis is parallel.

3.3 Interaction of Infinite Symmetry Elements

Infinite symmetry elements interact with one another and produce new symmetry elements, just as finite symmetry elements do. Moreover, the presence of the symmetry element with a translational component (screw axis or glide plane) assumes the presence of the full translation vector as seen in Figs. 3.1 and 3.2. Unlike finite symmetry, symmetry elements in a lattice do not have to cross, although they may have a common point or a line. For example, two planes can be parallel to one another. In this case, the resulting third symmetry element is a translation vector perpendicular to the planes with translation (t) twice the length of the interplanar distance (d), as illustrated in Fig. 3.3.

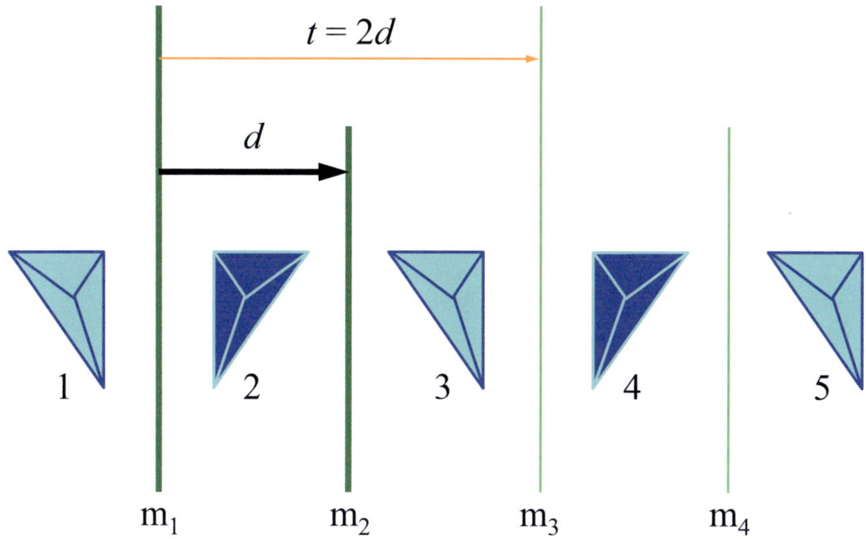

Fig. 3.3 Example illustrating how two parallel mirror planes with the interplanar distance *d* result in the translation $t = 2d$ perpendicular to both planes. Starting from planes m_1 and m_2 (*thick lines*) and pyramid 1, we obtain pyramid 2 by reflecting 1 in m_1. Pyramids 3 and 4 are obtained by reflecting pyramids 2 and 1, respectively, in m_2. Thus, plane m_3 (*thin line*) appears between pyramids 3 and 4. Pyramid 5 is obtained from pyramid 3 by reflecting it in m_1 and then resulting pyramid (not shown) in m_2, which leads to plane m_4 (*thin line*) and so on. It is easy to see that pyramids 1, 3, 5, and so on, as well as pyramids 2, 4, and so on are symmetrically equivalent to one another via the action of the translation $t = 2d$ (*thin orange arrow*)

Another example (Fig. 3.4) illustrates the result of the interaction between glide plane, b, and center of inversion. We begin from the symmetry elements shown using thick lines and marked in Fig. 3.4 as b and $\overline{1}$, and pyramid A. Note, that the two symmetry elements do not share a common point. Pyramid A is converted by the glide plane into B; B into A′; and so on. The center of inversion converts A into C and so on. All objects depicted in Fig. 3.4 can be obtained using the two original symmetry elements (keep in mind that although only 12 pyramids are shown in this figure, their number is in reality infinite). The derived symmetry element that converts B into C is the 2_1 screw axis. Additional derived symmetry elements are translation **b** that transforms A into A′, translation **a**, other glide planes, screw axes, and centers of inversion, as marked in Fig. 3.4. Thus, the infinite structure is produced and its unit cell is highlighted light yellow in the figure. It is easy to see that in this infinite structure there are an infinite number of symmetry elements, even though we started with just two of them: the glide plane, b, and the center of inversion, $\overline{1}$.

Note that only those symmetry elements which intersect the asymmetric part of the unit cell are independent; exactly in the same way as only those atoms that are

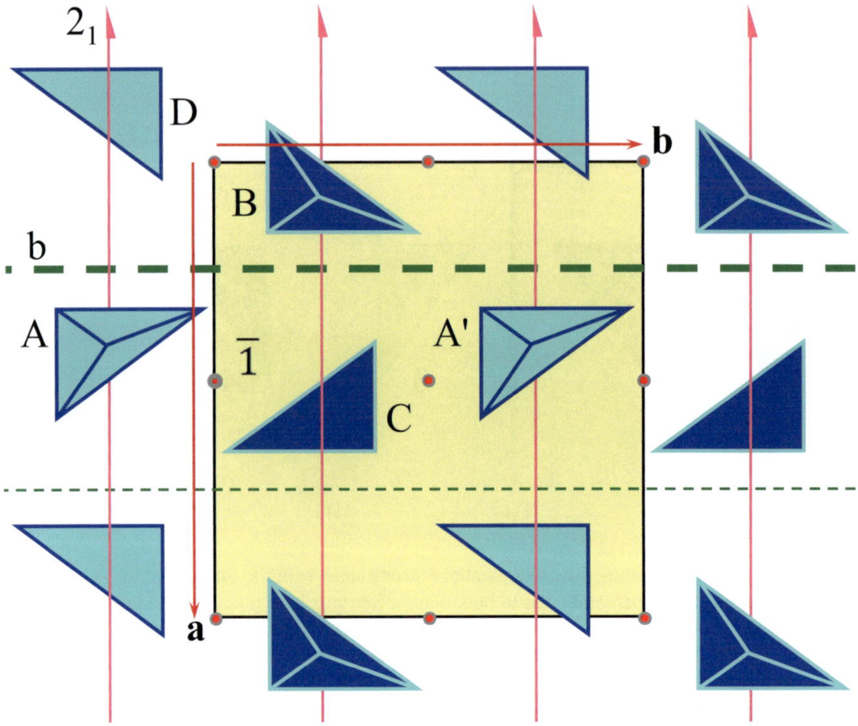

Fig. 3.4 Example illustrating the result of interaction between the glide plane, b, and the center of inversion, $\bar{1}$. The original symmetry elements are shown using *thick lines* and the derivatives using *thin lines*. The resulting unit cell is *highlighted*

located in the asymmetric part of the unit cell are independent (see Fig. 2.2).[2] Once the locations of independent atoms and symmetry elements in the unit cell are known, the whole crystal can be easily reconstructed. Symmetry elements interact with basis translations, which are symmetry elements themselves. Hence, the three noncoplanar vectors that form a three-dimensional lattice interact with symmetry elements and distribute them together with their "subelements" (e.g., a sixfold axis contains both two- and threefold axes as its subelements) in the specific order in the unit cell and in the entire lattice.

The distribution of the inversion centers, twofold screw axes, and glide planes in the unit cell can be seen in Fig. 3.4: these symmetry elements are repeated along the basis translation vectors every half of the full translation. This is true for triclinic, monoclinic, and orthorhombic crystal systems, that is, crystal systems with low symmetry. Similar arrangements of symmetry elements in tetragonal, trigonal, and hexagonal crystal systems are shown for a primitive lattice in Fig. 3.5, where only

[2] Strict definition can be given as follows: two objects (atoms, molecules, or symmetry elements) are symmetrically independent if there is no symmetry operation that converts one into another.

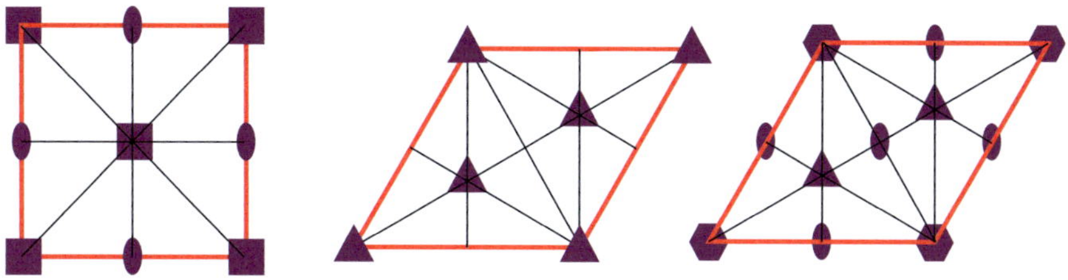

Fig. 3.5 Characteristic distribution of the two-, three-, four-, and sixfold rotation axes parallel to the unique axis in the unit cell in tetragonal (*left*), trigonal (*center*), and hexagonal (*right*) crystal systems as the result of their interaction with lattice translations

rotation axes perpendicular to the plane of the projection were taken into account. The distribution of other symmetry elements, including the infinite ones in primitive or centered lattices follows the same path, but types of axes and their order may be different.

The orientation and placement of crystallographic axes in a cubic crystal system are more complex and it is difficult to illustrate the same in a simple drawing. However, projections of rotation axes in cubic symmetry along the unit cell edges resemble tetragonal symmetry, while projections along the body diagonals of the cube resemble trigonal symmetry. The positions of symmetry elements (especially finite) in the unit cell are important, since they are directly related to special site positions, which are discussed later in Sects. 3.5 and 3.6.

3.4 Crystallographic Space Groups

Similar to finite symmetry elements, which can be combined into point groups (see Sect. 2.9), various combinations of the same symmetry elements plus allowed translations, while obeying the rules described in Sect. 2.11, result in the so-called crystallographic space groups. As follows from the differences in their names, symmetry elements in a space group are spread over the space of an infinite object, contrary to point groups, where all symmetry elements have at least one common point. Therefore, each point of a continuous object can be moved in a periodic fashion through space by the action of symmetry elements that form a space-group symmetry, whereas at least one point of the object remains unmoved by the action of symmetry elements that belong to a point-group symmetry.

Given the limitations on the allowed rotations and translations (see Sects. 2.4, 3.1, and 3.2), there is a total of 230 three-dimensional crystallographic space groups,[3]

[3] When vector directions (e.g., magnetic moments) are included into consideration, the number of space groups increases dramatically. Thus, a total of 1651 dichromatic (black and white) symmetry groups are used to treat symmetry of magnetically ordered structures. See A.V. Shubnikov and

which were derived and systematized independently by Fedorov[4] and Schönflies.[5] A complete list of space groups is found in Table 3.3, where they are arranged according to seven crystal systems (see Sect. 2.7) and 32 crystallographic point groups (see Sect. 2.9).

3.4.1 Relationships Between Point Groups and Space Groups

The symbols of 230 crystallographic space groups, which are shown in Table 3.3, are known as short international or short Hermann–Mauguin[6] symbols. They are based on the symbols of the corresponding point groups. The orientation of symmetry elements with respect to the three major crystallographic axes in a space group is the same as in the parent point group (see Table 2.7) and it depends on the position of the element in the symbol. The rules that govern space-group symbolic are quite simple:

- The international crystallographic space-group symbols begin with a capital letter designating Bravais lattice, i.e., P, A, B, C, I, F, or R (see Tables 2.12 and 2.13).
- The point-group symbol, in which rotation axes and mirror planes can be substituted with allowed screw axes or glide planes, respectively, is added as the second part of the international space-group symbol.
- In some cases, the second and the third symmetry elements of the point group can be switched, for example, point-group symmetry $\bar{4}$m2 produces two different space groups without introducing glide planes and/or screw axes in the space-group symbol, that is, P$\bar{4}$2m and P$\bar{4}$m2.

These rules not only reflect how the symbol of the space-group symmetry is constructed, but they also establish one of the possible ways to derive all 230 crystallographic space groups:

- First, consider the Bravais lattices (see Table 2.13) and point groups (see Table 2.7) which are allowed in a given crystal system.

V.A. Koptsik, Symmetry in science and art, Plenum Press (1974), https://www.springer.com/gp/book/9781468420692, for a brief description of color symmetry groups.

[4]Evgraf Stepanovich Fedorov (1853–1919) applied the theory of finite groups to crystallography to derive 230 space groups in 1891.

[5]Arthur Moritz Schönflies (1853–1928) derived 230 space groups independently of E.S. Fedorov in 1891. See https://www.iucr.org/publ/50yearsofxraydiffraction/full-text/schoenflies for a brief biography.

[6]Named after Carl Hermann (1898–1961) and Charles-Victor Mauguin (1878–1958) who developed the system of symbols. C. Hermann's and C.-V. Mauguin's biographies written by P.P. Ewald and F. Wyart, respectively, may be found in a book *50 years of X-ray Diffraction*, edited by P.P. Ewald. The book is available online at IUCr: https://www.iucr.org/publ/50yearsofxraydiffraction

Table 3.3 The 230 crystallographic space groups arranged by seven crystal systems and 32 crystallographic point groups, as listed in the International Tables for Crystallography, vol. A. Centrosymmetric groups are in **bold**, noncentrosymmetric groups that do not invert an object are in *italic*, and the remaining are noncentrosymmetric groups that invert an object (contain inversion axis or mirror plane)

Crystal system (total number of space groups)	Point group	Space groups based on a given point-group symmetry (the superscript indicates the space group number as adopted in the International Tables for Crystallography, vol. A)
Triclinic (2)	1	$P1^{1}$
	$\bar{\mathbf{1}}$	$\mathbf{P\bar{1}}^{2}$
Monoclinic (13)	*2*	$P2^{3}$, $P2_1{}^{4}$, $C2^{5}$
	m	Pm^{6}, Pc^{7}, Cm^{8}, Cc^{9}
	2/m	$\mathbf{P2/m}^{10}$, $\mathbf{P2_1/m}^{11}$, $\mathbf{C2/m}^{12}$, $\mathbf{P2/c}^{13}$, $\mathbf{P2_1/c}^{14}$, $\mathbf{C2/c}^{15}$
Orthorhombic (59)	*222*	$P222^{16}$, $P222_1{}^{17}$, $P2_12_12^{18}$, $P2_12_12_1{}^{19}$, $C222_1{}^{20}$, $C222^{21}$, $F222^{22}$, $I222^{23}$, $I2_12_12_1{}^{24}$
	mm2	$\mathrm{Pmm2}^{25}$, $\mathrm{Pmc2_1}^{26}$, $\mathrm{Pcc2}^{27}$, $\mathrm{Pma2}^{28}$, $\mathrm{Pca2_1}^{29}$, $\mathrm{Pnc2}^{30}$, $\mathrm{Pmn2_1}^{31}$, $\mathrm{Pba2}^{32}$, $\mathrm{Pna2_1}^{33}$, $\mathrm{Pnn2}^{34}$, $\mathrm{Cmm2}^{35}$, $\mathrm{Cmc2_1}^{36}$, $\mathrm{Ccc2}^{37}$, $\mathrm{Amm2}^{38}$, $\mathrm{Abm2}^{39}$, $\mathrm{Ama2}^{40}$, $\mathrm{Aba2}^{41}$, $\mathrm{Fmm2}^{42}$, $\mathrm{Fdd2}^{43}$, $\mathrm{Imm2}^{44}$, $\mathrm{Iba2}^{45}$, $\mathrm{Ima2}^{46}$
	mmm	$\mathbf{Pmmm}^{47}$, $\mathbf{Pnnn}^{48}$, $\mathbf{Pccm}^{49}$, $\mathbf{Pban}^{50}$, $\mathbf{Pmma}^{51}$, $\mathbf{Pnna}^{52}$, $\mathbf{Pmna}^{53}$, $\mathbf{Pcca}^{54}$, $\mathbf{Pbam}^{55}$, $\mathbf{Pccn}^{56}$, $\mathbf{Pbcm}^{57}$, $\mathbf{Pnnm}^{58}$, $\mathbf{Pmmn}^{59}$, $\mathbf{Pbcn}^{60}$, $\mathbf{Pbca}^{61}$, $\mathbf{Pnma}^{62}$, $\mathbf{Cmcm}^{63}$, $\mathbf{Cmca}^{64}$, $\mathbf{Cmmm}^{65}$, $\mathbf{Cccm}^{66}$, $\mathbf{Cmma}^{67}$, $\mathbf{Ccca}^{68}$, $\mathbf{Fmmm}^{69}$, $\mathbf{Fddd}^{70}$, $\mathbf{Immm}^{71}$, $\mathbf{Ibam}^{72}$, $\mathbf{Ibca}^{73}$, $\mathbf{Imma}^{74}$
Tetragonal (68)	*4*	$P4^{75}$, $P4_1{}^{76}$, $P4_2{}^{77}$, $P4_3{}^{78}$, $I4^{79}$, $I4_1{}^{80}$
	$\bar{4}$	$\mathrm{P\bar{4}}^{81}$, $\mathrm{I\bar{4}}^{82}$
	4/m	$\mathbf{P4/m}^{83}$, $\mathbf{P4_2/m}^{84}$, $\mathbf{P4/n}^{85}$, $\mathbf{P4_2/n}^{86}$, $\mathbf{I4/m}^{87}$, $\mathbf{I4_1/a}^{88}$
	422	$P422^{89}$, $P42_12^{90}$, $P4_122^{91}$, $P4_12_12^{92}$, $P4_222^{93}$, $P4_22_12^{94}$, $P4_322^{95}$, $P4_32_12^{96}$, $I422^{97}$, $I4_122^{98}$
	4mm	$\mathrm{P4mm}^{99}$, $\mathrm{P4bm}^{100}$, $\mathrm{P4_2cm}^{101}$, $\mathrm{P4_2nm}^{102}$, $\mathrm{P4cc}^{103}$, $\mathrm{P4nc}^{104}$, $\mathrm{P4_2mc}^{105}$, $\mathrm{P4_2bc}^{106}$, $\mathrm{I4mm}^{107}$, $\mathrm{I4cm}^{108}$, $\mathrm{I4_1md}^{109}$, $\mathrm{I4_1cd}^{110}$
	$\bar{4}$m2	$\mathrm{P\bar{4}2m}^{111}$, $\mathrm{P\bar{4}2c}^{112}$, $\mathrm{P\bar{4}2_1m}^{113}$, $\mathrm{P\bar{4}2_1c}^{114}$, $\mathrm{P\bar{4}m2}^{115}$, $\mathrm{P\bar{4}c2}^{116}$, $\mathrm{P\bar{4}b2}^{117}$, $\mathrm{P4n2}^{118}$, $\mathrm{I\bar{4}m2}^{119}$, $\mathrm{I\bar{4}c2}^{120}$, $\mathrm{I\bar{4}2m}^{121}$, $\mathrm{I\bar{4}2d}^{122}$
	4/mmm	$\mathbf{P4/mmm}^{123}$, $\mathbf{P4/mcc}^{124}$, $\mathbf{P4/nbm}^{125}$, $\mathbf{P4/nnc}^{126}$, $\mathbf{P4/mbm}^{127}$, $\mathbf{P4/mnc}^{128}$, $\mathbf{P4/nmm}^{129}$, $\mathbf{P4/ncc}^{130}$, $\mathbf{P4_2/mmc}^{131}$, $\mathbf{P4_2/mcm}^{132}$, $\mathbf{P4_2/nbc}^{133}$, $\mathbf{P4_2/nnm}^{134}$, $\mathbf{P4_2/mbc}^{135}$, $\mathbf{P4_2/mnm}^{136}$, $\mathbf{P4_2/nmc}^{137}$, $\mathbf{P4_2/ncm}^{138}$, $\mathbf{I4/mmm}^{139}$, $\mathbf{I4/mcm}^{140}$, $\mathbf{I4_1/amd}^{141}$, $\mathbf{I4_1/acd}^{142}$
Trigonal (25)	*3*	$P3^{143}$, $P3_1{}^{144}$, $P3_2{}^{145}$, $R3^{146}$
	$\bar{\mathbf{3}}$	$\mathbf{P\bar{3}}^{147}$, $\mathbf{R\bar{3}}^{148}$
	32	$P312^{149}$, $P321^{150}$, $P3_112^{151}$, $P3_121^{152}$, $P3_212^{153}$, $P3_221^{154}$, $R32^{155}$
	3m	$\mathrm{P3m1}^{156}$, $\mathrm{P31m}^{157}$, $\mathrm{P3c1}^{158}$, $\mathrm{P31c}^{159}$, $\mathrm{R3m}^{160}$, $\mathrm{R3c}^{161}$
	$\bar{\mathbf{3}}$**m**	$\mathbf{P\bar{3}1m}^{162}$, $\mathbf{P\bar{3}1c}^{163}$, $\mathbf{P\bar{3}m1}^{164}$, $\mathbf{P\bar{3}c1}^{165}$, $\mathbf{R\bar{3}m}^{166}$, $\mathbf{R\bar{3}c}^{167}$
Hexagonal (27)	*6*	$P6^{168}$, $P6_1{}^{169}$, $P6_5{}^{170}$, $P6_2{}^{171}$, $P6_4{}^{172}$, $P6_3{}^{173}$

(continued)

Table 3.3 (continued)

Crystal system (total number of space groups)	Point group	Space groups based on a given point-group symmetry (the superscript indicates the space group number as adopted in the International Tables for Crystallography, vol. A)
	$\bar{6}$	$P\bar{6}^{174}$
	6/m	**$P6/m^{175}$, $P6_3/m^{176}$**
	622	$P622^{177}$, $P6_122^{178}$, $P6_522^{179}$, $P6_222^{180}$, $P6_422^{181}$, $P6_322^{182}$
	6mm	$P6mm^{183}$, $P6cc^{184}$, $P6_3cm^{185}$, $P6_3mc^{186}$
	$\bar{6}m2$	$P\bar{6}m2^{187}$, $P\bar{6}c2^{188}$, $P\bar{6}2m^{189}$, $P\bar{6}2c^{190}$
	6/mmm	**$P6/mmm^{191}$, $P6/mcc^{192}$, $P6_3/mcm^{193}$, $P6_3/mmc^{194}$**
Cubic (36)	23	$P23^{195}$, $F23^{196}$, $I23^{197}$, $P2_13^{198}$, $I2_13^{199}$
	$m\bar{3}$	**$Pm\bar{3}^{200}$, $Pn\bar{3}^{201}$, $Fm\bar{3}^{202}$, $Fd\bar{3}^{203}$, $Im\bar{3}^{204}$, $Pa\bar{3}^{205}$, $Ia\bar{3}^{206}$**
	432	$P432^{207}$, $P4_232^{208}$, $F432^{209}$, $F4_132^{210}$, $I432^{211}$, $P4_332^{212}$, $P4_132^{213}$, $I4_132^{214}$
	$\bar{4}3m$	$P\bar{4}3m^{215}$, $F\bar{4}3m^{216}$, $I\bar{4}3m^{217}$, $P\bar{4}3n^{218}$, $F\bar{4}3c^{219}$, $I\bar{4}3d^{220}$
	$m\bar{3}m$	**$Pm\bar{3}m^{221}$, $Pn\bar{3}n^{222}$, $Pm\bar{3}n^{223}$, $Pn\bar{3}m^{224}$, $Fm\bar{3}m^{225}$, $Fm\bar{3}c^{226}$, $Fd\bar{3}m^{227}$, $Fd\bar{3}c^{228}$, $Im\bar{3}m^{229}$, $Ia\bar{3}d^{230}$**

- Second, consider all permissible substitutions of mirror planes and/or rotation axes with glide planes and screw axes, respectively. In this step, it is essential to examine whether the substitution results in a new group or not, or whether the resulting group can be reduced to the existing one by a permutation of the unit cell edges.

For example, think about the monoclinic point group m in the standard setting, where m is perpendicular to **b** (Table 2.7). According to Table 2.13, the following Bravais lattices are allowed in the monoclinic crystal system: P and C. There is only one finite symmetry element (mirror plane m) to be considered for replacement with glide planes (a, b, c, n, and d):

- The first and obvious choice of the crystallographic space group is Pm.
- By replacing m with a, we obtain new space group, Pa.
- Replacing m with b is prohibited since the plane b cannot be perpendicular to **b**.
- By replacing m with c, we obtain space group Pc. This space-group symmetry is identical to Pa, which is achieved by switching **a** and **c** because glide plane, a, translates a point by half of full translation along **a**, and glide plane, c, translates a point by half of full translation along **c**. Pc is a standard choice (see Table 3.3).
- By replacing m with n, we obtain space group Pn. Similar to Pa, this space group can be converted into Pc when the following transformation is applied to the unit cell vectors: $\mathbf{a_{new}} = -\mathbf{a_{old}}$, $\mathbf{b_{new}} = \mathbf{b_{old}}$, $\mathbf{c_{new}} = \mathbf{a_{old}} + \mathbf{c_{old}}$.
- Glide plane, d, is incompatible with the primitive Bravais lattice.

– By repeating the same process in combination with the base-centered lattice, C, two new space-groups symmetry, Cm and Cc, can be obtained.

Therefore, the following four monoclinic crystallographic space groups (Pm, Pc, Cm, and Cc) result from a single monoclinic point group (m) after considering all possible translations in three dimensions.

3.4.2 Full International Symbols of Crystallographic Space Groups

The 230 crystallographic groups listed in Table 3.3 are given in the so-called standard orientation (or setting), which includes proper selection of both the coordinate system and origin. However, there exist a number of publications in the scientific literature, where space-group symbols are different from those provided in Table 3.3. Despite being different, these symbols refer to one of the same 230 crystallographic space groups but using a nonstandard setting or even using a nonstandard choice of the coordinate system. These ambiguities primarily occur because of the following:

– The crystal structure was solved using a nonstandard setting, since most of the modern crystallographic software enables minor deviations from the standard, and the results were published as they were obtained, without converting them to a conventional orientation. It is worth noting that many but not all technical journals allow certain deviations from crystallographic standards.
– The crystal structure contains some specific molecules, blocks, layers, or chains of atoms or molecules, which may be easily visualized or represented using space-group symmetry in a nonstandard setting.

Deviations from the standard are most often observed in the monoclinic crystal system, because there are many different ways that result in a nonstandard setting in this crystal system. This uncertainty is even reflected in the International Tables for Crystallography, where there are two different settings in the monoclinic crystal system. When the unique twofold axis is parallel to **b** (i.e., to *Y*-axis), this setting is considered a standard choice, but when it is parallel to **c** (or to *Z*-axis), this is an allowed alternative setting. In addition, the unique twofold axis can be chosen to be parallel to **a** (i.e., to *X*-axis), which is considered a nonstandard setting.

To reflect or to emphasize a nonstandard choice of space-group symmetry, the so-called full international or full Hermann–Mauguin symbols can be used. In the full symbol, both the rotation axes parallel to the specific direction (see Table 2.7) and planes perpendicular to them (if any) are specified for each of the three positions. If the space group has no symmetry element parallel or perpendicular to a given direction, the symbol of the onefold rotation axis is placed in the corresponding position. For example, $P12_1/a1$ and $P112_1/a$, both refer to the same space-group symmetry where the orientation of the crystallographic axes, *Y* and *Z*, is switched.

Full international symbols may also be used to designate space-group symmetry in a standard setting, such as when the short notation Pmmm is replaced by the full Hermann–Mauguin notation P2/m2/m2/m. The full symbol in this case emphasizes the presence of the three mutually perpendicular twofold axes and mirror planes that are parallel and perpendicular, respectively, to the three major crystallographic axes.

Thus, one of the most commonly observed in both natural and synthetic crystals, space-group symmetry $P2_1/c$ (standard notation with the unique twofold screw axis parallel to *Y*, and $\alpha = \gamma = 90°$ and $\beta \neq 90°$) can be listed as follows:

- $P2_1/b11$ and $P2_1/c11$, when the unique twofold screw axis is chosen to be parallel to *X*. Both symbols represent nonstandard settings of this group, in which $\alpha \neq 90°$.
- $P12_1/c1$ (or short $P2_1/c$, which is standard) and $P12_1/a1$ (or short $P2_1/a$, which is also standard except for the choice of the glide plane), when the unique twofold screw axis is chosen to be parallel to *Y*.
- $P112_1/a$ and $P112_1/b$, when the unique twofold screw axis is chosen to be parallel to *Z* (both represent alternative setting, where the unique twofold screw axis is parallel to *Y*, and therefore, $\gamma \neq 90°$).
- $P2_1/n11$, $P12_1/n1$ (or short $P2_1/n$) or $P112_1/n$, when the unique twofold screw axis is chosen to be parallel to *X*, *Y*, or *Z*, respectively, but the selection of the diagonal glide plane, n, represents a deviation from the standard. The alternative $P2_1/n$ setting is routinely used if it results in the monoclinic angle β closer to 90° than in the standard $P2_1/c$.

Since the selection of the crystallographic coordinate system is not unique, conventionally, a right-handed set[7] of basis vectors **a, b, c** is chosen in compliance with Table 2.11 and in a way that the combination of symmetry elements in the space group is best visualized. The selection of the conventional origin is usually more complicated, but in general the origin of the coordinate system is selected at the center of inversion, if it is present, or at the point with the highest site symmetry, if there is no center of inversion in the group.

Additional information about each of the 230 three-dimensional crystallographic space groups can be found in the International Tables for Crystallography, vol. A. It includes their symbols, diagrams of all symmetry elements present in the group together with their orientation with respect to crystallographic axes, the origin of the coordinate system, and more, for both the conventional and alternative (if any) settings. The format of the International Tables for Crystallography and some relevant issues are briefly discussed in Sect. 3.5.

[7] In a right-handed set, the positive directions of basis vectors **c**, **a**, **b** are chosen from the middle of the palm of the right hand toward the ends of thumb, index, and middle fingers, respectively.

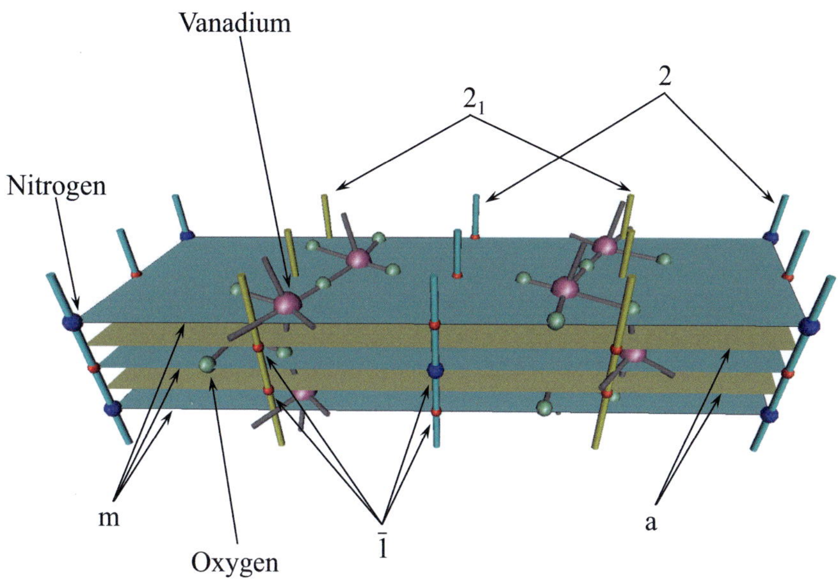

Fig. 3.6 The three-dimensional view of the crystal structure of *tma*V_8O_{20} (vanadium oxide layers and nitrogen atoms from *tma* molecules; the whole *tma* molecules are not shown for clarity) with added symmetry elements. The space-group symmetry of the material is C2/m

3.4.3 Visualization of Space-Group Symmetry in Three Dimensions

An excellent way to achieve a better understanding of both the infinite symmetry elements and how they are combined in crystallographic space groups is to use three-dimensional illustrations. Thus, the crystal structure of the vanadium oxide layer in *tma*V_8O_{20}[8] (*tma* is shorthand for tetramethylammonium, $[N(CH_3)_4]^+$) and symmetry elements in its unit cell are shown in Fig. 3.6.

The drawing of the *tma*V_8O_{20} crystal structure was created using the General Structure Analysis System (GSAS),[9] and the symmetry elements were added later manually. The following color scheme is used:

- Twofold rotation axes and mirror planes are shown in blue.
- Glide planes, a, and 2_1 screw axes are shown in yellow.
- Inversion centers are represented as small red spheres.

---

[8]T. Chirayil, P.Y. Zavalij, M.S. Whittingham, Synthesis and characterization of a new vanadium oxide, *tma*V_8O_{20}, J. Mater. Chem. **7**, 2193 (1997), https://doi.org/10.1039/A704624C

[9]A.C. Larson and R.B. Von Dreele, General Structure Analysis System (GSAS), Los Alamos National Laboratory Report, LAUR 86-748 (2004) available at https://www.ncnr.nist.gov/xtal/software/gsas.html

– The inversion center in the origin of coordinates (0, 0, 0), which is occupied by nitrogen atom is shown as a blue sphere.
– Vanadium and oxygen atoms are shown as pink and green spheres, respectively.

3.4.4 Space Groups in Nature

In conclusion of this section, it may be interesting to note that not all of the 230 crystallographic space groups have been found in real crystalline materials. For example, space groups $P4_222$ and P432 are not found in the Inorganic Crystal Structure Database (ICSD),[10] which at the end of 2020 contained crystallographic data of 200,000+ inorganic compounds. The frequency of the occurrence of various space groups is far from uniform and it varies for different classes of materials.

Organic compounds predominantly crystallize in the low-symmetry crystal systems: 95% of known organic crystal structures exhibit orthorhombic or lower symmetry. Specifically, nearly 80% of organic crystal structures fall into just five space groups: $P2_1/c$ (34.0%), $P\bar{1}$ (25.2%), C2/c (8.2%), $P2_12_12_1$ (7.0%), and $P2_1$ (5.2%), with the first two groups encompassing more than half of all organic compounds.[11] In contrast, the majority of inorganic compounds crystallize in space groups with orthorhombic or higher symmetry, exhibiting a much wider distribution across different space groups. In order of decreasing frequency, these groups are: Pnma, $P2_1/c$, $Fm\bar{3}m$, $Fd\bar{3}m$, I4/mmm, $P6_3/mmc$, $P\bar{1}$, C2/c, C2/m, and $Pm\bar{3}m$. These ten crystallographic space groups account for approximately 35% of inorganic structures, with 25% of the structures concentrated in the first six groups.[12]

3.5 International Tables for Crystallography

An example of how each of the 230 three-dimensional crystallographic space groups is listed in the International Tables for Crystallography[13] is shown in Table 3.4. There are 12 fields in Table 3.4; each of them contains the following information:

1. *Header*: Provides the short international (Hermann–Mauguin) space-group symmetry symbol (***Cmm*2**) followed by the Schönflies symbol of the same

[10] The Inorganic Crystal Structure Database (ICSD) Web Version 4.5.0 (Nov. 30, 2020), (FIZ Karlsruhe, Germany) https://icsd.fiz-karlsruhe.de/

[11] Space group statistics from CSD database as of Jan. 1, 2024 according to CCDC website https://www.ccdc.cam.ac.uk/

[12] Based on statistics from the ICSD website https://icsd.fiz-karlsruhe.de/ collected on Oct. 8, 2024.

[13] Format varies in earlier editions of the International Tables for Crystallography; here, we follow the format adopted in volume A since the 1983 edition.

Table 3.4 The space-group symmetry Cmm2 as it is listed in the International Tables for Crystallography, vol. A

Field	Content					
(1)	***Cmm*2**	C_{2v}^{11}	***mm*2**		**Orthorhombic**	
(2)	**No 35**	***Cmm*2**			**Patterson symmetry *Cmmm***	
(3)	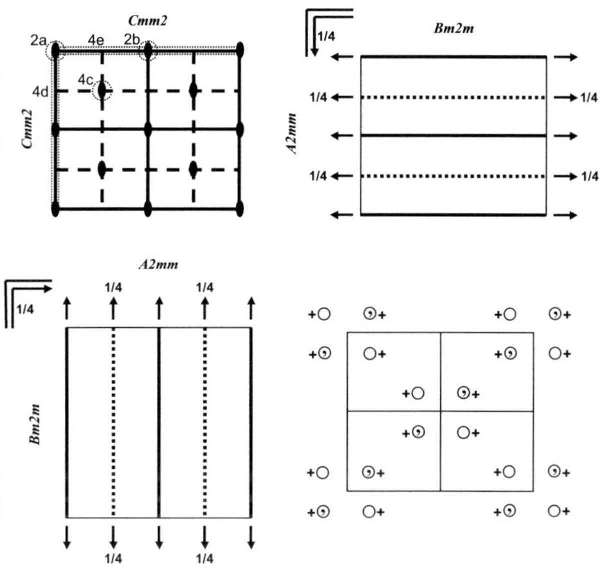					
(4)	**Origin** on mm2					
(5)	**Asymmetric unit** $0 \le x \le \frac{1}{4}$; $0 \le y \le \frac{1}{2}$; $0 \le z \le 1$					
(6)	**Symmetry operations**					
	For (0,0,0)+ set					
	(1) 1	(2) 2 0,0,*z*		(3) *m* *x*,0,*z*		(4) *m* 0,*y*,*z*
	For (½,½,0)+ set					
	(1) *t*(½,½,0)	(2) 2 ¼,¼,*z*		(3) *a* *x*,¼,*z*		(4) *b* ¼,*y*,*z*
(7)	**Generators selected** (1); *t*(1,0,0); *t*(0,1,0); *t*(0,0,1); *t*(½,½,0); (2); (3)					
(8)	**Positions**					
	Multiplicity, Wyckoff letter, Site symmetry		Coordinates (0,0,0)+ (½,½,0)+		Reflection conditions General	
	8 *f* 1	(1) *x*,*y*,*z*	(2) $\bar{x},\bar{y},z$	(3) $x,\bar{y},z$	(4) $\bar{x},y,z$	*hkl*: $h + k = 2n$
						0*kl*: $k = 2n$
						*h*0*l*: $h = 2n$
						*hk*0: $h + k = 2n$
						*h*00: $h = 2n$
						0*k*0: $k = 2n$
						Special: as above, plus

(continued)

Table 3.4 (continued)

Field	Content					
	4 *e* *m*	0, *y*, *z*		$0, \bar{y}, z$		No extra conditions
	4 *d* *m*	*x*, 0, *z*		$\bar{x}, 0, z$		No extra conditions
	4 *c* 2	¼, ¼, *z*		¼, ¾, *z*		*hkl*: *h* = 2*n*
	2 *b* *mm*2	0, ½, *z*				No extra conditions
	2 *a* *mm*2	0, 0, *z*				No extra conditions
(9)	**Symmetry of special projections**					
	Along [001] *c*2*mm*		Along [100] *p*1*m*1		Along [010] *p*11*m*	
	a′ = **a** **b′** = **b**		**a′** = ½b **b′** = **c**		**a′** = **c** **b′** = ½**a**	
	Origin at 0,0,*z*		Origin at *x*,0,0		Origin at 0,*y*,0	
(10)	**Maximal nonisomorphic subgroups**					
	I	[2] *C*1*m*1 (*Cm*, 8)		(1; 3)+		
		[2] *Cm*11 (*Cm*, 8)		(1; 4)+		
		[2] *C*112 (*P*2, 3)		(1; 2)+		
	IIa	[2] *Pba*2 (32)		1; 2; (3; 4) + (½,½,0)		
		[2] *Pbm*2 (*Pma*2, 28)		1; 3; (2; 4) + (½,½,0)		
		[2] *Pma*2 (28)		1; 4; (2; 3) + (½,½,0)		
		[2] *Pmm*2 (25)		1; 2; 3; 4		
	IIb	[2] *Ima*2 (**c′** = 2**c**) (46); [2] *Ibm*2 (**c′** = 2**c**) (*Ima*2, 46); [2] *Iba*2 (**c′** = 2**c**) (45); [2] *Imm*2 (**c′** = 2**c**) (44); [2] *Ccc*2 (**c′** = 2**c**) (37); [2] $Cmc2_1$ (**c** **′** = 2**c**) (36); [2] $Ccm2_1$ (**c′** = 2**c**) ($Cmc2_1$, 36);				
(11)	**Maximal isomorphic subgroups of lowest index**					
	IIc	[2] *Cmm*2 (**c’** = 2**c**) (35); [3] *Cmm*2 (**a’** = 3**a** or **b’** = 3**b**) (35);				
(12)	**Minimal nonisomorphic supergroups**					
	I	[2] *Cmmm* (65); [2] *Cmme* (67); [2] *P*4*mm* (99); [2] *P*4*bm* (100); [2] $P4_2cm$ (101);				
		[2] $P4_2nm$ (102); [2] $P\bar{4}2m$ (111); [2] $P\bar{4}2_1m$ (113); [3] *P*6*mm* (183)				
	II	[2] *Fmm*2 (42); [2] *Pmm*2 (**a′** = ½**a, b′** = ½**b**) (25)				

space group (C_{2v}^{11}, these symbols are not discussed in this book[14]), the corresponding point-group symmetry symbol (***mm*2**) and the name of the crystal system (**Orthorhombic**).

2. *Second header*: Includes the sequential number of the space-group symmetry (**35**) followed by the full international (Hermann–Mauguin) symbol (***Cmm*2**; in this case, the full symbol is the same as the short one) and Patterson symmetry (***Cmmm***). Although the Patterson function is discussed in Sect. 10.2.1, it is worth noting here that the Patterson symmetry is derived from the symmetry of

[14] In Schönflies notation, C_{Nv} designates point group with an N-fold rotation axis and a mirror plane parallel to this axis. Hence, C_{2v} is point group mm2 written in Schönflies notation. The superscript ([11]) added to the point group symbol is the sequential number in the series of space groups based on this particular point group (see row #7 in Table 3.3, where Cmm2 is No. 11 in this row). Hence, international space group symbols Pmn2 and $Cmc2_1$ become C_{2v}^{10} and C_{2v}^{12}, respectively, in Schönflies notation.

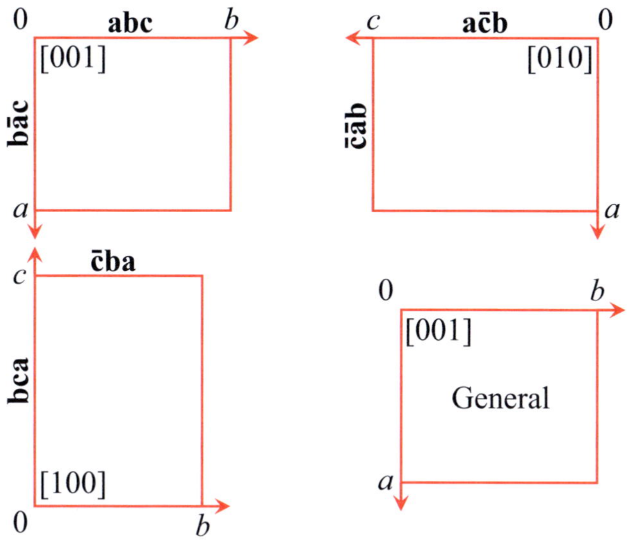

Fig. 3.7 Schematic representation of orthorhombic space-group symmetry diagrams showing the origin of coordinates (0), the labels of the axes (a, b, and c), and the projection directions ([001], [010], and [100]). The box marked "General" is a general position diagram (see the description of fields "Symmetry operations" and "Positions" in the numbered list in this section). The triplets (abc and so on) indicate settings symbols that result in the corresponding space-group symbols. Each settings symbol is obtained by a permutation of the unit cell axes, e.g., cab is the cyclic permutation of abc: a′ = c, b′ = a, c′ = b. It should be noted that a space-group symbol is invariant to the changes of the sign (direction) of the axes

the space group by replacing all screw axes and glide planes with the corresponding finite symmetry elements, and by adding a center of inversion, if it is not present in the space group.

3. *Space-group symmetry diagrams* are shown as one or several (up to 3) orthogonal projections along different unit cell axes. The projection direction is perpendicular to the plane of the figure. Projection directions, orientations of axes, and selection of the origin of the coordinates are dependent on the crystal system, and a full description can be found in the International Tables for Crystallography, vol. A. The schematic for the orthorhombic crystal system is shown in Fig. 3.7. If the horizontal symmetry element is elevated above the plane of the projection by a fraction of the translation along the projection direction other than 0 or ½, its elevation is shown as the corresponding fraction, that is, $^1/_8$, ¼, and so on. An additional diagram shows symmetrically equivalent points (both inside and in the immediate vicinity of one unit cell) that are related to one another by the symmetry elements present in the orientation matching the first diagram.
4. *Origin of the unit cell* is given as the site symmetry and its location, if necessary. In the example shown in Table 3.4, the origin of the unit cell is located on **mm2**, that is, on the twofold axis, which coincides with the line where the two

perpendicular mirror planes intersect. In this example, the origin can be chosen arbitrarily on the Z-axis since there are no symmetry elements with a fixed z-coordinate in the space group Cmm2.

5. *Asymmetric unit* represents the fraction of the unit cell, which generally contains symmetrically inequivalent points. It is delineated by the elementary parallelepiped specified in terms of fractions of the corresponding unit cell edges (for example, $\mathbf{0 \leq x \leq 1/4}$, $\mathbf{0 \leq y \leq 1/2}$, and $\mathbf{0 \leq z \leq 1}$) and (if necessary) by including supplementary restrictions, e.g., $x \leq (1 + y)/2$; $y \leq x/2$ as in the space group P$\bar{3}$m1.
6. *Symmetry operations*: For each point with coordinates x, y, z in the general position, the symmetry operation which transforms this point into symmetrical equivalent is listed together with its sequential number. The term "general position" applies to any point in the unit cell that is not located on any of the finite symmetry elements present in the group. Symmetry operations in the nonprimitive space groups are divided into sets. The sets are arranged based on the translation vector(s) corresponding to Bravais lattice centering, which are added to the coordinates of each point in the unit cell. The first set is always the **(0,0,0)** + set, which represents the primitive basis that is enough to describe the same but primitive space-group symmetry (in this case, it is Pmm2). The second set in Table 3.4 is (½,½,0) + set, which accounts for the presence of the base-centered lattice, C. Each symmetry operation in addition to its sequential number (in parenthesis) lists the nature of the operation and its location (t stands for translations). For example, t(½,½,0) stands for the translation of the point by ½ along both a (or X) and b (or Y); the notation a (x,¼,z) specifies glide plane, a, perpendicular to Y and intersecting Y at y = ¼.
7. *Generators selected* specify the minimum set of symmetry operations, including translations that are needed to generate the space-group symmetry. They begin with identity—symmetry operation (1), then translations (the first three represent full translations along the three major crystallographic axes and the fourth is the translation reflecting the presence of the base-centered Bravais lattice, C) followed by the numbers of symmetry operations from the first set in the previous field. Hence, the space group Cmm2 can be generated by using symmetry operations No. 1, No. 2 (the twofold axis coinciding with Z), and No. 3 (the mirror plane perpendicular to *Y* and intersecting *Y* at $y = 0$) in addition to four translations.
8. *Positions*: This field contains standardized information about possible locations (or sites) that can accommodate points (or atoms) in the unit cell and corresponding reflection conditions. Reflection conditions show the limitations on the possible combinations of Miller indices that are imposed by the symmetry of the space group and these are discussed in Sects. 9.3 and 9.4. Each record in this field corresponds to one site and each site is listed starting with the multiplicity of the site position (integer numbers, **8**, **4**, **4**, …, **2**) followed by

Wyckoff[15] letter (**f**, **e**, **d**, …, **a**), site symmetry (**1**, **m..**, **.m.**, …, **mm2**), and coordinate triplets of the symmetrically equivalent atoms. The multiplicity of the site position is the total number of symmetrically equivalent atoms that will appear in one unit cell as the result of having an atom with the coordinates corresponding to any of the listed triplets. For example, the multiplicity of the second site is 4 (Wyckoff notation is 4e). The *x*-coordinate of any independent atom in this site is fixed at 0, while *y*- and *z*-coordinates may vary between 0 and 1. Assume that $y = 0.15$ and $z = 0.31$. The complete list of four symmetrically equivalent atoms in this position is obtained as follows:

Atom1 $(0, y, z)$ plus (0, 0, 0): $x = 0$, $y = 0.15$, $z = 0.31$;
Atom2 $(0, \bar{y}, z)$ plus (0, 0, 0): $x = 0$, $y = 0.85$,[16] $z = 0.31$;
Atom3 $(0, y, z)$ plus (½,½, 0): $x = 0.5$, $y = 0.65$, $z = 0.31$;
Atom4 $(0, \bar{y}, z)$ plus (½,½, 0): $x = 0.5$, $y = 0.35$, $z = 0.31$.

Site positions in this field are arranged according to their multiplicities (from the highest to the lowest) and according to site symmetry (from the lowest to the highest). Wyckoff letters are assigned to site positions starting with "a" for the site with the lowest multiplicity and the highest symmetry. The coordinate triplets for the site with the highest multiplicity and the lowest symmetry (general position) are listed with the numbers of symmetry operations that generate this atom (as they appear in field "Symmetry operations").

9. *Symmetry of special projections* is usually given along *X*, *Y*, and *Z* axes, and in crystal systems with higher symmetry also along diagonals together with the axes and the origin of the projected unit cell. These projections correspond to two-dimensional crystallographic groups.
10. - (12) The last three fields entitled "Maximal nonisomorphic subgroups," "Maximal isomorphic subgroups of lowest index," and "Minimal nonisomorphic supergroups" list the closest subgroups and supergroups, their axes, and other relevant information. The discussion of these fields goes beyond the scope of this book, and the International Tables for Crystallography, vol. A should be consulted for further details.

[15] Ralph W.G. Wyckoff (1987–1994), the American crystallographer who in 1922 published a book *The Analytical Expression of the Results of the Theory of Space Groups*, which contained tables with the positional coordinates, both general and special, permitted by the combination of symmetry elements in the group, thus forming the basis for the description of 230 space groups in The International Tables for Crystallography. Also see the obituary published in Acta Cryst. A**51**, 649 (1995), https://doi.org/10.1107/S0108767395006374

[16] Any negative coordinate (i.e. $y = -0.15$) may be converted into a positive coordinate by adding a full translation along the same axis, i.e. $y = -0.15 + 1 = 0.85$.

3.6 Equivalent Positions (Sites)

As briefly mentioned in Sect. 3.5, equivalent positions (or sites) that are listed in the field "Positions" in Table 3.4 for each crystallographic space group represent sets of symmetrically equivalent points found in one unit cell. All equivalent points in one site are obtained from an initial point by applying all symmetry operations that are present in the unit cell. The fractional coordinates (coordinate triplet) of the initial (or independent) point are usually marked as x, y, z.

3.6.1 General and Special Equivalent Positions

The equivalent position is called *general* when the initial point is *not* located on any of the finite symmetry elements, that is, not on those elements that convert the point into itself, if they are present in the group. The general equivalent position has the highest multiplicity, and every one of the 230 space groups has one and only one general site. However, since the only limitation on the possible combination of values of x, y, and z in the coordinate triplet is imposed by geometrical constraints that prevent neighboring atoms from overlapping with one another, multiple sets of atoms occupying the general site with different coordinate triplets of independent atoms are possible in many crystal structures. An atom in the general equivalent position in any of the 230 three-dimensional space groups always has three positional degrees of freedom, that is, each of the three coordinates may vary independently.

When a point (or an atom) is placed on a finite symmetry element that converts the point into itself, the multiplicity of the site is reduced by an integer factor when compared to the multiplicity of the general site. Since different finite symmetry elements may be present in the same space-group symmetry, the total number of different "nongeneral" sites (they are called *special* sites or special equivalent positions) may exceed one. Contrary to a general equivalent position, one, two, or all three coordinates are constrained in every atom occupying a special equivalent position.

Both the multiplicity and Wyckoff letter combined together are often used as the name of the equivalent position. Sometimes when crystallographic data are published, the coordinates of all independent atoms are given in reference to equivalent positions they occupy. For example, if in a hypothetical crystal structure, nickel atoms occupy the site 4c in the space group Cmm2, they can be listed as "Ni in 4c, z = 0.1102," which indicates that there are a total of four nickel atoms in the unit cell and one of them has the coordinates $x = 1/4$, $y = 1/4$, $z = 0.1102$. The coordinates of the remaining three nickel atoms are easily determined from the coordinates of all symmetrically equivalent points in the position 4c, see field No. 8 "Positions" in Table 3.4.

3.6.2 Special Sites with Points Located on Mirror Planes

The first example of a special position was considered earlier (see the description of field "Positions" in Table 3.4), when we analyzed the coordinates of four symmetrically equivalent atoms located in the mirror plane in the space group Cmm2. Both this site and the corresponding mirror plane are marked as 4e in the first diagram in Table 3.4. A different special position on a different mirror plane in this space-group symmetry is marked as 4d in the same figure. Two additional examples are provided in Fig. 3.8.

In the tetragonal crystal system (Fig. 3.8, left), diagonal mirror planes perpendicular to the *XY* plane are possible. When an atom is placed in the special position on this plane, its coordinates are restricted in a way that $y = x$, while x and z may assume any values. This relationship is usually emphasized by specifying the corresponding coordinate triplet as x, x, z. Similarly, in the hexagonal crystal system (Fig. 3.8, right), multiple diagonal mirror planes perpendicular to the *XY* plane are possible. When an atom is placed in the special position on the mirror plane nearest to the *Y*-axis, its coordinates are restricted such that $y = 2x$, but x and z vary freely. The corresponding coordinate triplet becomes x, $2x$, z. Considering the two remaining mirror planes and moving clockwise, the point that belongs to the next mirror plane will have coordinate triplet x, x, z (i.e., the same as in the case of the diagonal mirror plane in the tetragonal crystal system), and for the third mirror plane, the triplet becomes x, $\frac{1}{2}x$, z (which is the same as $2x$, x, z).

Therefore, any atom in a special position where it belongs to a mirror plane has two positional degrees of freedom. Only two of the three coordinates vary independently, whereas the third is restricted to either a constant value, for example, positions 4e and 4d in the space group Cmm2, or it is constrained to be proportional to one of the two free coordinates (Fig. 3.8). Similar to a general position, any special position on the mirror plane can accommodate many independent atoms provided they do not overlap.

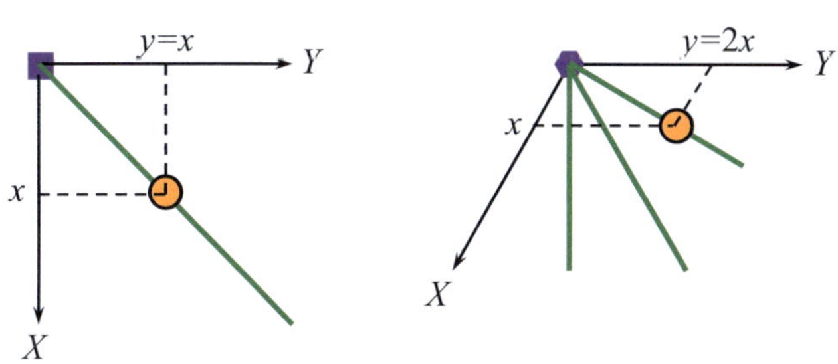

Fig. 3.8 Constraints imposed on the coordinates of atoms located on diagonal mirror planes (*thick solid lines*) in tetragonal (*left*) and hexagonal (*right*) crystal systems

3.6.3 Special Sites with Points Located on Rotation and Inversions Axes

Both the rotation and inversion axes can also be the source of special positions. Consider, for example, the site 2a (Table 3.4) where atoms are accommodated by the twofold rotation axis that follows the line at which two mutually perpendicular mirror planes intersect. In this case, two of the three coordinates in the triplet are fixed ($x = 0$ and $y = 0$), while the third coordinate (z) may assume any value. A similar special position is represented by the site 4c (Table 3.4), where the twofold rotation axis is parallel to Z and coincides with the line at which two mutually perpendicular glide planes (a and b) intersect. In this position, the two coordinates (x and y) are fixed at $x = y = ¼$ and z varies. Note that the multiplicities of these two sites (2a and 4c) are different, as they are defined by the total number of symmetrically equivalent points in the given space-group symmetry.

Twofold axes can also be parallel to face diagonal(s) in tetragonal, hexagonal, and cubic crystal systems, and one example for a cubic crystal system is shown in Fig. 3.9, left. The coordinates of a point (open circle) located on the twofold axis that coincides with the diagonal of a square face of the cube are constrained at x, x, 0. As we already know, there are four threefold rotation or inversion axes coinciding with the body diagonals of the cube in the cubic crystal system. Also shown in Fig. 3.9 on the left is one of these threefold rotation axes, where the coordinates of a point located in the special position defined by the orientation of this crystallographic axis (filled circle) are constrained at x, x, x.

Therefore, the number of positional degrees of freedom is further reduced to only one independent coordinate in special positions where atoms are located on rotation or inversion crystallographic axes. Similar to both the general position and special

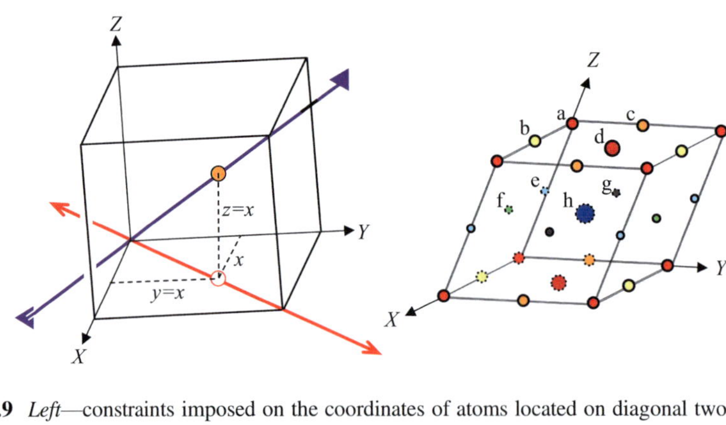

Fig. 3.9 *Left*—constraints imposed on the coordinates of atoms located on diagonal twofold and threefold axes in cubic crystal system. *Right*—the distribution of the inversion centers in the triclinic space-group symmetry $P\bar{1}$. Eight independent centers are labeled from "a" through "h." Inversion centers that are equivalent to one another are marked using symbols of the same size and color. The "invisible" centers are drawn using dotted lines

sites on mirror planes, any special equivalent position on a rotation or inversion axis can accommodate many independent atoms (geometrical constraints are always applicable).

3.6.4 Special Sites with Points Located on Centers of Inversion

Atoms can also reside on centers of inversion. Since there is no inversion center in the space group Cmm2, which was considered in Table 3.4, we turn our attention to the distribution of the centers of inversion in the unit cell that belongs to the triclinic space group symmetry $P\bar{1}$ (Fig. 3.9, right). Similar to the distribution of rotation axes shown in Fig. 3.5, this is a characteristic distribution of the inversion centers in a primitive centrosymmetric group.

Thus, there are no positional degrees of freedom available to an atom occupying any special position created by the presence of the center of inversion. Further, unlike in the cases when atoms are located on the mirror planes and rotation or inversion axes, only one atom can occupy any single special position on the center of inversion. Note that all of the inversion centers in the $P\bar{1}$ group are independent yielding a total of eight independent special positions, while in other space groups of higher symmetry some or all of these centers may be symmetrically related, and the number of such special positions may be reduced to as low as one.

3.7 Additional Reading

1. C. Giacovazzo, H.L. Monaco, G. Artioli, D. Viterbo, M. Milaneso, G. Gilli, P. Gilli, G. Zanotti, G. Ferraris, and M. Catti, Fundamentals of crystallography, 3rd edition, C. Giacovazzo, Ed., IUCr/Oxford University Press (2011). https://doi.org/10.1093/acprof:oso/9780199573653.001.0001
2. D. Schwarzenbach, Crystallography, Wiley, New York (1996). https://www.wiley.com/en-us/Crystallography-p-9780471955986
3. C. Hammond, The basics of crystallography and diffraction, 4th edition. IUCr/Oxford University Press (2015). https://doi.org/10.1093/acprof:oso/9780198738671.001.0001
4. D. Farmer, Groups and symmetry: A Guide to Discovering Mathematics (Mathematical World, Vol. 5) 1st Edition, ISBN: 0821804502, Amer. Math. Soc. (1995). https://bookstore.ams.org/mawrld-5
5. International Tables for Crystallography, Vol. A, Space-group symmetry, M. I. Aroyo, Ed., Second online edition, (2016). https://it.iucr.org/A/

6. International Tables for Crystallography. 6th edition, Teaching edition, Crystallographic symmetry, M. I. Aroyo, Ed., Wiley (2021). https://www.wiley.com/en-us/-p-9780470974223
7. IUCr Commission on Crystallographic teaching. Teaching pamphlets. https://www.iucr.org/education/pamphlets

3.8 Problems

1. Consider space-group symmetry Fdd2. Without using the International Tables for Crystallography, establish the following: (a) the crystal system; (b) the corresponding point-group symmetry; (c) the corresponding Laue class; (d) the relationships between the unit cell dimensions; and (e) explain the space-group symbol.
2. Consider orthorhombic space-group symmetry Pnma

   4.1. On what point-group symmetry is this space group based upon?
   4.2. When the *X* and *Y* directions are switched, one obtains a different setting of the same space group—Pmnb. In as few sentences as possible, yet with details, explain why not Pmna. If you think a sketch will help, supplement your explanations with a figure.
   4.3. Considering all possible permutations of *X*, *Y*, and *Z* axes, find all valid nonstandard settings of this space group.

3. Consider orthorhombic space-group symmetry P2mb illustrated in Fig. 3.10. Similar to as this is done in the International Tables for Crystallography, list multiplicities and coordinates of all symmetrically equivalent points for each independent site position (general and special, if any) in this space-group symmetry and briefly explain your answers.

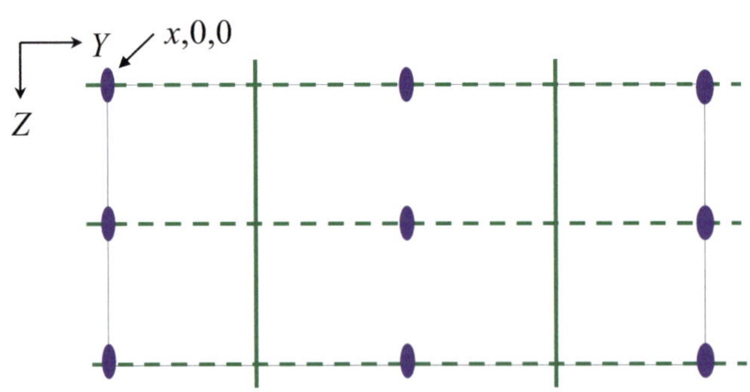

Fig. 3.10 Combination of symmetry elements in one unit cell (thin lines) for space-group symmetry P2mb. The origin of coordinates is on the twofold axis marked by arrow

4. Consider independent atoms with the following coordinates in the space-group symmetry C2/m: Atom1: $x = 0.15$, $y = 0.0$, $z = 0.33$; Atom2: $x = 0.5$, $y = 0.11$, $z = 0.5$; and Atom3: $x = 0.25$, $y = 0.25$, $z = 0.25$. Using the International Tables for Crystallography, carry out the following tasks:

   4.1. Apply the coordinates and centering vectors listed for the general equivalent position to generate all symmetrically equivalent atoms from the three listed independent atoms (the total in each case should be the same as the multiplicity of the general position).
   4.2. Find atoms with equal coordinate triplets (remember that the difference by a full translation in one, two, or three directions refers to the same atom) and cross them out. The total number of atoms left is the multiplicity of the corresponding special position.
   4.3. Establish both the multiplicity and the Wyckoff notation of special position for each of the three listed independent atoms.
   4.4. To which symmetry element(s), if any, do the independent atoms belong?
   4.5. Which of the three original independent atoms occupies the general equivalent position?

5. The crystal structure of a material is described in space-group symmetry $P6_3/mmc$ with the following atomic coordinates:

Atom	x	y	z
Ba1	0	0	0.25
Ba2	0.3333	0.6667	0.9110
Ni	0	0	0
Sb	0.3333	0.6667	0.1510
O1	0.4816	−0.0368	0.25
O2	0.1685	0.3370	0.4169

   Using the International Tables for Crystallography, describe every atom in terms of the multiplicities and Wyckoff letters of their site positions and establish the content of the unit cell, the simplest chemical formula, and the number of formula units[17] (Z) per unit cell.

6. Using a sheet of rectilinear graphing paper, draw a projection along the c-axis of one unit cell of the crystal structure of an RX_2 material that has the AlB_2-type crystal structure. The space-group symmetry is P6/mmm. Assume that $a = c = 4$ Å. Atoms are located in the unit cell as follows: R in 1(a) and X in 2 (d); both sites are fully occupied. Radius of atom R is larger than that of atom X by ~30%. Prepare the drawing by hand and make sure that it is to scale, including relative sizes of atoms (radii of circles). Mark elevations of all atoms above the plane of the projections as their fractional coordinates along c accurate to 1/100.

[17] Usually, a formula unit corresponds to the simplest chemical formula or to the stoichiometry of the molecule of a material.

Chapter 4
Formalization of Symmetry

So far we used both geometrical and verbal tools to describe symmetry elements (for example, plane, axis, center, and translation) and operations, such as reflection, rotation, inversion, and shift. This is quite convenient when the sole purpose of this description is to understand the concepts of symmetry. However, it becomes difficult and time-consuming when these tools are used to manipulate symmetry, for example, to generate all possible symmetry operations to complete a group. Therefore, two other methods are usually employed:

– The first one is symbolic, and it is used to simplify written descriptions of symmetry.
– The second method is algebraic, and it is very convenient in manipulating symmetry.

We begin with the symbolic description of symmetry operations, which is based on the fact that the action of any symmetry operation or any combination of symmetry operations can be described by the coordinates of the resulting point(s). In this section, we assume that the original or unique object has coordinates x, y, z.

4.1 Symbolic Representation of Symmetry

4.1.1 Finite Symmetry Operations

Consider a mirror plane that is perpendicular to the Z-axis and intersects with this axis at the origin ($z = 0$). This plane reflects objects leaving their x and y coordinates unchanged but the z-coordinate of the original object is inverted and becomes -z (may also be written with a bar over the letter, that is, $\bar{z}$, as in Table 3.4) after the reflection operation is performed. Therefore, the symbolic description of this mirror reflection operation is x, y, $-z$, or $x, y, \bar{z}$.

P. Zavalij, V. Pecharsky, *Fundamentals of Powder Diffraction and Structural Characterization of Materials*, https://doi.org/10.1007/978-3-031-91504-8_4

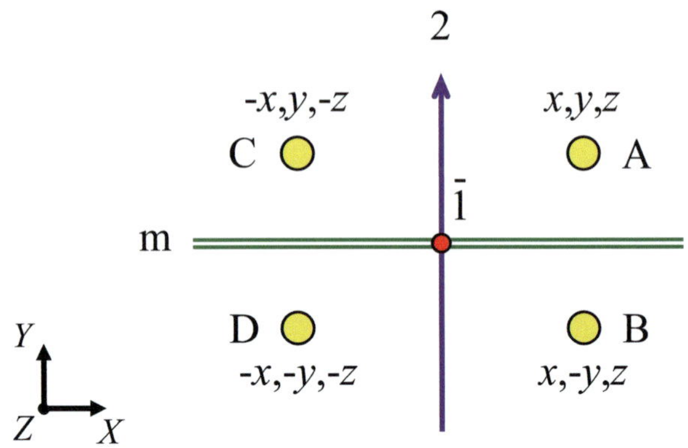

Fig. 4.1 Symmetry elements and symbolic description of symmetry operations in the point-group symmetry 2/m. Here, the origin of the coordinate system coincides with the center of inversion

Another example to consider is the four symmetry operations in the point-group symmetry 2/m as depicted in Fig. 4.1. The coordinates of four symmetrically equivalent points in Fig. 4.1 fully characterize all symmetry operations that form the group. Beginning from Point A:

- The onefold rotation converts the object into itself (A → A), which results in x, y, z.
- The mirror plane perpendicular to the Y-axis inverts y leaving x and z unchanged (A → B), which results in x, $-y$, z.
- The twofold rotation axis parallel to Y inverts both x and z leaving y unchanged (A → C), which results in $-x$, y, $-z$.
- The center of symmetry inverts all three coordinates of the object in the point with coordinates 0,0,0 (A → D), which results in $-x$, $-y$, $-z$.

Therefore, zero, one, two, or all three coordinates change their signs, but this only holds for symmetry elements of the first- and second-order when they are aligned with one of the three major crystallographic axes. Symmetry operations describing both diagonal symmetry elements and symmetry elements with the higher order (i.e., three-, four-, and sixfold rotations) may cause permutations and more complex relationships between the coordinates. For example:

- Reflection in a diagonal mirror plane may be symbolically described as y, x, z.[1]
- Rotations around the sixfold rotation axis parallel to Z result in x, y, z; $x - y$, x, z; y, $-x + y$, z; $-x$, $-y$, z; $-x + y$, $-x$, z; and $-y$, $x - y$, z.[2]
- Rotations around the threefold rotation axis along the body diagonal of a cube in the [111] direction are described by x, y, z; z, x, y; and y, z, x.

[1] In a tetragonal, hexagonal, or cubic crystallographic basis, where the plane dissects the positive directions of X- and Y-axes.

[2] In a hexagonal crystallographic basis, where X- and Y-axes form a 120° angle between them and Z-axis is perpendicular to both X and Y.

4.1.2 *Infinite Symmetry Operations*

All examples considered so far illustrate symmetry elements that traverse the origin of coordinates and do not have translations. When symmetry elements do not intersect the origin (0,0,0) or include translations, for example, glide planes and screw axes, their symbolic description includes fractions of full translations along the corresponding crystallographic axes. For example:

- Reflection in the mirror plane perpendicular to *Z* that intersects the *Z*-axis at $z = 0.25$ is described as x, y, $\frac{1}{2} - z$ (or x, y, $-z + \frac{1}{2}$).
- Rotation around and corresponding translation along the twofold screw axis, which coincides with *Y* results in $-x$, $\frac{1}{2} + y$, $-z$.
- Reflection in the glide plane, n, perpendicular to *X* and intersecting *X* at $x = 0.25$ is described symbolically as $\frac{1}{2}-x$, $\frac{1}{2} + y$, $\frac{1}{2} + z$.
- The nonprimitive translation in the base-centered unit cell C yields $x + \frac{1}{2}$, $y + \frac{1}{2}$, z.

This description formalizes symmetry operations by using the coordinates of the resulting points and, therefore, it is broadly used to represent both symmetry operations and equivalent positions in the International Tables for Crystallography (see Table 3.4). The symbolic description of symmetry operations, however, is not formal enough to enable easy manipulations involving crystallographic symmetry operations.

4.2 Algebraic Treatment of Symmetry Operations

Earlier (see Fig. 2.3) we established that there are four simple symmetry operations, namely: rotation, reflection, inversion, and translation. Among the four, reflection in a mirror plane may be represented as a complex symmetry element—twofold inversion axis—which includes simultaneous twofold rotation and inversion. Therefore, in order to minimize the number of simple symmetry operations, we begin with rotation, inversion, and translation, noting that complex operations can be described as simultaneous applications of these three simple transformations.

Algebraic description of symmetry operations is based on the following simple notion. Consider a point in a three-dimensional coordinate system with any (not necessarily orthogonal) basis, which has coordinates x, y, z. This point can be conveniently represented by the coordinates of the end of the vector, which begins in the origin of the coordinates 0,0,0 and ends at x, y, z. Thus, one only needs to specify the coordinates of the end of this vector in order to fully characterize the location of the point. Any symmetrical transformation of the point, therefore, can be described by the change in either or both the orientation and the origin of this vector.

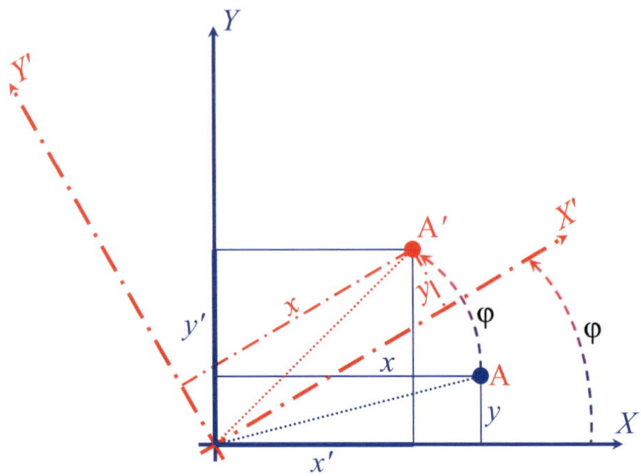

Fig. 4.2 Cartesian bases *XYZ* and *X′Y′Z′* (both *Z* and *Z′* are perpendicular to the plane of the projection) in which the coordinates of the point are invariant to the rotation around *Z* by angle φ

4.2.1 Transformation of Coordinates of a Point

Consider point A with coordinates *x*, *y*, *z* in a Cartesian[3] basis *XYZ*. Also, consider point A′ with coordinates *x′*, *y′*, *z′* in the same basis, which is obtained from Point A by rotating it around *Z* by angle φ. It is worth noting that since orientations of rotation axes in crystallography are restricted, for example, see Table 2.7, we may limit our analysis to rotations about one of the basis axes.

As shown in Fig. 4.2, it is possible to select a different Cartesian basis, *X′Y′Z′*, which is related to the original basis, *XYZ*, by the identical rotation around *Z* and in which the coordinates of the point A′ are *x*, *y*, *z*, that is, they are invariant to this transformation of coordinates. From the schematic shown in Fig. 4.2, it is easy to establish that the rotational relationships between the coordinate triplets *x*, *y*, *z* and *x′*, *y′*, *z′* in the original basis *XYZ* are given as

$$\begin{aligned} x' &= x\cos\varphi - y\sin\varphi \\ y' &= x\sin\varphi + y\cos\varphi \\ z' &= z \end{aligned} \tag{4.1}$$

Equation 4.1 is known as linear transformation of coordinates on the plane and can be written in matrix notation as shown below:

[3]Cartesian coordinate system (or basis) is the orthogonal system with $a = b = c = 1$ and $\alpha = \beta = \gamma = 90°$.

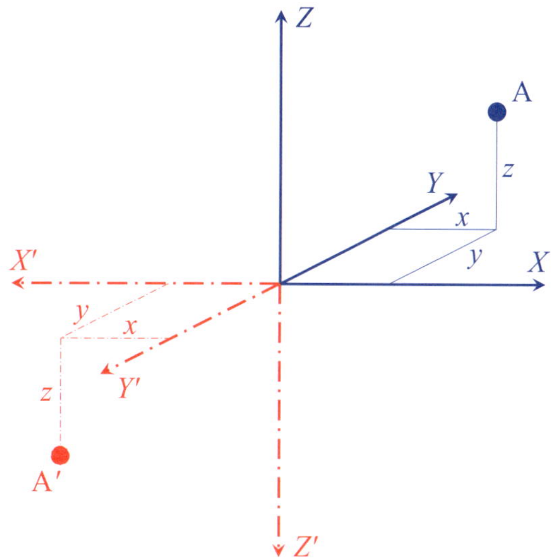

Fig. 4.3 Cartesian bases XYZ and $X'Y'Z'$ in which the coordinates of the point are invariant to the inversion through the origin of coordinates

$$\begin{pmatrix} x' \\ y' \\ z' \end{pmatrix} = \begin{pmatrix} \cos\varphi & -\sin\varphi & 0 \\ \sin\varphi & \cos\varphi & 0 \\ 0 & 0 & 1 \end{pmatrix} \begin{pmatrix} x \\ y \\ z \end{pmatrix} \tag{4.2}$$

When two points in the same Cartesian basis are related to one another via inversion through the origin of coordinates, then the coordinates of the inverted point are invariant with respect to a second Cartesian basis where the directions of all axes are reversed, as shown in Fig. 4.3.

Hence, the inversion through the origin of coordinates may be represented algebraically as

$$\begin{aligned} x' &= -x \\ y' &= -y \\ z' &= -z \end{aligned} \quad \text{or} \quad \begin{pmatrix} x' \\ y' \\ z' \end{pmatrix} = \begin{pmatrix} -1 & 0 & 0 \\ 0 & -1 & 0 \\ 0 & 0 & -1 \end{pmatrix} \begin{pmatrix} x \\ y \\ z \end{pmatrix} \tag{4.3}$$

When the two points are related to one another by roto-inversion, the resulting linear transformation is a combination of Eqs. (4.1) and (4.2) with Eq. (4.3):

$$\begin{aligned} x' &= -x\cos\varphi + y\sin\varphi \\ y' &= -x\sin\varphi - y\cos\varphi \\ z' &= -z \end{aligned} \tag{4.4}$$

and

$$\begin{pmatrix} x' \\ y' \\ z' \end{pmatrix} = \begin{pmatrix} -\cos\varphi & \sin\varphi & 0 \\ -\sin\varphi & -\cos\varphi & 0 \\ 0 & 0 & -1 \end{pmatrix} \begin{pmatrix} x \\ y \\ z \end{pmatrix} \quad (4.5)$$

Matrices in Eqs. (4.2) and (4.5) are related to one another simply by changing the sign of each element of the corresponding matrix. Considering Fig. 4.2 and Eq. (4.2), it is easy to see that when a point is rotated around the *Y*-axis, the corresponding transformation of coordinates is given by

$$\begin{aligned} x' &= x\cos\varphi - z\sin\varphi \\ y' &= y \\ z' &= x\sin\varphi + z\cos\varphi \end{aligned} \quad \text{or} \quad \begin{pmatrix} x' \\ y' \\ z' \end{pmatrix} = \begin{pmatrix} \cos\varphi & 0 & -\sin\varphi \\ 0 & 1 & 0 \\ \sin\varphi & 0 & \cos\varphi \end{pmatrix} \begin{pmatrix} x \\ y \\ z \end{pmatrix} \quad (4.6)$$

and for the rotation around *X* it becomes

$$\begin{aligned} x' &= x \\ y' &= y\cos\varphi - z\sin\varphi \\ z' &= y\sin\varphi + z\cos\varphi \end{aligned} \quad \text{or} \quad \begin{pmatrix} x' \\ y' \\ z' \end{pmatrix} = \begin{pmatrix} 1 & 0 & 0 \\ 0 & cos\,\varphi & -sin\,\varphi \\ 0 & sin\,\varphi & cos\,\phi \end{pmatrix} \begin{pmatrix} x \\ y \\ z \end{pmatrix} \quad (4.7)$$

A noteworthy property of matrices found in Eqs. (4.2), (4.3), and (4.5–4.7) is their unimodularity—the determinant of every matrix is equal to 1 or −1 for the rotation and inversion (or roto-inversion) operations, respectively, which is shown for the rotation around *Z* in Eq. (4.8).

$$\det\begin{pmatrix} \cos\varphi & -\sin\varphi & 0 \\ \sin\varphi & \cos\varphi & 0 \\ 0 & 0 & 1 \end{pmatrix} = \cos^2\varphi + \sin^2\varphi = 1 \quad (4.8)$$

Because of the restrictions imposed on the values of the rotation angles (see Table 2.3), sinφ and cosφ in a Cartesian basis are 0, 1, or −1 for one-, two-, and fourfold rotations, and they are $\pm 1/2$ or $\pm\sqrt{3}/2$ for three- and sixfold rotations. However, when the same rotational transformations are considered in the appropriate crystallographic coordinate system,[4] all matrix elements become equal to 0, −1, or 1. This simplicity (and undeniably, beauty) of the matrix representation of symmetry operations is the result of restrictions imposed by the three-dimensional periodicity of crystal lattice. The presence of rotational symmetry of any other order (e.g., fivefold rotation) will result in the noninteger values of the elements of the corresponding matrices in three dimensions.

When two points in the same Cartesian basis are related to one another by a translation, then the coordinates of the second point are invariant with respect to a different Cartesian basis, in which the orientations of the axes remain the same as in

[4]The angles between *X*-, *Y*-, and *Z*-axes are identical to α, β, and γ in the corresponding crystal system (see Table 2.10).

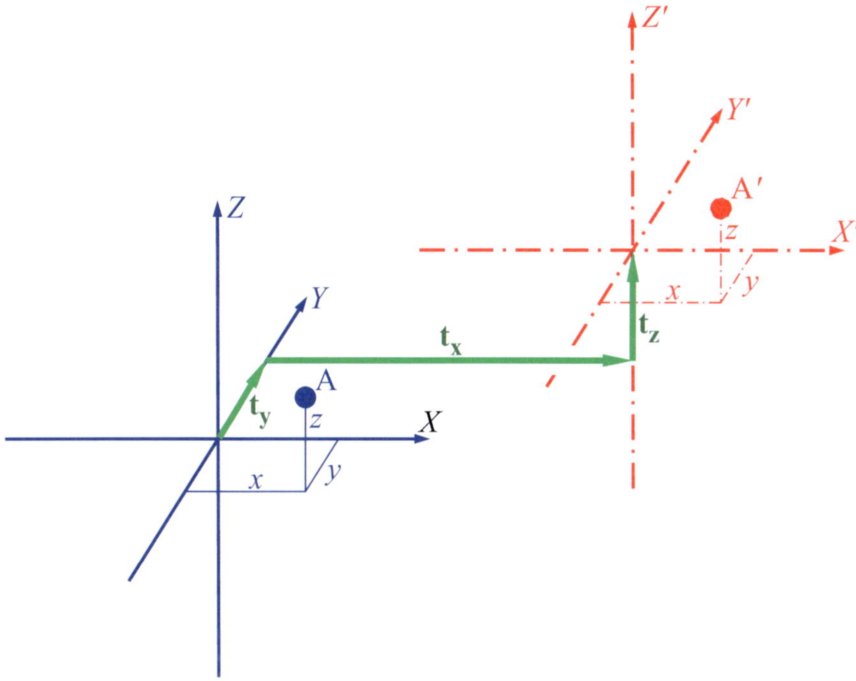

Fig. 4.4 Cartesian bases *XYZ* and *X′Y′Z′* in which the coordinates of the point are invariant to translations along three noncoplanar vectors $\mathbf{t_x}$, $\mathbf{t_y}$, and $\mathbf{t_z}$

the first basis but its origin is shifted along the three noncoplanar vectors, $\mathbf{t_x}$, $\mathbf{t_y}$, and $\mathbf{t_z}$, as shown in Fig. 4.4.

Thus, considering Fig. 4.4, the coordinates, x', y', z', of the point A′ in the original basis *XYZ* are given as

$$\begin{aligned} x' &= x + t_x \\ y' &= y + t_y \\ z' &= z + t_z \end{aligned} \tag{4.9}$$

or in matrix notation

$$\begin{pmatrix} x' \\ y' \\ z' \end{pmatrix} = \begin{pmatrix} x \\ y \\ z \end{pmatrix} + \begin{pmatrix} t_x \\ t_y \\ t_z \end{pmatrix} \tag{4.10}$$

where t_x, t_y, and t_z are the lengths of vectors $\mathbf{t_x}$, $\mathbf{t_y}$, and $\mathbf{t_z}$, respectively.

To generalize the results obtained in this section, we now consider two points, A and A′, with coordinates x, y, z and x', y', z', respectively, in the same Cartesian basis *XYZ*. An unrestricted transformation of A into A′ can be carried out first, by applying the corresponding rotation (and/or inversion) and second, by applying the corresponding translational transformation of the coordinates. For example, assuming that rotation occurs around the Z-axis, and considering Eqs. (4.1), (4.2), (4.9), and (4.10), the relationships between x, y, z, and x', y', z' are given as follows:

$$\begin{aligned} x' &= x\cos\varphi - y\sin\varphi + t_x \\ y' &= x\sin\varphi + y\cos\varphi + t_y \\ z' &= z + t_z \end{aligned} \tag{4.11}$$

or in matrix notation

$$\begin{pmatrix} x' \\ y' \\ z' \end{pmatrix} = \begin{pmatrix} \cos\varphi & -\sin\varphi & 0 \\ \sin\varphi & \cos\varphi & 0 \\ 0 & 0 & 1 \end{pmatrix} \begin{pmatrix} x \\ y \\ z \end{pmatrix} + \begin{pmatrix} t_x \\ t_y \\ t_z \end{pmatrix} \tag{4.12}$$

4.2.2 Rotational Transformations of Vectors

Any change in the orientation of a vector representing a point without changing the length and the position of the origin of the vector (see Fig. 4.5) can be described as a new vector, which is a product of a specific square matrix and the original vector. As we established in Sect. 4.2.1 (see Eqs. (4.2), (4.3), and (4.5–4.7)), in three dimensions the matrix has three rows and three columns, the vector has three rows, and the resulting product is shown in general form in Eq. (4.13).

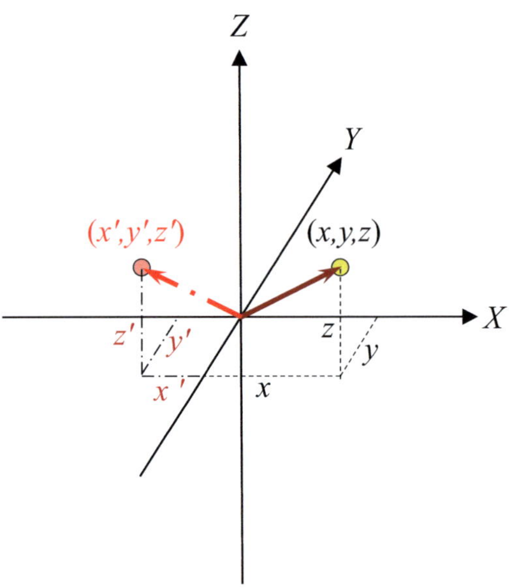

Fig. 4.5 The point with the coordinates x, y, z is represented by the vector (x, y, z) and the second, symmetrically equivalent, point with the coordinates x', y', z' is represented by the vector (x', y', z'), which has the same length as the original vector (x, y, z)

$$\begin{pmatrix} x' \\ y' \\ z' \end{pmatrix} = \begin{pmatrix} r_{11} & r_{12} & r_{13} \\ r_{21} & r_{22} & r_{23} \\ r_{31} & r_{32} & r_{33} \end{pmatrix} \begin{pmatrix} x \\ y \\ z \end{pmatrix} = \begin{pmatrix} r_{11}x + r_{12}y + r_{13}z \\ r_{21}x + r_{22}y + r_{23}z \\ r_{31}x + r_{32}y + r_{33}z \end{pmatrix} \tag{4.13}$$

Using the following notations

$$\mathbf{X} = \begin{pmatrix} x \\ y \\ z \end{pmatrix}, \quad \mathbf{R} = \begin{pmatrix} r_{11} & r_{12} & r_{13} \\ r_{21} & r_{22} & r_{23} \\ r_{31} & r_{32} & r_{33} \end{pmatrix} \text{ and } \mathbf{X}' = \begin{pmatrix} x' \\ y' \\ z' \end{pmatrix} \tag{4.14}$$

Equation 4.13 becomes

$$\mathbf{X}' = \mathbf{RX} \tag{4.15}$$

The matrix **R** is called rotation matrix, since the changing of the orientation of the vector without altering its length and without moving its origin away from the origin of the coordinate system is achieved by various rotational transformations of the vector, for example, inversion through a point, rotation around an axis, or reflection in a plane.

4.2.3 Translational Transformations of Vectors

When only the change of the length of the vector is involved, it is usually much more convenient to move the origin of the coordinate system in a way that the length and the orientation of the new vector remain the same in the new basis as those of the original vector in the original basis (see Fig. 4.4). This is achieved by translating the old origin of the coordinate system by (t_1, t_2, and t_3) along X, Y, and Z, respectively, which is equivalent to adding two vectors as shown in Eq. (4.16).

$$\begin{pmatrix} x' \\ y' \\ z' \end{pmatrix} = \begin{pmatrix} x \\ y \\ z \end{pmatrix} + \begin{pmatrix} t_1 \\ t_2 \\ t_3 \end{pmatrix} = \begin{pmatrix} x + t_1 \\ y + t_2 \\ z + t_3 \end{pmatrix} \tag{4.16}$$

Introducing the following notation in addition to Eq. (4.14)

$$\mathbf{T} = \begin{pmatrix} t_1 \\ t_2 \\ t_3 \end{pmatrix} \tag{4.17}$$

the short form of Eq. (4.16) becomes

$$\mathbf{X}' = \mathbf{X} + \mathbf{T} \tag{4.18}$$

4.2.4 Combined Symmetrical Transformations of Vectors

Unrestricted transformations of a vector can, therefore, be represented using a sequence of matrix–vector transformations by first evaluating the product of the rotation matrix and the original vector as shown in Eq. (4.13), and second, evaluating the sum of the obtained vector and the corresponding translation vector, as shown in Eq. (4.16). The combined unrestricted transformation is, therefore, represented in the expanded form using Eq. (4.19) or using the compact form as shown in Eq. (4.20).

$$\begin{pmatrix} x' \\ y' \\ z' \end{pmatrix} = \begin{pmatrix} r_{11} & r_{12} & r_{13} \\ r_{21} & r_{22} & r_{23} \\ r_{31} & r_{32} & r_{33} \end{pmatrix} \begin{pmatrix} x \\ y \\ z \end{pmatrix} + \begin{pmatrix} t_1 \\ t_2 \\ t_3 \end{pmatrix} = \begin{pmatrix} r_{11}x + r_{12}y + r_{13}z + t_1 \\ r_{21}x + r_{22}y + r_{23}z + t_2 \\ r_{31}x + r_{32}y + r_{33}z + t_3 \end{pmatrix} \tag{4.19}$$

$$\mathbf{X}' = \mathbf{RX} + \mathbf{T} \tag{4.20}$$

It is practically obvious that simultaneously or separately acting rotations (either proper or improper) and translations can be described using the combined transformations of vectors as defined by Eqs. (4.19) and (4.20). When finite symmetry elements intersect with the origin of coordinates, the respective translational part in Eqs. (4.19) and (4.20) is 0, 0, 0, and when the symmetry operation is a simple translation, the corresponding rotational part becomes identity, **E**, where

$$\mathbf{E} = \begin{pmatrix} 1 & 0 & 0 \\ 0 & 1 & 0 \\ 0 & 0 & 1 \end{pmatrix} \tag{4.21}$$

At this point, the validity of Eqs. (4.19) and (4.20) has been established when rotations were performed around an axis intersecting the origin of coordinates. We now establish their validity in the general case by considering vector **X** and a symmetry operation that includes both the rotational part, **R**, and translational part, **T**. Assume that the symmetry operation is applied in a crystallographic basis where the rotation axis is shifted from the origin of coordinates by a vector $\Delta\mathbf{T}$.

First, we select a new basis in which the rotation axis intersects with the origin of coordinates. This is equivalent to changing the coordinates of the original vector from **X** to **x**, where

$$\mathbf{x} = \mathbf{X} + \Delta\mathbf{T} \tag{4.22}$$

According to Eq. (4.20), the symmetry transformation in the new basis results in vector $\mathbf{x}'$, where

$$\mathbf{x}' = \mathbf{R}\mathbf{x} + \mathbf{T} = \mathbf{R}(\mathbf{X} + \Delta\mathbf{T}) + \mathbf{T} = \mathbf{R}\mathbf{X} + (\mathbf{R}\Delta\mathbf{T} + \mathbf{T}) \tag{4.23}$$

After switching back to the original basis by applying negative translation $-\Delta\mathbf{T}$ to the right-hand part of Eq. (4.23), we obtain the coordinates of the symmetrically equivalent vector $\mathbf{X}'$ in the original basis as

$$\mathbf{X}' = \mathbf{R}\mathbf{X} + (\mathbf{R}\Delta\mathbf{T} + \mathbf{T}) - \Delta\mathbf{T} = \mathbf{R}\mathbf{X} + [(\mathbf{R} - \mathbf{E})\Delta\mathbf{T} + \mathbf{T}] \tag{4.24}$$

where $\mathbf{E}$ is identity matrix. Noting that $(\mathbf{R} - \mathbf{E})\Delta\mathbf{T} + \mathbf{T} = \mathbf{T}$, that is, it is the translational part that reflects the shift of the rotation axis from the origin of coordinates in addition to the conventional translational part, $\mathbf{t}$, Eq. (4.24) becomes identical to Eq. (4.20). Thus, Eqs. (4.19) and (4.20) are valid for any crystallographic transformation in three dimensions.

It may be useful to briefly summarize the main properties of rotation transformation matrices, $\mathbf{R}$, some of which were already mentioned earlier:

- In conventional crystallography matrix, elements may accept only the following values:

$$\mathbf{R}_{ij} = 0, 1 \text{ or } -1 \tag{4.25}$$

- $\mathbf{R}$ is unimodular:

$$\det(\mathbf{R}) = 1 \text{ or } -1 \tag{4.26}$$

- The inverse rotation matrix is the same as the direct matrix, that is, $\mathbf{R}^{-1} = \mathbf{R}$ when the rotation order is 1 or 2; otherwise, $\mathbf{R}^{-1}$ represents rotation in the opposite direction.
- An order of an axis represented by the operation $\mathbf{R}$ can be determined in two steps:

1. In the case of an inversion axis, when $\det(\mathbf{R}') = 1$, the matrix should be converted to a simple rotation by multiplying its elements by -1:

$$\mathbf{R} = \mathbf{R}'\det(\mathbf{R}') \tag{4.27}$$

2. The order of the axis (N) is determined from the number of multiplications of the matrix $\mathbf{R}$ by itself that are required to obtain the identity matrix, $\mathbf{E}$:

$$\prod_{i=1}^{N} \mathbf{R} = \mathbf{E} \tag{4.28}$$

4.2.5 Augmentation of Matrices

For convenience, the 3×3 rotation matrix and the corresponding 3×1 translation vector may be combined into a single augmented matrix which has four rows and four columns. This matrix is shown in Fig. 4.6 together with the augmented vector to which the transformation is applied. The augmentation of the vector is required to ensure the compatibility with the matrix during their multiplication.

It is easy to verify that the product of the augmented matrix **A** and the augmented vector **V** results in the vector **V**′, which is the same as the vector **X**′ (Eqs. (4.19) and (4.20)) plus additional 1 as the fourth element of the vector as shown in Eq. (4.29). Therefore, instead of specifying rotational and translational parts separately, they can be combined into a single augmented matrix:

$$\mathbf{V}' = \begin{pmatrix} r_{11} & r_{12} & r_{13} & t_1 \\ r_{21} & r_{22} & r_{23} & t_2 \\ r_{31} & r_{32} & r_{33} & t_3 \\ 0 & 0 & 0 & 1 \end{pmatrix} \begin{pmatrix} x \\ y \\ z \\ 1 \end{pmatrix} = \begin{pmatrix} r_{11}x + r_{12}y + r_{13}z + t_1 \\ r_{21}x + r_{22}y + r_{23}z + t_2 \\ r_{31}x + r_{32}y + r_{33}z + t_3 \\ 0 + 0 + 0 + 1 \end{pmatrix} \tag{4.29}$$

4.2.6 Algebraic Representation of Crystallographic Symmetry

Considering conventional crystallographic symmetry, the elements representing rotations, that is, r_{ij}, are restricted to 1, 0, or −1, and the elements representing translations, that is, t_i in Eq. (4.29), are restricted to ±½, ±1/3, ±¼, ±1/6, ±1/8,

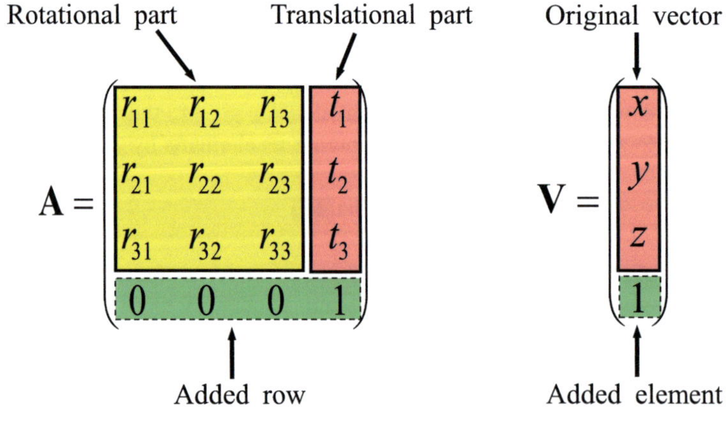

Fig. 4.6 The augmented 4×4 matrix, which combines both the rotational and translational parts as indicated, respectively, by yellow and pink boxes, and the added row highlighted by the green box (*left*). The corresponding modification of the original vector to ensure their compatibility in the matrix–vector multiplication (*right*)

including their integer multiples, and 0. In this way, all possible transformations of atoms by symmetry operations are represented by the multiplication of matrices and vectors.

Therefore, symmetrical transformations in the crystal are formalized as algebraic (matrix–vector) operations—an extremely important feature used in all crystallographic calculations in computer software. The partial list of symmetry elements along with the corresponding augmented matrices that are used to represent symmetry operations included in each symmetry element is provided in Tables 4.1 and 4.2. For a complete list, consult the International Tables for Crystallography, Vol. A.

Even though the column (Table 4.1) and the row (Table 4.2) labeled "First symmetry operation" seem redundant, their presence highlights the fact that each symmetry element contains onefold rotation or identity operation. Further, as easily seen from these tables, fourfold rotation axes also contain the twofold symmetry operation (e.g., the third symmetry operation for 4 || *Z* is the same as the second symmetry operation for 2 || *Z*) and sixfold rotation axes contain both three- and twofold rotations (e.g., second and third symmetry operations for 6 || *Z* are identical to second and third for 3 || *Z*, while the fourth symmetry operation for 6 || *Z* is identical to the second for 2 || *Z*).

4.2.7 Interaction of Symmetry Operations

We now consider how the two interacting symmetry operations produce a third symmetry operation, similar to how it was described in Sect. 2.5 but now in terms of their algebraic representation. Assume that two symmetry operations, which are given by the two augmented matrices $\mathbf{A^1}$ and $\mathbf{A^2}$, are applied in sequence to a point, coordinates of which are represented by the augmented vector $\mathbf{V}$. Taking into account Eq. (4.29), but written in a short form, the first symmetry operation will result in the vector $\mathbf{V^1}$ given as

$$\mathbf{V}^1 = \mathbf{A}^1\mathbf{V} \tag{4.30}$$

The second symmetry operation applied to the vector $\mathbf{V^1}$ will result in the third vector, $\mathbf{V^2}$ as follows:

$$\mathbf{V}^2 = \mathbf{A}^2\mathbf{V}^1 = \mathbf{A}^2\left(\mathbf{A}^1\mathbf{V}\right) \tag{4.31}$$

Recalling that the associative law holds for symmetry operations and for symmetry groups (see Sects. 2.5 and 2.6), Eq. (4.31) can be rewritten as:

$$\mathbf{V}^2 = \left(\mathbf{A}^2\mathbf{A}^1\right)\mathbf{V} \tag{4.32}$$

Table 4.1 Selected symmetry elements, their orientation and corresponding symmetry operations in the algebraic form as augmented matrices (see Fig. 4.6)

Symmetry element and orientation	Transformed coordinates	First symmetry operation	Second symmetry operation	Third symmetry operation	Fourth symmetry operation
1 (any)	x, y, z	$\begin{pmatrix} 1 & 0 & 0 & 0 \\ 0 & 1 & 0 & 0 \\ 0 & 0 & 1 & 0 \\ 0 & 0 & 0 & 1 \end{pmatrix}$	None	None	None
$\bar{1}$ at 0,0,0	$x, y, z;$ $-x, -y, -z$	$\begin{pmatrix} 1 & 0 & 0 & 0 \\ 0 & 1 & 0 & 0 \\ 0 & 0 & 1 & 0 \\ 0 & 0 & 0 & 1 \end{pmatrix}$	$\begin{pmatrix} -1 & 0 & 0 & 0 \\ 0 & -1 & 0 & 0 \\ 0 & 0 & -1 & 0 \\ 0 & 0 & 0 & 1 \end{pmatrix}$	None	None
$\bar{1}$ at ¼,¼,¼	$x, y, z;$ $½ - x, ½ - y, ½ - z$	$\begin{pmatrix} 1 & 0 & 0 & 0 \\ 0 & 1 & 0 & 0 \\ 0 & 0 & 1 & 0 \\ 0 & 0 & 0 & 1 \end{pmatrix}$	$\begin{pmatrix} -1 & 0 & 0 & 1/2 \\ 0 & -1 & 0 & 1/2 \\ 0 & 0 & -1 & 1/2 \\ 0 & 0 & 0 & 1 \end{pmatrix}$	None	None
m $\perp X$ at $X = 0$	$x, y, z;$ $-x, y, z$	$\begin{pmatrix} 1 & 0 & 0 & 0 \\ 0 & 1 & 0 & 0 \\ 0 & 0 & 1 & 0 \\ 0 & 0 & 0 & 1 \end{pmatrix}$	$\begin{pmatrix} -1 & 0 & 0 & 0 \\ 0 & 1 & 0 & 0 \\ 0 & 0 & 1 & 0 \\ 0 & 0 & 0 & 1 \end{pmatrix}$	None	None
m $\perp Y$ at $Y = 0$	$x, y, z;$ $x, -y, z$	$\begin{pmatrix} 1 & 0 & 0 & 0 \\ 0 & 1 & 0 & 0 \\ 0 & 0 & 1 & 0 \\ 0 & 0 & 0 & 1 \end{pmatrix}$	$\begin{pmatrix} 1 & 0 & 0 & 0 \\ 0 & -1 & 0 & 0 \\ 0 & 0 & 1 & 0 \\ 0 & 0 & 0 & 1 \end{pmatrix}$	None	None
m $\perp Z$ at $Z = 0$	$x, y, z;$ $x, y, -z$	$\begin{pmatrix} 1 & 0 & 0 & 0 \\ 0 & 1 & 0 & 0 \\ 0 & 0 & 1 & 0 \\ 0 & 0 & 0 & 1 \end{pmatrix}$	$\begin{pmatrix} 1 & 0 & 0 & 0 \\ 0 & 1 & 0 & 0 \\ 0 & 0 & -1 & 0 \\ 0 & 0 & 0 & 1 \end{pmatrix}$	None	None

m $\perp X$ at $X = ¼$	x, y, z; ½ $- x, y, z$	$\begin{pmatrix} 1 & 0 & 0 & 0 \\ 0 & 1 & 0 & 0 \\ 0 & 0 & 1 & 0 \\ 0 & 0 & 0 & 1 \end{pmatrix}$	$\begin{pmatrix} -1 & 0 & 0 & 1/2 \\ 0 & 1 & 0 & 0 \\ 0 & 0 & 1 & 0 \\ 0 & 0 & 0 & 1 \end{pmatrix}$	None	None
m $\perp X$ at $X = ½$	x, y, z; $1 - x, y, z$	$\begin{pmatrix} 1 & 0 & 0 & 0 \\ 0 & 1 & 0 & 0 \\ 0 & 0 & 1 & 0 \\ 0 & 0 & 0 & 1 \end{pmatrix}$	$\begin{pmatrix} -1 & 0 & 0 & 1 \\ 0 & 1 & 0 & 0 \\ 0 & 0 & 1 & 0 \\ 0 & 0 & 0 & 1 \end{pmatrix}$	None	None
b $\perp X$ at $X = 0$	x, y, z; $-x$, ½ $+ y, z$	$\begin{pmatrix} 1 & 0 & 0 & 0 \\ 0 & 1 & 0 & 0 \\ 0 & 0 & 1 & 0 \\ 0 & 0 & 0 & 1 \end{pmatrix}$	$\begin{pmatrix} -1 & 0 & 0 & 0 \\ 0 & 1 & 0 & 1/2 \\ 0 & 0 & 1 & 0 \\ 0 & 0 & 0 & 1 \end{pmatrix}$	None	None
b $\perp X$ at $X = ¼$	x, y, z; ½ $- x$, ½ $+ y, z$	$\begin{pmatrix} 1 & 0 & 0 & 0 \\ 0 & 1 & 0 & 0 \\ 0 & 0 & 1 & 0 \\ 0 & 0 & 0 & 1 \end{pmatrix}$	$\begin{pmatrix} -1 & 0 & 0 & 1/2 \\ 0 & 1 & 0 & 1/2 \\ 0 & 0 & 1 & 0 \\ 0 & 0 & 0 & 1 \end{pmatrix}$	None	None
c $\perp X$ at $X = 0$	x, y, z; $-x, y$, ½ $+ z$	$\begin{pmatrix} 1 & 0 & 0 & 0 \\ 0 & 1 & 0 & 0 \\ 0 & 0 & 1 & 0 \\ 0 & 0 & 0 & 1 \end{pmatrix}$	$\begin{pmatrix} -1 & 0 & 0 & 0 \\ 0 & 1 & 0 & 0 \\ 0 & 0 & 1 & 1/2 \\ 0 & 0 & 0 & 1 \end{pmatrix}$	None	None
n $\perp X$ at $X = 0$	x, y, z; $-x$, ½ $+ y$, ½ $+ z$	$\begin{pmatrix} 1 & 0 & 0 & 0 \\ 0 & 1 & 0 & 0 \\ 0 & 0 & 1 & 0 \\ 0 & 0 & 0 & 1 \end{pmatrix}$	$\begin{pmatrix} -1 & 0 & 0 & 0 \\ 0 & 1 & 0 & 1/2 \\ 0 & 0 & 1 & 1/2 \\ 0 & 0 & 0 & 1 \end{pmatrix}$	None	None

(continued)

Table 4.1 (continued)

Symmetry element and orientation	Transformed coordinates	First symmetry operation	Second symmetry operation	Third symmetry operation	Fourth symmetry operation
2 ‖ X through 0,0,0	$x, y, z;$ $x, -y, -z$	$\begin{pmatrix}1&0&0&0\\0&1&0&0\\0&0&1&0\\0&0&0&1\end{pmatrix}$	$\begin{pmatrix}1&0&0&0\\0&-1&0&0\\0&0&-1&0\\0&0&0&1\end{pmatrix}$	None	None
2 ‖ Y through 0,0,0	$x, y, z;$ $-x, y, -z$	$\begin{pmatrix}1&0&0&0\\0&1&0&0\\0&0&1&0\\0&0&0&1\end{pmatrix}$	$\begin{pmatrix}-1&0&0&0\\0&1&0&0\\0&0&-1&0\\0&0&0&1\end{pmatrix}$	None	None
2 ‖ Z through 0,0,0	$x, y, z;$ $-x, -y, z$	$\begin{pmatrix}1&0&0&0\\0&1&0&0\\0&0&1&0\\0&0&0&1\end{pmatrix}$	$\begin{pmatrix}-1&0&0&0\\0&-1&0&0\\0&0&1&0\\0&0&0&1\end{pmatrix}$	None	None
2_1 ‖ X through 0,0,0	$x, y, z;$ $½ + x, -y, -z$	$\begin{pmatrix}1&0&0&0\\0&1&0&0\\0&0&1&0\\0&0&0&1\end{pmatrix}$	$\begin{pmatrix}1&0&0&1/2\\0&-1&0&0\\0&0&-1&0\\0&0&0&1\end{pmatrix}$	None	None
3 ‖ Z through 0,0,0	$x, y, z;$ $-y, x-y, z;$ $y-x, -x, z$	$\begin{pmatrix}1&0&0&0\\0&1&0&0\\0&0&1&0\\0&0&0&1\end{pmatrix}$	$\begin{pmatrix}0&-1&0&0\\1&-1&0&0\\0&0&1&0\\0&0&0&1\end{pmatrix}$	$\begin{pmatrix}-1&1&0&0\\-1&0&0&0\\0&0&1&0\\0&0&0&1\end{pmatrix}$	None
3_1 ‖ Z through 0,0,0	$x, y, z;$ $-y, x-y, 1/3 + z;$ $y-x, -x, 2/3 + z$	$\begin{pmatrix}1&0&0&0\\0&1&0&0\\0&0&1&0\\0&0&0&1\end{pmatrix}$	$\begin{pmatrix}0&-1&0&0\\1&-1&0&0\\0&0&1&1/3\\0&0&0&1\end{pmatrix}$	$\begin{pmatrix}-1&1&0&0\\-1&0&0&0\\0&0&1&2/3\\0&0&0&1\end{pmatrix}$	None

4 ‖ Z through 0,0,0	x, y, z; $-y, x, z$; $-x, -y, z$; $y, -x, z$	$\begin{pmatrix} 1 & 0 & 0 & 0 \\ 0 & 1 & 0 & 0 \\ 0 & 0 & 1 & 0 \\ 0 & 0 & 0 & 1 \end{pmatrix}$	$\begin{pmatrix} 0 & -1 & 0 & 0 \\ 1 & 0 & 0 & 0 \\ 0 & 0 & 1 & 0 \\ 0 & 0 & 0 & 1 \end{pmatrix}$	$\begin{pmatrix} -1 & 0 & 0 & 0 \\ 0 & -1 & 0 & 0 \\ 0 & 0 & 1 & 0 \\ 0 & 0 & 0 & 1 \end{pmatrix}$	$\begin{pmatrix} 0 & 1 & 0 & 0 \\ -1 & 0 & 0 & 0 \\ 0 & 0 & 1 & 0 \\ 0 & 0 & 0 & 1 \end{pmatrix}$
4_1 ‖ Z through 0,0,0	x, y, z; $-y, x, \frac{1}{4} + z$; $-x, -y, \frac{1}{2} + z$; $y, -x, 3/4 + z$	$\begin{pmatrix} 1 & 0 & 0 & 0 \\ 0 & 1 & 0 & 0 \\ 0 & 0 & 1 & 0 \\ 0 & 0 & 0 & 1 \end{pmatrix}$	$\begin{pmatrix} 0 & -1 & 0 & 0 \\ 1 & 0 & 0 & 0 \\ 0 & 0 & 1 & 1/4 \\ 0 & 0 & 0 & 1 \end{pmatrix}$	$\begin{pmatrix} -1 & 0 & 0 & 0 \\ 0 & 1 & 0 & 0 \\ 0 & 0 & 1 & 1/2 \\ 0 & 0 & 0 & 1 \end{pmatrix}$	$\begin{pmatrix} 0 & 1 & 0 & 0 \\ -1 & 0 & 0 & 0 \\ 0 & 0 & 1 & 3/4 \\ 0 & 0 & 0 & 1 \end{pmatrix}$

Table 4.2 Selected symmetry elements in trigonal and hexagonal crystal systems, their orientation and corresponding symmetry operations in the algebraic form as augmented matrices (see Fig. 4.6)

Symmetry element and orientation	6 ∥ Z through 0,0,0	3 ∥ Z through 0,0,0	6 ∥ Z through 0,0,0	6_1 ∥ Z through 0,0,0	6_3 ∥ Z through 0,0,0
First symmetry operation	x, y, z $\begin{pmatrix} 1 & 0 & 0 & 0 \\ 0 & 1 & 0 & 0 \\ 0 & 0 & 1 & 0 \\ 0 & 0 & 0 & 1 \end{pmatrix}$	x, y, z $\begin{pmatrix} 1 & 0 & 0 & 0 \\ 0 & 1 & 0 & 0 \\ 0 & 0 & 1 & 0 \\ 0 & 0 & 0 & 1 \end{pmatrix}$	x, y, z $\begin{pmatrix} 1 & 0 & 0 & 0 \\ 0 & 1 & 0 & 0 \\ 0 & 0 & 1 & 0 \\ 0 & 0 & 0 & 1 \end{pmatrix}$	x, y, z $\begin{pmatrix} 1 & 0 & 0 & 0 \\ 0 & 1 & 0 & 0 \\ 0 & 0 & 1 & 0 \\ 0 & 0 & 0 & 1 \end{pmatrix}$	x, y, z $\begin{pmatrix} 1 & 0 & 0 & 0 \\ 0 & 1 & 0 & 0 \\ 0 & 0 & 1 & 0 \\ 0 & 0 & 0 & 1 \end{pmatrix}$
Second symmetry operation	$-y, x-y, z$ $\begin{pmatrix} 0 & -1 & 0 & 0 \\ 1 & -1 & 0 & 0 \\ 0 & 0 & 1 & 0 \\ 0 & 0 & 0 & 1 \end{pmatrix}$	$-y, x-y, z$ $\begin{pmatrix} 0 & -1 & 0 & 0 \\ 1 & -1 & 0 & 0 \\ 0 & 0 & 1 & 0 \\ 0 & 0 & 0 & 1 \end{pmatrix}$	$-y, x-y, z$ $\begin{pmatrix} 0 & -1 & 0 & 0 \\ 1 & -1 & 0 & 0 \\ 0 & 0 & 1 & 0 \\ 0 & 0 & 0 & 1 \end{pmatrix}$	$-y, x-y, 1/3+z$ $\begin{pmatrix} 0 & -1 & 0 & 0 \\ 1 & -1 & 0 & 0 \\ 0 & 0 & 1 & 1/3 \\ 0 & 0 & 0 & 1 \end{pmatrix}$	$-y, x-y, z$ $\begin{pmatrix} 0 & -1 & 0 & 0 \\ 1 & -1 & 0 & 0 \\ 0 & 0 & 1 & 0 \\ 0 & 0 & 0 & 1 \end{pmatrix}$
Third symmetry operation	$-x+y, -x, z$ $\begin{pmatrix} -1 & 1 & 0 & 0 \\ -1 & 0 & 0 & 0 \\ 0 & 0 & 1 & 0 \\ 0 & 0 & 0 & 1 \end{pmatrix}$	$-x+y, -x, z$ $\begin{pmatrix} -1 & 1 & 0 & 0 \\ -1 & 0 & 0 & 0 \\ 0 & 0 & 1 & 0 \\ 0 & 0 & 0 & 1 \end{pmatrix}$	$-x+y, -x, z$ $\begin{pmatrix} -1 & 1 & 0 & 0 \\ -1 & 0 & 0 & 0 \\ 0 & 0 & 1 & 0 \\ 0 & 0 & 0 & 1 \end{pmatrix}$	$-x+y, -x, 2/3+z$ $\begin{pmatrix} -1 & 1 & 0 & 0 \\ -1 & 0 & 0 & 0 \\ 0 & 0 & 1 & 2/3 \\ 0 & 0 & 0 & 1 \end{pmatrix}$	$-x+y, -x, z$ $\begin{pmatrix} -1 & 1 & 0 & 0 \\ -1 & 0 & 0 & 0 \\ 0 & 0 & 1 & 0 \\ 0 & 0 & 0 & 1 \end{pmatrix}$
Fourth symmetry operation	$-x, -y, z$ $\begin{pmatrix} -1 & 0 & 0 & 0 \\ 0 & -1 & 0 & 0 \\ 0 & 0 & 1 & 0 \\ 0 & 0 & 0 & 1 \end{pmatrix}$	$-x, -y, -z$ $\begin{pmatrix} -1 & 0 & 0 & 0 \\ 0 & -1 & 0 & 0 \\ 0 & 0 & -1 & 0 \\ 0 & 0 & 0 & 1 \end{pmatrix}$	$x, y, -z$ $\begin{pmatrix} 1 & 0 & 0 & 0 \\ 0 & 1 & 0 & 0 \\ 0 & 0 & -1 & 0 \\ 0 & 0 & 0 & 1 \end{pmatrix}$	$-x, -y, ½+z$ $\begin{pmatrix} -1 & 0 & 0 & 0 \\ 0 & -1 & 0 & 0 \\ 0 & 0 & 1 & 1/2 \\ 0 & 0 & 0 & 1 \end{pmatrix}$	$-x, -y, ½+z$ $\begin{pmatrix} -1 & 0 & 0 & 0 \\ 0 & -1 & 0 & 0 \\ 0 & 0 & 1 & 1/2 \\ 0 & 0 & 0 & 1 \end{pmatrix}$

Fifth symmetry operation	$x-y, x, z$ $\begin{pmatrix} 1 & -1 & 0 & 0 \\ 1 & 0 & 0 & 0 \\ 0 & 0 & 1 & 0 \\ 0 & 0 & 0 & 1 \end{pmatrix}$	$x-y, x, -z$ $\begin{pmatrix} 1 & -1 & 0 & 0 \\ 1 & 0 & 0 & 0 \\ 0 & 0 & -1 & 0 \\ 0 & 0 & 0 & 1 \end{pmatrix}$	$-x+y, -x, -z$ $\begin{pmatrix} -1 & 1 & 0 & 0 \\ -1 & 0 & 0 & 0 \\ 0 & 0 & -1 & 0 \\ 0 & 0 & 0 & 1 \end{pmatrix}$	$x-y, x, 1/6+z$ $\begin{pmatrix} 1 & -1 & 0 & 0 \\ 1 & 0 & 0 & 0 \\ 0 & 0 & 1 & 1/6 \\ 0 & 0 & 0 & 1 \end{pmatrix}$	$x-y, x, \frac{1}{2}+z$ $\begin{pmatrix} 1 & -1 & 0 & 0 \\ 1 & 0 & 0 & 0 \\ 0 & 0 & 1 & 1/2 \\ 0 & 0 & 0 & 1 \end{pmatrix}$
Sixth symmetry operation	$y, -x+y, z$ $\begin{pmatrix} 0 & 1 & 0 & 0 \\ -1 & 1 & 0 & 0 \\ 0 & 0 & 1 & 0 \\ 0 & 0 & 0 & 1 \end{pmatrix}$	$y, -x+y, -z$ $\begin{pmatrix} 0 & 1 & 0 & 0 \\ -1 & 1 & 0 & 0 \\ 0 & 0 & -1 & 0 \\ 0 & 0 & 0 & 1 \end{pmatrix}$	$-y, x-y, -z$ $\begin{pmatrix} 0 & -1 & 0 & 0 \\ 1 & -1 & 0 & 0 \\ 0 & 0 & -1 & 0 \\ 0 & 0 & 0 & 1 \end{pmatrix}$	$y, -x+y, 5/6+z$ $\begin{pmatrix} 0 & 1 & 0 & 0 \\ -1 & 1 & 0 & 0 \\ 0 & 0 & 1 & 5/6 \\ 0 & 0 & 0 & 1 \end{pmatrix}$	$y, -x+y, \frac{1}{2}+z$ $\begin{pmatrix} 0 & 1 & 0 & 0 \\ -1 & 1 & 0 & 0 \\ 0 & 0 & 1 & 1/2 \\ 0 & 0 & 0 & 1 \end{pmatrix}$

It follows from Eq. (4.32) and from our earlier consideration of interactions between symmetry elements, finding which symmetry operation appears as the result of consecutive application of any two symmetry operations is reduced to calculating the product of the corresponding augmented matrices. To illustrate how it is done in practice, consider Fig. 2.12 and assume that the twofold axis is parallel to Y. The corresponding symmetry operations, $\mathbf{A^1}$ and $\mathbf{A^2}$, are (Table 4.1):

$$\mathbf{A^1} = \begin{pmatrix} 1 & 0 & 0 & 0 \\ 0 & 1 & 0 & 0 \\ 0 & 0 & 1 & 0 \\ 0 & 0 & 0 & 1 \end{pmatrix} \quad \text{and} \quad \mathbf{A}^2 = \begin{pmatrix} -1 & 0 & 0 & 0 \\ 0 & 1 & 0 & 0 \\ 0 & 0 & -1 & 0 \\ 0 & 0 & 0 & 1 \end{pmatrix} \tag{4.33}$$

The presence of the center of inversion introduces one additional nonidentity symmetry operation, $\mathbf{A^3}$

$$\mathbf{A}^3 = \begin{pmatrix} -1 & 0 & 0 & 0 \\ 0 & -1 & 0 & 0 \\ 0 & 0 & -1 & 0 \\ 0 & 0 & 0 & 1 \end{pmatrix} \tag{4.34}$$

It is easy to see that the product of $\mathbf{A^2}$ and $\mathbf{A^3}$ is the fourth symmetry operation, $\mathbf{A^4}$, which is nothing else but the mirror reflection in the plane, which is perpendicular to Y and passes through the origin of coordinates:

$$\mathbf{A}^4 = \mathbf{A}^2\mathbf{A}^3 = \begin{pmatrix} 1 & 0 & 0 & 0 \\ 0 & -1 & 0 & 0 \\ 0 & 0 & 1 & 0 \\ 0 & 0 & 0 & 1 \end{pmatrix} \tag{4.35}$$

All other products between the four matrices do not result in new symmetry operations, that is,

$$\mathbf{A}^1\mathbf{A}^1 = \mathbf{A}^1;\ \ \mathbf{A}^1\mathbf{A}^2 = \mathbf{A}^2;\ \ \mathbf{A}^1\mathbf{A}^3 = \mathbf{A}^3;\ \ \mathbf{A}^1\mathbf{A}^4 = \mathbf{A}^4$$

$$\mathbf{A}^2\mathbf{A}^2 = \mathbf{A}^1;\ \ \mathbf{A}^2\mathbf{A}^3 = \mathbf{A}^4;\ \ \mathbf{A}^2\mathbf{A}^4 = \mathbf{A}^3$$

$$\mathbf{A}^3\mathbf{A}^3 = \mathbf{A}^1;\ \ \mathbf{A}^3\mathbf{A}^4 = \mathbf{A}^2$$

$$\mathbf{A}^4\mathbf{A}^4 = \mathbf{A}^1$$

By now we know quite well that this combination of symmetry operations corresponds to point-group symmetry 2/m (also see Fig. 4.1).

4.3 Additional Reading

1. C. Giacovazzo, H.L. Monaco, G. Artioli, D. Viterbo, M. Milaneso, G. Gilli, P. Gilli, G. Zanotti, G. Ferraris, and M. Catti, Fundamentals of crystallography, 3rd edition, C. Giacovazzo, Ed., IUCr/Oxford University Press (2011). https://doi.org/10.1093/acprof:oso/9780199573653.001.0001
2. D. Schwarzenbach, Crystallography, Wiley, New York (1996). https://www.wiley.com/en-us/Crystallography-p-9780471955986
3. C. Hammond, The basics of crystallography and diffraction, 4th edition. IUCr/Oxford University Press (2015). https://doi.org/10.1093/acprof:oso/9780198738671.001.0001
4. D. Farmer, Groups and symmetry: A Guide to Discovering Mathematics (Mathematical World, Vol. 5) 1st Edition, ISBN: 0821804502, Amer. Math. Soc. (1995). https://bookstore.ams.org/mawrld-5
5. International Tables for Crystallography, Vol. A, Space-group symmetry, M. I. Aroyo, Ed., Second online edition, (2016). https://it.iucr.org/A/
6. International Tables for Crystallography. 6th edition, Teaching edition, Crystallographic symmetry, M. I. Aroyo, Ed., Wiley (2021). https://www.wiley.com/en-us/-p-9780470974223
7. E. Prince, Mathematical techniques in crystallography and materials science. Third Edition, Springer, Berlin (2004). https://www.springer.com/gp/book/9783540211112.
8. M. I. Aroyo, A. Kirov, C. Capillas, J. M. Perez-Mato, and H. Wondratschek. Bilbao Crystallographic Server II: Representations of crystallographic point groups and space groups. Acta Cryst. **A62**, 115 (2006). https://doi.org/10.1107/S0108767305040286
9. Bilbao Crystallographic Server: https://www.cryst.ehu.es/.

4.4 Problems

1. Two primitive orthorhombic space groups are based on the following symmetry operations:

$$\text{(a)}\quad \begin{pmatrix} -1 & 0 & 0 & 0 \\ 0 & 1 & 0 & 0 \\ 0 & 0 & 1 & 0 \\ 0 & 0 & 0 & 1 \end{pmatrix} \quad \begin{pmatrix} 1 & 0 & 0 & 0 \\ 0 & -1 & 0 & 0 \\ 0 & 0 & 1 & 0 \\ 0 & 0 & 0 & 1 \end{pmatrix} \quad \begin{pmatrix} 1 & 0 & 0 & \frac{1}{2} \\ 0 & 1 & 0 & 0 \\ 0 & 0 & -1 & 0 \\ 0 & 0 & 0 & 1 \end{pmatrix}$$

$$\text{(b)}\quad \begin{pmatrix} -1 & 0 & 0 & 0 \\ 0 & 1 & 0 & 0 \\ 0 & 0 & 1 & 1/2 \\ 0 & 0 & 0 & 1 \end{pmatrix} \quad \begin{pmatrix} 1 & 0 & 0 & 0 \\ 0 & -1 & 0 & 0 \\ 0 & 0 & 1 & 0 \\ 0 & 0 & 0 & 1 \end{pmatrix} \quad \begin{pmatrix} 1 & 0 & 0 & 0 \\ 0 & 1 & 0 & 0 \\ 0 & 0 & -1 & 0 \\ 0 & 0 & 0 & 1 \end{pmatrix}$$

1.1. Identify these symmetry operations and write the international symbols of the two space groups.
1.2. How can you describe the difference (if any) and/or similarity (if any) between these two space groups?

2. Two primitive orthorhombic space groups are based on the following symmetry operations:

(c) $\begin{pmatrix} 1 & 0 & 0 & 0 \\ 0 & -1 & 0 & 0 \\ 0 & 0 & -1 & 0 \\ 0 & 0 & 0 & 1 \end{pmatrix}$ $\begin{pmatrix} -1 & 0 & 0 & 0 \\ 0 & 1 & 0 & 1/2 \\ 0 & 0 & -1 & 0 \\ 0 & 0 & 0 & 1 \end{pmatrix}$ $\begin{pmatrix} -1 & 0 & 0 & 0 \\ 0 & -1 & 0 & 0 \\ 0 & 0 & 1 & 0 \\ 0 & 0 & 0 & 1 \end{pmatrix}$

(d) $\begin{pmatrix} 1 & 0 & 0 & 1/2 \\ 0 & -1 & 0 & 0 \\ 0 & 0 & -1 & 0 \\ 0 & 0 & 0 & 1 \end{pmatrix}$ $\begin{pmatrix} -1 & 0 & 0 & 0 \\ 0 & 1 & 0 & 0 \\ 0 & 0 & -1 & 0 \\ 0 & 0 & 0 & 1 \end{pmatrix}$ $\begin{pmatrix} -1 & 0 & 0 & 0 \\ 0 & -1 & 0 & 0 \\ 0 & 0 & 1 & 0 \\ 0 & 0 & 0 & 1 \end{pmatrix}$

1.3. Identify these symmetry operations and write the international symbols of the two space groups.
1.4. How can you describe the difference (if any) and/or similarity (if any) between these two space groups?

3. Solve problem No. 2 in Sect. 2.13. Find or derive symmetry operations (rotation matrices) for both planes assuming that *X*-axis is left to right across the paper, *Y*-axis is bottom to top along the paper, *Z*-axis is perpendicular to the paper, and the origin of coordinates is located on the line along which the two planes intersect. Confirm the solution of problem No. 2 in Sect. 2.13 algebraically by finding all derivative symmetry operations. Relate symmetry operations to the corresponding symmetry elements.

Chapter 5
Nonconventional Symmetry

Conventional crystallography was developed using the explicit assumption that crystalline objects maintain ideal periodicity in three dimensions. As a result, any ideal three-dimensional crystal structure can be described using a periodic lattice and one of the 230 crystallographic space groups (see Sect. 3.4). The overwhelming majority of both naturally occurring and synthetic crystalline solids are indeed nearly ideally periodic. Their diffraction patterns are perfectly periodic since Bragg peaks are only observed at the corresponding points of the reciprocal lattice, which reflects both the symmetry and three-dimensional periodicity of the crystal lattice. Long ago, the first aperiodic crystal was reported,[1] and the apparent absence of the three-dimensional periodicity of diffraction patterns was later found in a number of materials. One of the most prominent examples is the 1984 discovery of the fivefold symmetry in the diffraction pattern of rapidly cooled $Al_{0.86}Mn_{0.14}$ alloy.[2] Supported by many experimental observations, several approaches to describe the symmetry of aperiodic structures have been developed and successfully used to establish the crystal structure of these unusual materials.

Probably the most fruitful method has been suggested by P.M. de Wolff[3] in which more than three physical dimensions are used to represent the crystal lattice and thus to restore its periodicity in the so-called superspace. Then the resulting aperiodic diffraction pattern is simply a projection of the crystal lattice, which is periodic in the superspace, upon the physical space, which is three-dimensional. The diffraction pattern of an aperiodic crystal usually contains a subset of strong, that is, highly

[1] U. Dehlinger, Über die Verbreiterung der Debyelinien bei kaltbearbeiteten Metallen (On the broadening of Debye lines in cold-worked metals), Z. Kristallogr. **65**, 615 (1927). https://doi.org/10.1524/zkri.1927.65.1.615

[2] D. Shechtman, I. Blech, D. Gratias and J.W. Cahn, Metallic phase with long-range orientational order and no translational symmetry, Phys. Rev. Lett. **53**, 1951 (1984). https://doi.org/10.1103/PhysRevLett.53.1951

[3] P.M. de Wolff, The pseudo-symmetry of modulated crystal structures, Acta Cryst. A**30**, 777 (1974). https://doi.org/10.1107/S0567739474010710

P. Zavalij, V. Pecharsky, *Fundamentals of Powder Diffraction and Structural Characterization of Materials*, https://doi.org/10.1007/978-3-031-91504-8_5

intense diffraction peaks, which are called main peaks, and their indices are described using three integers corresponding to a standard three-dimensionally periodic crystal lattice. The subsets of the so-called satellite peaks are weaker and their indices include more than three integers to reflect the increased dimensionality of the superspace (see Footnote 1 in Chap. 14, p.440).

5.1 Commensurate Modulation

Consider the simplest case, when the periodicity of the crystal lattice is perturbed in one dimension by periodic deviations of atoms from their ideal positions. As shown in Fig. 5.1, the basic structure (the upper row of atoms, which is conventional and ideally periodic with the translation vector **a**) is perturbed, in other words, modulated by a periodic function, with the period $\lambda = 1/q$, where q is the magnitude of the modulation vector, **q**, and $\mathbf{q} = \alpha\mathbf{a}^*$. The amplitude of the modulation function is A. The resulting modulated structure (the lower row of atoms in Fig. 5.1) is obtained by shifting atoms from their ideal positions.[4] The relationship between the translation vector of the underlying periodic lattice and the period of the modulation function defines the directions, and its amplitude defines the extent of the shifts, as indicated by the red arrows in Fig. 5.1.

When the value of α is rational, this results in a commensurate modulation. Upon further examination of Fig. 5.1, it is easy to see that here $\lambda = {}^{8}/_{3}a$, and $\mathbf{q} = {}^{3}/_{8}\mathbf{a}^*$, and the modulation is commensurate ($\alpha = {}^{3}/_{8}$). In principle, in the case of commensurate

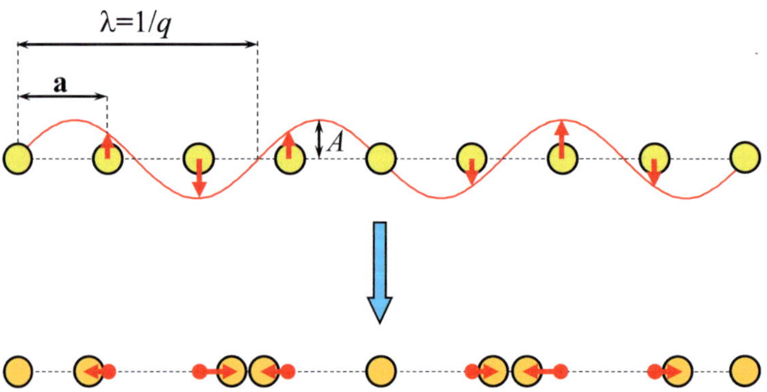

Fig. 5.1 The conventional, ideally periodic one-dimensional structure (yellow circles) and the corresponding modulation function with the period $\lambda = 1/q$, which is commensurate with **a**, and the amplitude A (*top*). The resulting commensurately modulated structure (*bottom*)

[4]In the cases shown in Figs. 5.1 and 5.2, the x-coordinate of each atom becomes $x^m = x + A \sin(2n\pi qa)$, where x^m and x correspond to the modulated and conventional periodic structures, respectively, and $n = \ldots, -2, -1, 0, 1, 2, \ldots$.

modulation, the "conventional" periodicity can be restored by selecting a much larger unit cell, which is often called a supercell. Considering the example shown in Fig. 5.1, the periodicity can be restored without introducing the perturbing function by choosing the unit cell with $\mathbf{a}_{\text{bottom}} = 8\mathbf{a}_{\text{top}}$, where "bottom" and "top" refer to the locations of the one-dimensional structures in the figure. However, atoms which are symmetrically related due to the presence of the modulation function (See Footnote 4) are no longer equivalent to one another in the enlarged unit cell, and therefore, the correct description using commensurate modulation is usually preferred.

5.2 Incommensurate Modulation

When α is irrational, the so-called incommensurate modulation occurs, and this is shown schematically in Fig. 5.2. The exact description of incommensurately modulated structure is impossible using only conventional crystallographic symmetry in the unit cell of any size smaller than the crystal. The periodicity of the structure can only be restored by using two different periodic functions. The first function is the conventional crystallographic translation, and the second one is a modulation function with a certain period, which is incommensurate with the corresponding translation, and a certain amplitude.

It may seem that by selecting a supercell large enough to match the basic structure (at least within experimental accuracy), we can always make modulated structure commensurate. Even so this supercell may be very well used to refine approximate structure, it does not make modulation commensurate. In order to determine the true nature of modulation, the unit cell and modulation vector should be measured at

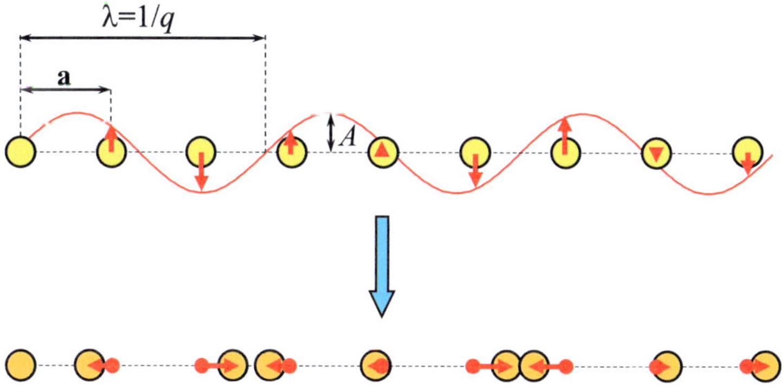

Fig. 5.2 The ideally periodic one-dimensional structure (yellow circles) and the corresponding modulation function (red line) with the period $\lambda = 1/q$, which is incommensurate with $\mathbf{a}$, and the amplitude A (*top*). The resulting incommensurately modulated structure (*bottom*)

different temperatures as the modulation vector remains constant in case of commensurate modulation but varies with temperature when modulation is incommensurate.

Modulation functions are most often modeled by Fourier series

$$u(x_i) = \sum_{n=1}^{m} A_n \sin(2\pi n x_i) + B_n \cos(2\pi n x_i) \tag{5.1}$$

but in more complex cases, Kernel functions, Bloch wave functions, or saw-tooth functions may be employed.

In addition to one-dimensional modulations, both two- and three-dimensional modulations are possible. Atomic parameters affected by modulations may be one or several of the following: positional (as shown in Figs. 5.1 and 5.2), occupancy, thermal displacement, and orientation of magnetic moments. The latter, that is, commensurately or incommensurately modulated orientations of magnetic moments are quite common in various magnetically ordered structures (for example, pure lanthanide metals such as Er and Ho), and both the value of the modulation vector and the amplitude of the modulation function often vary with temperature.

Symmetry of modulated structures is represented algebraically by using rotation matrices and translational vectors in an N-dimensional superspace, where N > 3. Additional dimensions are needed to describe the symmetry of the modulation functions. Thus, one-dimensional modulation is described in a four-dimensional superspace using 4×4 rotation matrices and 4×1 vectors (3 dimensions for a normal space plus one for the modulation function), while two- and three-dimensional modulations require 5×5 and 6×6 rotation matrices and 5×1 and 6×1 vectors, respectively.

The general form of a superspace symmetry operation may be given as follows:

$$\begin{pmatrix} R_E & 0 \\ R_M & R_I \end{pmatrix} \begin{pmatrix} t_E \\ t_I \end{pmatrix} \tag{5.2}$$

where R and t are, respectively, rotational and translational components; subscript E refers to the external (real space) dimensions, I refers to the internal (additional) dimension, and M is their common part.

Hence, modulated structures are described using the main lattice that reflects an average structure in a three-dimensional physical space with addition of one or more modulation waves. These modulation waves modify the main lattice and require addition of an appropriate number of dimensions to adequately describe both the symmetry and structure of a modulated crystal. Reciprocal lattice of a modulated structure, therefore, consists of main points and the so-called satellites, which correspond to the main three-dimensional structure and displacements due to modulation waves, respectively.

5.3 Composite Crystals

Composite structures have two or more substructures penetrating one another. For example, tunnels existing in a framework may be filled with atoms exhibiting different periodicity than that of the tunnel. When periodicity of the tunnel does not match periodicity of the filling atoms, this may result in additional one-dimensional modulation either commensurate or incommensurate. Another example is alternating packing of two or more distinctly different layers. One-dimensional modulation may result if there is a mismatch in one dimension of the layers, or periodicity of the whole structure may be complicated by a two-dimensional modulation if both independent dimensions of the neighboring layers do not match. Reciprocal lattices corresponding to composite structures, therefore, are superpositions of the reciprocal lattices of the substructures with satellites reflecting modulations induced by the mismatch.

Modulated structures maintain at least approximate translational symmetry in three-dimensional physical space and, if the modulations are weak, the average structure may still be represented by the main reciprocal lattice. The same is true for composite crystals in which the substructures can be represented independently by their own reciprocal lattices.

5.4 Symmetry of Modulated Structures

When one, two, or three independent modulation waves that correspond to one-, two-, or three-dimensional modulations are present, then four-, five-, or six-dimensional superspace is required for a complete description of their symmetry. Since the theory of modulated structure is relatively new, the symbols describing symmetry of superspace groups have been developed only recently.[5] Two different notations are commonly used at present. These are:

$$\text{the two} - \text{line symbol, for example, } B^{Pmna}_{s1\bar{1}}, \tag{5.3}$$

$$\text{or the one-line symbol, for example, } Pmna(0\tfrac{1}{2}\gamma)s00 \tag{5.4}$$

[5] T. Janssen, A. Jannar, A. Looijenga-Vos, P.M. de Wolf, Chap. 9.8 Incommensurate and commensurate modulated structures, in: International Tables for Crystallography, Vol. C, E. Prince, Ed., First online edition (2006). https://it.iucr.org/C/; A. Yamamoto, Crystallography of quasiperiodic crystals. Acta Cryst. **A52**, 509 (1996). https://doi.org/10.1107/S0108767396000967; a comprehensive list of superspace groups including symbols, symmetry operations, and reflection conditions can be found online at https://wcp2-ap.eng.hokudai.ac.jp/yamamoto/spgr.html

The two-line symbol begins with a capital letter denoting the Bravais-like symbol of the lattice in a basic space group.[6] It is followed by a superscript defining the basic three-dimensional space-group symmetry of the average structure and a subscript describing intrinsic translations[7] in the superspace for each symmetry element in the basic space-group symmetry. A separate axis is used for every symmetry operation listed in the basic space-group symbol. Sometimes, the two-line symbol is written in a single line. For instance, the two-line symbol shown in Eq. 5.3 becomes B : Pmna : s1$\overline{1}$ when written in a single line. The Bravais-like symbols for modulated lattices are different from those used in three-dimensional periodic lattices (see Table 2.13). Thus, the following rational lattice translation vectors of modulated structures are represented by the following letters:

(½, 0, 0), (0, ½, 0), and (0, 0, ½) by A, B, and C, respectively,
(1, 0, 0), (0, 1, 0), and (0, 0, 1) by L, M, and N, respectively,
(0, ½,½), (½,0, ½), and (½,½,0) by U, V, and W, respectively, and
(1/3,1/3,0) by R.

The intrinsic translations are coded as follows: s = ½, t = 1/3, q = ¼, and h = 1/6; if intrinsic translation is zero, it is denoted as 1 or $\overline{1}$.

Hence, the symbol given in Eq. 5.3 can be understood as follows:

- Bravais type B describes rational lattice translation vector (0, ½, 0), which, in this particular case, means that lattice vector **b** is doubled in the superspace compared to the basic structure.
- Pmna is space-group symmetry of the average (basic) structure.
- s1$\overline{1}$ denotes intrinsic translations along three dimensions for the symmetry operations corresponding to the symmetry elements listed in the basic space-group symbol, that is, m, n and a. We note that $\overline{1}$ reflects a negative diagonal element of the superspace symmetry operation, in other words $R_{44} = -1$.

The one-line symbol begins with a conventional notation of three-dimensional space-group symmetry of the average structure, followed by a modulation vector in parentheses, and intrinsic translations or symmetry elements for additional dimension(s). The latter are set to 0 if the intrinsic translations are zero, which is contrary to 1 or $\overline{1}$ found in the two-line symbol. Equation (5.4) describes the same superspace-group symmetry as Eq. (5.3). The main differences between the two notations are the presence or absence of Bravais-like notation and explicit listing of the modulation vector. Because of these differences, the two-line notation has been developed to directly identify symmetry operations, while the one-line symbol is easily deduced from diffraction data, since the modulation vector may be observed and measured directly from a three-dimensional diffraction pattern. It is important to realize that finding modulation vector using only powder diffraction data is a much more

---

[6] P. M. de Wolff, The Superspace groups for incommensurate crystal structures with a one-dimensional modulation. Acta Cryst. **A37**, 625 (1981). https://doi.org/10.1107/S0567739481001447

[7] Intrinsic translation is a part of a translational component of symmetry operation that is independent of the choice of the origin of coordinates.

complicated task when compared to finding the unit cell dimensions of a periodic lattice, simply because of the one-dimensional nature of the powder pattern.

Since composite structures may consist of two or more modulated substructures, two or more modulated group symbols in either of the mentioned notations are required to describe their full superspace symmetry.[8] For example:

$$\mathrm{R}_{111}^{\mathrm{P31c}} : \mathrm{P}_{1\mathrm{s}}^{\mathrm{R3m}}, \text{or P31c}\left(\frac{1}{3}\ \frac{1}{3}\ \gamma_1\right) : \mathrm{R3m}(00\gamma_2)0\mathrm{s} \tag{5.5}$$

It is worth mentioning that in some cases, the basic space groups are presented in nonstandard setting (e.g., triclinic body-centered lattice, $I\bar{1}$) in order for the substructures to match each other. Therefore, the following additional symbols are also used to designate Bravais-like types:

D, E, F, G, H, I, for (½, 1, 0), (½, 0, 1), (0, ½, 1), (1, ½, 0), (1, 0, ½), (0, 1, ½), and X, Y and Z for (0, 1, 1), (1, 0, 1) and (1, 1, 0).

The full list of superspace-group symbols for four-dimensional (3 + 1) modulated structures can be found in the International Tables for Crystallography.[9] The number of Bravais classes and symmetry groups increases rapidly with added dimensions, especially when all possible groups are included, not only those intrinsic to modulated structure. For example, there are 4783 nonisomorphic (excluding enantiomorphous pairs) four-dimensional groups, but only 370 of them are required for the description of aperiodic structures.[10] Because of increased dimensionality, the symbols of superspace groups became more and more complex, and their listings take more and more space. The list of aperiodic groups in up to six dimensions is available on A. Yamamoto's web page.[11]

5.5 Quasicrystals

The symmetry of quasicrystals can be represented by introducing a different perturbation function, which is based on the Fibonacci[12] numbers. An infinite Fibonacci sequence is derived from two numbers, 0 and 1, and is formed according to the following rule:

---

[8] A. Yamamoto, Unified setting and symbols of superspace groups for composite crystals. Acta Cryst. **A48**, 476 (1992). https://doi.org/10.1107/S010876739200014X

[9] International Tables for Crystallography, Vol. C, E. Prince, Ed., First online edition (2006), Table 9.8.3.5. https://it.iucr.org/C/,

[10] T. Janssen, J.L. Birman, F. Dénoyer, V.A. Koptsik, J.L. Verger-Gaugry, D. Weigel, A. Yamamoto, S.C. Abrahamsh, V. Kopsky, Report of a Subcommittee on the Nomenclature of n-Dimensional Crystallography. II. Symbols for arithmetic crystal classes, Bravais classes and space groups. Acta Cryst. **A58**, 605 (2002). https://doi.org/10.1107/S010876730201379X

[11] https://wcp2-ap.eng.hokudai.ac.jp/yamamoto/spgr.html

[12] Leonardo Pisano Fibonacci, a.k.a. Leonardo of Pizza (likely 1170–1250). Medieval Italian mathematician who in 1202 wrote *Liber abaci—The Book of the Abacus*, a.k.a. *The Book of*

$$F_{n+2} = F_{n+1} + F_n \tag{5.6}$$

This results in the series of numbers

0, 1, 1, 2, 3, 5, 8, 13, 21, … 5.7

Assume that we have a sequence of words containing letters L (for long distance or fragment) and S (for short distance or fragment), which are constructed by replacing each letter in the previous word using the following substitution rule: letter S is replaced by letter L, while letter L is replaced by the word LS. Starting from L as the first word, the infinite sequence of words is obtained, and the first six members of this sequence are shown in Fig. 5.3.

The frequency of occurrence of letters L and S in this sequence is represented in Table 5.1, and it is easy to recognize that they are identical to the consecutive members (F_{n+1} and F_n) of the Fibonacci series. The corresponding limit when the number of words, n, approaches infinity is the golden mean, τ

$$\tau = \lim_{n \to \infty} \left(\frac{F_{n+1}}{F_n} \right) = \frac{\sqrt{5}+1}{2} = 1.618. \tag{5.8}$$

The golden mean can also be represented as a continuous fraction, which contains only one number, 1, and therefore, it is sometimes referred to as the "most irrational" number.

$$\tau = \frac{1}{1 + \frac{1}{1 + \frac{1}{1 + \dots}}} \tag{5.9}$$

The perturbation (modulation) function used in the description of aperiodic structures is obtained by associating interatomic distances (or larger fragments in the crystal structure) with length ratio τ to 1 to letters L and S and the resulting modulation function is no longer a sinusoidal wave but is saw-tooth-like. It is worth noting that the periodicity in this simple one-dimensional case (Fig. 5.3 and Table 5.1) is absent, but the order is perfect: as soon as the law has been established, the "structure" of the series, that is, the location of S and L can be predicted at any point starting from the origin, or any other known location.

Unlike modulated structures, quasicrystals do not have even approximate translational symmetry in two- or three-dimensions of physical space and they can be described only in five- or six-dimensional superspace using 5×5 and 6×6 rotation matrices.

Calculation—in which he formulated the problem leading to the sequence of numbers 1, 1, 2, 3, 5, 8, 13, 21, 34, 55, … (without the first term, i.e., without 0): "How many pairs of rabbits can be produced in a year from one pair of rabbits assuming that every month each pair produces one new pair of rabbits, which becomes productive one month after birth?" See https://mathshistory.st-andrews.ac.uk/Biographies/Fibonacci/ for a brief biography.

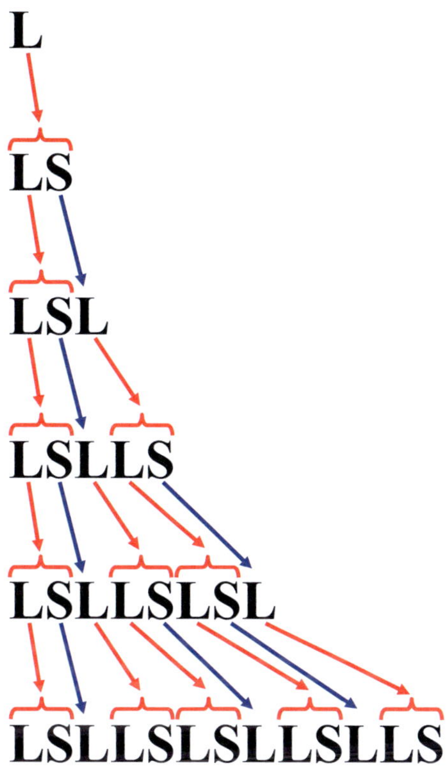

Fig. 5.3 The sequence of words containing quasiperiodic sequences of letters L and S based on the following substitution rule: S → L, and L → LS

Table 5.1 The frequency of occurrence of letters L (f_L) and S (f_S) in the infinite series of words based on the substitution rule S → L, and L → LS

n	Word	f_L	f_S
0	L	1	0
1	LS	1	1
2	LSL	2	1
3	LSLLS	3	2
4	LSLLSLSL	5	3
5	LSLLSLSLLSLLS	8	5
6	LSLLSLSLLSLLSLSLLSLSL	13	8
7	LSLLSLSLLSLLSLSLLSLSLLSLLSLSLLSLLS	21	13
…	…	…	…
		F_{n+1}	F_n

The more detailed description of the nonconventional symmetry goes beyond the scope of this book[13] as the application of powder diffraction for structural study of aperiodic crystals is quite complex. Interpretation of a powder diffraction pattern, which is a projection of reciprocal space on one dimension, presents challenges even for some conventional three-dimensionally periodic crystal structures. However, powder method may still be used successfully to study aperiodic structures when combined with other techniques, for example, electron diffraction.[14] Nevertheless, this chapter has been included here for completeness and to give the reader a flavor of the recent developments in the very old subject of crystallography.[15]

5.6 Symmetry of Magnetic Structures

The magnetic structure is defined as an ordered arrangement of nonzero magnetic spins or magnetic moments (Fig. 5.4a) within the crystal lattice. Determining the magnetic structure is essential for understanding and predicting the properties of magnetic materials. For instance, a magnetic structure with ordered and parallel magnetic moments (Fig. 5.4b) exhibits ferromagnetic properties. In contrast, an alternating orientation of magnetic moments (Fig. 5.4c) results in antiferromagnetic materials, or, if alternating magnetic moments have different lengths (Fig. 5.4d), ferrimagnetic materials. Only the ordered magnetic spins contribute to diffraction, a phenomenon observable exclusively with neutrons, which possess magnetic spin and interact with magnetic moments of atoms. Importantly, magnetic moments become ordered only below the ordering temperature: the Néel temperature (T_N) for antiferromagnetic materials or the Curie temperature (T_C) for ferromagnetic materials.

The terms *magnetic moment* and *magnetic spin* are often used interchangeably to describe the same phenomenon—a magnetic moment caused by the spinning of an unpaired electron. However, the symmetry operations are applied to the magnetic

---

[13] A more complete description of de Wolff's approach to treatment of various types of aperiodic crystals can be found in the International Tables for Crystallography, Vol. B, Second online edition, U. Shmueli, Ed., (2010). https://it.iucr.org/B/

[14] M. Dušek, V. Petříček, M. Wunschel, R.E. Dinnebier, S. van Smaalen, Refinement of modulated structures against X-ray powder diffraction data with JANA2000. J. Appl. Cryst. **34**, 398 (2001). https://doi.org/10.1107/S0021889801003302; H.A. Graetsch, Monoclinic $AlPO_4$ tridymite at 473 and 463 K from X-ray powder data, Acta Cryst. **C58**, 18 (2002). https://doi.org/10.1107/S0108270101015773

[15] The discovery of fivefold symmetry prompted the *ad*-interim Commission on Aperiodic Crystals of the International Union of Crystallography to change the definition of a crystal as a periodic three-dimensional arrangement of identical unit cells to the following: "…by 'crystal' we mean any solid having an essentially discrete diffraction diagram, and by 'aperiodic crystal' we mean any crystal in which three-dimensional lattice periodicity can be considered to be absent". International Union of Crystallography. Report of the Executive Committee for 1991, Acta Cryst. **A48**, 922–946 (1992). https://doi.org/10.1107/S0108767392008328

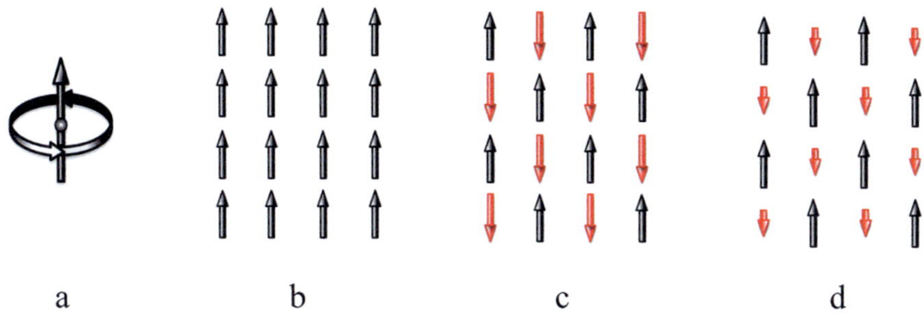

Fig. 5.4 Magnetic spin and the resulting magnetic moment are illustrated in (**a**), while the arrangement of magnetic moments is shown for ferromagnetic (**b**), antiferromagnetic (**c**), and ferrimagnetic (**d**) structures

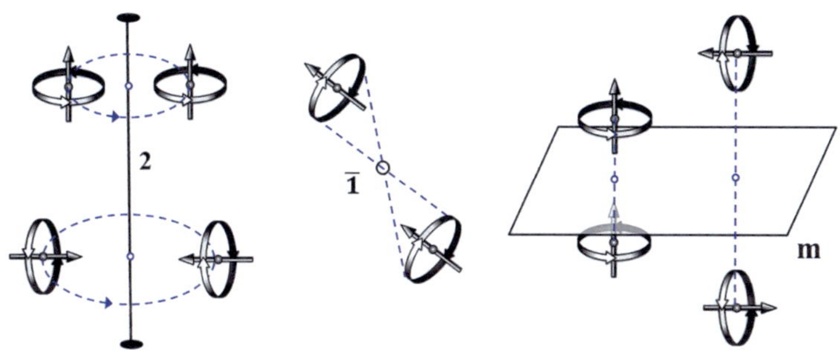

Fig. 5.5 Transformation of magnetic spins by a twofold rotation axis (2y), inversion center ($\bar{1}$), and mirror plane (m_y). The spin is represented as a *circle* with an *arrow* pointing in the spinning direction, while the magnetic moment is shown as a *vector* perpendicular to the circle

spin as a physical object, not to the magnetic moment, which is represented as a vector pointing toward the observer when viewed perpendicularly to a counterclockwise spinning circle (see Fig. 5.4a). Furthermore, the magnetic spin provides a clearer visualization of the geometry of transformations by basic symmetry elements—rotation, reflection, inversion, and translation—while the magnetic moment[16] is more convenient for symbolic and algebraic representations of symmetry.

Figure 5.5 illustrates the transformation of magnetic spins and magnetic moments in various orientations through rotation, reflection, and inversion. All symmetry operations transform the magnetic spin directly, regardless of its orientation, through rotation around axes, inversion at the center, and reflection in the plane. However, the transformation of the magnetic moment depends on both its orientation and the type of symmetry operation applied.

Rotating the spin also rotates the magnetic moment. In contrast, inverting the spin does not invert the moment but retains its direction. The effect of reflection on the

[16] A projection of the magnetic moment vector on the coordinate axes.

magnetic moment depends on its orientation: the moment perpendicular to the plane remains unchanged, while moments in other orientations have their direction reversed.

As discussed earlier, regular symmetry operations[17] that transform the atomic coordinates of a crystal structure also transform magnetic spins. However, magnetic structures can exhibit additional complex symmetry operations that reverse[18] the magnetic spin and, consequently, the magnetic moment. These complex operations are historically referred to as antisymmetry or antisymmetric operations. The concepts of antisymmetry and anti-identity were first introduced by Heesch in 1932.[19] Subsequently, the black-and-white symmetry theory was developed, and the black-and-white space groups containing antisymmetry were derived.[20] This symmetry, in addition to transforming the three-dimensional structure (atomic coordinates), also applies to a fourth parameter that can exist in two states, such as black-and-white color, reversed (−1) and not reversed (+1) magnetic spin, or similar properties. It is also referred to as two-color symmetry, magnetic symmetry, and is often associated with time-reversal symmetry.

Each regular symmetry operation, including identity and translation, has an antisymmetric analogue, denoted by prime symbol ($'$). This analogue is derived by multiplying the regular symmetry operation with the anti-identity ($1'$). The anti-identity is essentially an antisymmetric onefold rotation axes or an identity operation that performs no other action other than reversing the spin. For this reason, it is sometimes referred to as the spin-reversal operation. Equation (5.10) illustrates the multiplication of a mirror reflection operation (m), perpendicular to the Y-axis, by the anti-identity operation ($1'$), which reverses the magnetic spin. This results in the anti-reflection operation (m') or anti-mirror plane, expressed as $m \times 1' = m'$. Here, the magnetic symmetry operation is represented by a 4×4 matrix (M), with the magnetic component forming the fourth dimension. In this matrix, all magnetic elements in the fourth row and fourth column are zero, except for the diagonal element $M_{4,4}$, which equals 1 for regular symmetry transformations and −1 for spin-reversal or antisymmetry operations. Equation (5.11) symbolically represents the same transformation as Eq. (5.10) but records it in algebraic form. It is important to note that both notations indicate only the reversal (−1) of the spin or moment and not its actual orientation.

[17] Also called structural or nuclear symmetry.

[18] The term inverse or inversion is often used in this context. However, we will use the term reverse or reversal to distinguish magnetic moment transformation from the inversion symmetry operation centered at a point.

[19] Heesch, H. Über die vierdimensionalen Gruppen des dreidimensionalen Raumes. Z. Kristallogr. **73**, 325–345 (1930), https://doi.org/10.1524/zkri.1930.73.1.325

[20] See the brief history on the antisymmetry and black-and-white symmetry in: B.D. Litvin, Tables of crystallographic properties of magnetic space groups, Acta Cryst., **A64**, 419–424 (2008), https://doi.org/10.1107/S010876730800768X

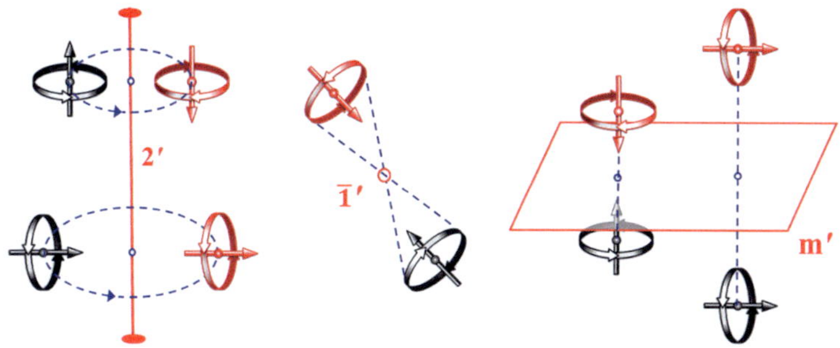

Fig. 5.6 Transformation of magnetic spins by a twofold anti-rotation axis ($2'_y$), anti-inversion center ($\overline{1}'$), and anti-mirror plane (m'_y). The antisymmetry elements, as along with the reversed spins and moments, are shown in *red*

$$\begin{pmatrix} 1 & 0 & 0 & 0 \\ 0 & -1 & 0 & 0 \\ 0 & 0 & 1 & 0 \\ 0 & 0 & 0 & 1 \end{pmatrix} \times \begin{pmatrix} 1 & 0 & 0 & 0 \\ 0 & 1 & 0 & 0 \\ 0 & 0 & 1 & 0 \\ 0 & 0 & 0 & -1 \end{pmatrix} = \begin{pmatrix} 1 & 0 & 0 & 0 \\ 0 & -1 & 0 & 0 \\ 0 & 0 & 1 & 0 \\ 0 & 0 & 0 & -1 \end{pmatrix} \tag{5.10}$$

$$m_y \times 1' = m'_y \quad \text{or} \quad (x, -y, z, 1) \times (x, y, z, -1) = (x, -y, z, -1) \tag{5.11}$$

The transformation of magnetic spin and moment in Fig. 5.6 is shown for the same objects as in Fig. 5.5, but using antisymmetry elements that reverse the spins and their corresponding moments, shown in red.

The interaction of magnetic symmetry operations follows the same rules as discussed in Sec. 4.2.7 for regular symmetry. The resulting symmetry operation can be derived by multiplying the corresponding 3×3 matrices, along with the additional multiplication of the $M_{4,4}$ diagonal elements, which indicates whether the operation reverses the magnetic spin. If both interacting operations are symmetric or both are antisymmetric, the resulting operation is symmetric ($1 \times 1 = 1$ and $-1 \times -1 = 1$). However, if the interacting operations are mixed, the resulting operation is antisymmetric ($1 \times -1 = -1$ and $-1 \times 1 = -1$). Thus, a twofold axis interacting with a perpendicular mirror plane and resulting in an inversion center can yield the following four point groups:

$$2 \times m = \overline{1} \rightarrow 2/m$$

$$2' \times m' = \overline{1} \rightarrow 2'/m'$$

$$2' \times m = \overline{1}' \rightarrow 2'/m$$

$$2 \times m' = \overline{1}' \rightarrow 2/m'$$

By analyzing the possible combinations of symmetric and antisymmetric operations in the 32 point groups, a total of 122 magnetic, or black-and-white, point groups can be deduced (Table 5.2). Fifty-eight of these are antisymmetric point groups, each containing at least one antisymmetric operation. Thirty-two of the

Table 5.2 Magnetic point group symmetry[a]

Point group[b] (uncolored)	Non-trivial magnetic point groups (black and white)				
1					
$\bar{1}$	$\bar{1}'$				
m	m'				
2	$2'$				
2/m	$2'/m$	$2/m'$	$2'/m'$		
222	$2'2'2$				
mm2	$m'm2'$	$m'm'2$			
mmm	$m'mm$	$m'm'm$	$m'm'm'$		
4	$4'$				
$\bar{4}$	$\bar{4}'$				
4/m	$4'/m$	$4/m'$	$4'/m'$		
422	$4'22'$	$42'2'$			
4mm	$4'm'm$	$4m'm'$			
$\bar{4}2m$	$\bar{4}'2'm$	$\bar{4}'2m'$	$\bar{4}2'm'$		
4/mmm	$4/m'mm$	$4'/mm'm$	$4'/m'm'm$	$4/mm'm'$	$4/m'm'm'$
3					
$\bar{3}$	$\bar{3}'$				
32	$32'$				
3m	$3m'$				
$\bar{3}m$	$\bar{3}'m$	$\bar{3}m'$	$\bar{3}'m'$		
6	$6'$				
$\bar{6}$	$\bar{6}'$				
6/m	$6'/m$	$6/m'$	$6'/m'$		
622	$6'22'$	$62'2'$			
6mm	$6'mm'$	$6m'm'$			
$\bar{6}m2$	$\bar{6}'m'2$	$\bar{6}'m2'$	$\bar{6}m'2'$		
6/mmm	$6/m'mm$	$6'/mmm'$	$6'/m'mm'$	$6/mm'm'$	$6/m'm'm'$
23					
$m\bar{3}$	$m'\bar{3}'$				
432	$4'32'$				
$\bar{4}3m$	$\bar{4}'3m'$				
$m\bar{3}m$	$m'\overline{3'}m$	$m\bar{3}m'$	$m'\bar{3}'m'$		

[a] https://www.cryst.ehu.es/cryst/mpoint.html

[b] 32 gray point groups are not listed as they have the same symbols as point groups with $1'$ added the end

original point groups transform magnetic spins without reversing them and are therefore referred to as uncolored (or colorless) point groups. Another 32 point groups, known as Heesch or gray groups, include individual anti-identity operations. The anti-identity operation, when used alone, reverses the magnetic spin without moving the atom. As a result, the reversed spin and the original spin cancel each other out, yielding a net magnetic moment of zero. Consequently, the 32 gray

Table 5.3 36 magnetic Bravais lattices[a]

Crystal system	Lattice type
Triclinic	P, P_{2s}
Monoclinic	P, P_{2a}, P_{2b}, P_C C, C_{2c}, C_P
Orthorhombic	P, P_{2a}, P_C, P_F C, C_{2c}, C_P, C_I F, F_C I, I_P
Tetragonal	P, P_{2c}, P_C, P_I I, I_P
Trigonal	R, R_R
Hexagonal	P, P_{2c}
Cubic	P, P_F F I, I_P

[a]45 lattices counting alternative orientations

groups, which contain both the anti-identity operation and the asymmetric analogue of each present symmetry operation, are used to describe the symmetry of non-magnetic or paramagnetic structures. For this reason, gray groups are sometimes referred to as paramagnetic groups.

Similar to the 230 structural space groups, magnetic space groups can be derived from the magnetic point groups by replacing rotation axes and mirror planes with screw axes and glide planes, as well as altering Bravais lattice types. However, magnetic space groups include many more lattice types for a given crystal system, as shown in Table 5.3. A new lattice type appears only when anti-translation operation is present.

For example, converting the base-centering translation 1/2**a** + 1/2**b** (Fig. 5.7, left) into anti-translation makes the magnetic lattice primitive (P), while the structural lattice remains base-center (C). The former is added as a subscript, yielding C_P magnetic Bravais lattice. Similarly, when two centering translations 1/2**a** + 1/2**c** and 1/2**b** + 1/2**c** (Fig. 5.7, middle) in the orthorhombic face-centered lattice (F) become anti-translations, the magnetic lattice becomes base-centered (C) and the lattice type symbol F_C. However, when a basic translation (**a**, **b**, or **c**) becomes antisymmetric translation, a new type of edge-centered magnetic lattice results as shown in Fig. 5.7, right. Here, translation **a** is magnetic anti-translation with a double magnetic vector $\mathbf{a}_{mag} = 2\mathbf{a}_{str}$. Such complex edge-centered lattice is labeled as P_{2a}.

All 36 possible magnetic Bravais lattices are listed in Table 5.3; their drawings could be found online elsewhere[21, 22]

[21] B.D. Litvin, Magnetic Group Tables. 1-, 2- and 3-Dimensional Magnetic Subperiodic Groups and Magnetic Space Groups. Part 1. Introduction. IUCr e-book, ISBN 978-0-9553602-2-0, 2014. https://doi.org/10.1107/9780955360220001

[22] A.S. Borovik-Romanov, H. Grimmer, M. Kenzelmann, Ch. 1.5. Magnetic properties, in: International Tables for Crystallography (2013), Vol. D: Physical properties of crystals. A. Authier,

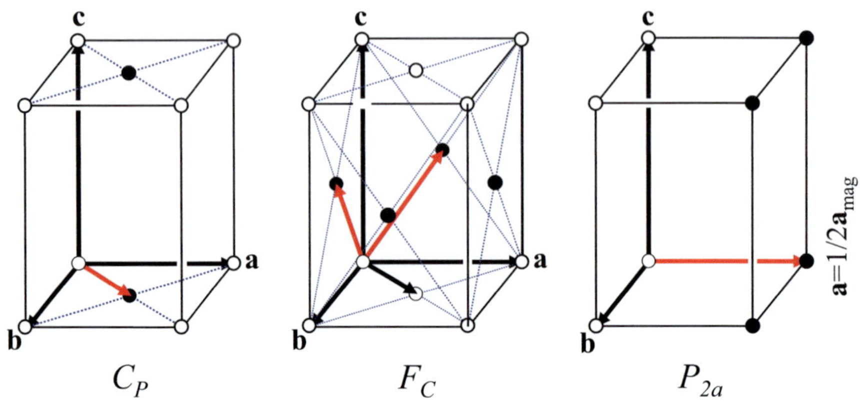

Fig. 5.7 Examples of magnetic Bravais lattices with anti-translation vectors shown in *red* (C_P: 1/2**a** + 1/2**b**, F_C: 1/2**a** + 1/2**c**, 1/2**b** + 1/2**c**, and P_{2a}: **a**) and lattice points with reversed spin shown in *black*

Table 5.4 Three types of magnetic space groups

Type I		230 uncolored space groups (no antisymmetry)
Type II		230 gray space groups containing the individual anti-identity operation
Type III	(a)	674 black-and-white space groups of the first kind, with primitive magnetic cell = primitive structural cell
	(b)	517 black-and-white groups of the second kind, containing anti-translation with primitive magnetic cell > primitive structural cell
Total		1651 magnetic space groups

[a] B.D. Litvin, Magnetic Group Tables. 1-, 2- and 3-Dimensional Magnetic Subperiodic Groups and Magnetic Space Groups. Part 1. Introduction (2013). IUCr e-book, ISBN 978-0-9553602-2-0. https://doi.org/10.1107/9780955360220001

As shown in Table 5.4, 1191 black-and-white groups are possible, of which 674 have the same structural and magnetic cell, containing only regular lattice types. The remaining 517 groups include anti-translation and therefore exhibit complex lattice types. Adding the 230 regular (uncolored) space groups and 230 gray space groups brings the total number of magnetic space groups to 1651.

Comprehensive information on magnetic space groups, presented in the format and spirit of the *International Tables* and compiled by D.B. Litvin, is available as an eBook.[23] Another excellent source of information on magnetic symmetry and its

Ed. Second online edition, ISBN: 978-1-118-76229-5, pp. 106-153. https://doi.org/10.1107/97809553602060000904

[23] B.D. Litvin, Magnetic Group Tables. 1-, 2- and 3-Dimensional Magnetic Subperiodic Groups and Magnetic Space Groups. Part 2. Tables of Magnetic Groups (2013). IUCr e-book, ISBN 978-0-9553602-2-0. https://doi.org/10.1107/9780955360220001

Table 5.5 Comparing symmetry operations for the general position in the $P2_1/m$ space group and its derivatives with different types of magnetic symmetry. Antisymmetry operations are shown in *red*, with anti-identity and anti-translation operations highlighted in **bold**

	$P2_1/m$		$P2_1/m1'$		$P2_1/m'$		$P_{2a}2_1/m$	
	Type I (uncolored)		Type II (gray)		Type IIIa (same cell)		Type IIIb (anti-translation)	
1	x,y,z,+1	[uvw]	x,y,z,+1	[0,0,0]	x,y,z,+1	[uvw]	x,y,z,+1	[uvw]
2	$\bar{x}$,y + 1/2,$\bar{z}$, +1	$[\bar{u}v\bar{w}]$	$\bar{x}$,y + 1/2,$\bar{z}$, +1	[0,0,0]	$\bar{x}$,y + 1/2,$\bar{z}$, +1	$[\bar{u}v\bar{w}]$	$\bar{x}$,y + 1/2,$\bar{z}$,+1	$[\bar{u}v\bar{w}]$
3	$\bar{x}$,$\bar{y}$,$\bar{z}$,+1	[uvw]	$\bar{x}$,$\bar{y}$,$\bar{z}$,+1	[0,0,0]	$\bar{x}$,$\bar{y}$,$\bar{z}$,−1	$[\bar{u}\bar{v}\bar{w}]$	$\bar{x}$,$\bar{y}$,$\bar{z}$,+1	[uvw]
4	x,$\bar{y}$ + 1/2,z, +1	$[\bar{u}v\bar{w}]$	x,$\bar{y}$ + 1/2,z, +1	[0,0,0]	x,$\bar{y}$ + 1/2, z,−1	$[u\bar{v}w]$	x,$\bar{y}$ + 1/2,z,+1	$[\bar{u}v\bar{w}]$
5			**x,y,z,−1**	**[0,0,0]**			**x + 1,y,z,−1**	$[\bar{u}\bar{v}\bar{w}]$
6			$\bar{x}$,y + 1/ 2,$\bar{z}$,−1	[0,0,0]			$\bar{x}$ + 1,y + 1/ 2,$\bar{z}$,−1	$[u\bar{v}w]$
7			$\bar{x}$,$\bar{y}$,$\bar{z}$,−1	[0,0,0]			$\bar{x}$ + 1$\bar{y}$,$\bar{z}$,−1	$[\bar{u}\bar{v}\bar{w}]$
8			x,$\bar{y}$ + 1/2, z,−1	[0,0,0]			x + 1,$\bar{y}$+1/2, z,−1	$[u\bar{v}w]$

applications is the BilBao Crystallographic Server.[24] Both sources provide *International Tables* style lists of symmetry transformations for the atomic coordinates, with the addition of transformations for magnetic spins and moments. The latter includes not only the fourth parameter, +1 and −1, indicating spin reversal, but also the symmetry operations performed on the vector of the magnetic moment. These are presented as the projections of the magnetic moment on the coordinate axes *u, v, w* or m_x, m_y, m_z in Litvin's eBook and the BilBao tables, respectively.

Table 5.5 illustrates the symmetry (and antisymmetry) transformations for a general site in $P2_1/m$ uncolored space group, the corresponding gray group, and two selected black-and-white groups of type III(a) and III(b). The symmetry operations are presented in symbolic form, as discussed earlier, with the fourth parameter equal to −1 for the spin-reversal operation and +1 otherwise.

The first column contains the regular (uncolored) $P2_1/m$ space group. The second column shows the gray group, $P2_1/m1'$, which includes the anti-identity operation (line 5). This operation creates antisymmetric duplicates of all symmetry operations and zeroes all magnetic vectors. The third column illustrates the $P2_1/m'$ black-and-white group of type III(a), which includes two antisymmetry operations (m' and $\bar{1}'$) and shares the same regular and magnetic cells. Two additional groups (not discussed here) involve different choices of antisymmetry operations: $P2_1'/m$ and $P2_1'/m'$. The fourth column shows type III(b) black-and-white group $P_{2a}2_1/m$, which contains the1,0,0 anti-translation vector (line 5). This vector doubles the lattice parameter *a* of the magnetic lattice but not the structural one. Note, however, that the symmetry operations provided here correspond to the original single cell. An alternative presentation of the same group, with symmetry operations in the double

[24] https://www.cryst.ehu.es

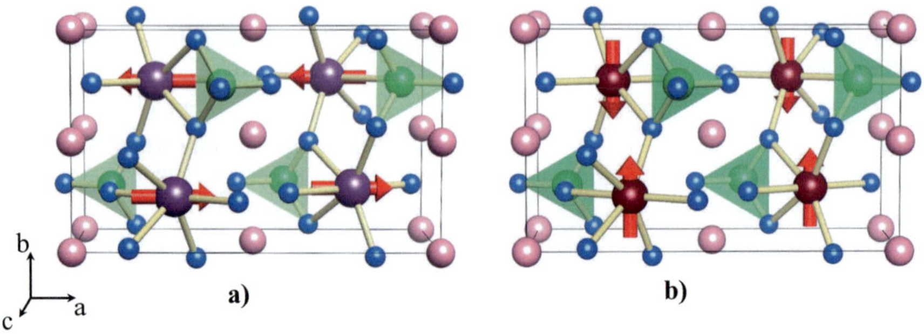

Fig. 5.8 Orientation of magnetic vectors in (**a**) Pn′m′a′ for $LiMnPO_4$ and (**b**) Pnma′ for $LiCoPO_4$. Li atoms are shown in *pink*, P atoms are *green balls*, and O atoms shown as *small blue balls*. Mn and Co atoms are represented by *large purple* and *dark red balls*, respectively (Many thanks to Dr. H.C. Mandujan, Dr. S. Gnewuch, and Dr. E.E. Rodriguez for their valuable discussions and for providing the data.)

cell, is indicated by changing the space-group symbol to P_a2_1/m. Both settings are used and can be generated using BilBao Crystallographic Server,[25] where they are referred to as the OG (Opechowski–Guccione) and BNS (Belov–Neronova–Smirnova) settings, respectively.

An interesting example of magnetic symmetry involves two antiferromagnetic compounds, $LiCoPO_4$ and $LiMnPO_4$, which share the same olivine nuclear structure type but exhibit two distinctly different magnetic structures, each described by a unique magnetic symmetry group: Pnma′ for $LiCoPO_4$ and Pn′m′a′ for $LiMnPO_4$.[26] The nuclear structure and the distinct orientations of the magnetic vectors in both structures are illustrated in Fig. 5.8.

Finally, magnetic symmetry not only showcases the beauty of symmetry and provides a framework for describing magnetic structures but also serves as a defining factor for magnetic properties. An excellent illustration of this is Ascher's trinity diagram[27,28,29] which illustrates the connection between magnetic symmetry and the associated magnetic properties, as shown in Fig. 5.9. The diagram features three intersecting circles representing ferromagnetic (M), ferroelectric (E), and ferrotoroidic (T) properties. The non-overlapping portion of each circle contains 14 groups, each allowing only one of the properties. The area where two circles

[25] https://www.cryst.ehu.es/#msgtop

[26] S. Gnewuch and E.E. Rodriguez. Distinguishing the Intrinsic Antiferromagnetism in Polycrystalline $LiCoPO_4$ and $LiMnPO_4$ Olivines. *Inorg. Chem.*, **59**, 5883–5895 (2020). https://doi.org/10.1021/acs.inorgchem.9b03545?ref=pdf

[27] E. Ascher, Kinetoelectric and Kinetomagnetic Effects in Crystals. Int. J. Magn. **5**, 287–295 (1979), https://www.edgar-ascher.ch/download/1970/1974-02.pdf

[28] V. Kopský, Crystallography and Magnetic Phenomena. Symmetry, **7**, 125–145 (2015), https://doi.org/10.3390/sym7010125

[29] H. Schmid, Some symmetry aspects of ferroics and single phase multiferroics. J. Phys.: Condens. Matter, **20**, 434231:1–24 (2008), https://doi.org/10.1088/0953-8984/20/43/434201

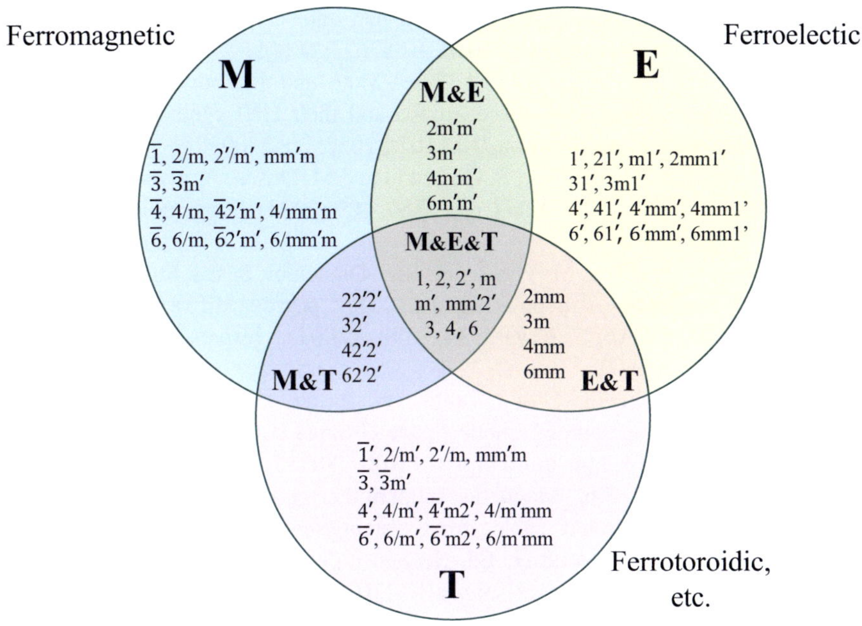

Fig. 5.9 Ascher's trinity diagram linking magnetic symmetry to physical properties

overlap contains four groups, each permitting a pair of properties, and where all three circles overlap, there are nine groups that allow all three properties. Thus, each circle contains 31 groups allowing each of the phenomena (See Footnote 27). Additionally, Schmidt (See Footnote 29) describes two similar trinity diagrams linking magnetic symmetry with higher-order tensors for magnetoelectricity and piezo effects.

5.7 Additional Reading

1. H.T. Stokes, B.J. Campbell, and S. van Smaalen, Generation of (3 + d)-dimensional superspace groups for describing the symmetry of modulated crystalline structures, Acta Cryst. **A67**, 45 (2011), https://doi.org/10.1107/S0108767310042297.
2. C. Janot, Quasicrystals. A primer, Clarendon Press, Oxford (1992), https://global.oup.com/academic/product/quasicrystals-9780199657407.
3. International Tables for Crystallography, Vol. A, Space-group symmetry, M. I. Aroyo, Ed., Second online edition, (2016), https://it.iucr.org/A/.
4. T. Janssen, G. Chapuis, M. de Boissieu, Aperiodic crystals: From modulated phases to quasicrystals. Oxford University Press, (2007), https://doi.org/10.1093/acprof:oso/9780198567776.001.0001.

5. A. Yamamoto, Crystallography of quasiperiodic crystals, Acta Cryst. **A52**, 509 (1996), https://doi.org/10.1107/S0108767396000967.
6. B.J. Campbell, H.T. Stokes, J.M. Perez-Mato and J. Rodriguez-Carvajal, A recapitulation of magnetic space groups and their UNI symbols, Acta Cryst. **B80**, 401 (2024), https://doi.org/10.1107/S2052520624008084.
7. M. S. Henriques, V. Petříček, S. Goswami and M. Dušek, Analysis of magnetic structures in JANA2020, Acta Cryst. **B80**, 409 (2024), https://doi.org/10.1107/S2052520624008163.
8. M. Dušek, V. Petříček, M. Wunschel, R.E. Dinnebier, S. van Smaalen, Refinement of modulated structures against X-ray powder diffraction data with JANA2000. J. Appl. Cryst. **34**, 398 (2001), https://doi.org/10.1107/S0021889801003302.
9. B.D. Litvin, Magnetic Group Tables. 1-, 2- and 3-Dimensional Magnetic Subperiodic Groups and Magnetic Space Groups. IUCr e-book, ISBN 978-0-9553602-2-0, 2014, https://doi.org/10.1107/9780955360220001.
10. D.B. Litvin, Ch. 3.6. Magnetic subperiodic groups and magnetic space groups, in: International Tables for Crystallography (2016). Vol. A: Space-group symmetry. A. Authier, Ed. Second online edition. ISBN: 978-0-470-68575-4, pp. 852-865, https://doi.org/10.1107/97809553602060000934.
11. A. S. Borovik-Romanov, H. Grimmer, M. Kenzelmann, Ch. 1.5. Magnetic properties, in: International Tables for Crystallography (2013), Vol. D: Physical properties of crystals. A. Authier, Ed. Second online edition, ISBN: 978-1-118-76229-5, pp. 106-153, https://doi.org/10.1107/97809553602060000904.
12. Qingzhen Huang, Powder diffraction of superconductors, Ch. 7.13.4 and A7.13.1, In: International Tables for Crystallography, Vol. H: Powder diffraction, C.J. Gilmore, J.A. Kaduk and H. Schenk, Eds. First online edition (2019), ISBN: 978-1-118-41628-0. https://doi.org/10.1107/97809553602060000115.
13. S.V. Gallego, E.S. Tasci, G. de la Flor, J.M. Perez-Mato, M.I. Aroyo, Magnetic symmetry in the Bilbao Crystallographic Server: a computer program to provide systematic absences of magnetic neutron diffraction. J. Appl. Cryst. **45**, 1236–1247 (2012), https://doi.org/10.1107/S0021889812042185.
14. J. Rodríguez-Carvaja, J. Villain, Magnetic structures. C.R.Physique, **20**, 770–802 (2019), https://doi.org/10.1016/j.crhy.2019.07.004.
15. D.E. Laughlin, M.A. Willard, M. E. McHenry, Magnetic Ordering: Some Structural Aspects. Cheminform, **31**(46), 1-17 (2000), https://doi.org/10.1002/chin.200046247, https://www.andrew.cmu.edu/user/dl0p/laughlin/pdf/252.pdf.
16. S. Gnewuch, E.E. Rodriguez, Distinguishing the Intrinsic Antiferromagnetism in Polycrystalline $LiCoPO_4$ and $LiMnPO_4$ Olivines. Inorg. Chem. **59**, 5883–5895 (2020), https://doi.org/10.1021/acs.inorgchem.9b03545.
17. V. Kopský, Crystallography and Magnetic Phenomena. Symmetry, **7**, 125–145 (2015), https://doi.org/10.3390/sym7010125.
18. E. Ascher, Kinetoelectric and Kinetomagnetic Effects in Crystals. Int. J. Magn. **5**, 287–295 (1979). https://www.edgar-ascher.ch/download/1970/1974-02.pdf.

19. H. Schmid, Some symmetry aspects of ferroics and single phase multiferroics. J. Phys.: Condens. Matter, **20**, 434231:1–24 (2008), https://doi.org/10.1088/0953-8984/20/43/434201.

5.8 Problems

1. When working with a modulated structure, a researcher finds that the value of the modulation vector **q** is (0, 0.172(1), 0). Is this one-, two-, or three-dimensional modulation? Is the modulation commensurate or incommensurate?
2. In another experiment, the modulation vector **q** is found to be (0.333(1), 0, 0.501 (2)). How many dimensions does a researcher need to fully describe symmetry of this structure? She wonders whether it is possible to treat this crystal structure in three dimensions. Please, help her in making the decision.
3. A quasicrystal has been described using a one-dimensional modulation function based on the Fibonacci series. Starting from the origin, how many long and short structural fragments will fit within the "period" No. 8.
4. What is allowed orientation of the mag moment for the atom located on the mirror plane and anti-mirror plane?
5. What is allowed orientation of the mag moment for the atom located on the twofold axis and twofold anti-rotation axis?
6. Describe transformation of the magnetic spin and moment by fourfold rotation and anti-rotation axis, similar to Figs. 5.2 and 5.3.
7. Describe transformation of the magnetic spin and moment by threefold rotation and anti-rotation, similar to Figs. 5.2 and 5.3.
8. Show and explain why for point group mm2 only two additional magnetic point groups are possible.
9. Show and explain why for point group mmm there are three possible magnetic groups.

Chapter 6
Properties, Sources, and Detection of Radiation

In the preceding five chapters, we introduced basic concepts of symmetry and discussed the structure of crystals in terms of three-dimensional periodic arrays of atoms and/or molecules, sometimes perturbed by various modulation functions. During this exploration, we implicitly assumed that this representation aligns with reality. Now, it is time to shift our perspective and contemplate how atoms or molecules can be observed—either directly or indirectly. Thus, we delve into understanding how it is possible to determine the crystal structure of a material and verify the concepts of crystallographic symmetry.

To embark on answering this question, consider the following mental experiment: picture yourself in a dark room next to this book. As human eyes are sensitive to visible light, you won't be able to see the book or read the words in total darkness (Fig. 6.1, left). Only when you turn on the light does the book become visible, and the information stored within becomes accessible (Fig. 6.1, right). The fundamental outcome of our experiment is that the book and its content can be observed by means of visible light after it has been scattered by the object (the book) and detected by the eyes.

In general, a source of rays and a suitable detector (such as the light bulb and the eye, respectively) are required to observe common objects. Atoms, however, are too small to be discerned using any visible light source, because atomic radii[1] range from a few tenths of an angström to a few angströms, and they are smaller than 1/1000 of the wavelengths present in visible light (from ~4000 to ~7000 Å). A suitable wavelength to observe individual atoms is that of X-rays. The latter are short-wavelength electromagnetic radiation discovered by W.C. Roentgen,[2] and they

[1] Atomic radius may be calculated self-consistently or it may be determined from experimental structural data. Effective size of an atom varies as a function of its environment and nature of chemical bonding. Several different scales—covalent, ionic, metallic, and Van der Waals radii—are commonly used in crystallography.

[2] Wilhelm Conrad Roentgen (1845–1923), a German physicist who on November 8, 1895, discovered X-rays and was awarded the first-ever Nobel Prize in Physics in 1901, "in recognition of the

P. Zavalij, V. Pecharsky, *Fundamentals of Powder Diffraction and Structural Characterization of Materials*, https://doi.org/10.1007/978-3-031-91504-8_6

Fig. 6.1 The illustration of an observer placed in the absolutely dark room with a book (*left*) and the same room with the light source producing visible rays of light (*right*)

have the wavelengths that are commensurate with both the atomic sizes and shortest interatomic distances.

It is worth mentioning Johann Puluj,[3] who conducted similar research years earlier,[4] although he is not as widely known as Wilhelm Roentgen. In 1881, Puluj invented a device called the "Puluj lamp,[5]" which emitted rays with properties very similar to X-rays. He even used it to take X-ray-like images of bones[6] but did not fully realize the potential of his discovery at the time, particularly in the field of medical imaging. Puluj's contributions to the X-ray technology are now recognized as an important precursor to Roentgen's discoveries.

Unfortunately, the index of refraction of X-rays is near unity for all materials and they cannot be focused by a lens in order to observe such small objects as atoms, as it

---

extraordinary services he has rendered by the discovery of the remarkable rays subsequently named after him." For more information about W.C. Roentgen, see his Nobel Prize entry: https://www.nobelprize.org/prizes/physics/1901/rontgen/facts/

[3]Johann Puluj (also known as Ivan Puluy) was a physicist of Ukrainian origin who, at that time, worked at University of Vienna and later at German Technical University in Prague. He played a significant role in the development of X-ray technology.

[4]J. Puluj, Strahlende Elektrodenmaterie, Akademie d. Wissenschaften Wien, **81**, 864–923 (1880). https://www.zobodat.at/pdf/SBAWW_81_2_0864-0923.pdf

[5]https://web.archive.org/web/20111205010856/http://www.uibk.ac.at/exphys/museum/html/tubes_1.html

[6]Emilie Těšínská, Johann Puluj (1845–1918): his career and the "invisible cathode rays," IUCr Newsletters, **28**(2), 26 (2020). https://www.iucr.org/news/newsletter/etc/articles?issue=148371&result_138339_result_page=26

is done by glass lenses in a visible light microscope or by magnetic lenses in an electron microscope. Thus, in general, X-rays cannot be used to image individual atoms directly.[7] However, as was first shown by Max von Laue in 1912 using a single crystal of hydrated copper sulfate ($CuSO_4{\cdot}5H_2O$), the periodicity of the crystal lattice allows atoms in a crystal to be observed with exceptionally high resolution and precision by means of X-ray diffraction. As we will see later, the diffraction pattern of a crystal is a transformation of an ordered atomic structure into reciprocal space, rather than a direct image of the former. Hence, the three-dimensional distribution of atoms in a lattice can be restored only after the diffraction pattern has been transformed back into direct space.

Particles in motion, such as neutrons and electrons, may be used as an alternative to X-rays. They produce images of crystal structures in reciprocal space because of their dual nature: as follows from quantum mechanics, waves behave as particles (e.g., photons), and particles (e.g., neutrons and electrons) behave as waves with wavelength λ determined by the de Broglie[8] equation:

$$\lambda = \frac{h}{mv} \tag{6.1}$$

where h is Planck's constant ($h = 6.626 \times 10^{-34}$ J s), m is the particle's rest mass, and v is the particle's velocity ($mv = p$, particle momentum).

For example, a neutron (rest mass, $m = 1.6749 \times 10^{-27}$ kg) moving at a constant velocity $v = 3000$ m/s will also behave as a wave with $\lambda = 1.319$ Å. Moreover, charged particles, for example, electrons, can be focused using magnetic lenses. Thus, modern high-resolution electron microscopes allow direct imaging of atomic structures (for the most part in two dimensions on a surface) with the resolution sufficient to distinguish individual atoms. Direct imaging methods, however, require sophisticated equipment and the accuracy in determining atomic positions is substantially lower than that possible by means of diffraction techniques.[9] Hence, direct visualization of a structure with atomic resolution is invaluable in certain applications, but the three-dimensional crystal structures are determined exclusively from diffraction data. For example, electron microscopy may be used to determine unit

[7]Direct imaging of atoms is feasible using X-ray holography, in which the wave after passing through a sample is mixed with a reference wave to recover phase information and produce three-dimensional interference patterns. See R. Fitzgerald, X-ray and γ-ray holography improve views of atoms in solids, Phys. Today **54**, 21 (2001). https://doi.org/10.1063/1.1372102

[8]Louis de Broglie (1892–1987) the French physicist who postulated the dual nature of the electron. In 1929, he was awarded the Nobel Prize in physics "for his discovery of the wave nature of electrons." See https://www.nobelprize.org/prizes/physics/1929/broglie/facts/ for details.

[9]Despite recent progress in the three-dimensional X-ray holography [e.g., see M. Tegze, G. Faigel, S. Marchesini, M. Belakhovsky, and A. I. Chumakov, Three-dimensional imaging of atoms with isotropic 0.5 Å resolution, Phys. Rev. Lett. **82**, 4847 (1999) https://doi.org/10.1103/PhysRevLett.82.4847], which in principle enables visualization of the atomic structure in three dimensions, its accuracy in determining coordinates of atoms and interatomic distances is much lower than possible by employing conventional diffraction methods.

cells or modulation vectors, both of which are valuable data that may be further employed in solving a crystal structure using diffraction methods and, specifically, powder diffraction.

Nearly immediately after their discovery, X-rays were put to use to study the internal structure of objects that are opaque to visible light but transparent to X-rays, for example, parts of a human body using radiography, which takes advantage of varying absorption: bones absorb X-rays stronger than surrounding tissues. It is interesting to note that the lack of understanding of their nature, which did not occur until 1912, did not prevent the introduction of X-rays into medicine and engineering. Today the nature and the properties of X-rays and other types of radiation are well-understood and they are briefly considered in this chapter.

6.1 Nature of X-Rays

Electromagnetic radiation is generated every time when electric charge accelerates or decelerates. It consists of transverse waves where electric (**E**) and magnetic (**H**) vectors are perpendicular to one another and to the propagation vector of the wave (**k**), see Fig. 6.2, top. The X-rays have wavelengths from ~0.1 to ~100 Å, which are

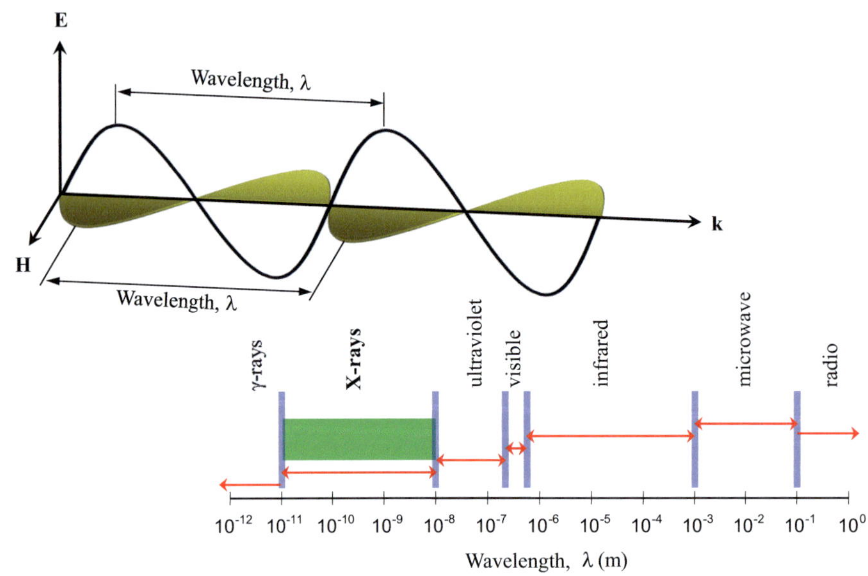

Fig. 6.2 *Top*—the schematic of the transverse electromagnetic wave in which electric (**E**) and magnetic (**H**) vectors are mutually perpendicular, and both are perpendicular to the direction of the propagation vector of the wave, **k**, where |**k**| = 1/λ. The wavelength, λ, is the distance between the two neighboring wave crests. *Bottom*—the spectrum of the electromagnetic waves. The range of typical X-ray wavelengths is shaded. The boundaries between different types of electromagnetic waves are diffuse

located between γ-radiation and ultraviolet rays as also shown in Fig. 6.2, bottom. The wavelengths, most commonly used in crystallography, range between ~0.5 and ~2.5 Å, as they are of the same order of magnitude as the shortest interatomic distances observed in both organic and inorganic materials. Additionally, these wavelengths can be easily produced in almost every research laboratory.

6.2 Production of X-Rays

X-rays are typically generated using two different methods or sources. The first method involves a device called an X-ray tube, where electromagnetic waves are produced through the impacts of high-energy electrons with a metal target. X-ray tubes are the simplest and most commonly utilized sources of X-rays in laboratories of any size. Therefore, an X-ray tube is often referred to as a laboratory or conventional X-ray source. Conventional X-ray sources, such as X-ray tubes, typically exhibit low efficiency, and their brightness is fundamentally limited by the thermal properties of the target material. Continuous cooling of the target material is necessary because nearly all kinetic energy of the accelerated electrons is converted into heat when they rapidly (and sometimes instantly) decelerate during impacts with a metal target.

The second is a much more advanced source of X-ray radiation—the synchrotron where high-energy electrons are confined in a storage ring. When they move in a circular orbit, electrons accelerate toward the center of the ring, thus emitting electromagnetic radiation. The synchrotron sources are extremely bright (or brilliant[10]) since every electron emits electromagnetic radiation, unlike in the X-ray tube, where the majority of electrons that reach the target convert their kinetic energy into heat. Hence, thermal losses are minimized and there is no target to cool. Their brightness is only limited by the flux of electrons in the high-energy beam. Today, the so-called third generation of synchrotrons is in operation, and their brilliance exceeds that of the conventional X-ray tube by nearly ten orders of magnitude.

Obviously, given the cost of both the construction and maintenance of a synchrotron source, owning one would be prohibitively expensive and inefficient for an average crystallographic laboratory. All synchrotron sources are multiple-user facilities, which are constructed and maintained using governmental support. For example, they are supported by the United States Department of Energy and National Science Foundation in the United States and by similar agencies in Europe, Japan, and other countries.

---

[10]The quality of synchrotron beams is typically characterized by brilliance, defined as brightness divided by the product of the source area (in mm^2) and a fraction of useful photon energy, known as bandwidth. (Refer to, for example, J. Als-Nielsen and D. McMorrow, Elements of modern X-ray physics, Second edition, Wiley, New York (2011). https://doi.org/10.1002/9781119998365).

In general, there is no principal difference in the diffraction phenomena using the synchrotron and conventional X-ray sources, except for the presence of several highly intense peaks with fixed wavelengths in the conventionally obtained X-ray spectrum and their absence, that is, the continuous distribution of photon energies when using synchrotron sources. Here and throughout this book, the X-rays from conventional sources are of concern, unless noted otherwise.

6.2.1 Continuous and Characteristic X-Ray Spectra

The X-ray spectrum, generated in a typical X-ray tube, is shown schematically in Fig. 6.3. It consists of several intense peaks, the so-called characteristic spectral lines, superimposed over a continuous background, known as the "white" radiation. The continuous part of the spectrum is generated by electrons decelerating rapidly and unpredictably—some instantaneously, other gradually—and the distribution of the wavelengths depends on the accelerating voltage but not on the nature of the anode material. White radiation, also known as *bremsstrahlung* (German for

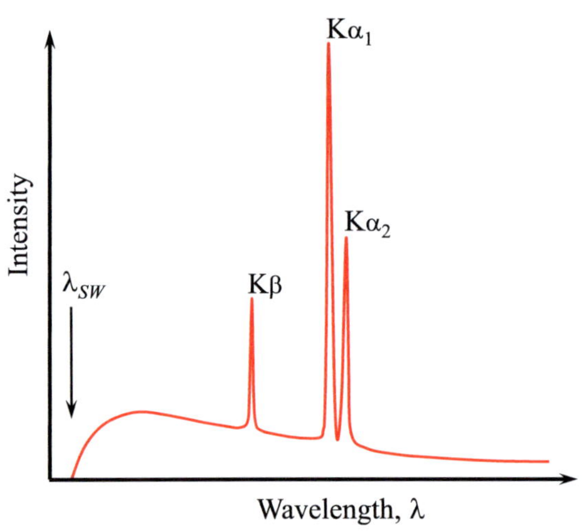

Fig. 6.3 The schematic of a typical X-ray emission spectrum, for clarity indicating only the presence of continuous background and three characteristic wavelengths: $K\alpha_1$, $K\alpha_2$, and $K\beta$, which have high intensities. The relative intensities of the three characteristic spectral lines are approximately to scale; however, the intensity of the continuous spectrum and the separation of the $K\alpha_1/K\alpha_2$ doublet are exaggerated. Fine structure of the $K\beta$ spectral line is not shown for clarity. The vertical arrow indicates the shortest possible wavelength of white radiation, λ_{SW}, as determined by Eq. (6.4)

"braking radiation"), is generally highly undesirable in X-ray diffraction analysis applications.[11]

While it is difficult to establish the exact distribution of the wavelengths in the white spectrum analytically, it is possible to establish the shortest wavelength that will appear in the continuous spectrum as a function of the accelerating voltage. Photons with the highest energy (i.e., rays with the shortest wavelength) are emitted by the electrons, which are stopped instantaneously by the target. In this case, the electron may transfer all of its kinetic energy

$$\frac{mv^2}{2} = eV \tag{6.2}$$

to a photon with the energy

$$h\nu = \frac{hc}{\lambda} \tag{6.3}$$

where m is the rest mass, v is the velocity, e is the charge of the electron (1.602×10^{-19} C), V is the accelerating voltage, c is the speed of light in vacuum (2.998×10^8 m/s), h is Planck's constant (6.626×10^{-34} J s), ν is the frequency, and λ is the wavelength of the wave associated with the energy of the photon.

After combining the right-hand parts of Eqs. (6.2) and (6.3) and solving with respect to λ, it is easy to obtain the equation relating the shortest possible wavelength (λ_{SW} in Å) and the accelerating voltage (in V).

$$\lambda_{SW} = \frac{1.240 \times 10^4}{V} \tag{6.4}$$

The three characteristic lines are quite intense and they result from the transitions of upper-level electrons in the atom core to vacant lower energy levels, from which an electron was ejected by the impact with an electron accelerated in the X-ray tube. The energy differences between various energy levels in an atom are element-specific, and therefore, each chemical element emits X-rays with a constant, that is, characteristic, distribution of wavelengths that appear due to excitations of core electrons by high-energy electrons bombarding the target, see Table 6.1. Obviously, before core electrons can be excited from their lower energy levels, the bombarding electrons must have energy, which is equal to or exceeds that of the energy difference between the two nearest lying levels of the target material.

[11] One exception is the so-called Laue technique, in which white radiation is employed to produce diffraction patterns from stationary single crystals, see Figs. 7.11 and 7.12.

Table 6.1 Characteristic wavelengths of five common anode materials and the K-absorption edges of suitable β-filter materials[a]

Anode material	Wavelength (Å)				β filter	K-absorption edge (Å)
	Kα[b]	Kα1	Kα2	Kβ		
Cr	2.291034	2.289726(3)	2.293651(3)	2.084881(4)	V	2.26921(2)
Fe	1.937352	1.936041(3)	1.939973(3)	1.756604(4)	Mn	1.896459(6)
Co	1.790276	1.788996(1)	1.792835(1)	1.620826(3)	Fe	1.743617(5)
Cu	1.541871	1.5405929(5)	1.5444274(5)	1.392234(6)	Ni	1.488140(4)
Mo	0.710747	0.7093172(4)	0.713607(12)	0.632887(13)	Nb Zr	0.653134(1) 0.688959(3)

[a] The wavelengths are taken from the International Tables for Crystallography, Vol. C, First online edition, E. Prince, Ed. (2006), https://it.iucr.org/Cb/. For details on absorption and filtering, see Sects. 8.6.5 and 11.2.2

[b] The weighted average value, calculated as $\lambda_{average} = (2\lambda_{K\alpha 1} + \lambda_{K\alpha 2})/3$

The transitions from L and M shells to the K shell, that is, L → K and M → K are designated as Kα and Kβ radiation,[12] respectively. Here K corresponds to the shell with principal quantum number n = 1, L to n = 2, and M to n = 3. The Kα component consists of two characteristic wavelengths designated as $K\alpha_1$ and $K\alpha_2$, which correspond to $2p_{1/2} \rightarrow 1s_{1/2}$ and $2p_{3/2} \rightarrow 1s_{1/2}$ transitions, respectively, where *s* and *p* refer to the corresponding orbitals. The subscripts ½ and 3⁄2 are equal to the total angular momentum quantum number, j.[13] The Kβ component also consists of several discrete spectral lines, the strongest being $K\beta_1$ and $K\beta_3$, which are so close to one another that they are practically indistinguishable in the X-ray spectra of many anode materials. There are more characteristic lines in the emission spectrum (e.g., Lα-γ and Mα-ξ); however, their intensities are much lower and their wavelengths are greater than those of Kα and Kβ. Therefore, they are not used in X-ray diffraction analysis and are not considered here.[14]

In addition to their wavelengths, the strongest characteristic spectral lines have different intensities: the intensity of $K\alpha_1$ exceeds that of $K\alpha_2$ by a factor of about 2, and the intensity of $K\alpha_{1,2}$ is approximately five times that of the intensity of the strongest Kβ line, although the latter ratio varies considerably with the atomic

[12] According to IUPAC [R. Jenkins, R. Manne, R. Robin, C. Senemaud, Nomenclature, symbols, units and their usage in spectrochemical analysis—VIII. Nomenclature system for X-ray spectroscopy (Recommendations 1991), Pure Appl. Chem. **63**, 735 (1991). https://doi.org/10.1351/pac199163050735] the old notations, e.g., Cu $K\alpha_1$ and Cu $K\beta_1$ should be substituted by the initial and final levels separated by a hyphen, e.g., Cu K–L_3 and Cu K–M_3, respectively. However, since the old notations remain in common use, they are retained throughout this book.

[13] $j = \ell \pm s$ when $\ell > 0$ and j = ½ when $\ell = 0$, where ℓ is the orbital, and s is the spin quantum numbers. Since ℓ adopts values 0, 1, 2, …, n − 1, which correspond to *s*, *p*, *d*, … orbitals and s = ±½, j is equal to ½ for *s* orbitals, ½ or 3⁄2 for *p* orbitals, and so on.

[14] Except for one experimental artifact shown later in Fig. 6.11, where two components present in the Lα characteristic spectrum of W (filament material contaminating Cu anode of a relatively old X-ray tube) are clearly recognizable in the diffraction pattern collected from the oriented single crystalline silicon wafer.

number. Spectral purity, that is, the availability of a single intense wavelength, is critical in most diffraction applications, and therefore, various monochromatization methods (see Sect. 11.2.2) are used to eliminate multiple wavelengths. Although the continuous X-ray emission spectrum does not result in distinct diffraction peaks from polycrystals, its presence increases the background noise, and therefore, white radiation must be minimized.

Typical anode materials that are used in X-ray tubes (Table 6.1) produce characteristic wavelengths between ~0.5 and ~2.3 Å. However, only two of them are used most commonly. These are Cu in powder and Mo in single-crystal diffractometry. Other anode materials can be used in special applications, for example, Ag anode (λ $K\alpha_1 = 0.5594218$ Å) can be used to increase the resolution of the atomic structure since using shorter wavelength broadens the range of $\sin\theta/\lambda$ over which diffracted intensity can be measured. Bragg peaks, however, are observed closer to each other and the resolution of the diffraction pattern may deteriorate. Conversely, Cr, Fe, or Co anodes may be used instead of a Cu anode in powder diffraction (or Cu anode instead of Mo anode in single-crystal diffractometry) to increase the resolution of the diffraction pattern (Bragg peaks are observed further apart), but the resolution of the atomic structure decreases.

6.2.2 *Conventional Sealed X-Ray Sources*

As noted earlier, the X-ray tube is a conventional laboratory source of X-rays. The two types of X-ray tubes in common use today are the sealed tube and the rotating anode tube. The sealed tube consists of a stationary anode coupled with a cathode, and both are placed inside a metal/glass, metal/ceramic, or all-metal container sealed under high vacuum. Common in the twentieth century, a metal/glass-sealed X-ray tube is illustrated in Fig. 6.4.

The X-ray tube assembly is a simple and maintenance-free device. However, the overall efficiency of an X-ray tube is very low—approximately 1% or less. Most of the energy supplied to the tube is converted into heat, and therefore, the anode must be continuously cooled with chilled water to avoid target meltdown. The input power to the sealed X-ray tube (~0.5–3 kW) is, therefore, limited by the tube's ability to dissipate heat, but the resultant energy of the usable X-ray beam is much lower than 1% of the input power because only a small fraction of the generated photons exits through each window. Additional losses occur during the monochromatization and collimation of the beam (see Sect. 11.2).

In the X-ray tube, electrons are emitted by the cathode, usually electrically heated tungsten filament, and they are accelerated toward the anode by a high electrostatic potential (30–60 kV) maintained between the cathode and the anode. The typical current in a sealed tube is between 10 and 50 mA. The X-rays are generated by the impacts of high-energy electrons with the metal target of a water-cooled anode, and they exit the tube through beryllium (Be) windows, as shown in Figs. 6.4 and 6.5.

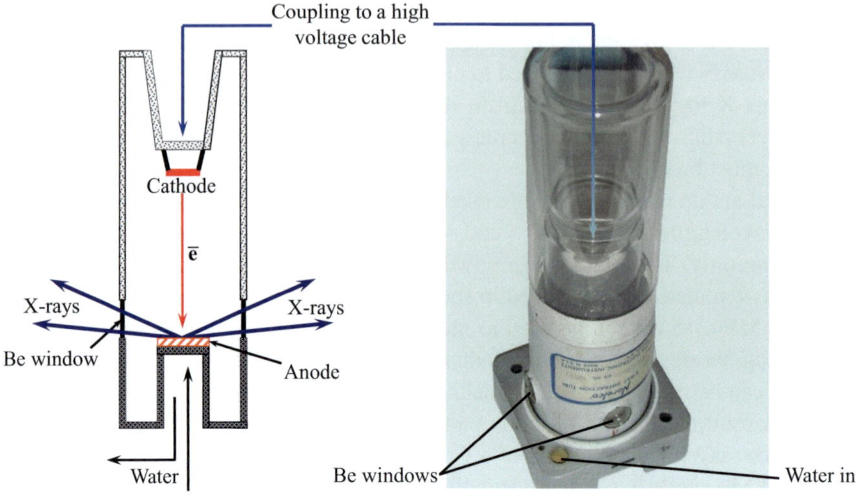

Fig. 6.4 The schematic (*left*) and the photograph (*right*) of the metal/glass-sealed X-ray tube. The bottom part of the tube is metallic and it contains the anode (high purity copper, which may be coated with a layer of a different metal, for example, Cr, Fe, Mo, etc., to produce a target other than copper), the windows (beryllium foil), and the cooling system. The top part of the tube contains the cathode (tungsten filament) and it is made from glass or ceramics, welded shut to the metal canister in order to maintain high vacuum inside the tube. The view of two additional windows (a total of four) and the "water out" outlet is obscured by the body of the tube (*right*). High voltage is supplied by a cable through a coupling located in the glass (or ceramic) part of the tube. Both the metallic can and the anode are grounded

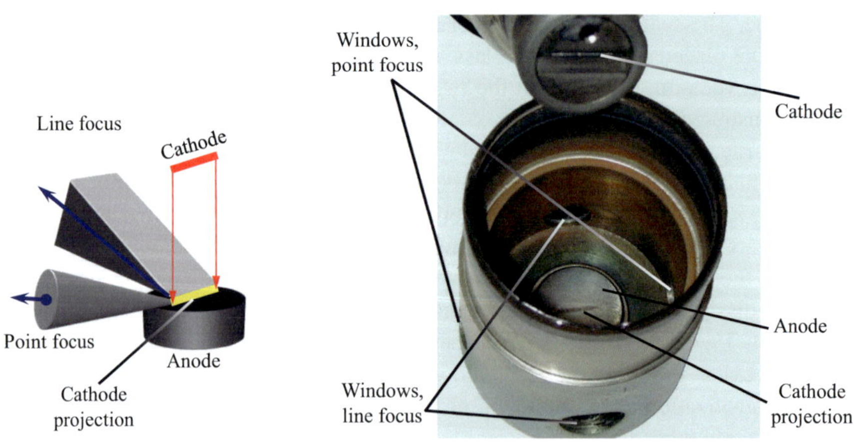

Fig. 6.5 The schematic explaining the appearance of two different geometries of the X-ray focus in a conventional sealed X-ray tube (*left*) and the disassembled tube (*right*). The photo on the right shows the metallic can with four beryllium windows, two of which correspond to line- and two to point-foci. The surface of the anode with the cathode projection is seen inside the can (*bottom, right*). What appears as a scratch on the surface of the anode is the damage from the high-intensity electron beam and a thin layer deposit of the cathode material (W), which occurred during the lifetime of the tube. The cathode assembly is shown on top, right

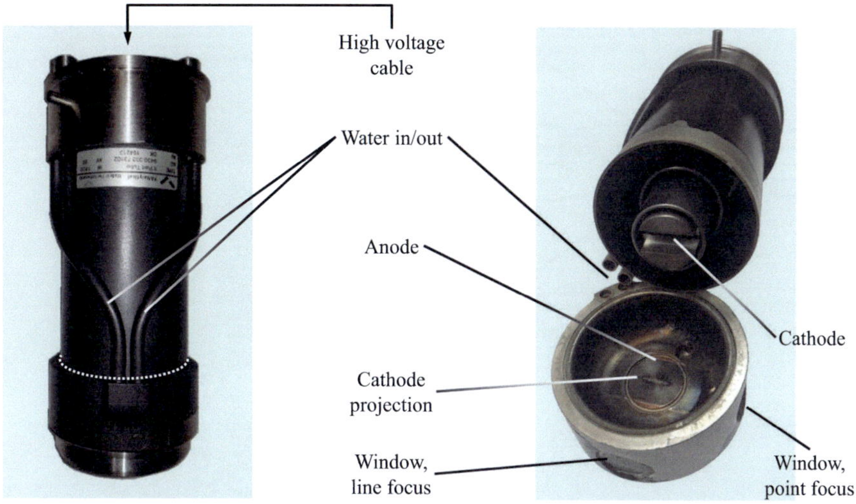

Fig. 6.6 Overall view of a sealed all-metal Malvern Panalytical X'Pert X-ray tube with copper anode rated to 60 kV and maximum input power of 1.8 kW(*left*). The same tube, sliced along the white dotted line seen on the left, illustrating the cathode, the anode coated with W film, the projection of the cathode visible as radiation damage on the surface of the anode, and two windows (*right*). Both the W coating and the radiation damage are the result of prolonged use

A classic metal/glass-sealed tube has four Be windows located 90° apart around the circumference of the cylindrical body. One pair of the opposite windows corresponds to a point-focused beam, which is mostly used in single-crystal diffraction, while the second pair of windows results in a line-focused beam, which is normally used in powder diffraction applications, see Fig. 6.5.

Given the geometry of the X-ray tube, the intensities of both the point- and line-focus beams are nearly identical, but their brightness is different: the point focus is brighter than the linear one. The use of the line focus in powder diffraction is justified by the need to maintain as many particles in the irradiated volume of the specimen as possible. The focus, that is, the projection of the cathode visible through beryllium windows at *ca* 5–6° angle is typically 0.1–0.2 mm wide[15] and 8–12 mm long. Similarly, point focus is employed in single-crystal diffraction because a typical size of the specimen is small (0.1–1 mm). Thus, high brightness of a point-focus beam enables one to achieve high scattered intensity in a single-crystal diffraction experiment.

Metal/glass-sealed X-ray tubes have been phased out by the manufacturers, and all-metal tubes, one of which is shown in Fig. 6.6, are common today. In addition to convenience in manufacturing, all-metal tubes are also more reliable in use as there is no danger that the glass part of the tube can be broken.

[15] The projection of the cathode on the anode surface is wider, 1–2 mm, see Fig. 11.7.

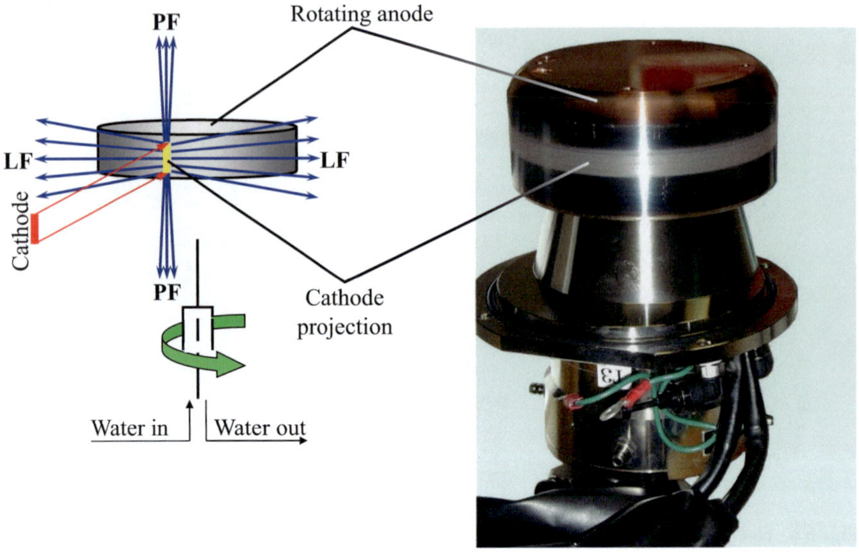

Fig. 6.7 The schematic (*left*) and the photograph (*right*) of the direct drive rotating anode assembly employed in a Rigaku TTRAX powder diffractometer. **PF** is point focus and **LF** is line focus. The trace seen on the anode surface on the right is surface damage caused by high-energy electrons bombarding the target and a thin layer deposit of the filament material (W), which occurred during anode operation

6.2.3 Rotating Anode X-Ray Sources

The low thermal efficiency of the sealed X-ray tube can be substantially improved by using a rotating anode X-ray source,[16] which is shown in Fig. 6.7. In this design, a massive disk-shaped anode is continuously rotated at a high speed while being cooled by a stream of chilled water. Both factors, that is, the anode mass (and therefore, the total area bombarded by high-energy electrons) and anode rotation, which constantly brings chilled metal into the impact zone, enable a routine increase in the X-ray tube input power to *ca* 15–18 kW and in some reported instances to 50–60 kW, that is, up to 20 times greater when compared to a standard sealed X-ray tube.

The resultant brightness of the X-ray beam increases proportionally to the input power; however, the lifetime of seals and bearings that operate in high vacuum is limited.[17] The considerable improvement in the incident beam brightness yields

[16] For more details on rotating anode X-ray sources, see W.C. Phillips, X-ray sources, Methods Enzymol. **114**, 300 (1985). https://doi.org/10.1016/0076-6879(85)14024-3, and references therein.

[17] In the laboratory of one of the authors (VKP), the direct drive rotating anode source manufactured by Rigaku has been in continuous operation (the anode is spinning and the X-rays are on 24 h/day, 7 days/week) for 20 years at the time of writing this book. The anode requires periodic refurbishing, which includes replacement of bearings and seals, and rebalancing approximately every 6 months.

much better diffraction patterns, especially when diffraction data are collected in conditions other than the ambient air (e.g., high or low temperature, high pressure, and others), which require additional shielding and windows for the X-rays to pass through, thus resulting in added intensity losses.

6.2.4 Novel Micro-focus X-Ray Sources

Over the past decade, the utilization of micro-focus sealed X-ray tubes represents a substantial breakthrough in X-ray equipment technology. These tubes are capable of generating a remarkably small focal spot, reaching dimensions as small as tens of microns or even less, by focusing an electron beam onto the target with the assistance of magnetic lenses.[18] Further enhancement of the quality and characteristics of the generated X-ray beam is achieved through the use of polycapillary optics or parabolic mirrors.

One notable advantage of micro-focus tubes is their substantially lower power requirements compared to conventional sealed tubes. For instance, a micro-focus tube might only require around 30 W (at 60 kV and 0.5 mA), whereas a conventional sealed tube might need as much as 2000 W (at 50 kV and 40 mA). This represents a reduction of two orders of magnitude in power consumption.

The lower power load has several benefits. First, micro-focus tubes generate very little heat, enabling air-cooling instead of more complex and energy-intensive cooling systems. Additionally, the reduced heat production contributes to longer lifetimes for these tubes. Despite the lower power requirements, micro-focus tubes are capable of producing brilliant X-ray beams comparable to those generated by rotating anode systems, see Sect. 6.2.2. These tubes typically find applications in cases where a point focus of the beam is desirable, such as in single-crystal diffraction (including protein diffraction), small-angle scattering, and pole figures. However, their utilization in conventional powder diffraction is limited due to the small cross-section of the beam.

The X-ray sources (tubes) discussed so far use solid metal targets, e.g., Cu and Mo anodes. However, a recent advancement is the introduction of novel metal-jet technology.[19] This technology employs a liquid-metal jet anode[20] and a micro-focused electron beam with a spot size on the liquid target ranging from 5 microns to a few tens of microns. The liquid-metal jet serves as both the target and heat conductor simultaneously.

---

[18] R. Zhou, X. Zhou, X. Li, Y. Cai, F. Liu, Study of the Microfocus X-Ray Tube Based on a Point-Like Target Used for Computed Tomography, Plos One, **11**(6), e0156224 (2016). https://doi.org/10.1371/journal.pone.0156224.g001

[19] MetalJet X-ray sources: https://www.excillum.com/products/metaljet/

[20] M. Otendala, T. Tuohimaa, U. Vogt, and H. M. Hertz, A 9 keV electron-impact liquid-gallium-jet x-ray source, Rev. Sci. Instr. **79**, 016102 (2008). https://doi.org/10.1063/1.2833838

The main limitation of solid anodes is the need to keep them below the melting point, which restricts their power loading capability. In the case of a liquid-metal source, the melting of the anode is not an issue as it is already in a liquid state. Consequently, the power load on the liquid anode can be significantly higher, for example, 250 W (at 60 kV and 4.2 mA). Two metals that are liquid at or near room temperature are gallium and indium. They produce X-ray wavelengths (λ Kα_1) of 1.340127(10) and 0.512125(5) Å, respectively, which are close to the wavelengths of Cu Kβ and Ag Kα from commonly used conventional tubes.

Ga-rich alloys are utilized when a longer wavelength is needed in powder diffraction and protein diffraction, while In-rich alloy yields a shorter wavelength more suitable for single-crystal diffraction. In addition, the metal-jet source can be optimized to generate both Ga and In X-ray beams, exiting through separate windows on opposite sides. This capability allows two different instruments to utilize the same metal-jet source. For example, a powder diffractometer or protein system can operate on the Ga wavelength side, while an instrument on the In wavelength side can collect data from single-crystal diffraction. The source based on metal-jet technology significantly outperforms conventional solid anode (including rotating anode) sources in all aspects, except for its high cost. On top of that, the power load capability and brilliance of the metal-jet source are at the level of first-generation synchrotrons but at a fraction of the cost of a synchrotron.

6.2.5 Synchrotron Radiation Sources

Synchrotron radiation sources were developed and successfully brought on line, beginning in the 1960s. They are the most powerful X-ray radiation sources today. Both the brilliance of the beam and the coherence of the generated electromagnetic waves are exceptionally high. The synchrotron output power exceeds that of the conventional X-ray tube by many orders of magnitude. Tremendous energies are stored in synchrotron rings (Fig. 6.8, left), where beams of accelerated electrons or positrons are moving in a circular orbit, controlled by a magnetic field, at relativistic velocities.

Electromagnetic radiation ranging from radiofrequency to short-wavelength X-ray region (Fig. 6.8, right) is produced due to the acceleration of charged particles toward the center of the ring. The X-ray beam is emitted in the direction, tangential to the electron/positron orbit.

Since there is no target to cool, the brilliance of the X-ray beam that can be achieved in synchrotrons is 4 (first-generation synchrotrons) to 12 and more (third- and fourth-generation synchrotrons) orders of magnitude higher than that from a conventional X-ray source. Moreover, given the size of the storage ring (hundreds of meters in diameter), the average synchrotron beam consists of weakly divergent beams that may be considered nearly parallel at distances typically used in powder diffraction (generally less than 1 m). This feature presents an additional advantage in powder diffraction applications since the instrumental resolution is also increased.

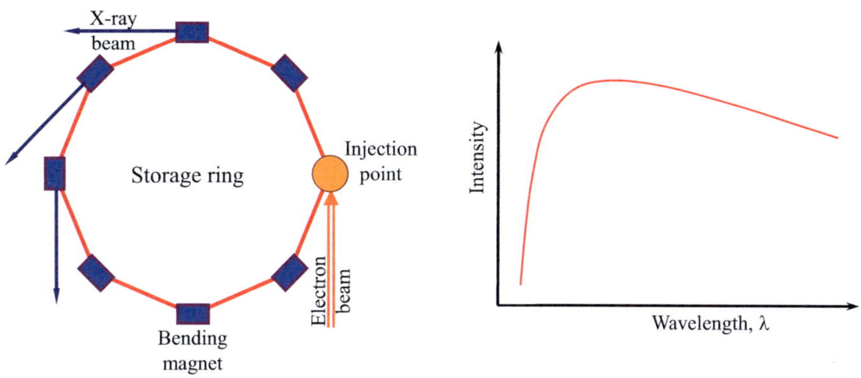

Fig. 6.8 Schematic diagram of a synchrotron illustrating X-ray radiation output from bending magnets that can be enhanced by wigglers (Wigglers and undulators are periodic magnet structures placed between the bending magnets. Wigglers are commonly used to increase photon flux and brightness, while undulators are employed to enhance photon flux in a desired range of energies, controlled by periodicity of the undulator. See, for example, H.N. Chapman, Fourth-generation light sources, IUCrJ. **10**, 246 (2023), https://doi.org/10.1107/S2052252523003585). Quasi-monochromatic X-rays can be obtained with undulators. Electrons must be periodically injected into the ring to replenish losses that occur during normal operation. Unlike in conventional X-ray sources, where both the long- and short-term stability of the incident photon beam are controlled by the stability of the power supply, the X-ray photon flux in a synchrotron changes with time: it decreases gradually due to electron losses and then periodically and sharply increases when electrons are injected into the ring

Another important advantage of the synchrotron radiation sources, in addition to the extremely high brilliance of the X-ray beam, is in the distribution of the beam intensity as a function of wavelength (Fig. 6.8, right). The high intensity, observed in a broad range of photon energies, allows for easy selection of nearly any desired wavelength. Further, the wavelength may be changed when needed and energy-dispersive experiments, in which the diffraction angle remains constant but the wavelength varies, can be conducted.

Thus, synchrotron radiation finds more and more use today, although its availability is restricted to the existing synchrotron sites.[21] However, some synchrotron sites are equipped with high-throughput automated powder diffractometers that are made available to a broad scientific community. For example, beamline 11-BM-B is equipped with 12 analyzers for high-speed, high-resolution collection of powder diffraction data and is available for rapid access using mail-in service.[22] Some of the well-known sites are the ALS—Advanced Light Source at Berkeley Lab,

[21] Room-size compact light sources based on inverse Compton scattering are under development. For additional details, see B. Günther, R. Gradl, C. Jud, E. Eggl, J. Huang, S. Kulpe, K. Achterhold, B. Gleich, M. Dierolf, F. Pfeiffer. The versatile X-ray beamline of the Munich Compact Light Source: design, instrumentation and applications. J. Synchrotron Rad., (2020) **27**, 1395–1414. https://doi.org/10.1107/S1600577520008309

[22] See https://www.aps.anl.gov/Beamlines/Directory/Details?beamline_id=90

APS—Advanced Photon Source at Argonne National Laboratory, NSLS—National Synchrotron Light Source at Brookhaven National Laboratory, SRS—Synchrotron Radiation Source at Daresbury Laboratory, ESRF—European Synchrotron Radiation Facility in Grenoble, and others.[23]

6.3 Other Types of Radiation

Other types of radiation that are commonly used in diffraction analysis are neutrons and electrons. The properties of both are compared with those of X-rays in Table 6.2.

Neutrons are usually produced in nuclear reactors; they have variable energy and therefore a white spectrum. Maximum flux of neutrons is usually obtained in an angstrom range of wavelengths. The main differences when compared to X-rays are as follows: (1) neutrons are scattered by nuclei, which are much smaller than electron clouds, and the scattering occurs on points; (2) scattering factors of elements remain constant over the whole range of Bragg angles; (3) scattering functions are not proportional to the atomic number and they are different for different isotopes of the same chemical element. Furthermore, since neutrons have spins, they interact with the unpaired electron spins (magnetic moments) and thus neutron diffraction data are commonly used to determine ordered magnetic structures. Other differences between neutrons and X-rays are nonessential in the general diffraction theory.

One of the biggest disadvantages of the conventional (reactor-generated) neutron sources is relatively low neutron flux at useful energies and weak interactions of neutrons with matter. Hence, a typical neutron experiment calls for 1–5 cm^3 of a material.[24] This problem is addressed in the new generation of highly intense pulsed (spallation) neutron sources.[25] In a spallation neutron source, bunches of protons are accelerated to high energies, and then released, bombarding a heavy metal target in short but extremely potent pulses. The collision of each proton with a heavy metal nucleus results in many expelled (spalled or knocked out) neutrons at various energies. The resultant highly intense (~10^2 times higher flux than in any conventional reactor) neutron beams have a nearly continuous energy spectrum, and they

[23] Web links to worldwide synchrotron and neutron facilities can be found at https://www.iucr.org/resources/other-directories/facilities

[24] This volume is a few orders of magnitude greater than needed for an X-ray diffraction experiment. Some of the third generation, high-flux neutron sources allow measurements of much smaller amounts, e.g., as little as 1 mm^3, but acceptable levels of the scattered intensity are generally achieved by sacrificing resolution.

[25] The most powerful operational pulsed neutron source is SNS—the Spallation Neutron Source—at the Oak Ridge National Laboratory (https://neutrons.ornl.gov/sns). The next is ISIS Neutron and Muon Source, which is located at the Rutherford Appleton Laboratory in the UK (https://www.isis.stfc.ac.uk/Pages/home.aspx). The European Spallation Source is under construction and is expected to become the world's most powerful neutron source in 2025 (https://europeanspallationsource.se/about)

Table 6.2 Comparison of three types of radiation used in powder diffraction

	X-rays (conv./synch.)	Neutrons	Electrons
Nature	Wave	Particle	Particle
Medium	Atmosphere	Atmosphere	High vacuum
Scattering by	Electron density	Nuclei and magnetic spins of electrons	Electrostatic potential
Scattering function	$f(s) \propto Z^{a}$	f is constant at all s	$f(s) \propto Z^{1/3,\ b}$
Wavelength range, λ	0.5–2.5/0.1–10 Å	~1 Å	0.02–0.05 Å
Wavelength selection	Fixed[c], Kα,β/variable	Variable[c]	Variable[c]
Focusing	None		Magnetic lenses
Lattice image	Reciprocal		Direct, reciprocal
Direct structure image	No		Yes
Applicable theory of diffraction	Kinematical		Dynamical
Use to determine atomic structure	Relatively simple		Very complex

[a] s—$\sin\theta/\lambda$, Z—atomic number, f—atomic (nuclear for neutrons) scattering function

[b] If unknown, electron scattering factor $f_e(s)$ may be derived from X-ray scattering factor $f_x(s)$ as $f_e(s) = k[Z - f_x(s)]/s^2$, where k is constant (Electron diffraction techniques, Vol. 1, J. M. Cowley, Ed., Oxford University Press, NY, 1992)

[c] According to Moseley's (Henry Gwyn Jeffreys Moseley (1887–1915). British physicist, who studied X-ray spectra of elements and discovered a systematic relationship between the atomic number of the element and the wavelength of characteristic radiation. A brief biography is available on Wikipedia at https://en.wikipedia.org/wiki/Henry_Moseley) law, X-ray characteristic frequency is $\nu = c/\lambda = C(Z - \sigma)^2$, while for neutrons and electrons $\lambda = h/mv = h(2mE)^{-1/2}$, where C and σ are constants, m is the mass, v is the velocity, and E is the kinetic energy of a particle

can be used in a variety of diffraction studies, mostly in the so-called time-of-flight (TOF) experiments. In the latter, the energy (and the wavelength) of the neutron that reaches the detector is calculated from the time it takes for a neutron to fly from the source to the specimen and then to the detector.

In addition to the direct imaging of crystal lattices (for example, in a high-resolution transmission electron microscope), electrons may be used in diffraction analysis. Despite the ease of the production of electrons by heating a filament in vacuum, electron diffraction is not as broadly used as X-ray diffraction. First, the experiments should be conducted in a high vacuum, which is inconvenient and may result in decomposition of some materials. Second, electrons strongly interact with materials. In addition to extremely thin samples, this requires the use of the dynamical theory of diffraction, thus making structure determination and refinement quite complex. Finally, the complexity and the cost of a high-resolution electron microscope usually considerably exceed those of a high-resolution powder diffractometer.

Neutron diffraction examples are discussed when deemed necessary, even though in this book we have no intention of covering the diffraction of neutrons (and electrons) at any significant depth. Interested readers can find more information on electron and neutron diffraction in some of the references provided at the end of this chapter.

6.4 Detection of X-Rays

The detector is an integral part of any diffraction analysis system, and its major role is to measure the intensity and, sometimes, the direction of the scattered beam. The detection is based on the ability of X-rays to interact with matter and to produce certain effects or signals, for example, to generate particles, waves, electrical current, etc., which can be easily registered. In other words, each photon entering the detector generates a specific event, better yet, a series of events that can be recognized, and from which the total photon count (intensity) can be determined. Obviously, the detector must be sensitive to X-rays (or in general to the radiation being detected) and should have an extended dynamic range and low background noise.[26]

6.4.1 Detector Efficiency, Linearity, Proportionality, and Resolution

An important characteristic of any detector is how efficiently it collects X-ray photons and then converts them into a measurable signal. Detector efficiency is determined by first, a fraction of X-ray photons that pass through the detector window (the higher, the better) and second, a fraction of photons that are absorbed by the detector and thus result in a series of detectable events (again, the higher, the better). The product of the two fractions, which is known as the absorption or quantum efficiency, should usually be between 0.5 and 1.

The efficiency of modern detectors is quite high, in contrast to the X-ray film, which requires multiple photons to activate a single grain of photon-sensitive silver halide. It is important to keep in mind that the efficiency depends on the type of the detector and it normally varies with the photon energy (wavelength) for the same type of the detector. The need for high efficiency is difficult to overestimate since

[26]For the purpose of this consideration, the dynamic range is the ability of the detector to count photons at both the low and high fluxes with the same effectiveness, and by the background noise we mean the events similar to those generated by the absorbed photons but occurring randomly and spontaneously in the detector without photons entering the detector.

every missed (i.e., not absorbed by the detector) photon is simply a lost photon. It is nearly impossible to account for the lost photons by any amplification method, no matter how far the amplification algorithm has been advanced.

The linearity of the detector is critical in obtaining correct intensity measurements (photon count). The detector is considered linear when there is a linear dependence between the photon flux (the number of photons entering through the detector window in 1 s) and the rate of signals generated by the detector (usually the number of voltage pulses) per second. In any detector, it takes some time to absorb a photon, convert it into a voltage pulse, register the pulse, and reset the detector to the initial state, that is, make it ready for the next operation. This time is usually known as the dead time of the detector—the time during which the detector remains inactive after it has just registered a photon.

The presence of the dead time always decreases the registered intensity. This effect, however, becomes substantial only at high photon fluxes. When the detector is incapable of counting every photon due to the dead time, some of them could be absorbed by the detector but remain unaccounted, that is, become lost photons. It is said that the detector becomes nonlinear under these conditions. Thus the linearity of the detector can be expressed as: (1) the maximum flux in photons per second that can be reliably counted (the higher the better); (2) the dead time (the shorter, the better); or (3) the percentage of the loss of linearity at certain high photon flux (the lower percentage, the better). The latter is compared for several different types of detectors in Table 6.3 along with other characteristics.

The proportionality of the detector determines how the size of the generated voltage pulse is related to the energy of the X-ray photon. Since X-ray photons produce a certain amount of events (ion pairs, photons of visible light, etc.), and each event requires certain energy, the number of events is generally proportional to the energy of the X-ray photon and therefore to the inverse of its wavelength. The amplitude of the generated signal is normally proportional to the number of these events, and thus, it is proportional to the X-ray photon energy, which could be used in pulse-height discrimination. Usually, the high proportionality of the detector enables one to achieve additional monochromatization of the X-ray beam in a straightforward fashion: during the registration, the signals that are too high or too

Table 6.3 Selected characteristics for Cu Kα radiation of the most common detectors

	Property				
Detector	Linearity loss at 40,000 cps[a]	Proportionality	Resolution for Cu Kα	Energy per event (eV)	No. of events[b]
Scintillation	<1%	Very good	45%	350	23
Proportional	<5%	Good, but fails at high photon flux	14%	26	310
Solid state	Up to 50%	Pileup in mid-range	2%	3.7	2200

[a] *cps* counts per second

[b] Approximate number of ion pairs or visible light photons resulting from a single X-ray photon assuming Cu Kα radiation with photon energy of about 8 keV

low and thus correspond to photons with exceedingly high or exceedingly low energies, respectively, are simply not counted.

Finally, the resolution of the detector characterizes its ability to resolve X-ray photons of different energies and wavelengths. The resolution (R) is defined as follows:

$$R(\%) = \frac{\sqrt{\delta V}}{V} 100\% \tag{6.5}$$

where V is the average height of the voltage pulse and δV is the spread of voltage pulses. The latter is also defined as the full width at half maximum of the pulse-height distribution in volts. The resolution for Cu Kα radiation for the main types of detectors is listed in Table 6.3. Thus, the resolution is a function of both the number of the events generated by a single photon and the energy required to generate the event, and it is critically dependent on how small is the spread in the number of events generated by different photons with identical energy. In other words, high resolution is only viable when every photon is absorbed completely, which is difficult to achieve when the absorbing medium is gaseous but is nearly ideal in the solid state simply due to the difference in their densities (see Sect. 8.6.5).

6.4.2 *Classification of Detectors*

Historically, the photographic film is the first and the oldest detector of X-rays, which was in use for many decades. Just as the visible light, the X-ray photons excite fine particles of silver halide when the film is exposed to X-rays. During the development, the exposed halide particles are converted into black metallic silver grains. Only the activated silver halide particles, that is, those that absorbed several X-ray photons (usually at least 3–5 photons), turn into metallic silver.

This type of detector is simple but is no longer in common use due to its low proportionality range and limited spatial and energy resolution. Moreover, the film development process is inconvenient and time-consuming. Finally, digitizing the information stored on the developed photographic film poses significant challenges.

However, over the last several decades, a new type of film called radiochromic film[27] has been developed. Unlike traditional films, it bypasses the development process and does not require the use of a dark room, as it is sensitive only to X-ray and not to visible light. It is best known by its trade name, Gafchromic™ film,[28] and is typically used for medical dosimetry, such as testing CT scanners. It can also be used, and sometimes is, for the alignment and calibration of X-ray diffractometers.

---

[27] https://en.wikipedia.org/wiki/Radiochromic_film

[28] http://www.gafchromic.com/gafchromic-film/index.asp, https://www.rpdinc.com/gafchromic-film-ebt4-8x10-25box-9593.html

In modern detectors, the signal, which is usually an electric current, is easily digitized and transferred to a computer for further processing and analysis. In general, detectors could be broadly divided into two categories: ratemeters and true counters. In a ratemeter, the readout is performed after hardware integration, resulting in an electrical current or a voltage signal that is proportional to the flux of photons entering the detector. True counters, however, count individual photons entering through the detector window and being absorbed by the detector.

Hence, the photographic film vaguely resembles a ratemeter because the intensity is extracted from the degree of darkening of the spots found on the film—the darker the spot, the higher the corresponding intensity, as a larger number of photons have been absorbed by the spot on the film surface. The three most commonly utilized types of X-ray detectors today are gas-proportional, scintillation, and solid-state detectors, all of which are true counters.

Yet another classification of detectors is based on whether the detector can resolve the location where the photon has been absorbed and thus detect the direction of the beam in space relative to the location of the specimen, in addition to counting the number of photons. Conventional gas-proportional, scintillation, and solid-state detectors do not support spatial resolution and are therefore known as point detectors. A point detector registers only the intensity of the diffracted beam, one point at a time. In other words, the readout of the detector corresponds to a specific value of the Bragg angle, determined by its position relative to both the sample and the incident beam. This is illustrated in Fig. 6.9 as the series of dots, each representing a single position of the detector and a single-point measurement of the intensity. Thus, to examine the distribution of the diffracted intensity as a function of Bragg angle

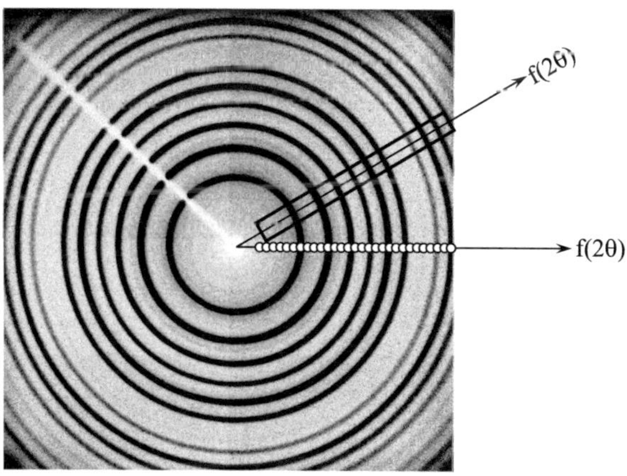

Fig. 6.9 The schematic explaining the difference between point (a set of *discrete dots*), line (represented by a *solid rectangle*), and area (depicted as *the entire picture*) detectors used in modern powder diffractometry. The faint trace extending from the center of the image to the upper left corner is the shade from the primary beam stop. The Bragg angle is zero at the center of the image and it increases along any radial line as shown by the *two arrows* (also see Fig. 8.4)

using a point detector, it is necessary to perform multiple-point measurements at varying Bragg angles.

Detectors that support spatial resolution are usually called position-sensitive detectors (PSD), as they can measure not only intensity (count the X-ray photons) of the diffracted beam but also the positions of the photons. Devices with the measured positions (points) arranged in a line are termed as 1D or line detectors, while those that provide resolution in two dimensions are known as 2D or area detectors.

Photographic film, for instance, is a typical example of an area detector because each point on the film can be characterized by two independent coordinates, and the entire film area is exposed simultaneously.

The following types of detectors are commonly used in powder diffractometry today: for 1D line detectors, they are solid-state strip and gas-proportional detectors. For 2D area detectors, options include charge-coupled devices (CCD), multi-wire proportional counter (MWPC), image plates (IPD), and others. The operation of a line detector is represented by the rectangle in Fig. 6.9, while area detectors measure the entire area of Fig. 6.9 simultaneously.

Both line and area detectors measure diffracted intensity at multiple points simultaneously, enabling a single measurement to resolve the diffraction pattern in one or two dimensions, respectively. This capability reduces measurement time by as much as one to two orders of magnitude.

6.4.3 Point Detectors

A typical *gas-proportional counter* detector usually consists of a cylindrical body filled with a mixture of gases (Xe mixed with some quench gas, usually CO_2, CH_4, or a halogen, to limit the discharge) and a central wire anode as shown schematically in Fig. 6.10 (left). High voltage is maintained between the cathode (the body of the counter) and the anode. When the X-ray photon enters through the window and is absorbed by the gas, it ionizes Xe atoms producing positively charged ions and electrons, that is, ion pairs (see Table 6.3). The resulting electrical current is

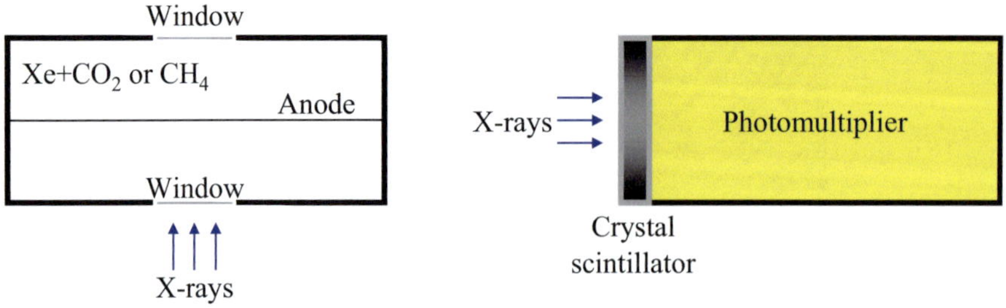

Fig. 6.10 The schematics of the gas-proportional counter (*left*) and the scintillation (*right*) detectors of X-rays

measured and the number of current pulses is proportional to the number of photons absorbed by the mixture of gases. The second window is typically added to allow nonabsorbed photons to exit, thereby reducing the X-ray fluorescence that may occur at the walls of the counter. In some cases, the cylinder can be filled by a mixture of gases under a pressure exceeding ambient level to enhance photon absorption and, therefore, improve photon detection by the detector.

Gas-proportional counters have relatively good resolution, so the heights of current pulses can be analyzed and discriminated to eliminate pulses that appear due to Kβ photons and due to low- and high-energy white radiation photons. The pulse-height discrimination is often used in combination with a β-filter to improve the elimination of the Kβ and white background photons.

The lifetime of a proportional detector is limited to about 2 years because a fraction of the gas, filling the detector, escapes through the windows, which are usually made from a thin and low-absorbing organic film to improve quantum efficiency. Another disadvantage of this type of the detector is its low effectiveness at high photon fluxes and with short-wavelength (high-energy) X-rays, such as Mo (the mass absorption coefficients of Xe are 299 and 38.2 cm^2/g for Cu Kα and Mo Kα radiation, respectively).

A typical *scintillation detector* employs a different principle for the detection of X-rays. It is constructed from a crystal scintillator coupled with a photomultiplier tube as shown in Fig. 6.10 (right). The X-ray photons, which are absorbed by the crystal, generate photons of blue light. After exiting the crystal, blue light photons are converted into electrons in a photomultiplier and amplified, and the resultant voltage pulses are registered as photon counts.

The crystal scintillator is usually made from cleaved, optically clear sodium iodide (NaI) activated with ~1% of Tl. The crystals are hydroscopic, and thus, they are usually sealed in a vacuum tight enclosure with a thin Be window in the front (X-rays entry window) and high-quality optical glass in the back (blue light photons exit window). The crystal is usually mounted on the photomultiplier tube using a viscous fluid that minimizes the refraction of blue light on the interface between the crystal and the photomultiplier.

Unfortunately, due to the relatively high energy of blue light photons, the X-ray wavelength resolution of the scintillation detector is quite low. Although the extremely short wavelengths (e.g., cosmic rays) can be discriminated and eliminated, the Kβ photons cannot be recognized and filtered out by the detector. Despite their low energy resolution, scintillation detectors are highly stable and effective, especially at high photon fluxes (see Table 6.3). They have very short dead time and therefore extended linear range. Because of this, scintillation detectors are by far the most commonly used detectors in the modern laboratory X-ray powder diffractometry.

Yet another physical phenomenon is used in *solid-state detectors* also called semiconductor detectors, which are manufactured from high-quality silicon or germanium single crystals doped with lithium and commonly known as Si(Li) or Ge(Li) solid-state detectors. The interaction of the X-ray photon with the crystal (detector) produces electron–hole pairs in quantities proportional to the energy of the photon divided by the energy needed to generate a single pair. The latter is quite low and amounts to approximately 3.7 eV for a Si-based detector. The electric potential difference applied across the crystal results in the photon-induced electric current, which is amplified and measured. The current is indeed proportional to the number of the generated electron–hole pairs.

In order to minimize noise and Li migration, the solid-state detector should be cooled, usually to about 80 K. This can be done using liquid nitrogen but it is quite inconvenient to have a cryogenic container mounted on the detector arm, especially since the container needs to be refilled every few days. Thus, solid-state detectors coupled with thermoelectric coolers have been developed and commercialized and successfully used in powder diffraction.

The substantial advantage of this type of detector is its high resolution at low temperature, even when compared with proportional counters (Table 6.3). Cooled solid-state detectors facilitate excellent filtration of both the undesirable Kβ and white radiation, thus resulting in a very low background without a significant loss of the intensity of the Kα_1/Kα_2 doublets. It is worth noting that even the highest quality, perfectly aligned monochromator decreases the characteristic Kα intensity by a factor of two or more. Thus, with the cooled solid-state detector the monochromator is no longer needed unless the extreme spectral purity is an issue. Moreover, modern solid-state detectors may be tuned to suppress the Kα doublet and register only Kβ energies. The latter is only about 20% as strong as the Kα component of the characteristics spectrum, but it has a single intense wavelength, which may be quite useful when working with complex powder diffraction patterns.

The major drawback of this type of detector, not counting the need for continuous cooling, is its relatively low linear range—only up to about 8×10^4 counts/s. This may result in experimental artifacts when measuring extremely strong Bragg peaks, similar to those shown in Fig. 6.11.

The inset of this figure shows sharp and random spikes peaking at ~1.2×10^5 counts instead of a smooth Bragg peak, due to the highly nonlinear response of the detector when photon flux exceeds 2–5×10^4 counts/s. Further, the photon energy resolution-based electronic monochromatization also fails. In addition to the (400) Bragg peak centered at $2\theta \cong 69.2°$, which is expected for this orientation of the single crystalline silicon wafer, three additional Bragg peaks are clearly recognizable in the measured diffraction pattern at $2\theta \cong 61.6°$, 65.9°, and 66.3°.

A simple analysis, based on the Braggs' equation, leads to the conclusion that these three peaks are all due to the reflections from the (400) crystallographic planes of Si, each representing a different wavelength in a characteristic spectrum of this particular X-ray tube. The first satellite Bragg peak ($2\theta \cong 61.6°$) corresponds to the incompletely suppressed Kβ component in the characteristic spectrum of the Cu anode. Obviously this occurs due to the extremely high intensity of the (400) peak from the nearly perfect single crystalline specimen. The two remaining Bragg peaks ($2\theta \cong 65.9°$ and 66.3°) are due to the presence of a W film deposited on the surface of

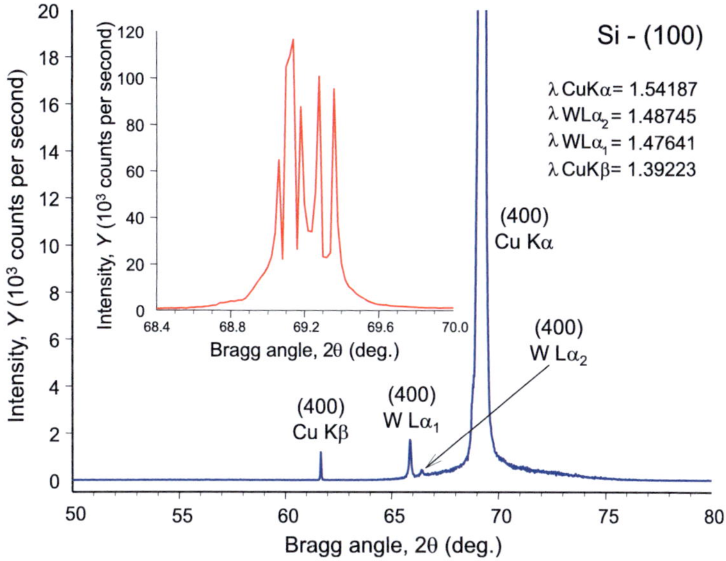

Fig. 6.11 The X-ray diffraction pattern collected from a single crystalline Si wafer in the reflecting position with the [100] direction perpendicular to its surface using a powder diffractometer equipped with the thermoelectric cooled Si(Li) solid-state detector. The inset shows multiple random spikes for the (400) Cu Kα peak when photon flux exceeds ~20,000 cps due to the nonlinearity of the detector. The additional Bragg peaks are marked with the corresponding characteristic wavelengths. (Experimental data courtesy of Dr. J.E. Snyder)

the anode (see Figs. 6.4 and 6.5) during the long-time operation of the X-ray tube.[29] These two peaks correspond, respectively, to the (400) Bragg reflections caused by W $L\alpha_1$ and $L\alpha_2$ characteristic lines. Another unusual feature of this diffraction pattern is the fact that the intensity corresponding to the very weak W $L\alpha_1$ peak is higher than that of the much stronger Cu Kβ peak. Indeed, this happens because the energy of the former is closer to the energy of Cu Kα photons. Since the detector is tuned to register Cu Kα photons, discriminating other photons with nearly the same energy is difficult to achieve. It is worth noting that if the same data would be collected using a brand new Cu-anode X-ray tube (which has no deposit from the W filament on the surface of the anode), only two characteristic Cu Bragg peaks would be visible.

6.4.4 Line (1D) Detectors

A *position-sensitive detector* (*PSD*), in addition to counting X-ray photons can detect the location of a photon absorption event and measure the intensity of the

[29]This particular X-ray tube was in service for more than a year and the total number of hours of operation was approaching 2000. The tube was regularly operated at ~75% of rated power.

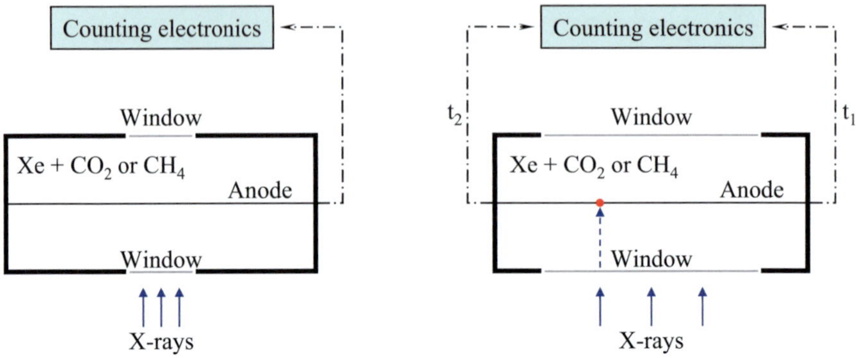

Fig. 6.12 The schematic compares the conventional gas-proportional detector (*left*), where the signal is collected on one side of the wire anode, with the position-sensitive detector (*right*), where signals are measured on both sides of the wire anode. The position of the electron discharge (represented by a red dot) is determined by the counting electronics based on the time difference between t1 and t2 when the two signals are recorded

diffracted beam in multiple points simultaneously, usually thousands. As a result, a powder diffraction experiment becomes much faster while maintaining nearly the same quality as that obtained using a standard gas-proportional counter.[30]

The basic principle of sensing the position of the photon absorption event by the *gas-proportional PSD* is based on the properties of the proportional counter. When an X-ray photon is absorbed, it creates Xe ion-electron pairs. The electrons accelerate along a path of minimum resistance toward the wire anode, where they are discharged, producing the electrical current pulse in the anode circuit. In a point detector, the amplitude of this pulse is measured at one end of a wire. Given the high speed of modern electronics, it is possible to measure the same signal at both ends of the wire anode. The time difference between these two measurements is used to determine the location of the discharge, provided the length of the wire anode is known, as illustrated schematically in Fig. 6.12.

The spatial resolution of the PSDs is not as high as that attainable with the precise positioning of point detectors. Nevertheless, it remains satisfactory (approaching about 0.01°) for conducting good-quality experiments. The minor loss of the resolution is a small price to pay for the ability to collect powder diffraction data across a wide range of Bragg angles simultaneously, which significantly reduces the duration of the experiment. Typically, this improvement translates from many hours when using a point detector to several minutes when using a position-sensitive detector.

Different models of the PSDs may vary in geometry, resolution, and Bragg angle range. Short linear PSDs cover a few degrees range (approximately 3–10°), while long curved PSDs may cover as much as 100–140° 2θ. The biggest advantage of

[30] A significant deterioration of the quality of X-ray powder diffraction data may occur when the studied specimen is highly fluorescent because it is impractical to monochromatize the diffracted beam when using line or area detectors. Also see Sect. 11.3.2.

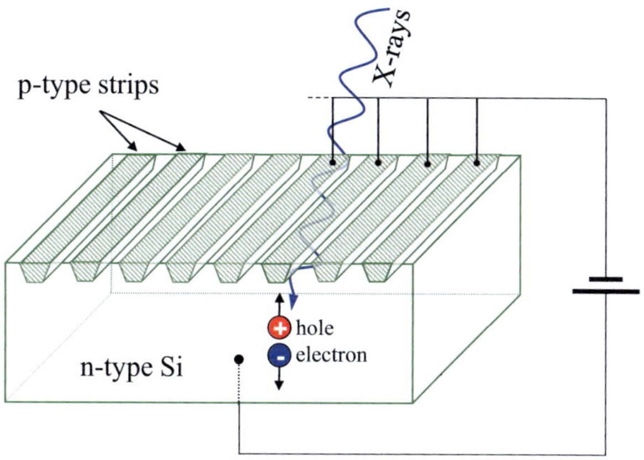

Fig. 6.13 The schematic of a strip detector

long-range PSDs is the considerable reduction in experiment time compared to short or medium range position-sensitive detectors. The disadvantage of long-range PSDs arises from often substantial differences in photon counting properties observed at different places along the detector, such as in the middle vs. the ends of its length. The large angular spread of long detectors also restricts the quality of X-ray focusing, usually resulting in the deterioration of the shape of Bragg peaks. A relevant discussion about the geometry of powder diffractometers equipped with PSDs can be found in Sect. 11.3.2.

Comparable to a gas-proportional PSD with the added functionality of detecting the location of photon absorption events, the *real-time multiple strip detector* (*RTMS*), also known as a silicon micro-strip detector, employs the principle used in a standard solid-state detector (see Sect. 6.4.3). Strip detectors are typically manufactured using a photolithographic process, where narrow p-doped strips are deposited on an n-type Si wafer. A simplified schematic of a strip detector is shown in Fig. 6.13. The electric potential difference applied to each strip across the wafer results in a photon-induced electric current, which is amplified and measured. Since both the holes and electrons created by each photon absorption event travel along the path of least resistance, the current measured individually for each strip provides information about the position where the X-ray photon was absorbed by the detector. The current remains proportional to the number of the generated electron–hole pairs, providing information about photon flux at a specific location on the detector. When a one-dimensional array of strips is replaced by two-dimensional arrays of dots, a 1D line strip detector transforms into a 2D area detector known as a silicon pixel detector.

6.4.5 *Area (2D) Detectors*

Area detectors record diffraction pattern in two dimensions simultaneously. Aside from photographic film, several types of electronic area detectors have advanced to a commercial status and are frequently used in modern X-ray powder diffraction analysis.

In a *charge-coupled device* (*CCD*) detector, X-ray photons are converted by a phosphor[31] into visible light, which is then captured using a CCD. This chip is similar to (or even the same as) those used in modern digital cameras. To reliably measure a large area, in some detector designs, the phosphor may be made several times larger than the chip, and then the generated visible light is demagnified to the size of the chip by using fiber optics, while in other designs several chips (e.g., a 2×2 or 3×3 chip arrays) are glued together. Similar to solid-state detectors, CCD chips are cooled with thermoelectric cooling device to reduce random (thermal) noise.

In an *image-plate detector* (IPD), X-ray photons are also captured by a phosphor.[32] However, the excited phosphor pixels are not immediately converted into the signals. Instead, the information is stored in the phosphor grains as a latent image, similar to the activation of silver halide particles in photographic film during exposure. After the data collection, the image is scanned (or "developed") by a laser, which deactivates pixels that releases the stored energy as blue light. Visible light photons are then detected conventionally by a photomultiplier tube, and the plate is reactivated by another laser. Image plates are integrating detectors capable of high counting rates and a wide dynamic range, but they have relatively long readout times.

Another type of area detector that sees more use in powder diffraction than other area detectors is the *multi-wire detector* (MWD), which operates on the same principle as gas-proportional counters and is therefore often referred to as a multi-wire proportional counter (MWPC). The MWD features two anodes composed of multi-wire grids, enabling the detection of X and Y positions where photons are absorbed, in addition to the count of absorbed photons. However, the MWD faces increasing competition from recently developed area detectors. One of these is the *micro-gap detector* related to MWD, which features high spatial resolution, very low noise levels, and a high dynamic range. Another is the *complementary metal-oxide-semiconductor* (CMOS) *detector*, which offers advantages such as shutter-free, dead-time-free continuous scanning.[33] CMOS sensors are also commonly used in digital cameras, smartphones, and other consumer electronics.

---

[31] A typical CCD phosphor is Tb^{3+} doped Gd_2O_2S, which converts X-ray photons into visible light photons.

[32] A typical image plate phosphor is Eu^{2+} doped BaFBr. When exposed to X-rays, Eu^{2+} oxidizes to Eu^{3+}. Thus produced electrons may either recombine with Eu^{3+} or they become trapped by F-vacancies in the crystal lattice of BaFBr. The trapped electrons may exist in this metastable state for a long time. They are released when exposed to a visible light and emit blue photons during recombination with Eu^{3+} ions, e.g., see K. Takahashi, K. Khoda, J. Miyahara, Y. Kanemitsu, K. Amitani, S. Shionoya, Mechanism of photostimulated luminescence in BaFX:Eu^{2+} (X = Cl, Br) phosphors, J. Luminesc. **31–32**, 266 (1984). https://doi.org/10.1016/0022-2313(84)90268-0

[33] See Table 2.1.6 with comparison of detectors properties in: International Tables of Crystallography, Vol. H,: Powder Diffraction, First Online Edition, C.J. Gilmore, J.A. Kaduk, H. Schenk, Ed. (2019). https://it.iucr.org/H/

Table 6.4 Qualitative comparison of the most common area detectors

Parameter	CCD	IPD	MWD	PAD
Active area size	Small	Large	Small	Large
Readout time	Medium	Long	Short	Near 0
Counting rate	High	High	Low	High
Dynamic range	Medium	High	Medium	High
Spatial resolution	High	Low	Low	High

Related to CMOS detectors but not exactly the same are novel *pixel array detectors* (PADs) ,[34] which have gained popularity over the last two decades and were initially developed for fourth-generation XFEL synchrotron beamlines.[35] They are known for their capability in single photon counting and accurate intensity determination across a very wide dynamic range. PAD detectors are referred to as pixel detectors because they consist of a semiconductor sensor layer that converts photons into electric signals. These signals are then counted by individual cells, where each cell (or pixel) functions as a detector itself. Pixel array detectors are developing very rapidly and in several directions with differences varying by function and application. The detectors most commonly used in X-ray crystallography are PILATUS by Dectris,[36] which employs hybrid pixel array (HPAD) technology, and PHOTON CPAD by Bruker,[37] which uses charge-integrating pixel array (CPAD) technology.

Table 6.4 provides a general comparison of the abovementioned four types of area detectors.

6.5 Additional Reading

1. International Tables for Crystallography, Vol. C, Mathematical, physical and chemical tables, First online edition, E. Prince, Ed. (2006). https://it.iucr.org/C/
2. International Tables for Crystallography, vol. H: Powder diffraction, First online edition, C.J. Gilmore, J.A. Kaduk, and H. Schenk, Eds. (2019). https://it.iucr.org/H/

[34] CMOS detectors are in fact a subset of pixel detectors using CMOS technology.

[35] H.T. Philipp, M. Hromalik, M. Tate, L. Koerner, S.M. Gruner, Pixel array detector for X-ray free electron laser experiments. Nuclear Instruments and Methods in Physics Research, A649, 67–69 (2011). https://doi.org/10.1016/j.nima.2010.11.189

[36] Brönnimann, C.; Trüb, P. (2018). "Hybrid pixel photon counting X-ray detectors for synchrotron radiation". In E Jaeschke; S Khan; JR Schneider; JB Hastings (eds.). *Synchrotron Light Sources and Free-electron Lasers*. Cham, Switzerland: Springer International. pp. 995–1027. https://doi.org/10.1007/978-3-319-14394-1_36

[37] B. Becker, J. Kaercher, M. Krug, S. Leo, T. Stuerzer, B. Weinhausen, R. Durst, A large area detector with indirect conversion, charge integration and photon counting operation. Proc. SPIE 11838, Hard X-Ray, Gamma-Ray, and Neutron Detector Physics XXIII, Vol. 11838, id. 118380N, 10 pp. (2021). https://doi.org/10.1117/12.2597029

3. B.D. Cullity S.R. Stock, Elements of X-Ray Diffraction, Third Edition, Pearson Education Limited (2013).
4. Electron diffraction techniques. Vol. 1, 2. J. Cowley, Ed., Oxford University Press. Oxford (1992). https://global.oup.com/academic/product/electron-diffraction-techniques-9780198555582
5. R. Jenkins and R.L. Snyder, Introduction to X-ray powder diffractometry. Wiley, New York (1996). https://www.wiley.com/en-us/Introduction+to+X+Ray+Powder+Diffractometry-p-9781118520925
6. J. Als-Nielsen and D. McMorrow, Elements of modern X-ray physics, Second edition, Wiley, New York (2011). https://doi.org/10.1002/9781119998365
7. A. Pifferi, X-ray sources, in: Encyclopedia of Condensed Matter Physics, Elsevier, pp. 323-329 (2005). https://doi.org/10.1016/B0-12-369401-9/00631-8
8. B. Becker, J. Kaercher, M. Krug, S. Leo, T. Stuerzer, B. Weinhausen, R. Durst, A large area detector with indirect conversion, charge integration and photon counting operation. Proceedings Volume 11838, Hard X-Ray, Gamma-Ray, and Neutron Detector Physics XXIII; 118380N (2021). https://doi.org/10.1117/12.2597029
9. J.P. Cline, M.H. Mendenhall, D. Black, D. Windover, and A. Henins, The Optics and Alignment of the Divergent Beam Laboratory X-ray Powder Diffractometer and its Calibration Using NIST Standard Reference Materials, J. Res. Natl. Inst. Stand. **120**, 173 (2015), https://doi.org/10.6028/jres.120.013
10. J.P. Cline, M.H. Mendenhall, J.J. Ritter, D. Black, A. Henins, J.E. Bonevich, P.S. Whitfield, and J.J. Filliben, The Certification of Standard Reference Material 1979: Powder Diffraction Line Profile Standard for Crystallite Size Analysis, J. Res. Natl. Inst. Stand. **125**, 125020 (2020), https://doi.org/10.6028/jres.125.020

6.6 Problems

1. A typical energy of electrons in a modern transition electron microscope is 300 keV. Calculate the corresponding wavelength of the electron beam assuming that the vacuum inside the microscope is ideal.
2. Calculate the energy (in keV) of the characteristic Cr $K_{\alpha 1}$ and Mo $K_{\alpha 1}$ radiation.
3. You are in charge of buying a new powder diffractometer for your company. The company is in business of manufacturing alumina (Al_2O_3)-based ceramics. The powder diffractometer is to become a workhorse instrument in the quality control department. Routine experiments will include collecting powder diffraction data from ceramic samples to analyze their structure and phase composition. High data collection speeds are critical because a typical daily number of samples to be analyzed using the new equipment is 100+. The following options are available from different vendors:

   (a) Sealed Cu X-ray tube, scintillation detector; the lowest cost.
   (b) Sealed Cu X-ray tube, solid-state detector; $10,000 more than the first option.

(c) Sealed Cu X-ray tube, curved position-sensitive detector; $25,000 more than the first option.
(d) What recommendation will you make to your boss without a fear of being fired during the first month after the delivery of the instrument?

Chapter 7
Fundamentals of Diffraction

When X-rays propagate through a substance, the occurrence of the following processes should be considered in the phenomenon of diffraction:

- Coherent scattering (Sect. 7.1), which produces beams with the same wavelength as the incident (primary) beam. In other words, the energy of the photons in a coherently scattered beam remains unchanged when compared to that in the primary beam.
- Incoherent (or Compton[1]) scattering, in which the wavelength of the scattered beam increases due to partial loss of photon energy in collisions with core electrons (the Compton effect).
- Absorption of the X-rays, see Sect. 8.6.5, in which some photons are dissipated in random directions due to scattering, and some photons lose their energy by ejecting electron(s) from an atom (i.e., ionization) and/or due to the photoelectric effect (i.e., X-ray fluorescence).

Incoherent scattering is not essential when the interaction of X-rays with crystal lattices is of concern, and it is generally neglected. When absorption becomes significant, it is usually considered as a separate effect. Thus, in the first approximation only coherent scattering results in the diffraction from periodic lattices and is considered in this chapter.

Generally, the interaction of X-rays (or any other type of radiation with the proper wavelength) with a crystal is multifaceted and complex, and there are two different levels of approximation—kinematical and dynamical theories of diffraction. In the *kinematical diffraction*, a beam scattered once is not allowed to be scattered again

[1] Arthur Holly Compton (1897–1962). The American physicist, best known for his discovery of the increase in wavelength of X-rays due to scattering of the incident radiation by free electrons—inelastic scattering of X-ray photons—known today as the Compton effect. With Charles Thomson Rees Wilson, Compton shared the Nobel Prize in physics in 1927 "for his discovery of the effect named after him." See https://www.nobelprize.org/prizes/physics/1927/compton/ for more information.

P. Zavalij, V. Pecharsky, *Fundamentals of Powder Diffraction and Structural Characterization of Materials*, https://doi.org/10.1007/978-3-031-91504-8_7

before it leaves the crystal. Thus, the kinematical theory of diffraction is based on the assumption that the interaction of the diffracted beam with the crystal is negligibly small. This requires the following postulations: (1) a crystal consists of individual mosaic blocks—crystallites[2]—which are slightly misaligned with respect to one another; (2) the size of the crystallites is small, and (3) the misalignment of the crystallites is large enough, so that the interaction of X-rays with matter at the length scale exceeding the size of mosaic blocks is negligible.

On the contrary, the theory of the *dynamical diffraction* accounts for scattering of the diffracted beam and other interactions of waves inside the crystal, and thus, the mathematical apparatus of the theory is quite complex. Dynamical effects become significant and the use of the theory of dynamical diffraction is justified only when the crystals are nearly perfect, or when there is an exceptionally strong interaction of the radiation with the material. In the majority of crystalline materials, however, dynamical effects are weak and they are usually noticeable only when precise single-crystal experiments are conducted. Even then, numerous dynamical effects (e.g., primary and/or secondary extinction, simultaneous diffraction, thermal diffuse scattering, and others) are usually applied as corrections to the kinematical diffraction model.

The kinematical approach is simple and adequately and accurately describes the diffraction of X-rays from mosaic crystals. This is especially true for polycrystalline materials where the size of crystallites is relatively small. Hence, the kinematical theory of diffraction is used in this chapter and throughout this book.

7.1 Scattering by Electrons, Atoms, and Lattices

It is well-known that when a wave interacts with and is scattered by a point object, the outcome of this interaction is a new wave, which spreads in all directions. If no energy loss occurs, the resultant wave has the same frequency as the incident (primary) wave and this process is known as elastic scattering. In three dimensions, the elastically scattered wave is spherical, with its origin in the point coinciding with the object as shown schematically in Fig. 7.1.

When two or more points are involved, they all produce spherical waves with the same λ, which interfere with each other simply by adding their amplitudes. If the two scattered waves with parallel propagation vectors are completely in-phase, the resulting wave has its amplitude doubled (Fig. 7.2, top), while the waves, which are completely out-of-phase, extinguish one another as shown in Fig. 7.2 (bottom).

The first case seen in Fig. 7.2 is called constructive interference and the second case is termed as destructive interference. Constructive interference, which occurs on

[2]Crystallite usually means a tiny single crystal (microcrystal). Each particle in a polycrystalline material usually consists of multiple crystallites that join together in different orientations. A small powder particle can be a single crystallite as well.

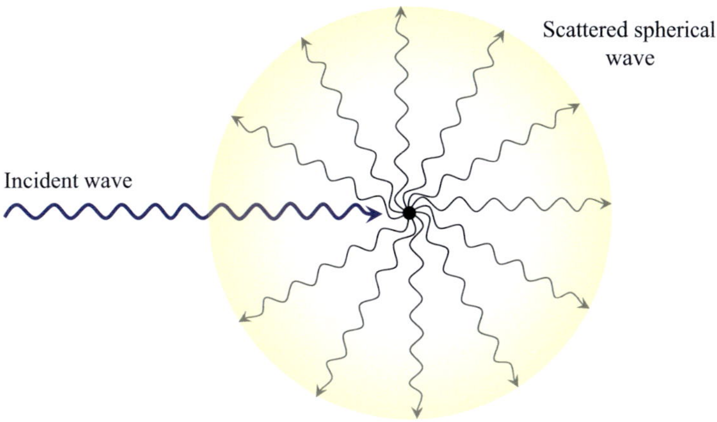

Fig. 7.1 The illustration depicts a spherical wave generated from the elastic scattering of an incident wave by a point object (represented as a filled dot *at the center*)

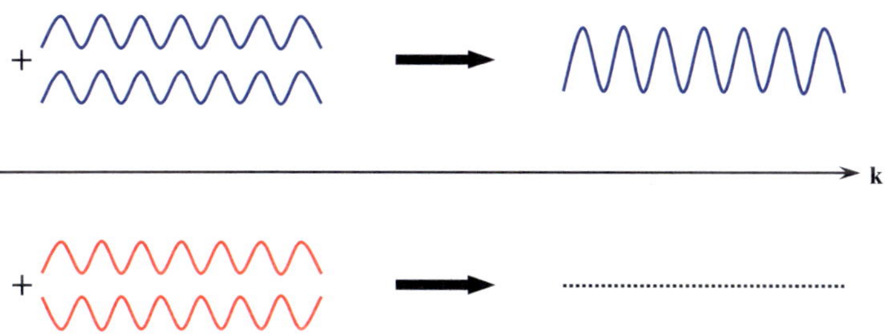

Fig. 7.2 The interaction between two waves with parallel propagation vectors (**k**) manifests in two limiting cases: constructive interference, where two in-phase waves combine to form a new wave with double the amplitude (*top*), and destructive interference, where two completely out-of-phase waves cancel each other out, resulting in zero amplitude for the resultant wave (*bottom*)

periodic arrays of points, increases the resultant wave amplitude by many orders of magnitude and this phenomenon is one of the cornerstones in the theory of diffraction.

Diffraction can be observed only when the wavelength is of the same order of magnitude as the repetitive distance between the scattering objects. Thus, for crystals, the wavelength should be in the same range as the shortest interatomic distances, that is, somewhere between ~0.5 and ~2.5 Å. This condition is fulfilled when using electromagnetic radiation, which within the mentioned range of wavelengths, are X-rays. It is important to note that X-rays scatter from electrons, so that the active scattering centers are not the nuclei, but the electrons, or more precisely the electron density, periodically distributed in the crystal lattice.

The other two types of radiation that can diffract from crystals are neutron and electron beams. Unlike X-rays, neutrons are scattered on the nuclei, while electrons, which have electric charge, interact with the electrostatic potential. Nuclei, their electronic shells (i.e., core electron density), and electrostatic potentials are all distributed similarly in the same crystal and their distribution is established by the crystal structure of the material. Thus, assuming a constant wavelength, the differences in the diffraction patterns when using various kinds of radiation are mainly in the intensities of the diffracted beams. The latter occurs because various types of radiation interact in their own way with different scattering centers. The X-rays are the simplest, most accessible, and by far the most commonly used waves in powder diffraction.

7.1.1 *Scattering by Electrons*

The origin of the electromagnetic wave elastically scattered by the electron can be better understood by recalling the fact that electrons are charged particles. Thus, an oscillating electric field (see Fig. 6.2) from the incident wave exerts a force on the electric charge (electron) forcing the electron to oscillate with the same frequency as the electric field component of the electromagnetic wave. The oscillating electron accelerates and decelerates in concert with the varying amplitude of the electric field vector and emits electromagnetic radiation, which spreads in all directions. In this respect, the elastically scattered X-ray beam is simply radiated by the oscillating electron; it has the same frequency and wavelength as the incident wave, and this type of scattering is also known as coherent scattering.[3]

For the sake of simplicity, we now consider electrons as stationary points and disregard the dependence of the scattered intensity[4] on the scattering angle.[5] Each

---

[3] It is worth noting that coherency of the electromagnetic wave elastically scattered by the electron establishes specific phase relationships between the incident and the scattered wave: their phases are different by π (i.e., scatterred wave is shifted with respect to the incident wave exactly by $\lambda/2$).

[4] The scattered (diffracted) X-ray intensity recorded by the detector is proportional to the amplitude squared.

[5] The absolute intensity of the X-ray wave coherently scattered by a single electron, I, is determined from the Thomson equation:

$$I = I_0 \frac{K}{r^2} \left(\frac{1 + \cos^2 2\theta}{2} \right)$$

where I_0 is the absolute intensity of the incident beam, K is constant ($K = 7.94 \times 10^{-30}$ m^2), r is the distance from the electron to the detector in m, and θ is the angle between the propagation vector of the incident wave and the direction of the scattered wave. It is worth noting that in a powder diffraction experiment, all prefactors in the right-hand side of the Thomson's equation are constant and can be omitted. The only variable part is, therefore, a function of the Bragg angle, θ. It emerges because the incident beam is generally unpolarized but the scattered beam is always partially polarized. This function, therefore, is called the polarization factor.

Thomson equation is named after sir Joseph John (J.J.) Thomson (1856–1940)—the British physicist who has been credited with the discovery of an electron. In 1906, he received the Nobel Prize in Physics "in recognition of the great merits of his theoretical and experimental investigations on the conduction of electricity by gases." See https://www.nobelprize.org/prizes/physics/1906/thomson/ for more information.

electron then interacts with the incident X-ray wave producing a spherical elastically scattered wave as shown in Fig. 7.1. Thus, the scattering of X-rays by a single electron yields an identical scattered intensity in every direction.

When more than one point is affected by the same incident wave, the overall scattered amplitude is a result of interference among multiple spherical waves. As established earlier (Fig. 7.2), the amplitude will vary depending on the difference in the phases of multiple waves with parallel propagation vectors but originating from different points.

The phase difference between these waves is also called the phase angle, φ. For example, diffraction from a row of five equally spaced points produces a pattern shown schemtically in Fig. 7.3, which depicts the intensity of the diffracted beam, *I*, as a function of the phase angle, φ. The major peaks (or diffraction maxima) in the

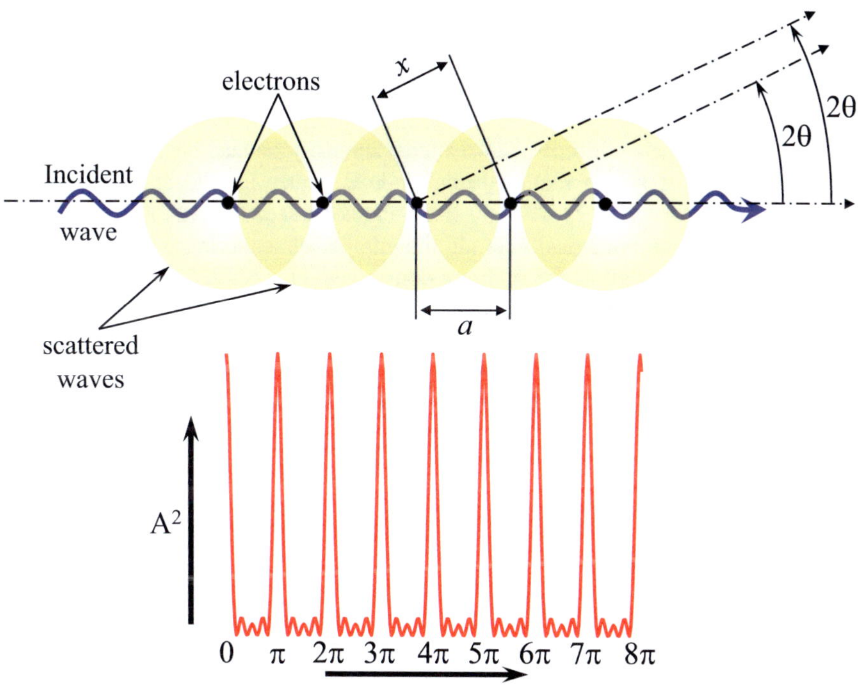

$$\varphi = 2\pi\frac{a-x}{\lambda} = 2\pi\frac{a(1-\cos 2\theta)}{\lambda} = 4\pi\frac{a\sin^2\theta}{\lambda}$$

Fig. 7.3 Top: five equally spaced points producing five spherical waves as a result of elastic scattering of the single incident wave. Bottom: the resultant scattered amplitude as a function of the phase angle, φ. In this geometry, the phase angle is a function of the spacing between the points, *a*, the wavelength of the incident beam, λ, and the scattering angle, 2θ. The relationship between the phase (φ) and scattering (θ) angles for the arrangement shown on *top* is easily derived by considering path difference (*a*-*x*) between any pair of neighboring waves, which have parallel propagation vectors

pattern are caused by the constructive interference, while the multiple smaller peaks are due to the superimposed waves, which have different phases but are not completely out-of-phase.

For a one-dimensional periodic structure, the intensity diffracted by the row of N equally spaced points is proportional to the so-called interference function, which is shown in Eq. (7.1).

$$I(\phi) \propto \frac{\sin^2 N\varphi}{\sin^2\varphi} \quad \text{or} \quad I(\varphi) \propto N^2, \quad \text{when } \varphi = h\pi, \text{and } h = \ldots, -2, -1, 0, 1, 2, \ldots \tag{7.1}$$

The example considered here illustrates scattering from only five points. When the number of equally spaced points increases, the major constructive peaks become sharper and more pronounced, while the minor peaks turn out to be less and less visible. The gradual change is illustrated in Fig. 7.4, where the resultant intensity from the rows of 5, 10, and 20 points is modeled as a function of the phase angle using Eq. (7.1).

When N approaches infinity, the scattered intensity pattern becomes a periodic delta function, that is, the scattered amplitude is nearly infinite at specific phase angles ($\varphi = h\pi$, $h = \ldots, -2, -1, 0, 1, 2, \ldots$) and is reduced to zero everywhere else. Since crystals contain practically an infinite number of scattering points, which are systematically arranged in three dimensions, they also should produce discrete diffraction patterns with sharp diffraction peaks observed only in specific directions. Just as in the one-dimensional case (Fig. 7.2), the directions of diffraction peaks (i.e., diffraction angles, 2θ) are directly related to the spacing between the diffracting points (i.e., lattice points, as established by the periodicity of the crystal) and the wavelength of the used radiation.

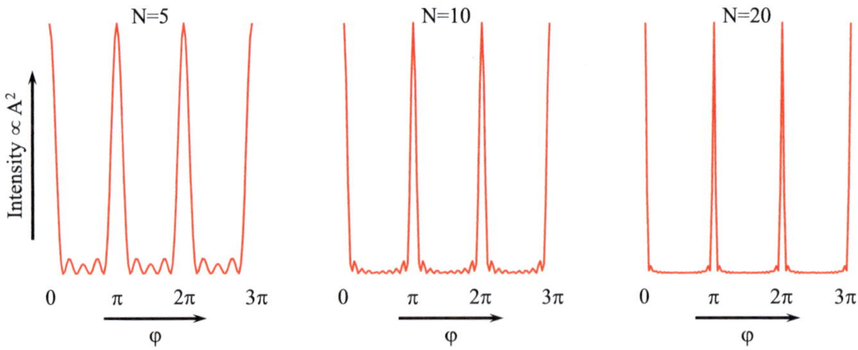

Fig. 7.4 The illustration of the changes in the diffraction pattern from a one-dimensional periodic arrangement of scattering points when the number of points (N) increases from 5 to 20. The horizontal scales are identical, but the vertical scales are normalized for the three plots

7.1.2 Scattering by Atoms and Atomic Scattering Factor

We now consider an atom instead of a stationary electron. The majority of atoms and ions consist of multiple electrons distributed around a nucleus as shown schematically in Fig. 7.5. It is easy to see that no path difference is introduced between the waves for the forward-scattered X-rays. Thus, intensity scattered in the direction of the propagation vector of the incident wavefront is proportional to the total number of core electrons, Z, in the atom. For any other angle, $2\theta > 0$, that is, when the propagation vector of the scattered waves, $\mathbf{k}''$, is different from the propagation vector of the incident waves, $\mathbf{k}$, the presence of core electrons results in the introduction of a certain path difference, δ, between the individual waves in the resultant wavefront.

The amplitude of the scattered beam is therefore a gradually decaying function of the scattered angle and it varies with φ and with θ. The intrinsic angular dependence of the X-ray amplitude scattered by an atom is called the atomic scattering function (or factor), f, and its behavior is shown in Fig. 7.6 (left) as a function of the phase angle.

Thus, when stationary, periodically arranged electrons are substituted by atoms, their diffraction pattern is the result of a superposition of the two functions, as shown in Fig. 7.6 (right). In other words, the amplitude squared of the diffraction pattern from a row of N atoms is a product of the interference function (Eq. 7.1) and the corresponding atomic scattering function squared, $f^2(\varphi)$:

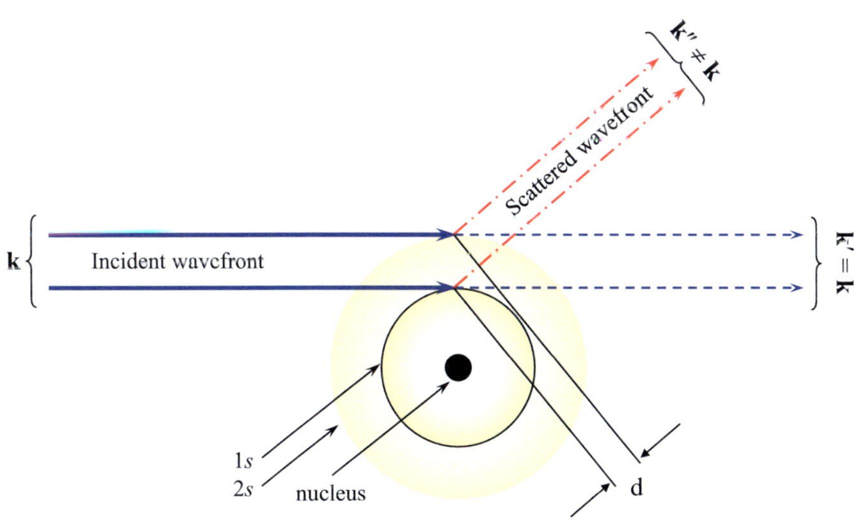

Fig. 7.5 The schematic of the elastic scattering of X-rays by s electrons illustrating the introduction of a path difference, δ, into the wavefront with a propagation vector $\mathbf{k}''$, when it is different from the propagation vector, $\mathbf{k}$, of the incident beam. The distribution of electrons in two s orbitals is determined from the corresponding wave functions

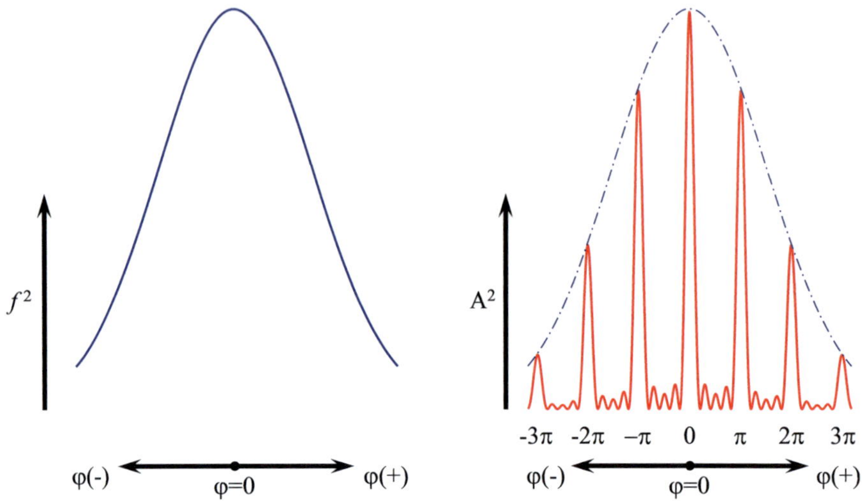

Fig. 7.6 The schematic showing the dependence of the intensity scattered by an atom, i.e., the atomic scattering factor, $f^2 \propto A^2$, as a function of the phase angle (left), and the resultant decrease of the intensity of the diffraction pattern from the row of five regularly spaced atoms, also as a function of the phase angle (right)

$$I(\varphi) \propto f^2(\varphi)\frac{\sin^2 N\varphi}{\sin^2\varphi} \tag{7.2}$$

It is worth noting that it is the radial distribution of core electrons in an atom, which is responsible for the reduction of the intensity when the diffraction angle increases. Thus, it is a specific feature observed in X-ray diffraction from ordered arrangements of atoms. If, for example, the diffraction of neutrons is of concern, they are scattered by nuclei, which may be considered as points. Hence, neutron scattering functions (factors) are independent of the diffraction angle and they remain constant for a given type of nuclei (also see Table 6.2).

7.1.3 Scattering by Lattices

The interference function in Eq. (7.2) describes a discontinuous distribution of the scattered intensity in the diffraction space.[6] Assuming an infinite number of points in a one-dimensional periodic structure ($N \rightarrow \infty$), the distribution of the scattered intensity is a periodic delta function (as mentioned earlier), and therefore, diffraction peaks occur only in specific points, which establish a one-dimensional lattice in the

[6]Diffraction space, in which diffraction peaks are arranged into a lattice, is identical to reciprocal space.

diffraction space. Hence, diffracted intensity is only significant at certain points, which are determined (also see Eq. 7.1 and Figs. 7.4 and 7.6) from

$$I(\varphi) \propto f^2(\varphi)\frac{\sin^2 N\varphi}{\sin^2\varphi} = f^2(\varphi)\frac{\sin^2 Nh\pi}{\sin^2 h\pi} \tag{7.3}$$

In three dimensions, a total of three integers (h, k and l)[7] are required to define the positions of intensity maxima in the diffraction space:

$$I(\varphi) \propto f^2(\varphi)\frac{\sin^2 N_1 h\pi}{\sin^2 h\pi}\frac{\sin^2 N_2 k\pi}{\sin^2 k\pi}\frac{\sin^2 N_3 l\pi}{\sin^2 l\pi} \tag{7.4}$$

where N_1, N_2, and N_3 are the total numbers of the identical atoms in the corresponding directions.

Conversely, when the unit cell contains more than one atom, the individual atomic scattering function $f(\varphi)$ should be replaced with scattering by the whole unit cell, since the latter is now the object that forms a periodic array. The scattering function of one unit cell, F, is called the structure factor or the structure amplitude. It accounts for scattering factors of all atoms in the unit cell together with other relevant atomic parameters. As a result, a diffraction pattern produced by a crystal lattice may be defined as

$$I(\varphi) \propto F^2(\varphi)\frac{\sin^2 U_1 h\pi}{\sin^2 h\pi}\frac{\sin^2 U_2 k\pi}{\sin^2 k\pi}\frac{\sin^2 U_3 l\pi}{\sin^2 l\pi} \tag{7.5}$$

where U_1, U_2, and U_3 are the numbers of the unit cells in the corresponding directions.

The phase angle is a function of lattice spacing (Fig. 7.3), which is a function of h, k, and l. As seen later (Sect. 9.1), the structure factor is also a function of the triplet of Miller indices (hkl). Hence, in general the intensities of discrete points (hkl) in the reciprocal space are given as

$$I(hkl) \propto F^2(hkl)\frac{\sin^2 U_1 h\pi}{\sin^2 h\pi}\frac{\sin^2 U_2 k\pi}{\sin^2 k\pi}\frac{\sin^2 U_3 l\pi}{\sin^2 l\pi} \tag{7.6}$$

The scattered intensity is nearly always measured in relative and not in absolute units, which necessarily introduces a proportionality coefficient, C. As we established earlier, when the phase angle is $n\pi$ (n is an integer), the corresponding interference functions in Eq. (7.6) are reduced to U_1^2, U_2^2, and U_3^2 and they become zero everywhere else. Hence, assuming that the volume of a crystalline material producing a diffraction pattern remains constant (this is always ensured in a properly

[7] The integers h, k, and l are identical to Miller indices.

arranged experiment), the proportionality coefficient C can be substituted by a scale factor $K = CU_1^2U_2^2U_3^2$.

In addition to the scale factor, intensity scattered by a lattice is also subject to different geometrical effects,[8] G, which are various functions of the diffraction angle, θ. All things considered, the intensity scattered by a lattice may be given by the following equation:

$$I(hkl) = K \times G(\theta) \times F^2(hkl) \tag{7.7}$$

This is a very general equation for intensity of the individual diffraction (Bragg) peaks observed in a diffraction pattern of a crystalline substance, and it is discussed in detail in Sect. 8.6, while the geometry of powder diffraction, that is, the directions in which discrete peaks can be observed, is discussed in the following two sections.

7.2 Geometry of Diffraction by Lattices

Both direct and reciprocal spaces may be used to understand the geometry of diffraction by a lattice. Direct-space concepts are intuitive, and therefore, we begin our consideration using physical space. Conversely, reciprocal space is extremely useful in the visualization of diffraction patterns in general and from powders in particular. In this section, therefore, we also show the relationships between geometrical concepts of diffraction in physical and reciprocal spaces.

7.2.1 Laue Equations

The geometry of diffraction from a lattice, or in other words the relationships between the directions of the incident and diffracted beams, was first given by Max von Laue in a form of three simultaneous equations, which are commonly known as Laue equations:

$$\begin{aligned} a(\cos\psi_1 - \cos\varphi_1) &= h\lambda \\ b(\cos\psi_2 - \cos\varphi_2) &= k\lambda \\ c(\cos\psi_3 - \cos\varphi_3) &= l\lambda \end{aligned} \tag{7.8}$$

Here, a, b, and c are the dimensions of the unit cell; ψ_{1-3} and φ_{1-3} are the angles that the incident and diffracted beams, respectively, form with the parallel rows of atoms in three independent directions; the three integer indices h, k, and l have the same

---

[8] One of these geometrical effects is the polarization factor introduced earlier in the Thomson's equation; see Sect. 7.1.1 and the corresponding Footnote 5 on page 158.

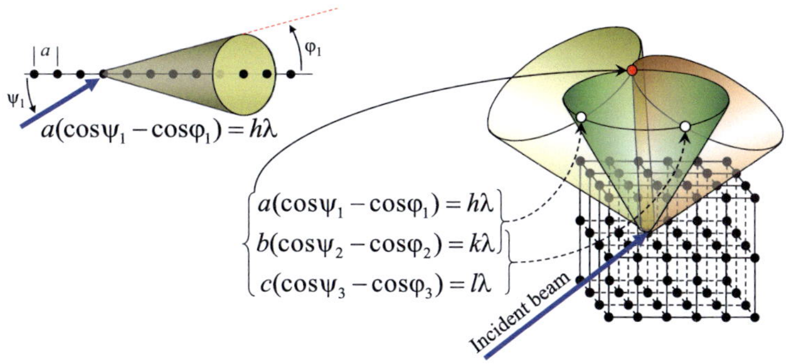

Fig. 7.7 Graphical illustration of Laue equations. A cone of diffracted beams, all forming the same angle φ_1 with a row of atoms, satisfying one Laue equation is shown on top left. Each of the three cones shown on bottom right also satisfies one of the three equations, while the intersecting cones satisfy either two or all three equations simultaneously as shown by arrows. A sharp diffraction peak is only observed in the direction of a point where three Laue equations are simultaneously satisfied

meaning as in Eqs. (7.6) and (7.7), that is, they are unique for each diffraction peak and define the position of the peak in the reciprocal space (also see Sect. 1.5), and λ is the wavelength of the used radiation. The cosines, $\cos\psi_i$ and $\cos\varphi_i$, are known as the direction cosines of the incident and diffracted beams, respectively. According to the formulation given by Laue, sharp diffraction peaks can only be observed when all three equations in (7.8) are satisfied simultaneously as illustrated in Fig. 7.7.

Laue equations once again indicate that a periodic lattice produces diffraction maxima at specific angles, which are defined by both the lattice repeat distances (a, b, c) and the wavelength (λ). Laue equations give the most general representation of a three-dimensional diffraction pattern and they may be used in the form of Eq. (7.8) to describe the geometry of diffraction from a single crystal.

7.2.2 Braggs' Law

More useful in powder diffraction is the law formulated by W.H. Bragg and W.L. Bragg (see Footnote 10 in Chap. 2, p. 42). It establishes certain relationships among the diffraction angle (Bragg angle), wavelength, and interplanar spacing.

According to the Braggs, diffraction from a crystalline sample can be explained and visualized by using a simple notion of mirror reflection of the incident X-ray beam from a series of crystallographic planes. As established earlier (Sect. 1.4.1), all planes with identical triplets of Miller indices are parallel to one another and they are equally spaced. Thus, each plane in a set (hkl) may be considered as a separate scattering object. The set is periodic in the direction perpendicular to the planes, and the repeat distance in this direction is equal to the interplanar distance d_{hkl}. Diffraction from a set of equally spaced objects is only possible at specific angle(s) as we

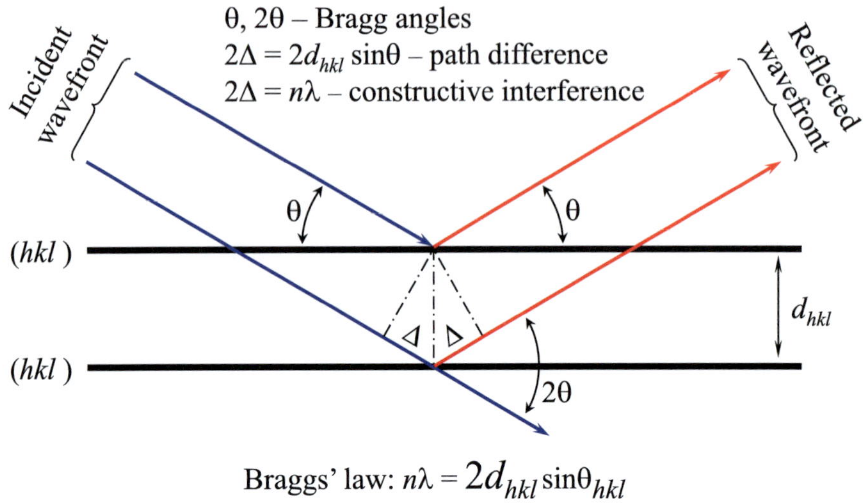

Fig. 7.8 Geometrical illustration of the Braggs' law

already saw in Sect. 7.1. The possible angles, θ, are established from the Braggs' law, which is derived geometrically in Fig. 7.8.

Consider an incident front of waves with parallel propagation vectors, which form an angle θ with the planes (*hkl*). In a mirror reflection, the reflected wavefront will also consist of parallel waves, which form the same angle θ with all planes. The path differences introduced between a pair of waves both before and after they are reflected by the neighboring planes, Δ, are determined by the interplanar distance as $\Delta = d_{hkl} \sin \theta_{hkl}$. The total path difference is 2Δ, and the constructive interference is observed when $2\Delta = n\lambda$, where *n* is an integer and λ is the wavelength of the incident wavefront. This simple geometrical analysis results in the Braggs' law:

$$2d_{hkl} \sin \theta_{hkl} = n\lambda \tag{7.9}$$

The integer *n* is known as the order of reflection. Its value is taken as 1 in all calculations, since orders higher than 1 ($n > 1$) can always be represented by first-order reflections ($n = 1$) from a set of different crystallographic planes with indices that are multiples of *n* because

$$d_{hkl} = n d_{nh,nk,nl} \tag{7.10}$$

and for any $n > 1$, Eq. (7.9) is simply transformed as follows:

$$2d_{hkl} \sin \theta_{hkl} = n\lambda \;\;\Rightarrow\;\; 2d_{nh,nk,nl} \sin \theta_{nh,nk,nl} = \lambda \tag{7.11}$$

7.2.3 Reciprocal Lattice and Ewald's Sphere

The best visual representation of the phenomenon of diffraction has been introduced by P.P. Ewald (see Footnote 9 in Chap. 1, p. 11). Consider an incident wave with a certain propagation vector, $\mathbf{k_0}$, and a wavelength, λ. If the length of $\mathbf{k_0}$ is selected as the inverse of the wavelength

$$|\mathbf{k_0}| = 1/\lambda \tag{7.12}$$

then the entire wave is fully characterized, and it is said that $\mathbf{k_0}$ is its wavevector. When the primary wave is scattered elastically, the wavelength remains constant. Thus, the scattered wave is characterized by a different wavevector, $\mathbf{k_1}$, which has the same length as $\mathbf{k_0}$:

$$|\mathbf{k_1}| = |\mathbf{k_0}| = 1/\lambda \tag{7.13}$$

The angle between $\mathbf{k_0}$ and $\mathbf{k_1}$ is 2θ (Fig. 7.9, left). We now overlap these two wavevectors with a reciprocal lattice (Fig. 7.9, right) such that the end of $\mathbf{k_0}$ coincides with the origin of the lattice. As shown by Ewald, diffraction in the direction of $\mathbf{k_1}$ occurs only when its end coincides with a point in the reciprocal lattice. Considering that $\mathbf{k_0}$ and $\mathbf{k_1}$ have identical lengths regardless of the direction of $\mathbf{k_1}$ (the direction of $\mathbf{k_0}$ is fixed by the origin of the reciprocal lattice), their ends are equidistant from a common point, and therefore, all possible orientations of $\mathbf{k_1}$ delineate a sphere in three dimensions. This sphere is called the Ewald's sphere,

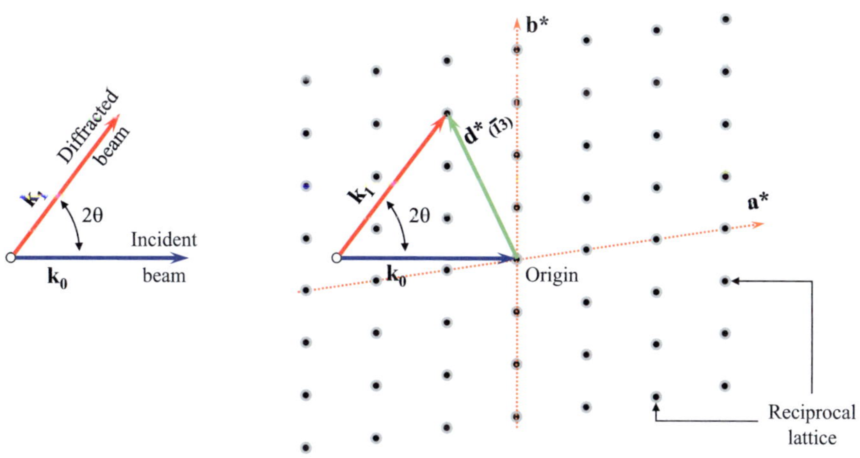

Fig. 7.9 The incident ($\mathbf{k_0}$) and diffracted ($\mathbf{k_1}$) wavevectors originating from a common point (*left*) and the same two vectors overlapped with the two-dimensional reciprocal lattice, which is based on the unit vectors **a*** and **b*** (*right*). The origin of the reciprocal lattice is chosen at the end of $\mathbf{k_0}$. When diffraction occurs from a point in the reciprocal lattice, e.g., the point ($\bar{1}3$), the corresponding reciprocal lattice vector $\mathbf{d}^*_{(hkl)}$ [e.g., $\mathbf{d}^*_{(\bar{1}3)}$] extends between the ends of $\mathbf{k_0}$ and $\mathbf{k_1}$

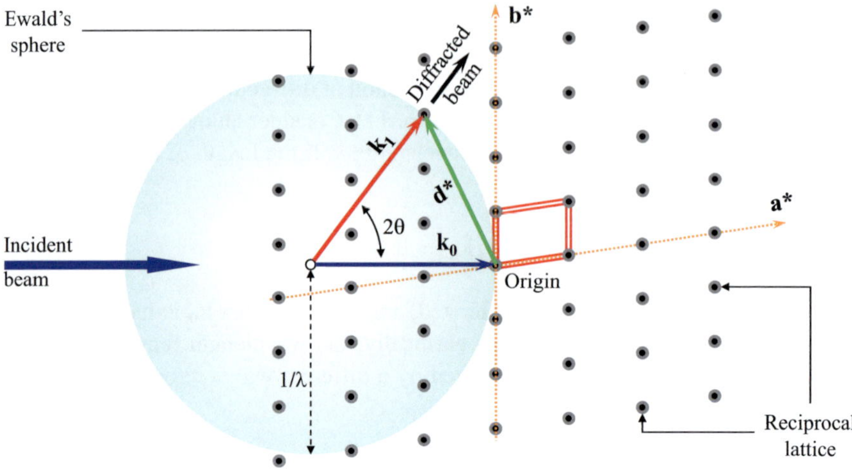

Fig. 7.10 The visualization of diffraction using the Ewald's sphere with radius 1/λ and the two-dimensional reciprocal lattice with unit vectors **a*** and **b***. The origin of the reciprocal lattice is located on the surface of the sphere at the end of $\mathbf{k_0}$. Diffraction can only be observed when a reciprocal lattice point, other than the origin, intersects with the surface of the Ewald's sphere [e.g., the point $(\bar{1}3)$]. The incident and the diffracted beam wavevectors, $\mathbf{k_0}$ and $\mathbf{k_1}$, respectively, have common origin in the center of the Ewald's sphere. The two wavevectors are identical in length, which is the radius of the sphere. The unit cell of the reciprocal lattice is shown using double lines

and it is shown schematically in Fig. 7.10. Obviously, the radius of the Ewald's sphere is the same as the length of $\mathbf{k_0}$; in other words, it is equal to 1/λ.

The simple geometrical arrangement of the reciprocal lattice, Ewald's sphere, and three vectors ($\mathbf{k_0}$, $\mathbf{k_1}$, and $\mathbf{d}^*_{hkl}$) in a straightforward and elegant fashion yields Braggs' equation. From both Figs. 7.9 and 7.10, it is clear that vector $\mathbf{k_1}$ is a sum of two vectors, $\mathbf{k_0}$ and $\mathbf{d}^*_{hkl}$:

$$\mathbf{k_1} = \mathbf{k_0} + \mathbf{d}^*_{hkl} \tag{7.14}$$

Its length is known (1/λ)[9] and its orientation with respect to the incident wavevector, that is, angle θ, is found from simple geometry after recalling that $|\boldsymbol{d}^*| = 1/d$:

$$|\boldsymbol{k}_1| \sin\theta = |\boldsymbol{k}_0| \sin\theta = \frac{1}{2}|\boldsymbol{d}^*| \quad \Rightarrow \quad 2d\sin\theta = \lambda \tag{7.15}$$

[9]The lengths of the propagation vectors $\mathbf{k_0}$ and $\mathbf{k_1}$ may also be defined in terms of their wavenumbers: $|\mathbf{k_0}| = |\mathbf{k_1}| = 2\pi/\lambda$. Equation (7.15) may then be rewritten as $|\mathbf{d^*}| = |\mathbf{Q}| = 4\pi\sin\theta/\lambda$, thus defining the so-called Q-vector, which is often used to represent diffraction data in synchrotron radiation experiments. It is also worth noting that since **d*** is a vector in reciprocal lattice, the value of sinθ/λ is independent of the wavelength.

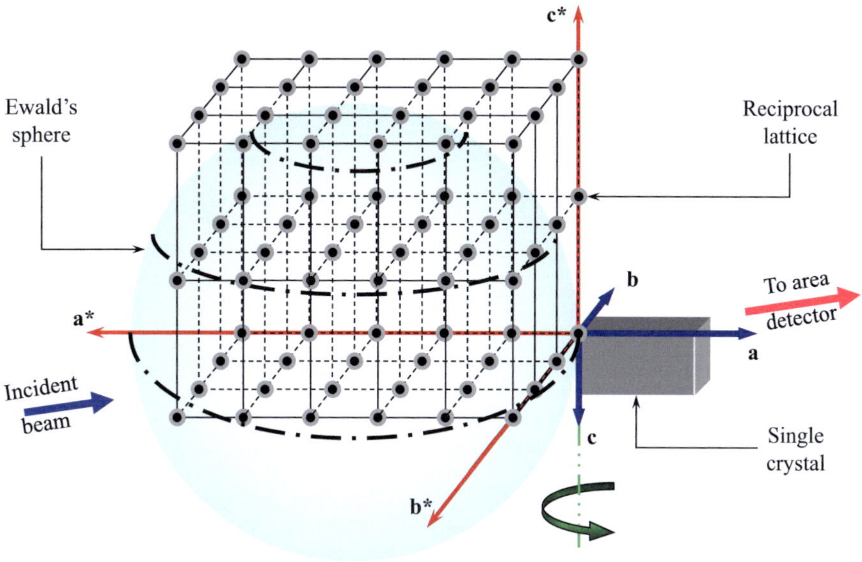

Fig. 7.11 The illustration of a single crystal showing the orientations of the basis vectors corresponding to both the direct (**a**, **b** and **c**) and reciprocal (**a***, **b*** and **c***) lattices and the Ewald's sphere. The reciprocal lattice is infinite in all directions but only one octant (where $h > 0$, $k > 0$ and $l > 0$) is shown for clarity

The Ewald's sphere and the reciprocal lattice are essential tools in the visualization of the three-dimensional diffraction patterns from single crystals, as illustrated in the next few paragraphs. They are also invaluable in the understanding of the geometry of diffraction from polycrystalline (powder) specimens, which is explained Chap. 8.

Consider a stationary single crystal, in which the orientation of basis vectors of the reciprocal lattice is established by the orientation of the corresponding crystallographic directions with respect to the external shape of the crystal, as shown in Fig. 7.11. Thus, when a randomly oriented single crystal is irradiated by monochromatic X-rays, only a few, if any (also see Fig. 7.10), points of the reciprocal lattice will coincide with the surface of the Ewald's sphere.[10] This occurs because first, the sphere has a constant radius determined by the wavelength, and second, the distribution of the reciprocal lattice points in three dimensions is fixed by both the lattice parameters and the orientation of the crystal. The resultant diffraction pattern may reveal just a few Bragg peaks, as shown in Fig. 7.12.

[10]When a stationary single crystal is irradiated by white, polychromatic X-rays, a single Ewald's sphere shown in Fig. 7.11 becomes a continuum of spheres. Different points of reciprocal lattice will then rest on surfaces of different Ewald's spheres, thus producing a much richer diffraction pattern. Thus technique is known as Laue technique, and it is most often employed for examination of symmetry and orientation of single crystals.

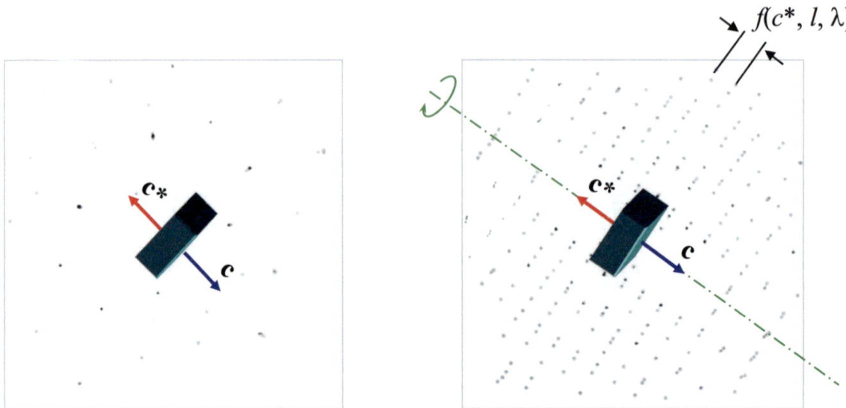

Fig. 7.12 The two-dimensional diffraction patterns from stationary (*left*) (The single crystal is triclinic $Pb_3F_5(NO_3)$: space group $P\bar{1}$, a = 7.3796(6), b = 12.1470(9), c = 16.855(1) Å, α = 100.460(2), β = 90.076(1), γ = 95.517(1)°, see: D.T. Tran, P.Y. Zavalij and S.R.J. Oliver, A cationic layered material for anion-exchange, J. Am. Chem. Soc. **124**, 3966 (2002), https://doi.org/10.1021/ja017333w) and rotating (*right*) (The single crystal is orthorhombic, $FePO_4{\cdot}2H_2O$: space group Pbca, a = 9.867(1), b = 10.097(1), c = 8.705(1) Å, see: Y. Song, P.Y. Zavalij, M. Suzuki, and M.S. Whittingham, New iron(III) phosphate phases: Crystal structure, electrochemical and magnetic properties, Inorg. Chem. **41**, 5778 (2002). https://doi.org/10.1021/ic025688q) single crystals recorded using a CCD detector. The incident wavevector is perpendicular to both the detector and the plane of the figure. The *dash-dotted* line on the *right* shows the rotation axis, which is collinear with **c***

Many more reciprocal lattice points are placed on the surface of the Ewald's sphere when the crystal is set in motion, for example, when it is rotated around an axis. The rotation of the crystal changes the orientation of the reciprocal lattice, but the origin of the latter remains aligned with the end of the incident wavevector. Hence, all reciprocal lattice points with $|\mathbf{d}^*| \leq 2/\lambda$ will coincide with the surface of the Ewald's sphere at different angular positions of the crystal. When the rotation axis is collinear with one of the crystallographic axes and is perpendicular to the incident beam, the reciprocal lattice points form planar intersections with the Ewald's sphere (Fig. 7.11, dash-dotted lines). The planes are parallel and equidistant, resulting in a diffraction pattern[11] similar to that illustrated in Fig. 7.12 (right).

[11] This type of the diffraction pattern enables one to determine the lattice parameter of the crystal in the direction along the axis of rotation. It is based on the following geometrical consideration: the distance between the planar cross-sections of the Ewald's sphere in Fig. 7.11 equals c^*; the corresponding diffraction peaks are grouped into lines (see Fig. 7.12, right), and the distance between the neighboring lines is a function of c^*, the distance from the crystal to the detector, l, and the wavelength, λ.

7.3 Additional Reading

1. International Tables for Crystallography, Vol. A-I published jointly with the International Union of Crystallography (IUCr) by Springer, https://it.iucr.org/
2. International Tables for Crystallography. Volume C: Mathematical, physical and chemical tables. Edited by T. R. Welberry. Second online edition (2021), https://doi.org/10.1107/97809553602060000117
3. International Tables for Crystallography. Volume H: Powder diffraction. Edited by C. J. Gilmore, J. A. Kaduk and H. Schenk. First online edition (2019), https://doi.org/10.1107/97809553602060000115
4. P. Scardi, S.J.L. Billinge, R. Nederd, and A. Cervellino, Celebrating 100 years of the Debye scattering equation, Acta Cryst. (2016). A72, 589 (2016), https://doi.org/10.1107/S2053273316015680
5. L. Gelisio and P. Scardi, 100 years of Debye's scattering equation, Acta Cryst. **A72**, 608 (2016), https://doi.org/10.1107/S2053273316014881
6. R.B. Neder and Th. Proffen, Teaching diffraction with the aid of computer simulations, J. Appl. Cryst. **29**, 727–735 (1996), https://doi.org/10.1107/S002188989600619X
7. Electron diffraction techniques. Vol. 1, 2. J. Cowley, Ed., Oxford University Press. Oxford (1992), https://global.oup.com/academic/product/electron-diffraction-techniques-9780198555582
8. R. Jenkins and R.L. Snyder, Introduction to X-ray powder diffractometry. Wiley, New York (1996), https://doi.org/10.1002/9781118520994
9. Modern powder diffraction. D.L Bish and J.E. Post, Eds. Reviews in Mineralogy, Vol. 20. Mineralogical Society of America, Washington, DC (1989), https://msaweb.org/volume-20-modern-powder-diffraction/

7.4 Problems

1. A student prepares a sample and collects a powder diffraction pattern on an instrument that is available in the laboratory overseen by his major professor. The student then takes the same sample to a different laboratory on campus and collects a second set of powder diffraction data. When he comes back to his office, he plots both patterns. The result is shown in Fig. 7.13. Analyze possible sources of the observed differences.
2. The following is the list of five longest interplanar distances possible in a crystal lattice of some material: 4.967, 3.215, 2.483, 2.212, and 1.607 Å. Calculate Bragg angles (2θ) at which Bragg reflections may be observed when using Cr $K_{\alpha 1}$ or Cu $K_{\alpha 1}$ radiation.

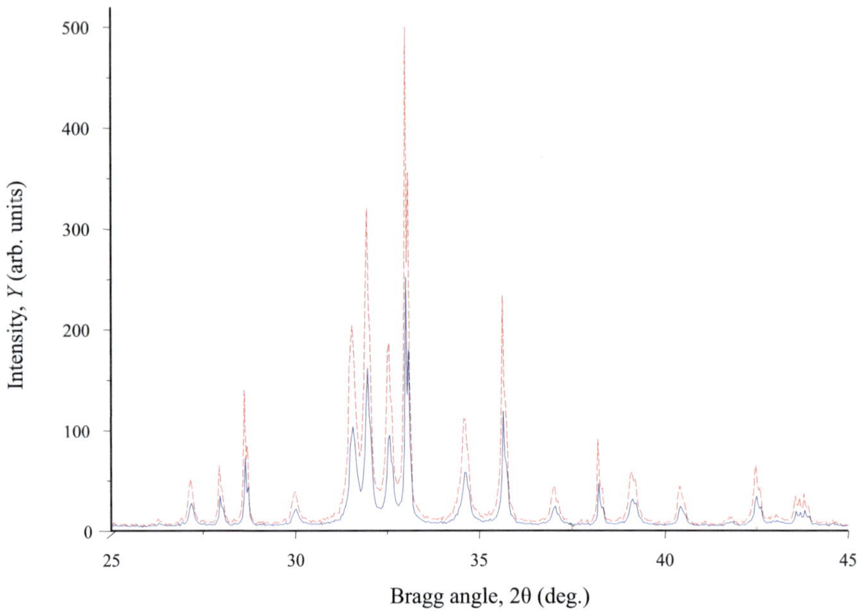

Fig. 7.13 Two powder diffraction patterns collected by a student using the same sample but two different powder diffractometers

3. Researcher finished collecting a powder diffraction pattern of an unknown crystalline substance. She used Cu K_α radiation, $\lambda = 1.54178$ Å. The first Bragg peak is observed at $2\theta = 9.76°$. Based on this information, she makes certain conclusions regarding the length of at least one of the three unit cell edges. What are these conclusions?

Chapter 8
The Powder Diffraction Pattern

The powder diffraction experiment is a cornerstone of basic materials characterization–diffraction analysis. This technique has been used for many decades with exceptional success to provide accurate information about the structure of materials. Although powder data typically lack the three-dimensionality of a diffraction image, the method's fundamental nature is evident from the fact that a powder diffraction pattern represents a one-dimensional projection of the three-dimensional reciprocal lattice of a crystal.[1] The quality of the powder diffraction pattern is usually limited by the nature and the energy of the available radiation, the resolution of the instrument, and the physical and chemical conditions of the specimen. Since many materials can only be prepared in a polycrystalline form, the powder diffraction experiment becomes the only realistic option for reliably determining the crystal structure of such materials.

Powder diffraction data are customarily recorded in the simplest possible fashion, where the scattered intensity is measured as a function of a single independent variable—the Bragg angle. What makes the powder diffraction experiment so powerful is that different structural features of a material have distinct effects on various parameters of its powder diffraction pattern. For example, the presence of a crystalline phase is manifested as a set of discrete intensity maxima—the Bragg reflections—each with a specific intensity and location. When atomic parameters, such as coordinates of atoms in the unit cell or populations of different sites in the lattice of the crystalline phase, are altered, this change affects relative intensities and/or positions of the Bragg peaks that correspond to this phase. When the changes are microscopic, such as when the grain size is reduced below a certain limit, or when the material has been strained or deformed, the shapes of Bragg peaks become

[1] Imaging of the reciprocal lattice in three dimensions is easily done in a single crystal diffraction experiment.

P. Zavalij, V. Pecharsky, *Fundamentals of Powder Diffraction and Structural Characterization of Materials*, https://doi.org/10.1007/978-3-031-91504-8_8

affected in addition to their intensities and positions. Hence, much of the structural information about the material is embedded in its powder diffraction pattern. When experimental data are properly collected and processed, a great deal of detail about a material's structure at different length scales, as well as its phase and chemical compositions, can be established.

8.1 Origin of the Powder Diffraction Pattern

As established in Sect. 7.2.3, the primary monochromatic beam is scattered in a particular direction, which is easily predicted using Ewald's representation (see Figs. 7.10–7.12). A similar, yet fundamentally different situation is observed in the case of diffraction from powders or from polycrystalline specimens, where multiple single crystals (crystallites or grains) are irradiated simultaneously by a monochromatic incident beam. When the number of grains in the irradiated volume is large and their orientations are completely random, the same is true for the reciprocal lattices associated with each crystallite. Consequently, the ends of the identical reciprocal lattice vectors, $\mathbf{d}^*_{hkl}$, become arranged on the surface of the Ewald's sphere in a circle perpendicular to the incident wavevector, $\mathbf{k_0}$. The corresponding scattered wavevectors, $\mathbf{k_1}$, are aligned along the surface of the cone, as shown in Fig. 8.1. The apex of the cone coincides with the center of Ewald's sphere, the cone axis is parallel to $\mathbf{k_0}$, and the solid cone angle is 4θ.

Assuming that the number of crystallites approaches infinity (the randomness of their orientations has been postulated in the previous paragraph), the density of the scattered wavevectors, $\mathbf{k_1}$, becomes constant on the surface of the cone. Consequently, the diffracted intensity will be uniform around the circumference of the cone base or, when measured by a planar area detector as shown in Fig. 8.1, around the ring, which the cone base forms with the plane of the detector. Similar rings but with different intensities and diameters will be formed by other independent reciprocal lattice vectors, and these are commonly known as the Debye rings.

The appearance of eight diffraction cones when polycrystalline copper powder is irradiated by the monochromatic Cu $K\alpha_1$ radiation is shown schematically in Fig. 8.2. All Bragg peaks, possible in the range $0° < 2\theta < 180°$, are also listed with the corresponding Miller indices and relative intensities in Table 8.1.

Assuming that the diffracted intensity is distributed evenly around the base of each cone (see the postulations made above), there is usually no need to measure the intensity of the entire Debye ring. Hence, in a conventional powder diffraction experiment, the measurements are performed only along a narrow cylinder (red) wrapping the Ewald's sphere around the equatorial plane, as shown in Fig. 8.2. Because of this, only one variable axis (2θ) is fundamentally required in powder diffractometry, yet the majority of instruments have two independently or jointly controlled axes. The latter is done due to a variety of reasons, such as the limits imposed by the geometry, more favorable focusing, advantage of keeping specimen

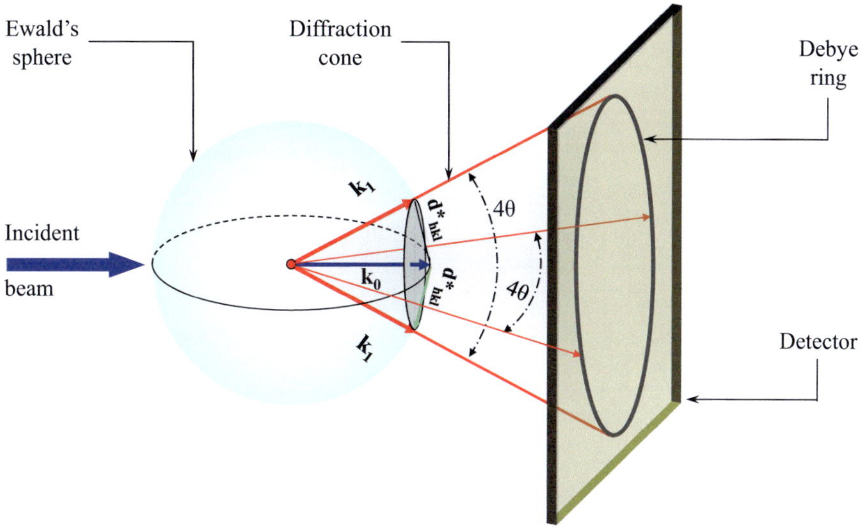

Fig. 8.1 The powder diffraction cone originates from the infinite number of completely randomly oriented, identical reciprocal lattice vectors, $\mathbf{d}^*_{hkl}$, which form a circle with their ends placed on the surface of Ewald's sphere. This arrangement produces the powder diffraction cone and the corresponding Debye (Petrus (Peter) Josephus Wilhelmus Debye (1884–1966), a Dutch physical chemist credited with numerous discoveries in physics and chemistry. Relevant to the subject of this book is his calculation of the effect of temperature on the scattered X-ray intensity (the Debye–Waller factor) and his work together with Paul Scherrer on the development of the powder diffraction method. In 1936, Peter Debye was awarded the Nobel Prize in Chemistry for "his contributions to our knowledge of molecular structure through his investigations on dipole moments and on the diffraction of X-rays and electrons in gases." For more information, see https://www.nobelprize.org/prizes/chemistry/1936/debye/) ring on the flat screen (film or area detector) (In this geometry (flat detector perpendicular to the incident beam placed behind the sample), it is fundamentally impossible to measure intensity scattered at $2\theta \geq 90°$. One alternative is to place the detector between the focal point of the X-ray tube and the sample; this enables to measure intensity scattered at $2\theta > 90°$. In either case, when diffraction occurs at $2\theta = 90°$, the measurement is impossible ($\tan 90° = \infty$). Furthermore when $2\theta \cong 90°$, the size of a flat detector becomes prohibitively large. For practical measurements, a flat detector may be tilted at any angle with respect to the propagation vector of the incident beam. Instead of a flat detector, a flexible image plate detector may be arranged as a cylinder with its axis perpendicular to the incident wavevector and traversing the location of the specimen, which facilitates simultaneous measurement of the entire powder diffraction pattern.). The detector is perpendicular to both the direction of the incident beam and the cone axis, and the radius of the Debye ring in this geometry is equal to SDD $\cdot\ \tan 2\theta$, where SDD is the sample-to-detector distance

horizontal, specific application, etc. More details about the geometry of modern powder diffractometers are given in Chap. 11.

In powder diffraction, the scattered intensity is customarily represented as a function of a single independent variable—Bragg angle (2θ)—as modeled in Fig. 8.3 for a polycrystalline copper. This type of the plot is standard and it is called the powder diffraction pattern or the histogram. In some instances, the diffracted intensity may be plotted versus the interplanar distance, d, the q-value ($q = 1/$

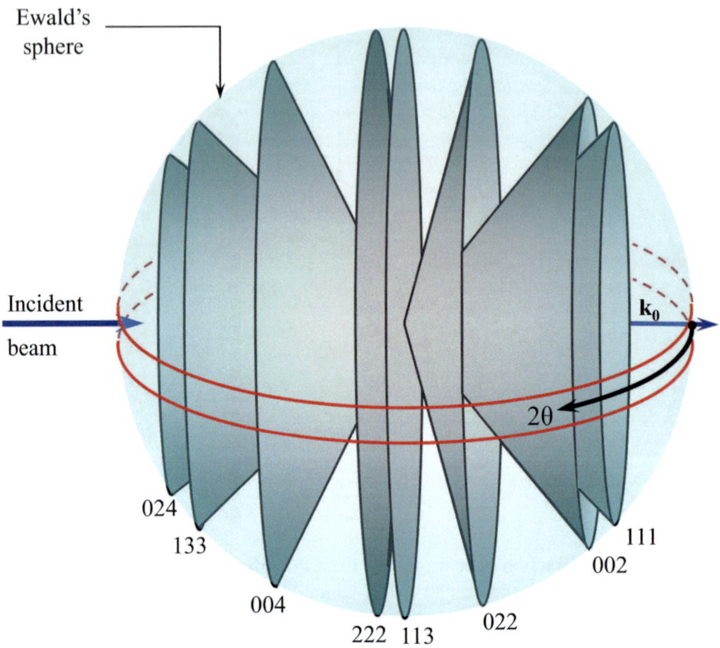

Fig. 8.2 The schematic of the powder diffraction cones produced by a polycrystalline copper sample using Cu Kα_1 radiation. The differences in the relative intensities of various Bragg peaks (diffraction cones) are not discriminated, and they may be found in Table 8.1. Each cone is marked with the corresponding triplet of Miller indices. The *red* cylinder around the equator depicts the portion of the Ewald's sphere scanned during the diffraction experiment using a point or 1D detector

Table 8.1 Bragg peaks from polycrystalline copper obtained using Cu Kα_1 radiation[a]

hkl	I/I_0	2θ (°)	*hkl*	I/I_0	2θ (°)
111	100	43.298	222	5	95.143
002	446	50.434	004	3	116.923
022	20	74.133	133	9	136.514
113	17	89.934	024	8	144.723

[a] The data are taken from the ICDD powder diffraction file, record No. 4-836: H.E. Swanson, E. Tatge, National Bureau of Standards (US), Circular **359**, 1 (1953)

$d^2 = d^{*2}$), or $\sin\theta/\lambda$ (or the Q-value, which is different from $\sin\theta/\lambda$ by a factor of 4π, see Footnote 9 in Chap. 7, p.168).

The scattered intensity is usually represented as the total number of the accumulated counts, counting rate (counts per second, cps), or in arbitrary units (a.u.). Regardless of which units are chosen to plot the intensity, the patterns are visually identical because the intensity scale remains linear and because the intensity measurements are normally relative, not absolute. In rare instances, the intensity is plotted as a common or a natural logarithm or a square root of the total number of

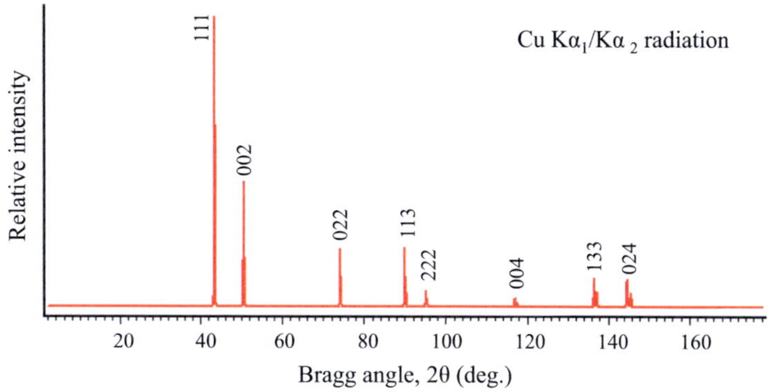

Fig. 8.3 The simulated powder diffraction pattern of copper (space group Fm$\bar{3}$m, $a = 3.615$ Å, Cu Kα_1/Kα_2 radiation, Cu atom in 4(a) position with $x = 0$, $y = 0$, $z = 0$)

the accumulated counts in order to better visualize both the strong and weak Bragg peaks on the same plot. The use of these two nonlinear intensity scales, however, always increases the visibility of the noise (i.e., highlights the presence of statistical counting errors). A few examples of the nonconventional representation of powder diffraction patterns are provided in Sect. 8.2.

In Chap. 7, we assumed that the diffracted intensity is observed as infinitely narrow diffraction maxima (delta functions). In reality, the Ewald's sphere has finite thickness due to wavelength aberrations, and reciprocal lattice points are far from infinitesimal shapeless points—they may be reasonably imagined as small diffuse spheres (do not forget that the reciprocal lattice itself is not real and it is nothing else than a useful mathematical concept). Therefore, Bragg peaks always have nonzero widths as functions of 2θ, which is illustrated quite well in Figs. 8.4 and 8.5 by the powder diffraction pattern of LaB_6.[2] The data were collected using Mo Kα radiation on a Bruker D8 Venture diffractometer equipped with a flat PHOTON III CPAD detector. This figure also serves as an excellent experimental confirmation of our conclusions made at the beginning of this section (e.g., see Figs. 8.1 and 8.2).

As shown in Fig. 8.4, when diffraction cones, produced by the LaB_6 powder, intersect with the flat detector, they create a set of concentric Debye rings. As in a typical powder diffractometer, only a narrow band has been scanned and the result of the integration is also shown in Fig. 8.4 as the scattered intensity versus tan2θ (note that the radial coordinate of the detector is tan2θ and not 2θ). The resultant diffraction pattern is shown in the standard format as relative intensity versus 2θ in Fig. 8.5, where each Bragg peak is labeled with the corresponding Miller indices. It is worth noting that the diffractometer used in this experiment is a single-crystal diffractometer, which was not designed to take full advantage of focusing of the scattered beam. As a result, the Bragg peaks shown in Fig. 8.5 are quite broad and the Kα_1/α_2 doublet is unresolved even when 2θ approaches 30°. As we will see in Chap. 12

[2]NIST standard reference material, SRM 660c (see https://www.nist.gov/srm)

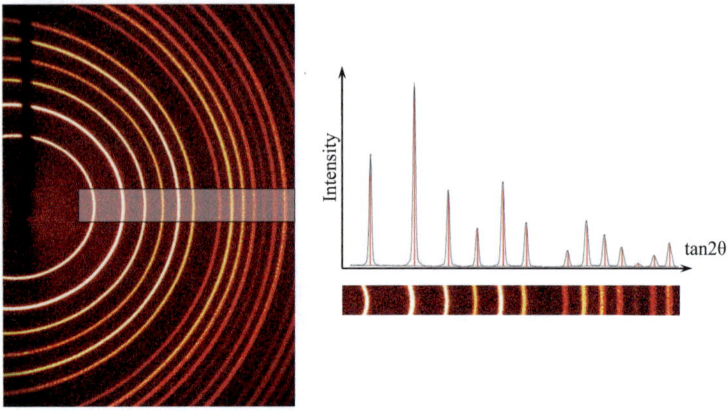

Fig. 8.4 *Left*—the X-ray diffraction pattern of a polycrystalline LaB_6 obtained using Mo Kα radiation and recorded using a flat CCD area detector (compare with Figs. 8.1 and 8.2). Measured intensity is proportional to the degree of darkening. The *dark shadow* extending from the center of the image to the *top* is the projection of the beam stop needed to protect the detector from being damaged by the high-intensity incident beam. The *light-shaded rectangular box* delineates the area in which the scattered intensity was integrated from the center of the image toward its right edge. *Right*—the resultant intensity as a function of tan2θ shown together with the area over which the integration has been carried out

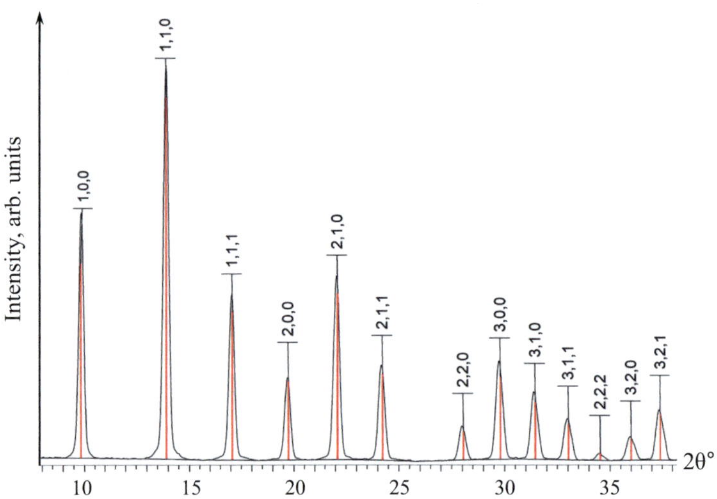

Fig. 8.5 The powder diffraction pattern of the polycrystalline LaB_6 as intensity versus 2θ obtained by the integration of the rectangular area from the two-dimensional diffraction pattern shown in Fig. 8.4

(e.g., see Fig. 12.21), a much better resolution is possible in high-resolution powder diffractometers, where the doublet becomes resolved at much lower Bragg angles.

8.2 Representation of Powder Diffraction Patterns

In a typical experiment, the intensity, diffracted by a polycrystalline sample, is measured as a function of Bragg angle, 2θ. Hence, powder diffraction patterns are usually plotted in the form of the measured intensity, *Y*, as the dependent variable versus the Bragg angle as the independent variable; see Figs. 8.6 (top) and 8.7a, c. In rare instances, for example, when there are just a few very intense Bragg peaks and all others are quite weak, or when it is necessary to directly compare diffraction patterns collected from the same material using different wavelengths, the scales of one or both axes may be modified for better viewing and easier comparison.

When the former is true (i.e., there are few extremely strong Bragg peaks while all others are weak), the vertical axis can be calibrated as a logarithm of intensity

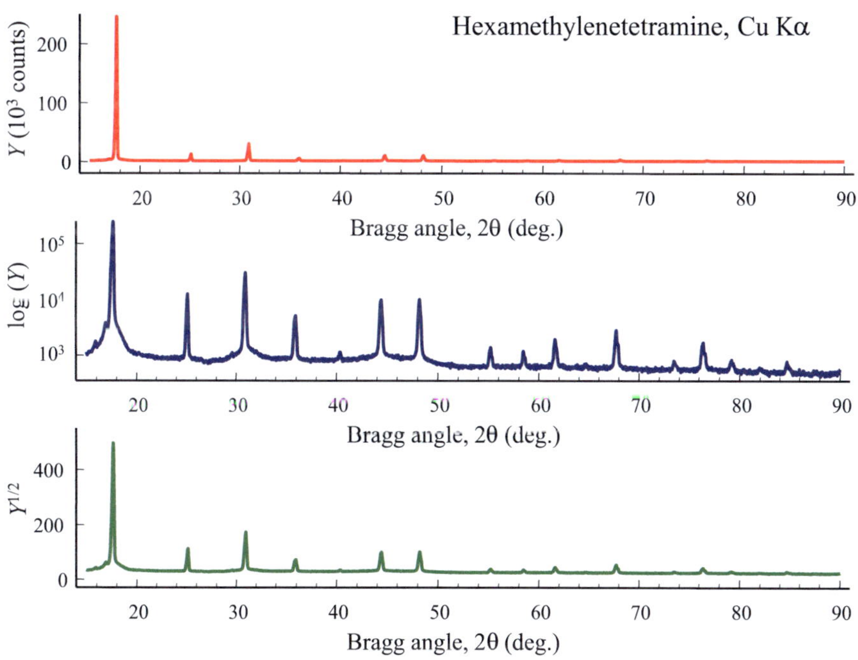

Fig. 8.6 The powder diffraction pattern (Powder diffraction data were collected on a Scintag XDS2000 powder diffractometer using Cu Kα radiation and cooled Ge(Li) solid state detector. The counting time was 10 s in every point; the data were collected with a 0.025° step of 2θ) of hexamethylenetetramine collected using Cu Kα radiation and plotted as the measured intensity in counts (*top*), common logarithm (*middle*), and square root (*bottom*) of the total number of registered photon counts versus 2θ

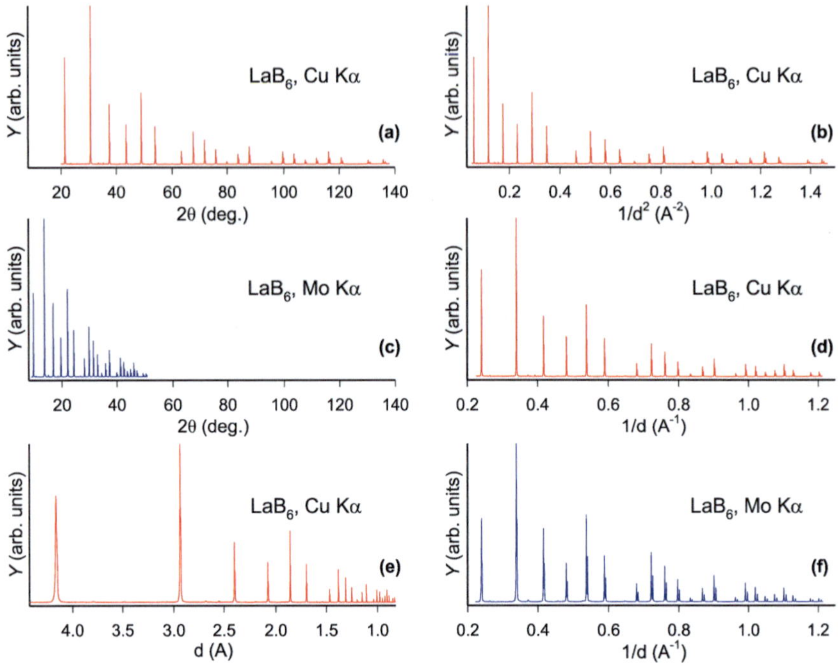

Fig. 8.7 Two powder diffraction patterns of LaB_6 collected using different wavelengths with the scattered intensity plotted versus different independent variables (Powder diffraction data were collected on a Scintag XDS2000 powder diffractometer using Cu Kα radiation and cooled Ge (Li) solid-state detector and on a Rigaku TTRAX rotating anode powder diffractometer using Mo Kα radiation with diffracted beam monochromator and scintillation detector. The data were collected with a 0.02° step of 2θ using Cu Kα and with a 0.01° step using Mo Kα radiations). Each plot contains the same number of Bragg peaks, which can be observed below $2\theta \cong 140°$ when using Cu Kα radiation

(Fig. 8.6, middle) or its square root (Fig. 8.6, bottom). This changes the scale and enables better visualization of the low-intensity features. In the example shown in Fig. 8.6, the middle (logarithmic) plot reveals all weak Bragg peaks in addition to the nonlinearity of the background and the details of the intensity distribution around the bases of the strongest peaks. The $Y^{1/2}$ scale is equivalent to the plot of statistical errors of the measured intensities (see Sect. 12.3.1), in addition to better visualization of weak Bragg peaks.

Various horizontal scales alternative to the Bragg angle, see Fig. 8.7, are usually wavelength-independent and their use is mostly dictated by special circumstances. For example, *d*-spacing (Fig. 8.7e) is most commonly used in the time-of-flight (TOF) experiments: according to Eq. (6.1), the wavelength is the inverse of the velocity of the particle (neutron). The time-of-flight from the specimen to the detector is therefore directly proportional to *d*. This scale, however, reduces the visual resolution in the low *d* range (equivalent to high Bragg angle range, see Eq. 7.9) when used in combination with X-ray diffraction data. In TOF experiments, the actual resolution of the diffraction pattern is reduced at low d, that is, at high neutron velocities.

The second scale is $1/d = 2 \sin \theta/\lambda$ (see Fig. 8.7d, f). It results in only slightly reduced resolution at high Bragg angles when compared to the 2θ scale. Recalling that $1/d = d^*$, this type of plot is a one-dimensional projection of the reciprocal lattice and it is best suited for direct comparison of powder diffraction data collected using different wavelengths. The similarity of these two diffraction patterns is especially impressive after comparing them when both are plotted versus 2θ (Fig. 8.7a, c).

The third is the q-values scale, where $q = 1/d^2 = 4\sin^2\theta/\lambda^2$, which provides the best resolution at high Bragg angles when compared to other wavelength-independent scales, see Fig. 8.7b. This scale results in the equally spaced Bragg peaks when the crystal system is cubic (see Sect. 14.6). In cases of lower symmetry crystal systems, only certain types of Bragg peaks are equally spaced along the q-axis, and in some instances, the q-scaled powder diffraction pattern may be used to assign indices and/or examine the relationships between the lattice parameters of the material with the unknown crystal structure.

8.3 Understanding of Powder Diffraction Patterns

The best way to appreciate and understand how structural information is encoded in a powder diffraction pattern is to consider the latter as a set of discrete diffraction peaks (Bragg reflections) superimposed over a continuous background. Although the background may be used to extract information about the crystallinity of the specimen and few other parameters about the material, we are concerned with the Bragg peaks and not with the background. In the majority of powder diffraction applications, the background is an inconvenience, which has to be dealt with, and generally every attempt is made to achieve its minimization during the experiment.

Disregarding the background, the structure of a typical powder diffraction pattern may be described by the following components: *positions*, *intensities*, and *shapes* of multiple Bragg reflections, for example, compare Figs. 8.6a and 8.7a. Each of the three components italicized here contains information about the crystal structure of the material, the properties of the specimen (sample), and the instrumental parameters, as shown in Table 8.2. Some of these parameters have a key role in defining a particular component of the powder diffraction pattern, while others result in various distortion(s), as also indicted in Table 8.2. It is worth noting that this table is not comprehensive and additional parameters may affect the positions, intensities, and shapes of Bragg peaks.

In addition to the influence of instrumental parameters, two types of crystallographic (structural) parameters define the makeup of every powder diffraction pattern. These are the unit cell dimensions and the atomic structure, which includes both the unit cell content and the spatial distribution of atoms within the unit cell. Thus, a powder diffraction pattern can be constructed (or simulated) as follows:

Table 8.2 Powder diffraction pattern as a function of various crystal structure, specimen, and instrumental parameters[a]

Pattern component	Crystal structure	Specimen property	Instrumental parameter
Peak position	**Unit cell parameters** (a, b, c, α, β, γ)	*Absorption* Porosity	**Radiation (wavelength)** *Instrument/sample alignment* Axial divergence of the beam
Peak intensity	**Atomic parameters** (x, y, z, B, etc.)	*Preferred orientation* Absorption *Porosity*	Geometry and configuration **Radiation (Lorentz, polarization)**
Peak shape	*Crystallinity* Disorder Defects	*Grain size* *Strain* *Stress*	**Radiation (spectral purity)** **Geometry** **Beam conditioning**

[a] Key parameters are shown in **bold**. Parameters that may have a significant influence are shown in *italic*

- Positions of Bragg peaks are established from the Braggs' law as a function of the wavelength and the interplanar distances, that is, d-spacing. The latter can be easily calculated from the known unit cell dimensions (Sect. 8.4). For instance, in the case of the orthorhombic crystal system, permissible Bragg angles can be found from

$$2\theta_{hkl} = 2\arcsin\left(\frac{\lambda}{2d_{hkl}}\right), \quad \text{where} \quad d_{hkl} = \left(\frac{h^2}{a^2}+\frac{k^2}{b^2}+\frac{l^2}{c^2}\right)^{-1/2} \tag{8.1}$$

  Since h, k, and l are integers, both the resultant d values and Bragg angles form arrays of discrete values for a given set of unit cell dimensions. Bragg angles are also dependent on the employed wavelength. The example of the discontinuous distribution of Bragg angles is shown using short vertical bars of equal length in Fig. 8.8a.
- As noted in Sect. 7.1, the intensity of diffraction maxima is a function of the periodicity of the scattering centers (unit cells), and therefore, the intensities can be calculated for individual Bragg peaks from the structural model. The latter requires the knowledge of the coordinates of atoms in the unit cell together with other relevant atomic and geometrical parameters. The influence of the varying intensity on the formation of the powder diffraction pattern is illustrated using the varying lengths of the bars in Fig. 8.8b—the longer the bar, the higher the intensity. Although not shown in Fig. 8.8, certain combinations of Miller indices may have zero or negligibly small intensity and, therefore, the corresponding Bragg reflections disappear or become unrecognizable in the diffraction pattern.
- The shape of Bragg peaks is usually represented by a bell-like function—the so-called peak-shape function. The latter is weakly dependent on the crystal structure and is the convolution of various individual functions, established by the instrumental parameters and to some extent by the properties of the specimen,

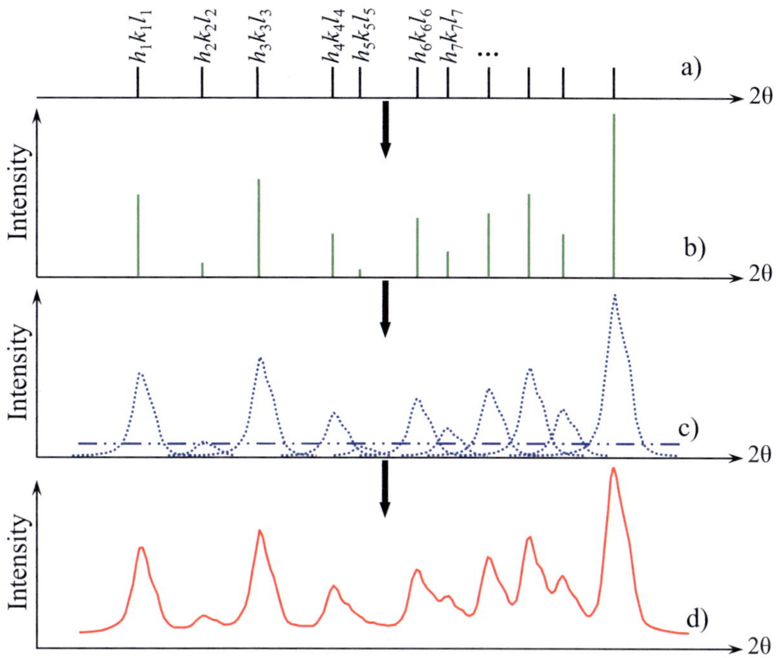

Fig. 8.8 The appearance of the powder diffraction pattern: (**a**) only Bragg peak positions (e.g., see Eq. 8.1) are represented by the *vertical bars* of equal length; (**b**) in addition to peak positions, their intensities are indicated by using the bars with variable lengths (the higher the intensity, the *longer the bar*); (**c**) peak shapes have been introduced by substituting individual intensities with appropriate peak-shape functions, and a constant background has been indicated by the *dash-double dotted line*; (**d**) the resultant powder diffraction pattern is the sum of all components shown separately in (**c**), i.e., discrete but partially overlapped peaks and continuous background

see Table 8.2. The shape of each peak can be modeled using instrumental and specimen characteristics, although in reality ab initio modeling is difficult and most often it is performed using various empirically selected peak-shape functions and parameters. If the radiation is not strictly monochromatic, that is, when both $K\alpha_1$ and $K\alpha_2$ components are present in the diffracted beam, the resultant peak should include contributions from both components as shown in Fig. 8.9. Thus, vertical bars with different lengths are replaced by the corresponding peak shapes, as shown in Fig. 8.8c. It should be noted that although the relative intensities of different Bragg reflections may be adequately represented by the lengths of the bars, this is no longer correct for peak heights: the bars are one-dimensional and have zero area, but peak area is a function of the full width at half maximum, which varies with Bragg angle. Individual peaks should have their areas proportional to intensities of individual Bragg reflections (see Sect. 8.6.1).

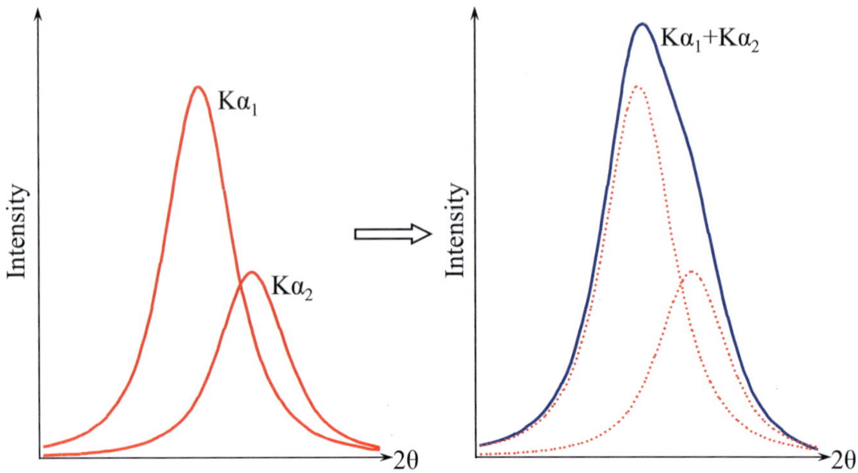

Fig. 8.9 The two individual functions corresponding to monochromatic $K\alpha_1$ and $K\alpha_2$ wavelengths (*left*) and the combined peak-shape function for a $K\alpha_1/K\alpha_2$ doublet as the sum of two peaks (*right*). Since both $K\alpha_1$ and $K\alpha_2$ peaks correspond to the same $\mathbf{d^*_{hkl}}$, their positions, θ_1 and θ_2, are related as $\sin\theta_1/\lambda K\alpha_1 = \sin\theta_2/\lambda K\alpha_2$ (see Eq. 7.9), while their areas (intensities) are related as approximately 2 to 1 (see Fig. 6.3)

– Finally, the resultant powder diffraction pattern is a sum of the individual peak-shape functions and a background function as illustrated in Fig. 8.8d, where the background function was assumed constant for clarity.

It is generally quite easy to simulate the powder diffraction pattern when the crystal structure of the material is known (the peak-shape parameters are empirical and the background, typical for a given instrument, may be measured). The inverse process, that is, the determination of the crystal structure from powder diffraction data is much more complex. First, individual Bragg peaks should be located on the pattern and both their positions and intensities determined by fitting to a certain peak-shape function, including the background. Second, peak positions are used to establish the unit cell symmetry, parameters, and content. Third, peak intensities are used to determine space-group symmetry and coordinates of atoms. Fourth, the entire diffraction pattern is used to refine all crystallographic and peak-shape function parameters, including the background. All these issues are discussed and illustrated beginning from Chap. 13.

8.4 Positions of Powder Diffraction Peaks

As discussed earlier in general terms, diffraction peaks appear at specific angles due to scattering by periodic lattices. Further, as shown by the Braggs and Ewald (Sect. 7.2), these angles are a discontinuous function of Miller indices, the interplanar distances (lengths of independent reciprocal lattice vectors), and the wavelength (radius of the Ewald's sphere). Therefore, both the unit cell dimensions and the wavelength are the two major factors that determine Bragg angles for the same combination of *h*, *k*, and *l*. As we will also see later, the observed peak positions can be distorted by instrumental and specimen parameters.

8.4.1 Peak Positions as a Function of Unit Cell Dimensions

The interplanar distance is a function of the unit cell parameters and Miller indices, *h*, *k*, and *l*, which fully describe every set of crystallographic planes. The corresponding formulae for the inverse square of the interplanar distance, $1/d^2$, are usually given separately for each crystal system,[3] as shown in Eqs. (8.2)–(8.7).

$$\text{Cubic}: \quad \frac{1}{d^2} = \frac{h^2 + k^2 + l^2}{a^2} \tag{8.2}$$

$$\text{Tetragonal}: \quad \frac{1}{d^2} = \frac{h^2 + k^2}{a^2} + \frac{l^2}{c^2} \tag{8.3}$$

$$\text{Hexagonal}: \quad \frac{1}{d^2} = \frac{4}{3}\frac{h^2 + hk + k^2}{a^2} + \frac{l^2}{c^2} \tag{8.4}$$

$$\text{Orthorhombic}: \quad \frac{1}{d^2} = \frac{h^2}{a^2} + \frac{k^2}{b^2} + \frac{l^2}{c^2} \tag{8.5}$$

$$\text{Monoclinic}: \quad \frac{1}{d^2} = \frac{h^2}{a^2 \sin^2\beta} + \frac{k^2}{b^2} + \frac{l^2}{c^2 \sin^2\beta} + \frac{2hl\cos\beta}{ac\sin^2\beta} \tag{8.6}$$

$$\begin{aligned} \text{Triclinic}: \quad \frac{1}{d^2} = \; & \left[\frac{h^2}{a^2\sin^2\alpha} + \frac{2kl}{bc}(\cos\beta\cos\gamma - \cos\alpha) + \right. \\ & \frac{k^2}{b^2\sin^2\beta} + \frac{2hl}{ac}(\cos\alpha\cos\gamma - \cos\beta) + \\ & \left. \frac{l^2}{c^2\sin^2\gamma} + \frac{2hk}{ab}(\cos\alpha\cos\beta - \cos\gamma)\right] / \\ & (1 - \cos^2\alpha - \cos^2\beta - \cos^2\gamma + 2\cos\alpha\cos\beta\cos\gamma) \end{aligned} \tag{8.7}$$

[3] Primitive rhombohedral lattices, i.e., when $a = b = c$ and $\alpha = \beta = \gamma \neq 90°$ are nearly always treated in the hexagonal basis with rhombohedral (R) lattice centering. In a primitive rhombohedral lattice:

$$\frac{1}{d^2} = \frac{(h^2 + k^2 + l^2)\sin^2\alpha + 2(hk + kl + hl)(\cos^2\alpha - \cos\alpha)}{a^2(1 - 3\cos^2\alpha + 2\cos^3\alpha)}$$

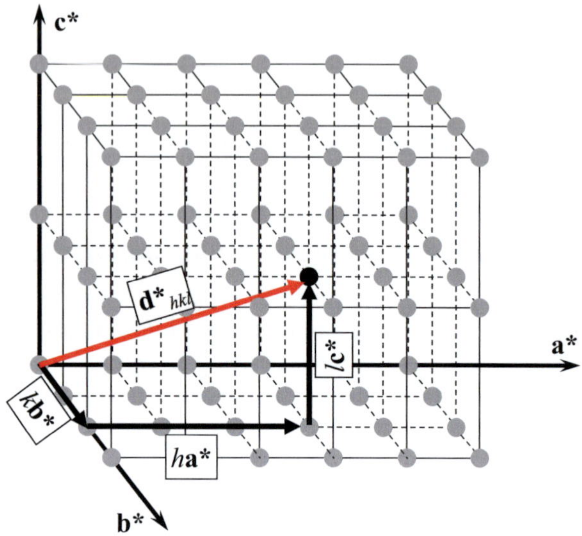

Fig. 8.10 The illustration of a reciprocal lattice vector, $\mathbf{d}^*_{hkl}$ (*red*), as a vectorial sum of three basis unit vectors, **a***, **b***, and **c*** multiplied by *h*, *k*, and *l*, respectively

The most complex formula is the one for the triclinic crystal system, in which a total of six independent parameters are required to describe the unit cell dimensions. Conversely, Eq. (8.7) is the most general, since Eqs. (8.2)–(8.6) are easily derived from it. For example, after introducing the corresponding relationships between the unit cell dimensions for the tetragonal crystal system (i.e., $a = b \neq c$, $\alpha = \beta = \gamma = 90°$) into Eq. (8.7), the latter is straightforwardly simplified to Eq. (8.3). Thus, the simplified formulae (8.2)–(8.6) are only useful in manual calculations, but when the list of possible *d*'s (or θ's) is generated using a computer program, it makes better sense to employ only the most general equation since obviously the resultant $1/d^2$ values are correct upon the substitution of the appropriate numerical values for *a*, *b*, *c*, α, β, and γ into Eq. (8.7).

The usefulness of the reciprocal lattice concept may be once again demonstrated here by illustrating how easily Eqs. (8.2)–(8.7) can be derived in the reciprocal space employing reciprocal lattice vectors. When the derivation is performed in the direct space, the geometrical considerations become quite complex.

Consider a reciprocal lattice as shown in Fig. 8.10. Any reciprocal lattice vector, $\mathbf{d}^*_{hkl}$, is a sum of three noncoplanar vectors (**a***, **b***, and **c*** are the unit vectors of the reciprocal lattice and *h*, *k*, and *l* are integers):

$$\mathbf{d}^*_{hkl} = h\mathbf{a}^* + k\mathbf{b}^* + l\mathbf{c}^* \tag{8.8}$$

For example, in the orthorhombic crystal system $\alpha^* = \beta^* = \gamma^* = 90°$. Hence, Eq. (8.8) is transformed into

$$\left(d^*_{hkl}\right)^2 = (ha^*)^2 + (kb^*)^2 + (lc^*)^2 \tag{8.9}$$

and Eq. (8.5) is obtained immediately because, in orthogonal coordinate system such as the orthorhombic system, $d^* = 1/d$, $a^* = 1/a$, $b^* = 1/b$, and $c^* = 1/c$.

In the triclinic crystal system, the equivalent of Eq. (8.9) is more complex

$$d^{*2} = h^2a^{*2} + k^2b^{*2} + l^2c^{*2} + 2hka^*b^*\cos\gamma^* + 2hla^*c^*\cos\beta^* + 2klb^*c^*\cos\alpha^* \tag{8.10}$$

but it becomes considerably more intuitive and easier to understand in terms of reciprocal lattice parameters than Eq. (8.7), which is given in terms of direct-space unit cell dimensions.

According to the Braggs' law (Eq. 7.9), the diffraction angle, θ_{hkl}, of a reflection from a series of lattice planes (*hkl*) is determined from the interplanar distance, d_{hkl}, and the wavelength, λ as:

$$\sin\theta_{hkl} = \frac{\lambda}{2d_{hkl}} \tag{8.11}$$

8.4.2 Other Factors Affecting Peak Positions

Equations (8.7) and (8.10) are exact assuming that both the powder diffractometer and the sample are ideal. In reality, various instrumental and specimen features may affect the observed positions of Bragg peaks. These factors are often known as systematic aberrations (distortions), and they are usually assembled into a single correction parameter, $\Delta 2\theta$. The latter is applied to the idealized Bragg angle, $2\theta_{calc}$, calculated from the unit cell dimensions and wavelengths using Eqs. (8.7) or (8.10) and (8.11), so that the experimentally observed Bragg angle, $2\theta_{obs}$, is given as

$$2\theta_{obs} = 2\theta_{calc} + \Delta 2\theta \tag{8.12}$$

For the most commonly used Bragg–Brentano focusing geometry (see Sect. 11.3), the overall correction is generally a sum of six factors:

$$\Delta 2\theta = \frac{p_1}{\tan 2\theta} + \frac{p_2}{\sin 2\theta} + \frac{p_3}{\tan\theta} + p_4\sin 2\theta + p_5\cos\theta + p_6 \tag{8.13}$$

The first two parameters, p_1 and p_2, account for the *axial divergence* of the incident beam (Sect. 11.2) and they can be expressed as

$$p_1 = -\frac{h^2K_1}{3R^2}; \quad p_2 = -\frac{h^2K_2}{3R^2} \tag{8.14}$$

where h is the length of the specimen parallel to the goniometer axis, R is the goniometer radius, and K_1 and K_2 are constants established by the collimator. Soller slits (see Sect. 11.2.1) usually minimize the axial divergence, and therefore, these two corrections are often neglected for practical purposes.

In addition to axial divergence, the first parameter (p_1) includes a shift that is due to peak asymmetry caused by other factors. One of these is the finite length of the receiving slit of the detector, which results in the measurement of a fixed length of an arc (see Fig. 8.4) rather than an infinitesimal point of the Debye ring. The curvature of the Debye ring increases[4] with the decreasing Bragg angle and the resultant increasing peak asymmetry cannot be corrected for by using Soller slits. This effect can be minimized by reducing the detector slit length, which however considerably lowers the measured intensity.

The third parameter, p_3, is given as

$$p_3 = -\frac{\alpha^2}{K_3} \tag{8.15}$$

where α is the *in-plane divergence* of the X-ray beam (see Sect. 11.2) and K_3 is a constant. This factor accounts for the zero curvature of flat samples, typically used in Bragg–Brentano goniometers. This geometry of the sample distorts the ideal focusing in which the curvature of the sample surface should vary with Bragg angle. The aberrations are generally insignificant, and they are usually neglected in routine powder diffraction experiments.

The fourth parameter is

$$p_4 = \frac{1}{2\mu_{eff}R} \tag{8.16}$$

where μ_{eff} is the *effective linear absorption* coefficient (see Sect. 8.6.5 and Eq. 8.51). This correction is known as the transparency-shift error, and it may play a role when examining thick (more than 50–100 μm) samples. The transparency-shift error is caused by the penetration of the beam into the sample, and the penetration depth is a function of Bragg angle. Usually, p_4 is the refined parameter since μ_{eff} is rarely known (both the porosity and the density of the powder sample are usually unknown). The transparency-shift error could be substantial for low-absorbing samples, for example, organic compounds, and it is usually negligible for highly

[4] Strictly speaking, the curvature of the Debye ring increases both below and above $2\theta = 90°$, e.g. see Fig. 8.2. Both the curvature and associated asymmetry become especially significant when $2\theta \leq \sim 20°$ and $2\theta \geq \sim 160°$. At low Bragg angles, this contribution to asymmetry results in the enhancement of the low angle slopes of Bragg peaks, while at high Bragg angles the asymmetry effect is opposite. Asymmetry at high Bragg angles is often neglected because the intensity of Bragg peaks is usually low due to a variety of geometrical and structural factors, which is discussed in Sects. 8.6 and 9.1. It is also worth mentioning that at $2\theta \cong 90°$ the contribution from p_1 becomes negligible because $\tan 2\theta \to \infty$.

absorbing specimens, that is, compounds containing heavy chemical elements. For low-absorbing materials, this shift can be reduced by using thin samples, however, doing so significantly decreases intensity at high Bragg angles. The latter is already small when a compound consists of light chemical elements due to their low X-ray scattering ability.

The fifth parameter characterizes *specimen displacement*, s, from the goniometer axis and it is expressed as

$$p_5 = -\frac{2s}{R} \tag{8.17}$$

where R is the radius of the goniometer. This correction may be substantial, especially when there is no good and easy way to control the exact position of the specimen surface.

The last parameter, p_6, is constant over the whole range of Bragg angles and the corresponding aberration usually arises due to improper setting(s) of zero angles for one or more diffractometer axes: detector and/or X-ray source. Hence, this distortion is called the *zero-shift* error. The zero-shift error can be easily minimized by proper alignment of the goniometer. However, in some cases, for example, in neutron powder diffraction, zero shift is practically unavoidable and, therefore, should be always accounted for.

Equations (8.13)–(8.17) define the most important factors affecting peak positions observed in a powder diffraction pattern, some of which combine several effects that have the same 2θ dependence. The latter, however, is not really important when parameters p_i are refined rather than modeled. A comprehensive analysis of errors in peak positions for a general case of focusing geometry (not only the Bragg–Brentano focusing geometry) may be found in the International Tables for Crystallography in Tables 5.2.4.1, 5.2.7.1, and 5.2.8.1 on pp. 494–498.[5]

In order to account for several different factors simultaneously, high accuracy of the experimental powder diffraction data is required, in addition to the availability of data in a broad range of Bragg angles. Even then, it may be difficult since p_4 and p_5 are strongly correlated and so is the zero-shift parameter, p_6. Generally, they cannot be distinguished from one another when only a small part of the diffraction pattern has been measured (e.g., below 60–70° 2θ). Thus, refinement of any single parameter (p_4, p_5 or p_6) gives similar results, that is, the satisfactory fit between the observed and calculated 2θ values. The problem is: how precise are the obtained unit cell parameters? If the wrong correction was taken into account, the resultant unit cell dimensions may be somewhat different from their true values. The best way to deal with the ambiguity of which correction to apply is to use an internal standard, which unfortunately contaminates the powder diffraction pattern with Bragg peaks of the standard material.

[5] International Tables for Crystallography, Vol. C, Mathematical, physical and chemical tables, First online edition, E. Prince, Ed. (2006). https://it.iucr.org/C/

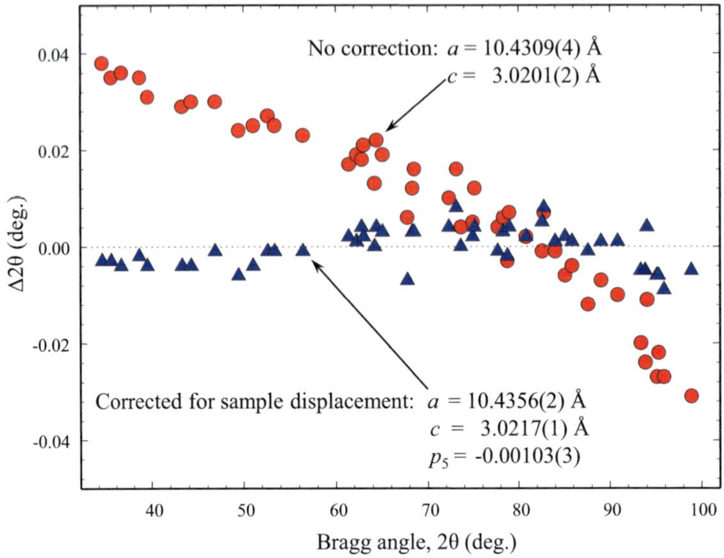

Fig. 8.11 The differences between the observed and calculated 2θ values plotted as a function of 2θ without (*red circles*) and with (*blue triangles*) specimen displacement correction. The corresponding values of the unit cell parameters and the specimen displacement parameter are indicated on the plot. The material belongs to the tetragonal crystal system

An example of how important the sample-displacement correction may become is shown in Fig. 8.11, where the differences between the observed and calculated Bragg angles are in the −0.03° to +0.04° range before correction (red circles). They fall within the −0.01° to +0.01° range (blue triangles) after refining the sample-displacement parameter along with the unit cell dimensions. Even though the difference in the unit cell dimensions obtained with and without the sample-displacement correction (Fig. 8.11) is not exceptionally large, it is still 10–20 times the least-squares standard deviations, that is, the differences in lattice parameters are statistically significant.

8.5 Shapes of Powder Diffraction Peaks

All but the simplest powder diffraction patterns are composed of more or less overlapped Bragg peaks due to the intrinsic one-dimensionality of the powder diffraction technique coupled with the usually large number of "visible" reciprocal lattice points, that is, those that have $d^*_{hkl} \leq 2/\lambda$, and the limited resolution of the instrument (e.g., see the model in Figs. 8.2 and 8.8d). Thus, processing of the data by fitting peak shapes to a suitable function is required in order to obtain both the positions and intensities of individual Bragg peaks. The same is also needed in structure refinement using the full-profile fitting approach—the Rietveld method.

The observed peak shapes are best described by the so-called peak-shape function (PSF), which is a convolution[6] of three different functions: instrumental broadening, Ω, wavelength dispersion, Λ, and specimen function, Ψ. Thus, PSF can be represented as follows:

$$PSF(\theta) = \Omega(\theta) \otimes \Lambda(\theta) \otimes \Psi(\theta) + b(\theta) \tag{8.18}$$

where b is the background function.

The instrumental function, Ω, depends on multiple geometrical parameters: the locations and geometry of the source, monochromator(s), slits, and specimen. The wavelength (spectral) dispersion function, Λ, accounts for the distribution of the wavelengths in the source and it varies depending on the nature of the source and the monochromatization technique. Finally, the specimen function, Ψ, originates from several effects. First is the dynamic scattering or deviations from the kinematical model. They result in a small but finite width, known as the Darwin[7] width, of the Bragg peaks. The second effect is determined by the physical properties of the specimen, such as crystallite (grain) size and microstrains. For instance, when the crystallites are small (typically smaller than ~1 μm) and/or they are strained, the resulting Bragg peak widths can increase substantially.

It is worth noting that unlike the instrumental and wavelength dispersion functions, the broadening effects introduced by the physical state of the specimen may be of interest in materials characterization. Thus, effects of the average crystallite size (τ) and microstrain (ε) on Bragg peak broadening (β, in radians) can be described in the first approximation as follows:

$$\beta = \frac{\lambda}{\tau \cdot \cos\theta} \tag{8.19}$$

and

$$\beta = k \cdot \varepsilon \cdot \tan\theta \tag{8.20}$$

---

[6] A convolution (⊗) of two functions, f and g, is defined as an integral

$$f(t) \otimes g(t) = \int_{-\infty}^{\infty} f(\tau)g(t-\tau)d\tau = \int_{-\infty}^{\infty} g(\tau)f(t-\tau)d\tau$$

which expresses the amount of overlap between the two functions as g is shifted over f. It effectively "blends" one function with another. The convolution is also known as "folding" (e.g. see E.W. Weisstein, Convolution, Eric Weisstein's world of mathematics, https://mathworld.wolfram.com/Convolution.html)

[7] Sir Charles Galton Darwin (1887–1962) the British physicist, who begun working with Ernest Rutherford and Niels Bohr, later using his mathematical skills to help Henry Moseley with his work on X-ray diffraction. A brief biography is available on Wikipedia at https://en.wikipedia.org/wiki/Charles_Galton_Darwin

where k is a constant that depends on the definition of a microstrain. It is important to note that β in Eqs. (8.19) and (8.20) is not the total width of a Bragg peak but it is an excess width, which is an addition to all instrumental contributions. The latter is usually established by measuring a standard material without microstrain and grain-size effects at the same experimental conditions.

In general, three different approaches to the description of peak shapes can be used. The first employs *empirical* peak-shape functions, which fit the profile without attempting to associate their parameters with physical quantities. The second is a *semi-empirical* approach that describes instrumental and wavelength dispersion functions using empirical functions, while specimen properties are modeled using realistic physical parameters. In the third, the so-called *fundamental parameters* approach,[8] all three components of the peak-shape function (Eq. 8.18) are modeled using rational physical quantities.

8.5.1 Peak-Shape Functions

Considering Figs. 8.8 and 8.9 and Eq. (8.18), the intensity, $Y(i)$, of the ith point ($1 \leq i \leq n$, where n is the total number of measured points) of the powder diffraction pattern, in the most general form is the sum of the contributions, y_k, from the m overlapped individual Bragg peaks ($1 \leq k \leq m$) and the background, $b(i)$.[9] Therefore, it can be described using the following expression:

$$Y(i) = b(i) + \sum_{k=1}^{m} I_k[y_k(x_k) + 0.5y_k(x_k + \Delta x_k)] \tag{8.21}$$

where I_k is the intensity of the kth Bragg reflection, $x_k = 2\theta_i - 2\theta_k$, and Δx_k is the difference between the Bragg angles of the $K\alpha_2$ and $K\alpha_1$ components in the doublet (if present). The presence of Bragg intensity as a multiplier in Eq. (8.21) enables one to introduce and analyze the behavior of different normalized functions independently of peak intensity, that is, assuming that the definite integral of a peak-shape function, calculated from negative to positive infinity, is unity in each case.

The four most commonly used empirical peak-shape functions (y) are as follows:

Gauss[10]:

[8] J. Bergmann, Contributions to evaluation and experimental design in the fields of X-ray powder diffractometry, Ph.D. thesis (in German), Dresden University for Technology (1984). See http://www.bgmn.de/methods.html for more information and other references.

[9] Several functions commonly used in approximating the background are discussed later, see Eqs. (13.1)–(13.6).

[10] Johann Carl Friedrich Gauss (1777–1885) was a German mathematician. A brief biography is available on Wikipedia, https://en.wikipedia.org/wiki/Carl_Friedrich_Gauss

$$y(x)=G(x)=\frac{C_G^{1/2}}{\sqrt{\pi}H}\exp\left(-C_G x^2\right) \tag{8.22}$$

Lorentz[11]:

$$y(x)=L(x)=\frac{C_L^{1/2}}{\pi H'}\left(1+C_L x^2\right)^{-1} \tag{8.23}$$

Pseudo-Voigt[12]:

$$y(x)=PV(x)=\eta\frac{C_G^{1/2}}{\sqrt{\pi}H}\exp\left(-C_G x^2\right)+(1-\eta)\frac{C_L^{1/2}}{\pi H'}\left(1+C_L x^2\right)^{-1} \tag{8.24}$$

Pearson-VII[13]:

$$y(x)=PVII(x)=\frac{\Gamma(\beta)}{\Gamma(\beta-1/2)}\frac{C_P^{1/2}}{\sqrt{\pi}H}\left(1+C_P x^2\right)^{-\beta} \tag{8.25}$$

where

- H and H' are the full widths at half maximum (often abbreviated as FWHM).
- $x=(2\theta_i-2\theta_k)/H_k$ is essentially the Bragg angle of the ith point in the powder diffraction pattern with its origin in the position of the kth peak divided by the peak's FWHM.
- $2\theta_i$ is the Bragg angle of the ith point of the powder diffraction pattern.
- $2\theta_k$ is the calculated (or ideal) Bragg angle of the kth Bragg reflection.
- $C_G=4\ln 2$, and $C_G^{1/2}/\sqrt{\pi}H$ is the normalization factor for the Gauss function such that $\int_{-\infty}^{\infty}G(x)dx=1$.
- $C_L=4$, and $C_L^{1/2}/\pi H'$ is the normalization factor for the Lorentz function such that $\int_{-\infty}^{\infty}L(x)dx=1$.

[11] Hendrik Antoon Lorentz (1853–1928) was a Dutch physicist best known for his contributions to the theory of electromagnetic radiation. In 1902, he shared the Nobel Prize in physics with Pieter Zeeman "in recognition of the extraordinary service they rendered by their researches into the influence of magnetism upon radiation phenomena." See https://www.nobelprize.org/prizes/physics/1902/lorentz/ for details.

[12] Named after Woldemar Voigt (1850–1919), the German physicist best known for his work in crystal physics. A brief biography is available on Wikipedia, http://en.wikipedia.org/wiki/Woldemar_Voigt

[13] Named after Karl Pearson (1857–1936), the British mathematician who derived several probability distribution functions known today as Pearson I to Pearson XII. A brief biography and a description of Pearson distributions are available at Wikipedia: https://en.wikipedia.org/wiki/Karl_Pearson and https://en.wikipedia.org/wiki/Pearson_distribution, respectively.

- $C_P = 4(2^{1/\beta} - 1)$, and $[\Gamma(\beta)/\Gamma(\beta - 1/2)]C_P^{1/2}/\sqrt{\pi}H$ is the normalization factor for the Pearson-VII function such that $\int_{-\infty}^{\infty} PVII(x)dx = 1$.
- $H = (U\tan^2\theta + V\tan\theta + W)^{1/2}$, which is known as Caglioti formula, is the full width at half maximum as a function of θ for Gauss, pseudo-Voigt, and Pearson-VII functions, and U, V and W are free variables.[14]
- $H' = X/\cos\theta + Y\tan\theta$ is the full width at half maximum as a function of θ for the Lorentz function, and X and Y are the free variables.
- $\eta = \eta_0 + \eta_1 2\theta + \eta_2 2\theta^2$, where $0 \le \eta \le 1$ is the pseudo-Voigt function mixing parameter, i.e., the fractional contribution of the Gauss function into the linear combination of Gauss and Lorentz functions, and η_0, η_1, and η_2 are the free variables.
- Γ is the gamma function.[15]
- $\beta = \beta_0 + \beta_1/2\theta + \beta_2/(2\theta)^2$ is the exponent as a function of Bragg angle in the Pearson-VII function, and β_0, β_1, and β_2 are the free variables.

The two simplest peak-shape functions (Eqs. 8.22 and 8.23) represent Gaussian and Lorentzian distributions, respectively, of the intensity in the Bragg peak. They are compared in Fig. 8.12, from which it is easy to see that the Lorentz function is sharp near its maximum but has long tails on each side near its base. Conversely, the Gauss function has no tails at the base but has a rounded maximum. Both functions are centrosymmetric, that is, $G(x) = G(-x)$ and $L(x) = L(-x)$.

The shapes of real Bragg peaks, which are the results of convoluting multiple instrumental and specimen functions (Eq. 8.18), are rarely described well by simple Gaussian or Lorentzian distributions, especially in X-ray diffraction. Usually, real peak shapes are located somewhere between the Gauss and Lorentz distributions and they can be better represented as the mixture of the two functions.[16] An ideal way would be to convolute the Gauss and Lorentz functions in different proportions. This convolution, however, is a complex procedure, which requires numerical integration every time one or several peak-shape function parameters change. Therefore, a much simpler linear combination of Gauss and Lorentz functions is used instead of a convolution, and it is usually known as the pseudo-Voigt function (Eq. 8.24). The Gaussian and Lorentzian are mixed in η to 1 − η ratio, so that the value of the mixing parameter, η, varies from 0 (pure Lorentz) to 1 (pure Gauss). Obviously, η has no

---

[14] G. Caglioti, A. Paoletti, and F.P. Ricci, Choice of collimators for a crystal spectrometer for neutron diffraction, Nucl. Instrum. Methods **3**, 223 (1958). https://doi.org/10.1016/0369-643X(58)90029-X

[15] Gamma function is defined as $\Gamma(z) = \int_0^{\infty} t^{z-1}e^t dt$, or recursively for a real argument as $\Gamma(z) = (z - 1)\Gamma(z - 1)$. It is nonexistent when $z = 0, -1, -2, \ldots$, and becomes $(z - 1)!$ when $z = 1, 2, 3, \ldots$. Gamma function is an extension of the factorial to complex and real arguments (e.g., see E.W. Weisstein, Gamma function, Eric Weisstein's world of mathematics, https://mathworld.wolfram.com/GammaFunction.html for more information).

[16] The most notable exception is the shape of peaks in neutron powder diffraction (apart from the time-of-flight data), which is typically close to the pure Gaussian distribution. Peak shapes in TOF experiments are usually described by a convolution of exponential and pseudo-Voigt functions.

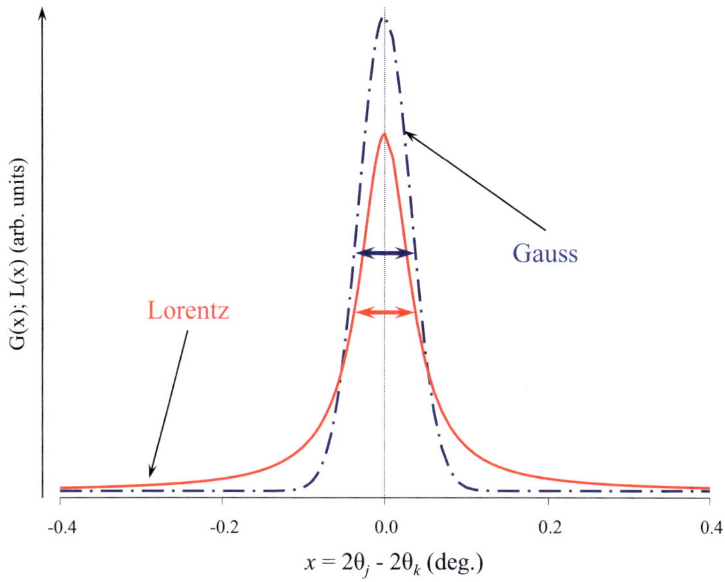

Fig. 8.12 *The illustration of* Gauss (*dash-dotted line*) and Lorentz (*solid line*) peak-shape functions. Both functions have been normalized to result in identical definite integrals ($\int_{-\infty}^{\infty} G(x)dx = \int_{-\infty}^{\infty} L(x)dx$) and full widths at half maximum (FWHM). The corresponding FWHMs are shown as thick horizontal arrows

physical meaning outside this range. When during refinement η becomes negative, this is usually called super-Lorentzian, and such an outcome points to an incorrect choice of the peak-shape function. Usually, Pearson-VII function should be used instead of pseudo-Voigt, see next paragraph.

The fourth commonly used peak-shape function is Pearson VII (Eq. 8.25). It is similar to Lorentz distribution, except that the exponent (β) varies in the Pearson-VII, while it remains constant ($\beta = 1$) in the Lorentz function. Pearson-VII provides an intensity distribution close to the pseudo-Voigt function: when the exponent, $\beta = 1$, it is identical to the Lorentz distribution, and when $\beta \cong 10$, Pearson-VII becomes nearly pure Gaussian. Thus, when the exponent is in the range $0.5 < \beta < 1$ or $\beta > 10$, the peak shape extends beyond Lorentz or Gauss functions, respectively, but these values of β are rarely observed in practice. An example of the X-ray powder diffraction profile fitting using Pearson-VII function is shown in Fig. 8.13. Both the pseudo-Voigt and Pearson-VII functions are also centrosymmetric.

The argument, x, in each of the four empirical functions establishes the location of peak maximum, which is obviously observed when $x = 0$ and $2\theta_i = 2\theta_k$. A second parameter, determining the value of the argument, is the full width at half maximum, H. The latter varies with 2θ and its dependence on the Bragg angle is most commonly represented by an empirical peak-broadening function, which has three free parameters U, V, and W (except for the pure Lorentzian, which usually has only two free parameters). Peak-broadening parameters are refined during the profile

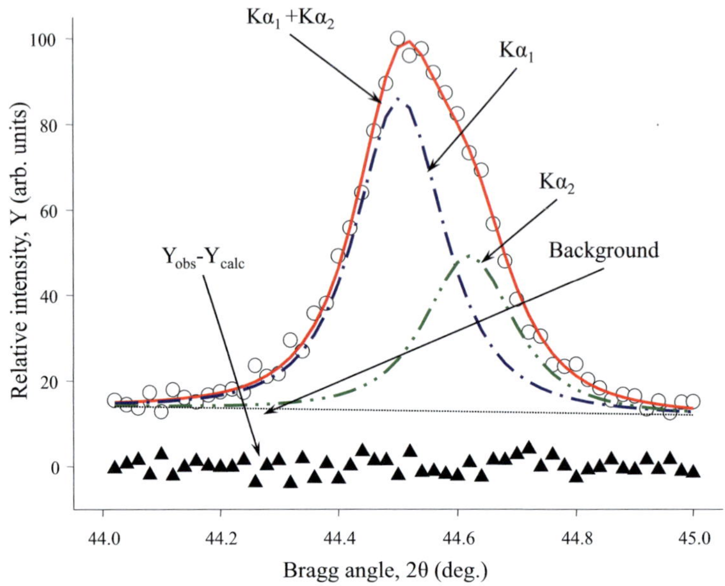

Fig. 8.13 The example of using Pearson-VII function to fit experimental data (*open circles*) representing a single Bragg peak containing $K\alpha_1$ and $K\alpha_2$ components

fitting. Hence, in the most general case, the peak full width at half maximum at a specific 2θ angle is represented as

$$H = \sqrt{U \tan^2\theta + V \tan\theta + W} \tag{8.26}$$

As an example, the experimentally observed behavior of FWHM for a standard reference material SRM-660 (LaB_6) is shown in Fig. 8.14 together with the corresponding interpolation using Eq. (8.26), both as functions of the Bragg angle, 2θ, rather than tanθ.

It is worth noting that the Lorentzian broadening function (H') parameters, X and Y, have the same dependence on Bragg angle as crystallite size- and microstrain-related broadening (compare Eqs. 8.19 and 8.20 with Eq. 8.23 and following explanation of notations). Therefore, when Bragg peaks are well-represented by Lorentz distribution, these physical characteristics of the specimen can be calculated from FWHM parameters after the instrumental and wavelength dispersion parts are subtracted.

The mixing coefficient, η for pseudo-Voigt function and the exponent, β for Pearson-VII function, generally vary for a particular powder diffraction pattern. Their behavior is typically modeled with a different empirical parabolic function of tanθ and 2θ, respectively, as follows from Eqs. (8.24) and (8.25). Peak shapes in the majority of routinely collected X-ray diffraction patterns are reasonably well-

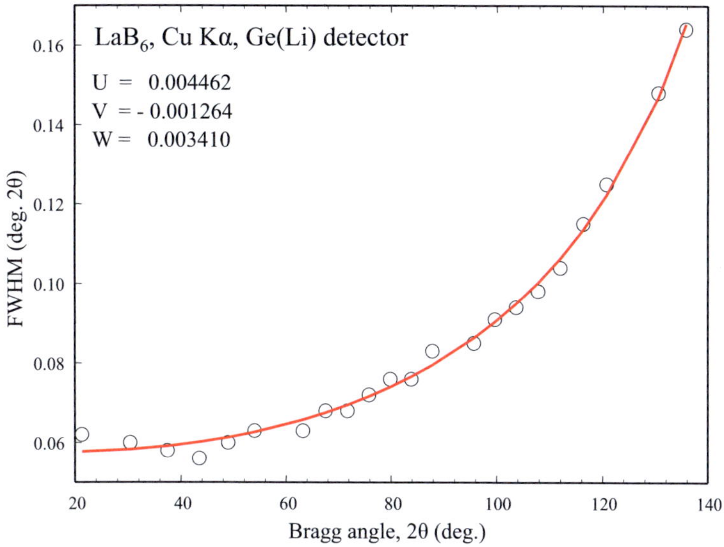

Fig. 8.14 Experimentally observed full width at half maximum of LaB_6 (*open circles*) as a function of 2θ. The solid line represents a least-squares fit and is calculated using Eq. (8.26) and $U = 0.004462$, $V = -0.001264$, and $W = 0.003410$

represented using pseudo-Voigt and/or Pearson-VII functions. Conversely, noticeable improvements in the experimental powder diffraction techniques, which occurred in the last decade, resulted in the availability of exceptionally precise and high-resolution data, especially when employing synchrotron radiation sources, where the use of these relatively simple functions is no longer justified. Furthermore, the ever-increasing computational power facilitates the development and utilization of advanced peak-shape functions, including those that extensively use numerical integration.

Most often, various modifications of the pseudo-Voigt function are employed to achieve improved precision, enhance the asymmetry approximation, account for the anisotropy of Bragg peak broadening, etc. For example, a total of four different functions (not counting those for the time-of-flight experiments) are employed in GSAS[17] or GSAS-II.[18]

The first function is the pure Gaussian (Eq. 8.22), which is suitable for neutron powder diffraction data.

[17] C.A. Larson and R.B. Von Dreele, GSAS: General structure analysis system. LAUR 86–748 (2004). The user manual and software are freely available at: https://subversion.xray.aps.anl.gov/trac/EXPGUI/wiki/GSASIntro

[18] B.H. Toby and R.B. Von Dreele, GSAS-II: the genesis of a modern open-source all purpose crystallography software package, **46**, 544 (2013). https://doi.org/10.1107/S0021889813003531

The second is a modified pseudo-Voigt (the so-called Thompson-modified pseudo-Voigt),[19] where the function itself remains identical to Eq. (8.24) but it employs a multi-term Simpson's integration introduced by C.J. Howard.[20] Its FWHM (H) and mixing (η) parameters are modeled as follows:

$$H = \left(\sum_{i=0}^{5} a_i H_G^{5-i} H_L^i \right)^{1/5} \tag{8.27}$$

$$\eta = \sum_{i=1}^{3} b_i \left(\frac{H_L}{H} \right)^i \tag{8.28}$$

where a_i and b_i are tabulated coefficients. Further,

$$H_G = 2\sigma\sqrt{2 \ln 2} \tag{8.29}$$

$$\sigma = \sqrt{U \tan^2\theta + V \tan\theta + W + P/\sin^2\theta} \tag{8.30}$$

$$H_L = (X + X_a \cos\phi)/\cos\theta + (Y + Y_a \cos\phi)\tan\theta \tag{8.31}$$

and H_G is the Gaussian full width at half maximum modified by an additional broadening parameter, P; H_L is Lorentzian full width at half maximum, which accounts for the anisotropic FWHM behavior by introducing two anisotropic broadening parameters, X_a (crystallite size) and Y_a (strain), and ϕ is the angle between a common anisotropy axis and the corresponding reciprocal lattice vector.

The major benefit achieved when using the modified pseudo-Voigt function is in the separation of FWHMs due to Gaussian and Lorentzian contributions to the peak-shape function. They represent two different effects contributing to the combined peak width, which are due to the instrumental (Gauss) and specimen (Lorentz) broadening. The specimen broadening parameters X and Y, being coefficients of $1/\cos\theta$ and $\tan\theta$, could be directly associated with the crystallite size and microstrain, respectively. Anisotropic broadening can be refined using two additional parameters, X_a and Y_a. The crystallite size (p) in Å can be obtained from these parameters as follows:

$$p_{iso} = p_{\perp} = \frac{180K\lambda}{\pi X} \quad \text{and} \quad p_{\parallel} = \frac{180K\lambda}{\pi(X + X_a)} \tag{8.32}$$

and microstrain (s) in percent as:

[19] P. Thompson, D.E. Cox, and J.B. Hastings, Rietveld refinement of Debye–Scherrer synchrotron X-ray data from $A1_2O_3$, J. Appl. Cryst. **20**, 79 (1987). https://doi.org/10.1107/S0021889887087090

[20] C.J. Howard, The approximation of asymmetric neutron powder diffraction peaks by sums of Gaussians, J. Appl. Cryst. **15**, 615 (1982). https://doi.org/10.1107/S0021889882012783

$$s_{iso} = s_{\perp} = \frac{\pi}{180}(Y - Y_{instr}) \cdot 100\% \quad \text{and} \quad s_{\parallel} = \frac{\pi}{180}(Y + Y_a - Y_{instr}) \cdot 100\% \tag{8.33}$$

where the subscript *iso* indicates isotropic parameters, ⊥ and || denote parameters that are perpendicular and parallel, respectively, to the anisotropy axis, K is the Scherrer constant,[21] and Y_{instr} is the instrumental part in the case of strain broadening.

The third function used in GSAS, is similar to the second function, as described in Eqs. (8.27)–(8.31). However, it fits real Bragg peak shapes better due to improved handling of asymmetry, which is treated in terms of axial divergence.[22] This function is formed by a convolution of pseudo-Voigt with the intersection of the diffraction cone and a finite receiving slit length using two geometrical parameters, S/L and D/L, where S and D are the sample and the detector slit dimensions in the direction parallel to the goniometer axis, and L is the goniometer radius. These two parameters can be measured experimentally or refined (after being suitably constrained because D and S are identical in a typical powder diffraction experiment) when low Bragg angle peaks are present. This peak-shape function also supports an empirical extension of microstrain anisotropy described by six parameters. The result is added to Y in the second part of Eq. (8.31) as $\gamma_L d^2$, where:

$$\gamma_L = \gamma_{11}h^2 + \gamma_{22}k^2 + \gamma_{33}l^2 + \gamma_{12}hk + \gamma_{13}hl + \gamma_{23}kl \tag{8.34}$$

Thus, the total number of parameters for this peak-shape function is 19.

The fourth function is also a modified pseudo-Voigt, and it accounts for anisotropic microstrain broadening as suggested by P. Stephens[23]:

$$H_S = = \left(\sum_{HKL} S_{HKL} h^H k^K l^L \right)^{1/2} \tag{8.35}$$

where S_{HKL} are coefficients, and H, K, and L (these are not the same as Miller indices *hkl*), represent permutations of positive integers restricted to H + K + L = 4. These coefficients are further restricted by Laue symmetry so that a total of 2 in the cubic crystal system to 15 coefficients in the triclinic crystal system may be used to describe strain broadening. The latter contributes to Gaussian and Lorentzian broadening by adding $\sigma_{S^2} d^4$ and $\gamma_S d^2$ to U and Y in Eqs. (8.26) and (8.31), respectively.

[21] K is known as the shape factor or Scherrer constant which varies in the range $0.89 < K < 1$ and usually $K = 0.9$ [H.P. Klug and L.E. Alexander, X-ray diffraction procedures for polycrystalline and amorphous materials, Second edition, John Wiley, NY (1974) p. 656].

[22] L.W. Finger, D.E. Cox, A.P. Jephcoat, A correction for powder diffraction peak asymmetry due to axial divergence, J. Appl. Cryst. **27**, 892 (1994). https://doi.org/10.1107/S0021889894004218

[23] P.W. Stephens, Phenomenological model of anisotropic peak broadening in powder diffraction, J. Appl. Cryst. **32**, 281 (1999). https://doi.org/10.1107/S0021889898006001

Here, $\sigma_S = (1 - \eta)H_S$, and $\gamma_S = \eta H_S$, where η is the pseudo-Voigt function mixing parameter as in Eq. (8.24).

Both the third and fourth functions describe asymmetric peaks much better than the first two and the simple pseudo-Voigt (Eq. 8.24), especially at low Bragg angles. The fourth function also serves as an excellent approximation of Bragg peaks, particularly in the presence of significant anisotropic broadening caused by microstrains. When the anisotropy is low, this function is similar to the third one but with a noticeably reduced number of free variables. Thus, the number of fitting parameters for the fourth function depends on the Laue class and it varies from 14 to 27. The number of free variables may be reduced further since the coefficients S_{HKL} have physical meaning, and some of them may be set to known predetermined values (for further details and examples, see the original paper, Footnote 23). The attractiveness of this model is that the anisotropy of microstrains can be visualized as the three-dimensional surface in reciprocal space with radial distances defined as:

$$D_S(hkl) = \frac{d^2}{C}\left(\sum_{HKL} S_{HKL} h^H k^K l^L\right)^{1/2} \tag{8.36}$$

In the modified pseudo-Voigt functions described earlier (Eqs. 8.27–8.31), both the Gaussian to Lorentzian mixing parameter (η, Eq. 8.27) and their individual contributions to the total peak width (*H*, Eq. 8.32) are tabulated. This feature may be used to lower the number of free parameters and to obtain more realistic peak-shape parameters that are due to the physical state of the specimen. Either or both may be achieved by using one of the following approaches:

- Employing a high-quality standard sample (e.g., LaB_6, see the footnote 2 in Chap. 8, p. 177) that has no measurable contributions from small crystallite size and microstrains, the peak-shape function parameters (*V*, *W*, and *P*), responsible for the instrumental and wavelength dispersion broadening, can be determined experimentally. These should remain constant during the following experiments when using different materials and thus should be kept fixed in future refinements. Obviously, the diffractometer configuration must be identical in the experiments conducted using both the standard and real samples. This method requires measuring a standard every time when any change in the experimental settings occurs, including replacement of the X-ray tube, selection of different divergence or receiving slits, monochromator geometry, filter, and other optical components.
- Taking advantage of the fundamental parameters approach, which is based on a comprehensive description of the experimental conditions and hardware configuration. It is developed quite well, and as a result, the corresponding peak-shape parameters may be computed and not necessarily refined. This technique requires realistic data about the experimental configuration, such as slit openings and heights, in-plane and axial divergences, monochromator characteristics, source and sample geometry and dimensions, and other data. Indeed, considerable effort is involved in order to obtain all required physical characteristics of the powder

Ω =

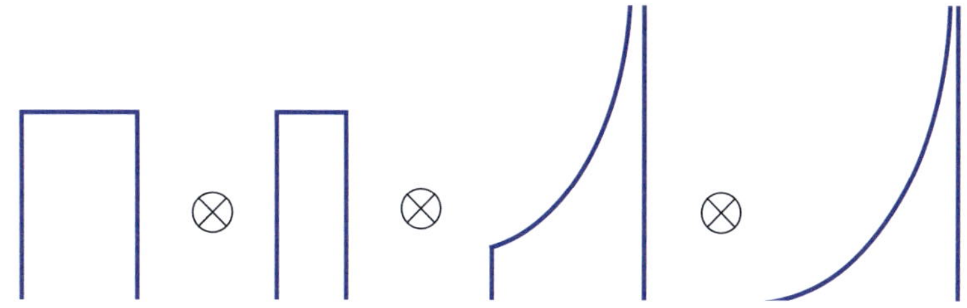

Fig. 8.15 Graphical representation of typical fundamental functions defining convoluted instrumental profile

diffractometer, the source, and the specimen. The resultant peak shape is then obtained as a convolution (Eq. 8.18) of the modeled instrumental function, Ω, wavelength distribution in the incident spectrum, Λ, and sample function, Ψ, with the pseudo-Voigt function.[24] For example, the instrumental function can be obtained by convolution of primitive (fundamental) functions describing effects of the corresponding instrumental characteristics on the peak shape as shown in Fig. 8.15. The fundamental parameters approach is implemented in software packages, such as TOPAS,[25] BGMN,[26] or as standalone tool for line profile analysis,[27] among others. More detailed information about both the technique and its implementation can be found in the corresponding references.[28]

[24] From this point of view, some applications of the modified pseudo-Voigt function (e.g., third and fourth peak-shape functions employed in GSAS) are in a way similar to the fundamental parameters approach as they use instrumental parameters to describe certain aspects of peak shape.

[25] A.A. Coelho, TOPAS and TOPAS-Academic: an optimization program integrating computer algebra and crystallographic objects written in C++, J. Appl. Cryst. 51, 210 (2018), https://doi.org/10.1107/S1600576718000183

TOPAS, Version 6.0, Bruker AXS, Karlsruhe, Germany (2017); A.A. Coeho, TOPAS Academic, Version 7, (2020), http://www.topas-academic.net/

[26] http://www.bgmn.de/

[27] M.H. Mendenhall, K. Mullen, and J.P. Cline, An Implementation of the Fundamental Parameters Approach for Analysis of X-ray Powder Diffraction Line Profiles, J. Res. Natl. Inst. Stand. **120**, 223 (2015), https://doi.org/10.6028/jres.120.014

[28] R.W. Cheary, A.A. Coelho, J.P. Cline, Fundamental parameters line profile fitting in laboratory diffractometers, J. Res. Natl. Inst. Stand. **109**, 1 (2004), https://nvlpubs.nist.gov/nistpubs/jres/109/1/j91che.pdf

R.W. Cheary and A. Coelho, A fundamental parameters approach to X-ray line-profile fitting, J. Appl. Cryst. **25**, 109 (1992), https://doi.org/10.1107/S0021889891010804

R.W. Cheary and A.A. Coelho, Axial divergence in a conventional X-ray powder diffractometer. II. Realization and evaluation in a fundamental-parameter profile fitting procedure, J. Appl. Cryst. **31**, 862 (1998), https://doi.org/10.1107/S0021889898006888

J. Bergmann, R. Kleeberg, A. Haase, and B. Breidenstein, Advanced fundamental parameters model for improved profile analysis, Mater. Sci. Forum **347**, 303 (2002), https://doi.org/10.4028/www.scientific.net/MSF.347-349.303, and references therein.

8.5.2 Peak Asymmetry

All peak-shape functions considered so far were centrosymmetric with respect to their arguments (x), which implies that both the low- and high-angle slopes of Bragg peaks have mirror symmetry with respect to a vertical line intersecting the peak maximum (e.g., see Fig. 8.12). In reality, Bragg peaks are asymmetric due to various instrumental factors, such as axial divergence and nonideal specimen geometry, and due to the nonzero curvature of the Debye rings (e.g., see Fig. 8.4), especially at low Bragg angles. The combined asymmetry effects usually result in the low angle sides of Bragg peaks being considerably broader than their high-angle sides, as illustrated schematically in Fig. 8.16. Peak asymmetry is usually strongly dependent on the Bragg angle and it is most prominently visible at low Bragg angles (2θ below ~20°–30°). At high Bragg angles, peak asymmetry may be barely visible but it is still present.

A proper configuration of the instrument and its alignment can substantially reduce peak asymmetry, but unfortunately, they cannot eliminate it completely. The major asymmetry contribution, which is caused by the axial divergence of the beam, can be successfully controlled by Soller slits, especially when they are used on both the incident and diffracted beam's sides. The length of the Soller slits is critical in handling both the axial divergence and asymmetry; however, the reduction of the axial divergence is usually accomplished at a sizeable loss of intensity.

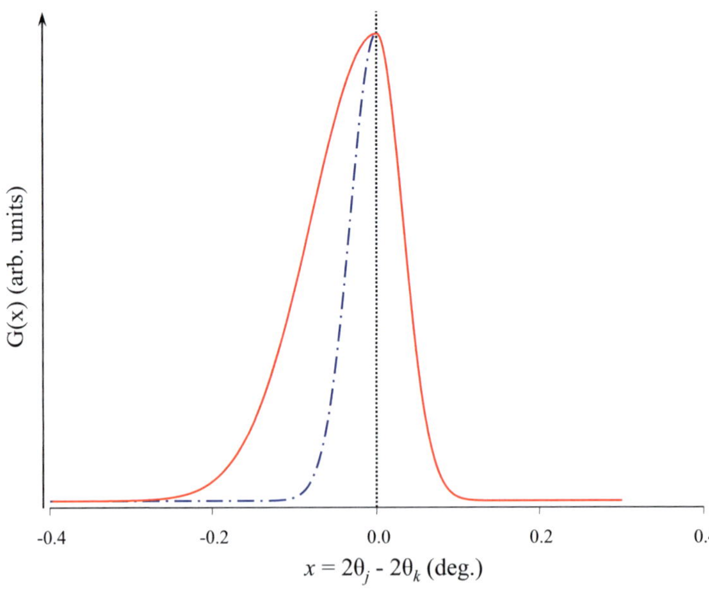

Fig. 8.16 The schematic illustrating the asymmetric Bragg peak (*solid line*) when compared with the symmetric peak composed of the dash-dotted line (*left slope*) and the solid line (*right slope*). Both peaks are modeled by the pure Gauss function (Eq. 8.22) using two different FWHMs on different sides of the peak maximum in the asymmetric case

Since asymmetry cannot be completely eliminated, it should be addressed in the profile-fitting procedure. Generally, there are three ways of treating the asymmetry of Bragg peaks, all achieved by various modifications of the selected peak-shape function:

– In the first method, the symmetry of a function is broken by introducing a multiplier, which increases the intensity on one side from the peak maximum (usually the low Bragg angle side) and decreases it on the opposite side. The same modification of intensities can also be achieved by introducing different peak widths on the opposite sides of the peak maximum, as has been done in Fig. 8.16. The following equation expresses the intensity correction, A, as a function of Bragg angle:

$$A(x_i) = 1 - \alpha \frac{z_i \times |z_i|}{\tan \theta} \tag{8.37}$$

In Eq. (8.37) α is a free variable, or the asymmetry parameter, which is refined during profile fitting and z_i is the distance from the maximum of the symmetric peak to the corresponding point of the peak profile, or $z_i = 2\theta_k - 2\theta_i$. This modification is applied separately to every individual Bragg peak, including $K\alpha_1$ and $K\alpha_2$ components. Since Eq. (8.37) is a simple intensity multiplier, it may be easily incorporated into any of the peak-shape functions considered earlier. In addition, in the case of the Pearson-VII function, asymmetry may be treated differently. It works nearly identical to Eq. (8.37) and all variables have the same meaning as in this equation but the expression itself is different:

$$A(x_i) = 1 + \alpha \frac{z_i^3}{\left(?H^2/C_P^{1/2} + z_i^2\right)^{3/2}} \tag{8.38}$$

where $C_P = 4(2^{1/\beta} - 1)$, see Eq. (8.25).

– Equations (8.37) and (8.38) are quite simple, but they are also far from the perfect in treating peak asymmetry, especially when high-quality powder diffraction data are available. Better results can be achieved by introducing the so-called split pseudo-Voigt or split Pearson-VII functions. Split functions employ two sets of peak-shape parameters (all or only some of them) separately applied to the opposite sides of each peak. For example, in a split Pearson-VII function, a different exponent β and its dependence on the Bragg angle may be used to model the low (left) and high (right) angle sides of the peak, while keeping the same FWHM parameters U, V, and W. This results in a total of nine peak-shape function parameters: U, V, W, $\beta^L{}_0$, $\beta^L{}_1$, $\beta^L{}_2$, $\beta^R{}_0$, $\beta^R{}_1$, and $\beta^R{}_2$, where superscripts ^{L} and ^{R} refer to parameters of the left and right sides of the peak, respectively (see Eq. 8.25 and the following explanation of notations). It is also possible to split the peak width (FWHM parameters), but then a total of 12 parameters should be refined, which is usually an overwhelming number of free variables for an average or even good-quality powder diffraction experiment.

– In some advanced implementations of the modified pseudo-Voigt function, an asymmetric peak can be constructed as a convolution of a symmetric peak shape and a certain asymmetric function, which can be either empirical or based on the real instrumental parameters. For example, as described in Sect. 8.5.1 and using the Simpson's multi-term integration rule, this convolution can be approximated using a sum of several (usually 3 or 5) symmetric Bragg peak profiles:

$$y(x)_{asym} = \sum_{i=1}^{n} g_i y(x)_{sym} \tag{8.39}$$

where n is the number of terms, $n = 3$ or 5; y_{sym} and y_{asym} are modeled symmetric and the resulting asymmetric peak-shape functions, respectively; and g_i are the coefficients describing Bragg angle dependence of the chosen asymmetry parameter. This approach is relatively complex but in the case of high accuracy data (e.g., precision X-ray or synchrotron powder diffraction), it adequately describes the observed asymmetry of Bragg peaks. An even more accurate method employs the modeling of asymmetry by using geometrical parameters responsible for axial divergence (see Sect. 8.5.1; Finger, Cox, and Jephcoat reference on page 171). Nevertheless, lower quality routine powder diffraction patterns to a large extent can be treated using the simpler Eq. (8.37).

8.6 Intensity of Powder Diffraction Peaks

Any powder diffraction pattern is composed of multiple Bragg peaks, which have different intensities in addition to varying positions and shapes. Numerous factors have either central or secondary roles in determining peak intensities. As briefly mentioned in Sect. 8.1 (Table 8.2), these factors can be grouped as: (1) *structural factors*, which are determined by the crystal structure; (2) *specimen factors* owing to its shape and size, preferred orientation, grain size and distribution, microstructure, and other parameters of the sample, and (3) *instrumental factors*, such as properties of radiation, type of focusing geometry, properties of the detector, slit, and/or monochromator geometry.

The two latter groups of factors may be viewed as secondary, so to say, they are less critical than the principal part defining the intensities of the individual diffraction peaks, which is the structural part.[29] Structural factors depend on the internal (or atomic) structure of the crystal, which is described by relative positions of atoms in the unit cell, their types, and other characteristics, such as thermal motion and population parameters. In this chapter, we consider secondary factors in addition to

[29] Some of the external factors, e.g., preferred orientation (see Sect. 8.6.6), may have a tremendous effect on the diffracted intensity. However, all secondary factors have similar or identical effects on the diffracted intensity, regardless of the crystal (atomic) structure of the material.

introducing the concept of the integrated intensity, while Chap. 9 is devoted to the major component of Bragg peak intensity—the structure factor.

8.6.1 Integrated Intensity

Consider the Bragg peak, which is shown in Fig. 8.17, and let us try to answer the question: which quantity most adequately describes its intensity, that is, what is the combined result of scattering from a series of crystallographic planes (*hkl*) or, which is the same, from the corresponding point in the reciprocal lattice? Is it the height of the peak (i.e., the *Y*-coordinate of the highest point)? Is it the area under the peak? Is it something else?

The value of the peak maximum (Y_{max}) is intuitively and often termed as its intensity. It can be easily measured and is indeed used in many applications where relative intensities are compared on a qualitative basis, such as when searching for a similar pattern in a powder diffraction database. This approach to measuring intensity can be considered semi-quantitative at best. However, it is unacceptable when accurate quantitative values are needed because both the instrumental and specimen factors may cause peak broadening, which can vary for Bragg peaks produced by diffraction from different hkl planes, even from the same crystalline material.

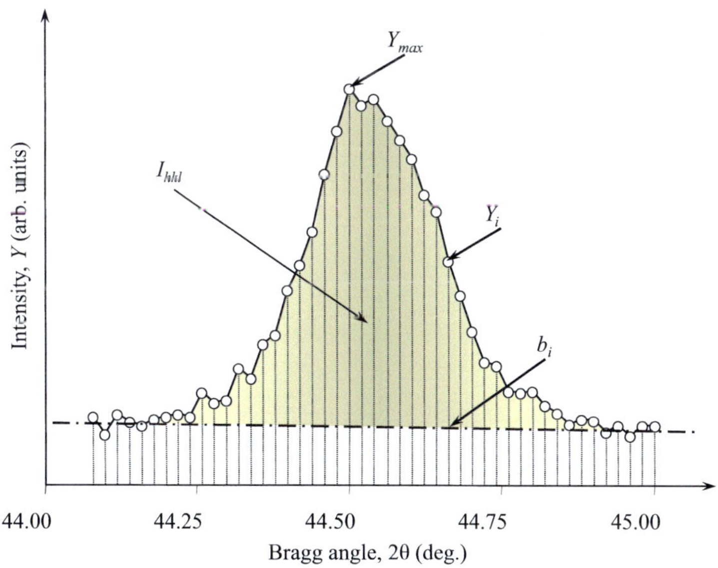

Fig. 8.17 The relationship between the measured shape (*the open circles* connected with *the solid lines*) and the integrated intensity (*shaded area*) of the Bragg peak. The background is shown using *the dash-dotted line*. The maximum measured intensity is indicated as Y_{max}. The measured intensities and the corresponding values of the background are indicated for one of the points as Y_i and b_i, respectively

Conversely, the area under the peak remains unchanged in most cases, even when substantial broadening, especially anisotropic, is present (see Eq. (8.21) indicating that Bragg peak intensity is a multiplier applied to the corresponding peak-shape function, which has unit area). Thus, the shaded area in Fig. 8.17 is known as the integrated intensity, and it represents the true intensity of Bragg peaks in powder diffraction.

The intensity, I_{hkl}, scattered by a reciprocal lattice point (*hkl*) corresponds to the integrated intensity of the matching Bragg peak. For simplicity, it is often called "intensity." In a powder diffraction experiment, the intensity is measured at different points of the powder pattern and is commonly referred as profile intensity. Profile intensity is usually labeled Y_i, where i is the sequential point number, starting from the first measured data point ($i = 1$).

Assuming that powder diffraction data were collected with a constant step in 2θ, the area of an individual peak may be calculated simply by adding the intensities (*Y*-coordinates) of all points measured within the range of the peak after the contribution from the background has been subtracted in every point. The background is shown as a nearly horizontal dash-dotted line in Fig. 8.17. The observed integrated intensity (I_{hkl}) of a Bragg peak (*hkl*) is, therefore, determined from numerical integration as:

$$\mathrm{I}_{hkl} = \sum_{i=1}^{j} \left(Y_i^{obs} - b_i \right) \tag{8.40}$$

where j is the total number of data points measured within the range of the peak.[30]

The integrated intensity is a function of the atomic structure, and it also depends on multiple factors, such as certain specimen and instrumental parameters. Considering Eq. (7.7) and after including necessary details, earlier grouped as "geometrical" effects, the calculated integrated intensity in powder diffraction is expressed as the following product:

$$I_{hkl} = K \times p_{hkl} \times L_\theta \times P_\theta \times A_\theta \times T_{hkl} \times E_{hkl} \times |F_{hkl}|^2 \tag{8.41}$$

where

[30] Strictly speaking, each Bragg peak begins and ends when its contribution becomes indistinguishable from that of the background. The determination of peak range at its base presents a challenging numerical problem since (1) diffracted intensity is always measured with a finite error; (2) it is nearly impossible to achieve zero background; and (3) Bragg peaks often overlap with one another. Thus, for all practical purposes, the beginning and the end of any Bragg peak (i.e., its width at the base) is usually assumed in terms of a certain number of full widths at half maximum to the left and to the right from peak maximum. For Bragg peaks, which are well represented by pure Gaussian distribution, the number of FWHMs can be limited to 2–3 on each side, while in the case of nearly Lorentzian distribution this number should be increased substantially (see Fig. 8.12). In some instances, the number of FWHMs can reach 10–20. It is also possible to define peak limits in terms of maximum intensity, for example, a peak extends only as far as profile intensity (Y_i) remains greater or equal than a certain small, predetermined fraction of the maximum intensity (Y_{max}).

- K is the scale factor, which is a multiplier required to normalize experimentally observed integrated intensities with absolute calculated intensities. Absolute calculated intensity is the total intensity scattered by the content of one unit cell in the direction (θ) defined by the length of the corresponding reciprocal lattice vector. Therefore, the scale factor is a constant for a given phase and it is determined by the number, spatial distribution, and states of the scattering centers (atoms) in the unit cell.
- p_{hkl} is the multiplicity factor, which is a multiplier which accounts for the presence of multiple symmetrically equivalent points in the reciprocal lattice, or in other words, the number of symmetrically equivalent reflections.
- L_θ is Lorentz factor defined by the geometry of diffraction.
- P_θ is the polarization factor, which accounts for a partial polarization of the scattered electromagnetic wave (see the footnote 5 and Thomson's equation in Sect. 7.1.1, page 158).
- A_θ is the absorption correction, which accounts for absorption of both the incident and diffracted beams and nonzero porosity of the powdered specimen.
- T_{hkl} is the preferred orientation factor, which accounts for possible deviations from a complete randomness in the distribution of grain orientations.
- E_{hkl} is the extinction multiplier accounting for deviations from the kinematical diffraction model. In powders, these are quite small and the extinction factor is nearly always neglected.
- F_{hkl} is the structure factor (or the structure amplitude) defined by the crystal structure of the material, including the coordinates and types of atoms, their distribution among different lattice sites, and their thermal motion.

The subscript *hkl* indicates that the multiplier depends on both the length and direction of the corresponding reciprocal lattice vector $\mathbf{d}^*_{hkl}$. Conversely, the subscript θ indicates that the corresponding parameter is only a function of Bragg angle, and thus, it only depends upon the length of the corresponding reciprocal lattice vector, d^*_{hkl}.

8.6.2 Scale Factor

As described earlier, the amplitude of the wave (and thus, the intensity, see Eqs. 7.1–7.7 and relevant discussion in Sect. 7.1) scattered in a specific direction by a crystal lattice is usually calculated for its symmetrically independent minimum—one unit cell. In order to compare the experimentally observed and the calculated intensities directly, it is necessary to measure the absolute value of the scattered intensity. This necessarily involves

- Measuring the absolute intensity of the incident beam exiting through the slits and reaching the sample.

- Precise account of inelastic and incoherent scattering, and absorption by the sample, sample holder, air, and other components of the system, such as windows of a sample attachment, if any.
- Measuring the portion of the diffracted intensity that passes through receiving slits, monochromator, and detector windows.
- Correction for efficiency of the detector, number of events generated by a single photon, detector proportionality, etc., all of which must be precise and reproducible.
- Knowledge of many other factors, such as the volume of the specimen which participates in scattering of the incident beam, the fraction of the irradiated volume which is responsible for scattering precisely in the direction of the receiving slit, and so on.

Since doing all this is impractical, the comparison of observed and calculated intensities is usually performed after normalizing the observed intensities with respect to the calculated ones using a scale factor. As long as all observed intensities are measured under nearly identical conditions—which is relatively easy to achieve—the scale factor remains constant for each phase and applies to the entire diffraction pattern.

Thus, scattered intensity is conventionally measured using an arbitrary relative scale, and the normalization is usually performed by analyzing all experimental and calculated intensities using a least-squares technique.[31] The scale factor is one of the variables in structure refinement and its correctness is critical in achieving the best agreement between the calculated and observed intensities.[32] Its value is also essential in quantitative analysis of multiple-phase mixtures.

8.6.3 Multiplicity Factor

As we established earlier, a powder diffraction pattern is one-dimensional, while the associated reciprocal lattice is three-dimensional. This results in scattering from multiple reciprocal lattice vectors that are related by symmetry operations and

[31] In certain applications, such as when the normalized structure factors should be calculated (see Sect. 10.2.2), an approximate scale factor must be determined before the crystal structure model is known. This can be achieved through various statistical approaches, taking into account that structure factor for the *000* reflection, denoted as F_{000}, is equal to the number of electrons in the unit cell. For further details, see A.J.C. Wilson, Determination of absolute from relative X-ray intensity data, Nature **150**, 151 (1942), https://doi.org/10.1038/150152a0. However, a detailed examination of this topic is beyond the scope of this book.

[32] The accuracy of the scale factor depends on several parameters. The most critical are: maintaining constant photon flux in the incident beam during measurements at any Bragg angle, ensuring that the volume of material producing scattered intensity remains unchanged, having a practically infinite number of crystallites with completely random orientations, accurately accounting for background, and properly considering X-ray absorption when applicable.

therefore occur at identical Bragg angles. Consider two points in a reciprocal lattice, $00l$ and $00\bar{l}$. By examining Eqs. (8.2) to (8.7), it becomes evident that, in any crystal system, $1/d^2_{00l} = 1/d^2_{00\bar{l}}$. Thus, Bragg reflections from these two reciprocal lattice points are observed at exactly the same Bragg angle.

Now consider the orthorhombic crystal system. Simple analysis of Eq. (8.5) indicates that the following groups of reciprocal lattice points will have identical reciprocal lattice vector lengths and thus are equivalent in terms of the corresponding Bragg angle:

$h00$ and $\bar{h}00$	2 equivalent points
$0k0$ and $0\bar{k}0$	2 equivalent points
$00l$ and $00\bar{l}$	2 equivalent points
$hk0$, $\bar{h}k0$, $h\bar{k}0$ and $\bar{h}\bar{k}0$	4 equivalent points
$h0l$, $\bar{h}0l$, $h0\bar{l}$ and $\bar{h}0\bar{l}$	4 equivalent points
$0kl$, $0\bar{k}l$, $0k\bar{l}$ and $0\bar{k}\bar{l}$	4 equivalent points
hkl, $\bar{h}kl$, $h\bar{k}l$, $hk\bar{l}$, $\bar{h}k\bar{l}$, $\bar{h}\bar{k}l$, $\bar{h}\bar{k}\bar{l}$ and $h\bar{k}\bar{l}$	8 equivalent points

Assuming that the symmetry of the structure is *mmm*, these equivalent reciprocal lattice points have the same intensity, in addition to the identical Bragg angles. Consequently, there is generally no need to calculate intensity separately for each reflection in a group of equivalents. It is sufficient to calculate it for one of the corresponding Bragg peaks and then multiply the calculated intensity by the number of the equivalents in the group, known as the multiplicity factor. The multiplicity factor for specific reflection hkl is, therefore, a number of reflections with different *hkl* indices generated by all possible symmetry operations. In the example considered here (orthorhombic crystal system with point-group symmetry mmm), the following multiplicity factors could be assigned to the following types of reciprocal lattice points:

$p_{hkl} = 2$ for $h00$, $0k0$ and $00l$
$p_{hkl} = 4$ for $hk0$, $0kl$ and $h0l$, and
$p_{hkl} = 8$ for hkl

Reciprocal lattices, and therefore diffraction patterns, are generally centrosymmetric, regardless of whether the corresponding direct lattices are centrosymmetric or not. Thus, pairs of reflections with the opposite signs of indices, (hkl) and ($\bar{h}\bar{k}\bar{l}$)—the so-called Friedel pairs—usually have equal intensity. However, they may differ in the presence of atoms that scatter anomalously (see Sect. 9.1.3). This phenomenon should be considered when evaluating multiplicity factors comprehensively. Relevant details associated with the effects of anomalous scattering on the multiplicity factor are discussed in Sect. 9.2.2.

8.6.4 Lorentz-Polarization Factor

The Lorentz factor accounts for two different geometrical effects and has two components. The first component is due to the finite size of reciprocal lattice points and the finite thickness of the Ewald sphere. The second component arises from the variable radii of the Debye rings. Both components are functions of θ.

Usually, the first component is derived by considering a reciprocal lattice rotating at a constant angular velocity around its origin. Under these conditions, various reciprocal lattice points are in contact with the surface of the Ewald sphere for different periods of time. Shorter reciprocal lattice vectors remain in contact with the sphere for longer periods compared to longer vectors. In powder diffraction, this contribution arises from the varying density of the equivalent reciprocal lattice points resting on the surface of the Ewald sphere, which is a function of d^*. It can be shown that the first component of the Lorentz factor is proportional to $1/\sin\theta$.

The second component accounts for a fixed length of the receiving slit. However, the portion of the Debye ring that passes through the slit depends on the radius of the ring (r_D), which is proportional to $\sin 2\theta$.[33] Since the scattered intensity is distributed evenly along the circumference of the ring, the intensity that reaches the detectors becomes inversely proportional to r_D and, therefore, directly proportional to $1/\sin 2\theta$.

The two factors combined result in the following proportionality

$$L \propto \frac{1}{\sin\theta \sin 2\theta} \tag{8.42}$$

which after recalling that $\sin 2\theta = 2\sin\theta\cos\theta$ and ignoring all constants (which are absorbed by the scale factor) becomes

$$L = \frac{1}{\cos\theta \sin^2\theta} \tag{8.43}$$

The polarization factor arises from partial polarization of the electromagnetic wave after scattering. Considering the orientation of the electric vector, the partially polarized beam can be represented by two components: one with its amplitude parallel ($A_{\parallel}$) to the goniometer axis and another with its amplitude perpendicular ($A_{\perp}$) to the same axis. The diffracted intensity is proportional to the square of the amplitude, and the two projections of the partially polarized beam on the diffracted wavevector are proportional to 1 for $(A_{\parallel})^2$ and $\cos^2 2\theta$ for $(A_{\perp})^2$. Thus, partial polarization after scattering yields the following overall factor (also see Thomson equation in the footnote 5 on page 158, Chap. 7):

[33] This proportionality holds as long as the distance between the specimen and the receiving slit of the detector remains constant at any Bragg angle.

$$P \propto \frac{1 + \cos^2 2\theta}{2} \tag{8.44}$$

When a monochromator is employed, it introduces additional polarization, which is accounted as:

$$P \propto \frac{1 - K + K \cdot \cos^2 2\theta \cdot \cos^2 2\theta_M}{2} \tag{8.45}$$

where $2\theta_M$ is the Bragg angle of the reflection from a monochromator (it is a constant for a fixed wavelength), and K is the fractional polarization of the beam. For neutrons, $K = 0$; for unpolarized and unmonochromatized characteristic X-ray radiation, $K = 0.5$ and $\cos 2\theta_M = 1$. For a monochromatic or synchrotron radiation, K should be established experimentally (i.e., measured) or refined.

The Lorentz and polarization contributions to the scattered intensity are nearly always combined together in a single Lorentz-polarization factor, which in the case when no monochromator is employed is given as:

$$LP = \frac{1 + \cos^2 2\theta}{\cos\theta \cdot \sin^2\theta} \tag{8.46}$$

or assuming $K = 0.5$ with a crystal monochromator

$$LP = \frac{1 + cos^2 2\theta \cos^2 2\theta}{\cos\theta \cdot \sin^2\theta} \tag{8.47}$$

Once again, all constant multipliers have been ignored in Eqs. (8.46) and (8.47). The Lorentz-polarization factor is strongly dependent on the Bragg angle as shown in Fig. 8.18. It is near its minimum between 80° and 120° 2θ and increases substantially at both low and high angles. Angles above approximately 150° 2θ are usually out of range in most routine powder diffraction experiments. As shown in Fig. 8.18, additional polarization caused by the presence of a monochromator results in a small change in the behavior of the Lorentz-polarization factor. However, it must be properly accounted for, especially when precision of diffraction data is high.

8.6.5 Absorption Factor

Absorption effects in powder diffraction depend on both the geometry and properties of the sample, as well as the focusing method. For example, when a flat sample is studied using the Bragg–Brentano technique, the scattered intensity is not affected by absorption as long as the specimen is highly impermeable, homogeneous, and thick enough so that the incident beam never penetrates all the way through the sample at any Bragg angle. In contrast, absorption by a thick flat sample in the

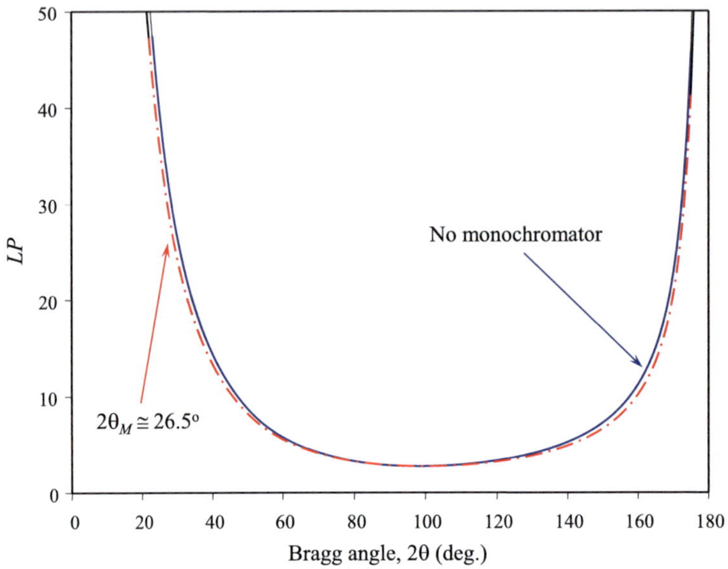

Fig. 8.18 Lorentz-polarization factor as a function of Bragg angle: the solid line represents calculation using Eq. (8.46) (no monochromator), and the dash-dotted line is calculated assuming graphite monochromator and Cu Kα radiation with $K = 0.5$ (Eq. 8.47)

transmission geometry significantly influences the scattered intensity, much more than if a thin sample of the same kind is under examination.

When X-rays penetrate into the matter, they are partially transmitted and partially absorbed. Thus, when an X-ray beam travels the infinitesimal distance, dx, its intensity is reduced by the infinitesimal fraction dI/I (Fig. 8.19a), which can be defined using the following differential equation:

$$\frac{dI}{I} = -\mu dx \tag{8.48}$$

where μ is the proportionality coefficient expressed in the units of the inverse distance, usually in cm^{-1}. This coefficient is also known as the linear absorption coefficient of a material.

The linear absorption coefficient of any chemical element is a function of the wavelength (photon energy). Both μ(λ) and μ(E) dependencies of Fe and Gd are shown in Fig. 8.20. In the range of wavelengths, which are of interest to powder diffraction, the μ(λ) functions consist of several continuous branches separated by abrupt changes in the absorption properties at certain element-specific wavelengths. The points, at which the discontinuities of the absorption coefficient occur, are called absorption edges. A single absorption edge—the K edge—is observed at the shortest wavelength, the next set of edges is called L, followed by the M edges, and so on. The number of absorption edges increases as the electronic structure of the

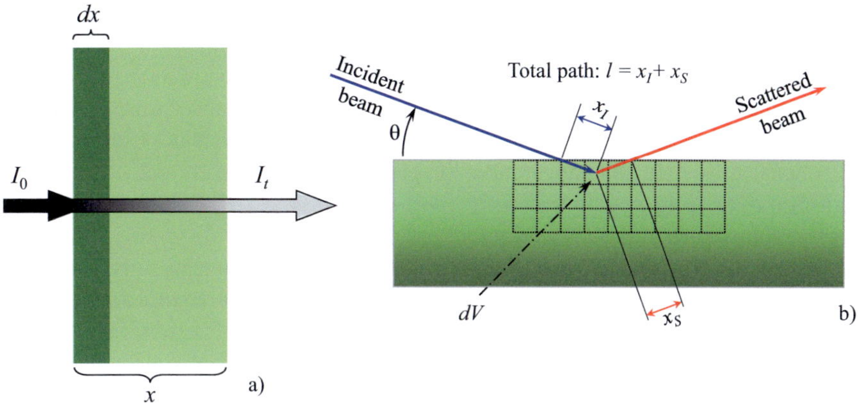

Fig. 8.19 Schematic explaining the phenomenon of absorption of X-rays by the matter (**a**) and the illustration of the derivation of Eq. (8.51) (**b**). The incident beam penetrates into the sample by the distance x_I before being scattered by the infinitesimal volume dV. The scattered beam traverses the distance x_S before exiting the sample. In the Bragg–Brentano geometry, $x_I = x_S$

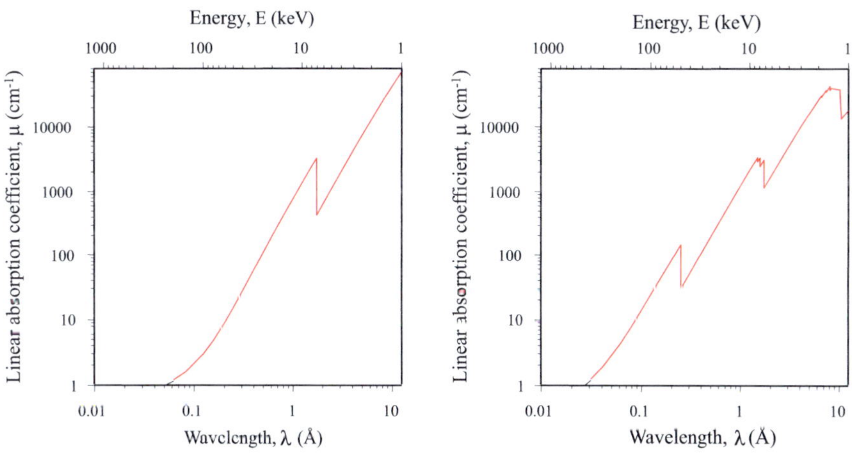

Fig. 8.20 The behavior of the linear absorption coefficients of the elemental iron (**a**) and gadolinium (**b**) as functions of the wavelength of the X-rays (bottom scale) and photon energy (top scale). The numerical data used to prepare both plots have been taken from the National Institute of Standards and Technology Physical Reference Data web page (https://www.nist.gov/pml/x-ray-mass-attenuation-coefficients)

element becomes more complex. For example, when $0 < \lambda$ B 13 Å, iron has a single K-absorption edge, but Gd in addition to the K edge has three L edges, followed by five M absorption edges. The linear absorption coefficient changes its value by a factor of 6–8 at the K edge; the relative changes become much smaller at the majority of L and M edges.

The continuous change of the liner absorption along each of the two branches is approximately defined as $\mu = kZ^3\lambda^3$, where Z is the atomic number of the chemical element and k is a constant, specific for each of the two continuous parts of the absorption function. The continuous branches correspond to the absorption occurring due to random scattering of photons by electrons, which is observed in all directions, thus reducing the number of photons in the transmitted beam in the direction of the propagation vector.

The appearance of the discontinuities is known as the true absorption, and it can be understood by considering Eq. (6.3). As the wavelength increases, the energy of the X-ray photons decreases and at a certain λ it matches the energies required to excite K electrons from their ground states for the K edge, L electrons for the L edges, M electrons for the M edges, and so on. This not only causes a rapid increase in the number of the absorbed photons but also results in the transitions of upper-level electrons to vacant K (L, M, ...) levels in the atoms of the absorber—a photoelectric effect, during which a fluorescent X-ray photon can be emitted in any direction. Both scattered and true absorption result in the reduction of the transmitted intensity, as defined by Eq. (8.48).

Absorption coefficients for all chemical elements are usually tabulated (see Table 8.3) in the form of mass absorption coefficients,[34] μ/ρ (the units are cm^2/g), instead of the linear absorption coefficients, μ.

The linear absorption coefficient of any material (solid, liquid, or gas) is then calculated as

$$\mu = \rho_m \sum_{i=1}^{n} w_i \left(\frac{\mu}{\rho}\right)_i \tag{8.49}$$

where w_i is the mass fraction of the chemical element in the material, $(\mu/\rho)_i$ is elemental mass absorption coefficient, and ρ_m is the density of the material. For example, the liner absorption coefficient of the stoichiometric mixture of gaseous hydrogen and oxygen (2 moles of H_2 per 1 mole of O_2) is only ~1/1200 of the linear absorption coefficient of water because the density of water is just over 1200 times greater than the density of the mixture of the two gases at atmospheric pressure and room temperature.

After integrating Eq. (8.48), the transmitted intensity (I_t) is easily calculated in terms of a fraction of the initial intensity of the beam (I_0):

$$I_t = I_0 \exp(-\mu x) \tag{8.50}$$

[34]This is reasonable because absorption of X-rays is proportional to the probability of a photon to encounter an atom when passing through matter. This probability is directly proportional to the number of atoms in the unit volume, i.e., to the density of the material.

Table 8.3 Mass absorption coefficients (in cm^2/g) of selected chemical elements for the commonly used anode materials[a]. The mass absorption coefficients of the best β-filter elements (see Sect. 11.2.2) for the corresponding anode materials are shaded

Element \ Anode	Cr		Fe		Cu		Mo	
	Kα	Kβ	Kα	Kβ	Kα	Kβ	Kα	Kβ
H	0.412	0.405	0.400	0.396	0.391	0.388	0.373	0.370
He	0.498	0.425	0.381	0.335	0.292	0.268	0.202	0.197
…								
C	15.0	11.2	8.99	6.68	4.51	3.33	0.576	0.458
N	24.7	18.6	14.9	11.0	7.44	5.48	0.845	0.645
O	37.8	28.4	22.8	17.0	11.5	8.42	1.22	0.908
…								
Sc	516	403	332	256	180	137	20.8	14.9
Ti	590	444	358	277	200	152	23.4	16.8
V	96.5	479	399	309	219	166	26	18.7
Cr	86.8	67.0	492	385	247	185	29.9	21.5
Mn	97.5	75.3	61.6	375	270	207	33.1	23.8
Fe	113	86.9	71.0	54.3	302	232	37.6	27.1

[a] Taken from the International Tables for Crystallography, Vol. C, Third Edition, E. Prince, Ed., published jointly with the International Union of Crystallography (IUCr) by Springer (2004)

Hence, intensity of the X-ray beam or other types of radiation is reduced due to absorption after passing through a layer of a material with a finite thickness. Now, consider Fig. 8.19b, where the incident beam is scattered by the infinitesimal volume dV of the flat sample in the reflection geometry. The total path of both the incident and diffracted beams through the sample is $l = x_I + x_S$. Thus, to calculate the effect of absorption in this and in any other geometry, it is necessary to perform the integration over the entire volume of the specimen which contributes to scattering.

Considering Eq. (8.50), the following integral equation expresses the reduction of the diffracted intensity, A, as the result of absorption[35]:

$$A = \frac{1}{V}\int_V \exp(-\mu_{\text{eff}} l)dV \tag{8.51}$$

It is important to recognize that an effective linear absorption coefficient, μ_{eff}, has been introduced into Eq. (8.51) to account for a lower density of dusted or packed powder when compared with the linear absorption coefficient, μ, of the bulk. The latter is usually used in diffraction from single crystals.

[35] In single crystal diffraction, absorption correction is usually applied to the observed intensities, and therefore, A is sometimes called the transmission factor, while the corresponding absorption correction is $A^* = 1/A$.

Equation (8.51) can be solved analytically for all geometries usually employed in powder diffraction.[36] For the most commonly used Bragg–Brentano focusing geometry, the two limiting cases are as follows:

- The material has a very high linear absorption coefficient or is thick enough that the transmission of the incident beam through the sample at any Bragg angle is negligible. In this case, the resultant absorption factor is constant and is typically neglected in Eq. (8.41), as it becomes part of the scale factor:

$$A = \frac{\mu_{\text{eff}}}{2} \tag{8.52}$$

- The material has low linear absorption or the sample is thin so that the incident beam is capable of penetrating all the way through the sample. The absorption correction in this case is a function of Bragg angle as shown in Eq. (8.53). Once again, the constant coefficient $1/2\mu_{eff}$ is omitted since it becomes a part of the scale factor:

$$A = \frac{1 - \exp(-2\mu_{\text{eff}} t / \sin\theta)}{2\mu_{\text{eff}}} \propto 1 - \exp(-2\mu_{\text{eff}} t / \sin\theta) \tag{8.53}$$

In Eq. (8.53), t is the sample thickness. Ignoring the absorption correction, especially when μ and/or t are small, meaning a weakly absorbing or thin sample, results in an underestimated calculated intensity at high Bragg angles. As a result, unphysical (negative) values of thermal displacement parameters are usually obtained.

The major difficulty in applying an absorption correction (Eq. 8.53) arises from usually unknown μ_{eff}. Obviously, the linear absorption coefficient, μ, can be easily calculated when the dimensions of the unit cell and its content are known (see Eq. 8.49), but it is applicable only for a fully dense sample. When a pulverized sample is used (and typically it is), μ_{eff} cannot be determined easily without measuring sample density. Often the combined parameter ($\mu_{\text{eff}}t$) can be refined or estimated and accounted in intensity calculations during Rietveld refinement (Sect. 15.7).

Another problem with pulverized samples is that their packing density varies as a function of the depth. This is known as the porosity effect, and for the Bragg–Brentano geometry, it may be expressed using two different approaches:

[36] Analytical integration necessarily assumes that μ_{eff} remains constant, even though the irradiated area of the specimen surface changes as a function of Bragg angle, as discussed in Chap. 12, Sects. 12.1.3 and 12.2.3. When diffraction from a single crystal is of concern, an analytical solution to this equation is rarely possible. Instead, it is usually integrated numerically using the known dimensions of the single crystal and the orientations of both the incident and diffracted beams with respect to crystallographic axes for each individual reflection *hkl*.

The first has been suggested by Pitschke et al.[37]

$$A = \frac{1 - a_1\left(1/\sin\theta - a_2/\sin^2\theta\right)}{1 - a_1(1 - a_2)} \tag{8.54}$$

and the second by Suortti[38]

$$A = \frac{a_1 + (1 - a_1)\exp(-a_2/\sin\theta)}{a_1 + (1 - a_1)\exp(-a_2)} \tag{8.55}$$

where a_1 and a_2 are two variables that can be refined. Both approximations also account for surface roughness as well as for absorption effects. They give practically identical results and the only difference is that the Suortti formula works better at low Bragg angles, according to Larson and Von Dreele.[39]

Approximations given in Eqs. (8.53)–(8.55) also account for some other effects that distort intensity, such as the improper size of the incident beam, which causes the beam to be broader than the sample at low Bragg angles. The refinement of the corresponding parameters may become unstable due to correlations with structural parameters, such as the scale factor and/or thermal displacement parameters of atoms. Therefore, any of these corrections should be introduced and refined with care.

8.6.6 Preferred Orientation

The conventional theory of powder diffraction assumes a completely random distribution of orientations among the infinite number of crystallites in a specimen used to produce a powder diffraction pattern. In other words, the same fraction of the specimen volume should be in the reflecting position for each Bragg reflection. Strictly speaking, this is possible only when the specimen contains an infinite number of crystallites. In practice, this can be only achieved when the number of crystallites is very large (usually in excess of 10^6–10^7 particles). Nonetheless, even when the number of crystallites approaches infinity, it does not necessarily mean that their orientations are completely random. The external shape of the crystallites also

[37] W. Pitschke, N. Mattern, and H. Hermann, Incorporation of microabsorption corrections into Rietveld analysis, Powder Diffraction **8**, 223 (1993). https://doi.org/10.1017/S0885715600019412

[38] P. Suortti, Effects of porosity and surface roughness on the X-ray intensity reflected from a powder specimen, J. Appl. Cryst. **5**, 325 (1972). https://doi.org/10.1107/S0021889872009707

[39] C.A. Larson and R.B. Von Dreele, General Structure Analysis System (GSAS), Los Alamos National Laboratory Report, LAUR 86–748 (2000). https://subversion.xray.aps.anl.gov/trac/EXPGUI/wiki/GSASIntro

plays an important role in achieving randomness of their orientations in addition to their number.

When the shapes of crystallites are isotropic, achieving a random distribution of their orientations is not problematic, and deviations from an ideal sample are usually negligible. However, quite often the shapes are anisotropic, for example, platelet-like or needle-like. This results in the distinctly nonrandom crystallite orientations due to natural preferences in packing of the anisotropic particles. This nonrandom particle orientation is called preferred orientation, and it can cause considerable distortions of the scattered intensity.

Preferred orientation effects are addressed by introducing the preferred orientation factor in Eq. (8.41) and/or by taking proper care in the preparation of the powdered specimen. The former can be quite difficult and even impossible when preferred orientation effects are severe. Therefore, every attempt should be made to physically increase randomness of particle distributions in the sample to be examined during a powder diffraction experiment. Sample preparation is discussed in Sect. 12.1, and in this section, we discuss the modeling of the preferred orientation using various functions that approximate the radial distribution of the crystallite orientations.

Consider two limiting anisotropic particle shapes: platelet-like and needle-like. When packed in a flat sample holder, platelets tend to align parallel to one another and to the sample surface.[40] Consequently, the number of platelets that are parallel or nearly parallel to the surface is much greater than those that are perpendicular or nearly perpendicular to it. In this case, a specific direction that is perpendicular to the flat sides of the crystallites is called the preferred orientation axis. This axis coincides with a reciprocal lattice vector $\mathbf{d}^{*T}_{hkl}$ that is normal to the flat side of each crystallite. Therefore, the intensity of reflections from reciprocal lattice points with vectors parallel to $\mathbf{d}^{*T}_{hkl}$ is larger than the intensity of reflections produced by any other point of the reciprocal lattice, with minimum for those with reciprocal lattice vectors perpendicular to $\mathbf{d}^{*T}_{hkl}$. This is simply because the distribution of their orientations is highly anisotropic. Preferred orientation in such cases is said to be uniaxial, and the preferred orientation axis is perpendicular to the surface of the flat specimen.

When packed into a flat sample holder, needle-like crystallites also tend to align parallel to the surface.[41] However, the preferred orientation axis, which in this case coincides with the elongated axes of the needles, is parallel to the sample surface. In addition to the nearly unrestricted distribution of needles' axes in the plane parallel to the sample surface (which becomes nearly ideally random when the sample spins around an axis perpendicular to its surface), each needle may be freely rotated around its longest direction. Hence, if the axis of the needle coincides, for example, with the vector $\mathbf{d}^{*T}_{hkl}$, then reflections from reciprocal lattice points with vectors

[40] Also see the schematic shown in Fig. 12.3.

[41] V.K. Pecharsky, L.G. Akselrud, and P.Y. Zavalij, On the method of accounting for preferred orientation (texture) in powder samples during the investigation of atomic structure of materials, Sov. Phys. Crystallogr. **32**, 514 (1987), WOS:A1987K313700008

parallel to $\mathbf{d}_{hkl}^{*T}$ are suppressed to a greater extent and reflections from reciprocal lattice points with vectors perpendicular to $\mathbf{d}_{hkl}^{*T}$ are strongly increased. This example describes the so-called in-plane preferred orientation.

In both cases, the most affected is the intensity of Bragg peaks that correspond to reciprocal lattice points with vectors parallel or perpendicular to $\mathbf{d}_{hkl}^{*T}$, while the effect on the intensity of other Bragg peaks is intermediate. Hence, the preferred orientation effect on the intensity of any reflection hkl can be described as a radial function of the angle ϕ_{hkl} between the corresponding vector $\mathbf{d}_{hkl}^{*T}$ and a specific $\mathbf{d}_{hkl}^{*T}$, which is the preferred orientation direction. The angle ϕ_{hkl} can be calculated from:

$$\cos\phi_{hkl} = \frac{\mathbf{d}_{hkl}^{*T} \cdot \mathbf{d}_{hkl}^{*}}{d_{hkl}^{*T} d_{hkl}^{*}} \tag{8.56}$$

where $\mathbf{d}_{hkl}^{*}$ is the reciprocal lattice vector corresponding to a Bragg peak hkl, and $\mathbf{d}_{hkl}^{*T}$ is the reciprocal lattice vector parallel to the preferred orientation axis. The numerator is a scalar product of the two vectors, and the denominator is a product of their lengths.

The simplest radial function that describes the anisotropic distribution of the preferred orientation factor as a function of angle ϕ_{hkl} is an ellipse (Fig. 8.21a), and the corresponding values of T_{hkl}, which are used in Eq. (8.41), can be calculated using the following expression (see Footnote 41):

$$T_{hkl} = \frac{1}{N} \sum_{i=1}^{N} \left[1 + \left(\tau^2 - 1\right) \cos^2\phi_{hkl}^{i}\right]^{-1/2} \tag{8.57}$$

In Eq. (8.57), the multiplier T_{hkl} is calculated as a sum over all N symmetrically equivalent reciprocal lattice points and τ is the preferred orientation parameter refined against experimental data. The magnitude of the preferred orientation parameter is defined as $T_{\perp}/T_{\parallel}$, where $T_{\perp}$ is the factor for Bragg peaks with reciprocal lattice vectors perpendicular, and $T_{\parallel}$ is the same for those which are parallel to the preferred orientation axis, respectively. In the case of the ellipsoidal preferred orientation function, $T_{\perp}/T_{\parallel}$ is equal to τ for the needles (in-plane preferred orientation) and $1/\tau$ for platelets (axial preferred orientation).

A different approach has been suggested by Dollase,[42] where the preferred orientation factor is represented by a more complex March–Dollase function:

$$T_{hkl} = \frac{1}{N} \sum_{i=1}^{N} \left(\tau^2 \cos^2\varphi_{hkl}^{i} + \frac{1}{\tau} \sin^2\varphi_{hkl}^{i}\right)^{-3/2} \tag{8.58}$$

[42] W.A. Dollase, Correction of intensities for preferred orientation in powder diffractometry: Application of the March model, J. Appl. Cryst. **19**, 267 (1986). https://doi.org/10.1107/S0021889886089458

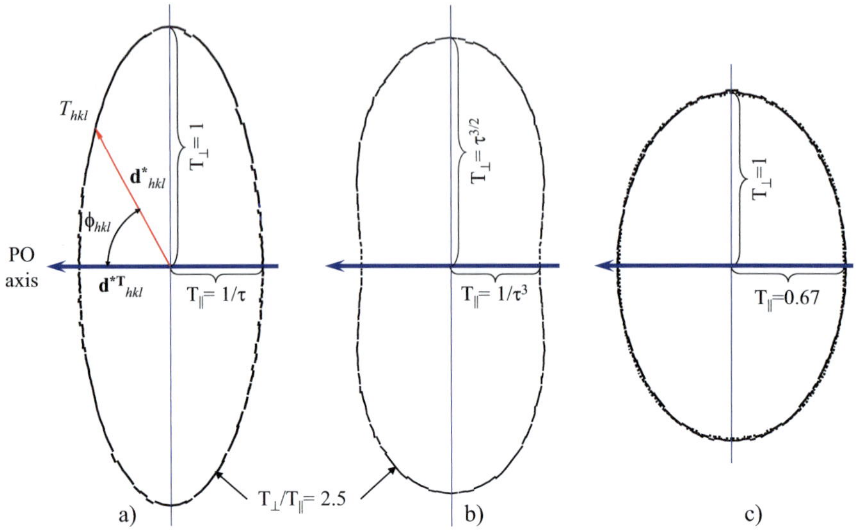

Fig. 8.21 Preferred orientation functions for needles represented by the ellipsoidal (**a**) and March-Dollase (**b**) functions with the magnitude $T_{\perp}/T_{\parallel} = 2.5$, and the two functions overlapped when $T_{\perp}/T_{\parallel} = 1.5$ (**c**). The two notations, $T_{\parallel}$ and $T_{\perp}$, refer to preferred orientation corrections in the directions parallel and perpendicular to the preferred orientation (PO) axis, respectively

Here, the preferred orientation magnitude $T_{\perp}/T_{\parallel}$ is $\tau^{4½}$ for needles and its inverse ($\tau^{-4½}$) is for plates. An example of the March–Dollase preferred orientation function for needles with magnitude $T_{\perp}/T_{\parallel} = 2.5$ is shown in Fig. 8.21b.

In both cases (Eqs. 8.57 and 8.58), the preferred orientation factor T_{hkl} is proportional to the probability of the reciprocal lattice point *hkl* being in the reflecting position (i.e., the probability of being located on the surface of the Ewald sphere). In other words, this factor is proportional to the number of crystallites with *hkl* planes parallel to the surface of the flat sample.

Both approaches work in a similar way. In the case of platelet-like particles, the function is stretched along $T_{\parallel}$ ($T_{\parallel} > T_{\perp}$), while in case of needles, it is stretched along $T_{\perp}$ ($T_{\parallel} < T_{\perp}$). Therefore, in both cases, $\tau < 1$ describes preferred orientation of the platelets, and $\tau > 1$ describes preferred orientation of the needles. Obviously, $\tau = 1$ corresponds to a completely random distribution of reciprocal lattice vectors and the corresponding radial distribution functions become a circle with unit radius (both Eqs. 8.57 and 8.58 result in $T_{hkl} = 1$ for any ϕ_{hkl}).

Both functions give practically the same result at low and moderate degrees of nonrandomness (i.e., at low preferred orientation contribution). The example with $T_{\perp}/T_{\parallel} = 1.5$ is shown in Fig. 8.21c, where the two functions (Eqs. 8.57 and 8.58) are nearly indistinguishable. Unfortunately, strong preferred orientation cannot be adequately approximated by either of these functions, and the best way around it is to reduce the preferred orientation by properly preparing the sample.

The platelets and needles discussed here are the two limiting but still the simplest possible cases. Particles may (and often do) have shapes of ribbons. These particles will pack the same way needles do—parallel to the sample surface but the ribbons will not be randomly oriented around their longest axes. Instead, they will tend to align their flat sides parallel to the sample surface. This case should be treated using two different preferred orientation functions simultaneously: one along the needle and one perpendicular to its flat surface. Thus, both types of functions (Eqs. 8.57 and 8.58) can be modified as follows:

$$T_{total} = k_0 + \sum_{i=1}^{N_a} k_i T_i \tag{8.59}$$

where T_{total} is the overall preferred orientation correction, N_a is the number of different preferred orientation axes, T_i is the preferred orientation correction for the ith axis, and k_i is the corresponding scale factor, which reflects the contribution of each axis. Here, k_0 is the portion of the sample not affected by preferred orientation at all. Equation (8.59) is sometimes used even when only one kind of the preferred orientation is present, thus giving the following very simple expression:

$$T_{total} = k + (1 - k) T_{hkl} \tag{8.60}$$

Yet another approach, which is based on the algorithm described by Bunge,[43] uses spherical harmonics expansion to deal with preferred orientation in three dimensions as a complex radial distribution:

$$T(h, y) = 1 + \sum_{l=2}^{L} \frac{4\pi}{2l + 1} \sum_{m=-l}^{l} \sum_{n=-l}^{l} C_l^{mn} k_l^m(h) k_l^n(y) \tag{8.61}$$

where

- h represents reflection, and y sample orientations
- L is the maximum order of a harmonic
- C_l^{mn} are harmonic coefficients
- $k(h)$ and $k(y)$ are harmonic factors as functions of reflection and sample orientations, respectively

The expression for harmonic factors is complex and is defined azimuthally by means of a Lagrange function. Sample orientation in routine powder diffraction experiment is fixed and so is the corresponding harmonic factor $k(y)$, which simplifies Eq. (8.61) to:

[43] H.-J. Bunge, Texture analysis in materials science, Butterworth, London (1982).

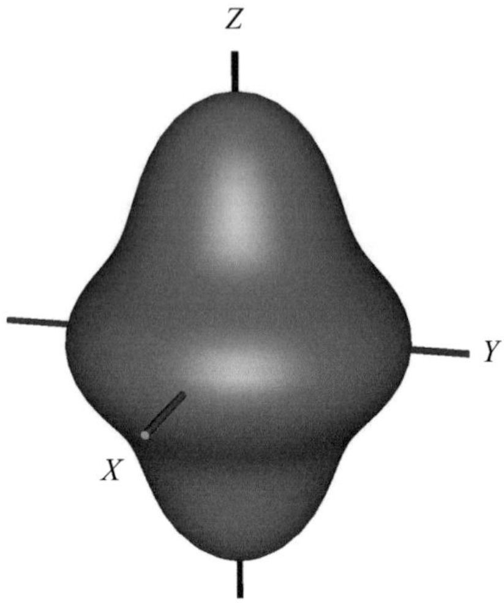

Fig. 8.22 The illustration of the complex distribution of reciprocal lattice vectors modeled using a spherical harmonic preferred orientation function for the (100) reflection

$$T(h) = 1 + \sum_{l=2}^{L} \frac{4\pi}{2l+1} \sum_{m=-l}^{l} C_l^m k_l^m(h) \tag{8.62}$$

The magnitude of the preferred orientation can be evaluated using the following function:

$$J = 1 + \sum_{l=2}^{L} \frac{1}{2l+1} \sum_{m=-l}^{l} \left|C_l^m\right|^2 \tag{8.63}$$

which is unity in the case of random orientation; otherwise, $J > 1$. When all grains are perfectly aligned (single crystal), the function (8.63) becomes infinity.

Only even orders are considered in Eqs. (8.62) and (8.63) due to the presence of the inversion center in the diffraction pattern. The number of harmonic coefficients C and terms $k(h)$ varies, depending on lattice symmetry and desired harmonic order L. Low symmetry results in multiple terms (triclinic has five terms for $L = 2$), and therefore, low orders (2 or 4) are usually sufficient. High symmetry requires fewer terms (e.g., cubic has only one term for $L = 4$), so higher orders may be required to adequately describe preferred orientation. The spherical harmonics approach is realized in GSAS.[44]

An example of the preferred orientation modeled using second- and fourth-order spherical harmonics in the orthorhombic crystal system [space group Cmc2_1, $C_2^0 =$ 0.17(1), $C_2^2 = 1.65(1)$, $C_4^0 = 0.17(2)$, $C_4^2 = 0.04(1)$, $C_4^4 = 0.56(2)$] is shown in Fig. 8.22.

[44] R. B. Von Dreele, Quantitative texture analysis by Rietveld refinement, J. Appl. Cryst. **30**, 517 (1997). https://doi.org/10.1107/S0021889897005918

Here the surface represents the probability of finding the reciprocal lattice point (100) in the diffractometer coordinate system assuming Bragg–Brentano focusing geometry. The *Z*-axis is perpendicular to the sample, and *X*- and *Y*-axes are located in the plane of the sample.

At present, the spherical harmonics approach is the most comprehensive method developed to account for preferred orientation effects. However, in routine experiments, it should be used with great care. The order of expansion should be increased gradually, and only as long as improvements are obvious, and the results make sense. Using an unnecessarily large number of harmonic coefficients may give excellent agreement with experimental data but can lead to overfitting and unrealistic models, resulting in incorrect structural and, especially, atomic displacement parameters. In its full form, Eq. (8.61) may be used in complex texture analysis, where powder diffraction data have been collected, not only as a function of Bragg angle 2θ but also at different orientations along Debye rings and with tilting the sample.

8.6.7 Extinction Factor

Extinction effects, which are dynamical in nature, may be noticeable in diffraction from nearly perfect and/or large mosaic crystals. Two types of extinction are generally recognized: primary, which occurs within the same crystallite, and secondary, which originates from multiple crystallites. Primary extinction is caused by back-reflection of the scattered wave into the crystal and it decreases the measured scattered intensity (Fig. 8.23, left). Furthermore, the re-reflected wave is usually out-of-phase with the incident wave, resulting in destructive interference and further lowering the intensity of the latter. Therefore, primary extinction lowers the observed intensity of very strong reflections from perfect crystals. Especially in

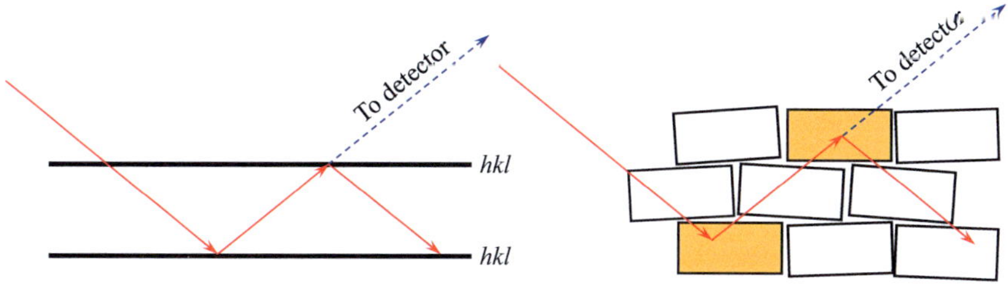

Fig. 8.23 The illustration of primary (*left*) and secondary (*right*) extinction effects, which reduce the intensity of strong reflections from perfect crystals and ideally mosaic crystals, respectively. The solid red lines indicate actual reflections paths, while the dashed blue lines indicate the expected paths that are partially suppressed by dynamical effects. The orange rectangles on the right indicate two different blocks of mosaic with identical orientations

powder diffraction, primary extinction effects are often smaller than experimental errors; however, when necessary, they may be included in Eq. (8.41) as[45]:

$$E_{hkl} = E_B \sin^2\theta + E_L \cos^2\theta \tag{8.64}$$

where E_B and E_L are Bragg ($2\theta = \pi$) and Laue ($2\theta = 0$) components, both defined as various functions of the extinction parameter, x, which is normally a refined variable:

$$x = (KN_c\lambda F_{hkl}D)^2 \tag{8.65}$$

In Eq. (8.65), K is the shape factor (it is unity for a cube of edge D, $K = 3/4$ for a sphere of diameter D, and $K = 8/3\pi$ for a cylinder of diameter D), λ is the wavelength, F_{hkl} is the calculated structure amplitude, and N_c is the number of unit cells per unit volume.

Secondary extinction (Fig. 8.23, right) occurs in a mosaic crystal when the beam, reflected from a crystallite, is re-reflected by a different block of the mosaic that happens to be in the diffracting position with respect to the scattered beam. This dynamical effect is observed in relatively large, nearly perfect mosaic crystals and reduces the measured intensities of strong Bragg reflections, similar to the primary extinction. It is not detected in diffraction from polycrystalline materials and therefore is always neglected.

8.7 Additional Reading

1. International Tables for Crystallography, Vol. A, Space-group symmetry, M. I. Aroyo, Ed., Second online edition (2016). https://it.iucr.org/A/
   International Tables for Crystallography, Vol. C, Mathematical, physical and chemical tables, First online edition, E. Prince, Ed. (2006). https://it.iucr.org/C/
   International Tables for Crystallography, Vol. A-I published jointly with the International Union of Crystallography (IUCr) by Springer, https://it.iucr.org/
2. R.B. Neder and Th. Proffen, Teaching diffraction with the aid of computer simulations, J. Appl. Cryst. **29**, 727 (1996). https://doi.org/10.1107/S002188989600619X
   Th. Proffen and R.B. Neder. Interactive tutorial about diffraction at https://web.pa.msu.edu/cmp/billinge-group/teaching/ibasic_r.html
3. Modern powder diffraction. D.L Bish and J.E. Post, Eds. Reviews in Mineralogy, Vol. 20. Mineralogical Society of America, Washington, DC (1989).

---

[45] T.M. Sabine, R.B. Von Dreele, and J.E. Jørgensen, Extinction in time-of-flight neutron powder diffractometry, Acta Cryst. **A44**, 374 (1988). https://doi.org/10.1107/S0108767388000625; T.M. Sabine, A reconciliation of extinction theories, Acta Cryst. **A44**, 368 (1988). https://doi.org/10.1107/S0108767388000613

4. R. Jenkins and R.L. Snyder, Introduction to X-ray powder diffractometry. Wiley, New York (1996).
5. T.M. Sabine, The flow of radiation in a polycrystalline material, in: The Rietveld method. IUCr monographs on crystallography 5, R.A. Young, Ed., Oxford University Press, Oxford (1993).
6. P. Suortti, Bragg reflection profile shape in X-ray powder diffraction patterns, in: The Rietveld method. IUCr monographs on crystallography 5, R.A. Young, Ed., Oxford University Press, Oxford (1993).
7. R. Delhez, T.H. de Keijser, J.I. Langford, D. Louër, E.J. Mittemeijer, and E.J. Sonneveld, Crystal imperfection broadening and peak shape in the Rietveld method, in: The Rietveld method. IUCr monographs on crystallography 5, R.-A. Young, Ed., Oxford University Press, Oxford (1993).
8. P.S. Whitfield, Spherical harmonics preferential orientation corrections and structure solution from powder diffraction data—a possible avenue of last resort, J. Appl. Cryst. **42**, 134 (2009), https://doi.org/10.1107/S0021889808041149

8.8 Problems

1. Vanadium oxide, V_2O_3, crystallizes in the space-group symmetry $R\bar{3}c$ with lattice parameters a = 4.954 Å and c = 14.00 Å. Calculate the interplanar spacing, *d*, and Bragg peak positions, 2θ, for the 104 (the strongest Bragg peak) and for the 012 (the lowest Bragg angle peak) reflections assuming Cu Kα_1 radiation with $\lambda = 1.5406$ Å.
2. A powder diffractometer ($R = 240$ mm) that you used recently is well maintained. However, the powder diffraction pattern you collected shows some systematic deviations between the calculated and observed positions of Bragg peaks. You plotted the corresponding deviations ($\Delta 2\theta = 2\theta_{obs} - 2\theta_{calc}$) versus different functions of the Bragg angle, which are shown in Fig. 8.24. Which factor is

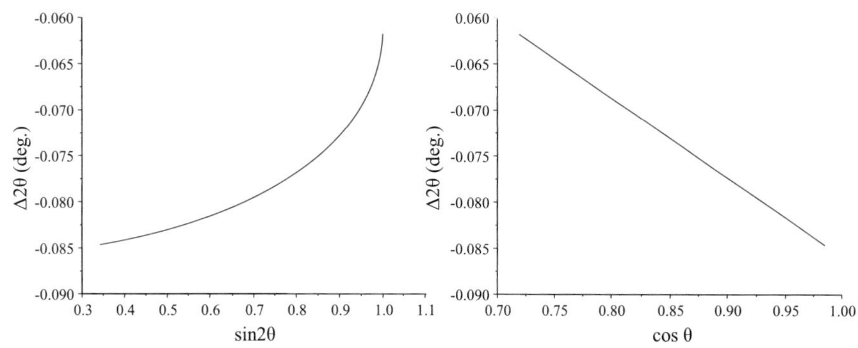

Fig. 8.24 Deviations between $2\theta_{obs} - 2\theta_{calc}$ plotted as functions of sin2θ (*left*) and cosθ (*right*). The estimated values of two data points on the straight line are (0.75, −0.0645) and (0.95, −0.082)

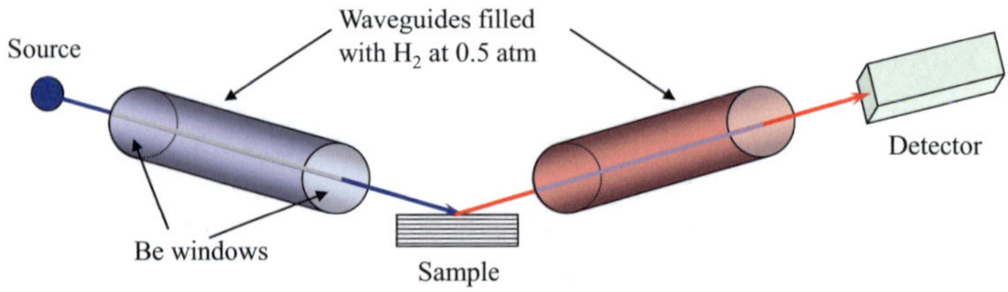

Fig. 8.25 The schematic of a modified powder diffractometer in which both the primary and scattered beams pass through the waveguides filled with hydrogen at $P_{H_2} = 0.5$ atm

primarily responsible for the observed deviations? Estimate the value of the corresponding physical parameter that systematically affects peak positions.

3. Using the refined sample-displacement correction p_5 from Fig. 8.11 and the goniometer ratios (sample-to-detector distance) of 217.5 mm, calculate the sample shift in micrometers. Is it possible to minimize this correction through better sample preparation?
4. What is the multiplicity factor for reflections *hkl* in the monoclinic crystal system in a standard setting? How about reflections *h*0*l* and 0*kl*? Do the multiplicity factors change for the same groups of Bragg reflections in the second allowed setting in the monoclinic crystal system (i.e., when the non 90° angle changes from β to Y)?
5. A powder diffractometer in your laboratory is equipped with a sealed X-ray tube, which has Cr anode. You need to design a β-filter to ensure that the intensity of the Kβ spectral line is less than 0.5% of the intensity of the $K\alpha_1$ part in the characteristic spectrum. Calculate the needed thickness of a foil made from the most appropriate element (which one?) and by how much the intensities of $K\alpha_1$ and Kβ lines are reduced after filtering.
6. Increasing the distance between the X-ray source and the sample and between the sample and the detector improves the resolution of a powder diffraction pattern. However, absorption of X-rays by air is nonnegligible, and increasing these distances usually results in the reduction of registered intensity. In order to minimize absorption by air, researcher considers installing two waveguides—tubes filled with hydrogen gas at 0.5 atm—one along the path of the incident beam and another along the path of the scattered beam. Each waveguide is capped with a pair of 0.1 mm thick Be windows. Schematic of the modifications is shown in Fig. 8.25.
   Assume that X-ray data are collected using Cu Kα radiation at room temperature (20 °C) and normal atmospheric pressure (1 atm) and that the length of each tube is 250 mm. Assume that air is 20 mass% oxygen and 80 mass% nitrogen.
   (a) Compute intensity gain, if any, after the installation of the tubes.
   (b) What is the required minimum length of each tube to gain intensity?
   Find all missing physical quantities using the textbook, relevant handbooks, and/or the web.

Chapter 9
Structure Factor

So far, we have discussed all the prefactors in Eq. (8.41), which depend on multiple parameters except for the crystal structure of the material. The only remaining term is the structure factor, $|F_{hkl}|^2$, which is the square of the absolute value of the structure amplitude, $\mathbf{F}_{hkl}$. This factor encompasses multiple contributions determined by the distribution of atoms in the unit cell and other structural features.

9.1 Structure Amplitude

When the unit cell contains only one atom, the resulting diffracted intensity is only a function of the scattering ability of this atom (see Sect. 7.1.2). However, when the unit cell contains multiple atoms with varying scattering abilities, the amplitude of the scattered wave is described by a complex function known as the structure amplitude:

$$\mathbf{F}(\mathbf{h}) = \sum_{j=1}^{n} g^j t^j(s) f^j(s) \exp\left(2\pi \mathbf{i}\mathbf{h}\cdot\mathbf{x}^j\right) \tag{9.1}$$

where

- $\mathbf{F}(\mathbf{h})$ is the structure amplitude of a Bragg reflection with indices *hkl*, represented as the vector $\mathbf{h}$ in three dimensions. The structure amplitude is often shown in vector form because it is a complex number
- n is the total number of atoms in the unit cell, including all symmetrically equivalent atoms
- s is $\sin\theta_{hkl}/\lambda$
- g^j is the population (or occupation) factor of the *j*th atom ($g^j = 1$ for a fully occupied site)

P. Zavalij, V. Pecharsky, *Fundamentals of Powder Diffraction and Structural Characterization of Materials*, https://doi.org/10.1007/978-3-031-91504-8_9

- t^j is the temperature factor, which describes the thermal motions of the *j*th atom
- $f^j(s)$ is the atomic scattering factor, which describes the interaction of the incident wave with a specific type of atom as a function of $\sin\theta/\lambda$ for X-rays or electrons. For neutrons, it is simply f^j, meaning it is independent of $\sin\theta/\lambda$
- $\mathbf{i} = \sqrt{-1}$
- $\mathbf{h} \cdot \mathbf{x}^j$ is a scalar product of two vectors: $\mathbf{h} = (h, k, l)$ and $\mathbf{x}^j = (x^j, y^j, z^j)$. The latter represents the fractional coordinates of the *j*th atom in the unit cell:

$$\mathbf{h} \cdot \mathbf{x}^j = (h \;\; k \;\; l) \times \begin{pmatrix} x^j \\ y^j \\ z^j \end{pmatrix} = hx^j + ky^j + lz^j \tag{9.2}$$

Taking into account Eq. (9.2), the previous Eq. (9.1) can be expanded as follows:

$$\mathbf{F}_{hkl} = \sum_{j=1}^{n} g^j t^j(s) f^j(s) \exp\left[2\pi\mathbf{i}\left(hx^j + ky^j + lz^j\right)\right] \tag{9.3}$$

9.1.1 Population Factor

Considering a randomly chosen unit cell, each available position with a specific coordinate triplet (x^j, y^j, z^j) may be occupied only by one atom (fractional population parameter $g^j = 1$) or left unoccupied ($g^j = 0$). Even very small crystals contain a nearly infinite number of unit cells. For example, a crystal in the form of a cube with 1 μm sides will contain 10^9 cubic unit cells with $a = 10$ Å. Diffraction is observed from all of the unit cells simultaneously. Hence, the resulting structure amplitude is normalized to a mean unit cell, representing the distribution of atoms averaged over the entire volume of the studied sample. In most compounds, each crystallite is fully ordered, and the content of every unit cell can be assumed to be identical throughout the whole crystal. Therefore, the population factor remains unity for every symmetrically independent atom. Occasionally, the population factor may be less than one but greater than zero. Some of the common reasons for this occurrence are briefly discussed next.

It is possible that in some of the unit cells, some atoms are missing. Thus, instead of complete occupancy, the corresponding site in an average unit cell will contain only a fraction ($0 < g^j < 1$) of the *j*th atom. In such cases, it is said that the lattice has defects and g^j smaller than 1 reflects the fraction of the unit cells where a specific position is occupied. Obviously, a fraction of the unit cells where the same position is empty complements g^j to unity and it is equal to $1 - g^j$.

A different situation arises when in an average unit cell atom *j* is located close to a finite symmetry element, such as a mirror plane. When a mirror plane produces atom j', symmetrically equivalent to *j*, the distance between j' and *j* can become unreasonably small, causing the core electrons of the two atoms overlap. However, in

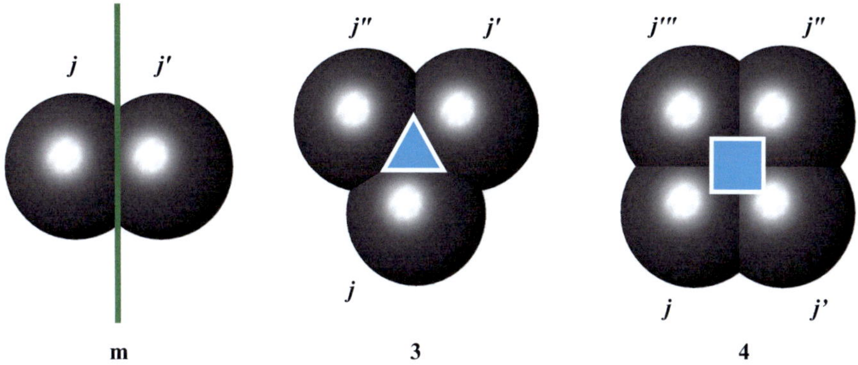

Fig. 9.1 The illustration of forbidden overlaps resulting from an atom being too close to a finite symmetry element: mirror plane (*left*), threefold rotation axis (*middle*), and fourfold rotation axis (*right*). Assuming there are no defects in the crystal lattice, these distributions require $g^j \leq ½, g^j \leq ⅓,$ and $g^j \leq ¼$, respectively

reality, atoms cannot be located at such a close distance (Fig. 9.1, left). This usually means that in some of the unit cells atom *j* is located on one side of the mirror plane, while in others it is positioned on the opposite side of the same plane. If this is the case, then in the absence of "conventional" defects, the population factors g^j and $g^{j'}$ are related as $g^j = 1 - g^{j'}$. However, due to mirror symmetry, g^j and $g^{j'}$ are equal, leading to $g^j = g^{j'}$.

In general, $g^j = 1/n$, where n is the multiplicity of the symmetry element causing the overlap of the corresponding atoms. When the symmetry element is a mirror plane, a twofold rotation axis, or a center of inversion, $n = 2$ and $g^j = 0.5$. For a threefold rotation axis, $n = 3$ and $g^j = ⅓$ and so on (Fig. 9.1, middle and right). When defects are present in addition to the overlap, g^j becomes less than $1/n$.

The condition illustrated in Fig. 9.1 usually occurs when the real symmetry of the unit cell (and the lattice) is lower than the symmetry of the average unit cell as detected from diffraction data. Therefore, the following should be taken into consideration:

- Distribution of atoms *j* and *j*′ in the lattice is random, leading to partial disorder in the crystal. In this case, the average cell can still be used to describe the average crystal structure, even if it has a higher symmetry than the real unit cell.
- Atoms *j* and *j*′ are distributed in the lattice in an ordered fashion. This usually indicates that an incorrect space-group symmetry was used to describe the crystal structure. Therefore, the symmetry should be lowered by modifying or removing those elements that cause atoms to overlap.[1]

[1] For example, overlaps due to a mirror plane may be avoided after doubling the unit cell dimension along the mirror plane and substituting it with a glide plane. When certain symmetry operations are excluded from the space-group symmetry, this may sometimes result in a substantial change, including switching to a lower symmetry crystal system. This approach ensures that atoms are no longer forced into unrealistically close proximity by the symmetry operations, thereby avoiding overlaps and providing a more accurate representation of the crystal structure.

Yet another possibility is the so-called statistical mixing of various atoms at the same site. This may occur when different unit cells contain different types of atoms in the same position.[2] In the most general case, the occupational disorder can be expressed as:

$$g^0 + \sum_{j=1}^{m} g^j = 1 \tag{9.4}$$

where g^0 represents the unoccupied fraction of the unit cells (i.e., defects), and m is the number of different types of atoms that occupy a given site position in different unit cells.

In all cases considered in this section, population factors can be refined, although some may be constrained by symmetry or must be restrained by other relationships. For example, when $g^0 = 0$ and $m = 3$ in Eq. (9.4), the following constraint applies: $g^3 = 1 - g^1 - g^2$. However, when more than two different chemical elements occupy the same site, their occupation factors often exhibit strong correlations. This correlation cannot be reliably refined, especially if the atomic numbers of the elements are similar. In such cases, the occupation factors should remain fixed at the value obtained from other sources, such as elemental analysis. They could be still refined simultaneously, but only when using several diffraction patterns obtained with different wavelengths specially chosen to increase difference in anomalous scattering of the chemical elements occupying the common site. Given the sensitivity of the population parameters to potentially overestimated symmetry or their correlation, it is crucial to analyze their values, especially if they converge into simple fractions, such as ½ or ⅓ (see Fig. 9.1), to ensure occupation factors make physical sense.

9.1.2 Temperature Factor (Atomic Displacement Parameters)

At any temperature higher than absolute zero, certain frequencies in a phonon spectrum of the crystal lattice are excited, resulting in atoms continuous oscillating about their equilibrium positions. To account for these vibrations, the so-called

[2] Strictly speaking, it should be referred to as "practically the same position," since different atoms would interact differently with their surroundings, leading to locally distinct environments unless their volumes and electronic properties are quite similar (e.g., Si and Ge, Fe^{3+} and Co^{3+}, and others). Consequently, in the majority of cases, their positions (coordinates) are at least slightly different. Conversely, the differences in the positions of these atoms can rarely be detected from diffraction experiment and are therefore neglected.

temperature factor is introduced into the general equation (Eq. 9.1) of the structure amplitude.

It is worth noting that according to a recommendation issued by the International Union of Crystallography (IUCr),[3] the corresponding parameters representing the temperature factor should be referred to as "atomic displacement parameters" (ADPs) rather than the commonly used "temperature parameters" or "thermal parameters." This change in terminology is based on the understanding that these parameters, when determined from X-ray diffraction experiment, reflect not only displacements caused by thermal motion but also other effects. These include deformation of the electron density around the atom due to chemical bonding, improperly or not accounted absorption, preferred orientation, porosity, and more. While these factors may influence the structure factor to a lesser degree than thermal motion, they are still significant enough to be considered.

Oscillatory motions of atoms may be quite complex, leading to several levels of approximations in expressing the temperature factor. In its simplest form, the temperature factor of the *j*th atom is represented as:

$$t^j = \exp\left(-B^j \frac{\sin^2\theta}{\lambda^2}\right) \tag{9.5}$$

where B^j is the displacement parameter of the *j*th atom, θ is the Bragg angle at which a specific reflection *hkl* is observed, and λ is the wavelength. This is the so-called isotropic approximation, which assumes equal probability of an atom to deviate in any direction regardless of its environment, treating atoms as diffuse spheres. In Eq. (9.5),

$$B^j = 8\pi^2 \left(\overline{u}^2\right)^j \tag{9.6}$$

$\left(\overline{u}^2\right)^j$ is the root mean square deviation in Å^2 of the *j*th atom from its equilibrium position (*x*, *y*, *z*). In crystallographic literature, it is also common to list isotropic atomic displacement parameters of atoms as $U = B/8\pi^2$ instead of *B*. This isotropic approximation is useful for simplifying calculations but may not fully capture the complexities of atomic displacements in all crystal structures.

Considering Eq. (9.6), isotropic displacement parameters are only physical when they are positive (also see Sect. 9.1.3). Depending on the nature of the material, they usually vary within relatively narrow ranges at room temperature. For inorganic ionic crystals and intermetallic compounds, the typical range of *B* values is approximately 0.5–1 Å^2; for other inorganic and many coordination compounds, *B* varies from 1 to 3 Å^2, while for organic and organometallic compounds and for solvent or

[3] K.N. Trueblood, H.-B. Bürgi, H. Burzlaff, J.D. Dunitz, C.M. Gramaccioli, H.H. Schulz, U. Shmueli, S.C. Abrahams, Atomic Displacement Parameter Nomenclature. Report of a Subcommittee on Atomic Displacement Parameter Nomenclature, Acta Cryst. **A52**, 770–781 (1996). https://doi.org/10.1107/S0108767396005697

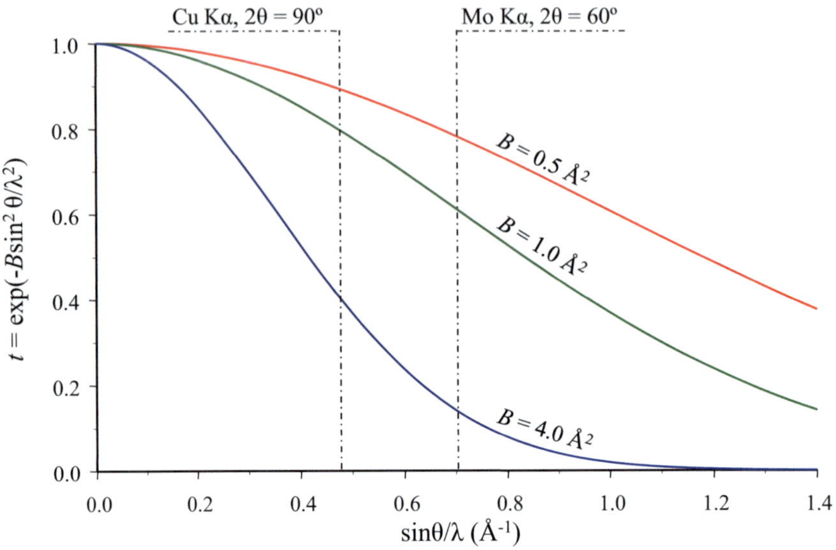

Fig. 9.2 Temperature factor as a function of $\sin\theta/\lambda$ for several different atomic displacement parameters: B = 0.5, 1.0, and 4.0 Å^2. The two vertical *dash-dotted lines* correspond to two commonly employed upper limits of the Bragg angle in diffraction experiments using Cu Kα and Mo Kα radiation

other intercalated nonbonded molecules or atoms, this range extends from 3 to 10 Å^2 or higher. As seen in Fig. 9.2, high value of the atomic displacement parameter results in a rapid decrease in the structure amplitude as the Bragg angle increases (also see Eq. 9.1).

As mentioned at the beginning of this section, the temperature factor absorbs other unaccounted or incorrectly accounted effects. The most critical are absorption, porosity, and other instrumental or sample effects (see Chap. 11), which systematically modify the diffracted intensity as a function of the Bragg angle. As a result, the *B* parameters of all atoms may become negative. If this occurs, the absorption correction should be reevaluated and re-refined, or the experiment should be repeated to minimize the deleterious instrumental influence on the distribution of intensities of Bragg peaks.

The next level of approximation accounts for the anisotropy of thermal motions in a harmonic approximation, describing atoms as ellipsoids, which can be represented in one of the three following forms, which are equivalent to one another:

$$t^j = \exp\left[-\left(\beta^j_{11}h^2 + \beta^j_{22}k^2 + \beta^j_{33}l^2 + 2\beta^j_{12}hk + 2\beta^j_{13}hl + 2\beta^j_{23}kl\right)\right] \quad (9.7)$$

$$t^j = \exp\left[-\frac{1}{4}\left(B^j_{11}h^2a^{*2} + B^j_{22}k^2b^{*2} + B^j_{33}l^2c^{*2} + 2B^j_{12}hka^*b^* + 2B^j_{13}hla^*c^* + 2B^j_{23}klb^*c^*\right)\right] \quad (9.8)$$

$$t^j = \exp\left[-2\pi^2\left(U_{11}^j h^2 a^{*2} + U_{22}^j k^2 b^{*2} + U_{33}^j l^2 c^{*2} + 2U_{12}^j hka^* b^* + 2U_{13}^j hla^* c^* + 2U_{23}^j klb^* c^*\right)\right] \tag{9.9}$$

where $\beta_{11}^{\mathrm{j}}\ldots\beta_{23}^{\mathrm{j}}$, $B_{11}^{\mathrm{j}}\ldots B_{23}^{\mathrm{j}}$, and $U_{11}^{\mathrm{j}}\ldots U_{23}^{\mathrm{j}}$ are the anisotropic atomic displacement parameters (six parameters[4] per atom).

As follows from Eqs. (9.8) and (9.9), the relationships between B_{ij} and U_{ij} are identical to that given in Eq. (9.6), and both are measured in Å^2. The β_{ij} parameters in Eq. (9.7) are dimensionless but may be easily converted into B_{ij} or U_{ij}. Very high-quality powder diffraction data are needed to obtain dependable anisotropic displacement parameters. Even then, they may be reliable only for those atoms that have large scattering factors (see Sect. 9.1.3).[5] Conversely, the refinement of anisotropic displacement parameters is essential when strongly scattering atoms are distinctly anisotropic.

The anisotropic displacement parameters can also be represented in a format of a tensor, $\mathbf{T_{ij}}$, which is a square matrix symmetrical with respect to its principal diagonal. For B_{ij}, it is given as:

$$\mathbf{B}_{ij} = \begin{pmatrix} B_{11} & B_{12} & B_{13} \\ B_{12} & B_{22} & B_{23} \\ B_{13} & B_{23} & B_{33} \end{pmatrix} \tag{9.10}$$

For other types of anisotropic displacement parameters (Eqs. 9.7 and 9.9), the matrices are identical except that the corresponding elements are β_{ij} and U_{ij}, respectively. The diagonal elements of the tensor, $\mathbf{T_{ii}}$ (i = 1, 2, 3), describe atomic displacements along three mutually perpendicular axes of the ellipsoid. Similar to the isotropic displacement parameter, these may not be negative and should have reasonable values at room temperature, as established by the nature of the material. Generally, the ratios between B_{ii} should not exceed 3–5, unless the large anisotropy can be explained.

All nine elements of the tensor $\mathbf{T_{ij}}$ establish the orientation of the ellipsoid in the coordinate basis of the crystal lattice. Therefore, any or all non-diagonal elements may be positive or negative, but certain relationships between them and the diagonal parameters should be observed, as shown in Eq. (9.11) for B_{ij}. If any of these relationships between the anisotropic displacement parameters is violated, then the set of parameters has no physical meaning.

[4]This is true for atoms located in the general site position (point symmetry 1), where all six parameters are independent from one another. In special positions, some or all of the anisotropic displacement parameters are constrained by symmetry. For example, B_{13} (β_{13}, U_{13}) and B_{23} (β_{23}, U_{23}) for an atom located on a mirror plane perpendicular to Z-axis are constrained to 0.

[5]More often than not, the anisotropic atomic displacement parameters determined from powder diffraction data affected by preferred orientation are incorrect and unphysical.

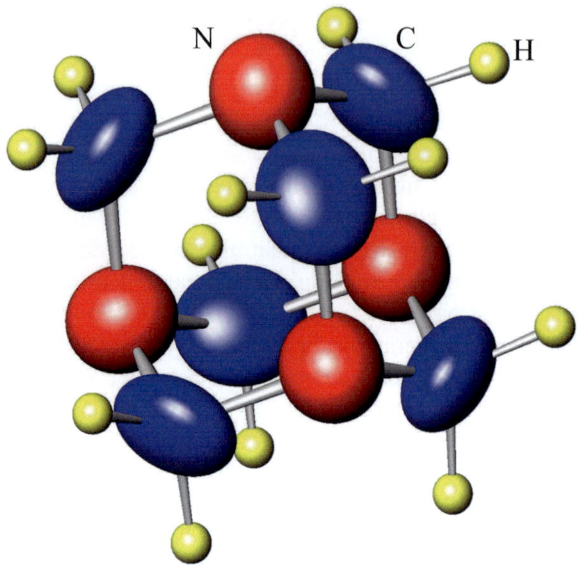

Fig. 9.3 The atomic displacement ellipsoids of carbon and nitrogen atoms shown at the 50% probability level for the hexamethylenetetramine molecule, as determined from powder diffraction data (see Problem 5 on page 779, Chap. 25). Hydrogen atoms were refined in the isotropic approximation (Eq. 9.5) and are shown as small spheres of an arbitrary radius

$$\begin{aligned} &B_{ii} > 0 \\ &B_{ii}B_{ij} > B_{ij}^2 \\ &B_{11}B_{22}B_{33} + B_{12}^2 B_{13}^2 B_{23}^2 > B_{11}B_{23}^2 + B_{22}B_{13}^2 + B_{33}B_{12}^2 \end{aligned} \tag{9.11}$$

The anisotropic displacement parameters can be visualized as ellipsoids (Fig. 9.3) that delineate the volume where atoms are located most of the time, typically at the 50% probability level. The magnitude of the anisotropy and the orientations of the ellipsoids may be used to validate the model of the crystal structure and the quality of refinement by comparing "thermal" motions of atoms with their bonding states. Because of this, when new structural data are published, the ellipsoid plot is usually required if the results are based on single-crystal diffraction data.

Yet another level of complexity in vibrational motion is taken into account by using the so-called anharmonic approximation of atomic displacement parameters. One commonly used approach is the cumulant expansion formalism suggested by Johnson,[6] in which the structure factor is given by the following general expression:

[6] C.K. Johnson, in: Thermal neutron diffraction: proceedings of the International Summer School at Harwell, 1–5 July 1968 on the accurate determination of neutron intensities and structure factors, pp. 132–160, B.T.M. Willis, Ed., London, Oxford University Press (1970).

$$\mathbf{F}(\mathbf{h}) = \sum_{j=1}^{n} g^j f^j(s) \exp\left(2\pi \mathbf{i}\mathbf{h}\cdot\mathbf{x}^j - \beta^j_{\mathrm{kl}} h_k h_l - \mathbf{i}\gamma^j_{\mathrm{klm}} h_k h_l h_m + \delta^j_{\mathrm{klmn}} h_k h_l h_m h_n - \cdots\right) \tag{9.12}$$

where

- β^j_{kl} are the anisotropic displacement parameters (see Eq. 9.7)[7]
- γ^j_{klm} are the third-order anharmonic displacement parameters
- δ^j_{klmn} are the fourth-order displacement parameters. The expansion in Eq. (9.12) may be continued to include fifth, sixth, and higher-order terms. The sign of the corresponding term is determined from the sign of $\mathbf{i}^p$, where p is the order of the anharmonic term
- k, l, m, n (p, q, etc., in the higher-order expansions) vary between 1 and 3
- h_{k}, h_{l}, h_{m}, and h_{n} (h_{p}, h_{q}, etc.) are the corresponding Miller indices ($h_1 = h$, $h_2 = k$, and $h_3 = l$, respectively)
- Other notations are identical to Eq. (9.1)

Displacement parameters are included in Eq. (9.12) with all possible permutations of indices. Thus, for a conventional anisotropic approximation after considering the diagonal symmetry of the corresponding tensor (Eq. 9.10)

$$\begin{aligned}&\beta_{11}hh + \beta_{22}kk + \beta_{33}ll + \beta_{12}hk + \beta_{21}kh + \beta_{13}hl + \beta_{31}lh + \beta_{23}kl + \beta_{32}lk = \\ &\beta_{11}h^2 + \beta_{22}k^2 + \beta_{33}l^2 + 2\beta_{12}hk + 2\beta_{13}hl + 2\beta_{23}kl\end{aligned} \tag{9.13}$$

the already known expression for the exponential factor in Eq. (9.7) is easily obtained. Following the same procedure, it is possible to show that in the case of the third-order anharmonic expansion, the number of the independent atomic displacement parameters is increased by 10 (γ_{111}, γ_{222}, γ_{333}, γ_{112}, γ_{113}, γ_{122}, γ_{133}, γ_{221}, γ_{223}, γ_{233}, and γ_{123}).[8] Similarly, the maximum number of parameters per atom in the fourth-order expansion is increased by 15 (δ_{1111}, δ_{2222}, δ_{3333}, δ_{1112}, δ_{1113}, δ_{1122}, δ_{1133}, δ_{1123}, δ_{1222}, δ_{1223}, δ_{1233}, δ_{1333}, δ_{2223}, δ_{2233}, and δ_{2333}), fifth order by 21, and so on.

A brief description of the anharmonic approximation is included here for completeness, as it is rarely feasible to obtain reasonable atomic displacement parameters

---

[7] In this treatment, the conventional harmonic anisotropic displacement model is a "second-order" approximation.

[8] Just as in the case of the conventional anisotropic approximation, the maximum number of displacement parameters is only realized for atoms located in the general site position (site symmetry 1). In special positions, some or all of the displacement parameters are constrained by symmetry. For example, γ_{333}, γ_{113}, γ_{223}, and γ_{123} for an atom located in the mirror plane perpendicular to Z-axis are constrained to 0. Furthermore, if an atom is located in the center of inversion, all parameters of the odd order anharmonic tensors (3, 5, etc.) are reduced to 0.

of this complexity from powder diffraction data. The total number of atomic displacement parameters for an atom in the fourth-order anharmonic approximation may reach 31 (6 anisotropic +10 third order +15 fourth order). The primary challenges preventing their determination in powder diffraction are uncertainty of the description of Bragg peak shapes, superposition of the reflections, nonideal models to account for preferred orientation, the inadequacy of accounting for porosity, etc.

9.1.3 Atomic Scattering Factor

As briefly mentioned earlier (see Sect. 7.1.2), the ability to scatter radiation varies depending on the type of an atom. Therefore, the general expression of the structure amplitude includes this factor as a multiplier (Eq. 9.1). For X-rays, the scattering power of various atoms and ions is proportional to the number of core electrons. This scattering power is measured using a relative scale normalized to the scattering ability of an isolated electron. The X-ray scattering factors depend on the radial distribution of the electron density around the nucleus and are also functions of the Bragg angle.

When neutrons are of concern, their coherent scattering by nuclei is independent of the Bragg angle, and the corresponding factors remain constant for any Bragg reflection. Scattering factors of different isotopes are represented in terms of coherent scattering lengths of a neutron and are expressed in femtometers (1 fm $= 10^{-15}$ m).

The best-known scattering factors for X-rays and scattering lengths for neutrons of all chemical elements are listed in the International Tables for Crystallography, vol. C. These values are provided for common isotopes and their naturally occurring mixtures (neutrons) and for neutral atoms and common ions (X-rays).

For practical computational purposes, the normal atomic scattering factors for X-rays as functions of Bragg angle are represented by the following exponential function:

$$f_0^j(\sin\theta/\lambda) = c_0^j + \sum_{i=1}^{4} a_i^j \exp\left(-b_i^j \sin\theta/\lambda\right) \tag{9.14}$$

Thus, scattering factors of various chemical elements and ions can be represented as functions of nine coefficients c_0, $a_1 - a_4$, $b_1 - b_4$, and $\sin\theta/\lambda$. These coefficients are also found in the International Tables for Crystallography, vol. C.

The X-ray atomic scattering factors f_0^j for several atoms and ions are shown in Fig. 9.4 as functions of $\sin\theta/\lambda$. As established earlier (see Sect. 7.1.2), the forward-scattered X-ray amplitude is always the result of a constructive interference and, therefore, at $\sin\theta/\lambda = 0$, the value of the scattering factor and the sum of five coefficients in Eq. (9.14) ($c_0 + a_1 + a_2 + a_3 + a_4$) are both equal to the number of electrons.

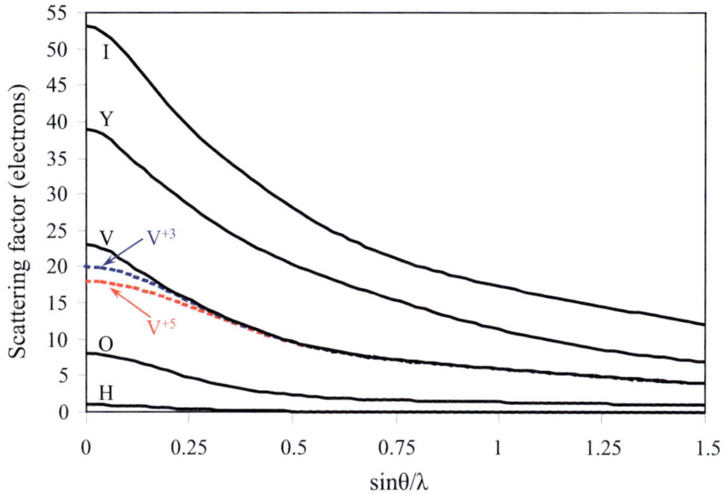

Fig. 9.4 Atomic scattering factors of H, O, V^{5+}, V^{3+}, V, Y, and I as functions of $\sin\theta/\lambda$, appearing in order from the bottom to the top. The scattering factors of vanadium ions become nearly identical to that of the neutral atom at $\sin\theta/\lambda \geq 0.25$, which corresponds to $2\theta \geq 45°$ (Cu Kα)

It is worth noting that the difference between scattering factors of neutral atoms and ions (compare the plots for V, V^{3+}, and V^{5+}) is substantial only at low Bragg angles (more precisely, at low $\sin\theta/\lambda$). Conversely, atoms in solids are rarely fully ionized even in their highest oxidation states. Therefore, atomic scattering factors for neutral atoms usually provide an adequate approximation and are commonly used in all calculations.

Considering Figs. 9.2 and 9.4, both the temperature and scattering factors decrease exponentially with $\sin\theta/\lambda$. Hence, when the displacement parameter of an individual atom becomes unphysical during a least-squares refinement, this usually means that the improper type of atom has been placed in the suspicious site. If the chemical element placed in a certain position has fewer electrons than it should, then during the refinement, its displacement parameter(s) become negative or much lower than those of the correctly placed atoms. This reduction in the displacement parameters(s) (Eqs. 9.5–9.9) offsets the inadequately low scattering factor (Eq. 9.14) by increasing the temperature factor. Similarly, when the site has more electrons than it should, the displacement parameter(s) of the corresponding atom become unusually high during the refinement. Similar effects are observed in neutron diffraction.

Anomalies in displacement parameter(s) are also observed when an atom is placed in a position that is empty or only partially occupied (i.e., the site has significant defects). In this case, displacement parameter(s) become extremely large, effectively reducing the contribution from this atom to the structure factor (see Fig. 9.2). All sites with unusually low or unusually high displacement parameter (s) should be tested by refining population parameter(s) while setting the displacement parameters at reasonable values to avoid possible and sometimes severe correlations. This refinement may reveal the nature of the chemical element located in a specific position by analyzing the number of electrons (or scattering length in neutron diffraction).

For example, assume that a Ni atom has been placed in a certain position and a least-squares refinement results in a large and positive isotropic displacement parameter for this site, while the displacement parameters of other atoms are in the "normal" range. After the population parameter of Ni has been refined while keeping its displacement parameter fixed at the average value of other "normal" atoms, the result is $g^{Ni} = 0.5$. Considering Fig. 9.4, the approximate number of electrons in this position is equal to $Z^{Ni}g^{Ni} = 28 \times 0.5 = 14$. Hence, the refinement result points to one of the following: (1) the site is occupied by Ni atoms at 50% population (50% of the unit cells have Ni in this position and in the remaining 50%, the same position is vacant) or (2) the site is occupied by an element that has approximately 14 electrons, such as Si. Depending on many other factors (e.g., precision of the experiment), the same position may also be occupied by Al (13 electrons), P (15 electrons), or by other neighboring atoms.

The final decision about the population of this "Ni" site can be made only when all available data about the material are carefully considered: (1) which chemical elements may be present in this sample or were used during its preparation, if known? (2) what are the results of the chemical analysis? (3) what is the gravimetric density of the material? and (4) what is the environment of this particular position? For example, an octahedral coordination usually points to Al, but not Si or P.

It should be emphasized that the calculation of the number of electrons described here is not exact since X-ray scattering factors are only directly proportional to the number of electrons at $2\theta = 0°$ and the proportionality becomes approximate at $\sin\theta/\lambda$. The population parameter(s) refinement results are usually more reliable in neutron diffraction because neutron scattering lengths are independent of the Bragg angle.

Normal X-ray scattering factor, f_0 (Eq. 9.14), describes the scattering ability of different atoms as a function of $\sin\theta/\lambda$ (or interplanar distance, d, since $\sin\theta/\lambda = ½d$) and, therefore, is wavelength-independent. This is only true for light chemical elements and relatively short wavelengths. Most atoms scatter X-rays anomalously and their scattering factors become functions of the wavelength.[9] Anomalous scattering is taken into account by including two additional parameters into the overall scattering factor of each chemical element in the following form:

$$f^j(s) = f_0^j(s) + \Delta f^{j\prime} + \boldsymbol{i}\Delta f^{j\prime\prime} \tag{9.15}$$

where

[9]Some isotopes also scatter neutrons anomalously. Footnote 9 on page 168, Chap. 7

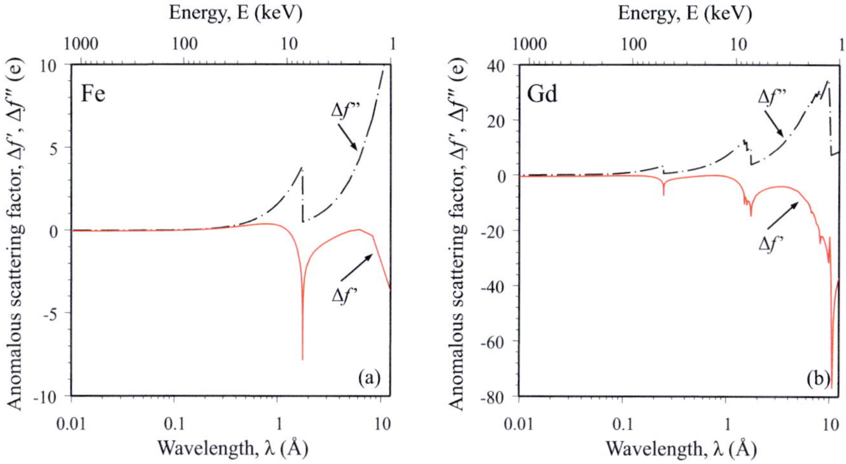

Fig. 9.5 Anomalous scattering factors, $\Delta f'$ and $\Delta f''$, of Fe and Gd as functions of wavelength (bottom scale) and photon energy (upper scale). Compare this figure with Fig. 8.20

– s is $\sin\theta/\lambda$
– f_0^j is the normal atomic scattering factor that depends only on the type of the scattering atom (number of electrons) and is a function of $\sin\theta/\lambda$
– $\Delta f^{j'}$ and $\Delta f^{j''}$ are the real and imaginary components, respectively, of the anomalous scattering factor and they depend on both the atom type and the wavelength

The anomalous scattering factors are also listed in the International Tables for Crystallography for all chemical elements and commonly used wavelengths of laboratory X-rays. They can be measured or calculated for any wavelength, which is important when using synchrotron radiation.[10] The anomalous scattering is usually at least an order of magnitude lower than normal scattering. Generally, the magnitude of the anomalous scattering factors is proportional to the wavelength and inversely proportional to the number of electrons in an atom. Anomalous scattering becomes strong and is at its maximum when the wavelength is near the corresponding absorption edge of an atom, as shown in Fig. 9.5. This effect may be used to discriminate atoms that have similar conventional scattering factors (e.g., Cu and Zn), which is most easily done using synchrotron radiation where the wavelength can be tuned at will.

[10]The most precise values are obtained from experimental measurements. When one of the two anomalous scattering coefficients is measured, the second can be relatively precisely calculated using the Kramers–Kronig equation. More details about anomalous scattering can be found in specialized literature, e.g., see J. Als-Nielsen and D. McMorrow, *Elements of modern X-ray physics*, Wiley, New York (2001).

9.1.4 Phase Angle

The structure amplitude, expressed earlier as a sum of exponents, can also be represented in a different format. Thus, by applying Euler's[11] formula:

$$e^{\mathbf{i}x} = \cos x + \mathbf{i}\sin x \tag{9.16}$$

Equation (9.3) becomes

$$\begin{aligned}\mathbf{F}_{hkl} = \quad & \sum_{j=1}^{n} g^j t^j(s) f^j(s) \cos\left[2\pi\left(hx^j + ky^j + lz^j\right)\right] + \\ & \mathbf{i}\sum_{j=1}^{n} g^j t^j(s) f^j(s) \sin\left[2\pi\left(hx^j + ky^j + lz^j\right)\right]\end{aligned} \tag{9.17}$$

The sums of cosines and sines in Eq. (9.17) signify the real (**A**) and imaginary (**B**) components of a complex number, respectively, which the structure amplitude indeed is. Hence, considering the notations introduced in Eqs. (9.1)–(9.17), it can be rewritten as:

$$\mathbf{F}(\mathbf{h}) = \mathbf{A}(\mathbf{h}) + \mathbf{iB}(\mathbf{h}) \tag{9.18}$$

Complex numbers can be represented as vectors in two dimensions with two mutually perpendicular axes: real and imaginary. Accordingly, the complex structure amplitude (Eq. 9.18) may be imagined as a vector **F**, and its real and imaginary components are the projections of this vector on the real and imaginary axes, respectively, as shown in Fig. 9.6, left. From simple geometry, the following relationships are true:

$$\begin{aligned}&|\mathbf{F}(\mathbf{h})|^2 = |\mathbf{A}(\mathbf{h})|^2 + |\mathbf{B}(\mathbf{h})|^2 = [|\mathbf{A}(\mathbf{h})| + \mathbf{i}|\mathbf{B}(\mathbf{h})|]\,[|\mathbf{A}(\mathbf{h})| - \mathbf{i}|\mathbf{B}(\mathbf{h})|] \\ &\text{and} \\ &\alpha(\mathbf{h}) = \arctan\left(\frac{|\mathbf{B}(\mathbf{h})|}{|\mathbf{A}(\mathbf{h})|}\right)\end{aligned} \tag{9.19}$$

where $|\mathbf{F}(\mathbf{h})|$, $|\mathbf{A}(\mathbf{h})|$, and $|\mathbf{B}(\mathbf{h})|$ are the lengths of the corresponding vectors (or their absolute values), and $\alpha(\mathbf{h})$ is the angle that vector $\mathbf{F}(\mathbf{h})$ makes with the positive direction of the real axis, also known as the phase angle (or the phase) of the structure amplitude. It is worth noting that since $\mathbf{A}(\mathbf{h})$ and $\mathbf{B}(\mathbf{h})$ are mutually perpendicular, the phase shift between the imaginary and real components of the structure amplitude is always $\pm\pi/2$.

[11] Leonhard Euler (1707–1783), the Swiss mathematician and physicist who made significant contributions to calculus, number theory, mechanics, optics, and astronomy. A brief biography is available on Wikipedia https://en.wikipedia.org/wiki/Leonhard_Euler

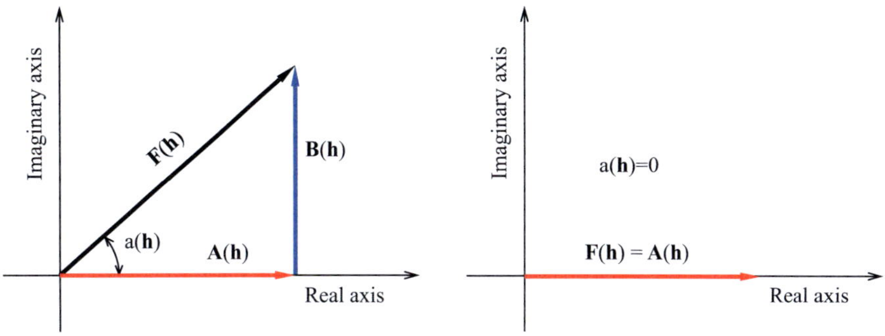

Fig. 9.6 The structure amplitude, **F**(**h**), shown as a vector representing a complex number with its real, **A**(**h**), and imaginary, **B**(**h**), components as projections on the real and imaginary axes, respectively, in the noncentrosymmetric (*left*) and centrosymmetric (*right*) structures. The imaginary component on the left is shifted from the origin of coordinates for clarity

As a result, the structure amplitude can be represented by its magnitude, |**F**(**h**)|, and phase angle, $\alpha(\mathbf{h})$, which varies between 0 and 2π. When the crystal structure is centrosymmetric (i.e., when it contains the center of inversion), each atom with coordinates (x, y, z) has a symmetrically equivalent atom with coordinates ($-x$, $-y$, $-z$). Thus, considering that $\cos(-\gamma) = \cos(\gamma)$ and $\sin(-\gamma) = -\sin(\gamma)$ and assuming that all atoms scatter normally, Eq. (9.17) can be simplified to

$$\mathbf{F}_{hkl} = \sum_{j=1}^{n} g^j t^j(s) f_0^j(s) \cos\left[2\pi\left(hx^j + ky^j + lz^j\right)\right] \tag{9.20}$$

where all sine terms are nullified. The structure amplitude, therefore, becomes a real number and it can be represented as a vector parallel to the real axis. When the phase angle is 0, **F**(**h**) has a positive direction (see Fig. 9.6, right); when the phase angle is π, **F**(**h**) has the opposite (negative) direction. In other words, in the presence of the center of inversion and in the absence of anomalously scattering atoms, the structure amplitude becomes a real quantity with a positive sign when $\alpha(\mathbf{h}) = 0$ and with a negative sign when $\alpha(\mathbf{h}) = \pi$ because its imaginary component is always zero.

When the anomalous scattering is present, the structure amplitude even for a centrosymmetric crystal becomes a complex number. This is shown in Eqs. (9.21) and (9.22). The first (general expression) is easily derived by combining Eqs. (9.15) and (9.17) and rearranging them to group both the real and imaginary components.

$$\mathbf{F}_{hkl} = \begin{pmatrix} \sum_{j=1}^{n} g^j t^j(s)\{f_0^j(s) + \Delta f^{j\prime}\} \cos\left[2\pi\left(hx^j + ky^j + lz^j\right)\right] - \\ \sum_{j=1}^{n} g^j t^j(s)\Delta f^{j\prime\prime} \sin\left[2\pi\left(hx^j + ky^j + lz^j\right)\right] \end{pmatrix} + \mathbf{i}\begin{pmatrix} \sum_{j=1}^{n} g^j t^j(s)\{f_0^j(s) + \Delta f^{j\prime}\} \sin\left[2\pi\left(hx^j + ky^j + lz^j\right)\right] + \\ \sum_{j=1}^{n} g^j t^j(s)\Delta f^{j\prime\prime} \cos\left[2\pi\left(hx^j + ky^j + lz^j\right)\right] \end{pmatrix} \tag{9.21}$$

The introduction of the center of inversion results in the cancellation of all sine terms in Eq. (9.21), and

$$\mathbf{F}_{hkl} = \sum_{j=1}^{n} g^j t^j(s)\{f_0^j(s) + \Delta f^{j\prime}\} \cos\left[2\pi\left(hx^j + ky^j + lz^j\right)\right] + \mathbf{i}\sum_{j=1}^{n} g^j t^j(s)\Delta f^{j\prime\prime} \cos\left[2\pi\left(hx^j + ky^j + lz^j\right)\right] \tag{9.22}$$

Recalling that usually $\Delta f^{j\prime\prime} \ll f_0^j + \Delta f^{j\prime}$, the absolute value of the imaginary component in Eq. (9.21) is smaller than that of the real part, and phase angles remain close to 0 or π in any centrosymmetric structure. In the noncentrosymmetric cases (Eq. 9.21), the phase angles may adopt any value between 0 and 2π.

Concluding this section, it is worth noting that the measured integrated intensity (Eq. 8.41) is proportional to the square of the structure amplitude. Thus, the relative absolute value of the structure amplitude is an easily measurable quantity—it can be obtained as the square root of intensity after dividing the latter by all known geometrical factors. The phase angle (or the sign in the case of normally scattering centrosymmetric structures), however, remains unknown. In other words, phases are lost and cannot be determined directly from either a powder or a single-crystal diffraction experiment. This creates the so-called phase problem in diffraction analysis, and its significance is discussed in Sect. 10.2.

9.2 Effects of Symmetry on the Structure Amplitude

Considering the most general expression of the structure amplitude, Eq. (9.21), its value is defined by the population and displacement parameters of all atoms present in the unit cell, their scattering ability, and coordinates. We know that the coordinates and other parameters of atoms in the unit cell are related by symmetry. As was shown in Chap. 4 (Eqs. 4.20 and 4.29), the coordinates of two symmetrically equivalent atoms ($\mathbf{x}$ and $\mathbf{x}'$) are related as

$$\mathbf{x}' = \mathbf{A} \cdot \mathbf{x} \tag{9.23}$$

where $\mathbf{x} = (x, y, z, 1)$, $\mathbf{x'} = (x', y', z', 1)$, and $\mathbf{A}$ is an augmented matrix representing a symmetry operation. By expanding Eq. (9.23), substituting the result to express the identical arguments of sine and cosine parts in Eq. (9.21) and regrouping it as shown in Eq. (9.24).

$$2\pi\mathbf{h} \cdot \mathbf{x}' = 2\pi \begin{bmatrix} x(hr_{11} + kr_{21} + lr_{31}) + y(hr_{12} + kr_{22} + lr_{32}) \\ +z(hr_{13} + kr_{23} + lr_{33}) + (ht_1 + kt_2 + lt_3) \end{bmatrix} \tag{9.24}$$

It is easy to see that the symmetry of the direct space (crystal structure) is carried over into the reciprocal space by modifying structure amplitudes. Thus, the contribution of two symmetrically equivalent atoms $\mathbf{x}$ and $\mathbf{x'}$ into the corresponding reciprocal lattice points, $\mathbf{h}$ and $\mathbf{h'}$, may be expressed as:

$$\mathbf{h}' = \mathbf{A}^{\mathrm{T}} \cdot \mathbf{h} \tag{9.25}$$

where $\mathbf{h}$ and $\mathbf{h'}$ are augmented reciprocal lattice vectors, and $\mathbf{A}^{\mathbf{T}}$ is obtained from $\mathbf{A}$ by transposing the rotational part of the augmented matrix:

$$\mathbf{A} = \begin{pmatrix} r_{11} & r_{12} & r_{13} & t_1 \\ r_{21} & r_{22} & r_{23} & t_2 \\ r_{31} & r_{32} & r_{33} & t_3 \\ 0 & 0 & 0 & 1 \end{pmatrix} \quad \text{and} \quad \mathbf{A}^{\mathbf{T}} = \begin{pmatrix} r_{11} & r_{21} & r_{31} & t_1 \\ r_{12} & r_{22} & r_{32} & t_2 \\ r_{13} & r_{23} & r_{33} & t_3 \\ 0 & 0 & 0 & 1 \end{pmatrix} \tag{9.26}$$

As a result of the symmetry transformation (Eqs. 9.25 and 9.26), the magnitude of the structure amplitude does not change,[12] and therefore, intensities scattered by symmetrically equivalent reciprocal lattice points preserve the point-group symmetry of the crystal.[13] However, phase angles preserve point-group symmetry only in groups with no other than Bravais lattice translations (the so-called symmorphic groups). For nonsymmorphic groups, phase angles of symmetrically related structure factors are dependent on the translational part of the symmetry operation as shown in Eq. (9.27).

$$\alpha\left(\mathbf{A}^{\mathbf{T}}\mathbf{h}\right) = \alpha(\mathbf{h}) - 2\pi\mathbf{h}\cdot\mathbf{t} \tag{9.27}$$

We note that, in noncentrosymmetric structures, transformations by the center of inversion (which is absent in the structure but is always present in the diffraction pattern) may affect both the amplitude and phase angle, see Sect. 9.2.1.

[12] J. Waser, Symmetry relations between structure factors. Acta Cryst. **8**, 595 (1955). https://doi.org/10.1107/S0365110X55001862

[13] Chap. 1.4 Symmetry in reciprocal space, by International Tables for Crystallography, Vol. B, Reciprocal space, U. Shmueli, Ed., Second online edition, (2010), https://it.iucr.org/B/

9.2.1 Friedel Pairs and Friedel's Law

We begin by considering the relationships between the structure amplitudes of two centrosymmetric reciprocal lattice points: (*hkl*) and $(\overline{hkl})$, the so-called Friedel (or Bijvoet) pair.[14] The analysis is relatively simple and is based on the known relationships between the three trigonometric functions:

$$\begin{aligned}&\cos(-x) = \cos(x), \quad \text{but}\\&\sin(-x) = -\sin(x) \quad \text{and} \quad \tan(-x) = -\tan(x)\end{aligned} \tag{9.28}$$

Regardless of whether the crystal structure is centrosymmetric or not, in the absence of the anomalous scattering, it directly follows from Eqs. (9.17)–(9.19) and (9.28) that

$$\begin{gathered}\mathbf{A}(\mathbf{h}) = \mathbf{A}(\overline{\mathbf{h}}), \quad \mathbf{B}(\mathbf{h}) = -\mathbf{B}(\overline{\mathbf{h}}), \quad \alpha(\mathbf{h}) = -\alpha(\overline{\mathbf{h}})\\|\mathbf{F}(\mathbf{h})| = \sqrt{|\mathbf{A}(\mathbf{h})|^2 + |\mathbf{B}(\mathbf{h})|^2} = \sqrt{|\mathbf{A}(\mathbf{h})|^2 + |\mathbf{B}(\overline{\mathbf{h}})|^2} = |\mathbf{F}(\overline{\mathbf{h}})|\\I(\mathbf{h}) = I(\overline{\mathbf{h}}) \propto |\mathbf{F}(\mathbf{h})|^2\end{gathered} \tag{9.29}$$

Equation (9.29) represents the algebraic formulation of Friedel's law which states that the absolute values of structure amplitudes and intensities are identical but the phase angles have opposite signs for Bragg reflections related to one another by the center of inversion. In another formulation, it states that the reciprocal space is always centrosymmetric in the absence of the anomalous scattering because $|\mathbf{F}(\mathbf{h})| = |\mathbf{F}(\overline{\mathbf{h}})|$. Friedel's law is illustrated in Fig. 9.7, left.

On the contrary, when anomalous scattering is substantial and considering Eq. (9.21) instead of Eq. (9.17), both the real and imaginary components of the structure amplitude are modified as follows:

[14] Georges Friedel (1865–1933), the French mineralogist and crystallographer who in 1913 formulated what is known today as Friedel's law, see Eq. (9.29). G. Friedel, Sur les symétries cristallines que peut révéler la diffraction des rayons X, C.R. Acad. Sci. Paris **157**, 1533 (1913); also see IUCr online dictionary of crystallography: https://dictionary.iucr.org/Friedel's_law and Wikipedia: https://en.wikipedia.org/wiki/Georges_Friedel. Johannes Martin Bijvoet (1892–1980), the Dutch chemist who developed a technique of establishing the absolute configuration of optically active molecules. J.M. Bijvoet, A.F. Peerdeman, and A.J. van Bommel, Determination of the absolute configuration of optically active compounds by means of X-rays, Nature (London) **168**, 271 (1951). https://doi.org/10.1038/168271a0. See a brief biography at: https://en.wikipedia.org/wiki/Johannes_Martin_Bijvoet

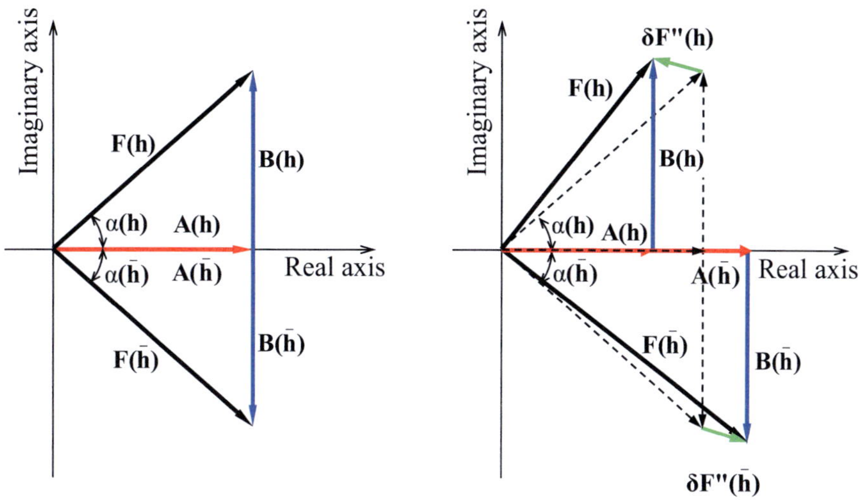

Fig. 9.7 The relationships between different components of the structure amplitude in a Friedel pair are shown for two cases: when all atoms scatter normally (*left*) and when anomalously scattering atoms are present in the crystal structure (*right*), assuming $\Delta f'$ is negative (which is typically the case) and $\Delta f''$ is positive (which is always the case)

$$\begin{aligned}\mathbf{A}(\mathbf{h}) &= \mathbf{A}^{\circ}(\mathbf{h}) + \delta\mathbf{A}''(\mathbf{h})\\ \mathbf{A}(\bar{\mathbf{h}}) &= \mathbf{A}^{\circ}(\mathbf{h}) - \delta\mathbf{A}''(\mathbf{h})\\ \mathbf{B}(\mathbf{h}) &= \mathbf{B}^{\circ}(\mathbf{h}) + \delta\mathbf{B}''(\mathbf{h})\\ \mathbf{B}(\bar{\mathbf{h}}) &= -\mathbf{B}^{\circ}(\mathbf{h}) + \delta\mathbf{B}''(\mathbf{h})\end{aligned} \tag{9.30}$$

where $\mathbf{A}^{\circ}$ and $\mathbf{B}^{\circ}$ represent components of the structure amplitude without anomalous scattering, and $\delta\mathbf{A}''$(h) and $\delta\mathbf{B}''$(h) are the projections of $\delta\mathbf{F}(\mathbf{h})''$ onto the real and imaginary axes, respectively. Note that because **A**(**h**) and $\delta\mathbf{B}''$ are functions of cos, they do not change the sign when **h** changes, whereas **B**(**h**) and $\delta\mathbf{A}''$ do change sign, as they are functions of sin. In Eq. (9.30), the factors **A** and **B** represent the real and imaginary parts of the structure amplitude, which are independent of the anomalous scattering. The terms $\delta\mathbf{A}''$ and $\delta\mathbf{B}''$ appear due to the effect introduced by the presence of the complex component of the anomalous scattering, $\Delta f''$, and are equal to $\Delta f''$ sin ($2\pi\mathbf{hx}$) and $\Delta f''$ cos ($2\pi\mathbf{hx}$), respectively. The combined effect of $\delta\mathbf{A}''$ and $\delta\mathbf{B}''$ results in the modification of the corresponding structure amplitude vectors by $\delta\mathbf{F}(\mathbf{h})''$ and $\delta\mathbf{F}(\bar{\mathbf{h}})''$. This breaks all equalities in Eq. (9.29) and is illustrated in Fig. 9.7 (right) and therefore the structure amplitudes, and therefore the intensities, of the Friedel pair are different:

$$\mathrm{I(h)} \approx |\mathrm{F(h)}|^2 = (\mathrm{A}^{\circ} + \delta\mathrm{A}'')^2 + (\mathrm{B}^{\circ} + \delta\mathrm{B}'')^2$$

$$\mathrm{I}(\overline{\mathrm{h}}) \approx \left|\mathrm{F}(\overline{\mathrm{h}})\right|^2 = (\mathrm{A}^\circ - \delta\mathrm{A}'')^2 + (-\mathrm{B}^\circ + \delta\mathrm{B}'')^2 \quad (9.31)$$

Therefore, $I(h) \neq I(\overline{h})$ and $|\mathrm{F(h)}|^2 \neq \left|\mathrm{F}(\overline{\mathrm{h}})\right|^2$, indicating that Friedel's law is violated in noncentrosymmetric crystal structures containing anomalously scattering atoms. We leave it to the reader to determine the difference $|F(h)|^2 - \left|F(\overline{h})\right|^2$ as a function of A°, B°, $\delta A''$, and $\delta B''$ using Eq. (9.31).

When the crystal structure is centrosymmetric (Eq. 9.22), Friedel's law (Eq. 9.29) becomes fully valid even in the presence of anomalously scattering atoms. This statement is easy to verify, and we leave this exercise to the reader.

9.2.2 Friedel's Law and Multiplicity Factor

Earlier, we considered the effect of lattice symmetry on the multiplicity factors in powder diffraction only in terms of the number of completely overlapping equivalent Bragg peaks (Sect. 8.6.3). Because Friedel's law is violated in noncentrosymmetric structures in the presence of anomalously scattering chemical elements, the evaluation of multiplicity factors should also include Eqs. (9.29) and (9.30). Thus, the multiplicity factor, p_{hkl}, may be comprehensively defined as the number of equivalent reflections (*hkl*) that satisfy Eq. (9.29), in addition to having identical lengths of the reciprocal lattice vectors calculated using Eq. (8.10) (or its equivalent for higher symmetry crystal systems).

Considering an orthorhombic crystal system, the equivalency relationships established earlier (Sect. 8.6.3) also satisfy Eq. (9.29). The following relationships hold:

$$|\mathbf{F}(hkl)| = |\mathbf{F}(\overline{h}kl)| = |\mathbf{F}(h\overline{k}l)| = |\mathbf{F}(hk\overline{l})| = |\mathbf{F}(\overline{h}k\overline{l})| = |\mathbf{F}(\overline{h}\,\overline{k}l)| = |\mathbf{F}(\overline{h}\,\overline{k}\,\overline{l})| = |\mathbf{F}(h\overline{k}\,\overline{l})|$$

when space-group symmetry contains a center of inversion or when there are no anomalously scattering atoms in the unit cell. The multiplicity factor in this case is $p = 8$ for all (*hkl*) reflections where none of the indices is 0. When the center of inversion is absent and when the anomalous scattering is considerable, the situation is different. For example, in the space group P222, the following two sets of structure amplitudes remain identical:

$$|\mathbf{F}(hkl)| = |\mathbf{F}(h\overline{k}\,\overline{l})| = |\mathbf{F}(\overline{h}k\overline{l})| = |\mathbf{F}(\overline{h}\,\overline{k}l)| \text{ and } |\mathbf{F}(\overline{h}\,\overline{k}\,\overline{l})| = |\mathbf{F}(hk\overline{l})| = |\mathbf{F}(h\overline{k}l)| = |\mathbf{F}(\overline{h}kl)|$$

but each set of four is different, and the multiplicity factor is reduced to $p = 4$ for each of the two groups of points in the reciprocal lattice.

9.3 Systematic Absences

As noted at the beginning of this section, crystallographic symmetry affects the structure amplitude and, consequently, the intensities of Bragg peaks. Translational symmetry, in particular, causes certain combinations of Miller indices to become extinct. This extinction occurs because symmetrical contributions to Eq. (9.24) result in the cancellation of relevant trigonometric factors in Eq. (9.21). Additionally, some combinations of indices are forbidden due to translational symmetry.

9.3.1 Lattice Centering

Consider a body-centered lattice in which every atom has a symmetrically equivalent atom shifted by (½, ½, ½). The two matrices **A** (Eq. 9.23) for every pair of the symmetrically identical atoms are

$$A_1 = \begin{pmatrix} 1 & 0 & 0 & 0 \\ 0 & 1 & 0 & 0 \\ 0 & 0 & 1 & 0 \\ 0 & 0 & 0 & 1 \end{pmatrix} \quad \text{and} \quad A_2 = \begin{pmatrix} 1 & 0 & 0 & 1/2 \\ 0 & 1 & 0 & 1/2 \\ 0 & 0 & 1 & 1/2 \\ 0 & 0 & 0 & 1 \end{pmatrix} \tag{9.32}$$

The corresponding arguments defined in Eq. (9.24) become as follows:

$$\begin{aligned} 2\pi(\mathbf{h}\cdot\mathbf{x})_1 &= 2\pi(hx + ky + lz) \\ 2\pi(\mathbf{h}\cdot\mathbf{x})_2 &= 2\pi(hx + ky + lz) + \pi(h + k + l) \end{aligned} \tag{9.33}$$

When these arguments are substituted into the most general equation (9.21) and the summation is carried over every pair of the symmetrically equivalent atoms, the resulting sums of the corresponding trigonometric factors are

$$\begin{aligned} \cos[2\pi(hx + ky + lz)] + \cos[2\pi(hx + ky + lz) + \pi(h + k + l)] &= 0 \\ \sin[2\pi(hx + ky + lz)] + \sin[2\pi(hx + ky + lz) + \pi(h + k + l)] &= 0 \end{aligned} \tag{9.34}$$

when $h + k + l = \pm1, \pm3, \pm5, \ldots = 2n + 1$ and

$$\begin{aligned} \cos[2\pi(hx+ky+lz)] + \cos[2\pi(hx+ky+lz) + \pi(h+k+l)] &= 2\cos[2\pi(hx+ky+lz)] \\ \sin[2\pi(hx+ky+lz)] + \sin[2\pi(hx+ky+lz) + \pi(h+k+l)] &= 2\sin[2\pi(hx+ky+lz)] \end{aligned} \tag{9.35}$$

when $h + k + l = 0, \pm2, \pm4, \ldots = 2n$, where n is an integer.

Table 9.1 Systematic absences caused by different Bravais lattices

Bravais lattice	Allowed reflections	Extinct (forbidden) reflections
P	All	None
I	$h + k + l = 2n$	$h + k + l = 2n + 1$
F	$h + k = 2n$ and $k + l = 2n$ and $h + l = 2n$[a]	$h + k = 2n + 1$ or $k + l = 2n + 1$ or $h + l = 2n + 1$
A	$k + l = 2n$	$k + l = 2n + 1$
B	$h + l = 2n$	$h + l = 2n + 1$
C	$h + k = 2n$	$h + k = 2n + 1$
R[b]	$-h + k + l = 3n$ (hexagonal basis)	$-h + k + l = 3n + 1$ and $3n + 2$
R[c]	$h - k + l = 3n$ (hexagonal basis)	$h - k + l = 3\text{n} + 1$ and $3n + 2$

[a] Alternative definition: all indices even ($h = 2n$ and $k = 2n$ and $l = 2n$) or all odd ($h = 2n + 1$ and $k = 2n + 1$ and $l = 2n + 1$)
[b] Standard setting
[c] Reverse setting

All prefactors in Eq. (9.21) (i.e., g, t, f_0, $\Delta f'$, and $\Delta f''$) are identical for the pairs of the symmetrically equivalent atoms. Consequently, Bragg reflections where the sums of all Miller indices are odd should have zero structure amplitude and zero intensity in any body-centered crystal structure. In other words, these reflections become extinct or absent and therefore cannot be observed as they are forbidden by symmetry.

This property, introduced by the presence of translational symmetry, is called the systematic absence (or the systematic extinction). Therefore, in a body-centered lattice, only Bragg reflections where the sums of all Miller indices are even (i.e., $h + k + l = 2n$, where $n = \pm 1, \pm 2, \pm 3, \ldots$) may have nonzero intensity and be observed. It is worth noting that some (but not all) of the Bragg reflections with $h + k + l = 2n$ may become extinct because their intensities are too low to be detected due to other factors, such as a specific distribution of atoms in the unit cell, which is not predetermined by symmetry.

Similar analyses can be easily performed for other types of Bravais lattices, and the resulting relationships between Miller indices are assembled in Table 9.1. As a self-exercise, try to derive the relationships between Miller indices of forbidden Bragg peaks for three different types of base-centered lattices: A, B, and C.

9.3.2 Glide Planes

We now consider a glide plane, n, perpendicular to the Y-axis. In the simplest case (i.e., when the plane also intersects the origin of coordinates), the equivalents of Eqs. (9.32)–(9.35) become

$$A_1 = \begin{pmatrix} 1 & 0 & 0 & 0 \\ 0 & 1 & 0 & 0 \\ 0 & 0 & 1 & 0 \\ 0 & 0 & 0 & 1 \end{pmatrix} \quad \text{and} \quad A_2 = \begin{pmatrix} 1 & 0 & 0 & 1/2 \\ 0 & -1 & 0 & 0 \\ 0 & 0 & 1 & 1/2 \\ 0 & 0 & 0 & 1 \end{pmatrix} \tag{9.36}$$

$$\begin{aligned} 2\pi(\mathbf{h}\cdot\mathbf{x})_1 &= 2\pi(hx + ky + lz) \\ 2\pi(\mathbf{h}\cdot\mathbf{x})_2 &= 2\pi(hx - ky + lz) + \pi(h + l) \end{aligned} \tag{9.37}$$

$$\begin{aligned} \cos[2\pi(hx + 0y + lz)] + \cos[2\pi(hx - 0y + lz) + \pi(h + l)] &= 0 \\ \sin[2\pi(hx + 0y + lz)] + \sin[2\pi(hx - 0y + lz) + \pi(h + l)] &= 0 \end{aligned} \tag{9.38}$$

when $h + l = \pm 1, \pm 3, \pm 5, \ldots = 2n + 1$ and $k = 0$, while

$$\begin{aligned} \cos[2\pi(hx + 0y + lz)] + \cos[2\pi(hx - 0y + lz) + \pi(h + l)] &= 2\cos[2\pi(hx + lz)] \\ \sin[2\pi(hx + 0y + lz)] + \sin[2\pi(hx - 0y + lz) + \pi(h + l)] &= 2\sin[2\pi(hx + lz)] \end{aligned} \tag{9.39}$$

when $h + l = 0, \pm 2, \pm 4, \ldots = 2n$ and $k = 0$. Note, that Eqs. (9.38) and (9.39) cannot be simplified when $k \neq 0$.

Hence, Bragg peaks ($h0l$) in which the sum of h and l is odd should have zero structure amplitude and zero intensity due to the presence of the glide plane n in this orientation. Only Bragg peaks in which the sum of these Miller indices is even (i.e., $h + l = 2n$) may have nonzero intensity and be observed. We note that the presence of this glide plane only affects Bragg peaks with $k = 0$ because of the change of the sign of ky in Eq. (9.37). Again, some (but not all) reflections $h0l$ with $h + l = 2n$ may be absent due to peculiarities of a crystal structure not associated with the presence of the glide plane n, but no peaks with $h + l = 2n + 1$ can be observed.

Similar analyses can be performed for other types of glide planes in different orientations, not necessarily traversing the origin of coordinates.[15] The relationships between allowed Miller indices for various glide planes in different orientations are shown in Table 9.2.

9.3.3 Screw Axes

Consider a screw axis, 2_1, which is parallel to the Y-axis. The equivalents of Eqs. (9.32)–(9.35) when this axis passes through the origin of the coordinate system are as follows:

[15] As an exercise, try to derive the relationships between Miller indices of the systematically absent Bragg peaks for a glide plane, a, perpendicular to the Y and then to the Z directions.

Table 9.2 Combinations of indices which are allowed by various glide planes

Glide plane	Orientation[a]	Reflection type	Allowed reflections[b]
a	(010)	$h0l$	$h = 2n$
	(001)	$hk0$	$h = 2n$
	(110)	hhl	$h = 2n$
b	(100)	$0kl$	$k = 2n$
	(001)	$hk0$	$k = 2n$
c	(100)	$0kl$	$l = 2n$
	(010)	$h0l$	$l = 2n$
	(110)	hhl	$l = 2n$
	(110)	hhl	$l = 2n$
d	(100)	$0kl$	$k + l = 4n$ $(k, l = 2n)$
	(010)	$h0l$	$h + l = 4n$ $(h, l = 2n)$
	(001)	$hk0$	$h + k = 4n$ $(h, k = 2n)$
	(110)	hhl	$2h + l = 4n$ $(l = 2n)$
n	(100)	$0kl$	$k + l = 2n$
	(010)	$h0l$	$h + l = 2n$
	(001)	$hk0$	$h + k = 2n$
	(110)	hhl	$l = 2n$

[a] The orientations are given as the crystallographic indices of the corresponding plane
[b] All other reflections are forbidden

$$A_1 = \begin{pmatrix} 1 & 0 & 0 & 0 \\ 0 & 1 & 0 & 0 \\ 0 & 0 & 1 & 0 \\ 0 & 0 & 0 & 1 \end{pmatrix} \quad \text{and} \quad A_2 = \begin{pmatrix} -1 & 0 & 0 & 0 \\ 0 & 1 & 0 & 1/2 \\ 0 & 0 & -1 & 0 \\ 0 & 0 & 0 & 1 \end{pmatrix} \tag{9.40}$$

$$\begin{aligned} 2\pi(\mathbf{h}\cdot\mathbf{x})_1 &= 2\pi(hx + ky + lz) \\ 2\pi(\mathbf{h}\cdot\mathbf{x})_2 &= 2\pi(-hx + ky - lz) + \pi k \end{aligned} \tag{9.41}$$

$$\begin{aligned} \cos[2\pi(0x + ky + 0z)] + \cos[2\pi(-0x + ky - 0z) + \pi k] = 0 \\ \sin[2\pi(0x + ky + 0z)] + \sin[2\pi(-0x + ky - 0z) + \pi k] = 0 \end{aligned} \tag{9.42}$$

when $k = \pm 1, \pm 3, \pm 5, \ldots = 2n + 1$, $h = 0$ and $l = 0$, while

$$\begin{aligned} \cos[2\pi(0x + ky + 0z)] + \cos[2\pi(-0x + ky - 0z) + \pi k] = 2\cos[2\pi ky] \\ \sin[2\pi(0x + ky + 0z)] + \sin[2\pi(-0x + ky - 0z) + \pi k] = 2\sin[2\pi ky] \end{aligned} \tag{9.43}$$

when $k = 0, \pm 2, \pm 4, \ldots = 2n$, $h = 0$, and $l = 0$. Similar to Eqs. (9.38) and (9.39), Eqs. (9.42) and (9.43) cannot be simplified as shown when either $h \neq 0$ or/and $l \neq 0$.

Hence, Bragg peaks $(0k0)$, in which k is odd, should have zero structure amplitude and zero intensity due to the presence of the 2_1 axis in this orientation. Only Bragg peaks in which k is even may have nonzero intensity and be observed. Similar to glide planes, the presence of screw axes only affects Bragg peaks with two Miller

Table 9.3 Combinations of indices, which are allowed by various screw axes

Screw axis	Orientation[a]	Reflection type	Allowed reflections[b]
2_1, 4_2	[100]	$h00$	$h = 2n$
	[010]	$0k0$	$k = 2n$
2_1, 4_2, 6_3	[001]	$00l$	$l = 2n$
2_1	[110]	$hh0$	$h = 2n$
4_1, 4_3	[100]	$h00$	$h = 4n$
	[010]	$0k0$	$k = 4n$
	[001]	$00l$	$l = 4n$
3_1, 3_2, 6_2, 6_4	[001]	$00l$	$l = 3n$
3_1, 3_2	[111]	hhh	$h = 3n$
6_1, 6_5	[001]	$00l$	$l = 6n$

[a] The orientations are given as the crystallographic directions
[b] All other reflections are forbidden

indices equal to 0 because of the change in the sign of two terms in Eq. (9.41) (hx and lz in this example).

Corresponding analyses may be performed for other types of screw axes in different orientations.[16] Again, the result remains identical whether the axis intersects the origin of coordinates or not. The relationships between Miller indices of Bragg peaks allowed by various screw axes in different orientations are shown in Table 9.3.

9.4 Space Groups and Systematic Absences

As shown in Sect. 9.3, the presence of translational symmetry causes the extinction of certain types of reflections. This property of infinite symmetry elements is used to determine possible space-group(s) symmetry from diffraction data by analyzing the Miller indices of the observed Bragg peaks. It is worth noting that only infinite symmetry elements cause systematic absences and, therefore, may be detected from this analysis. Finite symmetry elements, such as simple rotation and inversion axes, mirror planes, and center of inversion, produce no systematic absences and, therefore, are not distinguishable using this approach.

As is often the case in diffraction analysis, when the space-group symmetry is known, it is quite easy to predict which types of reflections can, and which cannot, be observed directly from the space-group symbol. For example, when the space-group symmetry is Pnma, only the following types of Bragg peaks may have nonzero intensity (as established by analyzing Tables 9.1, 9.2, and 9.3):

[16] As an exercise, try to derive the relationships between the Miller indices of the systematically absent Bragg peaks for a 4_1 screw axis parallel to the Z direction.

hkl: any (primitive Bravais lattice)
*hk*0: $h = 2n$ (glide plane a ⊥ *Z*)
*h*0*l*: any (mirror plane m ⊥ *Y*)
0*kl*: $k + l = 2n$ (glide plane n ⊥ *X*)
*h*00: $h = 2n$ (derived from *hk0*: $h = 2n$ because $k = 0$)
0*k*0: $k = 2n$ (derived from 0*kl*: $k + l = 2n$ because $l = 0$)
00*l*: $l = 2n$ (derived from 0*kl*: $k + l = 2n$ because $k = 0$)

When the space-group symmetry is unknown, that is, when reflection conditions are analyzed from diffraction data, the answer may not be unique. For example, the combination of systematic absences listed here also corresponds to a different space-group symmetry, $Pn2_1a$,[17] and, therefore, the two space groups are indistinguishable from a simple analysis of the present Bragg peaks.

All space groups that produce identical combinations of systematic absences are combined into the so-called diffraction groups. For example, space groups *P*31c and $P\bar{3}1c$ result in the same allowed reflection conditions: $l = 2n$ for both 0*kl* and 00*l* type reflections. They belong to the trigonal diffraction group P--c. The three hexagonal space-group symmetries—$P\bar{6}2c$, $P6_3mc$, and $P6_3/mmc$—also belong to the P--c diffraction group, but the crystal system is hexagonal. When powder diffraction data are used to analyze systematic absences, these two diffraction groups cannot be distinguished from one another due to the same symmetry of shapes of both the trigonal and hexagonal unit cells. The two are discernible using diffraction data from a single crystal, where the threefold rotation axis is different from the sixfold rotation axis. In powder diffraction, however, Bragg reflections with different intensities in the trigonal crystal system, such as *hkl* and *khl*, are observed at the same Bragg angle and, therefore, cannot be distinguished from each other.

Nonetheless, analysis of the systematic absences (the complete list is found in Tables 9.5, 9.6, 9.7, 9.8, 9.9, and 9.10) usually allows one to narrow the choice of space-group symmetry to just a few possibilities, and the actual symmetry of the material is usually established in the process of the complete determination of its crystal structure. Especially when powder diffraction data are used, it only makes sense to analyze low-angle Bragg peaks to minimize the potential influence of the nearly completely overlapped reflections with indices not related by symmetry. An example of space-group determination is shown in Table 9.4.

This powder diffraction pattern was indexed in the monoclinic crystal system with $a = 9.264$ Å, $b = 9.728$ Å, $c = 9.905$ Å, $\beta = 106.08°$. The sum $h + k$ is even for all observed Bragg peaks, which clearly points to a base-centered lattice C. The first four columns in Table 9.4 contain Miller indices and Bragg angles 2θ, calculated assuming a base-centered lattice C without applying any other conditions, that is, as in the space group C2/m. The next two columns list Bragg angles and integrated intensities of the observed peaks that correspond to the calculated Bragg angles.

[17] As an exercise, please verify that this space-group symmetry produces exactly the same combination of the allowed Miller indices of Bragg reflections as Pnma.

Table 9.4 Analysis of the observed combinations of indices in the monoclinic base-centered unit cell (E—even, O—odd). The entries corresponding to Bragg peaks that are theoretically possible but are not observed in the powder diffraction pattern are labeled as "unobserved". The observed Bragg reflections *h*0*l*, in which $l = 2n$, are shown in *bold* (zero is considered an even number)

h	*k*	*l*	$2\theta_{calc}$	$2\theta_{obs}$	I_{obs}	*h* + *k* for all *hkl*	*l* for *h*0*l*, 00*l*	*h* for *h*0*l*, *h*00	*k* for 0*k*0
0	0	1	9.227	Unobserved	E	O	E		
1	1	0	13.415	13.425	1000	E			
1	1	–1	14.672	14.675	259	E			
1	1	1	17.838	17.838	93	E			
0	2	0	18.168	18.172	112	E			E
0	**0**	**2**	**18.574**	**18.573**	**30**	**E**	**E**	**E**	
2	0	–1	19.466	Unobserved	E	O	E		
2	**0**	**0**	**19.876**	**19.869**	**104**	**E**	**E**	**E**	
0	2	1	20.433	20.428	352	E			
1	1	–2	20.619	20.622	86	E			
2	**0**	**–2**	**23.203**	**23.197**	**86**	**E**	**E**	**E**	
2	0	1	24.233	Unobserved	E	O	E		
1	1	2	25.179	25.186	4	E			
0	2	2	26.121	26.111	15	E			
2	2	–1	26.774	26.771	36	E			
2	2	0	27.079	27.074	112	E			
0	0	3	28.049	Unobserved	E	O	E		
1	1	–3	28.560	28.559	65	E			
1	3	0	29.234	29.237	37	E			
2	0	–3	29.632	Unobserved	E	O	E		
2	2	–2	29.652	Unobserved	E				
1	3	–1	29.854	29.853	35	E			
3	1	–1	30.306	30.321	38	E			
2	2	1	30.480	30.507	6	E			
2	**0**	**2**	**31.001**	**30.997**	**77**	**E**	**E**	**E**	
3	1	0	31.441	31.422	24	E			
1	3	1	31.591	31.576	27	E			

Table 9.5 Reflection conditions for the monoclinic crystal system (Laue class 2/m, unique axis *b*)

Reflection conditions			Diffraction symbol	Point group, space groups		
hkl, 0*kl*, *hk*0	*h*0*l*, *h*00, 00*l*	0*k*0		2	m	2/ m
			P1-1	**P121**[3]	**P1m1**[6]	**P12/m1**[10]
		k	$P12_11$	**$P12_11$**[4]		**$P12_1/m1$**[11]
	h		P1a1		P1a1[7]	P12/a1[13]
	h	*k*	$P12_1/a1$			$P12_1/a1$[14]
	l		P1c1		**P1c1**[7]	**P12/c1**[13]
	l	*k*	$P12_1/c1$			**$P12_1/c1$**[14]
	h + *l*		P1n1		P1n1[7]	P12/n1[13]
	h + *l*	*k*	$P12_1/n1$			$P12_1/n1$[14]
C	*h*	*k*	C1-1	**C121**[5]	**C1m1**[8]	**C12/m1**[12]
C	*h*, *l*	*k*	C1c1		**C1c1**[9]	**C12/c1**[15]
A	*l*	*k*	A1-1	A121[5]	A1m1[8]	A12/m1[12]
A	*h*, *l*	*k*	A1n1		A1n1[9]	A12/n1[15]
I	*h* + *l*	*k*	I1-1	I121[5]	I1m1[8]	I12/m1[12]
I	*h*, *l*	*k*	I1a1		I1a1[9]	I12/a1[15]

As follows from Table 9.4, there are only two possible diffraction groups for the monoclinic C-centered lattice. The second diffraction group, C1c1, differs from the first one, C1-1, by the presence of *h*0*l* reflections only with even *l*. As is easy to see from Table 9.4, none of the Bragg peaks *h*0*l* with $l = 2n + 1$ is observed, or in other words, the allowed reflections condition is $l = 2n$ for *h*0*l* (these are shown in bold). Other conditions are derived from the C-centering of the lattice. Thus, the following two space-groups are possible for the material: C1c1 and C12/c1. The latter group was confirmed during structure determination.

In general, the determination of the space-group symmetry or the list of possible space groups using Tables 9.5, 9.6, 9.7, 9.8, 9.9, and 9.10 should be performed as follows:

- Based on the symmetry of the unit cell shape, the proper table (crystal system) must be selected. Only in one case, when the unit cell is primitive with $a = b \neq c$, $\alpha = \beta = 90°$, and $\gamma = 120°$, both trigonal and hexagonal crystal systems should be analyzed.
- All reflections should first be checked for general systematic absences (conditions) caused by lattice centering, which are shown in the first column of each table. This should narrow the list of possible space groups.
- Subsequently, move to the next column in the corresponding table. Identify reflections that belong to a specified combination of Miller indices listed in the column header and analyze whether the extinction conditions are met. This should further narrow the list of possible space groups.

Table 9.6 Reflection conditions for the orthorhombic crystal system (Laue class mmm)

Reflection conditions							Diffraction symbol	Point group, space groups		
hkl	*0kl*	*h0l*	*hk0*	*h00*	*0k0*	*00l*		222	mm2 m2m 2mm	mmm
							P---	**P222**^{16}	**Pmm2**^{25}	**Pmmm**^{47}
									Pm2m^{25}	
									P2mm^{25}	
						l	P--2$_1$	**P222$_1$**^{17}		
					k		P-2$_1$-	P22$_1$2^{17}		
					k	*l*	P-2$_1$2$_1$	P22$_1$2$_1$^{18}		
				h			P2$_1$--	P2$_1$22^{17}		
				h		*l*	P2$_1$-2$_1$	P2$_1$22$_1$^{18}		
				h	*k*		P2$_1$2$_1$-	**P2$_1$2$_1$2**^{18}		
				h	*k*	*l*	P2$_1$2$_1$2$_1$	**P2$_1$2$_1$2$_1$**^{19}		
			h	*h*			P--a		Pm2a^{28}	
									P2$_1$ma^{26}	**Pmma**^{51}
			k		*k*		P--b		Pm2$_1$b^{26}	
									P2mb^{28}	Pmmb^{51}
			h + *k*	*h*	*k*		P--n		Pm2$_1$n^{31}	
									P2$_1$mn^{31}	**Pmmn**^{59}
		h		*h*			P-a-		Pma2^{28}	Pmam^{51}
									P2$_1$am^{26}	
		h	*h*	*h*			P-aa		P2aa^{27}	Pmaa^{49}
		h	*k*	*h*	*k*		P-ab		P2$_1$ab^{29}	Pmab^{57}
		h	*h* + *k*	*h*	*k*		P-an		P2an^{30}	Pman^{53}
		l				*l*	P-c-		**Pmc2$_1$**^{26}	
									P2cm^{28}	Pmcm^{51}
		l	*h*	*h*		*l*	P-ca		P2$_1$ca^{29}	Pmca^{57}
		l	*k*		*k*	*l*	P-cb		P2cb^{32}	Pmcb^{55}
		l	*h* + *k*	*h*	*k*	*l*	P-cn		P2$_1$cn^{33}	Pmcn^{62}
		h + *l*		*h*		*l*	P-n-		**Pmn2$_1$**^{31}	
									P2$_1$nm^{31}	Pmnm^{59}
		h + *l*	*h*	*h*		*l*	P-na		P2na^{30}	**Pmna**^{53}
		h + *l*	*k*	*h*	*k*	*l*	P-nb		P2$_1$nb^{33}	Pmnb^{62}
		h + *l*	*h* + *k*	*h*	*k*	*l*	P-nn		P2nn^{34}	Pmnn^{58}
	k				*k*		Pb--		Pbm2^{28}	
									Pb2$_1$m^{26}	Pbmm^{51}
	k		*h*	*h*	*k*		Pb-a		Pb2$_1$a^{29}	Pnma^{57}
	k		*k*		*k*		Pb-b		Pb2b^{27}	Pbmb^{49}
	k		*h* + *k*	*h*	*k*		Pb-n		Pb2n^{30}	Pbmn^{53}
	k	*h*		*h*	*k*		Pba-		**Pba2**^{32}	**Pbam**^{55}
	k	*h*	*h*	*h*	*k*		Pbaa			Pbaa^{54}
	k	*h*	*k*	*h*	*k*		Pbab			Pbab^{54}
	k	*h*	*h* + *k*	*h*	*k*		Pban			**Pban**^{50}
	k	*l*			*k*	*l*	Pbc-		Pbc2$_1$^{29}	**Pbcm**^{57}
	k	*l*	*h*	*h*	*k*	*l*	Pbca			**Pbca**^{61}
	k	*l*	*k*		*k*	*l*	Pbcb			Pbcb^{54}
	k	*l*	*h* + *k*	*h*	*k*	*l*	Pbcn			**Pbcn**^{60}
	k	*h* + *l*		*h*	*k*	*l*	Pbn-		Pbn2$_1$^{33}	Pbnm^{62}

(continued)

Table 9.6 (continued)

Reflection conditions							Diffraction symbol	Point group, space groups		
hkl	*0kl*	*h0l*	*hk0*	*h00*	*0k0*	*00l*		222	mm2 m2m 2mm	mmm
	k	*h* + *l*	*h*	*h*	*k*	*l*	Pbna			Pbna[60]
	k	*h* + *l*	*k*	*h*	*k*	*l*	Pbnb			Pbnb[56]
	k	*h* + *l*	*h* + *k*	*h*	*k*	*l*	Pbnn			Pbnn[52]
	l					*l*	Pc--		Pcm2$_1$[26] Pc2m[28]	Pcmm[51]
	l		*h*	*h*		*l*	Pc-a		Pc2a[32]	Pcma[55]
	l		*k*		*k*	*l*	Pc-b		Pc2$_1$b[29]	Pcmb[57]
	l		*h* + *k*	*h*	*k*	*l*	Pc-n		Pc2$_1$n[33]	Pcmn[62]
	l	*h*		*h*		*l*	Pca-		**Pca2$_1$**[29]	Pcam[57]
	l	*h*	*h*	*h*		*l*	Pcaa			Pcaa[54]
	l	*h*	*k*	*h*	*k*	*l*	Pcab			Pcab[61]
	l	*h*	*h* + *k*	*h*	*k*	*l*	Pcan			Pcan[60]
	l	*l*				*l*	Pcc-		**Pcc2**[27]	**Pccm**[49]
	l	*l*	*h*	*h*		*l*	Pcca			**Pcca**[54]
	l	*l*	*k*		*k*	*l*	Pccb			Pccb[54]
	l	*l*	*h* + *k*	*h*	*k*	*l*	Pccn			**Pccn**[56]
	l	*h* + *l*		*h*		*l*	Pcn-		Pcn2[30]	Pcnm[53]
	l	*h* + *l*	*h*	*h*		*l*	Pcna			Pcna[50]
	l	*h* + *l*	*k*	*h*	*k*	*l*	Pcnb			Pcnb[60]
	l	*h* + *l*	*h* + *k*	*h*	*k*	*l*	Pcnn			Pcnn[52]
	k + *l*				*k*	*l*	Pn--		Pnm2$_1$[31] Pn2$_1$m[31]	Pnmm[59]
	k + *l*		*h*	*h*	*k*	*l*	Pn-a		Pn2$_1$a[33]	**Pnma**[62]
	k + *l*		*k*		*k*	*l*	Pn-b		Pn2b[30]	Pnmb[53]
	k + *l*		*h* + *k*	*h*	*k*	*l*	Pn-n		Pn2n[34]	Pnmn[58]
	k + *l*	*h*		*h*	*k*	*l*	Pna-		**Pna2$_1$**[33]	Pnam[62]
	k + *l*	*h*	*h*	*h*	*k*	*l*	Pnaa			Pnaa[56]
	k + *l*	*h*	*k*	*h*	*k*	*l*	Pnab			Pnab[60]
	k + *l*	*h*	*h* + *k*	*h*	*k*	*l*	Pnan			Pnan[52]
	k + *l*	*l*			*k*	*l*	Pnc-		**Pnc2**[30]	Pncm[53]
	k + *l*	*l*	*h*	*h*	*k*	*l*	Pnca			Pnca[60]
	k + *l*	*l*	*k*		*k*	*l*	Pncb			Pncb[50]
	k + *l*	*l*	*h* + *k*	*h*	*k*	*l*	Pncn			Pncn[52]
	k + *l*	*h* + *l*		*h*	*k*	*l*	Pnn-		**Pnn2**[34]	**Pnnm**[58]
	k + *l*	*h* + *l*	*h*	*h*	*k*	*l*	Pnna			**Pnna**[52]
	k + *l*	*h* + *l*	*k*	*h*	*k*	*l*	Pnnb			Pnnb[52]
	k + *l*	*h* + *l*	*h* + *k*	*h*	*k*	*l*	Pnnn			**Pnnn**[48]
C	*k*	*h*	*h* + *k*	*h*	*k*		C---	**C222**[21]	**Cmm2**[35] Cm2m[38] C2mm[38]	**Cmmm**[65]
C	*k*	*h*	*h* + *k*	*h*	*k*	*l*	C--2$_1$	**C222$_1$**[20]		
C	*k*	*h*	*h*, *k*	*h*	*k*		C--(ab)		Cm2a[39] C2mb[39]	**Cmma**[67] Cmmb[67]
C	*k*	*h*, *l*	*h* + *k*	*h*	*k*	*l*	C-c-		**Cmc2$_1$**[36] C2cm[40]	**Cmcm**[63]
C	*k*	*h*, *l*	*h*, *k*	*h*	*k*	*l*	C-c(ab)		C2cb[41]	**Cmca**[64]
C	*k*, *l*	*h*	*h* + *k*	*h*	*k*	*l*	Cc--		Ccm2$_1$[36] Cc2m[40]	Ccmm[63]

(continued)

Table 9.6 (continued)

Reflection conditions							Diffraction symbol	Point group, space groups		
hkl	*0kl*	*h0l*	*hk0*	*h00*	*0k0*	*00l*		222	mm2 m2m 2mm	mmm
C	*k*, *l*	*h*	*h*, *k*	*h*	*k*	*l*	Cc-(ab)		Cc2a[41]	Ccmb[64]
C	*k*, *l*	*h*, *l*	*h* + *k*	*h*	*k*	*l*	Ccc-		**Ccc2**[37]	**Cccm**[66]
C	*k*, *l*	*h*, *l*	*h*, *k*	*h*	*k*	*l*	Ccc(ab)			**Ccca**[68] Cccb[68]
B	*l*	*h* + *k*	*h*	*h*		*l*	B---	B222[21]	Bmm2[38] Bm2m[35] B2mm[38]	Bmmm[65]
B	*l*	*h* + *l*	*h*	*h*	*k*	*l*	B-2_1-	B22$_1$2[20]		
B	*l*	*h* + *l*	*h*, *k*	*h*	*k*	*l*	B--b		Bm2$_1$b[36] B2mb[40]	Bmmb[63]
B	*l*	*h*, *l*	*h*	*h*		*l*	B-(ac)-		Bma2[39] B2cm[39]	Bmam[67] Bmcm[67]
B	*l*	*h*, *l*	*h*, *k*	*h*	*k*	*l*	B-(ac)b		B2cb[41]	Bmab[64]
B	*k*, *l*	*h* + *l*	*h*	*h*	*k*	*l*	Bb--		Bbm2[40] Bb2$_1$m[36]	Bbmm[63]
B	*k*, *l*	*h* + *l*	*h*, *k*	*h*	*k*	*l*	Bb-b		Bb2b[37]	Bbmb[66]
B	*k*, *l*	*h*, *l*	*h*	*h*	*k*	*l*	Bb(ac)-		Bba2[41]	Bbcm[64]
B	*k*, *l*	*h*, *l*	*h*, *k*	*h*	*k*	*l*	Bb(ac)b			Bbab[68] Bbcb[68]
A	*k* + *l*	*l*	*k*		*k*	*l*	A---	A222[21]	**Amm2**[38] Am2m[38] A2mm[35]	Ammm[65]
A	*k* + *l*	*l*	*k*	*h*	*k*	*l*	A2_1--	A2$_1$22[20]		
A	*k* + *l*	*l*	*h*, *k*	*h*	*k*	*l*	A--a		Am2a[40] A2$_1$ma[36]	Amma[63]
A	*k* + *l*	*h*, *l*	*k*	*h*	*k*	*l*	A-a-		**Ama2**[40] A2$_1$am[36]	Amam[63]
A	*k* + *l*	*h*, *l*	*h*, *k*	*h*	*k*	*l*	A-aa		A2aa[37]	Amaa[66]
A	*k*, *l*	*l*	*k*		*k*	*l*	A(bc)		**Abm2**[39] Ac2m[39]	Abmm[67] Acmm[67]
A	*k*, *l*	*l*	*h*, *k*	*h*	*k*	*l*	A(bc)-a		Ac2a[41]	Abma[64]
A	*k*, *l*	*h*, *l*	*k*	*h*	*k*	*l*	A(bc)a-		**Aba2**[41]	Acam[64]
A	*k*, *l*	*h*, *l*	*h*, *k*	*h*	*k*	*l*	A(bc)aa			Abaa[68] Acaa[68]
I	*k* + *l*	*h* + *l*	*h* + *k*	*h*	*k*	*l*	I---	**I222**[23] **I2$_1$2$_1$2$_1$**[24]	**Imm2**[44] Im2m[44] I2mm[44]	**Immm**[71]
I	*k* + *l*	*h* + *l*	*h*, *k*	*h*	*k*	*l*	I--(ab)		Im2a[46] I2mb[46]	**Imma**[74] Immb[74]
I	*k* + *l*	*h*, *l*	*h* + *k*	*h*	*k*	*l*	I-(ac)-		**Ima2**[46] I2cm[46]	Imam[74] Imcm[74]
I	*k* + *l*	*h*, *l*	*h*, *k*	*h*	*k*	*l*	I-cb		I2cb[45]	Imcb[72]
I	*k*, *l*	*h* + *l*	*h* + *k*	*h*	*k*	*l*	I(bc)--		Ibm2[46] Ic2m[46]	Ibmm[74] Icmm[74]
I	*k*, *l*	*h* + *l*	*k*, *l*	*h*	*k*	*l*	Ic-a		Ic2a[45]	Icma[72]
I	*k*, *l*	*h*, *l*	*h* + *k*	*h*	*k*	*l*	Iba-		**Iba2**[45]	**Ibam**[72]

(continued)

Table 9.6 (continued)

Reflection conditions							Diffraction symbol	Point group, space groups		
hkl	*0kl*	*h0l*	*hk0*	*h00*	*0k0*	*00l*		222	mm2 m2m 2mm	mmm
I	*k*, *l*	*h*, *l*	*h*, *k*	*h*	*k*	*l*	Ibca			**Ibca**[73]
F	*k* + *l*	*h* + *l*	*h* + *k*	*h*	*k*	*l*	F---	**F222**[22]	**Fmm2**[42] Fm2m[42] F2mm[42]	**Fmmm**[69]
F	*k*, *l*	[b]	[c]	h_4[d]	k_4[d]	l_4[d]	F-dd		F2dd[43]	
F	[a]	*h*, *l*	[c]	h_4[d]	k_4[d]	l_4[d]	Fd-d		Fd2d[43]	
F	[a]	[b]	*h*, *k*	h_4[d]	k_4[d]	l_4[d]	Fdd-		**Fdd2**[43]	
F	[a]	[b]	[c]	h_4[d]	k_4[d]	l_4[d]	Fddd			**Fddd**[70]

[a] $k + l = 4n$ and $k = 2n$, $l = 2n$
[b] $h + l = 4n$ and $h = 2n$, $l = 2n$
[c] $h + k = 4n$ and $h=2n$, $k=2n$
[d] h_4: $h = 4n$; k_4: $k = 4n$; l_4: $l = 4n$

- The last step should be repeated for all columns under the general header "reflection conditions." When finished, the list should be narrowed to a single line, that is, the corresponding diffraction group should be found for the known symmetry of the unit cell shape.
- In the case of monoclinic and orthorhombic crystal systems, it may be necessary to transform the unit cell to achieve a standard setting of the space-group symmetry. The space groups in standard settings are shown in the corresponding tables in bold. Usually, only a permutation of the coordinate axes is needed, but in some cases of the monoclinic crystal system, such as I-centering or the presence of the glide plane n, a more complex transformation of the unit cell vectors may be required. If necessary, the International Tables for Crystallography (vol. A) should be used as a reference on how to perform a transformation of unit cell vectors.

In Tables 9.5, 9.6, 9.7, 9.8, 9.9, and 9.10, reflection conditions as determined by the presence of translational symmetry are listed for all crystal systems, except triclinic. The latter has no translational symmetry, and therefore, the two possible space groups, P1 and P$\bar{1}$, belong to the same diffraction class, P1. If a cell showing reflection conditions is empty, there are no restrictions imposed on Miller indices, and any combination of them is allowed. Unless noted otherwise, the corresponding symbol(s) or expression(s) present in the cell indicate that the index or the combination is even. Monoclinic and orthorhombic space groups in their standard settings are shown in **bold**; trigonal, hexagonal, tetragonal, and cubic space groups are only shown in standard setting. Space groups with unique reflection conditions (i.e., those uniquely determined from the systematic absences) are shown in rectangles shaded

Table 9.7 Reflection conditions for the trigonal crystal system. There are no space groups with a unique set of reflection conditions

Reflection conditions					Laue class						
					$\bar{3}$	$\bar{3}$m1			$\bar{3}$1m		
					Point group, space groups						
hkl	*h*0*l*	*hhl*	00*l*	Diffraction symbol	3, $\bar{3}$	321	3m1	$\bar{3}$m1	312	31m	$\bar{3}$1m
				P---[a]	$P3^{143}$ $P\bar{3}^{147}$	$P321^{150}$	$P3m1^{156}$	$P\bar{3}m1^{164}$	$P312^{149}$	$P31m^{157}$	$P\bar{3}1m^{162}$
			l_3[b]	$P3_1$--[a]	$P3_1{}^{144}$ $P3_2{}^{145}$	$P3_121^{152}$ $P3_221^{154}$			$P3_112^{151}$ $P3_212^{153}$		
		l	*l*	P--c[a]						$P31c^{159}$	$P\bar{3}1c^{163}$
	l		*l*	P-c-[a]			$P3c1^{158}$	$P\bar{3}c1^{165}$			
R	[c]	l_3[b]	l_3[b]	R--	$R3^{146}$ $R\bar{3}^{148}$	$R32^{155}$	$R3m^{160}$	$R\bar{3}m^{166}$			
R	[c], *l*	l_3[b]	l_6[b]	R-c			$R3c^{161}$	$R\bar{3}c^{167}$			

[a] These diffraction groups have the same reflection conditions as the following hexagonal groups: P---, $P6_2$--, P--c, and P-c-. Therefore, they are indistinguishable from systematic absences using powder data

[b] l_3: $l = 3n$; l_6: $l = 6n$

[c] When R is $-h + k + l = 3n$, then $h + l = 3n$; when R is $h - k + l = 3n$, then $-h + l = 3n$

Table 9.8 Reflection conditions for the hexagonal crystal system. There are no space groups with a unique set of reflection conditions

Reflection conditions					Laue class				
					6/m		6/mmm		
					Point group, space groups				
hkl	*h*0*l*	*hhl*	00*l*	Diffraction symbol	6, $\bar{6}$	6/m	622, 6mm	$\bar{6}$2m, $\bar{6}$m2	6/mmm
				P---[a]	$P6^{168}$ $P\bar{6}^{174}$	$P6/m^{175}$	$P622^{177}$ $P6mm^{183}$	$P\bar{6}2m^{189}$ $P\bar{6}m2^{187}$	$P6/mmm^{191}$
			l	$P6_3$	$P6_3{}^{173}$	$P6_3/m^{176}$	$P6_322^{182}$		
			l_3[b]	$P6_2$--[a]	$P6_2{}^{171}$ $P6_4{}^{172}$		$P6_222^{180}$ $P6_422^{181}$		
			l_6[b]	$P6_1$--	$P6_1{}^{169}$ $P6_5{}^{170}$		$P6_122^{178}$ $P6_522^{179}$		
		l	*l*	P--c[a]			$P6_3mc^{186}$	$P\bar{6}2c^{190}$	$P6_3/mmc^{194}$
	l		*l*	P-c-[a]			$P6_3cm^{185}$	$P\bar{6}c2^{188}$	$P6_3/mcm^{193}$
	l	*l*	*l*	P-cc			$P6cc^{184}$		$P6/mcc^{192}$

[a] These diffraction groups have the same reflection conditions as the following trigonal groups: P---, $P3_1$--, P--c, and P-c-. Therefore, they are indistinguishable from systematic absences using powder data

[b] l_3: $l = 3n$; l_6: $l = 6n$

for a standard setting. Superscripts indicate the space-group number as listed in the International Tables for Crystallography, vol. A. The following abbreviations are employed to specify allowed combinations of indices due to lattice centering:

A: $k + l = 2n$; B: $h + l = 2n$; C: $h + k = 2n$
I: $h + k + l = 2n$
F: $h + k = 2n$ and $h + l = 2n$ and $k + l = 2n$
R: $-h + k + l = 3n$ or $h - k + l = 3n$

9.5 Additional Reading

1. International Tables for Crystallography, Vol. A, Space-group symmetry, M. I. Aroyo, Ed., Second online edition (2016). https://it.iucr.org/A/
International Tables for Crystallography, Vol. C, Mathematical, physical and chemical tables, First online edition, E. Prince, Ed. (2006). https://it.iucr.org/C/
International Tables for Crystallography, Vol. A-I published jointly with the International Union of Crystallography (IUCr) by Springer, https://it.iucr.org/

Table 9.9 Reflection conditions for the tetragonal crystal system

Reflection conditions							Diffraction symbol	Laue Class				
								4/*m*		4/*mmm*		
								Point group, space groups				
hkl	*hk*0	0*kl*	*hhl*	00*l*	0*k*0	*hh*0		4, $\bar{4}$	4/*m*	422, 4*mm*	$\bar{4}2m$, $\bar{4}m2$	4/*mmm*
							P---	P4[75] $P\bar{4}$[81]	P4/m[83]	P422[89] P4mm[99]	$P\bar{4}2m$[111] $P\bar{4}m2$[115]	P4/mmm[123]
					k		$P\text{-}2_1\text{-}$			$P42_12$[90]	$P\bar{4}2_1m$[113]	
				l			$P4_2\text{--}$	$P4_2$[77]	$P4_2/m$[84]	$P4_222$[93]		
				l	*k*		$P4_22_1\text{-}$			$P4_22_12$[94]		
				l_4[a]			$P4_1\text{--}$	$P4_1$[76] $P4_3$[78]		$P4_122$[91] $P4_322$[95]		
				l_4[a]	*k*		$P4_12_1\text{-}$			$P4_12_12$[92] $P4_32_12$[96]		
			l	*l*			P--c			$P4_2mc$[105]	$P\bar{4}2c$[112]	$P4_2/mmc$[131]
			l	*l*	*k*		$P\text{-}2_1c$				$P\bar{4}2_1c$[114]	
		k			*k*		P-b-			P4bm[100]	$P\bar{4}b2$[117]	P4/mbm[127]
		k	*l*	*l*	*k*		P-bc			$P4_2bc$[106]		$P4_2/mbc$[135]
		l		*l*			P-c-			$P4_2cm$[101]	$P\bar{4}c2$[116]	$P4_2/mcm$[132]
		l	*l*	*l*			P-cc			P4cc[103]		P4/mcc[124]
		k+*l*		*l*	*k*		P-n-			$P4_2nm$[102]	$P\bar{4}n2$[118]	$P4_2/mnm$[136]
		k+*l*	*l*	*l*	*k*		P-nc			P4nc[104]		P4/mnc[128]
	h+*k*				*k*		Pn--		P4/n[85]			P4/nmm[129]
	h+*k*			*l*	*k*		$P4_2/n\text{--}$		$P4_2/n$[86]			
	h+*k*		*l*	*l*	*k*		Pn-c					$P4_2/nmc$[137]
	h+*k*	*k*			*k*		Pnb-					P4/nbm[125]
	h+*k*	*k*	*l*	*l*	*k*		Pnbc					$P4_2/nbc$[133]
	h+*k*	*l*		*l*	*k*		Pnc-					$P4_2/ncm$[138]
	h+*k*	*l*	*l*	*l*	*k*		Pncc					P4/ncc[130]
	h+*k*	*k*+*l*		*l*	*k*		Pnn-					$P4_2/nnm$[134]
	h+*k*	*k*+*l*	*l*	*l*	*k*		Pnnc					P4/nnc[126]
I	*h*+*k*	*k*+*l*	*l*	*l*	*k*		I---	I4[79] $I\bar{4}$[82]	I4/m[87]	I422[97] I4mm[107]	$I\bar{4}2m$[121] $I\bar{4}m2$[119]	I4/mmm[139]
I	*h*+*k*	*k*+*l*	*l*	l_4[a]	*k*		$I4_1\text{--}$	$I4_1$[80]		$I4_122$[98]		
I	*h*+*k*	*k*+*l*	[b]	l_4[a]	*k*	*h*	I--d			$I4_1md$[109]	$I\bar{4}2d$[122]	
I	*h*+*k*	*k*,*l*	*l*	*l*	*k*		I-c-			I4cm[108]	$I\bar{4}c2$[120]	I4/mcm[140]
I	*h*+*k*	*k*,*l*	[b]	l_4[a]	*k*	*h*	I-cd			$I4_1cd$[110]		
I	*h*,*k*	*k*+*l*	*l*	l_4[a]	*k*		$I4_1/a\text{--}$		$I4_1/a$[88]			
I	*h*,*k*	*k*+*l*	[b]	l_4[a]	*k*	*h*	Ia-d					$I4_1/amd$[141]
I	*h*,*k*	*k*,*l*	[b]	l_4[a]	*k*	*h*	Iacd					$I4_1/acd$[142]

[a] l_4: $l = 4n$

[b] $2h + l = 4n$; $l = 2n$

2. Modern powder diffraction. D.L Bish and J.E. Post, Eds. Reviews in Mineralogy, Vol. 20. Mineralogical Society of America, Washington, DC (1989).
3. R. Jenkins and R.L. Snyder, Introduction to X-ray powder diffractometry. Wiley, New York (1996).

Table 9.10 Reflection conditions for the cubic crystal system

Reflection conditions				Diffraction symbol	Laue Class				
					$m\bar{3}$		$m\bar{3}m$		
					Point group				
hkl	*0kl*	*hhl*	*00l*		23	$m\bar{3}$	432	$\bar{4}3m$	$m\bar{3}m$
				P---	$P23^{195}$	$Pm\bar{3}^{200}$	$P432^{207}$	$P\bar{4}3m^{215}$	$Pm\bar{3}m^{221}$
			l	$P2_1$	$P2_13^{198}$				
				$P4_2$			$P4_232^{208}$		
			l_4[a]	$P4_1$			$P4_132^{213}$ $P4_332^{212}$		
		l	*l*	P--n				$P\bar{4}3n^{218}$	$Pm\bar{3}n^{223}$
	[b]		*l*	Pa--		$Pa\bar{3}^{205}$			
	k+l		*l*	Pn--		$Pn\bar{3}^{201}$			$Pn\bar{3}m^{224}$
	k+l	*l*	*l*	Pn-n					$Pn\bar{3}n^{222}$
I	*k+l*	*l*	*l*	I---	$I23^{197}$ $I2_13^{199}$	$Im\bar{3}^{204}$	$I432^{211}$	$I\bar{4}3m^{217}$	$Im\bar{3}m^{229}$
I	*k+l*	*l*	l_4[a]	$I4_1$--			$I4_132^{214}$		
I	*k+l*	[c], *l*	l_4[a]	I--d				$I\bar{4}3d^{220}$	
I	*k,l*	*l*	*l*	Ia--		$Ia\bar{3}^{206}$			
I	*k,l*	[c], *l*	l_4[a]	Ia-d					$Ia\bar{3}d^{230}$
F	*k,l*	*h+l*	*l*	F---	$F23^{196}$	$Fm\bar{3}^{202}$	$F432^{209}$	$F\bar{4}3m^{216}$	$Fm\bar{3}m^{225}$
F	*k,l*	*h+l*	l_4[a]	$F4_1$--			$F4_132^{210}$		
F	*k,l*	*h,l*	*l*	F--c				$F\bar{4}3c^{219}$	$Fm\bar{3}c^{226}$
F	[d], *k,l*	*h+l*	l_4[a]	Fd--		$Fd\bar{3}^{203}$			$Fd\bar{3}m^{227}$
F	[d], *k,l*	*h,l*	l_4[a]	Fd-c					$Fd\bar{3}c^{228}$

[a] l_4: $l = 4n$
[b] Conditions are for $0kl$: $k = 2n$; for $h0l$: $l = 2n$; for $hk0$: $h = 2n$
[c] $2h + l = 4n$
[d] $2k + l = 4n$

4. J. Als-Nielsen and D. McMorrow, Elements of modern X-ray physics, Wiley, New York (2001).
5. W. Clegg, Synchrotron chemical crystallography. J. Chem. Soc., Dalton Trans. **19**, 3223 (2000). https://doi.org/10.1039/B004136J

9.6 Problems

1. A student collected a powder diffraction pattern from an organometallic compound on a standard powder diffractometer equipped with a sealed Cu Kα X-ray tube. She noticed that scattered intensity decays rapidly and she cannot see any Bragg peaks beyond 2θ = 60°. Her goal is to have reliable intensities at 90° or higher of 2θ. She thinks for a minute and then calls the crystallography lab at her

university to schedule time on one of their units. The lab has three powder diffractometers, all equipped with Cu Kα X-ray tubes: a rotating anode unit operating at ambient environment and two sealed tube units, one with a cryogenic attachment (the lowest temperature is 77 K) and another with a furnace (the highest temperature is 1100 K). The student asked for time on which unit and why?

2. Consider Fig. 10.1 in Chap. 10, which shows powder diffraction patterns collected from the same material ($CeRhGe_3$) at room temperature ($T \approx 295$ K) using X-rays and at $T = 200$ K using neutrons. Setting aside differences between the intensities of individual Bragg peaks, the most obvious overall difference between the two sets of diffraction data is that diffracted intensity is only slightly suppressed toward high Bragg angles ($\sin\theta/\lambda$) in neutron diffraction, while it is considerably lower in the case of X-ray data. Can you explain why?
3. Establish which combinations of indices are allowed and which are forbidden in the space-group symmetry $Cmc2_1$. List the symmetry elements that cause each group of reflections to become extinct?
4. Powder diffraction pattern of a compound with an unknown crystal structure was indexed with the following unit cell parameters (shown approximately): $a = 10.34$ Å, $b = 6.02$ Å, $c = 4.70$ Å, $\alpha = 90°$, $\beta = 90°$, and $\gamma = 90°$. The list of all Bragg peaks observed from 15° to 60° 2θ is shown in Table 9.11. Analyze the systematic absences (if any) present in this powder diffraction pattern and suggest possible space groups for the material.
5. Powder diffraction pattern of a compound with an unknown crystal structure was indexed with the following unit cell parameters: $a = b = 4.07$ Å, $c = 16.3$ Å, $\alpha = \beta = 90°$, $\gamma = 120°$. The list of all Bragg peaks observed from 2° to 120° 2θ is

Table 9.11 List of Bragg peaks with their intensities and indices observed in a powder diffraction pattern of a material indexed in the following unit cell: $a = 10.34$ Å, $b = 6.02$ Å, $c = 4.70$ Å, $\alpha = 90°$, $\beta = 90°$, and $\gamma = 90°$

h	*k*	*l*	I/I$_0$	2θ°	*h*	*k*	*l*	I/I$_0$	2θ°
2	0	0	255	17.105	2	0	2	207	42.154
1	0	1	583	20.712	3	2	1	59	44.296
2	1	0	207	22.629	2	1	2	47	44.899
0	1	1	77	23.966	4	2	0	19	46.246
1	1	1	741	25.495	5	0	1	17	47.978
2	0	1	120	25.572	0	2	2	133	49.120
2	1	1	665	29.598	0	3	1	15	49.339
0	2	0	106	29.650	1	2	2	28	49.961
3	0	1	327	32.152	1	3	1	123	50.166
2	2	0	23	34.448	5	1	1	68	50.470
3	1	1	1000	35.518	2	2	2	332	52.407
1	2	1	317	36.451	4	0	2	332	52.407
4	1	0	204	37.813	2	3	1	132	52.595
1	0	2	116	39.242	4	1	2	160	54.879
2	2	1	139	39.550	6	1	0	195	55.369
4	0	1	169	39.723	3	3	1	241	56.506
1	1	2	207	42.154	4	3	0	143	58.128

Table 9.12 List of Bragg peaks with their intensities and indices observed in a powder diffraction pattern of a material indexed in the following unit cell: $a = b = 4.07$ Å, $c = 16.3$ Å, $\alpha = \beta = 90°$, and $\gamma = 120°$

h	*k*	*l*	I/I_0	2θ°	*h*	*k*	*l*	I/I_0	2θ°
0	0	3	10,000	18.541	0	1	−10	159	75.324
0	1	−1	8851	34.192	0	2	4	256	76.249
0	1	2	280	35.927	0	2	−5	62	79.293
0	0	6	156	37.629	0	0	12	122	80.452
0	1	−4	4599	42.243	0	1	11	159	82.405
0	1	5	817	46.517	0	2	7	572	87.346
0	1	−7	3157	56.682	1	1	9	1097	88.383
0	0	9	1389	57.905	0	2	−8	1045	92.316
1	1	0	3322	60.143	1	2	−1	534	100.328
0	1	8	5186	62.442	1	2	2	59	101.427
1	1	3	456	63.511	1	2	−4	164	105.431
0	2	1	832	71.064	0	0	15	200	107.679
0	2	−2	136	72.107	1	1	12	189	109.601
1	1	6	206	73.182	1	2	−7	439	117.229

shown in Table 9.12. Analyze the systematic absences (if any) present in this powder diffraction pattern and suggest possible space groups for the material.

6. At room temperature, the lanthanide material cerium (Ce) has a face-centered cubic crystal structure, which is known as γ-Ce. The space group is $Fm\bar{3}m$ and the lattice parameter $a = 5.161$ Å. When cooled below 77 K, it transforms to α-Ce, which also has a face-centered cubic crystal structure (space group $Fm\bar{3}m$) with the lattice parameter $a = 4.85$ Å.

   (a) Calculate Bragg angles (2θ) for all Bragg peaks that may be observed between 0° and 100° 2θ using Cu Kα radiation $\lambda = 1.54178$ Å for both α- and γ-Ce.
   (b) Sketch diagrams of both diffraction patterns indicating only the positions of possible diffraction peaks.
   (c) Discuss the differences between the two diffraction patterns (if any) that you expect to see when the actual diffraction patterns are collected.

Chapter 10
Solving the Crystal Structure

In Chaps. 7–9, we established that the diffraction pattern of a single crystal or a polycrystalline material is transformation of an ordered atomic structure into reciprocal space rather than a direct image of the structure itself. The three-dimensional arrangement of atoms in a lattice can be reconstructed only after the diffraction pattern has been transformed back into direct space. However, in powder diffraction, the situation is complicated by the fact that the diffraction pattern is a one-dimensional projection of three-dimensional reciprocal space. We have no intention of covering the comprehensive derivation of the relevant mathematical tools since it is mainly of interest to experts and can be found in many excellent books and reviews.[1] Therefore, in this chapter, we only briefly describe a general approach to solving the crystal structure.

10.1 Fourier Transformation

In crystallography, direct and reciprocal spaces are related to one another as forward and reverse Fourier[2] transformations. In three dimensions, these relationships can be represented by the following Fourier integrals:

---

[1] For example, see the International Tables for Crystallography, Vol. B, Reciprocal space, U. Shmueli, Ed., Second online edition (2010), https://it.iucr.org/B/; International Tables for Crystallography, Vol. C, Mathematical, physical and chemical tables, First online edition, E. Prince, Ed. (2006). https://it.iucr.org/C/

[2] Jean Baptiste Joseph Fourier (1768–1830) the French mathematician who is best known for introducing analysis of periodic functions in which a complex function is represented by a sum of simple sine and cosine functions. A brief biography is available on Wikipedia https://en.wikipedia.org/wiki/Joseph_Fourier. See Eric Weisstein, MathWorld, for more information about Fourier transformation: https://mathworld.wolfram.com/FourierTransform.html

P. Zavalij, V. Pecharsky, *Fundamentals of Powder Diffraction and Structural Characterization of Materials*, https://doi.org/10.1007/978-3-031-91504-8_10

$$\mathbf{\Phi}(\mathbf{h}) = \int_V \rho(\mathbf{x}) \exp[2\pi\mathbf{i}(\mathbf{h}\cdot\mathbf{x})] d^3\mathbf{x} \tag{10.1}$$

$$\rho(\mathbf{x}) = \int_{V*} \mathbf{\Phi}(\mathbf{h}) \exp[-2\pi\mathbf{i}(\mathbf{h}\cdot\mathbf{x})] d^3\mathbf{h} \tag{10.2}$$

where Eq. (10.2) is the forward (−**i**) and Eq. (10.1) is reverse (+**i**) Fourier transform. Here, **Φ**(**h**) is the function of the scattered amplitude defined in the reciprocal space and $\rho(\mathbf{x})$ is the corresponding function defined in the direct space. For example, $\rho(\mathbf{x})$ is the electron density when scattering of X-rays is of concern, or it is the nuclear density when scattering of neutrons on nuclei is considered. **h** and **x** are the coordinate vectors in reciprocal and direct spaces, respectively, while V^* and V are, correspondingly, reciprocal and direct-space volumes, and $\mathbf{i} = \sqrt{-1}$.

Both integrals do not require the assumption of periodicity and they can be used to calculate the scattered amplitude or the corresponding density function of any direct or reciprocal object, respectively. For example, Eq. (10.1) results in the atomic scattering factor, $f(\sin\theta/\lambda) \propto |\mathbf{\Phi}(\mathbf{h})|$, when the integration is performed for an isolated atom. In this case, $\rho(\mathbf{x})$ is the electron density distribution in the atom, which is usually obtained from quantum mechanics.

Considering a crystal, in which the electron density function is periodic, the integral in Eq. (10.1) can be substituted by a sum:

$$\mathbf{F}(\mathbf{h}) = V\sum_{\mathbf{x}} \rho(\mathbf{x}) \exp[2\pi\mathbf{i}(\mathbf{h}\cdot\mathbf{x})] \tag{10.3}$$

where **F**(**h**) is the structure amplitude at a reciprocal lattice point **h**, V is the volume of the unit cell of the direct lattice, and the summation is carried over all possible coordinate vectors, **x**, in the unit cell for a specific **h**.

When Eq. (10.3) is compared with Eq. (9.1), it is easy to see that the distribution of the electron density in the unit cell is modeled by n products, $g^j t^j(\sin\theta/\lambda) f^j(\sin\theta/\lambda)$, where g^j, t^j, and f^j are the population, temperature, and scattering factors of the jth atom, respectively, and the summation ranges over all atoms (from 1 to n) that are present in the unit cell.

A similar substitution of the forward Fourier integral (Eq. 10.2) results in the following sum, which enables one to calculate the distribution of the electron (or nuclear) density in the unit cell from the known structure amplitudes:

$$\rho(\mathbf{x}) = \frac{1}{V}\sum_{\mathbf{h}} \mathbf{F}(\mathbf{h}) \exp[-2\pi\mathbf{i}(\mathbf{h}\cdot\mathbf{x})] \tag{10.4}$$

Here, the summation is carried over all reciprocal lattice points, **h**, for a given coordinate vector, **x**. This last equation has exceptional practical importance as it allows one to convert the array of numbers—the observed structure amplitudes obtained from the experimentally measured intensities—into the image of the atomic structure represented as the distribution of the electron (or nuclear) density in the unit cell.

Thus, in expanded form, the value of the electron density at any point in the unit cell with coordinates x, y, and z ($0 \leq x \leq 1, 0 \leq y \leq 1$, and $0 \leq z \leq 1$) can be calculated using structure amplitudes obtained from an X-ray diffraction experiment as:

$$\rho_{xyz} = \frac{1}{V} \sum_{h=-\infty}^{h=+\infty} \sum_{k=-\infty}^{k=+\infty} \sum_{l=-\infty}^{l=+\infty} \mathbf{F}_{hkl} \exp[-2\pi \boldsymbol{i}(hx + ky + lz)] \tag{10.5}$$

where $\mathbf{F}_{hkl}$ are the structure amplitudes represented as complex numbers.

Equation (10.5) may be converted into a more practical form since only the absolute values of the structure amplitude $|\boldsymbol{F}_{hkl}^{obs}|$ are directly observable in a diffraction experiment. Thus,

$$\rho_{xyz} = \frac{1}{V} \sum_{h=-\infty}^{h=+\infty} \sum_{k=-\infty}^{k=+\infty} \sum_{l=-\infty}^{l=+\infty} |\mathbf{F}_{hkl}^{obs}| \cos[2\pi(hx + ky + lz) - \alpha_{hkl}] \tag{10.6}$$

where α_{hkl} is the phase angle of the reflection (hkl), see Eq. (9.19).

When the crystal structure is centrosymmetric and contains no anomalously scattering atoms, phase angles are fixed at $\alpha_{hkl} = 1$ or π and Eq. (10.6) is simplified to

$$\rho_{xyz} = \frac{1}{V} \sum_{h=-\infty}^{h=+\infty} \sum_{k=-\infty}^{k=+\infty} \sum_{l=-\infty}^{l=+\infty} s_{hkl} |\mathbf{F}_{hkl}^{obs}| \cos[2\pi(hx + ky + lz)] \tag{10.7}$$

where $s_{hkl} = 1$ or -1 for $\alpha_{hkl} = 1$ or π, respectively.

Considering Friedel's law, the summation in Eqs. (10.5)–(10.7) can be simplified by excluding the negative values of one of the indices and by changing the prefactor from $1/V$ to $2/V$ in order to keep the correct absolute values of ρ_{xyz}. Since no real experiment produces an infinite number of data points (structure amplitudes), the practical use of Eqs. (10.5)–(10.7) is accomplished by including all available data, and the summation is carried over from h_{min} to h_{max}, k_{min} to k_{max}, and l_{min} to l_{max}.

Equation (10.6) is most commonly used to calculate the distributions of the electron (nuclear) density in the unit cell, which are also known as Fourier maps, from X-ray (neutron) diffraction data. The locations of peaks on the Fourier map calculated using X-ray diffraction data represent the coordinates of atoms, while the electron density integrated over the range of the peak corresponds to the number of electrons in the atom. The major problem in using Eq. (10.6) is that only the absolute values of the structure amplitudes, $|\boldsymbol{F}_{hkl}^{obs}|$, are known directly from the experiment because they are obtained as square roots of the integrated intensities of the corresponding Bragg peaks (Eq. 8.41) after eliminating all prefactors. As mentioned earlier, the information about phase angles α_{hkl} (or signs, s_{hkl}, see Eq. 10.7) is missing and is not measurable directly.

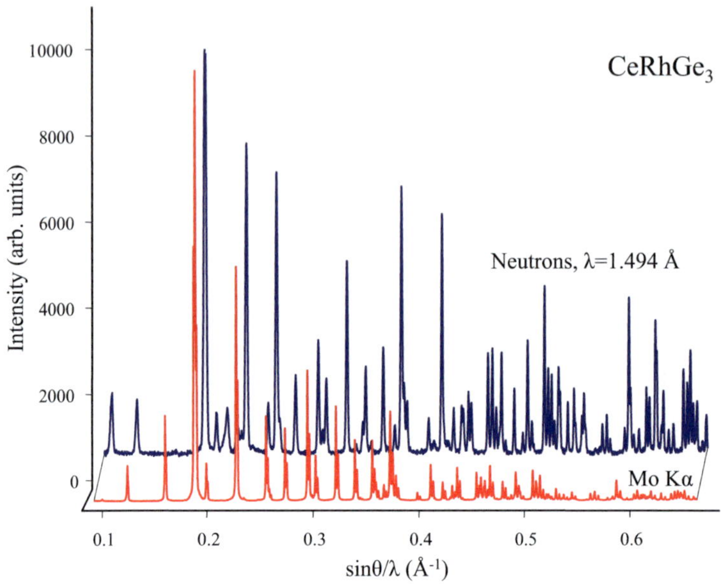

Fig. 10.1 Two powder diffraction patterns of the intermetallic compound $CeRhGe_3$. The *bottom plot* depicts X-ray data collected using Mo Kα radiation at room temperature, and the *top plot* displays neutron diffraction data collected at T = 200 K, using thermal neutrons with λ = 1.494 Å. (Neutron diffraction data courtesy of Dr. O. Zaharko)

When the distribution of atoms in the unit cell is known at least approximately, for example, when a model of the crystal structure exists, then the phases can be easily computed from Eqs. (9.17) to (9.19),[3] and all parameters in Eqs. (10.5)–(10.7) are defined. Examples of electron and nuclear density distributions in the unit cell of the intermetallic compound $CeRhGe_3$ calculated from Eq. (10.6) by using X-ray and neutron powder diffraction data (Fig. 10.1) are shown in Fig. 10.2. In both cases, phase angles were computed by employing Eqs. (9.17)–(9.19) and the known model of the crystal structure of the material. The individually observed absolute structure amplitudes were calculated from the observed integrated intensities using Eq. (8.41) after the overlapped Bragg peaks were deconvoluted.

Hence, in order to calculate a Fourier map and thus visualize the crystal structure from diffraction data, the phase angles of every Bragg reflection must somehow be established. There are various techniques that enable the recovery of phases or, in other words, enable the solution of the atomic structure. These methods are discussed in the following section.

When the crystal structure is unknown, nearly every available ab initio phase determination technique usually results in approximate phase angles and, therefore, instead of a complete atomic structure, only a partial model may be found from a

[3] Phase angles can be also determined using the so-called direct phase recovery techniques, see Sect. 10.2.2.

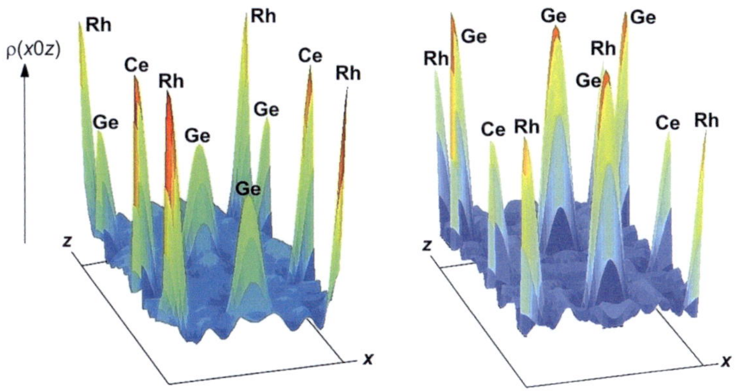

Fig. 10.2 The electron (*left*) and nuclear (*right*) density distributions in the $x0z$ plane of the unit cell of $CeRhGe_3$ calculated from X-ray and neutron powder diffraction data, respectively (Fig. 10.1). The contour of the unit cell is shown schematically as the rectangle under each Fourier map. The peaks correspond to various atoms located in this plane and are so labeled on the figure. The volumes of the peaks are proportional to the scattering ability of atoms: for X-ray, the scattering power decreases in the series Ce($58\overline{e}$) → Rh($45\overline{e}$) → Ge($32\overline{e}$); for neutrons, the coherent scattering lengths decrease in the reverse order: Ge(8.19 fm) → Rh(5.88 fm) → Ce(4.84 fm)

subsequently calculated Fourier map. Thus, in the remainder of this section, we briefly discuss how the initial model of the crystal structure can be improved and completed using Fourier transformations.

As soon as phase angles, α_{hkl}, are established at least approximately, they can be used in combination with available $\left|\boldsymbol{F}_{hkl}^{obs}\right|$ to compute a Fourier map and establish the distribution of electron or nuclear density in the unit cell. Even though the phases may be approximate (i.e., inexact), the values of $\left|\boldsymbol{F}_{hkl}^{obs}\right|$ are much more precise and, therefore, the Fourier map is usually more accurate than the model employed to calculate phase angles. Thus, a computed Fourier map can be used to improve and refine the model of the crystal structure by finding coordinates of atoms with higher precision (compared to the initial model) and by locating missing atoms.

If a crude initial model was used to determine approximate phases, some atoms might not appear on the map and these must be deleted because they were not confirmed as a result of Fourier transformation. The new or improved model is then used as the next level approximation to calculate the new set of phase angles. A subsequent Fourier map must be calculated using the new set of phases combined with experimental $\left|\boldsymbol{F}_{hkl}^{obs}\right|$.

This process may be repeated as many times as needed until all atoms in the unit cell are located and the subsequent Fourier map(s) do not improve the model. Equations (10.5)–(10.7) may be combined with a least-squares refinement using the observed data, resulting in a more accurate model of the crystal structure, including the positional and displacement parameters of the individual atoms already included in the model. The success in the solving of the crystal structure is critically dependent on both the accuracy of the initial model (initial set of phase angles) and the accuracy of the experimental structure amplitudes. Needless to say, when the precision of the latter is low, the initial model should be more detailed and precise.

The effectiveness of Fourier transformation in locating missing atoms may be improved by using the so-called difference Fourier map, which is defined as

$$\Delta\rho_{xyz} = \frac{1}{V}\sum_{h=-\infty}^{h=+\infty}\sum_{k=-\infty}^{k=+\infty}\sum_{l=-\infty}^{l=+\infty}\left(\mathbf{F}_{hkl}^{obs} - \mathbf{F}_{hkl}^{calc}\right)\exp[-2\pi\mathbf{i}(hx+ky+lz)] \quad (10.8)$$

which is equivalent to Eq. (10.5), except that the observed structure amplitudes are reduced by the complex structure amplitudes calculated from the existing model. The resulting Fourier map, therefore, will not contain "known" atoms and only the missing ones should become "visible." This approach works especially well when locating weakly scattering atoms in the presence of strong scatterers (e.g., when locating hydrogen atoms in organic and especially metal-organic compounds) because the difference Fourier map reveals fine details that may be otherwise hidden in the background of the normal (Eq. 10.5) map.

The difference electron density distribution also finds an interesting application when the fully refined model of the crystal structure is used to compute a Fourier transformation. Although it may seem that such a Fourier map should result in zero electron density throughout the unit cell (since the differences in Eq. 10.8 are expected to approach zero), this is true only if the electron shells of atoms in the crystal structure were not deformed. In reality, atoms do interact and form chemical bonds with their neighbors. This causes a redistribution of the electron density compared to isolated atoms, for which atomic scattering functions are known.

Thus, the difference Fourier map calculated using Eq. (10.8) for a complete and accurate model, using highly precise X-ray diffraction data reveals the so-called deformation electron density distribution. The latter is essentially a difference between the electron density in a real crystal, composed of chemically bound atoms, and individual isolated atoms or ions, which sometimes enables the visualization of excess electron density due to the formation of chemical bonds.[4] This technique has many restrictions and requirements, the major ones being: extremely high accuracy of the experimental diffraction data (including the accuracy of the observed structure amplitudes) and availability of data at high $\sin\theta/\lambda$ (often requiring low temperatures to achieve this). Unfortunately, powder diffraction fails to meet the first requirement, except for very simple structures, due to the intrinsic and often unavoidable overlapping of Bragg reflections.

Other types of Fourier transformations may also be calculated when the coefficients in Eq. (10.5) are modified or substituted. For example, when squared observed structure amplitudes $\left|\boldsymbol{F}_{hkl}^{obs}\right|^2$ (for which phase angles are not required!) or normalized structure amplitudes $\boldsymbol{E}_{hkl}^{obs}$ are used instead of $\boldsymbol{F}_{hkl}^{obs}$, the resultant maps usually provide

---

[4] See P. Coppens, X-ray charge densities and chemical bonding. IUCr Texts on Crystallography 4, Oxford University Press, Oxford, New York (1997).

means to solve the crystal structure (see Sect. 10.2). Different modifications of $\boldsymbol{F}_{hkl}^{obs}$ may reveal the distribution of the electrostatic potential and other properties of crystals.

10.2 Phase Problem

Despite the apparent simplicity with which a crystal structure can be restored by applying Fourier transformation to diffraction data (Eqs. 10.5–10.8), the fact that the phase, or the sign in the case of centrosymmetric structures, of the structure amplitude is lost creates the so-called phase problem. In the simplest case (Eq. 10.6), both the absolute values of the structure amplitudes and their phases (Eq. 9.19) are needed to locate atoms in the unit cell. The former are relatively easily determined from powder (Eq. 8.41) or single-crystal diffraction data, but the latter cannot be obtained experimentally.

Determination of the crystal structure of an unknown material is generally far from a straightforward procedure, especially when only powder diffraction data are available. It is truly a problem-solving process and not a simple refinement. The latter is a technique that improves structural parameters of the approximately or partially known model, usually by using a least-squares minimization against available diffraction data, and often may be fully automated. It is worth noting that the least-squares method is inapplicable to the ab initio structure solution because the phases of the structure amplitudes are unknown. Thus, during the crystal structure solving process, phase angles, which have been lost, must be recovered using suitable numerical technique.

A large variety of methods, developed with the specific goal to solve the crystal structure from diffraction data, can be divided into two major groups. The first group entails techniques that are applicable in direct space by constructing a model of the crystal structure from considerations other than the available array of structure amplitudes. These include:

- Purely geometrical modeling in the case of simple inorganic structures
- Examining various ways of packing and differences in conformations of molecules with known geometry when dealing with molecular structures
- Finding analogies with closely related compounds, such as isostructural series of intermetallic compounds and partially isostructural host frameworks in various intercalates
- Using a range of minimization methods, including quantum chemical, energy, entropy, and geometry optimizations, and other recently developed advanced techniques

When one or more models are constructed, they are tested against experimental diffraction data. Often, some of these approaches are combined, but they always stem from the requirement that the generated model must make physical, chemical, and crystallographic sense. Thus, their successful utilization requires a certain level of experience and knowledge of how different classes of crystals are built, e.g., what to expect in terms of coordination and bond lengths for a particular material based solely on its chemical composition. Direct-space modeling approaches will be discussed, to some extent, in Sect. 15.1 and Chap. 24.

The second group of methods utilizes an experimental array of diffraction data, specifically the absolute values of structure amplitudes, to provide initial clues about the crystal structure of a material. These methods are applicable in reciprocal space. The first of the two reciprocal space methods reviewed in this section is the Patterson technique, which is best known for its applications as the "heavy-atom method."

Furthermore, as we will see later, even though the phase angles of Bragg reflections are not directly observed or measured, they usually have certain relationships with one another and with the absolute values of structure amplitudes. This property supports a second reciprocal space approach: the so-called direct phase determination techniques or simply "direct methods." The latter are always referred to in the plural form because they are based on several fundamental principles and typically incorporate various algorithms combined together.

Needless to say, the crystal structure determined using any of the reciprocal space methods should also be reasonable from physical, chemical, and crystallographic perspectives.

10.2.1 Patterson Technique

As suggested by Patterson[5,6] in 1934, the complex coefficients in the forward Fourier transformation (Eqs. 10.2 and 10.5) may be substituted by the squares of structure amplitudes, which are real. Therefore, no information about phase angles is required to calculate the distribution of the following density function in the unit cell:

$$P_{uvw} = \frac{2}{V} \sum_{h=-\infty}^{h=+\infty} \sum_{k=-\infty}^{k=+\infty} \sum_{l=-\infty}^{l=+\infty} \left|\mathbf{F}_{hkl}^{obs}\right|^2 \cos[2\pi(hu + kv + lw)] \tag{10.9}$$

Here, the multiplier two appears because only one-half of reciprocal space is used in the summation; thus, the validity of Friedel's law is implicitly assumed.

[5] Arthur Lindo Patterson (1902–1966), the British-born physicist is best known for the introduction the function named after him. These two papers have been written during Patterson's tenure at the Massachusetts Institute of Technology. In 1980, the American Crystallographic Association has established the A.L. Patterson award "To recognize and encourage outstanding research in the structure of matter by diffraction methods." See https:/www.amercrystalassn.org/awards for details.

[6] A.L. Patterson, A Fourier series method for the determination of the components of the interatomic distances in crystals, Phys. Rev. **46**, 372 (1934). https://doi.org/10.1103/PhysRev.46.372

The resultant function, unfortunately, does not reveal the distribution of atoms in the unit cell directly, but it represents the distribution of interatomic vectors, all of which begin in a common point—the origin of the unit cell. Thus, P_{uvw} is often called the function of interatomic vectors and is also known as the Patterson function of the F^2-Fourier series. The corresponding vector density distribution in the unit cell is known as the Patterson map.

The interpretation of the Patterson function is based on a specific property of Fourier transformation (denoted as $\Im[\ldots]$) when it is applied to convolutions ($\otimes$) of functions:

$$\begin{aligned} \Im[f(x) \otimes g(x)] &= \Im[f(x)]\Im[g(x)] \\ \Im[f(x)g(x)] &= \Im[f(x)] \otimes \Im[g(x)] \end{aligned} \qquad (10.10)$$

As follows from Eq. (10.10), the multiplication of functions in the reciprocal space (e.g., structure amplitudes) results in a convolution of functions (e.g., electron or nuclear density) in the direct space and vice versa. Since Eq. (10.9) contains the structure amplitude multiplied by itself, the resultant Patterson function, P_{uvw}, represents a self-convolution of the electron (nuclear) density. Hence, it may be described as follows:

$$P_{uvw} = \int_V \rho_{x,y,z}\rho_{x-u,y-v,z-w}dV \qquad (10.11)$$

In this equation, P_{uvw} at every point (u, v, w) inside the unit cell is calculated as the sum (integral) of product of the electron (or nuclear) density at two points separated by a vector (u, v, w).

For simplicity, assume that the distribution of electron or nuclear density in the unit cell is discrete rather than continuous and is zero everywhere except for the locations of atoms, which are viewed as dimensionless points (see Fig. 10.2, which illustrates that both electron and nuclear densities decrease rapidly away from the centers of atoms). Under this assumption, the result of Eq. (10.11) is a set of peaks originating in (0, 0, 0) and ending at (u, v, w) with heights (more precisely with peak volumes since atoms are not dimensionless points), H_{ij}, given as:

$$H_{ij} \propto Z_i Z_j \qquad (10.12)$$

where Z_i and Z_j are the number of electrons (or scattering lengths) of the ith and jth atoms that are connected by a vector (u, v, w) defined as $\boldsymbol{u} = \pm[\boldsymbol{x}_i - \boldsymbol{x}_j]$.

According to this interpretation of Eq. (10.11), the value of the Patterson function is zero at any other point in the unit cell. An example of an idealized Patterson function corresponding to a simple two-dimensional structure containing a total of four atoms in the unit cell is shown in Fig. 10.3.

Thus, since a Patterson map contains peaks, which are related to the real distribution of atoms in the unit cell, it is possible to establish both the coordinates of

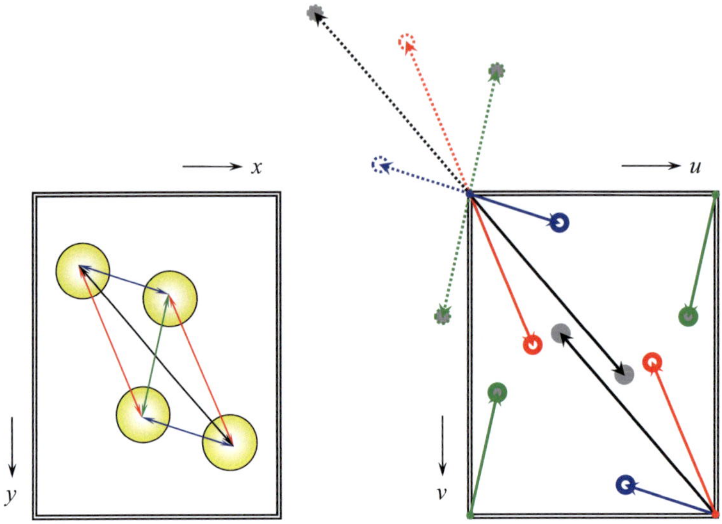

Fig. 10.3 The relationships between the distribution of atoms (*left*) and the Patterson function (*right*) in a two-dimensional unit cell. All possible interatomic vectors are drawn on the left. On the right, they are brought to a common origin (*upper left corner*) of the unit cell. Vectors that are outside the unit cell are shown using dotted lines. The content of one unit cell in the Patterson space (*right*) is shown using solid lines. *Blue* and *red circles* depict a twofold increase in the height of the corresponding peaks of the Patterson function when compared with those marked using *gray* and *green circles*, which occurs due to a complete overlap of vectors coinciding with the parallel sides of the parallelogram of atoms on the *left*

atoms and their scattering power by analyzing coordinates and heights of Patterson peaks. Unfortunately, the analysis of the distribution of the interatomic vector density function is sometimes easier said than done due to the presence of several complicating factors.

The first difficulty is that Patterson peaks are usually broader than electron (or nuclear) density peaks, which is the result of convolution (Eq. 10.11). The second complication is that the total number of Patterson peaks in the unit cell equals to $n(n-1)$, where n is the total number of atoms in the unit cell (see Fig. 10.3 where four atoms shown on the left produce 12 Patterson peaks shown on the right, four pairs of which are completely overlapped). The third difficulty is derivative of the first two and it arises from the overlapping (often quite substantial) of different interatomic vectors.

An example of the distribution of the interatomic vectors density function in the $u0w$ plane of $CeRhGe_3$ is illustrated in Fig. 10.4. When compared with the electron and nuclear density distributions (Fig. 10.2), there are many more peaks in the two Patterson maps. Similar to the results shown in Fig. 10.2, both Patterson functions are nearly identical except for the distribution of peak intensities, which is expected due to the differences in the scattering ability of Ce, Rh, and Ge when using X-rays and neutrons.

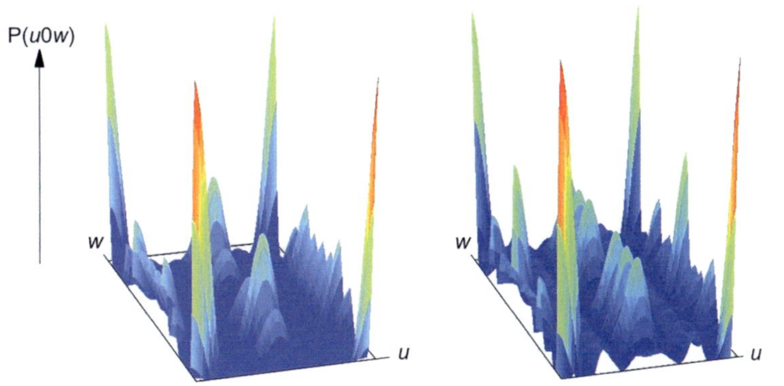

Fig. 10.4 Patterson functions calculated in the $u0w$ plane using Eq. (10.9) and employing experimental X-ray (*left*) and neutron (*right*) powder diffraction data shown in Fig. 10.1. The strongest peak in any Patterson function is always observed at (0, 0, 0) because the origins of all vectors coincide with the origin of coordinates. In this particular example, since the real crystal structure contains an atom at (0, 0, 0) (see Fig. 10.2), some of the peaks on the Patterson map correspond to the actual locations of atoms (i.e., $\mathbf{u}^j = \mathbf{x}^j \pm 0$). The contour of the unit cell is shown schematically as the rectangle beneath each Patterson map

The complicating factors mentioned here reduce the resolution of the Patterson map, which may make it extremely difficult or impossible to recover the atomic structure, especially in cases of complex crystal structures containing many atoms with nearly equal atomic numbers, such as organic compounds. Conversely, when only a few atoms in the unit cell have much stronger scattering ability than the rest, identifying Patterson peaks corresponding to these strong scatterers is relatively easy According to Eq. (10.12), these peaks are stronger than all others, except for the peak at the origin. The latter is used to scale the Patterson map since by definition (Eq. 10.11), its magnitude is the sum of the squared scattering factors of all atoms in the unit cell; in X-ray diffraction, it becomes the sum of squared atomic numbers.

The application of the Patterson technique to locate strongly scattering atoms is often called the heavy-atom method. This name arises from the fact that heavy atoms scatter X-rays more effectively, and the Patterson technique is most often applied to analyze X-ray diffraction data. This method allows constructing of a partial structure model consisting of "heavy" atoms only, which primarily define phase angles of all reflections (see Eq. 9.21). The "heavy atoms-only" model can be relatively easily completed using sequential Fourier syntheses (either standard, Eq. 10.6, and/or difference, Eq. 10.8), usually enhanced by a least-squares refinement of all found atoms.

The analysis of the Patterson function requires extensive use of symmetry. Consider all possible interatomic vectors (calculated as $\mathbf{u}_{ij} = \pm[\mathbf{x}_i - \mathbf{x}_j]$) originating from an atom in the general site position of the space group $P2_1/m$, which are listed in Table 10.1. Only three of the vectors (shaded in the first row of the table) are

Table 10.1 Interatomic vectors (shown in bold) produced by an atom located in the general site position in the monoclinic crystal system, space group P2$_1$/m

Symmetry element	Symmetry operation	*x,y,z*	-*x*,½+*y*,-*z*	-*x*,-*y*,-*z*	*x*,½-*y*,*z*
1	*x,y,z*	0,0,0	**2*x*, ½, 2*z***	**2*x*, 2*y*, 2*z***	**0, -½+2*y*, 0**
2$_1$	-*x*,½+*y*,-*z*	**-2*x*, ½, -2*z***	0,0,0	**0, ½+2*y*, 0**	**-2*x*, 2*y*, -2*z***
$\bar{1}$	-*x*,-*y*,-*z*	**-2*x*, -2*y*, -2*z***	**0, - ½-2*y*, 0**	0,0,0	**-2*x*, -½, -2*z***
m	*x*,½-*y*,*z*	**0, ½-2*y*, 0**	**2*x*, -2*y*, 2*z***	**2*x*, ½, 2*z***	0,0,0

unique, and the relationships between them are established by the combination of symmetry elements in the unit cell.

Thus, a vector 2*x*, ½, 2*z* is the result of a 2_1 screw axis parallel to the *b*-axis, and when this vector is correctly identified on the Patterson map, the two coordinates *x* and *z* of the corresponding atom are found. A second vector 0, −½ + 2*y*, 0 is due to a mirror plane and it yields the missing coordinate *y*. Finally, a vector 2*x*, 2*y*, 2*z* is due to a center of inversion, which in this case could be used to confirm all three coordinates.

When a structure contains only a single independent heavy atom, the solution is nearly always trivial. It may be possible to manually solve a structure with two to four independent heavy atoms, although the task becomes much more challenging. Solving a structure from a Patterson map for more complex crystal structures is typically performed using computer programs.

A more detailed analysis of Table 10.1 indicates that the symmetry of the Patterson function is different from that of the original crystal structure. A comprehensive examination of Patterson symmetry exceeds the scope of this book. However, it can be shown that symmetry elements with a translational component present in the crystal structure space are transformed into conforming finite symmetry elements in the Patterson function space, except for the lattice translations, which are preserved. Thus, screw axes become rotation axes, and glide planes are transformed into mirror planes. Moreover, a center of inversion is always added to the symmetry of the Patterson function. For example, the space group P2$_1$/m discussed here results in Patterson symmetry corresponding to space group P2/m, I$\bar{4}$2m is transformed into I4/mmm, Fdd2 turns into Fmmm, and so on. A complete list of Patterson function symmetry for all space groups can be found in the International Tables for Crystallography, Vol. A.

10.2.2 *Direct Methods*

In this approach, the phase angles of reflections are derived directly from the observed structure amplitudes through mathematical relationships between the intensities and indices of the reflections. These relationships are based on the following postulations:

- The electron density is nonnegative anywhere in the unit cell, that is, $\rho_{xyz} \geq 0$ for all x, y, z.
- The atomic structure is composed of nearly spherical atoms spread almost evenly throughout the unit cell volume.

These two general properties of the electron density result in special relationships between the phase angles of triplets of reflections, which have arithmetically (but not symmetrically) related indices. The triplets of related reflections are defined as follows:

$$\begin{array}{lll} \text{First reflection,} & \mathbf{h}: & h, k, l \\ \text{Second reflection,} & \mathbf{h}': & h', k', l \\ \text{Third reflection,} & \mathbf{h}-\mathbf{h}': & h-h',\ k-k',\ l-l' \end{array} \tag{10.13}$$

The phase relationships within a triplet are not strict, and their probability depends on the magnitude of the associated structure amplitudes. These amplitudes are scaled and normalized to reduce their dependence on the atomic scattering factors and vibrational motions, as both reduce the structure amplitude exponentially at high $\sin\theta/\lambda$; see Eq. (9.5) and Figs. 9.2 and 9.4. The normalized structure factor is commonly denoted as E_{hkl}, and it is calculated from the conventional structure factor as follows:

$$E_{hkl} = \frac{|\boldsymbol{F}_{hkl}|}{\left\langle F_{\text{exp}}^2 \right\rangle^{1/2}} \tag{10.14}$$

where the expected average value of the structure factor, $\left\langle F_{\text{exp}}^2 \right\rangle$, is estimated as:

$$F_{\text{exp}}^2 = \sum_{j=1}^{n} f_j^2(s) \tag{10.15}$$

and $f_j(s)$ is the atomic scattering factor of the jth atom; s is $\sin\theta/\lambda$ of the reflection (hkl).

In the *centrosymmetric* structures, the relationships between the signs of the reflections forming a triplet (Eq. 10.13) are described by the Sayre equation[7]:

$$\mathrm{s}_{\mathbf{h}} \approx \mathrm{s}_{\mathbf{h}'} \mathrm{s}_{\mathbf{h}-\mathbf{h}'} \tag{10.16}$$

[7] D. Sayre, The squaring method: a new method for phase determination, Acta Cryst. **5**, 60 (1952). https://doi.org/10.1107/S0365110X52000137

where s = +1 when $\alpha_{hkl} = 0$ (positive F_{hkl}), s = −1 when $\alpha_{hkl} = \pi$ (negative F_{hkl}), and the symbol ≈ has a meaning of "probably equal"; for example, $s_{123} \approx s_{102}s_{021}$.

The probability of this sign relationship in the triplet is defined as:

$$P_+ = ½ + ½ \tanh\left[N^{-1/2}|E_{\mathbf{h}}|E_{\mathbf{h}'}E_{\mathbf{h}-\mathbf{h}'}\right] \tag{10.17}$$

where P_+ is the probability of the positive sign for a reflection **h**, and:

$$N^{-1/2} = \sigma_3/\sigma_2^{3/2} \quad \text{and} \quad \sigma_m = \sum_{i=1}^{n} Z_i^m \tag{10.18}$$

Here, Z is the atomic number and n is the number of atoms in the unit cell.

In the noncentrosymmetric structures, reflection phases in the triplet are in the following relationship, also given by Sayre (Footnote 7):

$$\alpha_{\mathbf{h}} \approx \alpha_{\mathbf{h}'} + \alpha_{\mathbf{h}-\mathbf{h}'} \tag{10.19}$$

where the symbol ≈ also means "probably equal" and the probability of this relationship for the phase $\alpha_{\mathbf{h}}$ is defined by Cochran as[8]:

$$P_{\alpha_{\mathbf{h}}} = \frac{1}{2\pi I_0\left(k_{\mathbf{h},\mathbf{h}'}\right)} \exp\left[k_{\mathbf{h},\mathbf{h}'} \cos\left(\alpha_{\mathbf{h}'} + \alpha_{\mathbf{h}-\mathbf{h}'}\right)\right] \tag{10.20}$$

where I_0 is Bessel function, and $k_{\mathbf{h},\mathbf{h}'} = 2N^{-1/2}|E_{\mathbf{h}}E_{\mathbf{h}'}E_{\mathbf{h}-\mathbf{h}'}|$.

In order to generate phase angles using either of the two Sayre equations (10.16) or (10.19), some reflections with known phases are needed, as both equations define one phase from two others that are known. There are several sources of reflection phases at this stage. First, certain phase angles may be set at arbitrary values to fix the origin, and in the case of noncentrosymmetric structures, one additional phase with an arbitrary value is needed to fix the enantiomorph. Second, phase angles of several strong reflections are selected for permutations. For example, in the centrosymmetric case, a total of four reflections would have 2^4 possible combinations of signs, resulting in 16 different sets of possible phase angles.

For each permutation, the phases of all other reflections are generated using the Sayre equations. Thus, direct methods always result in more than one array of phases, reducing the problem to selecting the correct solution, if one exists. Several different figures of merit and/or their combinations have been developed and are used to evaluate the probability and relationships between phases. Consequently, the solutions are sorted according to their probability, from the highest to the lowest. Each solution is analyzed and evaluated, starting from the one that is most probable.

[8] W. Cochran, Relations between the phases of structure factors, Acta Cryst. **8**, 473 (1955). https://doi.org/10.1107/S0365110X55001485

Usually, each reflection forms more than one triplet, and each of the triplets may be used for phase determination (estimation). To employ all triplets and obtain the best agreement between phase angles resulting from different triplets, Karle and Hauptman[9,10] introduced a general expression for phase determination from triplets. This relationship is known today as the tangent formula:

$$\tan \alpha_{\mathbf{h}} \cong \frac{\sum_{\mathbf{h}} |E_{\mathbf{h}'}||E_{\mathbf{h}-\mathbf{h}'}| \sin(\alpha_{\mathbf{h}'} + \alpha_{\mathbf{h}-\mathbf{h}'})}{\sum_{\mathbf{h}} |E_{\mathbf{h}'}||E_{\mathbf{h}-\mathbf{h}'}| \cos(\alpha_{\mathbf{h}'} + \alpha_{\mathbf{h}-\mathbf{h}'})} \tag{10.21}$$

where the sums include all triplets in which the reflection in question, **h**, is involved.

Finally, the generated phase angles are used in a forward Fourier transformation combined with the normalized structure amplitudes E_{hkl}:

$$\rho_{xyz} = \frac{1}{V} \sum_{h=-\infty}^{h=+\infty} \sum_{k=-\infty}^{k=+\infty} \sum_{l=-\infty}^{l=+\infty} E_{hkl} \cos\left[2\pi i\left(hx + ky + lz - \alpha_{hkl}^{direct}\right)\right] \tag{10.22}$$

The resulting density distribution is called the E-map. It is quite similar to a conventional Fourier map representing electron density, but E-maps are usually sharper because the normalized structure amplitudes, E_{hkl}, are corrected for the effects of atomic scattering and thermal displacements.

A generic algorithm of employing direct methods to solve a structure can be summarized in the following steps:

1. Observed structure amplitudes are scaled, and normalized structure amplitudes are calculated using Eq. (10.14). Reflections with large normalized structure amplitudes (a standard cut-off is $E_{hkl} \geq E_{min} = 1.2$) are selected for phase determination and refinement.
2. Reflections to fix the origin and the enantiomorph (if needed) are selected. There are special requirements for the combinations of indices of these reflections, which are established by space-group symmetry.
3. Several of the strongest reflections (those that have the largest E_{hkl}) are chosen for phase permutations. In addition to being the strongest, these reflections should have combinations of indices that result in as many triplets as possible with all reflections selected in Step 1.
4. Phases are assigned to all, or as many as possible, reflections selected in Step 1 using the Sayre equations (Eq. 10.16 or 10.19) and the probability relationships (Eq. 10.17 or 10.20).

[9] Jerome Karle, US crystallographer, and Herbert A. Hauptman, US mathematician, laid a foundation toward the development of modern direct phase determination techniques. They won the 1985 Nobel Prize in Chemistry "for their outstanding achievements in the development of direct methods for the determination of crystal structures"—https://www.nobelprize.org/prizes/chemistry/1985/

[10] J. Karle and H. Hauptman, A theory of phase determination for the four types of non-centrosymmetric space groups, Acta Cryst. **9**, 635 (1956). https://doi.org/10.1107/S0365110X56001741

5. The best-possible agreement between phases is obtained using a least-squares refinement in combination with the tangent formula (Eq. 10.21).
6. Figures of merit are calculated, and the resultant sets of phase angles are sorted from the most probable to the least probable solution.
7. An E-map (Eq. 10.22) is computed for the most probable solution. The peaks are located on the map, and a partial or complete model of the crystal structure is created.
8. The obtained model of the crystal structure is analyzed with respect to common chemical and crystallographic sense: are all bond distances, angles, coordination polyhedra, etc., reasonable? If yes, move to Step 9. If no, return to Step 7 using the next best solution.
9. The model of the crystal structure is verified and completed by computing phases for all available (conventional) structure amplitudes using the current structural model (Eq. 9.19) and successive calculation of Fourier (Eq. 10.6) and/or difference Fourier maps (Eq. 10.8). Once all atoms are located, the complete structure is refined using the least-squares technique against all available diffraction data.
10. If no solution is found, Step 2 should be repeated with different parameters and lists of reflections in the starting sets. It may be necessary to expand or reduce the list of reflections under consideration by changing the cut-off value of E_{min} from the standard value of 1.2.

10.2.3 Structure Solution from Powder Diffraction Data

Solving the crystal structure using either heavy atom or direct techniques is not always straightforward, even when well-resolved and highly accurate diffraction data from a single crystal are available. In powder diffraction, the intrinsic exact and incidental approximate overlap of multiple Bragg peaks complicates the process. This overlap can become especially severe as the unit cell volume and structural complexity increase.

Thus, there is a fundamental difference between the accuracy of structure amplitudes obtained from single-crystal and powder diffraction data. The former are always resolved, and there is only one reflection (combination of indices), *hkl*, per Bragg peak, whereas in the latter some reflections may be fully or partially overlapped.[11] The intensities of individual reflections *hkl* may still be recovered

[11] A similar overlap may occur even in single crystal data when merohedral twinning results in partial or full overlap of some Bragg reflections, while non-merohedral twinning results in a complete overlap of all reflections. Yet, even in cases of this severity, single crystal data contain more unbiased information because the measured diffraction pattern is three-dimensional, and also because every powder pattern is, in a way, always "merohedrally twinned" for all non-centrosymmetric structures as well as low symmetry trigonal, hexagonal, tetragonal, and cubic point groups, regardless of whether the crystallites are twinned or not.

from powder diffraction data but their accuracy is critically dependent on both the degree of the overlap and the quality of the pattern. However, the absolute or near absolute overlapping of some reflections makes it impossible to obtain the individual intensities regardless of the quality of data, and only the combined total intensity is known (e.g., reflections 43*l* and 05*l*, which are not symmetrically equivalent, in both the cubic and tetragonal crystal systems).

As established earlier, individual intensities (or structure amplitudes) of Bragg reflections are needed in order to solve the crystal structure using direct or Patterson methods. In direct methods, accurate normalized structure amplitudes are required to generate phase angles and to evaluate their probabilities. In Patterson methods, accurate structure amplitudes result in higher accuracy and resolution of the Patterson map.

Thus, when reflections overlap to a degree where the individual intensities can no longer be considered reliable, they may be dealt with using two different approaches:

- *Discarding low-accuracy reflections*: Reflections with low accuracy in individual intensities, typically those that are completely or nearly completely overlapped, are discarded. This approach is most effective when using direct methods for structure solution, as substantial errors, even in some normalized structure amplitudes can impact the phase angles of many other reflections.
- *Dividing total intensity*: The total intensity of the diffraction peak is divided equally among the individual reflections, represented as $I_{\text{total}} = \Sigma I_i$. An alternative "blind" division accounts for the multiplicity factors of different Bragg reflections, so that $I_{\text{total}} = \Sigma m_i I_i$, where m_i is the multiplicity factor of the ith reflection, depending on symmetry and combination of indices (see Sects. 8.6.3 and 9.2.2). No clear preference can be given to any method of intensity division, as each is quite arbitrary. This approach of handling the overlapped intensities, rather than discarding them, is most beneficial in the Patterson method.

Even when the crystal structure is partially solved, the individual intensities are still needed to complete the structure by means of calculating Fourier or difference Fourier maps. Obviously, the result of Fourier transformations is affected by the accuracy in the absolute values of the structure amplitudes in addition to the precision of their phase angles (Eqs. 10.6 and 10.8).[12] Considering fully overlapped Bragg reflections, the situation with prorating individual intensities becomes different compared to when the crystal structure is completely unknown: at this point, the calculated intensities of all individual reflections are known with the accuracy of the current structural model. These calculated intensities may be, and usually are, used to divide the total intensity of the peak between all overlapped reflections proportionally to their calculated intensities (see Eq. 15.7 in Sect. 15.4).

The division of intensities of the overlapped Bragg reflections is critical only when they are needed to calculate Patterson, Fourier, or E-map(s). There is no need

[12] It turns out that precisely known phase angles are more critical in determining the coordinates of atoms in the unit cell than the corresponding absolute values of individual structure factors.

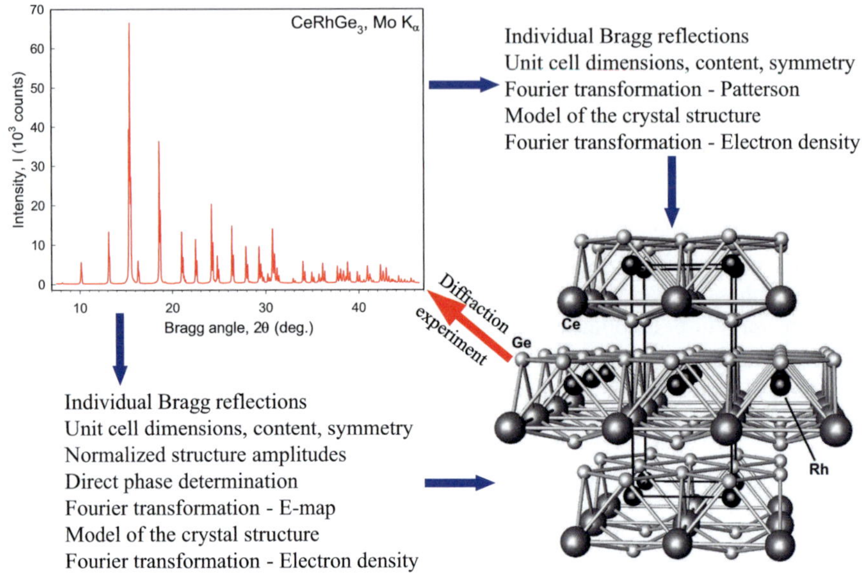

Fig. 10.5 The schematic illustrating how the powder diffraction pattern (*top left*) can be transformed into the image of the crystal structure in three dimensions (*bottom right*). Both the diffraction pattern and the model of the crystal structure represent the intermetallic compound $CeRhGe_3$

for their separation during a least-squares refinement of structural parameters because each point of the diffraction profile is simply taken as a sum of contributions from multiple Bragg reflections.[13]

We conclude this section where we began this chapter by repeating the following statement: "The diffraction pattern of a crystal is a transformation of an ordered atomic structure into reciprocal space rather than a direct image of the former, and the three-dimensional distribution of atoms in a lattice can be restored only after the diffraction pattern has been transformed back into direct space." We now are able to support this statement with Fig. 10.5, which illustrates how this can be performed in practice.

The very existence of the powder diffraction pattern, which is an experimentally measurable function of the crystal structure and other parameters of the specimen convoluted with various instrumental functions, has been made possible by the commensurability of properties of X-rays and neutrons with the properties and structure of solids. As in any experiment, the quality of structural information, which may be obtained via different pathways (two possibilities are illustrated in Fig. 10.5 as two series of required steps), is directly proportional to the quality of experimental data. The latter is usually achieved in a thoroughly planned and well-

[13]The least-squares refinement of structural parameters employing full profile powder diffraction data is discussed beginning with Chap. 15.

executed experiment as detailed in Chap. 12. Similarly, each of the data processing steps, which were described in this chapter and are summarized in Fig. 10.5, requires knowledge, experience, and careful execution. We will describe these steps in Chaps. 13–15, followed by numerous examples starting from Chap. 16.

10.3 Total Scattering Analysis Using Pair-Distribution Function

So far, we have only been concerned with diffraction from ideally periodic structures with a well-defined long-range order. However, diffraction, or more precisely, interference of scattered waves, occurs when either or both long- or short-range order is missing or is approximate. In other words, diffraction can also be observed from low crystallinity and amorphous solids, as well as from nanocrystalline materials. These materials normally produce diffraction patterns that have broad peaks, sometimes very broad halos over the instrumental background, instead of sharp diffraction maxima.

There are two kinds of low crystallinity or amorphous solids. Consider a normal crystalline material whose crystallinity is reduced to some extent by decreasing particle or grain size and/or by introducing local strains. This will reduce long-range order by reducing the length scale over which translational periodicity is present and will lead to broadening of Bragg peaks. Regardless of how broad the peaks are, the diffraction pattern will still resemble that of the original crystalline material. Even when the broadening is severe and only a few humps are visually evident in the pattern, these humps will still coincide with the locations of the strongest Bragg peaks from a crystalline phase. Hence, in this case, it is correct to say that lattice and order are still there, but a significant disorder is also present. Another kind of amorphous solid, or solid in the so-called glassy state, does not have, nor has ever had, nor is derived from, lattice and long-range order. Therefore, broad peaks from a glassy phase do not match the locations of crystalline Bragg peaks.

An important difference between these two kinds of amorphous solids is directly related to the scope of this book. The crystal structure of the first kind of materials, which have a distorted long-range order and can be described by a distorted lattice, can still be confirmed by the same methods that are standard for crystalline materials, i.e., the Rietveld method (see Chap. 15). Obviously, relevant peak-broadening corrections must be applied when doing so. When significant, these corrections may be quite inaccurate, and a structural model can only be estimated at the qualitative or atomic connectivity level. It is also clear that this approach cannot be used when dealing with glassy materials, such as polymers and low-crystallinity macromolecular solids, and a variety of nanomaterials, such as nanoparticles, nanotubes, and others.

In contrast to discrete diffraction patterns of crystalline solids that consist of well-defined Bragg peaks, which correspond to points of the reciprocal lattice intersecting the Ewald's sphere (see Sect. 8.1), diffraction patterns from low crystallinity or amorphous solids are treated as continuous. The former requires that the crystal lattice must be established first before any structure determination can be performed, which may be challenging enough even when dealing with a well-ordered material (see Chap. 14). Ab initio determination of the lattice becomes a nearly impossible task when periodicity is present on a short-length scale, simply because precision in the determination of Bragg angles becomes very low. On the contrary, treatment of a continuous diffraction pattern, which is also known in modern literature as the total scattering analysis, does not require any knowledge about the lattice.

Recently, total scattering analysis techniques, which are based on a pair-distribution function (PDF), have been under intense development for structural studies of low crystallinity and amorphous solids. The name—total scattering—reflects the fact that all points of the powder diffraction pattern are analyzed, even those that in conventional powder diffraction are considered useless, such as the background. Another name—pair distribution—places emphasis on pairs of atoms, or on the distribution of interatomic vectors,[14] which is directly calculable from a continuous pattern and is used for the comparison with the pattern calculated from a structural model and/or for improving the model. This approach is often referred to as "beyond Bragg…,[15]" "beyond crystallography…,[16]" or "underneath the Bragg peaks." The latter was adopted by Egami and Billinge as the title of a book,[17] which is recommend to reader interested in a comprehensive coverage of the total scattering analysis techniques and their applications. Here, only a brief description of the main principles of the method is given.

The use of the continuous diffraction pattern instead of the discrete Bragg peaks requires replacing the diffraction vectors **h** in Eqs. (10.1) and (10.2) with the vector **Q** (see Eq. 7.15 and the corresponding footnote 9 on page 168, Chap. 7). In the absence of a lattice, or when only a heavily distorted long-range order is present, the atomic coordinate vectors **x** are replaced with the interatomic vectors **r**. Thus, the X-ray intensity (in electrons) scattered by an array of atoms present in either a

[14] Here, interatomic vectors are identical to those considered above during the discussion of the Patterson function (see Sect. 10.2.1). The difference is that in the total scattering analysis, the experimentally observed distribution of the interatomic vectors is a one-dimensional direct space projection of the three-dimensional function of interatomic vectors.

[15] V. Petkov, P.Y. Zavalij, S. Lutta, M.S. Whittingham, V. Parvanov, S. Shastri, Structure beyond Bragg: Study of V_2O_5 nanotubes, Phys. Rev. B **69**, 085410 (2004). https://doi.org/10.1103/PhysRevB.69.085410

[16] S.L.J. Billinge, M.G. Kanatzidis, Beyond crystallography: the study of disorder, nanocrystallinity and crystallographically challenged materials with pair distribution functions, Chem. Comm. 749 (2004). https://doi.org/10.1039/B309577K

[17] T. Egami and S.J.L. Billinge, Underneath the Bragg peaks. Structural analysis of complex materials. Pergamon (2003). https://doi.org/10.1016/S1369-7021(03)00635-7

crystalline or amorphous solid as a function of interatomic distance can be represented as was given by Debye[18] back in 1915:

$$I(Q) = \sum_{i=1}^{N} \sum_{j=1}^{N} f_i f_j \frac{\sin(Qr_{ij})}{Qr_{ij}} \tag{10.23}$$

where $Q = |\mathbf{Q}| = 4\pi \sin(\theta)/\lambda$ is the magnitude of the scattering vector, $r_{ij} = |\mathbf{r}_{ij}|$ is the interatomic distance between atoms i and j, and f_i and f_j are the atomic scattering factors of atoms i and j.

This equation can be easily used to simulate a continuous powder diffraction pattern as a function of either Q or 2θ. To obtain the pattern, the sum over all interatomic distances has to be calculated. However, since many similar distances usually exist within the structure, it is more efficient to group all similar distances first. Then, one distance from each group can be used with an appropriate multiplier.

Instead of $I(Q)$ in Eq. (10.23), a different function, $G(r)$, is used in total scattering analysis. This function is known as the pair-distribution function (or reduced pair-distribution function) and can be calculated by performing a Fourier transformation of the experimental data, as shown in Eqs. (10.24) and (10.25). Therefore, it is a direct-space function:

$$G(r) = \frac{2}{\pi} \int_0^{\infty} Q \times [S(Q) - 1] \times \sin(Qr) dQ \tag{10.24}$$

Here, $S(Q)$ is the total scattering structure function, and $F(Q) = Q \times [S(Q) - 1]$ is the reduced structural function, which is calculated directly from the experimental powder pattern as follows:

$$F(Q) = Q \times \frac{I^{coh}(Q) - \sum c_i |f_i(Q)|^2}{\sum c_i |f_i(Q)|^2} \tag{10.25}$$

where $I^{coh}(Q)$ is the experimental intensity corrected for background and other effects, such as scattering by a sample holder, and normalized by the photon or neutron flux and the number of atoms in the sample, and c_i and f_i are the concentration of atoms and the atomic scattering factor, respectively.

The reduced pair-distribution function $G(r)$ can be represented as follows[19]:

$$G(r) = 4\pi r[\rho(r) - \rho_0] = 4\pi r \rho_0 [g(r) - 1] \tag{10.26}$$

[18] P. Debye, Dispersion of Röntgen rays. Ann. Phys. **46**, 809 (1915). https://doi.org/10.1002/andp.19153510606

[19] Note that when the second (the rightmost) form of Eq. (10.26) is used, $g(r)$ is the pair distribution function, while $G(r)$ becomes the reduced pair distribution function, as illustrated in Egami and Billinge's book.

where ρ_0 is the average atomic number density of the material, and $\rho(r)$ is the atomic pair density. The latter can be expressed as:

$$\rho(r) = \frac{1}{4\pi r^2 N} \sum_i^N \sum_{j \neq i}^N \frac{f_i f_j}{f^2} \delta(r - r_{ij}) \tag{10.27}$$

Here, $\delta(x)$ is the delta function, which equals one when $x = 0$ and zero everywhere else. Combining Eqs. (10.26) and (10.27) yields the following equation for the calculated $G(r)$ function:

$$G_{\text{calc}}(r) = \frac{1}{rN} \sum_i^N \sum_{j \neq i}^N \left[\frac{f_i f_j}{f^2} \delta(r - r_{ij}) \right] - 4\pi r \rho_0 \tag{10.28}$$

Thus, the calculated (Eq. 10.28) and experimental (Eq. 10.24) $G(r)$ functions can be used to compare the simulated and observed diffraction data and to improve the structural model.[20]

Let us consider the meaning of $G(r)$ and $\rho(r)$ functions. The latter is simpler as its physical meaning directly follows from Eq. (10.27). It can be described as the mean weighted[21] density of atoms[22] at a radial distance r from an atom at the origin. The atomic density function, $\rho(r)$, is nothing but the one-dimensional projection of the modified interatomic vector or Patterson function (see Sect. 10.2.1, Eq. 10.9). The $\rho(r)$ assumes only positive values, while the $G(r)$ function, being derived from $\rho(r)$ (Eq. 10.26), oscillates around the horizontal axis as depicted in Fig. 10.6, yet it features the same distribution of peaks. The physical meaning of the $G(r)$ is less intuitive than $\rho(r)$; nonetheless, it is used in total scattering analysis due to the following reasons:

- $G(r)$ can be directly obtained from raw experimental data (Eqs. 10.24 and 10.25). It can also be easily calculated from a structural model (Eq. 10.28).
- The difference plot $G_{\text{obs}}(r) - G_{\text{calc}}(r)$ has the same significance at all r values, which makes visual analysis of the model very easy. For example, large differences at high r but small differences at low r mean that the model describes the real structure best at short distances.
- Oscillations of $G(r)$ provide direct evidence of the degree of ordering. For example, the same amplitude or the same order of magnitude oscillations seen over a large range of r indicates the presence of a long-range order (Fig. 10.6,

[20] C.L. Farrow, P. Juhas, J.W. Liu, D. Bryndin, E.S. Bozin, J. Bloch, Th. Proffen, S.J.L. Billinge, PDFfit2 and PDFgui: computer programs for studying nanostructure in crystals, J. Phys.: Condens. Matter. **19**, 335219 (2007). https://doi.org/10.1088/0953-8984/19/33/335219. Software application DIFFpy (PDFfit and PDFgui) for PDF-based refinement along with a manual and a tutorial is available at https://www.diffpy.org

[21] Weighted by the atomic scattering factors.

[22] The density of atoms is the number of atoms per unit area of a sphere with radius r ($A = 4\pi r^2$).

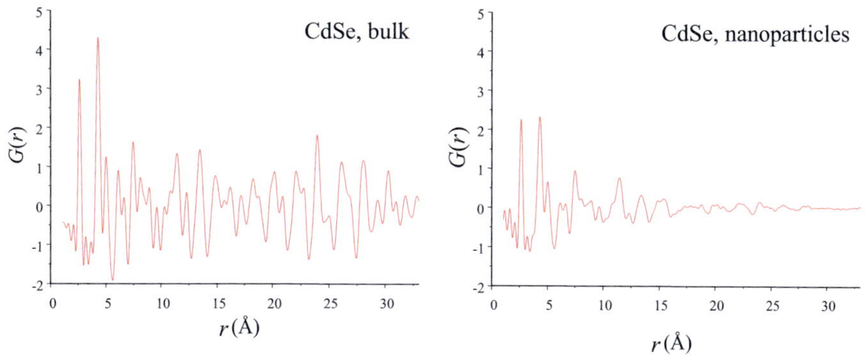

Fig. 10.6 Experimental pair-distribution function of CdSe (For more information on the CdSe refinement, see: S.K. Pradhan, Z.T. Deng, F. Tang, C. Wang, Y. Ren, P. Moeck and V. Petkov, "3D structure of CdX (X = Se, Te) nanocrystals by total X-ray diffraction", J. Appl. Phys. **102**, 044304 (2007). https://doi.org/10.1063/1.2767615) from bulk material (*left*) and 30 Å nanoparticles (*right*). Numerical data used to produce these plots were taken from the DIFFpy tutorial

left), while a rapidly diminishing amplitude above a certain value of r indicated the length scale of the order (such as the average size of nanoparticles) as seen in Fig. 10.6 (right).

– Finally, $G(r)$ is less affected by a finite observable range of Q of any experimental pattern.

Improvement of the model (structure refinement) in total scattering analysis is carried out in a similar fashion to Rietveld refinement, which is described in Chap. 15, except that the pair-distribution function $G(r)$ is used instead of a raw profile during the minimization. The minimized function (compare with Eq. 15.28 in Sect. 15.7.1) is

$$\Phi = \sum_{i=1}^{N} w_i \left(G^{\text{calc}}(r_i) - G^{\text{obs}}(r_i) \right)^2 \tag{10.29}$$

The relative simplicity of the refinement using the PDF approach, which may be considered even simpler than the Rietveld refinement, is based on the fact that no preliminary knowledge of the unit cell is required. However, the technique still requires a structural model, which is often a major challenge to create, especially when the lattice and long-range order are missing. Models are commonly generated from a parent periodic structure and then optimized using, for example, reverse Monte Carlo simulations.[23]

[23]T. Proffen, R.B. Neder, DISCUS, a program for diffuse scattering and defect structure simulations, J. Appl. Cryst. **30**, 171 (1997). https://doi.org/10.1107/S002188989600934X; T. Proffen, R.B. Neder DISCUS, a program for diffuse scattering and defect structure simulations—Update, J. Appl. Cryst. **32**, 838 (1999). https://doi.org/10.1107/S0021889899003039

Both the quality and range (maximum available values of Q) of experimental data for the total scattering analysis must be much higher compared to typical sets of data used in the Rietveld refinement. The quality (accuracy of the measured intensities) is important because every point of the pattern counts, and both random and systematic errors add up at every point of the pair-distribution function. The availability of data at high Q is critical because truncation results in termination ripples after Fourier transformation. Therefore, it is extremely important to have experimental patterns of the highest quality (high counting statistics) over the maximum possible Q range. This is easily achievable when using synchrotron radiation. Several synchrotron sources have specialized total scattering beamlines that allow for measuring high-quality powder diffraction patterns with very short wavelengths.[24] It is also possible to use laboratory diffractometers equipped with X-ray tubes producing low characteristic wavelengths, for example, tubes with Mo and Ag anodes.

Recently, the total scattering analysis has proven to be quite successful not only in studying amorphous and low-crystallinity materials, such as glasses, polymers, liquid crystals, and solids with a substantial degree of structural disorder and quasicrystals,[25] but also in the emerging field of nanomaterials.[26] The latter actually has provided great momentum to major developments of this old (see Eq. 10.23), yet nontraditional method. It is worthwhile to note that in addition to modeling structures of solids, the PDF-related methods also facilitate the examination of other properties—those that influence diffraction patterns—such as grain size and shape, microstrain, stacking faults, and other properties.

10.4 Additional Reading

1. International Tables for Crystallography, Vol. A, Space-group symmetry, M. I. Aroyo, Ed., Second online edition (2016). https://it.iucr.org/A/
   International Tables for Crystallography, Vol. C, Mathematical, physical and chemical tables, First online edition, E. Prince, Ed. (2006). https://it.iucr.org/C/
   International Tables for Crystallography, Vol. A-I. All volumes are published jointly with the International Union of Crystallography (IUCr) by Springer. https://it.iucr.org/

[24]For example, see the 11-ID-B beamline at APS (https://www.aps.anl.gov/Beamlines/Directory), which is specialized in high-energy X-ray diffraction and pair distribution function (PDF) and allows X-ray wavelengths as low as 0.1 Å.

[25]S. Brühne, E. Uhrig, K.D. Luther, W. Assmus, M. Brunelli, A.S. Masadeh, S.L. Billinge, PDF from X-ray powder diffraction for nanometer-scale atomic structure analysis of quasicrystalline alloys, Z. Kristallogr. **220**, 962 (2005). https://doi.org/10.1524/zkri.2005.220.11_2005.962

[26]S.K. Pradhan, Y. Mao, S.S. Wong, P. Chupas, V. Petkov, Atomic-scale structure of nanosized titania and titanate: particles, wires, and tubes. Chem. Mater. **19**, 6180 (2007). https://doi.org/10.1021/cm7019069

2. P. Coppens, X-ray charge densities and chemical bonding. IUCr Texts on Crystallography 4, Oxford University Press, Oxford (1997).
3. C. Giacovazzo, Direct phasing in crystallography: fundamentals and applications. IUCr monographs on crystallography 8, Oxford University Press, Oxford (1998).
4. T. Egami and S.J.L. Billinge, Underneath the Bragg peaks. Structural analysis of complex materials. Pergamon Materials Series. Pergamon, Amsterdam (2003).
5. R.B. Neder and T. Proffen, Fitting of nano particle structures to powder diffraction pattern using DISCUS, p. 49 in: CPD Newsletter "2D Powder Diffraction", Issue 32 (2005). https://www.iucr.org/__data/assets/pdf_file/0003/21657/CPD32.pdf

10.5 Problems

1. A material crystallizes in space-group symmetry Cmmm. After deconvoluting a powder diffraction pattern collected from this material, a student computes the Patterson function. He finds that two strongest peaks in the Patterson function have different coordinates (0, 0, 0 and ½, ½, 0) but identical heights. Is this result expected, or was there some kind of error in his computations?
2. A total of 16 reflections have been chosen as a basis set for a direct phase determination attempt. Knowing that the crystal structure is centrosymmetric and none of the atoms scattering anomalously, calculate how many different unrestricted combinations of phases in the basis set are tested when computations are completed.

Chapter 11
Powder Diffractometry

The powder diffraction experiment is the cornerstone of a truly fundamental materials characterization technique—diffraction analysis—and it has been used for many decades with exceptional success to provide accurate information about the structure of materials. Although powder data usually lack the three-dimensionality of a diffraction image, the fundamental nature of the method is easily appreciated from the fact that each powder diffraction pattern represents a one-dimensional projection of the three-dimensional reciprocal lattice of a crystal.[1] The quality of the powder diffraction pattern is usually limited by the nature and energy of the available radiation, the resolution of the instrument, and the physical and chemical conditions of the specimen. Since many materials can only be prepared in a polycrystalline form, the powder diffraction experiment becomes the only realistic option for a reliable determination of the crystal structure of such materials.

Powder diffraction data are customarily recorded in virtually the simplest possible fashion, where the scattered intensity is measured as a function of a single independent variable—the Bragg angle. What makes the powder diffraction experiment so powerful is that different structural features of a material have different effects on various parameters of its powder diffraction pattern. For example, the presence of a crystalline phase is manifested as a set of discrete intensity maxima—the Bragg reflections—each with a specific intensity and location. When atomic parameters, for example, coordinates of atoms in the unit cell or populations of different sites in the lattice of the crystalline phase, are altered, this change affects relative intensities and/or positions of the Bragg peaks that correspond to this phase. When the changes are microscopic, for example, when the grain size is reduced to below a certain limit or when the material has been strained or deformed, the shapes of Bragg peaks

[1] We remind you once again that imaging of the reciprocal lattice in three dimensions is easily doable in a single-crystal diffraction experiment.

P. Zavalij, V. Pecharsky, *Fundamentals of Powder Diffraction and Structural Characterization of Materials*, https://doi.org/10.1007/978-3-031-91504-8_11

become affected in addition to their intensities and positions. Hence, much of the structural information about the material is embedded into its powder diffraction pattern, and when experimental data are properly collected and processed, a great deal of detail about a material's structure at different length scales, as well as its phase and chemical compositions can be established.

11.1 Brief History of the Powder Diffraction Method

The X-ray powder diffraction method dates back to Debye and Scherrer,[2] who were the first to observe diffraction from LiF powder and succeeded in solving its crystal structure. Later, Hull[3] suggested and Hanawalt, Rinn, and Frevel[4] formalized the approach enabling one to identify crystalline substances based on their powder diffraction patterns. Since that time, the powder diffraction method has enjoyed enormous respect in both academia and industry as a technique that allows one to readily identify the substance both in a pure form and in a mixture, in addition to its ability to provide information about the crystal structure (or the absence of crystallinity) of an unknown powder.

In the early days of the method, powder diffraction data were recorded on X-ray film in a variety of cameras. Using film, the resulting diffraction pattern is usually observed as a series of elliptically distorted narrow concentric ring segments (Fig. 11.1), where each ring corresponds to one or more Bragg peaks. Multiple Bragg peaks may be convoluted into a poorly resolved or completely unresolved single ring due to the limitations imposed by the one-dimensionality of the technique and by the resolution of both the film and the instrument, for example, the Debye–Scherrer camera (Fig. 11.2).

From the locations of Debye rings on the film, along with their varying intensity (degree of darkening), it is possible to identify the material and to establish its crystal structure. Given the analogue nature of the film, it is nearly as easy to grasp the overall "structure" of the diffraction pattern, as it is difficult to convert it into a digital

[2]P. Debye, and P. Scherrer, Interferenzen an regellos orientierten Teilchen in Röntgenlight, Phys. Z. **17**, 277 (1916). https://eudml.org/doc/58947, https://eudml.org/doc/58948. Paul Scherrer (1890–1969) was a Swiss physicist, best known for developing the powder diffraction method together with Peter Debye. Paul Scherrer Institute (PSI) located in Villigen, Switzerland, has been named after him in 1988.

[3]A.W. Hull, A new method of chemical analysis, J. Am. Chem. Soc. **41**, 1168 (1919). https://doi.org/10.1021/ja02229a003. Albert W. Hull (1880–1966), the American physicist credited with the invention of the magnetron, became interested in X-ray diffraction after a colloquium given by Sir W.H. Bragg. A.W. Hull then proceeded with experiments and solved the crystal structure of elemental iron. Hull's autobiography can be found in the book *50 Years of X-ray Diffraction*, P.P. Ewald, Ed. (IUCr, 1962, 1999). The electronic version of Hull's autobiography is available at https://www.iucr.org/__data/assets/pdf_file/0015/771/hull.pdf

[4]J.D. Hanawalt, H.W. Rinn, and L.K. Frevel, Chemical analysis by X-ray diffraction, Ind. Eng. Chem. Anal. **10**, 457 (1938). https://doi.org/10.1021/ac50125a001

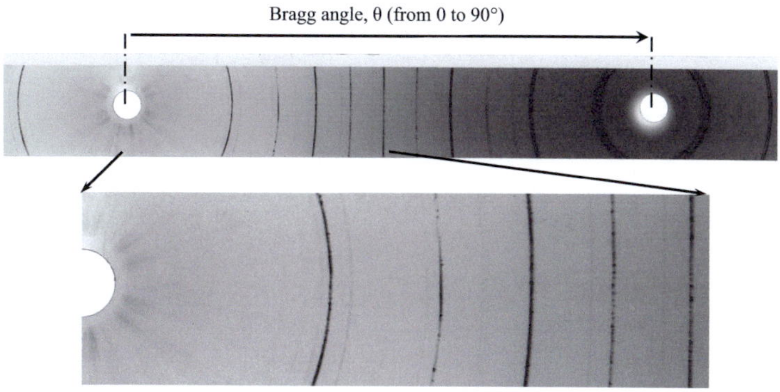

Fig. 11.1 Film with the X-ray diffraction pattern of the polycrystalline LuAu, recorded in a Debye–Scherrer camera using Cu Kα radiation. Bragg peaks appear as concentric ring segments with varying darkness and curvature. The spottiness of some rings, clearly visible at low Bragg angles in the expanded view, indicates an insufficient number of grains in the irradiated volume of the sample. This result was achieved by annealing the needle at 900 °C to promote grain growth. (Film courtesy of Dr. Karl A. Gschneidner, Jr.)

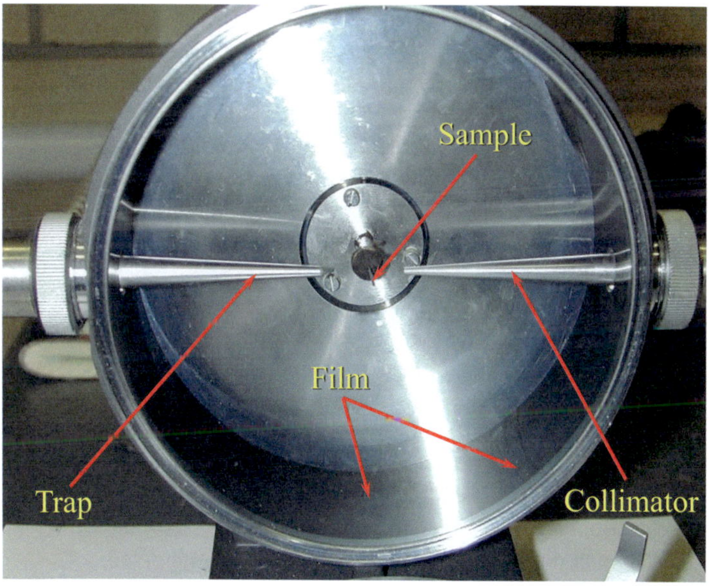

Fig. 11.2 Debye–Scherrer camera without a cover, including a cylindrical sample, collimator, incident beam trap, and the location of the X-ray film

format. Considerable effort is usually required to measure both the Bragg angles and diffracted intensities with high precision.

Cameras for X-ray powder diffraction are relatively simple yet precise optical instruments and require a dark room for loading and developing the X-ray film.

Fig. 11.3 Two Debye–Scherrer cameras with covers, which have been loaded with X-ray film and are installed on the X-ray generator, ready for collecting powder diffraction data

Debye–Scherrer cameras (Fig. 11.2) were the most commonly used in crystallographic laboratories, and many remain on display today. Debye–Scherrer cameras are exceptionally reliable and nearly maintenance-free devices. When loaded with both the sample and the film, the entire diffraction pattern is recorded simultaneously in a single exposure (Fig. 11.3). The typical time to register one powder diffraction pattern on film ranges from 1 to 3 h, depending on the brightness of the source, radius of the camera, the crystallinity of the specimen, and the sensitivity of the film.[5]

Powder diffraction data today are almost exclusively collected using much more sophisticated analytical instruments—powder diffractometers (Fig. 11.4). A powder diffractometer furnishes fully digitized experimental data in the form of diffracted intensity as a numerical function of Bragg angle (see Fig. 11.5). By their nature, powder diffractometer data are exceptionally well-suited for computerized processing. They usually provide accurate information about the structure of materials, especially when coupled with Rietveld analysis,[6] in which subtle anomalies of Bragg peak shapes are used in addition to the integrated intensities of Bragg reflections to extract important information about structural details.

Considering Fig. 11.5, the resolution of powder diffraction data collected using a powder diffractometer is usually much better than that achieved with X-ray film data. This is illustrated in an expanded view, where two closely located Bragg peaks (at $2\theta \cong 81.38°$ and $81.60°$) are easily recognizable. However, when comparing

[5] It may take as long as 12–24 h, especially when using low energy X-rays (e.g. Cr Kα radiation) in combination with a highly absorbing powder and a large camera radius.

[6] The Rietveld method is considered in Sect. 15.7.

Fig. 11.4 The overall view of a powder diffractometer. On the *right*, the radiation enclosure is opened to expose the goniometer, which rests on top of the high-voltage power supply. The computer on the *left* is used to manage and control data collection and to carry out preliminary processing of the data, such as conversion from a software-specific binary to ASCII format

Fig. 11.5 with Fig. 11.1, it is evident that X-ray film data are pseudo two-dimensional, as they allow examination of the distribution of intensity along Debye rings in addition to the distribution of intensity as a function of Bragg angle. The inherent one-dimensionality of conventional powder diffractometer data implies that the experimentalist should be fully aware of potential pitfalls of the technique, often related to improper preparation of the sample and/or with improper selection of data collection parameters and conditions.

Despite their undeniable historical significance, X-ray film data are seldom employed in practical powder diffraction analysis today, and in this book, we will focus solely on powder diffractometry. Interested reader is referred to several excellent texts dedicated to the analysis of film data, for example, those written by Azaroff and Buerger,[7] Lipson and Steeple,[8] Klug and Alexander,[9] and Cullity.[10]

---

[7] L.V. Azaroff, and M.J. Buerger, The powder method in X-ray crystallography, McGraw-Hill, New York (1958).

[8] H. Lipson, and H. Steeple, Interpretation of X-ray powder diffraction patterns, Macmillan, London/St Martin's Press, New York (1970).

[9] H.P. Klug, and L.E. Alexander, X-ray diffraction procedures for polycrystalline and amorphous materials, Second edition. John Wiley, New York (1974).

[10] B.D. Cullity S.R. Stock, Elements of X-Ray Diffraction, Third Edition, Pearson Education Limited (2013).

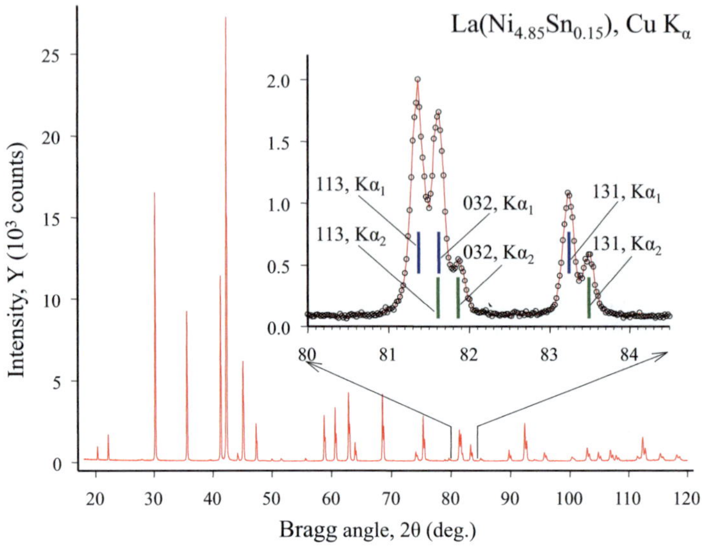

Fig. 11.5 The X-ray powder diffraction pattern of polycrystalline $LaNi_{4.85}Sn_{0.15}$ recorded on a Rigaku TTRAX rotating anode powder diffractometer using Cu Kα radiation. The expanded view shows three Bragg peaks between 80° and 84.5° 2θ. Each of the Bragg peaks consists of nearly resolved $K\alpha_1/K\alpha_2$ doublets. The $K\alpha_2$ component of the strongest Bragg peak (113) at $2\theta \cong 81.6°$ is completely overlapped with the $K\alpha_1$ component of the second Bragg peak (032) at $2\theta \cong 81.6°$. The $K\alpha_1$ component of 113 peak at $2\theta \cong 81.38°$, the sum of the 113 $K\alpha_2$ and 032 $K\alpha_1$ components at $2\theta \cong 81.6°$, and the 032 $K\alpha_2$ component at $2\theta \cong 81.82°$ are well-resolved. Every point in the inset represents a single experimental measurement, $Y(2\theta_i)$. The lines connecting the data points are guides for the eye

The last few decades of the twentieth century transformed the powder diffraction experiment from a technique familiar to a few into one of the most broadly practicable analytical diffraction experiments. This shift was particularly due to the availability of a much greater variety of radiation sources. Sealed and rotating anode X-ray tubes were supplemented by intense neutron and brilliant synchrotron radiation sources. Without doubt, the accessibility of both neutron and synchrotron radiation sources started a revolution in powder diffraction, especially with respect to previously unimaginable kinds of information that can be extracted from a one-dimensional projection of the three-dimensional reciprocal lattice of a crystal. Yet, powder diffraction fundamentals remain the same, no matter the brilliance of the source of particles or X-ray photons employed to produce diffraction peaks, and no matter how basic or advanced the method used to record the powder diffraction data is.

The conventional analytical powder diffractometer has been, and continues to be, a workhorse for thousands of researchers, both experienced and those who are at the beginning of their careers in science and industry. Accordingly, this chapter is illustrated with many examples obtained using standard analytical instruments. Needless to say, using a more advanced radiation source to study the phenomenon of powder diffraction from the same quality specimen will only result in a better (i.e., more accurate) set of experimental data.

11.2 Beam Conditioning in Powder Diffractometry

As mentioned in Sect. 11.1, beginning approximately in the 1970s, powder cameras and X-ray film were steadily replaced by automated analytical instruments, namely powder diffractometers. Despite a large variety of both commercial and unique apparatus found in analytical laboratories worldwide, nearly all of them share many common characteristics dictated by the properties of X-rays.[11] Since standard X-ray tubes produce divergent beams, most high-resolution powder diffractometers use self-focusing geometries, to improve both the diffracted intensity and the resolution of the instrument. This is usually achieved by highly precise X-ray optics, which are incorporated into the critical part of powder diffractometer hardware—the goniometer—through thorough alignment of the latter.

The most common features of focusing optics in powder diffractometry are summarized in Fig. 11.6. Focusing powder diffractometers usually operate in the so-called θ–2θ or θ–θ scanning regimes (or scanning modes), where the incident and diffracted beams both form the same angle θ with the surface of a flat sample, while the diffracted beam forms a 2θ angle with the incident beam (also see Fig. 11.5). The directions of the beams are shown by arrows in Fig. 11.6 and in other schematics in this chapter.

The incident beam passes through at least two slits before reaching the sample. The Soller slit limits the divergence of the incident beam in the direction perpendicular to the plane in which the diffracted intensity is measured, also known as out-of-plane or axial divergence. Axial divergence of the incident beam is not shown in Fig. 11.6. The divergence slit establishes the in-plane aperture of the incident beam and the in-plane divergence. Since the sample is irradiated by the divergent incident beam, the diffracted beam converges (self-focuses) at the receiving slit, which is located at the same distance from the center of the sample as the focal point of the source. These two distances remain constant at any Bragg angle, and both the focal point of the X-ray source and the receiving slit of the detector are located on the circumference of an imaginary circle, known as the goniometer circle. The radius of the goniometer circle is identical to the goniometer radius.

The diffracted beam passes through the second Soller slit before reaching the detector when no monochromator is employed (Fig. 11.6, left), or it is reflected by a crystal-monochromator on its path to the detector (Fig. 11.6, right). An additional

[11] In this book, we are predominantly concerned with the so-called Bragg–Brentano focusing geometry. Other types of geometries, such as those using Seemann–Bohlin and Guinier geometries, will not be considered here because their use in the determination and refinement of crystal structure from powder diffraction data is limited when compared to the Bragg–Brentano technique.

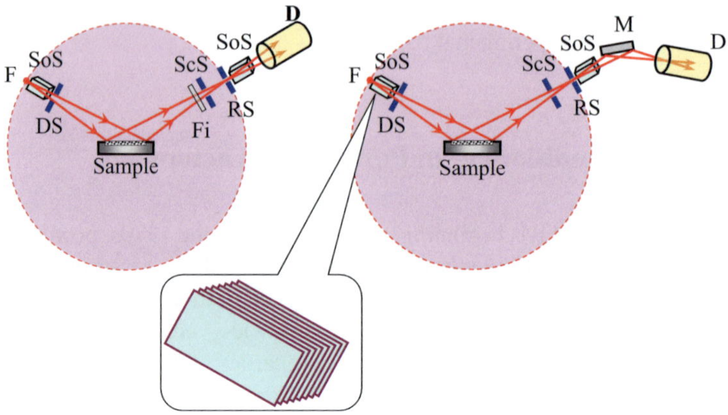

Fig. 11.6 Typical focusing optics employed in modern powder diffractometry: without (*left*) and with (*right*) the diffracted beam monochromator. *F* focus of the X-ray source, *SoS* Soller slits, *DS* divergence slit, *Fi* β-filter, *ScS* scatter slit (optional), *RS* receiving slit, *M* monochromator, *D* detector. In each case, both the focal point of the source and the receiving slit are equidistant from the common goniometer axis, which coincides with the center of the sample. These points are located on the surface of a cylinder known as the goniometer circle, shown using a dashed line. The Soller slit, seen in the expanded view, consists of multiple thin parallel plates that limit the divergence of the beam in the direction perpendicular to the plate surfaces (and to the plane of the figure) usually to within 1–5°. Both schematics are not to scale. Also, see Figs. 11.16, 11.17, 11.22, 11.23, and 11.24

scatter slit, located before the receiving slit, can be employed to reduce the background. The Soller slit on the diffracted beam side can be placed between the scatter and receiving slits.

The diffracted beam is monochromatized using a β-filter (Fig. 11.6, left) or a crystal monochromator (Fig. 11.6, right). Sometimes, the monochromatization geometries shown in Fig. 11.6 are reversed, so that the incident beam rather than the diffracted beam is monochromatized using either a β-filter or a crystal monochromator. The monochromatization of the diffracted beam is advantageous in that fluorescent X-rays (which may be quite intense in some combinations of samples and photon energies, e.g., see Sect. 8.6.5, and/or Fig. 11.25) can be suppressed, thus reducing the background.

The common optical features described here can be realized in various ways in the actual hardware designs of powder diffractometer goniostats. Thus, goniometers differ from one another by:

- The orientation of both the goniometer axis and specimen surface (or specimen axis) with respect to the horizon, i.e., they may be located in a vertical or horizontal plane.
- Diffraction geometry—reflection or transmission—when scattered intensity is registered after the reflection from or after the transmission through the sample, respectively.

– Motions of the goniometer arms, i.e., according to which arms of the goniometer are movable and which are stationary.

Both the polychromatic nature and the angular divergence of the primary X-ray beam generated using either a sealed or rotating anode X-ray tube (see Sect. 6.2 and Figs. 6.4, 6.5, 6.6, and 6.7) result in complex diffraction patterns when X-rays are employed in the "as-produced" condition. This occurs because (1) white radiation causes a high background; (2) the presence of the three intense characteristic lines ($K\alpha_1$, $K\alpha_2$, and $K\beta$) in the spectrum results in three Bragg peaks from each set of crystallographic planes (i.e., from each point in the reciprocal space); and (3) the angular divergence in all directions yields broad and asymmetric Bragg peaks. Thus, the incident X-ray beam needs to be modified (conditioned) to improve the quality of the powder diffraction pattern.

Angular divergence (dispersion) can be reduced by collimation, the process of selecting electromagnetic waves with parallel or nearly parallel propagation vectors. The undesirable satellite wavelengths can be removed by various monochromatization approaches, which convert polychromatic radiation into a single wavelength ($K\alpha_1$ or $K\beta$ in conventional X-ray sources, or narrow bandwidth using a synchrotron source) or at least into a double wavelength ($K\alpha_1$ plus $K\alpha_2$) beam. The $K\alpha_1$ plus $K\alpha_2$ doublet wavelength is, by and large, the only acceptable combination of polychromatic X-rays commonly used today.

The main problem in both collimation and monochromatization is not just how to reduce both the angular and wavelength (energy) dispersions, but how to do so with minimal loss of intensity (photon flux) of both the incident and diffracted beams. Needless to say, different diffraction methods require different degree of collimation and monochromatization. Thus, high-resolution or low-angle scattering applications usually require parallel and narrow beams containing a single wavelength, while routinely used powder diffractometers have less strict requirements. For example, the commonly used Bragg–Brentano focusing technique was specifically developed to work well with slightly divergent incident beams, and this focusing method will not produce the best results when coupled with a perfectly parallel incident beam.

11.2.1 Collimation

The simplest collimation can be achieved by placing a slit between the X-ray source and the sample, as shown in Fig. 11.7, top left. The angular divergence of the collimated beam is established by the dimensions of the source, as well as the size and the placement of the slit. This slit is called the divergence slit (DS in Fig. 11.6), and in the majority of powder diffractometers, the placement of the divergence slit is fixed at a certain distance from the X-ray tube focus.

Considering the geometry shown in Fig. 11.7 and assuming that the distance between the slit and the tube focus is much larger than the slit opening, the angular divergence of the collimated beam, i.e., the angle α, is given as

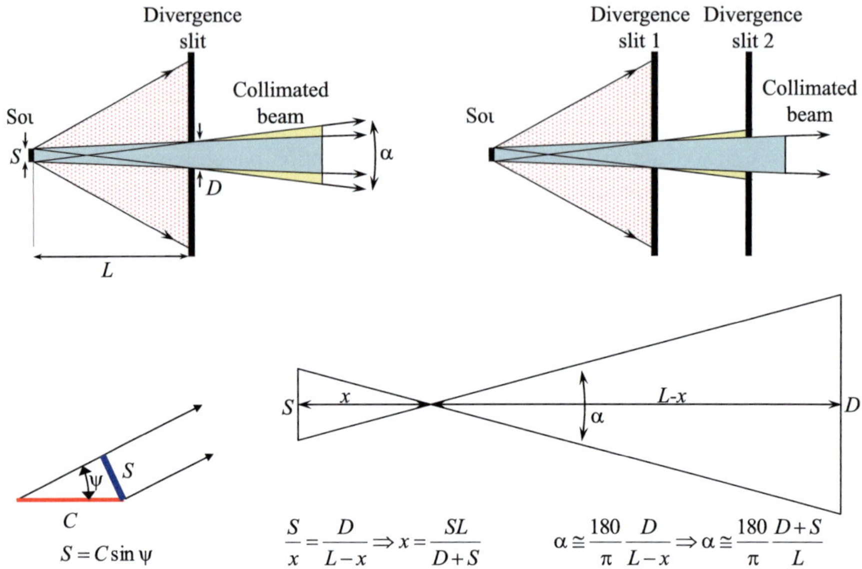

Fig. 11.7 The schematic showing collimation of the incident X-ray beam by using a single divergence slit (*top, left*) or coupled divergence slits (*top, right*). The schematic on the bottom left illustrates the size of the source (S) when the projection of the cathode (C) is viewed at a take-off angle, ψ (For a line focus, a typical width of the cathode projection $C = 1$ mm; a typical take-off angle $\psi = 6°$. Hence, a typical size of the source in powder diffraction is $S \cong 0.1$ mm). Equation (11.1) is derived on the *bottom, right*

$$\alpha(^\circ) \cong \frac{180}{\pi}\frac{D+S}{L} \tag{11.1}$$

where D is the divergence slit opening in mm, S is the width of the focus of the X-ray tube (in mm) visible at a take-off angle ψ, and L is the distance between the tube focus and the slit in mm. Since S and L are usually fixed, the variable beam divergence is customarily achieved by varying the slit opening, D. A second divergence slit can be placed further on the way of the beam to provide additional collimation as shown in Fig. 11.7, top right.

Both collimation methods shown in Fig. 11.7 are commonly used in powder diffractometers for routine powder diffraction experiments.[12] However, high-resolution and low-angle scattering diffractometers require better and therefore more complex collimation, which to some extent overlaps with the

[12]Note that the two-slit configuration, in which the top and the bottom edges of the focal spot lie on the same lines as top and bottom edges of both slits, respectively, provides a more even distribution of intensities in the incident beam when compared to the one slit configuration.

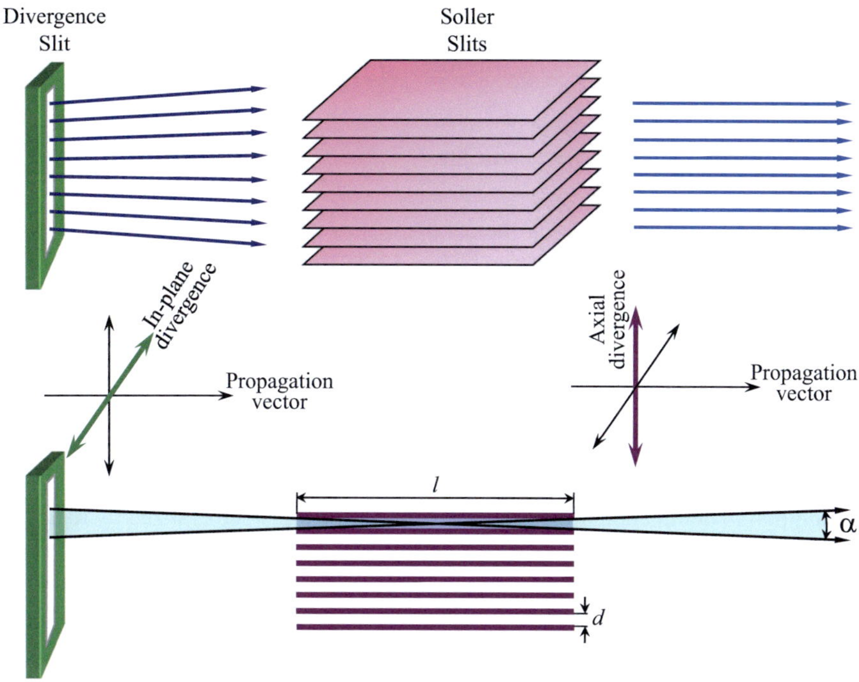

Fig. 11.8 The schematic showing how the X-ray beam is collimated by using both the divergence and Soller slits (*top*). The beam, collimated in-plane by the divergence slit, is further collimated axially by the Soller slits. The coordinates in the middle of the drawing indicate the corresponding directions. The bottom part of the figure illustrates the analogy of Eq. (11.2) with Eq. (11.1)

monochromatization described in Sect. 11.2.2, but otherwise is not considered in this book.[13]

Divergence slits reduce angular dispersion of the incident X-ray beam in the plane perpendicular to the goniometer axis; in other words, they are used to control the in-plane divergence. Angular divergence in the direction parallel to the goniometer, known as axial divergence, is controlled by the Soller slits. Soller slits are usually manufactured from a set of parallel, equally spaced thin metal plates as shown in Fig. 11.8 (top).

Each pair of the neighboring plates functions like a regular divergence slit. The major differences in the design of Soller slits, compared to divergence slits, lie in the requirement to minimally affect the in-plane size of the beam, regardless of the distance between the plates. This is done to maximize the intensity of the incident beam and maintain the high quality of powder diffraction patterns. Soller slits, installed on both the primary and diffracted beam paths, substantially reduce axial

[13] Advanced collimation and monochromatization techniques are described in: D.K. Bowen and B.K. Tanner, High resolution X-ray diffractometry and topography, Taylor & Francis, London/Bristol, PA (1998).

divergence of the X-ray beam and asymmetry (see Sect. 8.5.2) of Bragg peaks. The axial divergence of the beam collimated by Soller slits can be estimated in the same way as for a single divergence slit:

$$\alpha(^\circ) \cong \frac{180}{\pi}\frac{2d}{l} \tag{11.2}$$

where d is the distance between the parallel plates in mm and l is the plate length (also in mm) as shown at the bottom of Fig. 11.8. Thus, the axial divergence can be reduced or increased by varying either or both the length of the plates (l) and the distance between them (d).

11.2.2 Monochromatization

In addition to collimation, the X-ray beam should be monochromatized by reducing the intensity of white radiation and by eliminating the undesirable characteristic wavelengths from the X-ray spectrum, leaving only a single usable wavelength. As noted earlier, when conventional X-ray sources are employed, the two wavelength beams ($K\alpha_1$ plus $K\alpha_2$) are acceptable because the complete removal of the $K\alpha_2$ component substantially reduces the intensity of the incident beam and therefore increases the time of the experiment needed to obtain high-quality X-ray diffraction data. When synchrotron sources are used, the monochromatization process selects a single wavelength from a continuous X-ray spectrum. The most common methods utilized in the instrumental monochromatization of X-ray beams are as follows:

- Using a β-filter (conventional X-ray sources only)
- Using diffraction from a crystal monochromator (any source, including neutrons)
- Pulse-height selection using a proportional counter (X-rays)
- Energy resolution using a solid-state detector (X-rays)

The monochromatization using a β-filter employs the presence of the K-absorption edge (see Fig. 8.20) to selectively absorb Kβ radiation and transmit the $K\alpha_1$ and $K\alpha_2$ parts of the X-ray spectrum, as shown in Fig. 11.9. Thus, a properly selected β-filter material has its K-absorption edge below the wavelength of the $K\alpha_1$ characteristic line and just above the wavelength of the Kβ line.[14]

[14] Better monochromatization can be achieved using the so-called balanced filters. The first filter material is chosen to suppress Kβ and other short wavelengths. The second filter material is selected such that its K absorption edge is located above the wavelength of the $K\alpha_2$ component and the thickness of each material is selected to equally suppress the Kβ intensity. As a result, the balanced filter reduces the intensity of both short and long wavelengths. Balanced filters are seldom used today because the resulting intensity loss is usually higher when compared with a well-aligned crystal monochromator.

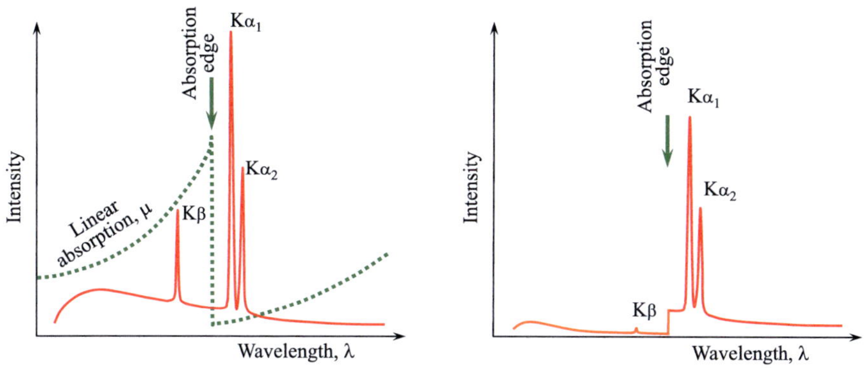

Fig. 11.9 *Left*—the schematic of the X-ray emission spectrum shown as the solid line overlapped with the schematic of the μ(λ) function of the properly selected β-filter material (*dotted line*). *Right*—the resultant distribution of intensity after filtering as a function of the wavelength

The general rule for choosing the β-filter is to use a material rich in a chemical element that is one atomic number less than the anode material in the periodic table. This ensures the location of the K-absorption edge between the $K\alpha_1$ and Kβ lines. For heavy anode materials (e.g., Mo), this rule can be extended to elements two atomic numbers below the anode material. A list of β-filter elements suitable for the commonly used anode materials can be found in Table 8.3. Thus, for a Cu anode, a foil made from Ni will work best as the β-filter, while for Mo radiation both Nb and Zr are good β-filters. The former material (Zr) is more often used in practice because its K-absorption edge is closer to the wavelength of the $K\alpha_1$ line.

The major disadvantages of β-filters are: (1) they are incapable of complete elimination of the Kβ intensity and (2) they leave a considerable amount of white radiation after filtering. Further, any β-filter also reduces all intensities in the spectrum at wavelengths higher than the corresponding absorption edge (see Fig. 11.9). This results in the reduction of the intensity of the $K\alpha_{1,2}$ components, although obviously the latter are reduced by a much smaller factor than the intensity of the Kβ line. Fundamentally, the use of a β-filter improves $I_{K\alpha}/I_{K\beta}$ and $I_{K\alpha}/I_{white}$ ratios and this improvement is proportional to the thickness of the filter. For example, assume that the ratio of the intensities of the $K\alpha_1$ and Kβ spectral lines in the unfiltered spectrum is approximately 5:1. After passing through a properly designed β-filter, the ratio becomes approximately 100–500 to 1.

Considering Eq. (8.50) and the as-produced ratio of intensities between $K\alpha_1$ and Kβ lines (5:1), the filtered $I_{K\alpha_1}/I_{K\beta}$ intensity ratio can be expressed as follows:

$$\frac{I_{K\alpha_1}}{I_{K\beta}} = 5\frac{\exp(-\mu_\alpha t)}{\exp(-\mu_\beta t)} \tag{11.3}$$

where μ_α and μ_β are the linear absorption coefficients of the filter material for $\lambda_{K\alpha}$ and $\lambda_{K\beta}$, respectively, and t is the thickness of the β-filter. This equation can also be used

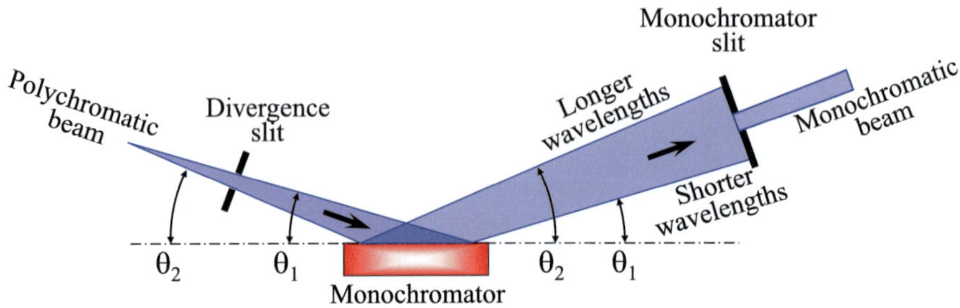

Fig. 11.10 The schematic explaining the principle of monochromatization using a single-crystal monochromator. Generally, $\theta^M \cong (\theta_2 + \theta_1)/2$. The directions of the propagation vectors are indicated by arrows

to calculate the filter thickness, t, which is necessary to obtain the desired $I_{K\alpha}/I_{K\beta}$ ratio. Thus, depending on the specific need and the application, a compromise is made between the purity of the spectrum and the intensity of the Kα radiation.

A different, more complex, but much improved monochromatization approach takes advantage of diffraction from a high-quality single crystal, precisely positioned with respect to the X-ray propagation vector. Examples of commonly used crystal monochromator materials include pyrolytic graphite, Si, Ge, and LiCl.

A nearly perfect single crystal is placed at a specific angle (θ^M) with respect to the primary or the diffracted X-ray beams and, according to the Braggs' law (Eq. 7.9), only discrete wavelengths can be transmitted at this angle. Assuming that $n = 1$, the single transmitted wavelength, λ_t, is a function of the corresponding interplanar distance of the crystal, d^M_{hkl}, and θ^M. In reality, even the best crystal monochromators allow some wavelength dispersion in the transmitted beam because of the unavoidable imperfections.

$$2d^M_{hkl} \sin \theta^M = n\lambda_t \tag{11.4}$$

The general principle of operation of a single-crystal monochromator is illustrated in Fig. 11.10. The X-rays in the divergent beam that reach the monochromator form slightly different angles (ranging from θ_1 to θ_2) with the crystal. They are reflected from the set of crystallographic planes, *hkl*, which are parallel to the crystal surface.

According to Braggs' law (Eq. 7.9), both the incident and diffracted beams with identical wavelengths should form the same angle with the surface of the crystal. Hence, each wavelength, λ_i, is diffracted at a particular angle $\theta_1 \leq \theta_i \leq \theta_2$ which yields an uneven spatial distribution of wavelengths in the beam reflected by the crystal. In effect, the shorter wavelengths are grouped at the low Bragg angles (they are darker in Fig. 11.10) and the longer wavelengths are observed at the high Bragg angles (they are lighter in Fig. 11.10). For example, the Kβ line falls into the low Bragg angle range and the $K\alpha_1/K\alpha_2$ doublet falls into the high angle range. Either option can be easily selected using a narrow slit properly installed in the path of X-rays reflected by a crystal monochromator.

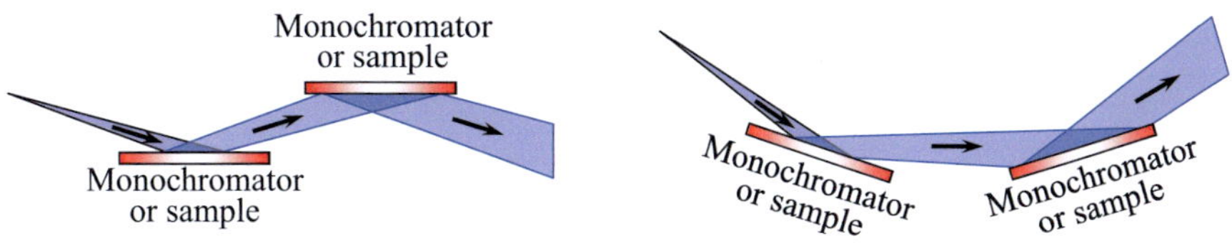

Fig. 11.11 Parallel (*left*) and angular (*right*) arrangements of two crystal monochromators or the sample and the crystal monochromator commonly used to improve the monochromatization of the resultant X-ray beam

Even though the diffracted beam is not perfectly monochromatic at any specific angle due to various imperfections (defects, distortions, stresses, etc.) present in the crystal monochromator, the separation of Kα and Kβ wavelengths is large enough to allow easy elimination of the Kβ and nearly all white X-rays during the monochromatization. However, the separation of Kα_1 and Kα_2 wavelengths requires more than the simple arrangement shown in Fig. 11.10. The wavelength (photon energy) resolution can be improved by using diffraction from two or more monochromators placed in sequence. We note that the diffraction from the sample, in the first approximation, works as a preliminary monochromator. Thus, the distribution of the wavelengths may be assumed as uniform in the whole range of incident angles (from θ_1 to θ_2 in Fig. 11.10), but this is no longer the case in the beam reflected by the first monochromator or by the sample.

The two configurations shown in Fig. 11.11—parallel and angular—are commonly used to improve the monochromatization of the beam. Both diffracting plates can be crystal monochromators, or one can be the sample while another is the monochromator. The latter arrangement is especially common in powder diffractometry.

Two parallel plates (Fig. 11.11, left) remove the Kβ line completely. The Kα_1 and Kα_2 doublet can be separated at specific conditions when the first plate is used as a monochromator but at a cost of the reduced beam intensity. When the second plate is used as monochromator, the resultant intensity is higher, but the Kα_1/Kα_2 doublet is nearly impossible to separate. This layout also improves the collimation of the beam. The second arrangement (Fig. 11.11, right) is more efficient. Angular orientation of the two crystals separates Kα_1 and Kα_2 lines quite well due to the larger spatial difference between the two wavelengths after diffraction from the second plate. Yet again, this is usually done at a considerable intensity loss penalty.

The three most common geometries of a crystal monochromator and a sample, used in powder diffraction, are illustrated in Fig. 11.12 and their characteristics are compared in Table 11.1. Diffracted beam monochromators (Fig. 11.12a, b) have a relatively high-intensity output (or in other words have low-intensity losses) but do not separate the Kα_1 and Kα_2 doublet. The removal of both the Kβ component and white radiation is excellent. The latter is especially important when the sample is strongly fluorescent. Diffracted beam monochromators are, by far, more commonly used than the primary beam monochromators. The advantage of the angular configuration (Fig. 11.12b) lies in the lower torque exerted on the detector arm compared

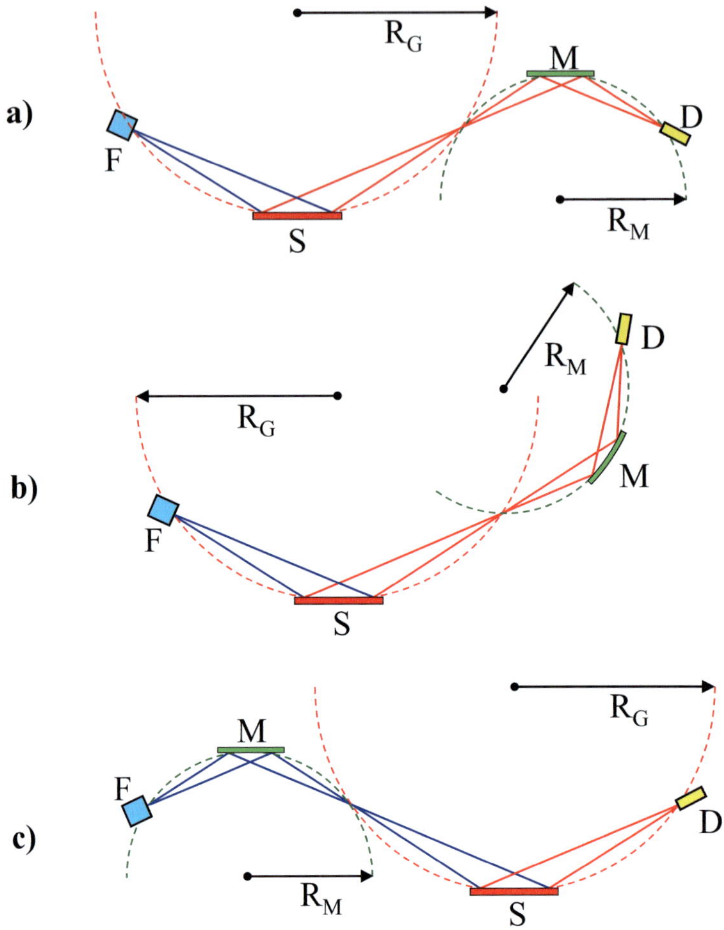

Fig. 11.12 The three different monochromator/sample geometries used in powder diffraction: (**a**) flat *diffracted* beam monochromator, parallel arrangement; (**b**) curved *diffracted* beam monochromator, angular arrangement; and (**c**) flat *primary* beam monochromator, parallel arrangement. *F* focus of the X-ray source, *S* sample, *M* crystal monochromator, *D* detector, R_M radius of the monochromator focusing circle, R_G radius of the goniometer focusing circle

with the linear configuration. However, the crystal surface must be curved to match the radius of the monochromator focusing circle for optimal results, and this setup imposes more restrictions on the maximum 2θ angle that could be measured.

The most important advantage of the primary beam monochromator (Figs. 11.12c and 11.13) is the possibility to separate the $K\alpha_1/K\alpha_2$ doublet,[15] which is especially

[15]This is usually achieved by using curved Johansson-type crystal monochromators, first described by T. Johansson, Über ein neuartiges, genau fokussierendes Röntgenspektrometer, Z. Physik **82**, 507 (1933). https://doi.org/10.1007/BF01342254

Table 11.1 Comparison of the three common monochromator/sample configurations used in powder diffractometry

Property	a[a]	b	c
Position in the beam	Diffracted	Diffracted	Primary
Orientation to the sample	Parallel	Angular	Parallel
Shape of the crystal	Flat/curved	Curved	Flat/curved
Intensity loss	Low	Low	Low/high
$K\alpha_1$ and $K\alpha_2$ separation	No	No	No/complete
Alignment	Simple	Simple	Simple/complex
Additional torque on the detector arm	High	Low	None
Use with area detectors	Impossible	Impossible	Possible

[a] Labeling of the columns corresponds to Fig. 11.12

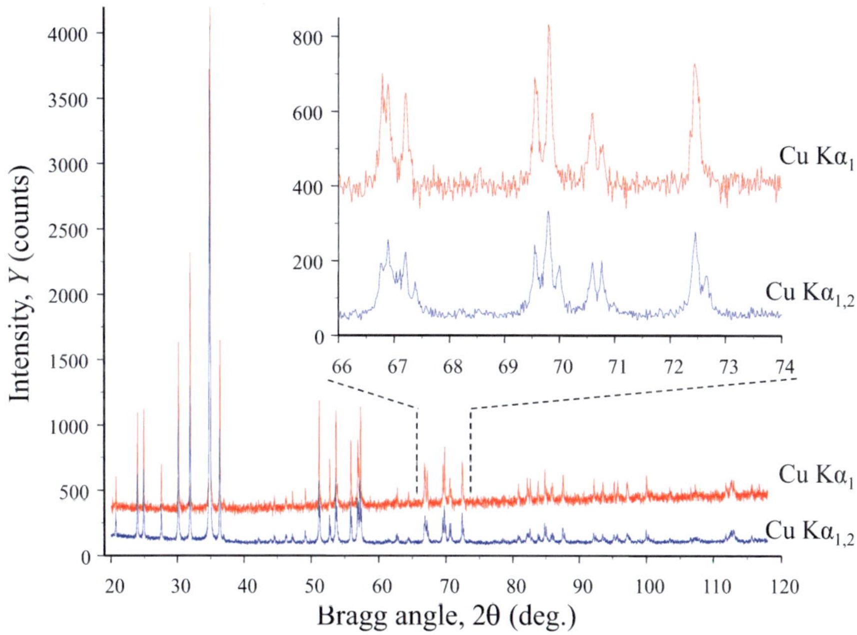

Fig. 11.13 Powder diffraction patterns collected from the same $Gd_5(Si_{1.5}Ge_{1.5})$ specimen using Cu $K\alpha$ radiation. The patterns shown in *blue* were recorded using a strip detector and a diffracted beam monochromator (the $K\beta$ component of the characteristic spectrum was eliminated, but the $K\alpha_2$ component was not). The patterns in *red* have been recorded using the primary beam Johansson monochromator, which results in purely monochromatic Cu $K\alpha_1$ radiation. The increased background in the top patterns is due to high fluorescence of Gd. (Data are courtesy of Dr. Y. Mudryk)

important when working with complex diffraction patterns where the maximum resolution of Bragg peaks is critical. A primary beam monochromator is the only type of geometry that can be used with area detectors (see Sect. 6.4.4). The main disadvantages are relatively high-intensity losses, the need for precise alignment, and potential for high fluorescent background in some combinations of materials and photon energies.

Monochromatization by pulse-height selection using a proportional detector is based on the fact that the signal generated in the detector is generally proportional to the energy of the absorbed X-ray photon and therefore inversely proportional to its wavelength (see Sect. 6.4.3). Monochromatization by pulse-height selection is far from ideal and it does not even result in the complete elimination of the Kβ intensity. Nevertheless, its use substantially improves the quality of powder diffraction data by reducing the background noise, especially when combined with a β-filter.

Cooled solid-state detectors find more and more use as both detectors and monochromators because of their extremely high sensitivity to the energy of absorbed photons, which enables precise energy discrimination. Thus, the detector can be tuned to register wavelengths only within certain limits, for example, $K\alpha_1$ and $K\alpha_2$ energies, or in some of the most recent applications, only Kβ component. The latter is one of the ways to obtain single wavelength X-rays without using crystal monochromators and, therefore, without a substantial loss of intensity except for the fact that the intensity of the Kβ line is only ~1/5 of the intensity of the $K\alpha_1$ line. Another advantage of this approach is a clean registered diffraction pattern because the majority of white radiation is also eliminated. The disadvantages of this monochromatization technique arise from the limitations intrinsic to solid-state detection (see Table 6.3 and Fig. 6.11).

11.3 Principles of Goniometer Design in Powder Diffractometry

As shown in Fig. 11.14 (left), when the examined specimen is parallel to the horizon, in a horizontal goniometer design, it has the obvious advantage that no special care is needed to hold the powder in the sample holder—the powder is simply held by gravity. The disadvantage of this design is that the detector arm and the X-ray source move in a vertical plane, necessitating powerful stepping motors and precise

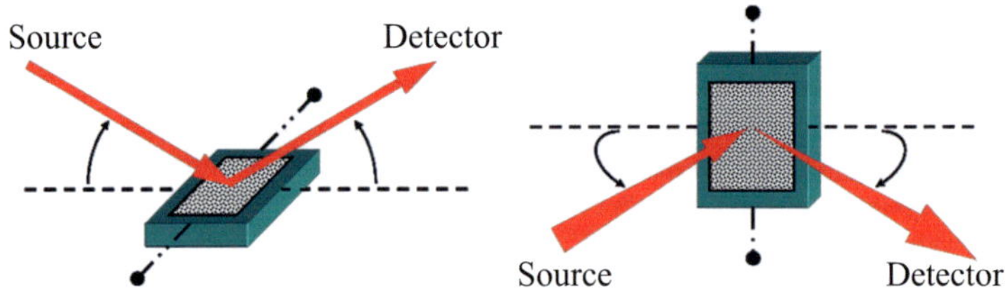

Fig. 11.14 Horizontal (*left*) and vertical (*right*) orientations of a flat sample, along with the goniometer axis, which is depicted using a *dash-double-dotted line* with *small circles* at the ends. The *dashed line* indicates the location of the optical axis, which connects the focus of the X-ray tube, the receiving slit, and the sample center

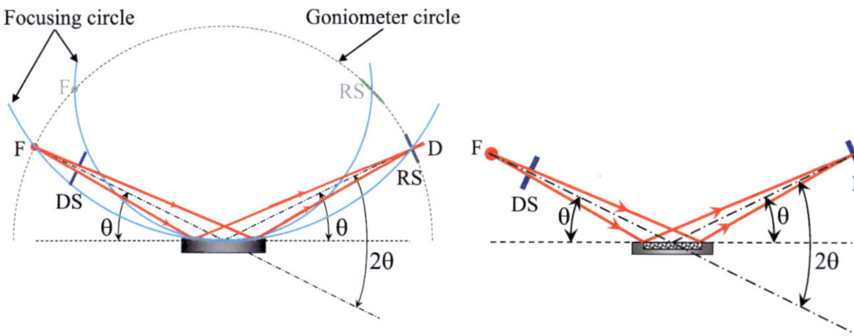

Fig. 11.15 The schematic of the ideal focusing geometry (*left*) and its common modification known as the Bragg–Brentano geometry using a flat sample (*right*) when the self-focused diffracted beam is registered by the detector after reflection from the sample. *F* focus of the X-ray source, *D* detector, θ Bragg angle, *DS* divergence slit, *RS* receiving slit. To achieve ideal focusing of the reflected divergent beam, the curvature of the sample must coincide with the circumference of the focusing circle as indicated by two *circles* of different radius shown in *blue* on the *left*. This is impractical because the curvature of the specimen becomes a function of the Bragg angle, and therefore, flat specimens are employed instead

counterbalancing to control heavy goniometer arms with the required precision, typically within 1/1000 of a degree.

Conversely, the simplicity of the goniometer arms motion in a horizontal plane, when the sample is located in a vertical plane (Fig. 11.14, right), is offset by the need of more complicated sample preparation to ensure that it stays in place and does not fall of. This is usually achieved by side packing the sample holder or by mixing a powder with a binder (e.g., X-ray amorphous and chemically inert petroleum jelly, oil, grease, or varnish), which typically increase preferred orientation or background, respectively (see Sect. 12.1 for more details on sample preparation).

The reflection geometry fully leverages the focusing of the diffracted beam, as shown in Fig. 11.15. Commonly known as the Bragg–Brentano geometry or focusing method, this approach results in both high-resolution and high diffracted intensity. Moreover, the Bragg–Brentano setup allows for relatively straightforward sample preparation, and when with the horizontal goniometer axis, the sample can even accommodate liquid state.

A disadvantage of the Bragg–Brentano geometry, aside from its susceptibility to preferred orientation, is the difficulty in preparing a sample with sufficient thickness to ensure complete opacity to X-rays. This is especially true when examining weakly absorbing materials, such as molecular compounds containing only light elements carbon, nitrogen, oxygen, and hydrogen.[16]

The two commonly employed transmission geometries are shown in Fig. 11.16. Powder diffraction in the transmission mode can be observed from both flat and

[16]Low absorption may affect positions of Bragg peaks due to transparency shift (Chap. 8, Sect. 8.4.2) and/or systematically distort scattered intensity (Chap. 8, Sect. 8.6.5).

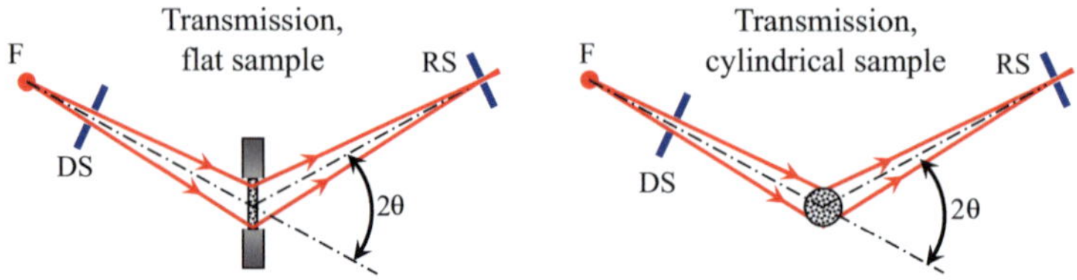

Fig. 11.16 Transmission geometry in the case of flat (*left*) and cylindrical (*right*) samples. *F* focus of the X-ray source, *DS* divergence slit, *RS* receiving slit, θ Bragg angle

cylindrical samples. Flat samples usually require small amounts of material; however, preparing a high-quality, uniformly dense specimen can be difficult, and these may be strongly affected by preferred orientation.

Cylindrical samples, commonly used in the Debye–Scherrer cameras (Fig. 11.2), are also employed in powder diffractometry. Like flat transmission samples, the cylindrical specimen geometry requires only small amounts of powder. This form is least susceptible to the nonrandom distribution of particle orientations, including preferred orientation effects, as long as the cylinder is spinning during data collection.

Both flat and cylindrical transmission samples are commonly used in combination with position-sensitive or image-plate detectors. The major disadvantage of the transmission geometry arises from the fact that self-focusing of the diffracted beam is not as precise as in the Bragg–Brentano geometry. Hence, laboratory instruments employing transmission geometry usually have lower resolution when compared to those operating in the reflection geometry. It is worth noting that imprecise self-focusing is generally not an issue when using synchrotron X-ray sources, which produce nearly parallel X-ray beams.

Powder diffractometer goniostats can be constructed in a way that both the detector and the sample revolve around a common goniometer axis in a synchronized fashion, or the sample is stationary, but both the detector and the X-ray source arms are synchronized, as shown in Fig. 11.17. When cylindrical samples are employed, the synchronization of the goniometer arms is generally unnecessary, only the detector arm (if any, e.g., see Fig. 11.23) needs to be rotated.

The schematic of a goniostat, which realizes horizontal Bragg–Brentano focusing geometry with both the detector and source arms in the synchronized rotation about a common horizontal goniometer axis, is shown in Fig. 11.18. Both arms of the goniometer revolve in the vertical plane, and this geometry of a powder diffractometer is in the most common use today.

Some powder diffractometers, particularly those which are used for routine analysis of multiple samples of the same kind, can be equipped with multiple sample changers, typically accommodating 4–12 specimens. This allows for straightforward software control of the data collection process across a series of samples, enabling better automation as datasets can be collected without operator intervention, such as overnight or during weekends. Multiple sample changers are common in powder

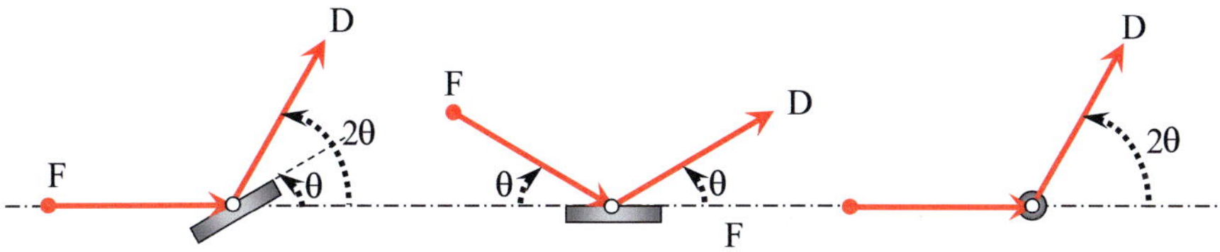

Fig. 11.17 Synchronization of the goniometer arms: the X-ray source is stationary, while the sample and the detector rotations are synchronized to fulfill the θ–2θ requirement (*left*); the sample is stationary while the source and the detector arms are synchronized to realize the θ–θ condition (*middle*)—this geometry is in common use at present; only the detector arm revolves around the goniometer axis in the case of a cylindrical sample (*right*). *F* focus of the X-ray tube indicating the position of the X-ray source arm, *D* detector arm, *θ* Bragg angle. The common goniometer axis (which is perpendicular to the plane of the projection) around which the rotations are synchronized is shown as the open circle in each of the three drawings. The location of the optical axis is shown as the dash-dotted line

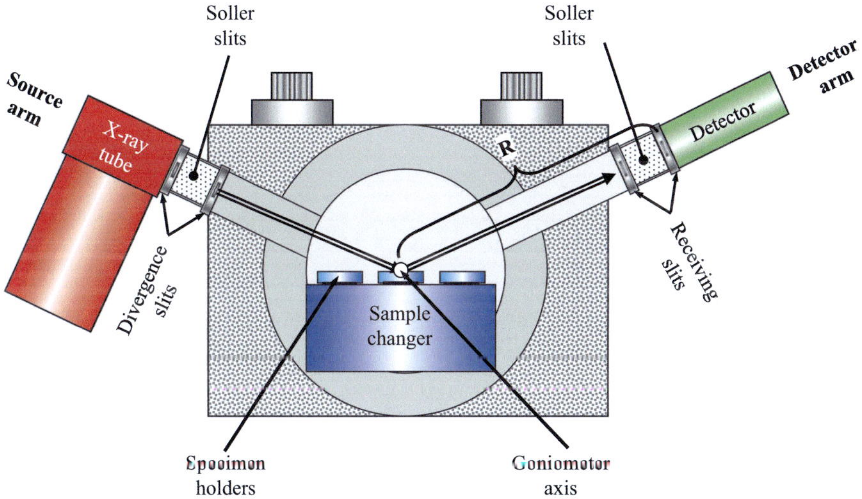

Fig. 11.18 The schematic of a goniostat of a powder diffractometer with the horizontal axis and synchronized rotations of both the source and detector arms. *R* is the radius of the goniometer

X-ray diffractometers used in analytical laboratories for quality control purposes. For example, they are ideal for analyzing multiple samples to ensure that a product meets the required properties, especially when these properties critically depend on the structure or phase composition of materials produced in different batches or at various stages of the production process.

11.3.1 Goniostats with Strip and Point Detectors

Photographs of several different powder diffractometer goniostats equipped with strip and point detectors are shown in Figs. 11.19, 11.21, and 11.22, respectively. The first example (Fig. 11.19) is a Bragg–Brentano θ–2θ goniometer, in which the X-ray source housing is rigidly mounted on the goniometer frame and remains stationary during data collection. The θ–2θ mode is achieved by synchronizing the rotations of the sample holder and the detector at a 1:2 ratio around the common horizontal goniometer axis, as illustrated schematically in Fig. 11.17, left.

The variable slit box located between the X-ray source and the sample (Fig. 11.19, left) contains divergence slit, which controls the aperture and the divergence of the incident beam in the vertical plane. Soller slit, which limits the divergence of the incident beam in the horizontal plane, is located just before the incident beam variable slit box. The sample holder here is a sample spinner attachment.

The second variable slit box is located on the detector arm between the sample and the detector. In this setup, the slit functions as a scatter slit, rather than a receiving slit. It is followed by another Soller slit positioned just before the detector.

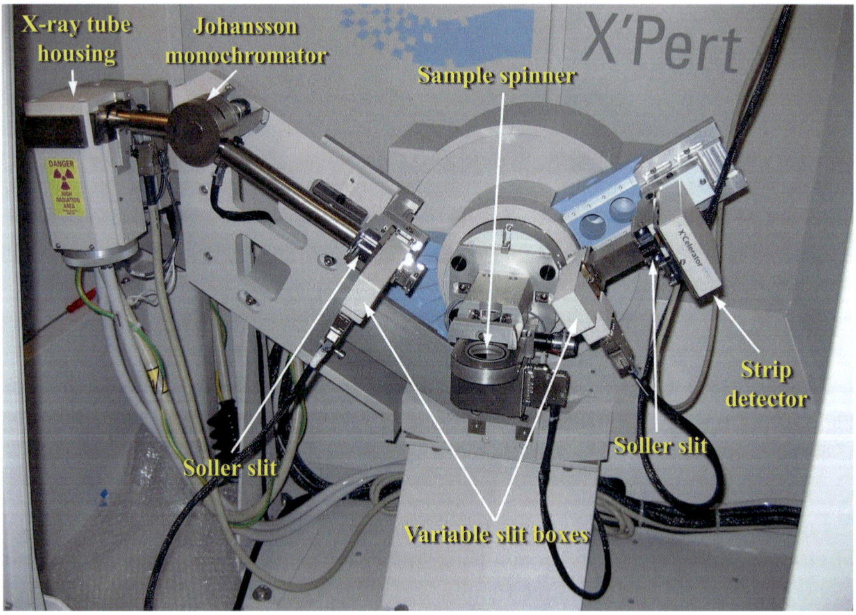

Fig. 11.19 The overall view of the goniostat of the Malvern Panalytical X'Pert powder diffractometer. This diffractometer has the horizontal goniometer axis, stationary X-ray source, and synchronized rotations of both the detector arm and sample holder. The goniometer is equipped with a Johansson-type primary beam monochromator, which eliminates Cu Kα_2 or Co Kα_2 wavelengths in addition to Kβ components of the characteristic spectrum, and a solid-state real-time multiple strip detector, which accelerates data collection

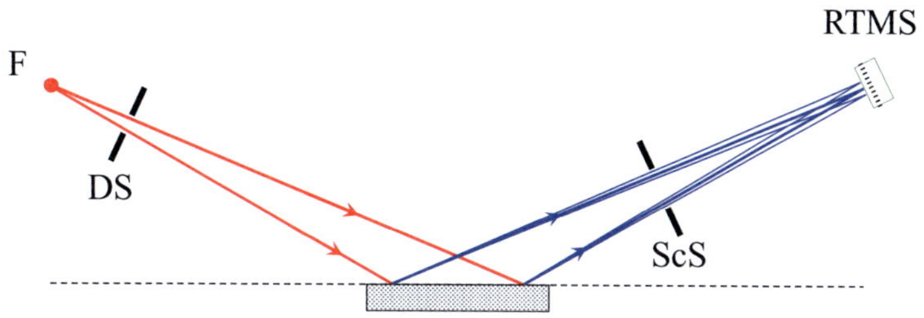

Fig. 11.20 The schematic of focusing in the Bragg–Brentano geometry when the real-time multiple strip detector (see Fig. 6.13) is utilized. *F* focus of the X-ray source, *DS* divergence slit, *ScS* scatter slit, *RTMS* strip detector

The detector here is a real-time multiple strip detector (trade name X'Celerator), which speeds up data collection by measuring about 2° of 2θ simultaneously as shown schematically in Fig. 11.20. Hence, as the Bragg angle changes, the strip detector registers intensity scattered over a small range of Bragg angles rather than at a single angular position. A narrow range of the registered Bragg angles allows for monochromatization of the scattered beam using a crystal monochromator, even though the intensity loss is greater when compared with a well-aligned monochromator and a point detector.

Another example of a goniostat with the Bragg–Brentano geometry, a horizontal axis, and two movable arms is shown in Fig. 11.21. This is a θ–θ goniometer, where the rotations of both arms are synchronized to move at the same speed in opposite directions, while the sample surface remains horizontal. This diffractometer is equipped with a real-time multiple strip detector (trade name LynxEye) (Fig. 11.21, right). On the detector side, it features a combination of a scatter slit, a Soller slit located between the sample and the detector without a receiving slit, and an optional β-filter. This configuration also includes a 9-sample changer that allows for spinning a sample in a horizontal plane.

The strip detector is usually employed for fast data collection since it simultaneously records approximately 3.6° of 2θ range in a standard setup with a 217½ mm goniometer radius. A typical experiment lasts less than 1 h, yet it provides high-quality patterns. A 5–10 min long experiment is usually sufficient for phase identification, and an hour-long experiment is adequate for precise unit cell determination and quantitative analysis and in many cases is good enough for solving crystal structure and Rietveld refinement. Although the speed of this detector is comparable to area detectors, the resolution is close to that of point detectors.

A different goniostat with the horizontal orientation of the specimen and Bragg–Brentano geometry is shown in Fig. 11.22. The X-ray tube housing is mounted on the movable arm, and both the X-ray source and the detector can be rotated in a synchronized fashion about the common horizontal goniometer axis (also see Fig. 11.17, middle).

The X-ray source in this example is a rotating anode X-ray tube. Another distinct feature of this goniometer is the presence of variable divergence (Fig. 11.22, left) and

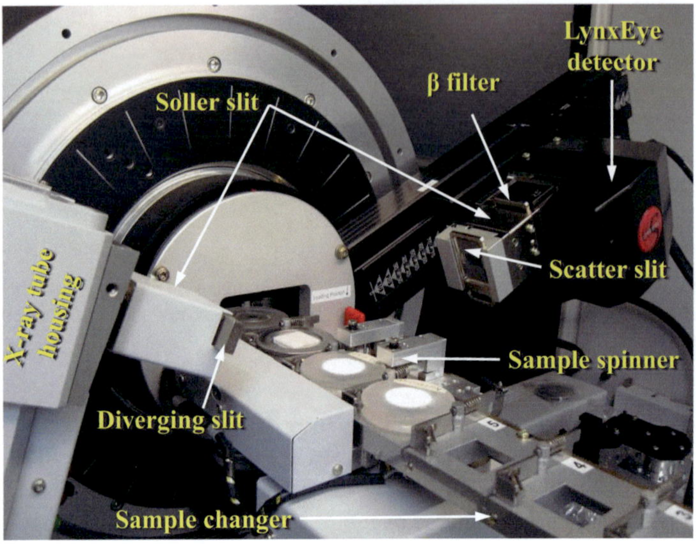

Fig. 11.21 The overall view of the Bruker D8 Advance powder diffractometer. It has a horizontal goniometer axis and the so-called θ–θ geometry—a synchronized rotation of both the X-ray source and the detector arms. This diffractometer is equipped with Cu Kα radiation from sealed tube, 2.5° Soller slits on both incident and diffracted beams, a Ni β-filter, a 9-samples changer with spinner, and a LynxEye multiple strip detector for fast data collection, which measures 3.6° 2θ (usually 180 data points) simultaneously while providing high-resolution data

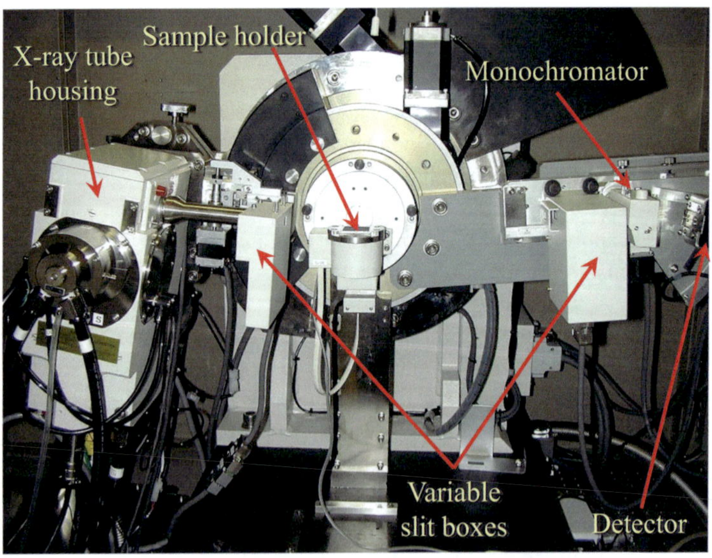

Fig. 11.22 The overall view of the goniostat of the Rigaku TTRAX rotating anode powder diffractometer with the horizontal goniometer axis and synchronized rotations of both the X-ray source and detector arms. This θ–θ goniometer is equipped with variable divergence, scatter and receiving slits, curved crystal monochromator, and scintillation detector

scatter and receiving (Fig. 11.22, right) slits. This combination of slits enables one to maintain the irradiated area of the studied specimen constant at any Bragg angle, which may be useful in some applications. Both variable slit boxes also contain a set of Soller slits each to control the divergence of both the incident and diffracted beams in the horizontal plane.

The curved crystal-monochromator is positioned between the receiving slit box and the detector (also see Fig. 11.6, right). This goniometer is shown with the specimen spinning sample holder attachment, which enables continuous spinning of the sample during data collection to achieve better particle orientations averaging, thereby reducing preferred orientation effects.

The goniometer shown in Fig. 11.22 is equipped with a scintillation detector, which has high linearity (within 1% in excess of ~10^5 counts/s), which is important with the high brightness of the incident beam produced by a rotating anode X-ray tube. Only a small part of the scintillation detector is visible in Fig. 11.22. The massive counterweight to balance the heavy X-ray tube housing is seen in the background of the photograph as the dark segment (top right).

Overall, powder diffractometers with point detectors offer the best resolution of the resulting powder diffraction data. Although the instrumental resolution improves with an increasing goniometer radius, the intensity of the diffracted beam unfortunately decreases due to the divergence of the incident beam produced by an analytical X-ray tube, as well as absorption by the air. As a result, typical goniometer radii range from approximately 150–300 mm.

11.3.2 Goniostats with Area Detectors

The schematic of a powder diffractometer goniostat utilizing transmission geometry with a cylindrical specimen and a curved position-sensitive detector (PSD) is shown in Fig. 11.23. When using a curved position-sensitive detector covering a long (from ~0° to ~90–140°) 2θ range, generally there is no need to rotate the detector arm and only sample spinning is required to improve particle orientations averaging and minimize preferred orientation.

The greatest advantage of this geometry is in the speed of data collection: the entire diffraction pattern can be recorded in as little as a few seconds because the diffracted intensity in the whole range of Bragg angles covered by the circumference of the curved position-sensitive detector is registered simultaneously. The downside is that it is impossible to monochromatize the diffracted beam effectively, which results in the increased background, particularly when the sample is strongly fluorescent. Another difficulty may occur in the interpretation of powder diffraction data collected using the geometry shown in Fig. 11.23 because of the lower resolution of curved PSDs and increased widths of Bragg peaks when compared with point or strip detectors.

In principle, curved position-sensitive detector can be replaced by a linear position-sensitive detector covering segments 3–10° (2θ) wide. This approach is

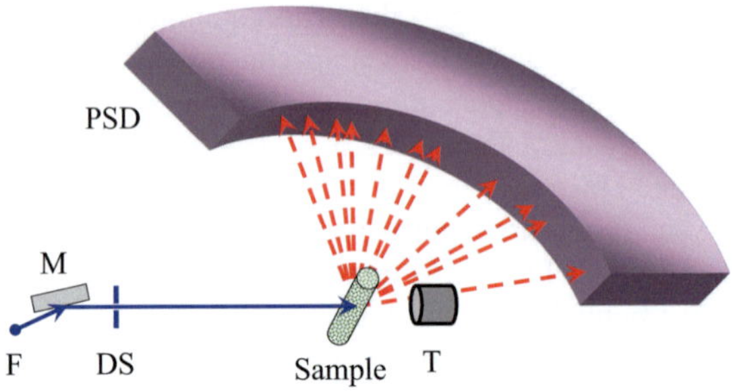

Fig. 11.23 The schematic of a powder diffractometer cylindrical sample in the transmission mode and a curved position-sensitive detector (PSD). *Solid arrows* indicate the incident beam and *broken arrows* indicate the diffracted beams pathways. *F* focal point of the X-ray source, *M* monochromator, *DS* divergence slit, *T* incident beam trap

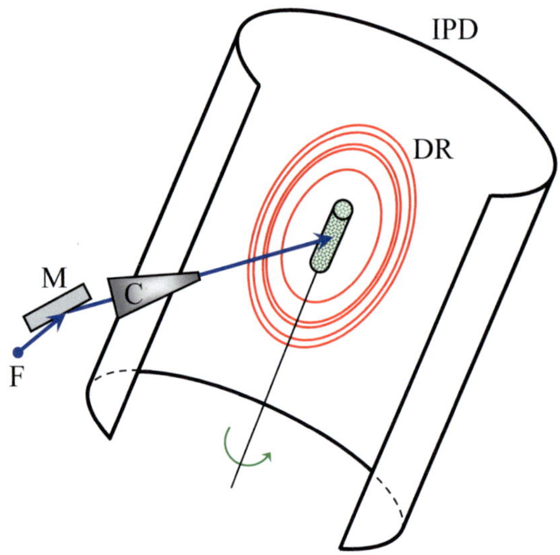

Fig. 11.24 The schematic of a powder diffractometer with the vertical goniometer axis, cylindrical sample in the transmission mode, and image-plate detector (IPD). Blue arrows show the incident beam path. *Red rings* indicate intercepts of Debye cones with the IPD. *F* focal point of the X-ray source, *M* monochromator, *C* collimator, *DR* Debye rings

similar to what is achieved when using a strip detector. It considerably increases resolution and decreases Bragg peak widths, but the problem of the enhanced background remains.

For a time, image-plate detectors (IPD) became popular in powder diffractometry (Fig. 11.24). The monochromatized and collimated beam passes through a cylindrical or flat sample, and the diffracted beams are registered by the image-plate detector in nearly all directions simultaneously. Because of the size of the detector, which can be made as large as necessary, the entire circumference of the Debye rings—or at least significant portion of them—is typically registered, rather than just a small sector as is the case with other powder diffraction geometries considered earlier.

The use of image-plate detectors restores the pseudo-two-dimensionality of the X-ray powder diffraction pattern, as in film-based registration. Experimental

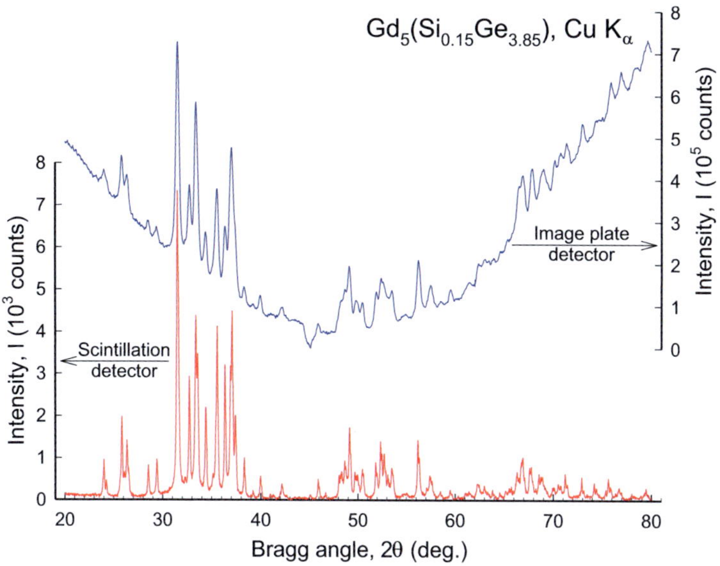

Fig. 11.25 Comparison of the X-ray diffraction data collected from the same powder using a point detector with a diffracted beam monochromator in Bragg–Brentano geometry (shown in *red*, left-hand intensity scale) versus an image-plate detector with an incident beam monochromator (shown in *blue*, right-hand intensity scale) in the transmission geometry from a cylindrical sample. The extremely high and nonlinear background observed with the image-plate detector is due to the strong X-ray fluorescence of Gd atoms interacting with Cu Kα radiation. The widths of the Bragg peaks with the image-plate detector are increased, resulting in low resolution of the data. Despite the nearly two orders of magnitude higher photon count, the best peak-to-background ratio for the image-plate detector data is 3:1, compared to nearly 70:1 for the point detector data

diffraction data can be collected at high speeds, nearly identical to those achievable with curved position-sensitive detectors. Further, the incident beam can be collimated into a small area and it is fundamentally possible to examine powder diffraction from just a few crystalline grains or even from a single grain, provided all possible grain orientations with respect to the incident beam have been arranged by properly varying the orientation of the specimen. The issues encountered in today's image-plate detector-equipped powder diffractometers resemble those of curved position-sensitive detectors: high background and relatively low resolution (see Fig. 11.25).

When the two powder diffraction patterns (Fig. 11.25) obtained from the same material are compared, the data collected using the Bragg–Brentano geometry and point detector with the diffracted beam crystal monochromator are definitely more useful in structural analysis than the set of data collected using an image-plate detector and an incident beam crystal monochromator. The dataset collected using an image-plate detector has insufficient quality due to the unfavorable coincidence of conditions. First, the crystal structure of the material is complex (a total of ~300 Bragg peaks are possible in the range shown in Fig. 11.25: $20° \leq 2\theta \leq 80°$). Second,

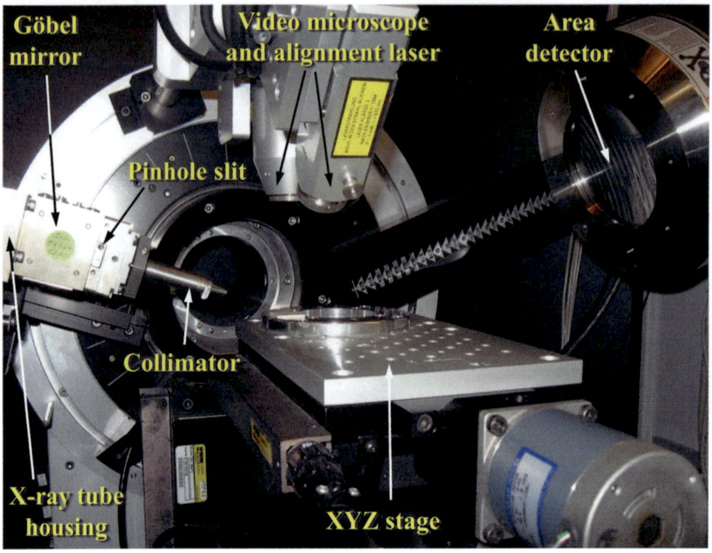

Fig. 11.26 Overall view of the Bruker D8/C2 Discover goniometer configured for powder diffractometry with GADDS (General Area Detector Diffraction System). The goniometer axis is horizontal, with independent rotations of both the X-ray source and detector arms. This goniometer is equipped with a Göbel mirror that produces a parallel beam, an XYZ sample stage, a video camera with a laser alignment system, and an area detector, specifically, the HiStar multi-wire gas-proportional detector

the powder contains more than 70 wt.% of Gd, a chemical element that scatters anomalously (see Fig. 9.5) and therefore produces a strong fluorescent background when using Cu Kα radiation.

Another kind of an area detector is a multi-wire or similar detector (see Sect. 6.4.4). An example of a powder diffractometer with such a detector is shown in Fig. 11.26. This system is a θ–θ diffractometer with two independently movable arms; however, this is not a Bragg–Brentano system. The X-ray tube (*the left arm*) is equipped with a Göbel mirror, pinhole slit, and collimator,[17] which transform the divergent incident beam into a bright parallel beam. Simultaneously, the X-ray beam is monochromatized by eliminating the Kβ component and the white portion of the spectrum. The pinhole slit and collimator produce a point beam typically ranging from several tenths of a millimeter to 1 mm in diameter. A line focus can also be generated when the pinhole slit and collimator are replaced with a standard divergence slit and Soller slit (neither is shown in Fig. 11.26). In this configuration, the multi-wire detector (trade name HiStar)[18] is positioned on the right arm at the

[17] See https://www.azom.com/details.asp?ArticleID=741 for more information about Göbel X-ray mirrors.

[18] Recently, it was upgraded with the more advanced Vantec-500 detector based on Bruker's MikroGap™ technology.

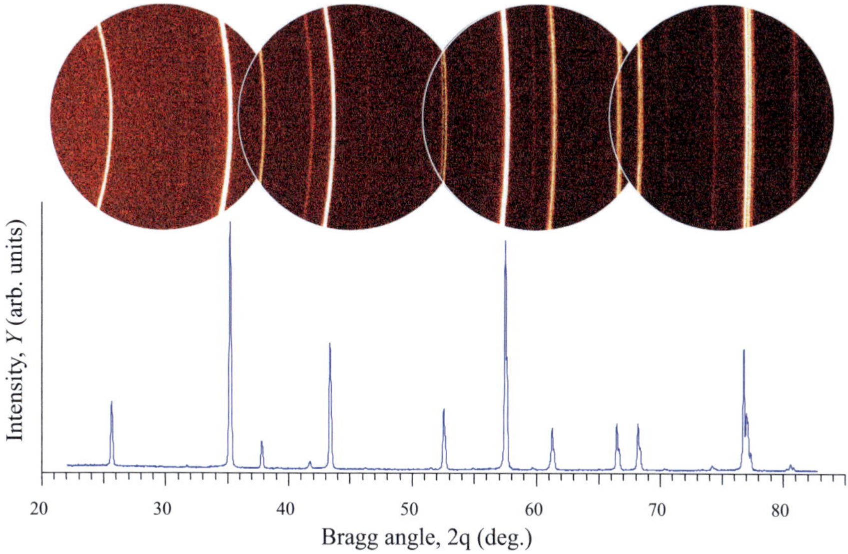

Fig. 11.27 Examples of two-dimensional diffraction images (*top*) from a corundum plate (SRM 1976) collected on D8/C2 Discover-GADDS powder diffractometer using Cu Kα radiation, Göbel mirror, HiStar detector at 30 cm with the resolution of 1024 × 1024 pixels. The plot at the *bottom* shows the corresponding one-dimensional pattern after the integration of the diffracted intensities along the Debye rings

maximum possible distance of 300 mm from the goniometer axis to achieve the highest resolution.

The two-dimensional diffraction image is usually recorded with 1024 × 1024 pixels resolution, but other resolutions (512 × 512 or 2048 × 2048) can be used when necessary. The specimen is placed on the XYZ sample stage, and its position is aligned using a video camera coupled with a laser located above the sample stage (see Fig. 11.26). The camera and the laser are aligned in such a way that their axes intersect exactly at the goniometer axis, and this is used to precisely position the height (Z-coordinate) of a specimen to minimize sample-displacement errors and to select the desired spot for measurement. One of the biggest advantages of this configuration is that it can accommodate samples of various shapes and sizes or even large bulk material (e.g., a rock or a manufactured part). Multiple locations can then be examined by focusing on specific spots and recording X-ray powder diffraction patterns.

As-recorded two-dimensional images, an example of which is shown at top of Fig. 11.27, are analyzed and, if needed, integrated and merged to produce a one-dimensional powder diffraction pattern, as shown at the bottom of Fig. 11.27. Depending on the detector-to-sample distance, it takes three to six images to cover a maximum possible range of Bragg angles. It is worth noting that data collection with the area detector is usually performed at specific angles of the X-ray source arm (θ_1) and the detector arm (θ_2). The obtained two-dimensional images are then converted (integrated) into one-dimensional patterns, which are subsequently merged to produce a single diffraction pattern. The merging conditions are set so that the images partially overlap, as shown in Fig. 11.27. Integration along the rings both increases the intensity and minimizes the effects related to nonideal particle orientation averaging. The latter can be further improved by oscillating the specimen in the plane of diffraction.

Thus, both the linear position-sensitive and area detectors (such as image plate, multi-wire, and others) are used in specialized applications of powder diffraction, including in situ studies of phase transformations, microdiffraction, and local nondestructive analysis. However, their use in high-precision determination of the crystal structure of materials is limited.

In conclusion, it is important to note that despite its long history, powder diffractometry remains a rapidly developing field, particularly in terms of instrumentation. Both position-sensitive and area detectors, made routine by exceptional technological advancements in high-speed electronics and tremendous computing power, have significantly accelerated the powder diffraction experiment. Moreover, X-ray mirrors and capillaries have successfully entered the market, enabling nearly parallel X-ray beams in analytical laboratory instruments—not just with synchrotron radiation sources. While it is difficult to predict how advanced powder diffraction instruments will become in the near future, the essence of a quality powder diffraction experiment will remain the same: the best powder diffraction data will always require high-precision and the best-possible resolution and minimum background.

11.4 Nonambient Powder Diffractometry

Many materials undergo polymorphic transformations (changes in their crystal structure) as temperature, pressure, and other thermodynamic parameters vary. The relative simplicity of the powder diffraction experiment makes this technique well-suited for in situ examination of the crystal structure of materials under nonambient conditions. The two thermodynamic parameters most commonly varied in powder diffraction studies are temperature and pressure. Here, we will briefly describe the basic principles used in the design of powder diffraction experiments at variable temperature and pressure Additionally, we will mention a less common approach that allows, in addition to temperature, for examining the effect of a magnetic field on the crystallography of magnetic materials.

11.4.1 Variable Temperature Powder Diffractometry

As shown in Fig. 11.28 (left), to manipulate the temperature of the specimen (or any other thermodynamic variable), the sample must be enclosed in a controlled environment. For temperatures exceeding ambient, this environment is typically a

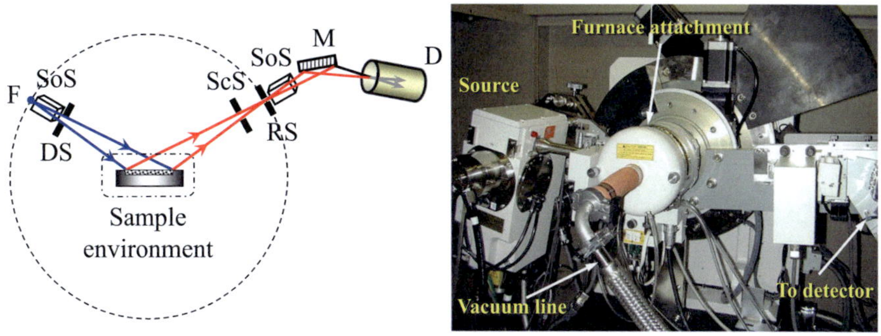

Fig. 11.28 The schematic of the optics of a powder diffractometer with diffracted beam monochromator and a sample in a controlled environment (*left*) and a photograph of the Rigaku TTRAX goniometer with a furnace attachment (*right*)

furnace (see Fig. 11.28, right), while for temperatures below ambient, a cryostat is used. The keys to a successful controlled-temperature experiment include protecting the sample from the environmental factors, such as preventing oxidation at elevated temperature or ice formation at cryogenic conditions, ensuring temperature stability and avoiding temperature gradient(s) across the specimen, and protecting the goniometer from nonambient temperatures. All this must be done while minimizing absorption of both the incident and scattered beams and avoiding increased background scattering.

Protection of the sample nearly always means surrounding it by a vacuum tight enclosure and keeping it in vacuum or in a controlled atmosphere.[19] Hence, the enclosure should normally be outfitted with low-absorbing windows that will transmit the incident beam to the sample and the scattered beam to the detector. For X-ray experiments, the windows can be made from beryllium; however, due to its toxicity, modern controlled-temperature powder diffractometer attachments typically use Mylar[20] for the windows. In neutron scattering, the windows are often made from vanadium, which is nearly transparent at thermal neutron energies. Despite the excellent heat resistance of Mylar®, when experiments are performed at high temperatures, keeping the sample in a vacuum is often the only way to prevent windows burn through due to high rates of convective heat exchange. Consequently, most furnaces are water-cooled in order to protect both the attachment body and the goniometer from damage caused by excessive heat.

Normally, temperature is controlled by fully automated temperature controllers, which are relatively inexpensive and quite accurate instruments. Temperature

[19] In some cryogenic applications, it is possible to control sample temperature by flowing dry, cold nitrogen gas over the sample and surrounding it with a cone of dry, room-temperature gas (typically nitrogen or air), thereby preventing the formation of ice. This method is more commonly used in single-crystal rather than in powder diffraction, as the specimen size in the latter is generally much larger than that in the former.

[20] Mylar® is a strong heat resistance polyester film invented by DuPont in 1952.

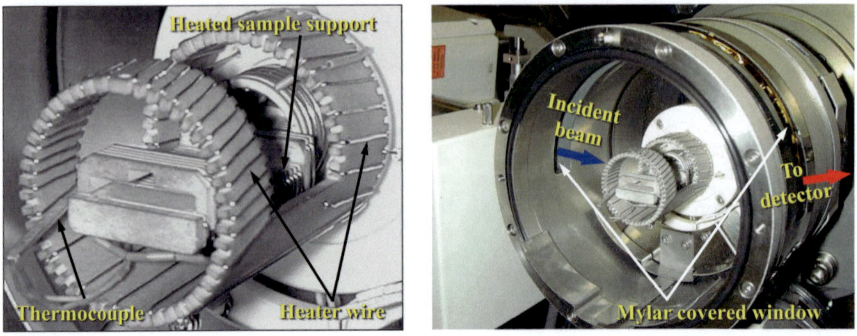

Fig. 11.29 Sample heater (*left*) and the overall view of the high-temperature furnace attachment with the cover removed to expose the inside (*right*)

uniformity is achieved by making a heater much larger than the sample, and by surrounding the sample by the heating element as much as possible, thus combining both the direct and radiative heat exchange, see Fig. 11.29. Obviously, the heating element must be designed so it does not obstruct the optical pathways. In cryogenic applications, the powder may be cooled either by a close cycle refrigerator—cryocooler—or by using a continuous flow cryostat. In the former case, the sample holder is usually attached to a cold finger by a flexible heat link to avoid vibrations. Sample holders must be good heat conductors and inert. For high-temperature applications, sample holders are usually made from platinum, and for low temperature experiments, they are made from copper.

In addition to studying phase transformations, which are accompanied by rearrangements of both the locations and intensities of Bragg peaks and are therefore easily detectable, temperature-dependent powder diffractometry is also useful for examining subtle structural effects, for example, thermal expansion. Thermal expansion is conventionally studied by dilatometry—the measurement of changes of the external dimensions of a solid. In polycrystalline solid, changes of shape reflect the underlying changes of lattice dimensions and, therefore, are directly observable in a powder diffraction experiment. Moreover, unlike dilatometry, which requires single crystals and often multiple experiments to determine anisotropy of thermal expansion, powder diffraction data provide all the necessary information in a single experiment using polycrystalline material, as can be seen in Fig. 11.30.

11.4.2 Principles of Variable Pressure Powder Diffractometry

Structural changes under pressure are commonly examined using a diamond anvil cell, shown in Fig. 11.31. The incident beam must pass through a diamond before reaching the sample, and the diffracted beam must also pass through a layer of the opposing diamond anvil. Considering the limited range of Bragg angles available for

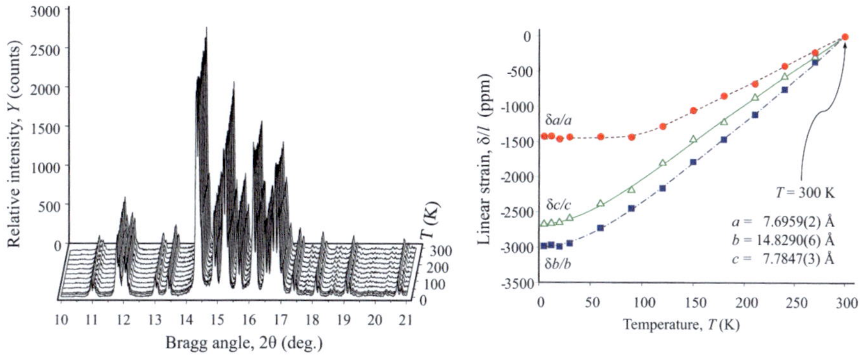

Fig. 11.30 A series of powder diffraction patterns of Gd_5Ge_4 collected between 5 and 300 K (*left*) and anisotropic linear thermal expansion along the three independent crystallographic direction of the orthorhombic lattice of the compound computed from the powder diffraction data (*right*) (Ya. Mudryk, A. P. Holm, K. A. Gschneidner, Jr., and V. K. Pecharsky, Crystal structure—magnetic property relationships of Gd_5Ge_4 examined by in situ X-ray powder diffraction, Phys. Rev. B **72**, 064442, 2005). The data were collected between 7° and 42° 2θ using equipment seen in Fig. 11.32, but for clarity only small fragments are shown on the *left*. Error bars on the right are smaller than the size of the data points

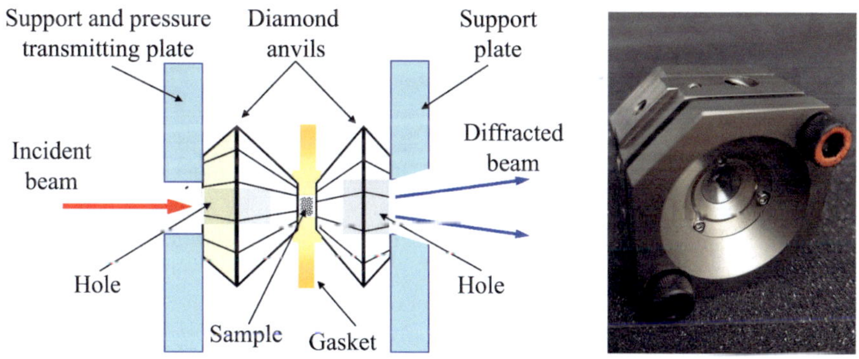

Fig. 11.31 The schematic of a diamond anvil cell (*left*), commonly known as Bridgman (Percy Williams Bridgman (1882–1961) was an American physicist best known for his work with high pressures. He is also credited with developing a technique for growing large single crystals (the Bridgman method). He won the 1946 Nobel Prize in Physics "for the invention of an apparatus to produce extremely high pressures, and for the discoveries he made therewith in the field of high pressure physics." See https://www.nobelprize.org/prizes/physics/1946/bridgman/) anvils, and an actual Diacell Bragg-(S) Plus (From Almax-easyLab, see https://www.almax-easylab.com/product/diacell-bragg-s-plus/) (*right*), which uses conical Boehler-Almax anvils for high-pressure X-ray research. The sample is placed inside an opening in a metal gasket, typically 250 μm in diameter. The diamonds have hollow centers to reduce X-ray absorption. The anvils are pressurized either by a screw or a hydraulic press. The pressure-transmitting medium is usually alcohol, such as methanol, or fluorinert (Fluorinert, a family of perfluorinated liquids trademarked by 3M. The liquids are chemically inert, electrically insulating, and good heat conductors)

examination, diamond anvil cells find most common use in powder diffraction experiments using short wavelengths, specifically high-energy synchrotron sources. Another reason for using short wavelengths is their lower absorption by the diamond anvils. Typical pressures achievable in diamond anvil cells range from a few kbar to over 1 Mbar. Normally, the sample is mixed with some kind of pressure standard. Diamond anvil cells can be coupled with temperature-controlled environment, most commonly in the cryogenic regime. For more details, we refer the interested reader to a recent review about high-pressure diffractometry (mostly single-crystal method).[21]

11.4.3 Powder Diffractometry in High-Magnetic Fields

Magnetic ordering of solids nearly always results in shape change, commonly known as spontaneous magnetostriction.[22] Furthermore, when ferromagnetically ordered solids are magnetized, the so-called forced magnetostriction leads to further shape changes. Obviously, these shape changes reflect intrinsic structural effects. Moreover, some materials may exhibit crystallographic variations that are far beyond a simple lattice expansion in a specific direction as a result of changes in their magnetic structure. Magnetic fields can induce significant structural rearrangements, such as polymorphic transformations. Therefore, understanding the underlying atomistic mechanisms is of considerable importance.

The idea of varying magnetic field around a specimen in a powder diffraction experiment is actually quite simple, as shown in Fig. 11.32 (left). All one has to do is to surround the specimen with two coils and energize them as needed. In practice, however, the instrumentation is quite complex, because in order to reach substantial field values, one must employ superconducting coils, which must be cooled to the temperature of liquid helium. This results in heavy and bulky equipment, as shown in Fig. 11.32 (right). In addition, considering that magnetic ordering most commonly occurs at temperatures below ambient, high-magnetic field attachments are normally outfitted by some kind of cryogenic control of sample temperature.

Magnetic field effects are most commonly studied using neutron scattering, but recently, the technique has been successfully demonstrated in laboratory powder diffraction experiments.[23] Having a high-magnetic field and accompanying field gradients in close vicinity to moving electrons (e.g., in the X-ray tube or in the detector) and to components of goniometer that are ferromagnetic at room

[21] A. Katrusiak, High-pressure crystallography, Acta Cryst. **A64**, 135 (2008). https://doi.org/10.1107/S0108767307061181

[22] A. Lindbaum and M. Rotter, Spontaneous magnetoelastic effects in gadolinium compounds, in, K.H.J. Buschow, Ed., Handbook of Magnetic Materials (Elsevier Science, Amsterdam, 2002), vol. 14, p. 307.

[23] A.P. Holm, V.K. Pecharsky, K.A. Gschneidner, Jr., R. Rink, and M.N. Jirmanus, X-ray powder diffractometer for in situ structural studies in magnetic fields from 0 to 35 kOe between 2.2 and 315 K, Rev. Sci. Instr. **75**, 1081 (2004). https://doi.org/10.1063/1.1667253

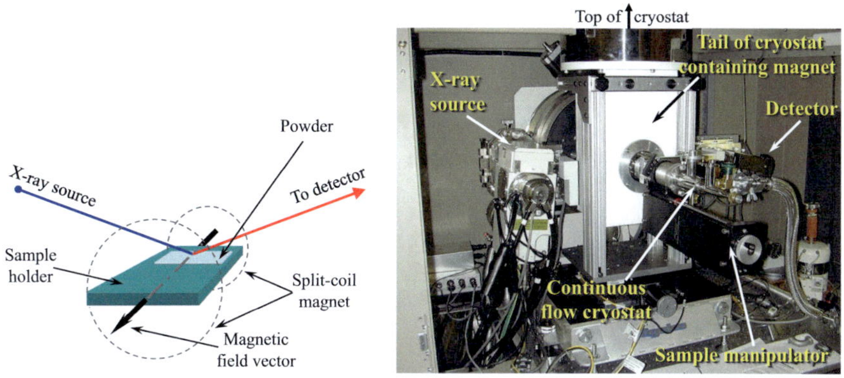

Fig. 11.32 A simplified schematic of the sample holder surrounded by a split coil magnet and the X-ray beam path from the source through the split coil to the detector (*left*), and the view of the bottom part of a cryostat containing a split coil superconducting magnet mounted on a Rigaku TTRAX powder diffractometer (*right*). The radiation safety enclosure is opened to expose the details of the instrument

temperature requires a careful control of the experiment and additional shielding. We refer the interested reader to the original paper, which describes many of the potential pitfalls, as well as the benefits of powder diffraction experiments in high-magnetic fields.

One example of useful structural information that can be obtained using in situ powder diffractometry in high-magnetic fields is illustrated in Fig. 11.33. As follows from the temperature-dependent powder diffraction data shown in Fig. 11.30, the Gd_5Ge_4 compound has no polymorphic transformations over the temperature range from 5 to 300 K in a zero magnetic field. However, when the sample is magnetized at low temperature (the temperature of the experiment illustrated in Fig. 11.33 was 29 K), magnetic field exceeding 15 kOe induces a martensitic-like phase transformation in this material. The field-induced transformation proceeds via shear displacements of the neighboring slabs of atoms during which the slabs shift by nearly a quarter of an angström in opposite directions, as indicated by arrows placed near the slabs in Fig. 11.33, but the symmetry of the crystal lattice remains unchanged (space group is Pnma for both the high-magnetic field and low-magnetic field polymorphs). Unit cell dimensions change by as much as 1.9% (19,000 ppm). Large shear displacements of the slabs result in substantial changes of the interslab distances as shown for a few of them in Fig. 11.33, but the average change of the intraslab distances is only a few percent. When Gd_5Ge_4 is demagnetized, the low field crystal structure is recovered exhibiting a narrow hysteresis.

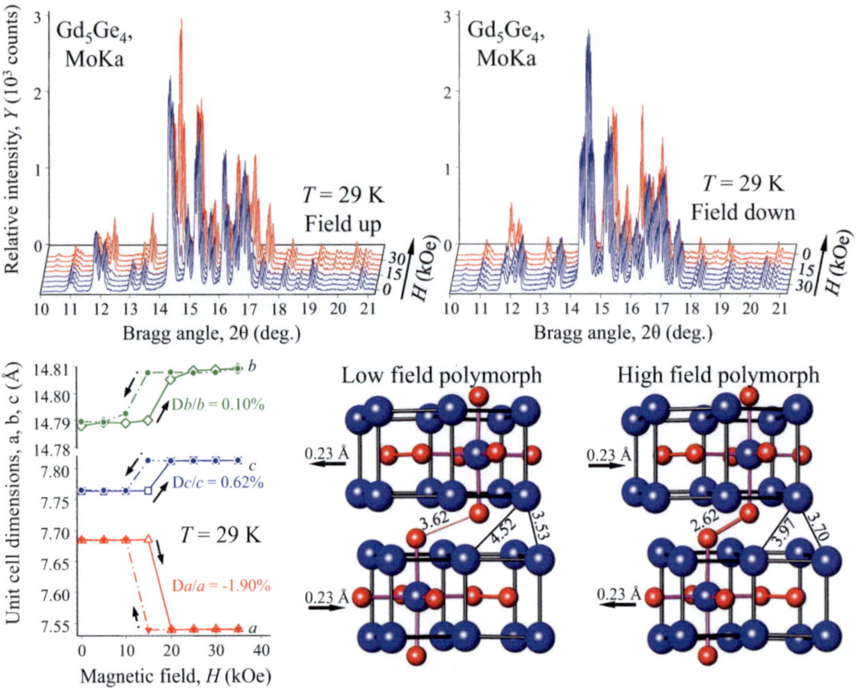

Fig. 11.33 (*Top*) Two series of powder diffraction patterns of Gd_5Ge_4 collected isothermally at $T = 29$ K with the magnetic field varying between 0 and 35 kOe. (*Bottom*) Unit cell dimensions as functions of magnetic field and details of the field-induced crystallographic phase transformation in the compound (V. K. Pecharsky, A. P. Holm, K. A. Gschneidner, Jr., and R. Rink, Massive magnetic-field-induced structural transformation in Gd_5Ge_4 and nature of the giant magnetocaloric effect, Phys. Rev. Lett. **91**, 197204 (2003). https://doi.org/10.1103/PhysRevLett.91.197204). The data were collected between 7° and 42° 2θ using equipment shown in Fig. 11.32, but only small fragments are shown for clarity

11.5 Additional Reading

1. International Tables for Crystallography, Vol. A, Space-group symmetry, M. I. Aroyo, Ed., Second online edition, (2016). https://it.iucr.org/A/
   International Tables for Crystallography, Vol. B, Reciprocal space, U. Shmueli, Ed., Second online edition, (2010). https://it.iucr.org/B/
   International Tables for Crystallography, Vol. C, Mathematical, physical and chemical tables, First online edition, E. Prince, Ed. (2006). https://it.iucr.org/C/
   International Tables for Crystallography, Vol. A-I published jointly with the International Union of Crystallography (IUCr) by Springer. https://it.iucr.org/
2. R. Jenkins and R.L. Snyder, Introduction to X-ray powder diffractometry. Wiley, New York (1996).
3. A. Katrusiak, High-pressure crystallography, Acta Cryst. **A64**, 135 (2008), https://doi.org/10.1107/S0108767307061181

4. J. Als-Nielsen and D. McMorrow, Elements of modern X-ray physics, Wiley, New York (2001).
5. H. Ehrenberg, 2D Powder Diffraction: In situ powder diffraction—Electric Fields, p. 7 in: CPD Newsletter "2D Powder Diffraction", Issue 32 (2005), available at https://www.iucr.org/__data/assets/pdf_file/0003/21657/CPD32.pdf

11.6 Problems

1. There are 25 plates in a Soller slit. Axial size of the incident beam when it exits the slit is 12 mm. Calculate the length of the plates along the X-ray beam (l) if the slit results in the axial divergence of the beam, $\alpha = 2.5°$. Neglect the thickness of the plates.
2. A crystal monochromator is made from high-quality pyrolytic graphite (space group $P6_3/mmc$, a = 2.464, c = 6.711 Å). Assume that this crystal is used to suppress the $K\alpha_2$ spectral line of Cu $K\alpha$ radiation by using the reflection from (002) planes and that the crystal is cleaved parallel to the (001) plane. Estimate the linear separation (δ, in mm) between the centers of two Bragg peaks ($K\alpha_1$ and $K\alpha_2$) at 200 mm distance after the reflection from the crystal. Assuming that the crystal is nearly ideal, calculate angle θ which the incident beam should form with the surface of the crystal for best result.
3. Two powder diffraction patterns (Fig. 11.34) were collected from the same material. The first experiment was conducted at elevated temperature and the

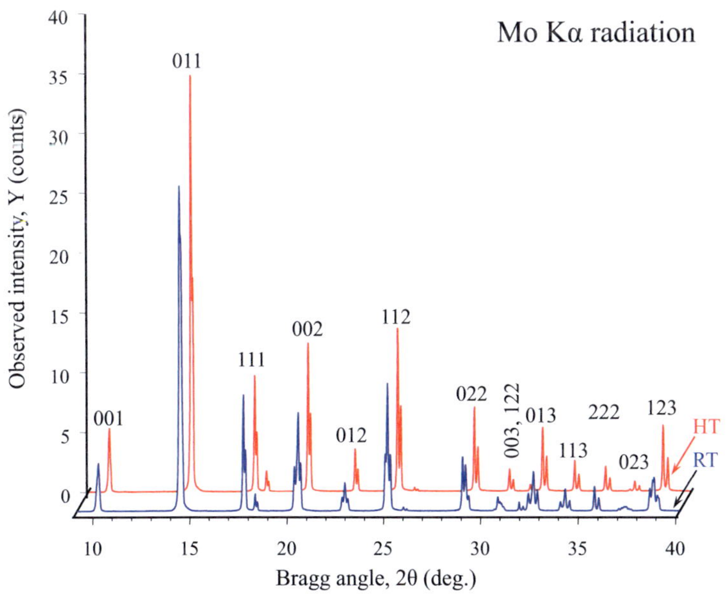

Fig. 11.34 Powder diffraction patterns of a material collected at high temperature (HT) and room temperature (RT) using Mo $K\alpha$ radiation

second at room temperature. The high-temperature crystal structure is cubic with the indices of Bragg reflections as marked. Note that a small amount of an impurity phase is also present, as can be concluded from three weak unindexed Bragg peaks observed at approximately 18°, 26°, and 32° of 2θ. What happens when the sample is cooled to room temperature? What can you tell about the crystal structure of the majority phase at room temperature?

Chapter 12
Collecting Quality Powder Diffraction Data

Many factors affect the quality of the powder diffraction data (e.g., see Figs. 11.1, 11.5, and 11.25), and the state of the specimen used in a powder diffraction experiment is one of them. Additionally, several data acquisition parameters must be carefully selected. Here we consider issues related to both specimen preparation and the selection of instrument-related parameters to achieve the highest possible quality of the resulting powder diffraction pattern.

12.1 Sample Preparation

The importance of proper sample preparation cannot be overstated, especially since it is entirely within the control of the operator conducting the experiment. Poorly prepared samples will inevitably produce unusable experimental data, necessitating a repeat of the entire process and wasting both time and resources. In contrast, while preparing a high-quality sample for powder diffraction may take, the time invested is well worth it.

12.1.1 Powder Requirements and Powder Preparation

The true powder diffraction pattern can only be obtained from a specimen containing an infinite number of individual particles realizing an infinite number of orientations within the irradiated volume (e.g., see Fig. 8.1). In other words, the particles in the specimen should have a completely random distribution of crystallographic orientations of grains or crystallites with respect to one another. Clearly, this ideal situation is impossible to achieve. However, if one considers a 10 mm diameter and 0.1 mm deep cylindrical sample holder filled with 50 μm diameter spherical particles, it is easy to estimate that it will hold nearly 9×10^4 of such particles at

P. Zavalij, V. Pecharsky, *Fundamentals of Powder Diffraction and Structural Characterization of Materials*, https://doi.org/10.1007/978-3-031-91504-8_12

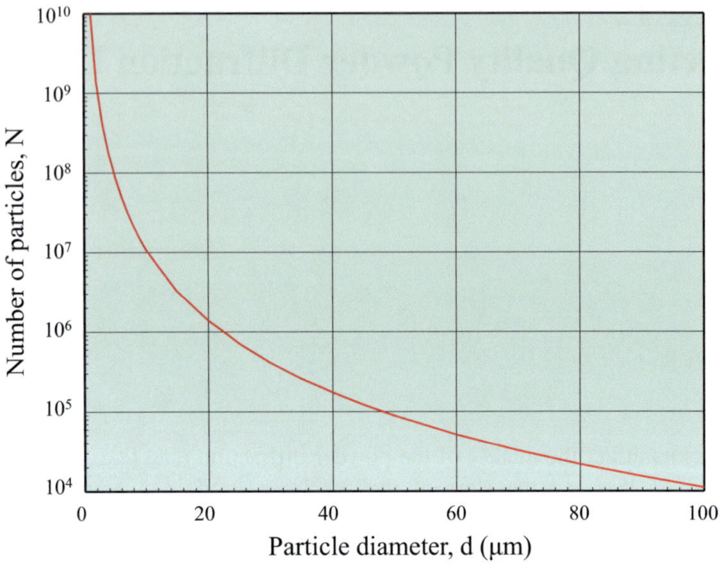

Fig. 12.1 The total number of spherical particles in a disk-like specimen with a diameter of 10 mm and a depth of 0.1 mm, as a function of particle size assuming close packing of the spheres

74.05% packing density by assuming close packing of the spheres (Fig. 12.1). When the particle size is reduced to 30 μm, the same volume will contain ~4 × 10^5 particles, and when the particles are 10 μm in diameter, it will take a total of ~1 × 10^7 different particles to fill the same volume. These are large numbers, which may be considered sufficient to approximate the infinite quantity of particles required for collecting powder diffraction data.

It is, therefore, obvious that a nearly infinite number of particles in a specimen is easily achieved by reducing average particle size. Another very effective approach to increase both the number of particles in the irradiated volume and the randomness of their orientations is to spin the specimen continuously during data collection.[1]

The majority of materials, which are routinely examined by powder diffraction, are initially in a state unsuitable for the straightforward preparation of the specimen. Unless the material is already in the form of a fine powder with the average particle size between 10 and 50 μm, its particle size should be reduced.

The most common approach to reducing particle size is to grind the substance using a mortar and pestle, though mechanical mills also do a fine job (Fig. 12.2). Both the former and the latter are available in a variety of sizes and materials. Mortars and pestles are usually made from agate or ceramics. Agate is typically used to grind hard materials (e.g., minerals or metallic alloys), while ceramic equipment is suitable to grind soft inorganic and molecular compounds. Mechanical mills require

[1] Sample spinning is the simplest way to better randomize particle orientations. Whenever possible, it should be employed during data collection.

Fig. 12.2 The tools most commonly used in sample preparation for powder diffraction experiments: agate mortar and pestle (*left*) and ball-mill vials made from hardened steel (*middle*) and agate (*right*)

vials and balls, which are usually made from hardened steel, tungsten carbide, or agate.

Depending on the nature of the material, grinding—whether manual or mechanical—may take anywhere from several seconds to several minutes. Prolonged grinding should not be relied upon, as it usually results in the creation of an excessive surface (its structure is different from the bulk) and in the agglomeration of small particles. When aggregation effects are severe, adding a chemically inert liquid, which does not dissolve the material under examination, may help to prevent excessive conglomeration. In some cases, prolonged mechanical grinding may cause serious degradation in the crystallinity of the material. If the crystallinity is reduced due to induced stresses and micro-deformations, it can be restored by a brief (10–30 min) heat treatment at temperatures near 1/3 of the melting temperature of the material on the absolute thermodynamic temperature scale.

A mortar and pestle are normally the only grinding option when the quantity of the material is limited, while mechanical mills usually require much larger amount of material to begin with. It is important to keep the milling equipment clean to prevent cross-contamination of samples. One of the best ways to clean a mortar and pestle before grinding a new sample is to use a 50:50 (by volume) mixture of nitric acid (HNO_3) and water. When using a hardened steel or a tungsten carbide ball mill, the vial should be cleaned using a blaster with fine (<5 μm) Al_2O_3 powder, and the previously used balls should normally be discarded. Agate vials and balls can be cleaned using a 50:50 mixture of water and nitric acid. At the very least, flashing with a low-boiling point organic solvent (e.g., acetone or alcohol), wiping, and drying the grinding equipment should always be performed.

One of the biggest concerns when using mechanical grinding in a ball mill is the possibility of contamination of the sample with the vial (balls) material. This is particularly true when hard materials are subject to mechanical processing. To ensure that no contamination has occurred, it is advisable to perform a chemical analysis of the material before and after ball milling. When a mortar and pestle are used to grind hard materials, excessive tapping on large pieces of a sample to break them should be avoided as much as possible since small chips of agate or ceramic may cross-contaminate the resulting powder.

Some materials, for example, metallic alloys and polymers, may be quite ductile. Grinding them using a mortar and pestle or a ball mill is nearly always unsuccessful, and the powders may be produced by filing. Needless to say, the file should be clean (preferably a brand new tool) to prevent cross-contamination. It is often the case that filings of ductile metallic alloys are contaminated by small particles broken off the file. The latter can be easily removed from the produced powder by using a strong permanent magnet (e.g., $Nd_2Fe_{14}B$ or $SmCo_5$) provided the powder of interest is paramagnetic or diamagnetic at room temperature. Powders produced by filing usually must be heat-treated before preparing a specimen for a powder diffraction experiment to relieve the processing-induced stresses.[2]

Regardless of the method employed to produce fine particles, the resultant powder should be screened using appropriate size sieve(s). The most commonly used sieves have openings from 25 to 75 μm. It is also important to ensure that the sieve is clean before sifting the powder under examination to eliminate cross-contamination. Sieves may be cleaned using a pressurized gas (e.g., nitrogen or helium from a high-pressure cylinder), and/or they can be washed in a low-boiling point solvent (e.g., acetone or alcohol) before drying by a high-pressure gas. It may be problematic to sift powders of low-density materials, but every effort should be made to do so. Sifting not only eliminates large particles from the powder, but it also helps to break down agglomerates that may have formed during grinding.

12.1.2 Powder Mounting

As mentioned at the beginning of this section, another important requirement imposed on a high-quality powder sample is the realization of the infinite number of possible orientations of the particles with respect to one another, in other words, complete randomness in their orientations. The reduction of particle size is the necessary but not sufficient condition to achieve this. In reality, nearly ideal randomness in particle orientations is only feasible with a large number of particles, which have spherical or nearly spherical (isotropic) shapes. In many cases, grinding or milling produces particles with far from isotropic shapes; therefore, special precautions should be taken when mounting powders on sample holders. The most severe cases of nonrandom particle orientation distributions are expected when platelet-like or needle-like particles are produced by grinding, see Fig. 12.3.

When powder particles have thin platelet-like shapes, they will tend to agglomerate, aligning their flat surfaces nearly parallel to one another (Fig. 12.3, left). As a result, the orientations of platelets are randomized via rotations about a common axis

[2]Since one-third of the absolute melting temperature is usually sufficient for an effective stress-relief, some materials may self-anneal during the filing. One of the examples is lead (Pb), which is a ductile metal and has melting temperature 601 K. Lead powder self-anneals at room temperature (~298 K), thus producing sharp Bragg peaks in the as-filed state.

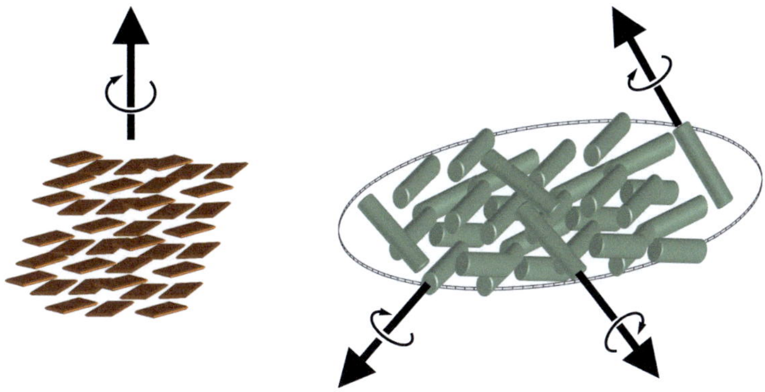

Fig. 12.3 The two limiting cases of nonrandom particle orientation distributions due to distinctly anisotropic particle shapes: platelet-like (*left*) and needle-like (*right*) particles. The arrows indicate the directions around which the particles may rotate freely

normal to their largest faces, and such samples are expected to have a uniaxial preferred orientation (or texture).

When particles are in the form of thin needles (Fig. 12.3, right), the orientations of the axes of the needles are naturally confined to the plane of the sample. Further, each needle has an additional rotational degree of freedom, which is rotation about its longest axis, and such samples are expected to have an in-plane preferred orientation. Even more complex preferred orientation effects may occur when elongated particles are flat or ribbon-like. Then ribbons may align in the plane of the sample, and because they are flat, their surfaces may arrange parallel to the sample surface as well. This is the case of two combined preferred orientation axes: one is along the axis of the ribbon, and the second is perpendicular to the flat face.

Regardless of the type of preferred orientation present in the sample, it will systematically affect the diffracted intensities in its own way. In severe cases, nothing more than lattice parameters (if any) can be determined from highly textured powder diffraction data since it is impossible to precisely account for the changes in the diffracted intensity caused by the exceedingly nonrandom distribution of particle orientations.

Depending on the diffraction geometry, proper sample preparation techniques are different. We begin with flat samples used in combination with the Bragg–Brentano focusing geometry and horizontal goniometer axis, since it is the most common geometry employed today. A few of these sample holders are shown in Fig. 12.4. The two types of sample holders most frequently used in the Bragg–Brentano method are also shown in Fig. 12.5. They can be made of metal, plastic, or glass. The difference between the two is that the holder on the left has a shallow cylindrical or rectangular cavity (usually ¼ to 1 mm deep) to accommodate the powder (also see Fig. 12.4c). The container on the right has a rough spot on its surface (also see Fig. 12.4b). Rough surface is best created using a sand blaster with hard particles

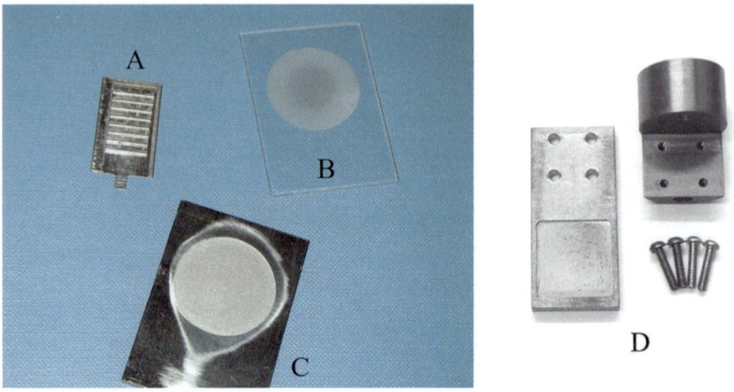

Fig. 12.4 Examples of sample holders used with Bragg–Brentano geometry. Part (**a**) is platinum sample holder used in high-temperature powder diffraction experiments (see Fig. 11.29). The fins seen in (**a**) are designed to improve heat conductivity and uniformity of powder temperature. Part (**b**) is glass slide with a round area of about 1 inch in diameter sand blasted to improve randomization of particle orientations. Part (**c**) is stainless steel sample holder with a ~0.5 mm deep cavity. Part (**d**) is copper sample holder with a rectangular cavity to accommodate powder for low-temperature high-magnetic field experiments (see Fig. 11.32). The four screws shown in (**d**) are used to attach the sample holder to a cold finger

Fig. 12.5 Examples of flat sample holders used in powder diffractometers with reflection geometry. The holder on the *left* has a cavity, which is filled with the powder. The holder on the *right* has a rough area, which accommodates a thin layer of the powder

10–50 μm in diameter. When the roughness of the sample surface is of the same order as particle size, this helps to diminish the preferred orientation effects.

It is worth mentioning that when background is of concern, for example, when only a small amount of a low-absorbing sample is available, low-background sample holders may be used. These holders (or only the inserts) are made from single crystals cut in such a way that no Bragg reflections can occur in a whole range of Bragg angles. For example, silicon single crystal plate cut parallel to the (510) planes or quartz single crystal cut at 6° with respect to (001) planes will contribute no Bragg peaks when used in a standard Bragg–Brentano diffractometer.

When a relatively large quantity of powder is available, the sample for X-ray diffraction can be prepared by filling the volume of the cavity (Fig. 12.5, left) with the dry powder. The excess powder should be removed from the surface of the sample holder by a single sweep with a razor blade (Fig. 12.6) or by an edge of a glass slide. Under normal circumstances, the powder should never be compacted inside the hole using a smooth flat surface, for example, the flat side of a glass slide. If the powder is packed by pressing against a smooth flat surface, this will inevitably

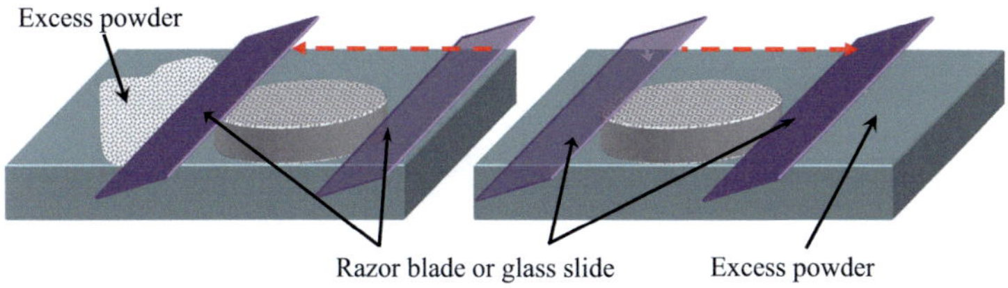

Fig. 12.6 Excess powder is removed from the top of the sample holder with a single sweep by an edge of a razor blade or a glass slide. The direction of the sweep depends on the physical properties of the powder and is usually established by trial-and-error

rearrange particle orientations in the specimen causing strong preferred orientation (see Fig. 12.3). Compacting is only forgivable if particles in the powder are nearly spherical, which is quite rare. If necessary, the powder inside the hole may be compressed by gently pressing it against a rough surface with a roughness that matches the average particle size, especially when the material is lightweight.[3] In the worst-case scenario, the edge of a glass slide can be used for this purpose.

Though filling the specimen holder with powder produces the best-quality flat samples for powder diffraction experiments, the procedure requires some experience, and a uniform distribution of particle orientations may be difficult to achieve, especially when working with light, fluffy powders. It may be helpful, therefore, to prepare a viscous suspension of powder using chemically inert, low-boiling temperature liquid, which does not dissolve the material and then pour the suspension into the hole. Excess mixture is then removed by a single cut with a razor blade, and the remaining solvent should be slowly evaporated before installing the sample on a diffractometer.

Another method of reducing the preferred orientation while mounting the powder into the sample holder is the so-called side or back filling (Fig. 12.7). It requires a special or modified sample holder with an opening on the side or on the back. The front of the holder is covered during packing, and the surface facing the powder should have nearly the same roughness as the average particle size. Using this technique, the powder can be packed better and with lower preferred orientation on the front surface of the sample that is irradiated, which is the most critical part of the specimen in X-ray powder diffraction due to the limited penetration depth of X-rays.

An effective way of avoiding preferred orientation is spraying the fine powder suspended in a quick drying polymer solution.[4] Small droplets spheroidize before the solution dries in-flight and form the tiny solid spheres containing only a few

[3] Flat surfaces with varying roughness may be created by sandblasting a set of glass slides using particles of different diameter. A quick solution may be achieved by gluing a piece of sandpaper with appropriate roughness on a glass slide.

[4] S. Hillier. Spray drying for X-ray powder diffraction specimen preparation, p. 7 in: CPD Newsletter "Powder Diffraction in mining and minerals", Issue 27 (2002) available at https://www.iucr.org/__data/assets/pdf_file/0017/21626/cpd27.pdf

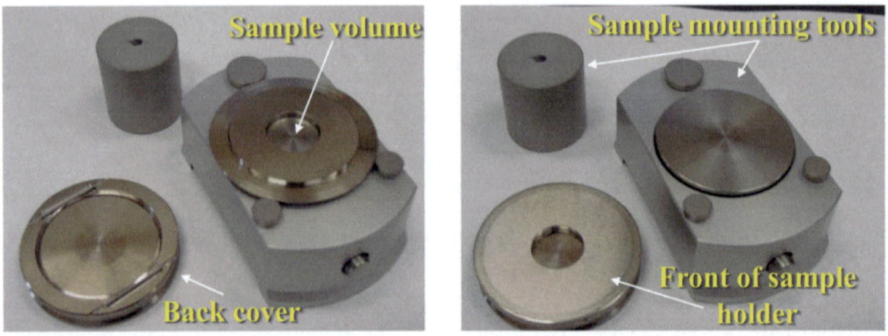

Fig. 12.7 Tools used for back filling a sample holder with powder (sold by Malvern Panalytical). On the *left*, the front side of the sample holder is clamped on a support. After the sample volume is filled from the back side and packed, the front side of the holder couples to the back cover, and the sample holder is released from the clamping tool, as seen on the *right*

particles embedded in each droplet.[5] This method removes preferred orientation nearly completely because the resulting particles are spherical and thus maintain random orientations during mounting. It is, however, complex and introduces a substantial amount of a polymer, which increases background noise, thus reducing the overall quality of the resultant powder diffraction pattern.

Good-quality specimens with minimal preferred orientation effects can be prepared by dusting the ground powders through a sieve directly on a sample holder, thus covering the rough spot. This is the only feasible option when using sample holders without the hole to accommodate the powder (e.g., see Fig. 12.4b). It is best to cover the sample holder with a specially made mask, which is removed when the dusting is completed (Fig. 12.8).

Usually, it takes several passes of dusting the powder on top of the sample holder surface to ensure complete and uniform coverage of the area, which is irradiated during the X-ray diffraction experiment. It is also possible to apply a thin layer of oil, grease, or slow-drying glue to the appropriate spot on the sample holder surface and then dust the powder on top. After dusting can be complete, the excess powder is easily removed by flipping the sample holder or by gently blowing air over the sample surface. The downside of using fluid to hold the powder on the surface is that the powder will remain contaminated with the fluid after the experiment is finished. Another fact to keep in mind when preparing samples by dusting is that it is nearly impossible to create a smooth surface of the specimen: pressing a sample holder

[5]Excellent powdered samples (see Fig. 12.16) can be prepared by using the high-pressure gas atomization (HPGA) technique. HPGA involves melting a material of interest and then spraying the melt through a nozzle employing a high-pressure nonreactive gas (e.g. nitrogen, argon, or helium). Liquid droplets (usually between ~10 and ~100 μm in diameter) spheroidize and then rapidly solidify in-flight maintaining nearly spherical form. The resulting powders may require brief homogenization and/or recrystallization heat treatment before they may be employed to collect powder diffraction data. HPGA-prepared powders are not embedded into polymer shells; however, this technique requires large amount of a starting material and is cost-ineffective in routine diffraction studies.

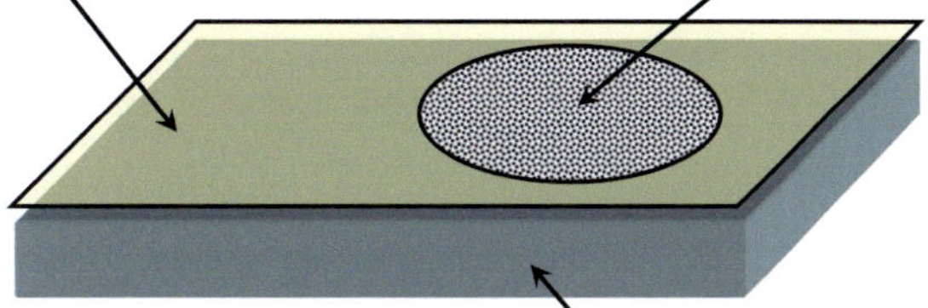

Fig. 12.8 Sample for powder diffraction prepared by dusting ground powder on the sample holder covered with a mask

against a flat surface is not an option as it usually induces strong preferred orientation effects.

The most common approach to prepare flat samples for transmission geometry is by dusting the powder on an X-ray transparent film covered with a thin layer of slow-drying varnish or glue and letting it dry before installing the sample on the goniometer. Obviously, both the film and dry varnish/glue should not be crystalline.

Cylindrical specimens are usually prepared by sinking a thin glass capillary into a liquid binder (e.g., purified petroleum jelly or liquid varnish) and then by dipping the capillary into a pile of loose powder. Alternatively, powder can be mixed with oil to a consistency of thick slurry and then the capillary is simply dipped into the mixture. In both cases, the capillary may need to be exposed to the powder several times to ensure complete and uniform coverage of its surface.

In many instances, especially when the studied powder is air- or moisture-sensitive, it can be placed inside a low-absorbing glass capillary (e.g., a borosilicate glass, or a polymer capillary such as Kapton[6]), after which the capillary is sealed. Filling capillaries with powders is usually a tedious process and it requires larger diameter capillaries than those usually used for surface coverage. However, a significant advantage of using the capillary is the near-complete removal of preferred orientation by spinning the capillary.

12.1.3 Sample Size

Several additional issues must be considered when preparing samples for X-ray powder diffraction experiment. One is the length of a flat sample, *L*, along the optical axis of the goniometer in the Bragg–Brentano geometry. It should be large enough relative to the specific beam divergence so that, at any Bragg angle during data collection, the X-ray beam projection on the sample surface does not exceed the specimen's length. Referring to Fig. 12.9 and assuming that the angular divergence

[6] Kapton® is a polyimide polymer developed by DuPont. It remains stable between ~4 and 700 K.

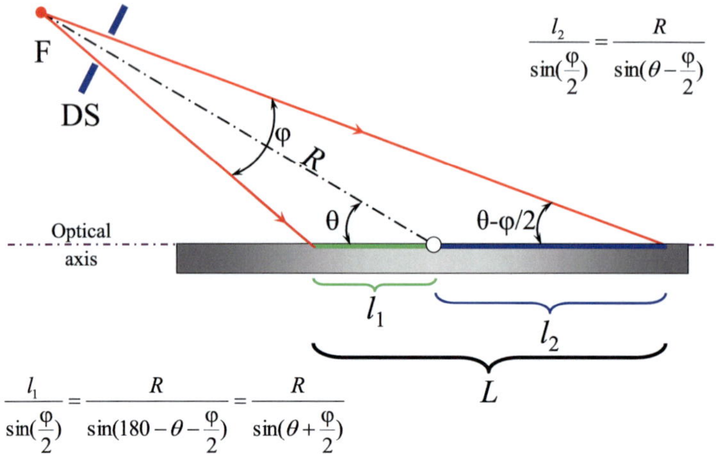

Fig. 12.9 The length of the incident beam projection, *L*, on the surface of the flat sample in Bragg–Brentano geometry. *F* focal point of the X-ray source; *DS* divergence slit; *R* goniometer radius; φ angular divergence of the incident beam; θ Bragg angle. The location of the goniometer axis is indicated by the *open circle*

of the beam is φ, it is easy to derive the relationship for the varying irradiated length, *L*, as a function of φ, goniometer radius, *R*, and Bragg angle, θ:

$$L = l_1 + l_2 = \frac{R\sin\left(\frac{\varphi}{2}\right)}{\sin\left(\theta + \frac{\varphi}{2}\right)} + \frac{R\sin\left(\frac{\varphi}{2}\right)}{\sin\left(\theta - \frac{\varphi}{2}\right)} \cong \frac{\varphi R}{\sin\theta} \tag{12.1}$$

In Eq. (12.1), φ is the angular divergence of the incident beam and *R* is the goniometer radius in mm. When the angular divergence of the incident beam is small ($\varphi \leq \sim 1°$) and $\theta \geq \sim 5°$, the approximation shown in Eq. (12.1) on the right may be used, where ϕ is the angular divergence of the incident beam in radians, and *R* is in millimeters.

Based on Eq. (12.1), the distance l_2, becomes critical at low Bragg angles, wide divergence slit apertures, and large goniometer radii (Fig. 12.10). Practical sample sizes in powder diffraction are usually kept below 25 mm in length, and when low Bragg angle data are desired, the appropriate aperture of the incident beam should be selected to avoid the situation when the size of the X-ray spot exceeds the size of the sample. If it occurs, the measured intensities will be underestimated at low Bragg angles because the sample will be illuminated by only a fraction of the incident beam compared to the illumination at high Bragg angles. Analytical accounting of this effect is difficult due to inhomogeneous distribution of intensity (photon flux) in the cross-section of the incident beam.

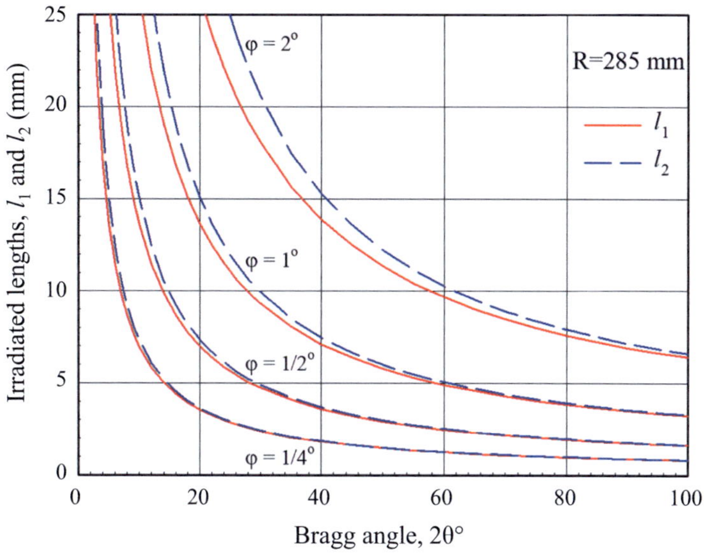

Fig. 12.10 Irradiated lengths, l_1 and l_2, of the flat specimen in the Bragg–Brentano geometry as functions of Bragg angle calculated using Eq. (12.1) for different angular divergences of the incident beam assuming goniometer radius, $R = 285$ mm

12.1.4 Thickness and Uniformity

The second important factor is the absorption of X-rays by the sample. In the Bragg–Brentano geometry, the sample should be completely opaque to X-rays. Assuming that the absorption of 99.9% of the incident beam intensity represents complete opacity, then the beam intensity should be reduced by a factor of 1000 and the following equation can be written (also see Eq. 8.50):

$$\frac{I_t}{I_0} = \exp(-2\mu_{\text{eff}}l) = 10^{-3} \qquad (12.2)$$

where

- I_0 and I_t are intensities of the incident and transmitted beams, respectively;
- μ_{eff} is the effective linear absorption coefficient (in mm^{-1}), which is specific for each sample and it should also account for the porosity of the powder;
- l is the beam path (in mm) through the sample and it is related to the sample thickness, t, as $l = t/\sin\theta$.

After solving Eq. (12.2) with respect to t, the minimum sample thickness (in mm) can be estimated from

$$t \cong \frac{3.45}{\mu_{\mathrm{eff}}} \sin\theta_{\max} \tag{12.3}$$

In Eq. (12.3), $\theta_{\max}$ represents the maximum Bragg angle to be measured during the experiment. Hence, it is usually easy to prepare a satisfactory Bragg–Brentano specimen from dense metallic alloys containing elements heavier than Al since they have large linear absorption coefficients, but it may be difficult to prepare a specimen of sufficient thickness from materials with low density and/or containing light atoms (C, N, O, and H), for example, organic compounds.

Conversely, transmission geometry requires that the sample absorbs minimally. This is usually not a problem if the studied specimen is a molecular substance. However, when the material is a dense alloy or intermetallic compound containing heavy elements, preparing a high-quality specimen for transmission powder diffraction can be problematic. With flat transmission samples, the best approach is to arrange no more than a single layer of particles mounted on the film. When cylindrical specimens are employed, the radius of the capillary should be reduced to a practical minimum. Unfortunately, these measures usually reduce the number of particles in the irradiated volume, and the quality of the resulting diffraction pattern deteriorates.

The third important issue, which is generic to any type of specimen for powder diffraction except cylindrical, is the uniformity of the sample. A portion of the specimen that scatters the incident beam changes as a function of Bragg angle (e.g., see Fig. 12.10 and Eqs. (12.1) and (12.5)). Thus, when the packing density of the powder is not uniform, this will result in random changes of the number of particles in the irradiated volume and in the measured diffracted intensity that are nearly impossible to account for. Achieving a uniform packing density is challenging because compacting the powder to improve uniformity typically increases preferred orientation, which has different but equally detrimental effects on the resulting relative intensities. Effects of sample nonuniformity are best reduced by rapidly spinning the sample during data collection. As mentioned earlier, every effort should be made to collect powder diffraction data from a spinning sample, since it also considerably improves both the number of particles and randomness of their orientations in the irradiated volume.

12.1.5 Sample Positioning

No matter how much time has been spent on the sample preparation and how good the resulting specimen is, it always needs to be properly positioned on the goniometer. Consider, for example, Fig. 12.11, which shows the effect of sample displacement in the Bragg–Brentano and transmission geometries.

When the sample is properly aligned (i.e., when its surface coincides with the goniometer axis, as shown by the dashed image of the flat sample), the correct Bragg

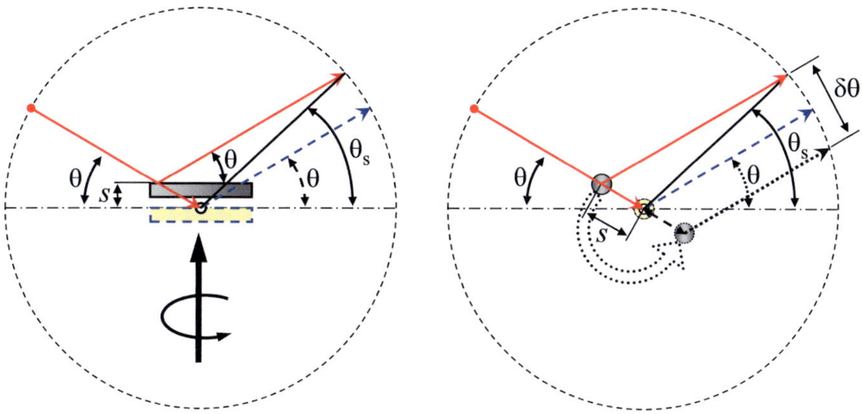

Fig. 12.11 The effect of sample displacement by distance *s* from the goniometer axis on the measured Bragg angle, θ_s, in the Bragg–Brentano geometry (*left*) and in the transmission geometry (*right*). The goniometer axis is indicated by the small *open circle* in the center of the drawings. The optical axis is shown as the *dash-dotted line*. The ideal location of the sample is shown as the *light-shaded dashed rectangle* on the *left* and as the *light-shaded dashed circle* in the center of the drawing on the *right*

angle (θ) is measured, provided the sample is completely opaque (see Sect. 8.4.2 and Eq. 8.16). However, when the sample is displaced by the distance s from the goniometer axis, this displacement results in a different measured Bragg angle, θ_s, even though both the incident and diffracted beams form the same angle θ with the sample surface. Further, if sample displacement is severe, the focusing of the diffracted beam is no longer precise and this will result in the loss of the resolution. The latter issue becomes particularly important when using goniometers with small radii.

Similar errors in the measured Bragg angle may occur in the transmission geometry (see Fig. 12.11, right). Moreover, when a cylindrical sample continuously deviates from the goniometer axis in a circular fashion during spinning, considerable Bragg peak broadening, $\delta\theta$, may be observed, as also indicated in Fig. 12.11 (right).

Errors in the registered Bragg angles associated with the nonideal positioning of the sample are usually not as severe when compared to those observed in intensity measurements due to improper sample preparation. They can be nearly completely eliminated by maintaining the goniometer properly aligned. Furthermore, sample positioning errors in Bragg angles are systematic, and they can be accounted for analytically based on the known geometry of a powder diffractometer. For example, sample-displacement error in the Bragg–Brentano geometry (Fig. 12.11, left) is

$$\theta_s - \theta = \frac{s\cos\theta}{R} \quad (12.4)$$

In Eq. (12.4), s is sample displacement, R is goniometer radius, and the resulting difference in Bragg angles, $\theta_s - \theta$, is in radians (compare Eq. (12.4) with Eqs. (8.13)

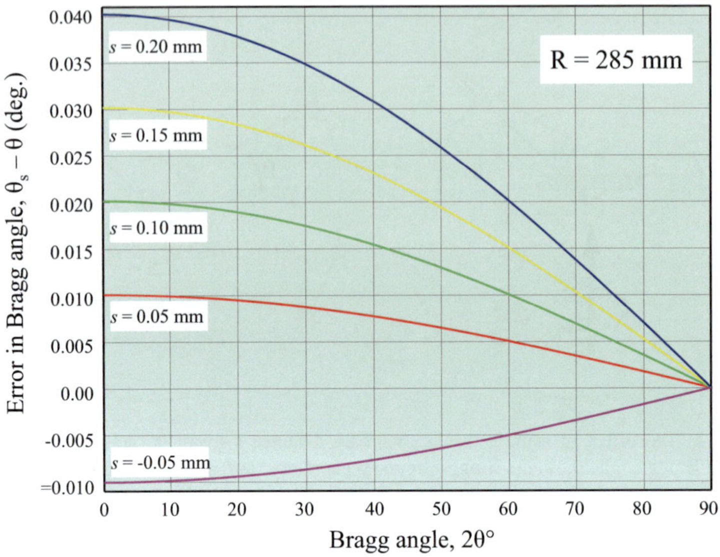

Fig. 12.12 The effect of sample displacement, *s*, on the observed Bragg angles calculated from Eq. (12.4) assuming Bragg–Brentano geometry and goniometer radius $R = 285$ mm. θ_s is the observed Bragg angle, and θ is the Bragg angle in the absence of sample displacement

and (8.17)). This error is commonly observed due to the varying sample thickness (especially when the sample was prepared by dusting) and the varying sample transparency. It is worth noting that the displacement, *s*, can be refined together with lattice parameters, provided low Bragg angle peaks are included in the refinement since they are the most sensitive to *s*, as shown in Fig. 12.12.

12.1.6 Effects of Sample Preparation on Powder Diffraction Data

In summary, preparing a high-quality specimen for powder diffraction can be more challenging than it appears. This task demands both experience and creativity. The negative impact of an improperly prepared sample is illustrated in the following three figures, shown in Rietveld format. These figures present experimental powder diffraction data collected in the Bragg–Brentano geometry from two different specimens, both prepared from the same material, which was intentionally left as coarse powder.

In the first example (Fig. 12.13), the sample holder, which had a cylindrical hole, 25 mm in diameter and 1 mm deep to hold the powder, was filled completely with the coarsely ground, unscreened powder using the technique shown in Fig. 12.6. The aperture of the incident beam was selected so that the length ($L = l_1 + l_2$, see

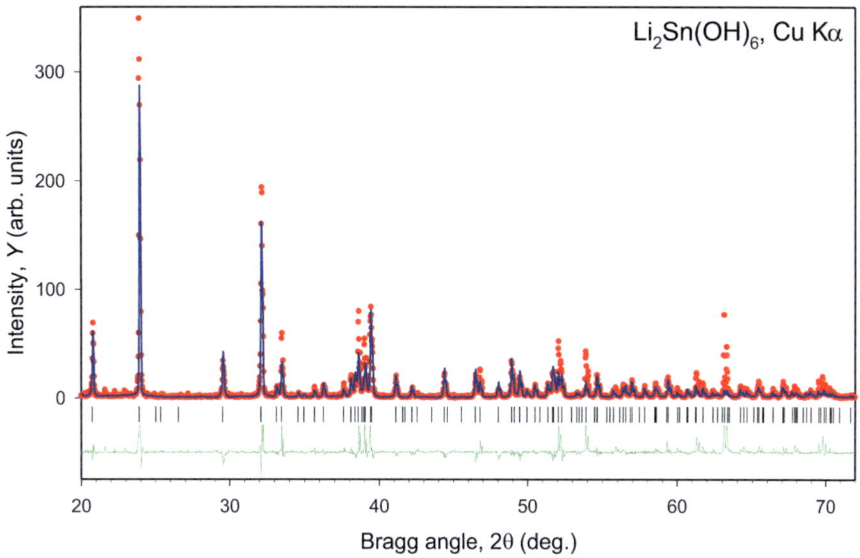

Fig. 12.13 Coarsely ground powder and properly selected incident beam aperture. Experimental data are shown using *small red circles*; the calculated diffraction pattern is shown using *solid blue lines*. The difference Y_{obs}—Y_{calc} is shown at the *bottom* of the plot in green, and the calculated positions of Bragg peaks are marked using *vertical bars*. The data were collected without spinning the sample

Figs. 12.9 and 12.10) of the irradiated area was ~20 mm at $2\theta = 20°$. Assuming that the calculated diffraction pattern in Fig. 12.13 accurately represents the correct distribution of relative intensities, it is evident that some of the observed diffraction peaks at certain Bragg angles are much stronger than expected. These anomalies are associated with the nonideal specimen, which contained coarse grains. As a result, the total number of particles in the irradiated volume was far from infinite, and/or their orientations were not random.

To illustrate the origin of these random intensity spikes, it is worthwhile recalling the spottiness of the Debye rings seen in the film in Fig. 11.1 and modeled for the first four strongest peaks from this film in Fig. 12.14. Assume that a coarse-grained LuAu specimen is under examination using a powder diffractometer. Contrary to the film data, a much smaller fraction of the Debye ring passes through the receiving slit and, therefore, is registered by the detector in a conventional powder diffractometer. Two possible traces of the receiving slit traversing Debye rings during data collection are indicated by the two bars in Fig. 12.14. The sequence of high- and low-intensity spots varies from one Debye ring to another along these two different directions. This variation leads to random, significant, and unpredictable fluctuations in the registered intensity, determined by a specific distribution of particles in the sample.

We note that the actual path of the detector slit remains constant, as shown by the solid elongated rectangle in Fig. 12.14. However, when the particles in the specimen

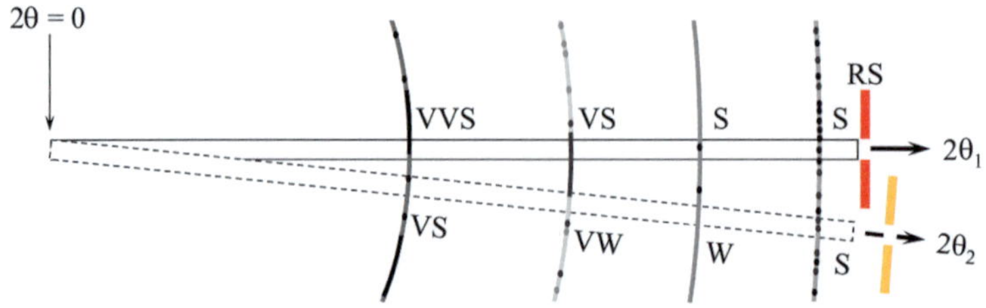

Fig. 12.14 The model showing the spottiness of the first four distinct Debye rings in the X-ray film with the powder diffraction pattern of LuAu from Fig. 11.1. The relative intensities of Bragg peaks vary considerably in the two directions shown as elongated rectangles (each direction represents the possible trace of the receiving slit traversing the Debye rings during data collection). The character notations in the figure characterize the expected measured intensity of Bragg peaks as follows: *VVS* very very strong; *VS* very strong; *S* strong; *W* weak; *VW* very weak. RS is the receiving slit

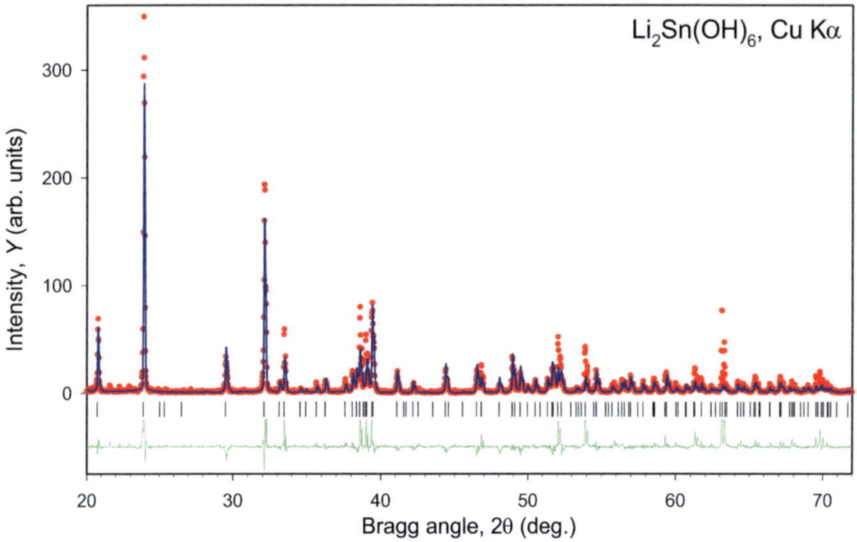

Fig. 12.15 Coarsely ground powder and an improper incident beam aperture. The data were collected without spinning the sample. Notations are identical to Fig. 12.13

are rearranged, the distribution of spots along each Debye ring changes, effectively altering the apparent path of the detector's receiving slit, as indicated by the dashed elongated rectangle. In conventional powder diffractometry, it is nearly impossible to detect these random and significant intensity spikes or dips from the visual analysis of the collected data. However, they can be easily identified through a simple visual analysis of the film or area detector data.

In the second example (Fig. 12.15), the same coarse powder was placed in a 0.2 mm deep sample holder and the powder covered a smaller area (10×5 mm^2). The aperture of the diffractometer was left identical to the first experiment. After

Fig. 12.16 High-quality, nearly spherical powder prepared by high-pressure gas atomization from the melt and proper sample length, *L*. The X-ray powder diffraction data were collected from a continuously spinning sample (20 mm diameter and 1 mm deep) prepared as shown in Fig. 12.6. Notations are the same as in Fig. 12.13. The powder contains a small fraction of a second phase, which is identified by the second row of vertical bars shifted downward. The inset shows the scanning electron microscopy image of the powder morphology. (Powder courtesy of Dr. I.E. Anderson)

comparison of Fig. 12.15 with Fig. 12.13, it is easy to see that the relative intensities of peaks at low Bragg angles are clearly reduced. This happened because the projection of the incident X-ray beam at low Bragg angles (~20 mm) exceeded the sample length (10 mm), and since only a fraction of the incident beam energy was scattered by the sample, the diffracted intensity at low Bragg angles was correspondingly reduced. In addition, the reduction of the number of particles in the irradiated sample volume (smaller area and smaller depth) further exacerbated intensity spikes at random Bragg angles. Clearly, the presence of fewer particles worsens the randomness in the distribution of their orientations.

Conversely, when the specimen is nearly ideal—meaning the particles are small and nearly spherical—and the aperture of the diffracted beam is compatible with the sample size, highly accurate powder diffraction data can be obtained, as shown in Fig. 12.16. Assuming that the calculated diffraction pattern here also represents the correct diffracted intensity, the difference between the observed and calculated diffracted intensities is minimal across the entire range of Bragg angles with no spikes or dips at random angles, as was seen in Figs. 12.13 and 12.15.

The particles in the specimen that was used to collect the data illustrated in Fig. 12.16 consisted of multiple cellular grains with an average diameter of 1 μm—the result of rapid solidification. The powder was subjected to a brief homogenization heat treatment at 950 °C (5 min), which was sufficient to even out the distribution of Ni and Sn in the crystal lattice of the major phase. The powder was screened through a 10 μm sieve after the heat treatment, and only particles smaller than ~10 μm in diameter were used to prepare a flat specimen for data collection employing the Bragg–Brentano focusing geometry. The data were collected using a scintillation detector with monochromatization of the diffracted beam by a curved graphite single-crystal monochromator. The peak-to-background ratio for the strongest Bragg reflection at $2\theta \cong 42°$ was ~180:1.

12.2 Data Acquisition

Once the sample for a powder diffraction experiment has been properly prepared, the next crucial step is acquiring experimental diffraction data. This step is essential for obtaining reliable results. Many of the data acquisition variables can be adjusted to improve or degrade the quality of the dataset. These include instrumental parameters such as the wavelength (energy) of the X-rays, their monochromatization, apertures of the goniometer optics, power settings, and data collection parameters like scanning mode, scan range, step size, and counting time.

12.2.1 Wavelength

The wavelength of the X-rays can usually be freely selected only when using synchrotron radiation. In conventional laboratory settings, changing the X-ray wavelength typically requires replacing the X-ray tube, followed by tube aging and realignment of the goniometer. These procedures should be performed by trained personnel and are rarely undertaken for a single sample. Therefore, the selection of the X-ray tube (and anode) type is generally based on practical considerations, the type of materials typically studied, and the purpose of the powder diffraction examination.

The most typical wavelength selection in powder diffractometry is with a copper (Cu) anode, although other anode materials may be used. These include chromium (Cr), iron (Fe), cobalt (Co), and molybdenum (Mo) anodes. The corresponding characteristic wavelengths for these anode materials are listed in Table 6.1. Long wavelengths (Cr, Fe, Co) are preferred when the accuracy of lattice parameters is of greatest concern since it is possible to measure intensity scattered at high Bragg angles.[7] Short wavelength (Mo) can be used to examine a large volume of the reciprocal lattice. However, since powder diffraction data are one-dimensional, the resolution may be too low for complex crystal structures and/or materials with low crystallinity when excessive Bragg peak overlap makes the diffraction pattern extremely difficult to analyze and process.[8]

Another important consideration in selecting X-ray wavelengths for a powder diffraction experiment is whether or not the studied material contains chemical elements with one of their absorption edges located just above the used characteristic wavelength. For example, the K-absorption edge of Co is ~1.61 Å, while the strongest $K\alpha_1/K\alpha_2$ spectral lines of copper have wavelengths ~1.54 Å. Hence, nearly all $K\alpha_1/K\alpha_2$ characteristic Cu radiation is absorbed by Co, and this type of X-ray tube is hardly suitable for X-ray powder diffraction analysis of Co-based materials. The same is true for Fe-based materials because the K-absorption edge of Fe is ~1.74 Å. Conversely, Co and Fe anodes are well-suited for collecting powder diffraction data from Co- and Fe-containing substances using Co $K\alpha_1/K\alpha_2$ and Fe $K\alpha_1/K\alpha_2$ doublets.

Certain combinations of characteristic wavelengths and chemical elements may cause considerable X-ray fluorescence (e.g., see Fig. 11.25). This phenomenon is similar to the fluorescence in the visible spectrum and it originates from the fact that electrons in an atom can be excited and removed from their ground states by energy transfer from photons of sufficient energy. Electrons from higher energy levels produce fluorescent X-rays when they lower their energy by occupying the formed vacancies. Fluorescent radiation is dissipated in all directions and it usually results in an increase in the background. Chemical elements with strong true absorption generally produce strong X-ray fluorescent background.

12.2.2 *Monochromatization*

Either the diffracted or incident beam should be monochromatized during data collection. When conventional X-ray sources are employed, additional peaks due to the presence of weaker spectral lines with various energies in the characteristic spectrum "contaminate" the diffraction picture and increase peak overlap. The most critical unwanted wavelength for any anode material is $K\beta_1$ (see Table 6.1). When a continuous (synchrotron or neutron) spectrum is available, it would be impossible to

[7] See Sect. 14.12 for details about various factors affecting precision of the unit cell dimensions.

[8] Short wavelength radiation may adversely affect resolution of the data collected using goniometers with small radii (usually under ~200 mm). When a large radius goniometer (usually 250 mm or greater) is employed, the reduction of the wavelength from Cu- to Mo-anode has little effect on pattern resolution because the widths of Bragg peaks decrease with the decreasing wavelength. Regardless of the goniometer radius, the use of short wavelengths requires a more precise alignment when compared to long wavelengths.

see discrete Bragg reflections as a function of Bragg angle without monochromatization, as directly follows from Bragg's equation.

The simplest monochromatization tool that can be used in powder diffractometry is a β-filter (also see Sect. 11.2.2) . Filter materials have their K-absorption edges just above the wavelength of the strongest Kβ spectral line of the anode of the X-ray tube, and their performance is based on how completely they absorb the characteristic impurity wavelengths and how well they transmit the desired parts of the characteristic X-ray spectrum. For example, to eliminate (absorb) nearly all the β-component of a copper anode ($\lambda \cong 1.39$ Å) but to transmit most of the α-component ($\lambda \cong 1.54$ Å), the filter should be made from Ni (the K-absorption edge of Ni is ~1.49 Å). This results in nearly an eightfold difference in the linear absorption coefficients of Ni for Kβ and Kα parts of the characteristic copper spectrum.

Various β-filters are most often used to monochromatize the diffracted beam (e.g., see Fig. 11.6, left), but sometimes they are used to eliminate Kβ radiation from the incident beam in conventional X-ray sources. The advantages of β-filters are in their simplicity and low cost. The disadvantages include: (1) incomplete monochromatization because a small fraction of Kβ spectral line intensity always remains in the X-ray beam; (2) the intensity of the Kα spectral line is reduced by a factor of two or more, and (3) the effectiveness of a β-filter is low for white X-rays above Kα and it rapidly decreases below Kβ. Therefore, β-filters are nearly helpless in eliminating the background, especially when the latter is enhanced by X-ray fluorescence (the filter itself fluoresces due to the true Kβ absorption).

The most common monochromatization option used in modern powder diffractometry is by means of crystal monochromators (see Fig. 11.6, right and Sect. 11. 2.2). Monochromators transmit only specific, narrowly selected wavelengths. As follows from the Bragg equation ($n\lambda = 2d_m \sin \theta_m$), for a constant interplanar distance, d_m, only one wavelength, λ, is being transmitted at a given monochromator angle, θ_m, assuming that $n = 1$. A great variety of crystal monochromators are used in practice, but the best results are usually obtained using curved crystal monochromators, as shown in Fig. 12.17 since they achieve the most precise focusing of the X-ray beam and therefore lower the intensity losses.

A well-aligned monochromator usually leaves only $K\alpha_1$ and $K\alpha_2$ characteristic wavelengths and considerably reduces background when it is used to monochromatize the diffracted beam. In fact, a diffracted beam monochromator is very effective in nearly complete elimination of even severe fluorescence (see Fig. 11.25). In some instances, high-quality curved monochromators can be used to eliminate the $K\alpha_2$ component in combination with the relatively large focusing distance and narrow monochromator slit (see Fig. 11.19). Spectral purity of X-rays is a definite advantage of this monochromatization approach. Further, if necessary, the monochromator angle, θ_m, can be selected to eliminate the $K\alpha_1/\alpha_2$ doublet and to leave only the $K\beta$ part, which is truly monochromatic in light anode materials, including the most commonly used Cu anode.

On a downside, high-quality crystal monochromators are relatively expensive, and they also reduce the intensity of characteristic X-rays by a factor of two to three. It is worth noting that when the monochromator is made from a low-quality single

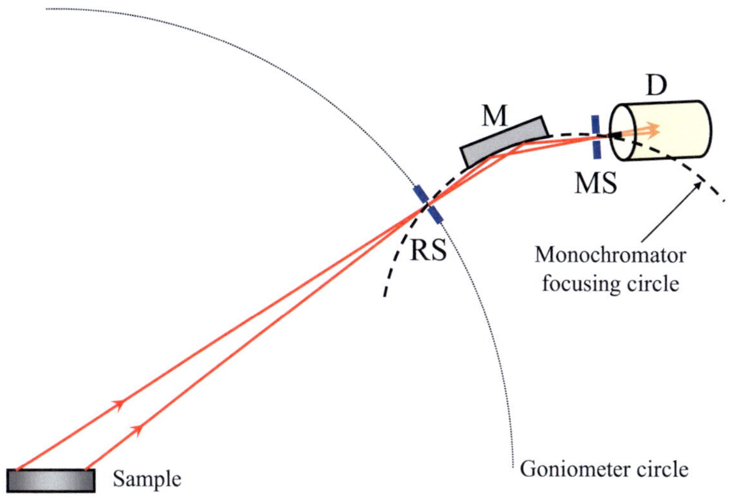

Fig. 12.17 The schematic of monochromatization of the diffracted beam using a curved crystal monochromator. *RS* receiving slit; *M* curved monochromator; *MS* monochromator scatter slit; *D* detector

crystal or when it is improperly aligned, the resulting reduction of the transmitted intensity may be severe. Crystal monochromators are susceptible to radiation damage, which is especially true for primary beam monochromators. As a result, the quality of a single crystal deteriorates with time, and primary beam monochromator crystals should be replaced after a prolonged use.

A third monochromatization option, such as energy-dispersive solid-state detectors, has become available quite recently. By adjusting the energy window, these detectors can be set to register only the specific energy of X-ray photons, thereby achieving monochromatization electronically (Sects. 6.4.3 and 11.2.2). The most important advantage of this approach is in the virtual absence of intensity loss. However, a significant downside is that the detector typically requires continuous cooling. Additionally, if the powder diffractometer is used in special applications, such as examining diffraction patterns from large single crystals, very strong diffracted intensity at different wavelengths could be accidentally registered.[9]

All monochromatization options discussed here have been used successfully in powder diffractometry. With point detectors, which register diffracted intensity at a specific angle, one point at a time (or a small range of Bragg angles when using a short strip detector), either or both the incident and diffracted beam can be monochromatized. When position-sensitive or image-plate detectors are used, the only feasible option is to use a β-filter or a crystal monochromator to achieve monochromatization of the incident beam. As shown in Fig. 11.25, for some

[9] An example or registering Cu Kβ and W Lα wavelengths is found in Fig. 6.11.

materials the background becomes too high, which makes diffraction data nearly useless in the determination of the atomic parameters of the material.

12.2.3 Incident Beam Aperture

The aperture of the incident beam should be selected to match the diffraction geometry and sample size. In the commonly used Bragg–Brentano focusing geometry, the key requirement is that, at any Bragg angle, the length of the incident beam' projection does not exceed the length of the sample (see Figs. 12.9 and 12.10 and Eq. 12.1). This is accomplished by properly selecting the divergence slit (see Fig. 11.6). When the slits are calibrated in degrees, the use of Eq. (12.1) to determine the proper size of the divergence slit is straightforward. However, in most common commercial diffractometers, slits are calibrated in millimeters and their angular divergence can be estimated (also see Eq. 11.1) using the following simple relationship, which assumes the infinitesimal X-ray tube focus: [10]

$$\varphi \cong \frac{57.3\delta}{r} \tag{12.5}$$

In Eq. (12.5), φ is the angular divergence of the slit in degrees, δ is the slit opening in mm, and r is the distance from the X-ray tube focus to the divergence slit in millimeters. It may be useful to check that the correct aperture of the incident beam has been selected by mounting a fluorescent screen on the sample holder at the lowest Bragg angle that is examined during the experiment. Alternatively, it is always possible to select a slit narrower than the acceptable maximum to ensure that the projection of the incident beam fits within the sample length along the optical axis of the goniometer at the minimum Bragg angle. Unnecessary reduction of the aperture of the incident beam, however, results in the proportional reduction in the diffracted intensity, see Fig. 12.19.

In the transmission geometry, the requirements are different. When a flat transmission sample is used, the aperture of the incident beam is defined by the largest Bragg angle of interest, because at $\theta = 0$ the sample is perpendicular to the incident beam (and not parallel, as in the Bragg–Brentano geometry). Equation (12.1) then becomes the following (with the same notations as in Eq. (12.1)):

$$L \cong \frac{\varphi R}{\sin(90 - \theta)} \tag{12.6}$$

[10] As established in Chap. 11 (see Fig. 11.7), this is a valid approximation because the typical projection of the 1 mm wide line focus of the X-ray tube, visible at a small take-off angle (usually 5–6°), results in the source size on the order of 0.1 mm.

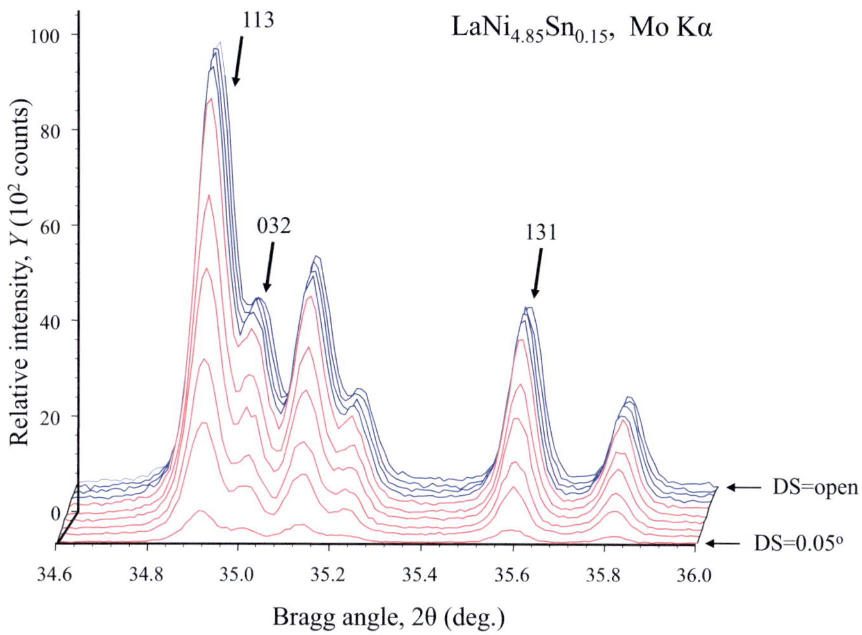

Fig. 12.18 The set of X-ray powder diffraction patterns collected from the nearly spherical $LaNi_{4.85}Sn_{0.15}$ powder (see Fig. 12.16, inset) on a Rigaku TTRAX rotating anode powder diffractometer using Mo Kα radiation. Goniometer radius $R = 285$ mm; receiving slit RS = 0.03°; flat specimen diameter $d = 20$ mm. Incident beam apertures were 0.05°, 0.17°, 0.25°, 0.38°, 0.5°, 0.75°, 1°, 1.5°, 2°, and completely opened (~5°), respectively. An automatic variable scatter slit was used to reduce the background. The data were collected with a fixed step $\Delta 2\theta = 0.01°$, and the sample was continuously spun during the data collection

The opposite constraint on the relationships between the incident beam size and sample size should be followed when a cylindrical transmission sample is under examination: the projection of the beam should be large enough to irradiate the whole sample at any position it may occupy during spinning. This includes a small precession around the goniometer axis. If this requirement is not followed, the inhomogeneities in the powder packing density may cause sporadic changes in measured intensities.[11]

The effects of the incident beam aperture on both the intensity of the diffracted beam and the resolution of the goniometer in the Bragg–Brentano geometry are shown in Figs. 12.18 and 12.19. Bragg peak intensities increase rapidly and nearly linearly as a function of the incident beam aperture as long as the opening of the divergence slit keeps the incident beam in check and the total length of its projection (L, Eq. 12.1) remains shorter than the diameter of the specimen (20 mm), as shown in Fig. 12.20.

[11] The homogeneity of an incident beam may become an issue when a cylindrical specimen experiences large amplitude oscillations around the goniometer axis during spinning.

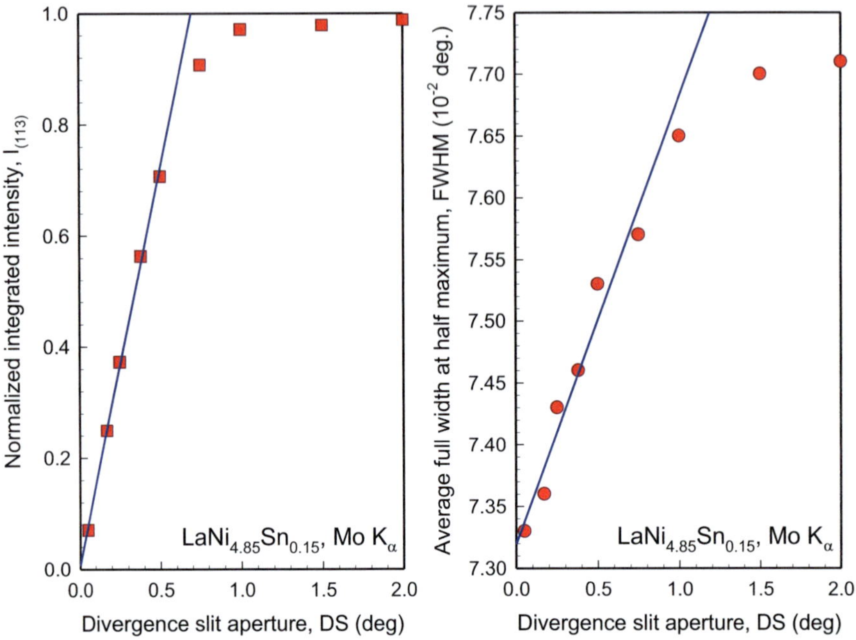

Fig. 12.19 The normalized integrated intensity of the strongest peak (*left*) and the average full width at half maximum (FWHM, *right*) of the three Bragg peaks shown in Fig. 12.18. The set of X-ray powder diffraction patterns collected from the nearly spherical LaNi4.85Sn0.15 powder (see Fig. 12.16, inset) on a Rigaku TTRAX rotating anode powder diffractometer using Mo Kα radiation. Goniometer radius $R = 285$ mm; receiving slit RS $= 0.03°$; flat specimen diameter $d = 20$ mm. Incident beam apertures were 0.05°, 0.17°, 0.25°, 0.38°, 0.5°, 0.75°, 1°, 1.5°, 2°, and completely opened (~5°), respectively. An automatic variable scatter slit was used to reduce the background. The data were collected with a fixed step $\Delta 2\theta = 0.01°$, and the sample was continuously spun during the data collection., both as functions of the divergence slit aperture. The intensity of the strongest Bragg peak was normalized with respect to its value when the divergence slit was completely opened (DS $\cong$ 5°). The corresponding values at the completely opened divergence slit are not shown to clarify the behavior at low apertures

Bragg peak intensity continues to increase as long as the incident beam's aperture is small enough and the incident beam's projection is fully contained within the sample boundaries. The background level follows a similar trend to Bragg peak intensities, as shown in Fig. 12.18. When the aperture of the divergence slit reaches and exceeds 1.5°, no further increase in the diffracted intensity is observed. This is consistent with Eq. (12.1) and Figs. 12.19 and 12.20, which indicate that for this goniometer the projection of the incident beam exceeds the sample ($d = 20$ mm) at $2\theta \cong 35°$ when DS aperture (φ) is between 1.0 and 1.5°.[12] Predictably, the linear

[12]The behavior of L as a function of DS (φ) is nearly linear for a constant θ when φ is small, which is apparent from the second part of Eq. (12.1). The deviation from linearity observed in Fig. 12.19, left, when DS is between 0.5 and 1°, i.e., when the projection of the incident beam remains within

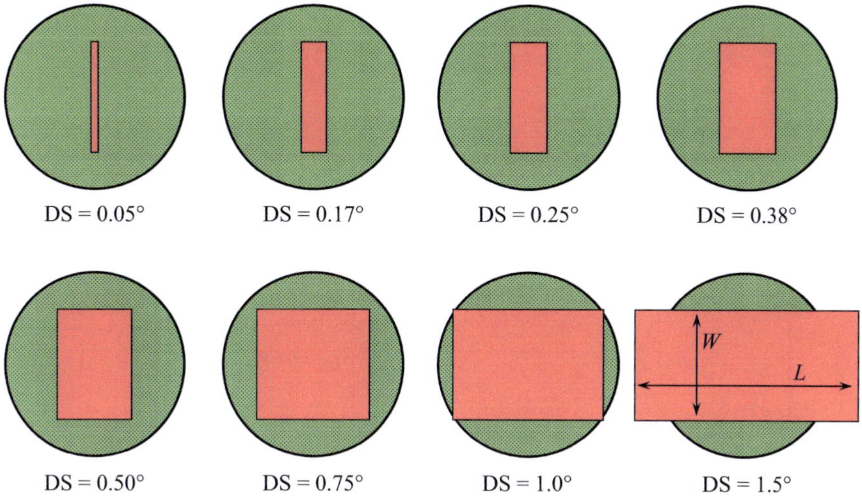

Fig. 12.20 The projections of the incident beam (*pink rectangles*) on the sample surface (*green circles*) at $2\theta \cong 35°$ modeled using Eq. (12.1) (φ = DS), goniometer radius $R = 285$ mm, and a constant width, $W = 12$ mm. Specimen diameter is $d = 20$ mm

behavior of the scattered intensity when DS $\leq 0.5°$ extrapolates to $I \rightarrow 0$ with DS $\rightarrow 0$.[13]

The varying incident beam aperture has minimal effect on the resolution of the instrument due to excellent focusing. As shown in Fig. 12.19 (right), the average full width at half maximum (FWHM) increases from ~0.073 to ~0.077° (i.e., only by ~5%) when the divergence slit aperture increases from 0.05° to completely opened (i.e., by as much as ~10,000%). The dependence of the FWHM on the slit opening saturates at wide apertures, which is consistent with the full illumination of the specimen when $\varphi_{DS} > 1°$.

In addition to controlling the aperture of the incident beam in the plane perpendicular to the goniometer axis, its divergence in the plane parallel to the goniometer axis should be controlled by Soller slits (see Fig. 11.6) . This is done to reduce the asymmetric broadening of Bragg peaks, which may be very strong at low Bragg angles and small goniometer radii.

The effect of varying the axial divergence of the incident beam is shown in Fig. 12.21, where one low- and one middle-Bragg angle peaks were measured with

---

the sample boundaries as seen in Figure 12.20, is associated with the inhomogeneity of the incident beam. Its intensity as a function of L becomes nonlinear at large divergence slit openings with the maximum in the center of the projection, gradually and non-linearly decreasing toward its ends. The distribution of intensity in the incident beam as a function of L is source-dependent and, if necessary, it may be measured experimentally.

[13]This is expected assuming the ideal homogeneity of both the incident beam and the sample packing density. The former is true for small divergence slit openings, and the latter is true for the used sample, which was prepared from the nearly spherical particles.

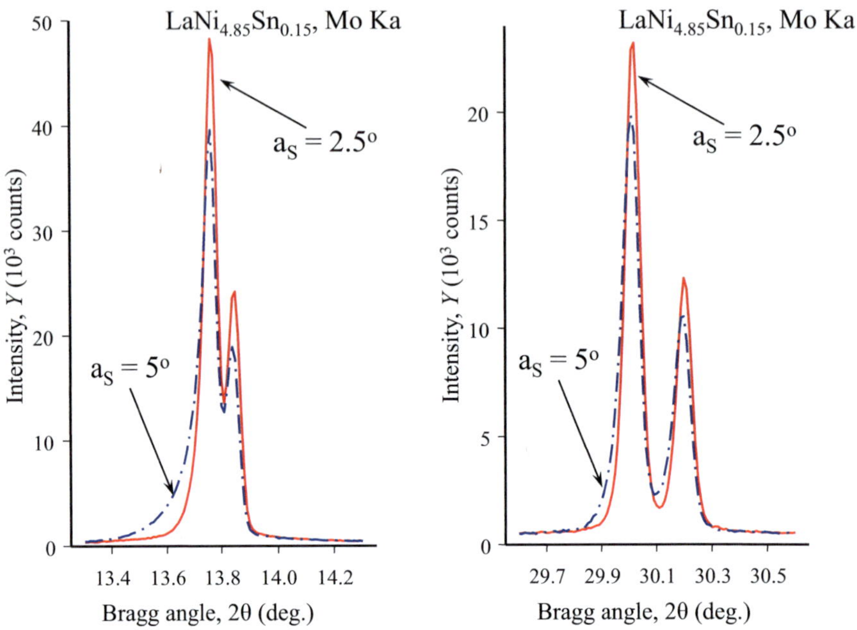

Fig. 12.21 Two individual Bragg peaks observed in the diffraction pattern of the $LaNi_{4.85}Sn_{0.15}$ powder (see Fig. 12.16, inset). The data were collected on a Rigaku TTRAX rotating anode powder diffractometer using Mo Kα radiation and two different sets of Soller slits controlling the axial divergence of the incident beam: $\alpha_S = 2.5°$ (*solid red lines*) and $\alpha_S = 5°$ (*dash-dotted blue lines*). The Soller slit controlling the axial divergence of the diffracted beam was constant, $\alpha_D = 2.5°$. The goniometer radius $R = 285$ mm; receiving slit RS = 0.03°; flat specimen diameter $d = 20$ mm; incident beam aperture DS = 0.17°. An automatic variable scatter slit was employed to reduce the background. The data were collected with a fixed step $\Delta 2\theta = 0.01°$, and the sample was continuously spun during the data collection

different Soller slits. Obviously, large axial divergence results in the appearance of a broad tail extending toward low Bragg angles and in somewhat reduced resolution of the instrument due to the related change in the peak shape caused by broadening of both the base and FWHM. As the Bragg angle increases, the peak asymmetry becomes less obvious but is still present.

12.2.4 Diffracted Beam Aperture

The aperture of the receiving slit (Fig. 12.22) is as important as that on the incident beam side. The receiving slit should be selected as small as reasonably possible to improve the resolution of the instrument. While the size of the receiving slit does not affect the measurements at specific Bragg angles (as does an improperly selected divergence slit), inevitably any reduction in the receiving slit opening results in a

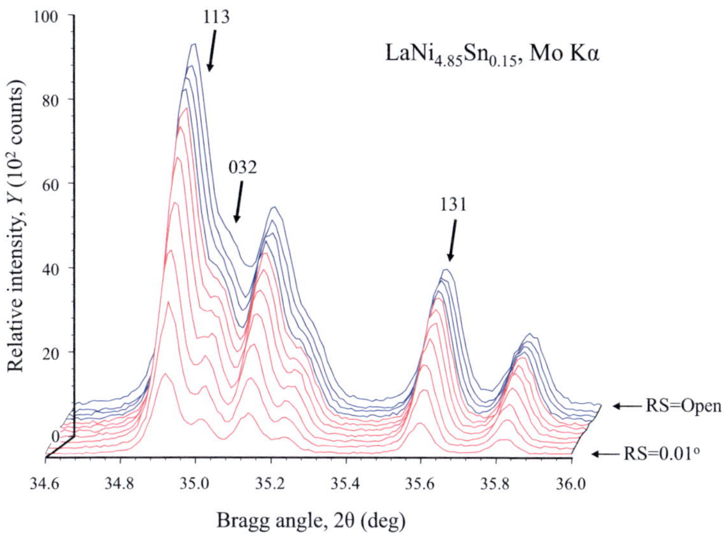

Fig. 12.22 The set of X-ray powder diffraction patterns collected from the $LaNi_{4.85}Sn_{0.15}$ powder (see the inset in Fig. 12.16) on a Rigaku TTRAX powder diffractometer using Mo Kα radiation. Goniometer radius, $R = 285$ mm; divergence slit, DS = 0.5°; flat specimen diameter, $d = 20$ mm. Diffracted beam apertures were 0.01°, 0.02°, 0.03°, 0.04°, 0.05°, 0.06°, 0.07°, 0.08°, 0.1°, 0.12°, and completely opened (~1°), respectively. An automatic variable scatter slit was used to reduce the background. The data were collected with a fixed step $\Delta 2\theta = 0.01°$, and the sample was continuously spun during the data collection

proportional reduction of the intensity registered by the detector at all Bragg angles. The balance between the needed intensity and resolution may be difficult to find ab initio, and it is best to perform several quick scans of a narrow Bragg angle range with varying receiving slit and make a selection where both the resolution and registered intensity are satisfactory. It is advisable to perform quick scans in the range of Bragg angles that includes the strongest Bragg peaks.

The influence of the diffracted beam aperture on both the diffracted intensity and goniometer resolution in the Bragg–Brentano geometry is shown in Figs. 12.22 and 12.23, respectively. Bragg peak intensities increase rapidly and almost linearly as a function of the receiving slit opening when the receiving slit is narrow. However, as the slit widens, the increase becomes slow and nonlinear, with a clear tendency toward saturation. In contrast, the background increases steadily as a function of the receiving slit aperture.

Unlike the incident beam aperture (Fig. 12.19), the varying receiving slit opening has strong influence on the resolution of the instrument. When the receiving slit aperture increases, the $K\alpha_1$ components of the two Bragg peaks, 113 and 032, which are relatively well-resolved when RS $\leq$ 0.04°, are observed as a main peak and a shoulder when RS $\geq$ 0.07°. As shown in Fig. 12.23, this is due to the considerable change in the FWHM, which increases from ~0.073° for RS = 0.01° to ~0.104°

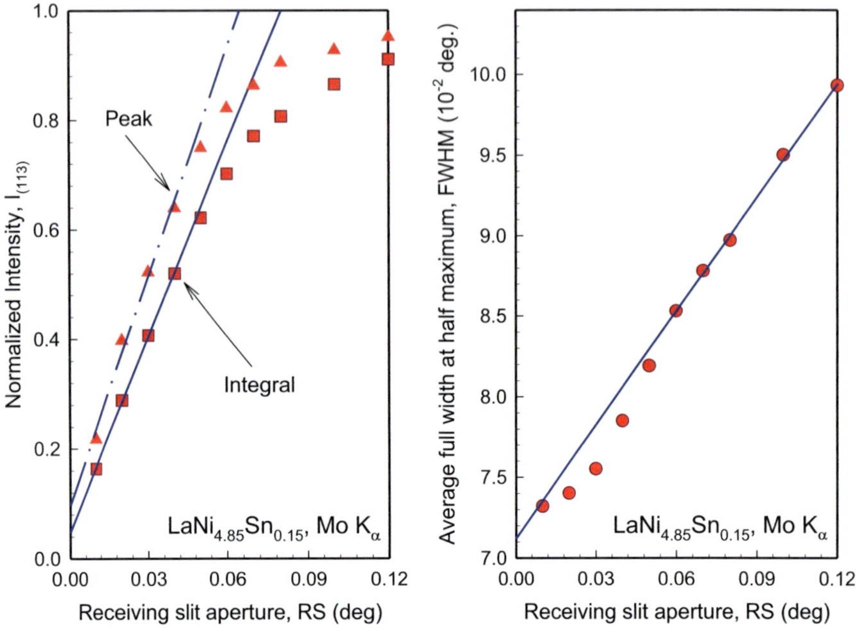

Fig. 12.23 The intensity of the strongest peak (*left*) and the average full width at half maximum (*right*) of the three Bragg peaks shown in Fig. 12.22, as functions of the receiving slit aperture. Both the integrated and peak intensities were normalized with respect to their values at the completely opened receiving slit (RS ≅ 1°). The corresponding values at the completely opened receiving slit are not shown to clarify the behavior at low apertures

Fig. 12.24 Examples of proper (ScS) and improper (ScS') selection of scatter slit aperture. *DS* divergence slits; *RS* receiving slits

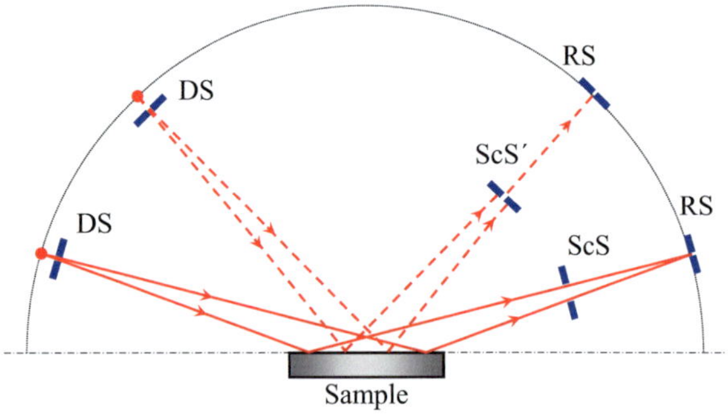

when the receiving slit is in a completely opened position, so that peak broadening exceeds 40%.

As noted earlier, a scatter slit can be used to reduce the background noise before it reaches the detector. The aperture of the scatter slit should be selected to enable unobstructed passage of the monochromatic diffracted beam at any Bragg angle, see Fig. 12.24. In this example, the scatter slit ScS is wide enough to transmit the beam without affecting its intensity. On the contrary, the scatter slit ScS' is too narrow, and only a fraction of the diffracted intensity will reach the detector.

The best practical way to select the aperture of the scatter slit manually is to find a well-resolved and strong Bragg peak at low angles and after setting the detector at the peak maximum find the minimum opening of the scatter slit that does not reduce the intensity of the peak without moving the detector arm. Some goniometer designs enable continuous variation of the scatter slit aperture, and if this is the case, it is advisable to expand the scatter slit gap by a few percent after finding the minimum noninterfering opening. Another option is to move (if possible) the slit by a few millimeters toward the receiving slit. It is worth noting that a slightly wider scatter slit aperture is preferable to a narrower one, as an improperly selected scatter slit can lead to a reduction in measured intensity.

12.2.5 Variable Aperture

Some of the modern powder diffractometers may be equipped with the variable divergence, scatter, and receiving slits (e.g., see Fig. 11.19—The overall view of the goniostat of the Malvern Panalytical X'Pert powder diffractometer. This diffractometer has the horizontal goniometer axis, stationary X-ray source, and synchronized rotations of both the detector arm and sample holder. The goniometer is equipped with a Johansson-type primary beam monochromator, which eliminates Cu Kα2 or Co Kα2 wavelengths in addition to Kβ components of the characteristic spectrum, and a solid-state real-time multiple strip detector, which accelerates data collection—and Fig. 11.22). This enables slit selection at the software level for any aperture, and in addition, it is possible to vary the aperture of the incident beam continuously during data collection. Thus, the variable slit option facilitates experiments where the length of the irradiated area of the flat sample can be kept constant as shown in Fig. 12.25. This may be particularly useful when it is impossible to spin the sample during data collection (e.g., when using high or low temperature or high-pressure attachments) to improve particle orientations averaging. Another benefit of using variable slits is an increased intensity at high Bragg angles, which are naturally

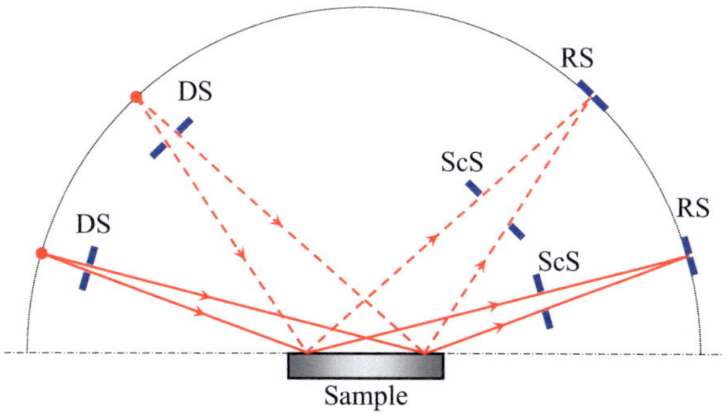

Fig. 12.25 The schematic of goniometer optics during data collection using variable divergence and scatter slits apertures, which enables a constant irradiated area on the sample at any Bragg angle. *DS* divergence slit; *ScS* scatter slit; *RS* receiving slit

suppressed due to various specimen and instrumental factors, such as X-ray atomic scattering (Sect. 9.1.3), atomic displacements (Sect. 9.1.2), and Lorentz-polarization (Sect. 8.6.4).

The resultant diffraction pattern is then numerically processed to convert the measured profile intensities to a constant incident beam divergence, which is a standard in most crystallographic software. An added benefit when using variable slits is the availability of the software control over the aperture of the scatter slit. The opening can be varied automatically and continuously, providing optimal background reduction at all Bragg angles.

However, it is important to note that correcting the measured intensity for the variable aperture of the incident beam introduces additional errors into the experimental data. If precise intensities are of greatest concern, the use of a variable divergence slit option should be avoided. These errors can become particularly severe if the diffractometer is misaligned, the slit is located too close to the X-ray tube's focus, or the incident beam is strongly inhomogeneous. Conversely, maintaining a constant irradiated area eliminates errors caused by sample packing inhomogeneity, such as differences in packing density between the center and the edges of a flat specimen.

An alternative to the continuously variable aperture methods is measuring low and high Bragg angle regions separately, using different fixed divergence slits. The low Bragg angle region is measured with narrow slits, while the high Bragg angle region is measured with wide slits to improve the intensity at high angles. This approach makes it difficult to treat the entire powder diffraction pattern simultaneously in many common applications, such as peak search and phase identification. However, most software programs designed for crystal structure refinement allow the processing of multiple diffraction patterns, making the multiple divergence slits method acceptable.[14]

12.2.6 Power Settings

Before selecting power settings, it is important to know the power ratings of both the high-voltage generator and the X-ray tube. The two adjustable parameters are accelerating voltage and tube current, whose product (accelerating voltage in kV and tube current in mA) determines the output power of the generator and the input

[14] When relative diffracted intensities are of greatest concern, the multiple divergence slits technique is preferred over the continuously varying divergence slit method. The two largest contributions to intensity correction errors when slit opening varies arise from both the imprecise mechanical control of the slit opening and from the inhomogeneity of the incident beam. The latter becomes especially severe at large apertures (see Figs. 12.18–12.20 and relevant text). Neither of the errors can be easily accounted for. In the former (multiple divergence slits), these issues become irrelevant because multiple diffraction patterns are employed in calculations with their own scale factors (see Sect. 16.5 and Eq. 16.3 on p. 548).

power of the X-ray source in watts. Typically, the maximum power is limited by the X-ray tube's power rating.[15]

The accelerating voltage should typically be set at or slightly above the threshold of the most efficient generation of the anode material's characteristic emission spectrum. For example, the optimal voltage for exciting characteristic radiation is about 45 kV for a Cu anode and 80 kV for a Mo anode. However, most commercial high-voltage generators do not operate above 60 kV, which means the intensity distribution between white and characteristic radiation for a Mo anode is not optimum.

The second parameter, tube current, should be set as high as possible without exceeding the allowable power rating of either the tube or generator, since the intensity of the characteristic radiation in the incident beam is proportional to the tube current. Additionally, the standard lifetime of most X-ray tubes (several thousand hours) can be significantly extended if the tube is operated at 75% or less of its rated power.[16]

12.2.7 *Classification of Powder Diffraction Experiments*

Provided all instrumental parameters discussed here have been properly selected, the next step in acquiring high-quality experimental data is to choose the scanning mode, scan range, data collection step, and counting time. The scanning mode is applicable to both point detectors and short linear or curved position-sensitive detectors. When using long curved position-sensitive detectors and/or image plates, the scanning mode generally becomes irrelevant, since there is no need to move the detector and the entire diffraction pattern (or a large part of it) is recorded at once, to using X-ray films. Most data acquisition settings depend on both the type of powder diffraction experiment and the specific information to be gathered from the data. Based on the counting time, powder diffraction experiments using point detectors can be broadly classified as fast, overnight, or weekend-long experiments.[17]

Fast experiments are usually conducted in the time frame from several minutes to several hours depending on the brightness of the incident beam (i.e., whether the X-ray source is a rotating anode or a sealed X-ray tube) and the crystallinity of the

---

[15] Note that only brief general guidance on the selecting power setting is provided here. Users must refer to the generator and X-ray tube instructions, as well as the standard operating procedures developed for the specific instrument and laboratory.

[16] For a properly aged X-ray tube, the aging procedure is described by the manufacturer.

[17] Note that this classification was originally used for point detectors measuring only one point at a time. Nowadays, most modern diffractometers are equipped with position-sensitive detectors (PSDs) that can simultaneously measure at least a hundred to several hundred points. Therefore, PSD experiments can be classified as fast (5–15 min), medium (20–40 min), and slow (1–2 h), corresponding to fast, overnight, and weekend-long experiments using a point detector, respectively. Nonetheless, we will retain the latter classification for the time being.

material. Fast powder diffraction experiments give the experimentalist a general idea about the complexity of a diffraction pattern and data collected in a fast experiment are hardly ever useful beyond the verification of the selected instrumental parameters for longer experiment(s) and/or for simple phase identification purposes. Yet, an hour-long experiment collected from a nearly perfect sample may be suitable for indexing and accurate refinement of the unit cell parameters.

Overnight experiments normally take from several hours to a whole day or night. It makes sense to run these experiments overnight, since modern powder diffractometers are completely automated. Overnight powder diffraction experiments usually provide good-quality diffraction patterns, which are satisfactory for further numerical processing that requires precise peak positions and their intensities.

Weekend-long experiments normally take more than a day. They are conducted when precise peak shapes are desired, for example, when microstructural properties are analyzed based on the anomalies of Bragg peak shapes, or when the crystal structure of the material is determined and refined by using the Rietveld method (see Sect. 15.7).

12.2.8 Step Scan

The scanning mode defines how the detector and the X-ray source, or the detector and the sample arms move during data collection (this depends on the goniometer design, see Figs. 11.22 and 11.19—The overall view of the goniostat of the Malvern Panalytical X'Pert powder diffractometer. This diffractometer has the horizontal goniometer axis, stationary X-ray source, and synchronized rotations of both the detector arm and sample holder. The goniometer is equipped with a Johansson-type primary beam monochromator, which eliminates Cu Kα2 or Co Kα2 wavelengths in addition to Kβ components of the characteristic spectrum, and a solid-state real-time multiple strip detector, which accelerates data collection., respectively): are they in an intermittent (stepping) or in a continuous motion. When the goniometer arms are in an intermittent motion, the diffracted intensity is measured when both the X-ray source (sample) and the detector are at rest,[18] and this is usually referred as the step-scan mode.

A generic algorithm of data collection in the step-scanning mode is shown in the form of a flowchart in Fig. 12.26. It includes the following sequence of events:

- Movable goniometer arms advance to their initial (or next) positions.
- Diffracted intensity is measured for a certain time (counting time) in this configuration.

[18] In this context, "sample at rest" means that the angle between the sample surface and the incident beam is constant. The specimen, however, normally continues to spin about an axis, which is normal to its surface for better particle averaging and minimization of other errors associated with preferred orientation and sample inhomogeneity.

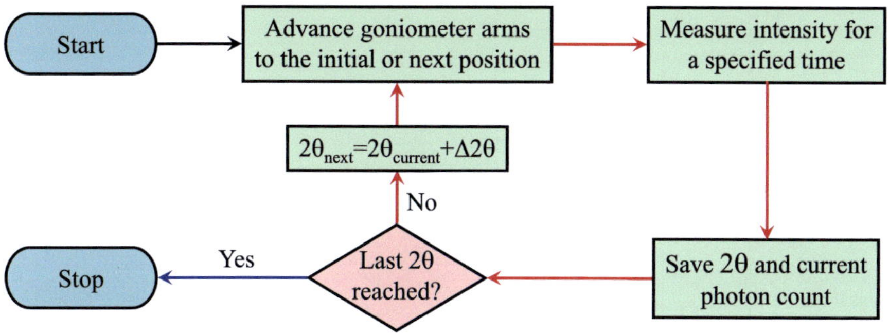

Fig. 12.26 The flowchart visualizing a generic step-scan data acquisition algorithm. The main loop is highlighted by using red arrows

```
Instrument specific header
Bragg angle 1, Number of accumulated counts 1[, Error 1]
Bragg angle 2, Number of accumulated counts 2[, Error 2]
Bragg angle 3, Number of accumulated counts 3[, Error 3]
..., ...[, ...]
Bragg angle N, Number of accumulated counts N[, Error N]
```

Fig. 12.27 Example of powder diffraction data file format. Optional experimental errors in the measured intensity are shown in square brackets

- Current Bragg angle and intensity count are saved when the counting time expires.
- Computer analyzes whether the last Bragg angle value has been reached:
  - If the answer is no, the loop is repeated by advancing goniometer arms to the next position, which is calculated by adding a fixed step to the current value of 2θ.
  - If the answer is yes, the data collection is finished.

The resulting powder diffraction pattern usually consists of intensity data as a numerical function of Bragg angle, as shown in Fig. 12.27.

The two most important parameters of the step-scan experiment, defined by the user, are the size of the step in terms of Bragg angle, $\Delta 2\theta$, and counting time, *t*. The step size is constant throughout the entire experiment, and it is usually selected between 0.01 and 0.05° of 2θ. Using Cu Kα radiation, the value $\Delta 2\theta = 0.02°$ is quite standard. When using Mo Kα radiation, step size should be generally reduced to $\Delta 2\theta = 0.01°$. High-energy synchrotron radiation may require steps as low as 0.001°. Neutron powder diffraction data are usually collected with 0.05–0.1° 2θ steps.

When Bragg peaks are extremely broad, and when conducting fast experiments, larger step sizes for analytical X-ray diffractometers are acceptable. On the contrary, when the crystallinity of the examined material is high and Bragg peaks are narrow,

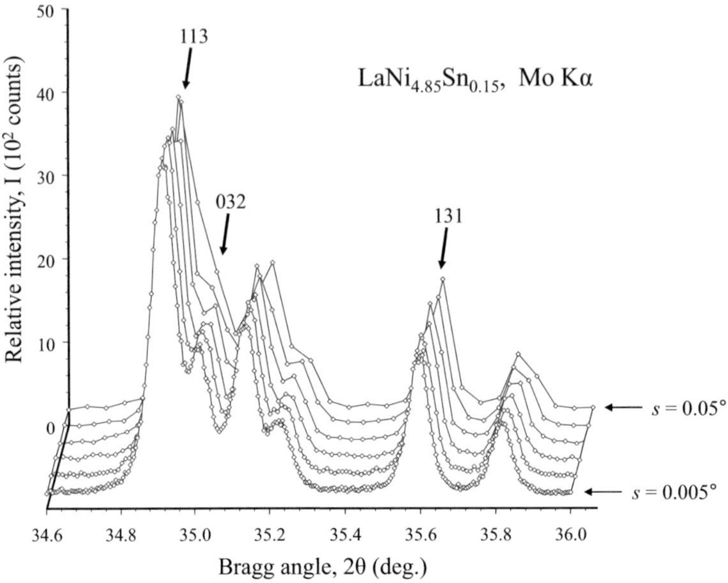

Fig. 12.28 A set of X-ray powder diffraction patterns collected from the $LaNi_{4.85}Sn_{0.15}$ powder (see the inset in Fig. 12.16) on a Rigaku TTRAX rotating anode powder using Mo Kα radiation. Goniometer radius, $R = 285$ mm; divergence slit, DS = 0.5°; receiving slit, RS = 0.03°; flat specimen diameter, $d = 20$ mm. Step-scan mode with steps 0.005°, 0.01°, 0.02°, 0.03°, 0.04°, and 0.05°. An automatic variable scatter slit was used to reduce the background

smaller step sizes should be employed. A rule of thumb for the selection of the step size is that at least 8–12 data points should be measured for well-resolved peaks within one full width at half maximum. Even lower $\Delta 2\theta$ values are used when precise Bragg peak shape data are needed for structure refinement and microstructure determination. When selecting $\Delta 2\theta$, it is important to remember that small step sizes improve resolution, which obviously cannot exceed the limits imposed by the goniometer optics, while large step sizes reduce the duration of the experiment.

The effect of the varying step size on the resulting powder diffraction data is shown in Fig. 12.28. The resolution of the two closely located Bragg peaks, 113 and 032, is nearly completely lost when $\Delta 2\theta$ was 0.04 and 0.05°. The loss of the resolution in this case is neither associated with the intrinsic changes of the peak shape nor with the increase of the full width at half maximum, as is the case when the aperture of the receiving slit varies (see Figs. 12.22 and 12.23). It is simply the result of missing the details of the distribution of the diffracted intensity as a function of Bragg angle because the size of the step exceeded the opening of the receiving slit.

Counting time establishes the length of time that the goniometer arms are held in fixed positions during which the diffracted intensity is measured. Unlike step size, which is constant in the overwhelming majority of powder diffraction experiments, counting time may be a constant or variable for each collected data point. The constant counting time is selected in most experiments, as it provides correct relative intensity measurements without additional data processing. Variable counting time can be employed when especially precise information about weak diffraction peaks is required.

A typical way of selecting the variable counting time is to specify the number of photon counts to be accumulated at every goniometer arm position. Correct relative intensities are then obtained by scaling each intensity data point to a fixed counting time, for example, 1 s. The variable counting time approach is rarely used in practical powder diffractometry because data collection process takes an exceedingly long time when the background is low.

12.2.9 Continuous Scan

The continuous scanning mode involves the uninterrupted movement of the goniometer arms at a constant speed, with intensity readings periodically saved at specific intervals ($\Delta 2\theta$) of the Bragg angle. According to a generic algorithm of this scanning mode (Fig. 12.29), the detector begins registering photons as soon as the goniometer arms are set in motion at a constant angular velocity (scan rate), starting from the selected initial positions. Accumulation of counts continues until the predetermined sampling interval, $\Delta 2\theta$, has been scanned. As soon as this condition is met, the accumulated intensity count is saved together with the median angle of the scanned range. The photon count is then reset to zero, and a new cycle of counts accumulation begins. The process is repeated until the final $\Delta 2\theta$ interval has been scanned.

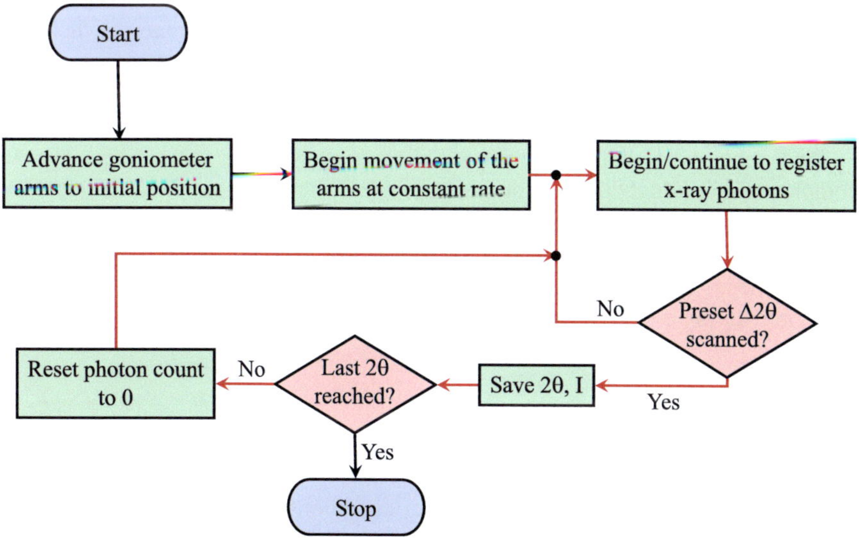

Fig. 12.29 The flowchart visualizing a generic algorithm of continuous scan data acquisition. Main loops are highlighted by using red arrows

Hence, a continuous scan produces data nearly identical to those collected by a step scan, so powder diffraction data are saved in the format shown in Fig. 12.27. The only difference is that the intensity is not given for a fixed detector position, but for a median Bragg angle in the scanned interval. To minimize the introduction of a small, but systematic error, an intensity measurement during continuous scanning always begins from $2\theta = 2\theta_{median} - \Delta 2\theta/2$, where $2\theta_{median}$ is the median Bragg angle saved in the data file. For example, the diffracted intensity at the Bragg angle $2\theta = 10°$ with a sampling interval $\Delta 2\theta = 0.02°$ is the result of accumulating an X-ray photon count during continuous scanning from 9.99 to 10.01° of 2θ.

The two most important parameters in a continuous scan, which are defined by the user, are the sampling interval (step), *s*, and the angular velocity (scan rate), *r*, which are typically measured in degrees 2θ and degrees 2θ/min, respectively. The sampling step is equivalent to the step size in the step-scan mode. Everything said about the size of the step in Sect. 12.2.8, therefore, applies to the sampling step during the continuous scan. The two parameters—counting time, *t*, in the step scan and the scan rate, *r*, in the continuous scan—are related to one another as follows:

$$t = \frac{60s}{r} \tag{12.7}$$

where *t* is in seconds, *s* is in degrees, and *r* is in degrees/min. Thus, a continuous scan with the rate $r = 0.1$ deg./min and with the sampling step $s = 0.02°$ is equivalent to a step scan with the same step and counting time 12 s/step. When the sampling step is reduced at a constant scan rate, this is equivalent to the proportional reduction of counting time and vice versa.

In modern diffractometers, both scanning modes result in nearly identical quality of experimental data. A step scan is usually considered as the one with less significant positioning errors, which could be important in experiments where the maximum lattice parameter precision is essential. Continuous scans are most often used for fast experiments, while step scans are usually employed in overnight or weekend-long experiments.

Predictably, when counting times are short ($t \ll 1$ s), step scans take longer to complete (the required time may be easily doubled when compared with the identical quality continuous scans). This occurs because no intensity is measured when the goniometer arms move to the next position (compare the flowchart from Fig. 12.26 with that from Fig. 12.29). The difference in the time of the experiment becomes negligible during overnight or weekend-long experiments.

12.2.10 Scan Range

Regardless of the scanning mode and other conditions, the range of Bragg angles within which the diffracted intensity is measured is also crucial. This range is typically specified by the user as the start and end Bragg angles, $2\theta_s$ and $2\theta_e$, respectively. The start angle should be a few degrees before the first observable Bragg peak to ensure the measurement of a sufficient number of background data points. If an unknown powder is examined, the beginning of the scan range should be selected at the lowest possible Bragg angle, which is allowed by the geometry of a sample holder, because some sample holders shield all or a fraction of the incident beam when $0 \leq 2\theta \leq {\sim}2$ to 5°. It is important to keep in mind that at very low Bragg angles, the background caused by a fraction of the divergent incident beam reaching the detector may be very high but most powder diffractometers will provide reliable data at Bragg angles as low as 2–4° 2θ. Even lower Bragg angles may be examined provided both the divergence and receiving slits are extremely narrow to eliminate strong contribution from the incident beam.

No powder diffraction experiment should be started at $2\theta_s = 0°$, since the extremely high intensity of the incident beam at full tube power may damage the detector even when the narrowest slits have been used. The most dependable way to determine $2\theta_s$ for an overnight or a weekend-long experiment is to perform a quick scan at 5–10°/min beginning at the minimum allowable Bragg angle and ending at $2\theta_e \cong 30°$–40°. Based on this result, the correct $2\theta_s$ may be properly selected.[19]

The selection of the end angle is usually based on the following factors. First, each goniometer has certain physical limits on the maximum allowed Bragg angle, which are dependent on hardware design. These are established by the manufacturer to ensure that the detector arm does not collide with the X-ray tube housing during data collection. The majority of modern powder diffractometers operating in θ–*2θ* (or θ–θ) modes can reach $2\theta_e$ as high as 140–160°.

Other considerations are: what is the purpose of powder diffraction data, which wavelength is used, and what is the nature of the examined material? Thus, when employing Cu Kα radiation:

- The end Bragg angle of 50–70° is usually sufficient for evaluation of crystallinity or for phase identification purposes.
- The end angle should be selected as high as possible when experimental data are used for precise unit cell or crystal structure refinement. This usually can be established by a quick scan (5–10°/min) with $2\theta_s \cong 50$–80° and $2\theta_e$ near the physical limit of the goniometer, and the proper $2\theta_e$ is selected a few degrees higher than the last distinguishable Bragg peak.
- Materials containing only light elements (e.g., organic compounds) scatter X-rays well at room temperature only at low Bragg angles (usually less than 50–70° 2θ when Cu Kα radiation is employed); therefore, it makes no sense to measure diffraction data at higher Bragg angles.

When using short wavelengths, the diffraction pattern is compressed to lower Bragg angles when compared to that collected using long wavelengths, which directly

[19] Reminder: once the $2\theta_s$ has been found, the divergence slit for the complete experiment should be selected based on the actual size of the prepared specimen as was discussed in Sects. 12.1.3 and 12.2.3.

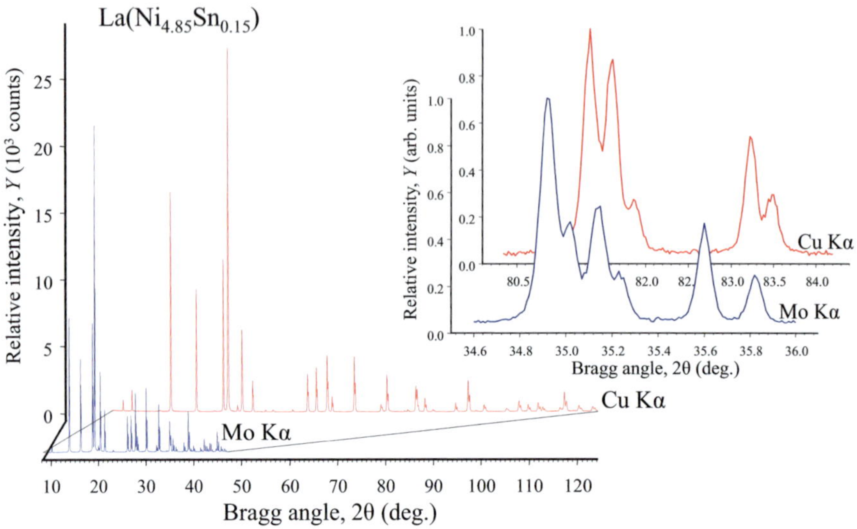

Fig. 12.30 Two X-ray powder diffraction patterns collected from the same flat specimen, 20 mm in diameter and 1 mm deep cavity, filled with nearly spherical $LaNi_{4.85}Sn_{0.15}$ powder (see the inset in Fig. 12.16) on a Rigaku TTRAX rotating anode powder diffractometer using Mo Kα and Cu Kα radiations. Goniometer radius $R = 285$ mm. For the experiment using Mo Kα radiation: divergence slit, DS = 0.5°; receiving slit, RS = 0.03°; step-scan mode, $\Delta 2\theta = 0.01°$. For the experiment using Cu Kα radiation: divergence slit, DS = 0.875°; receiving slit, RS = 0.03°; step-scan mode, $\Delta 2\theta = 0.02°$. An automatic variable scatter slit was used to reduce the background in both experiments. The inset shows the expanded view of the three Bragg peaks observed between ~34.6 and 36.0° 2θ using Mo Kα radiation and the same peaks observed between ~80.3 and 84.2° 2θ using Cu Kα radiation. The resolution is preserved because the corresponding FWHMs are ~0.075 and 0.172° for Mo and Cu Kα radiations, respectively

follows from the Braggs' law. For example, when X-ray diffraction data were obtained using Cu Kα radiation at $2\theta_e \cong 120°$, this is identical to $2\theta_e \cong 47°$ when using Mo Kα radiation. An example of two equivalent sets of diffraction data collected from the same powdered specimen using Cu Kα and Mo Kα radiation is shown in Fig. 12.30.

It may appear that when short wavelengths are used to collect powder diffraction data, this should result in the reduced resolution, thus creating potential problems during data processing. As shown in the inset of Fig. 12.30, this is not the case because the Bragg peaks observed using shorter wavelength radiation are sharper than those observed using longer wavelengths, provided the instrumental resolution remains constant, and the data collection step size has been appropriately reduced. For example, the full widths at half maximum of the three Bragg peaks shown in the inset in Fig. 12.30 decreases from ~0.172 to ~0.075° 2θ for Cu and Mo Kα radiations, respectively. Further, the resolution of $K\alpha_1/K\alpha_2$ doublet is considerably improved for Mo $K\alpha_{1,2}$ radiation because the separation of the doublet here is nearly 3 × FWHM, and it is only ~1.6 × FWHM for Cu $K\alpha_{1,2}$ radiation.

The use of short wavelengths somewhat improves the scattered intensity because reflections shift to low Bragg angles where the Lorentz-polarization factor is high.[20] Conversely, the total flux of low energy photons is usually higher than that of high-energy photons at the same X-ray tube power settings. Another advantage of using high-energy X-rays is because peak broadening at low angles is less susceptible to a variety of factors, such as crystallite size, strain, and some instrumental influences. The largest drawback is the inevitable increase in peak asymmetry and associated loss of resolution.

12.3 Quality of Experimental Data

The quality of powder diffraction data can be easily recognized visually from a plot of diffracted intensity as a function of Bragg angle, as depicted in Fig. 12.31. The two sets of data shown in this figure were collected from the same specimen, prepared from the nearly ideal $LaNi_{4.85}Sn_{0.15}$ powder (see the electron micrograph

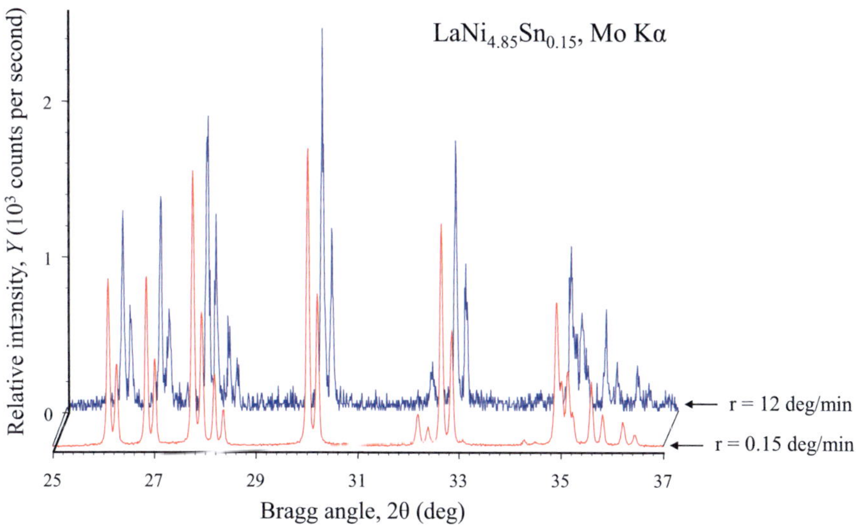

Fig. 12.31 The example of different quality X-ray powder diffraction data. Only thin lines connecting observed data points are shown to emphasize the difference in data quality. Both experiments were carried out using the same powdered specimen and the same powder diffractometer with rotating anode X-ray source (Fig. 11.22) but with different scan rates

[20] Intensity gain due to Lorentz-polarization factor (see Sect. 8.6.4) is partially offset by the requirement of reduced divergence slit opening (see Sects. 12.1.3 and 12.2.3), provided all other things remain constant, including the brightness of the incident beam.

in the inset of Fig. 12.16), using the same diffractometer in a continuous scan mode with sampling interval $s = 0.01°$, employing Mo Kα radiation. One of the datasets, which is smooth and appears to be of high quality, was collected with a scan rate $r = 0.15°$/min. This scan rate is equivalent to a counting time $s = 4$ s in each point (see Eq. 12.7). The second, noisy, powder diffraction pattern was collected with a scan rate $r = 12°$/min, which is equivalent to a counting time $s = 0.05$ s per point, and it gives the impression of lower quality data. In Fig. 12.31, it is evident that strong noise makes the weak Bragg peaks nearly unrecognizable (e.g., compare the range of Bragg angles between 33 and 35° 2θ). Additionally, the noise gives rise to false anomalies that resemble nonexistent weak Bragg peaks (e.g., see the range between 36.6 and 37° 2θ).

Strict numerical characterization of the quality of powder diffraction data is possible in addition to visual analysis. This section briefly considers the most important issues related to the proper selection of data acquisition parameters to ensure the consistent reliability of the powder diffraction experiment, as illustrated in Fig. 12.16. It can be implicitly assumed that the errors associated with positioning of the goniometer arms are negligible; in other words, the measured Bragg angles are precise. This is usually true for new goniometers but may not be the case for goniostats that have been in service for many years (20+ years) and have worn-out gear mechanisms.

If positioning errors are not negligible, the goniometer should be repaired or replaced, as it is intrinsically impossible to obtain high-quality powder diffraction data using a goniometer, which produces either systematic or random (and therefore, uncontrollable) errors in Bragg angles. Periodic testing of positioning errors of the goniometer should be done by measuring one or several standard materials available from the National Institute of Standards and Technology (NIST).[21] The most commonly used standards are silicon (SRM 640b), corundum (SRM 676), and LaB_6 (SRM 660c).

12.3.1 *Quality of Intensity Measurements*

Regardless of the selection of the scan mode (see Sects. 12.2.8 and 12.2.9), the intensity recorded at each Bragg angle is measured during a certain period of time—counting time—which is one of the multiple user-specified parameters in the data acquisition process. As clearly seen in Fig. 12.31, the importance of the counting time parameter is difficult to overestimate since it directly influences the accuracy of the measured diffracted intensity and the overall quality of diffraction data.

In powder diffraction, X-ray photons or neutrons (in X-ray and neutron diffraction experiments, respectively) are registered by the detector as random events. The

[21] For a complete list of X-ray standards available from NIST, see https://www.nist.gov/srm

measured intensity is directly proportional to the number of counts; therefore, the accuracy of intensity measurements is governed by statistical principles. Although this discussion primarily refers to X-ray diffraction and photons, the same conclusions apply to neutron diffraction and neutron counts.

Assume that a total of N photon counts were registered by the detector. In this case, the spread, σ, is defined by the Poisson[22] probability distribution

$$\sigma = \sqrt{N} \tag{12.8}$$

The corresponding error, ε, in an individual measurement (i.e., in the number of counts registered at each Bragg angle) depends on the confidence level and is given as

$$\varepsilon = \frac{Q\sigma}{\mathrm{N}} = \frac{Q}{\sqrt{N}} \times 100\% \tag{12.9}$$

where Q = 0.67, 1.64, 2.59, and 3.09 for the 50, 90, 99, and 99.9% confidence levels, respectively. Thus, to achieve a 3% error in the intensity measurement at 50% confidence, it is necessary to register a total of ~500 photons, while to reach 3% error at 99% confidence, as many as ~7400 counts need to be accumulated. For a 1% error at the same confidence levels, ~4500 and ~67,000 photons must be registered by the detector. Equation (12.9) reflects only statistical counting errors, which are present even in the best-quality data. Other systematic and random errors, for example, those that appear due to the nonideal sample, may affect the resulting experimental data, and these errors were described in Sect. 12.1.

Consider the example shown in Table 12.1, which shows the effect of counting time on the statistical error of a single X-ray intensity measurement. When the counting time is limited to 1 s, a total of 100 photons are registered with the resulting ~16% error at the 90% confidence level. When counting time is increased to 25 s, the statistical error in the 2500 registered photons is reduced to 3.3%. It is worth noting that as follows from Eq. (12.9) and Table 12.1, when a twofold increase in the

Table 12.1 Effect of counting time on the statistical error during a single measurement

Photon flux (counts per second)	Counting time (s)	Number (N) of registered counts	Spread = $\sqrt{N}$ (counts)	Error at 90% confidence (%)
100	1	100	10	16.4
100	25	2500	50	3.3

[22] Siméon-Denis Poisson (1781–1840) was a French mathematician credited with numerous discoveries in mathematics, geometry, and physics. He is probably best known for Poisson's ratio, which is the relative contraction strain divided by the relative extension strain under uniaxial compression, and Poisson's noise, which is defined in Eq. (12.8). See https://en.wikipedia.org/wiki/Simeon_Poisson for a brief biography.

accuracy of the intensity measurement is desired, the counting time must be increased fourfold.

Obviously, the quality of intensity measurements in powder diffraction is inversely proportional to the statistical measurement errors; therefore, it is directly proportional to the square root of the total number of registered photons. Assuming constant brightness of the X-ray source, the most certain way to improve the quality of the diffraction data is to use a lower scanning rate or longer counting time in continuous and step-scan experiments, respectively.

Increased photon counts in principle can be achieved by using larger divergence and receiving slits, increasing the input power to the X-ray tube, using a position-sensitive or image-plate detector and some other modifications of various data acquisition and instrumental parameters. The side effects arising from improvements in counting statistics should also be considered. For example,

- Increasing the counting time translates into long (overnight or weekend-long) experiments that take more time to collect.
- Increasing the divergence slit may be incompatible with the sample size (see Sects. 12.1.3 and 12.2.3).
- Increasing the receiving slit decreases the resolution (see Sect. 12.2.4).
- Raising the input power is limited by the ability of an X-ray tube to dissipate heat and results in a reduced life of the tube.
- Using position-sensitive or image-plate detectors results in a lower resolution and higher background (see Fig. 11.25).

In extreme cases, the only feasible option may be to seek beam time at the nearest synchrotron.

The optimal counting time depends on the requirements dictated by the desired quality of diffraction data. For example, the International Centre for Diffraction Data (ICDD),[23] which maintains and distributes the most extensive database of powder diffraction data, has established the following criteria for submitting new experimental powder diffraction patterns to the Powder Diffraction File™:

- At least 50,000 counts total should be accumulated for peaks with relative intensity 50% or higher of the strongest observed Bragg peak.
- At least 5000 counts total should be accumulated for peaks with intensity 5% or higher of the strongest Bragg peak.

The ICDD requirements are quite strict, and they are established to control the quality of the database. In a typical powder diffraction experiment according to the classification introduced in Sect. 12.2.7 and assuming that an average diffraction pattern will consist of about 4000 data points, the counting time will vary from

[23] International Centre for Diffraction Data (ICDD®) is a nonprofit scientific organization dedicated to collecting, editing, publishing, and distributing powder diffraction data for the identification of crystalline materials. ICDD on the Web: https://www.icdd.com/

0.5 to 2 s for a fast experiment, from 6 to 10 s in an overnight experiment, and counting time will exceed 20 s during a weekend-long experiment. Hence, fast, overnight, and weekend-long step-scan experiments using a sealed X-ray tube source will usually provide adequate quality of the powder diffraction pattern for phase identification, unit cell parameters refinement, and crystal structure solution and refinement, respectively.

In addition to the statistical error in each individual data point, it is often desirable to have the means of describing the quality of the whole powder diffraction pattern using a single figure of merit. This can be done by using the observed residual, R_{obs},[24] which is given by Eq. (12.10).

$$R_{obs} = \sum\nolimits_{i=1}^{n} \sigma_i / \sum\nolimits_{i=1}^{n} N_i \times 100\% \tag{12.10}$$

The numerator in Eq. (12.10) represents photon counting errors (see Eq. 12.8) and the denominator corresponds to the total number of counts, both summed over the total number of the measured data points, n. For example, the two diffraction patterns shown in Fig. 12.31 are characterized by $R_{obs} = 3.59$ and 30.9% for the higher and the lower quality data, respectively. The range shown in this figure includes a total of 1201 data points and the total number of accumulated counts varies by nearly two orders of magnitude: ~6.4×10^5 and ~8×10^3, for the high- and poor-quality data, respectively. The R_{obs} figure of merit reflects the overall quality of the powder diffraction pattern and is comparable with other residuals (R-factors) commonly calculated to represent the quality of fitting in the Pawley, Le Bail, and Rietveld methods (see Eqs. 15.18–15.22 in Sect. 15.6.2): a low R_{obs} characterizes high-quality data, while a high R_{obs} corresponds to low-quality diffraction patterns.

12.3.2 Factors Affecting Resolution

The intrinsic one-dimensionality of the powder diffraction experiment implies that every attempt should be made to collect diffraction data with the highest possible resolution. As established in Sect. 12.2, among many factors affecting the resolution of experimental data, the most important are the goniometer radius, the wavelength of the radiation used, the receiving slit aperture, and the step size or sampling interval.

[24] This figure of merit is different from the so-called expected residual used to quantify the quality of experimental data in a Rietveld refinement. The latter is given as $R_{exp} = \left((n-p)/\sum w_i Y_i^2\right)^{1/2}$, see Eq. (15.21) on p. 477, where n is the total number of collected data points, p is the total number of least-squares parameters, and w_i and Y_i are the weight and the intensity, respectively, of the ith data point ($1 \le i \le n$), also see Chap. 16 onward.

The goniometer radius is generally fixed. However, since the angular resolution remains constant for a given condition of the specimen, linear resolution of the goniometer is inversely proportional to its radius. Therefore, when especially high resolution is an issue, X-ray diffraction data should be collected using the instrument with the largest available radius. It is important to remember that a large goniometer radius usually translates into a decreased diffracted intensity and the powder diffraction experiment will take a longer time to achieve an identical R_{obs}, when compared to that using the goniometer with a smaller radius.

Similarly, using lower energy X-rays (i.e., X-rays with longer wavelengths) will normally result in the improved resolution of the powder diffraction pattern but not $K\alpha_1/K\alpha_2$ doublets (see Fig. 12.30). The caveat in using long wavelengths is the reduction of the volume of the "visible" reciprocal lattice when compared to that using short wavelengths because only those reciprocal lattice points which have $d^*_{hkl} \leq 2/\lambda$ (e.g., see Fig. 7.10) can be positioned on the surface of the Ewald's sphere.[25] Further, as discussed earlier (see Sect. 12.2), the easy selection of the wavelength is only feasible when using a synchrotron X-ray source.

12.4 Additional Reading

1. R. Jenkins and R.L. Snyder, Introduction to X-ray powder diffractometry, Wiley, New York (1996).
2. W. Parrish and J.I. Langford, 2.3. Powder and related techniques, in: Part 2: Diffraction Geometry and its practical realization; International Tables for Crystallography, Vol. C, Mathematical, physical and chemical tables, First online edition, E. Prince, Ed. (2006). https://it.iucr.org/C/
3. Jens Als-Nielsen and Des McMorrow, Elements of modern X-ray physics, Wiley, New York (2001).
4. D. Louër, Laboratory X-ray powder diffraction, in: Structure determination from powder diffraction data. IUCr monographs on Crystallography 13, W.I.F. David, K. Shankland, L.B. McCusker, and Ch. Baerlocher, Eds., Oxford University Press, Oxford (2002).
5. P.W. Stephens, D. E. Cox, and A. N. Fitch, Synchrotron radiation powder diffraction, in: Structure determination from powder diffraction data. IUCr monographs on Crystallography 13, W.I.F. David, K. Shankland, L.B. McCusker, and Ch. Baerlocher, Eds., Oxford University Press, Oxford (2002).

[25] An additional reduction of the number of experimentally measurable reciprocal lattice points occurs due to a physical limit of any goniometer, which makes it impossible to collect data at Bragg angles exceeding ~150° 2θ.

6. R. J. Hill and I. C. Madsen, Sample preparation, instrument selection and data, in: Structure determination from powder diffraction data. IUCr monographs on Crystallography 13, W.I.F. David, K. Shankland, L.B. McCusker, and Ch. Baerlocher, Eds., Oxford University Press, Oxford (2002).
7. R. M. Ibberson and W. I. F. David, Neutron powder diffraction, in: Structure determination from powder diffraction data. IUCr monographs on Crystallography 13, W.I.F. David, K. Shankland, L.B. McCusker, and Ch. Baerlocher, Eds., Oxford University Press, Oxford (2002).
8. R.J. Hill, Data collection strategies: fitting the experiment to the need, in: The Rietveld method. IUCr monographs on crystallography 5, R.A. Young, Ed., Oxford University Press, Oxford (1993).
9. D.K. Bowen and B.K. Tanner. High resolution X-ray diffractometry and topography. T.J. International Ltd., Padstow, United Kingdom (1998).
10. Modern powder diffraction. Reviews in mineralogy, v. 20, D.L. Bish and J.E. Post, Eds., Mineralogical Society of America, Washington, DC (1989).
11. L.V. Azaroff and M.J. Buerger, The powder method in X-ray crystallography, McGraw-Hill, New York (1958).
12. H. Lipson and H. Steeple, Interpretation of X-ray powder diffraction patterns, Macmillan, London/St. Martin's Press, New York (1970).
13. H.P. Klug and L.E. Alexander, X-ray diffraction procedures for polycrystalline and amorphous materials, Second Edition, Wiley, New York (1974).
14. B.D. Cullity S.R. Stock, Elements of X-Ray Diffraction, Third Edition, Pearson Education Limited (2013).
15. Accuracy in powder diffraction II. Proceedings of the International Conference, May 26–29, 1992. E. Prince and J.K. Stalick, Eds., National Institute of Standards and Technology, Washington, DC (1992).
16. Accuracy in powder diffraction. Proceedings of a Symposium on Accuracy in Powder Diffraction, June 11–15, 1979. S. Block and C.R. Hubbard, Eds., National Bureau of Standards, Washington, DC (1980).
17. A laboratory manual for X-ray powder diffraction (computer file on a CD), L.J. Poppe, Ed., U.S. Geological Survey, Woods Hole, MA (2001).
18. A.J.C. Wilson, Mathematical theory of X-ray powder diffractometry, Gordon and Breach, New York (1963).
19. A. Wielders, Rob Delhez, X-ray powder diffraction on the red planet, p. 5; R. Delhez, L. Marinangeli, S. van der Gaast, Mars-XRD: the X-ray Diffractometer for rock and soil analysis on Mars in 2011, p.6; Charge coupled devices as X-ray detectors for XRD/XRF, p. 10 in: CPD Newsletter "Powder Diffraction on Mars, the Red Planet", Issue 30 (2005), available at https://www.iucr.org/__data/assets/pdf_file/0020/21656/CPD30.pdf.

12.5 Problems

1. Assume that you are to collect powder diffraction data from a powder with the purpose to establish and refine its crystal structure. Earlier, you have used the following powder diffractometer:

   (a) Sealed X-ray tube source, curved position-sensitive detector, the radius of its goniometer is 150 mm, and diffraction data are collected in the transmission mode using cylindrical specimens. You employed this equipment to characterize the phase purity of your materials.
   (b) In addition to this device, other departments at your university have the following powder diffractometer systems:
   (c) Sealed X-ray tube source, Bragg–Brentano goniometer, radius 185 mm, scintillation detector.
   (d) Rotating anode source, Bragg–Brentano goniometer, radius 285 mm, scintillation detector.
   (e) Sealed X-ray tube source, Bragg–Brentano goniometer, radius 250 mm, cooled solid-state detector.

   Establish the order in which you would call people in charge of the diffractometers to arrange for data collection and explain why.
2. Now assume that you have 20 different samples to characterize with respect to their phase composition. Further, each department charges $50.00 per hour for the use of their equipment (the money goes to a special account, which pays for a service contract and routine maintenance). Assuming the availability of the same diffractometers as in Problem 1, what would be the order on your calling list and why?
3. When you called the person in charge of the diffractometer D (see Problem 1), she told you that the goniometer axis is horizontal and that the X-ray source arm is stationary. Do you need to worry about mixing your powder with a binder?
4. As part of the preparation for your experiment described in Problem 1, you used a mortar and pestle to grind the sample. You were completely satisfied with the result since the powder appeared fine and homogeneous to your eyes, but when you discussed the process with your thesis advisor, he asked you to screen the powder through a 25 μm sieve. This discussion happened just before you were about to take off across campus since your allotted time on the powder diffractometer starts in 10 min. Describe your course of action and explain why?
5. You made a flat sample in the preparation for a highly precise powder diffraction experiment using diffractometer D (see Problem 1). The sample completely fills a cylindrical opening 25 mm in diameter. The lowest Bragg angle during the experiment $2\theta = 10°$. The set of divergence slits available on this instrument includes the following apertures: 0.05, 0.1, 0.25, 0.5, 0.75, 1, 2, and 5 mm. Knowing that the distance between the focus of the X-ray tube and the

divergence slit is 60 mm, select the most appropriate divergence slit to be used in your experiment and explain why?

6. You are having difficulties with making a flat sample for a high-precision powder diffraction experiment—the powder just will not spread evenly in a cylindrical hole ~20 mm diameter and 1 mm deep. You are considering the following options:

   (a) Compact the powder by pressing a glass slide against the surface of the sample.
   (b) Make a suspension in petroleum ether and pour it into the hole.
   (c) Backfill the holder while the front of the sample is pressed against a glass slide with a strip of rough sand paper glued to one of its surfaces.
   (d) Use a different sample holder and dust the powder on top.

   Arrange these options in the order which should result in the best-quality specimen for powder diffraction and explain why.

7. You were able to arrange time on the powder diffractometer C (see Problem 1). Several sample holders are available:

   (a) Powder fills a cylindrical hole 25 mm diameter and 1 mm deep.
   (b) Powder fills a cylindrical hole 20 mm diameter and 1 mm deep.
   (c) Powder fills a square hole $30 \times 30 \times 1$ mm^3.
   (d) Powder is dusted on top of a round rough spot 25 mm in diameter.

   Arrange these sample holders in the order from most to least suitable for a high-precision powder diffraction experiment and explain why. Assume you are working with a molecular compound.

8. Answer Problem 7 if you are working with an intermetallic compound containing a lanthanide element.
9. You are using Fe Kα radiation to collect powder diffraction data employing powder diffractometer C (see Problem 1). After several quick scans, you established that the receiving slit with the aperture of 0.03° results in both acceptable resolution and intensity. Bragg peaks appear to have a full width at half maximum between 0.4 and 0.5° 2θ. What is the largest allowable step during data collection and why?
10. You are about to perform a quick scan from 5 to 70° 2θ to verify the crystallinity of your material and to check whether the receiving slit is of adequate width or not. You are planning for the experiment which will take overall 10–15 min. Which scanning mode (step scan or continuous scan) is the best to accomplish the task and why?
11. In preparation for an overnight experiment, you performed a quick scan, which resulted in the powder diffraction pattern shown in Fig. 12.32. The receiving slit was narrow, RS = 0.01°. You are planning to double the receiving slit aperture and your goal is to measure the highest intensity data point with 1% or better error at the 99.9% confidence level during a step scan. Estimate the counting time parameter for the upcoming overnight experiment.

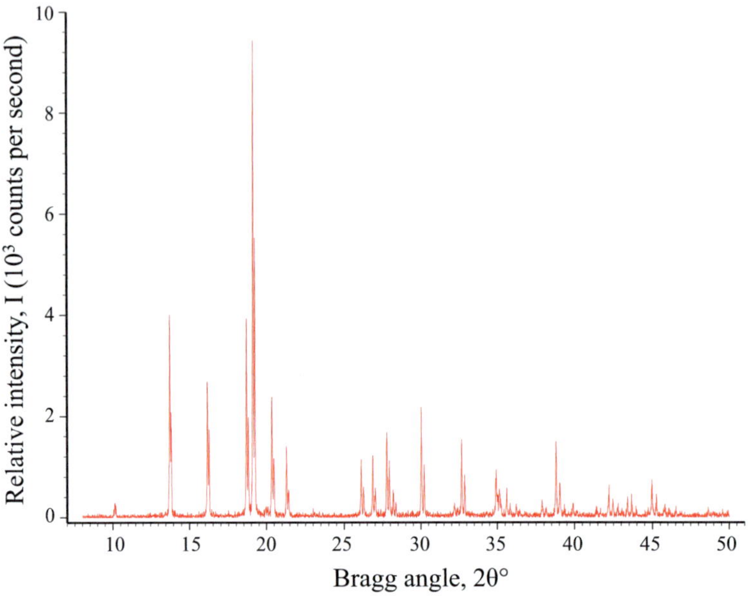

Fig. 12.32 The result of a quick scan experiment using a narrow receiving slit to determine correct data collection parameters for a future overnight experiment

12. Assume that the overnight experiment described in Problem 11 is to be performed using the continuous scanning mode from 8 to 60° 2θ with a sampling step $\Delta 2.\theta = 0.005°$. Estimate both the scanning rate and the time it will take to finish the experiment.

Chapter 13
Preliminary Data Processing and Phase Analysis

So far, we have considered the fundamentals of crystallographic symmetry, the phenomenon of diffraction from a crystal lattice, and the basics of a powder diffraction experiment. Familiarity with these broad subjects is essential for understanding how waves are scattered by crystalline matter, how structural information is encoded into a three-dimensional distribution of discrete intensity maxima, and how this information is convoluted with various instrumental and specimen-dependent functions when projected along one direction and measured as the scattered intensity Y versus the Bragg angle 2θ. We have already learned that this knowledge can be applied to the structural characterization of materials as it enables us to decode a one-dimensional snapshot of a reciprocal lattice and, therefore, to reconstruct a three-dimensional distribution of atoms in an infinite crystal lattice through forward Fourier transformation.

Our experience with applications of the powder method in diffraction analysis has been largely conceptual, and in the remainder of this book, we discuss key issues that arise during the processing and interpretation of powder diffraction data. Despite the apparent simplicity of one-dimensional diffraction patterns—observed as a series of constructive interference peaks, both resolved and partially or completely overlapped, created by elastically scattered waves and placed on top of nonlinear background noise—the complexity of their interpretation stems from the intricate processes involved in converting the underlying structure into the experimentally observed data. As a result, nearly every component of data processing in powder diffraction is computationally intensive.

The presence of symmetry (Chaps. 1–5) coupled with well-defined analytical relationships determining both the directions and intensities of scattered beams (Chaps. 7–9), in addition to known properties of both the specimen and instrument employed to obtain a powder diffraction pattern (Chaps. 11 and 12), makes it possible to develop both the general methodology and algorithm(s) suitable for automation. Given the amount of numerical data collected in a typical powder

P. Zavalij, V. Pecharsky, *Fundamentals of Powder Diffraction and Structural Characterization of Materials*, https://doi.org/10.1007/978-3-031-91504-8_13

diffraction experiment,[1] its interpretation and processing usually involve a broad use of computers.

We do not intend to comprehensively cover or evaluate any specific product among the large variety of available applications, whether freeware or commercial. Instead, we illustrate multifaceted aspects of data processing using only a few computer codes, while paying attention to both the capabilities and limitations of the powder diffraction method, as well as providing some examples of how the analysis of a powder diffraction pattern can be practically accomplished.

In this chapter, preliminary data treatment, phase identification, and quantification using database search-match were performed and illustrated with the DMSNT[2] and WinCSD[3] software packages. These programs were selected due to our familiarity with the software and the availability of numerous examples at the time. Although DMSNT and WinCSD are no longer available or supported, all examples have been retained, as they can be easily replicated with alternative software, provided the user understands the underlying process and functionality of the software. Typically, phase analysis and preliminary data treatment software is supplied by the instrument manufacturer. Examples include DIFFRAC.EVA (Bruker[4]), HighScore (Malvern Panalytical[5]), SmartLab Studio II (Rigaku[6]), and XRDanalysis (Anton Paar[7]). A notable standalone yet comprehensive software package is JADE (ICDD/MDI[8]).

13.1 Interpretation of Powder Diffraction Data

Given the nature of the powder diffraction method, the resultant experimental data can be employed to obtain and/or confirm the following information:

---

[1]For example, the range $10 \leq 2\theta \leq 90°$ scanned with a step $\Delta 2\theta = 0.02°$ results in 4001 measured data points.

[2]DMSNT—Data Management Software from Scintag Inc. (1999), later rebranded as WinXRD—Data Collection and Analysis Package from Thermo ARL. It is no longer supported. However, it is mentioned here because it was used to prepare some illustrations and examples in this text.

[3]L.G. Akselrud, P.Yu. Zavalii, Yu.N. Grin, V.K. Pecharsky, B. Baumgartner, E. Wolfel, "Use of the CSD program package for structure determination from powder data", Mater. Sci. Forum 133–136, 335 (1993), https://doi.org/10.4028/www.scientific.net/MSF.133-136.335; L. Akselrud, Yu. Grin, WinCSD: software package for crystallographic calculations (Version 4), J. Appl. Crystallogr. 47, 803 (2014). https://doi.org/10.1107/S1600576714001058

[4]https://www.bruker.com/en/products-and-solutions/diffractometers-and-x-ray-microscopes/x-ray-diffractometers/diffrac-suite-software/diffrac-eva.html

[5]https://www.malvernpanalytical.com/en/products/category/software/x-ray-diffraction-software/highscore

[6]https://rigaku.com/products/x-ray-diffraction-and-scattering/xrd/smartlab-studio-ii

[7]https://www.anton-paar.com/us-en/products/details/xrdanalysis/

[8]https://www.icdd.com/mdi-jade/

- Phase composition of a material, including both qualitative and quantitative analyses, coupled with searches of various databases.
- Indices of Bragg reflections, observed integrated intensities, and precise lattice parameters.
- Distribution of atoms in the unit cell (i.e., the crystal structure), either to verify that the material has one of the already known types of crystal structures or to solve it from first principles.[9]
- Precise structural details, including equilibrium positions of atoms in the unit cell, individual atomic displacement, and population parameters by employing the Rietveld method (see Sect. 15.7).
- Various microscopic structural characteristics of the specimen.

The first four items in this list represent the most common goals that are typically achieved during the characterization of polycrystalline materials using powder diffraction data. The results are often employed to establish and/or clarify relationships between crystal structures and properties of materials—knowledge that is critical in modern science and engineering. The most typical sequences of data processing steps (or phases) are, therefore, visualized in a general schematic depicted in Fig. 13.1. Correspondingly, these four major steps are the focus of the remainder of this book. As shown in Fig. 13.1, the three different quality levels of powder diffraction data are usually associated with the expected outcomes of the experiment.[10]

Thus, a fast experiment is typically suitable for evaluation of the specimen and phase identification, i.e., qualitative analysis. When necessary, this can be followed by an overnight experiment for indexing and accurate lattice parameter refinement and/or a weekend-long experiment for complete crystal structure determination and refinement. In some instances, such as when a specimen has exceptional quality and its crystal structure is known or very simple, all relevant parameters can be determined using data collected in an overnight experiment. Similarly, fast experiments may be suitable for accurate unit cell determination and semi-quantitative analysis, in addition to phase identification. There is also no requirement to first run a fast scan followed by overnight or weekend-long experiment. While doing this may be beneficial when using point detector, if the goal is full structure determination

[9] In the context of this book, structure solution from first principles (often referred to as the ab initio structure determination) means that all crystallographic data, including lattice parameters and symmetry and the distribution of atoms in the unit cell, are inferred from the analysis of the scattered intensity as a function of Bragg angle, collected during a powder diffraction experiment. Additional information, such as the gravimetric density of a material, its chemical composition, basic physical and chemical properties, may be used as well, when available.

[10] The classification "fast," "overnight," and "weekend" experiments is usually applied to laboratory powder diffractometers equipped with conventional sealed X-ray tube sources and point detectors. Obviously, when the brilliance of the available source increases dramatically, the time of the actual experiment will decrease. It is worth noting that since specialized beam time (e.g., a synchrotron source) is limited, this normally implies that the majority of samples should undergo a thorough preliminary examination using conventional X-ray sources.

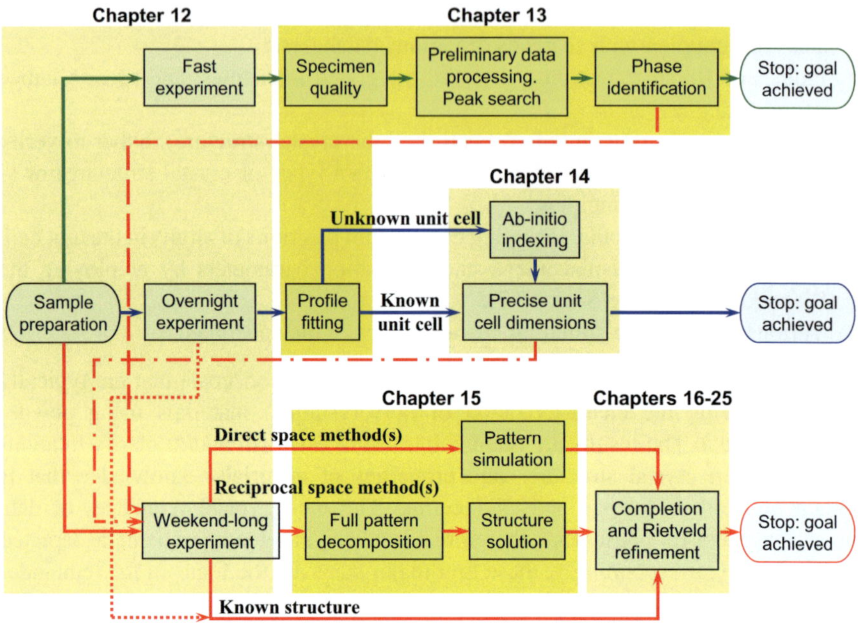

Fig. 13.1 The flowchart illustrates the common steps involved in structural characterization of materials by using the powder diffraction method. It always begins with the sample preparation, followed by a properly executed experiment, both of which are covered in Chaps. 11 and 12. This chapter discusses preliminary data processing and profile fitting, along with common issues related to phase identification and analysis. Unit cell determination, crystal structure solution, and refinement are addressed in Chaps. 14 and 15 onward. The flowchart highlights the typical applications for the three types of experiments, though any or all data processing steps may be applied to fast, overnight, and weekend-long experiments, depending on their quality and characterization goals

and/or accurate quantitative analysis and the sample is known to be good quality, a slow experiment (overnight or weekend-long) can be performed immediately, especially when using PSDs. In any case, one should use their judgment and experience to assess both the suitability of the experimental data and the reliability of the results.

When the sole goal of the experiment is to identify phases present in the polycrystalline material, it may be achieved in a fast experiment, which may take as little as 10–15 min depending on the quality of the specimen, brightness of the beam, and geometry of the instrument. Another important application of the rapidly collected data is visual examination of $Y(2\theta)$ to evaluate both the crystallinity of the specimen and the complexity of the pattern as illustrated by several distinct examples in Fig. 13.2.

Considering the three examples illustrated in Fig. 13.2, Specimen A is suitable for any kind of conventional structural characterization using powder diffraction. Further, judging from the relatively large separation between Bragg reflections at low

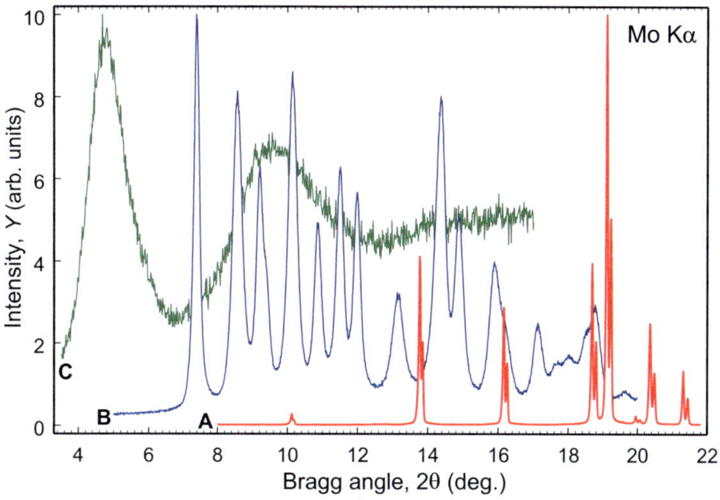

Fig. 13.2 Fragments of powder diffraction patterns collected from three different materials on a Rigaku TTRAX rotating anode powder diffractometer using Mo Kα radiation. *Pattern A* (red) represents a material with excellent crystallinity: narrow and sharp Bragg reflections with $K\alpha_1/K\alpha_2$ doublets becoming partially resolved at $2\theta \cong 14°$, low and nearly linear background. *Pattern B* (blue) is also a crystalline material; however, its crystallinity is poor (and/or grain sizes are extremely small, and/or the material has been strained), which is evident from broad but still distinct Bragg reflections; certain amount of an amorphous component may be present in this sample, judging from a minor nonlinearity of the background between ~6° and 20° 2θ. *Pattern C* (green) is collected from a material, which is clearly noncrystalline in a conventional sense: long-range order and periodicity are absent because no Bragg reflections have been observed. Provided Powder C is amorphous, broad halo(s) usually contain information about nearest neighbors in the coordination spheres of atoms. Each pattern has been scaled individually

angles, it is easy to conclude that the crystal structure of this material is quite simple.[11] Specimen B in the current state may or may not be suitable for the determination of its crystal structure. If possible, the material should be recrystallized or heat-treated, depending on its nature, to increase grain size and/or relieve strain. If successful, the widths of Bragg peaks could be reduced, and pattern resolution improved. Judging from the number of Bragg peaks observed at low angles and from the fact that the first Bragg reflection in Specimen B occurs at a lower angle compared with Specimen A, the crystal structure of the former is more complex than that of the latter.[12] Finally, Specimen C is not suitable for analyzing

[11] Pattern A was collected from a $LaNi_{4.85}Sn_{0.15}$ powder prepared by high-pressure gas atomization and then heat treated at 950 °C for 5 min. The compound has a hexagonal crystal structure with unit cell dimensions $a \cong 5.04$, $c \cong 4.01$ Å.

[12] Pattern B was collected from an anhydrous $FePO_4$ powder obtained by decomposition (holding at 80 °C in vacuum for 12 h) of the hydrothermally prepared dihydrate $FePO_4{\cdot}2H_2O$. The compound has a monoclinic crystal structure with unit cell dimensions $a \cong 5.48$, $b \cong 7.48$, $c \cong 8.05$ Å, $\beta \cong 95.7°$.

the long-range crystal structure, and the powder should be appropriately modified, if possible, to restore the crystallinity of the material.[13]

Provided the specimen produces a satisfactory diffraction pattern and in order to proceed with phase identification, both the positions and intensities of Bragg peaks should be determined. This is usually achieved by an automated peak-search procedure using the pattern, which has been processed to eliminate either or both the background and the $K\alpha_2$ components of the peaks. When the pattern is complex, automated peak-search procedures may result in numerous missed peaks, and then manual peak hunting or even semimanual profile fitting algorithms should be employed in order to obtain the list of Bragg reflections suitable for phase identification. The latter is accomplished by comparing the resulting list of Bragg reflections with one or more databases (e.g., the Powder Diffraction File™ maintained and distributed by the ICDD[14]) using different search-match algorithms.

If phase identification was the sole purpose of the experiment, and the obtained list of Bragg angles and intensities has no known match, the collected data can be used as a fingerprint of a new crystalline substance. The identity of the material should be confirmed using various experimental techniques (e.g., differential thermal analysis, spectroscopic and/or microscopic methods, and so on) to ensure that it is not a mixture of compounds. If phase identification was only the first step in a powder diffraction experiment, the next move is to proceed with the precise determination of lattice parameters. This is easier done when unit cell dimensions are known at least approximately, but they may also be established from first principles.

Since the highest possible accuracy of peak positions is essential, both the indexing of the pattern and lattice parameter refinement impose high demands on the quality of data, and overnight or weekend-long experiments should be performed. Profile fitting of powder data is typically conducted first, followed by assignment of indices to individual Bragg reflections. The latter can be accomplished using approximately known unit cell dimensions or in a process of the so-called ab initio indexing. Subsequent least-squares refinement usually yields precise unit cell dimensions. In some cases, when the suspected unit cell is quite accurate, lattice parameters can be obtained in a process known as the full-pattern decomposition.

When the unit cell is established and refined, it may mark the end of data processing, but it could also be just the beginning of a sequence, leading to a solution of the crystal structure from first principles. It is important to remember that the ab initio structure determination may be performed in two quite different ways. The first works in reciprocal space, where the intensities of Bragg reflections are employed to recover their phases by applying either heavy-atom (Patterson) or direct methods. Both phase angle recovery methods require highly accurate individual integrated

[13] Pattern C was collected from a molecular compound (Ph_3PCH_2COPh–Br) ball-milled for 1 h, which resulted in the complete loss of crystallinity. The appropriate processing in this case would be to slowly re-crystallize the compound using a suitable solvent and then gently grind the powder, if necessary, to prepare a specimen for a powder diffraction experiment.

[14] ICDD®—International Centre for Diffraction Data: https://www.icdd.com

intensities that may be obtained from exceptional quality data (usually a weekend-long experiment) via full-pattern decomposition. The second group of techniques involves simulating the crystal structure in real space, ranging from database searches and simple geometrical modeling to energy minimization and random structure generation. It may be necessary to employ multiple approaches, obtain a more accurate and better-resolved powder diffraction pattern, or examine different indexing solutions from the previous processing step before the crystal structure model is deemed acceptable. Once the crystal structure is solved, one final step—Rietveld refinement—is usually required to confirm and complete the structure.

The Rietveld method (see Sect. 15.7) is used both to finalize the crystal structure model—such as when it is necessary to locate a few missing atoms in the unit cell by coupling it with Fourier series calculations—and to confirm the crystal structure determination by refining positional and other relevant parameters of individual atoms, along with profile variables. The fully refined structural model must be both physically and chemically sound. If there are doubts, additional investigations should be carried out; they may include a better experiment and testing other feasible structural models in addition to employing various experimental techniques, such as chemical, thermogravimetric, spectroscopic, electron microscopic, neutron scattering, and others. When structure determination is completed, it should be followed by the calculation of bond lengths and bond angles, structure drawing, and preparation of the crystallographic data for publication in a suitable journal and/or in a database.

Referring once again to Fig. 13.1, the flowchart presented is quite generic and does not account for many other important applications of powder diffraction, such as microstructure analysis, determination of grain sizes and microstrain, low angle scattering, thin film structure analysis, texture analysis, and diffraction from polymers, fibers, liquids, and amorphous materials. Additionally, Rietveld refinement is widely used for quantitative phase and chemical analyses. The process illustrated in Fig. 13.1 should not be viewed as a rigid set of steps. Many more connections between the boxes can be drawn, depending largely on the type of the material, the complexity of its crystal structure, the quality of the experiment, the "noncrystallographic" knowledge about the material, and the major goal of the diffraction study.

13.2 Preliminary Data Processing

As we know (e.g., see Fig. 12.28), a typical powder diffraction pattern is collected in a form of scattered intensity as a numerical function of Bragg angle (also see Figs. 13.2 and 8.2–8.7). What is required in many applications of the method (e.g., phase analysis, unit cell dimensions, and structure solution) is a list of integrated intensities or equivalent absolute values of structure factors associated with the corresponding Miller indices of the reciprocal lattice points and the observed Bragg angles (e.g., see Table 8.1). Obviously, the availability of such a list eliminates the effects of multiple instrumental and specimen-related parameters from raw data, but its creation requires certain numerical processing of the pattern. Thus, raw powder diffraction data converted into the list of observed Bragg angles and intensities are known as the reduced or digitized powder diffraction patterns. It is worth noting that reduced patterns may be employed in crystal structure refinement but only in simple cases, when no substantial overlapping of Bragg reflections occurs.

In order to obtain a reduced pattern, scattered intensity maxima should be located and their relative intensities (either as peak height values or, most commonly, as peak areas) should be established together with the corresponding values of Bragg angles. Depending on the required accuracy and the availability of crystallographic information, for example, unit cell dimensions and symmetry, there are three common approaches to extracting intensities and positions of Bragg peaks from raw powder diffraction data:

- *Peak search*, which is unbiased by any kind of structural information. This approach is based on automatic recognition of Bragg peaks. The accuracy of peak positions, their intensities, and the completeness of the search are generally lower, especially when weak Bragg reflections are involved, and can vary, depending on both the algorithm used and quality of the raw data. A typical automatic peak search produces reduced patterns that are suitable for successful identification of phases when combined with an appropriate database search-match algorithm. This occurs because search-match utilities typically rely on only a few strong Bragg peaks, which are conclusively detected by most automatic peak-search algorithms. When the specimen is well-crystallized, the results may become suitable for a quantitative phase analysis, indexing, and even for lattice parameter refinement. However, the quality of a typical fast experiment is usually insufficient, and it may be necessary to conduct an overnight experiment. In many applications, an unbiased peak search requires additional preliminary processing of the data, including background subtraction, $K\alpha_2$ stripping, smoothing, and sometimes other corrections, such as Lorentz-polarization or conversion of variable slit data to a fixed slit scale. These additional processing steps may be necessary to improve the reliability of peak detection.
- *Profile fitting*, which is typically influenced by the user's decision or by the results of an automatic peak search whether a peak is present at a certain angle or not. Automatically found or manually marked Bragg peaks are fitted to experimental data using one of the many available peak-shape functions (Sect. 8.5.1) . The fit can be performed using the entire pattern at once or by splitting the pattern into several regions, which are processed separately. Free parameters in these fits usually include peak positions, their integrated intensity, and other relevant variables required to describe peak-shape functions. The latter may be peak-specific, or they may vary as certain analytical functions of Bragg angle and, therefore, become common for the entire pattern or only a selected range of Bragg angles. This approach is used when no information about the unit cell is available. It produces highly accurate observed peak positions ($2\theta^{obs}$) and integrated intensities (I^{obs}), provided the quality of experimental data is adequate. The obtained

reduced patterns may be used in a quantitative phase analysis, indexing from first principles and lattice parameter refinement, all with higher-than-average probability of success. Profile fitting results may be suitable for crystal structure determination; however, they can be problematic when diffraction patterns are exceedingly complex and contain numerous clusters of heavily overlapped Bragg reflections.

- *Full-pattern decomposition*, which is fundamentally biased by the chosen unit cell dimensions. It relies on fitting the whole powder pattern at once. In this method, positions of Bragg peaks are established from lattice parameters and symmetry. Only unit cell dimensions are refined, and the resulting peak positions are not "observed" per se, but rather they are calculated from the refined lattice parameters. Peak shapes are dependent on a few free variables in relevant analytical functions of Bragg angle, as was described in Sect. 8.5.1. The integrated intensities, however, are determined individually for each Bragg reflection. This method extracts quite reliable individual intensities and, hence, is a typical data processing step that precedes the structure solution from first principles, as discussed in Sect. 15.4 and illustrated in Chaps. 16–20. It is worth noting that the full-pattern decomposition approach is used increasingly often to obtain accurate lattice parameters when a powder pattern has been indexed but the crystal structure remains unknown or is not of interest for a specific application.Preliminary treatment of powder diffraction data and their conversion into reduced powder patterns for phase identification and a database search are nearly always included in data processing software suites, which are available with the purchase of a powder diffractometer. Perhaps this is the main reason explaining the lack of comparable freeware. We note that software developers use a range of data processing algorithms, and therefore, we will only be concerned with generic issues without getting into software-specific details, which may be found in the corresponding manuals.

13.2.1 Background

Background (e.g., see Figs. 13.2 and 13.3) is unavoidable in powder diffraction, and each powder pattern has a different level of background noise. The latter originates from inelastic scattering, scattering from air, sample holder and particle surfaces, X-ray fluorescence, incomplete monochromatization, detector noise, etc. As a result, the background must be accounted for, which is usually done by either subtracting it during preliminary processing of the data or adding its contribution (e.g., see Eq. (8.21)) to the calculated intensity, $Y(\theta)^{calc}$, during profile fitting.

Subtracting the background is considered mandatory during preliminary data processing before eliminating (stripping) $K\alpha_2$ contributions for a subsequent peak search. The background should be removed because $K\alpha_2$ stripping is based on the fixed $K\alpha_1$ to $K\alpha_2$ intensity ratio, which is 2:1 (see Sect. 6.2.1), in addition to Bragg angle splitting due to the difference in the wavelengths of these two components of

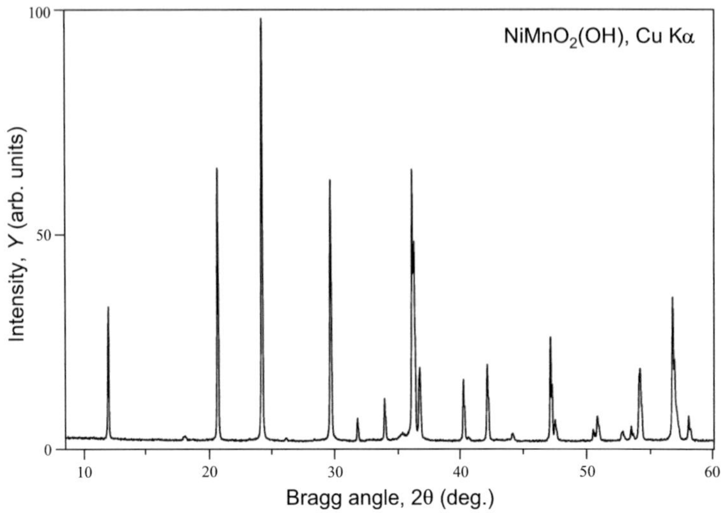

Fig. 13.3 Powder diffraction pattern of the orthorhombic $NiMnO_2(OH)$ collected on a Scintag XDS2000 powder diffractometer using Cu Kα radiation. In this example, the background noise contributes a few percent to the highest measured scattered intensity

the characteristic X-ray spectrum. The 2:1 intensity ratio includes elastic scattering on reciprocal lattice points, but not the background noise.

The background *should never be* subtracted prior to full-pattern decomposition and full profile-based Rietveld refinement.[15] In these cases, it is approximated using various analytical functions with coefficients, which are refined along with other parameters. Thus, computed background is then added to the intensity calculated as a function of Bragg angle.

In the majority of preliminary data processing algorithms, the background is handled in two different ways:

- Automatic, when the background points are chosen based on certain criteria and then these points are employed to approximate the background by means of certain smooth analytical function.
- Manual, when the user selects points that belong to the background (i.e., they do not belong to any of the visible Bragg peaks), which are then used in the same way as in automatic background removal.

[15] The only exception is the subtraction of a constant component of the background, i.e., the lowest observed intensity, to improve the visibility and enhance contributions from Bragg peaks, if required. Constant background subtraction can be described by the following analytical expression: $Y_i' = Y_i - min(Y_i) + 1$, where Y_i is the measured intensity, Y_i' is the intensity with constant background subtracted, unity is added to avoid zero intensity and a potential for a division by 0, and i varies from 1 to n, where n is the total number of collected data points. While doing so, one must remember that this subtraction will affect weights that are assigned to every data point during least-squares minimization (see Eqs. 14.60–14.62 in Sects. 14.12.1 and 15.5).

Both approaches have their advantages and disadvantages. Manual subtraction is slow but generally yields a more accurate background approximation, while an automatic algorithm is faster and easier but sometimes lacks the required accuracy. A typical background is enhanced at low Bragg angles. It has a broad minimum in the mid-angles range and then gradually increases toward high Bragg angles. Actual behavior, however, depends on many factors, such as the material of the sample holder, incident beam and monochromator geometry, sample chemistry and microstructure, and other parameters. In the powder diffraction pattern shown in Fig. 13.3, the background looks nearly linear but in reality, this is not the case as can be seen in Fig. 13.4, where the ordinate has been rescaled to reveal low-intensity details.

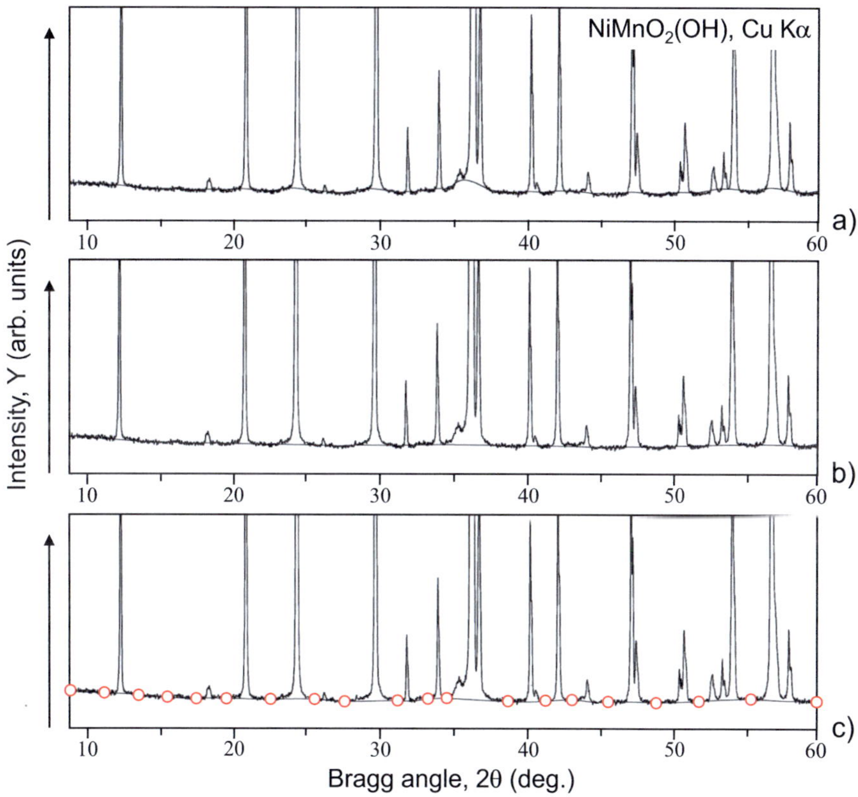

Fig. 13.4 The powder diffraction pattern from Fig. 13.3 shown with a different intensity scale. The background (*thin lines* at the *bottom* of each plot) has been approximated by means of (**a**) an automatic algorithm with the default box width (Width (in degrees 2θ) of the box (window) used in the Box Car Curve Fit method for background removal. The window determines how close the background curve follows the data. An underestimated box width results in humps under the peaks, as seen in (**a**). An overestimated box width translates in the background that does not follow the data. A proper box width (usually established experimentally) results in the background that follows the data closely without following the peaks, as seen in (**b**)) of 1.5°; (**b**) an automatic algorithm with a 4.0° box width; (**c**) background points selected manually as indicated by *small circles*

As seen in Fig. 13.4a, the background determined automatically using the default box width of 1.5° (see caption of Fig. 13.4) is far from the best choice in this case because greater than needed curvature results in accepting broad bases of the strongest peaks as the background. This distorts intensities of Bragg peaks, and therefore, Kα_2 stripping and other following treatments cannot be performed satisfactorily. In the next example (Fig. 13.4b), the background was also treated automatically, but this time with the box width increased to 4.0°. The result is a much better approximation but a careful examination of the bases of the strong peaks indicates that the background remains overestimated in these critical regions. Manual selection of points for a background approximation is shown in Fig. 13.4c, which appears to be the best choice for this particular powder diffraction pattern. Both examples b and c are suitable for a subsequent Kα_2 stripping. It is worth noting that neither the automatic nor the manual background subtraction works well when there is a substantial overlapping and clustering of Bragg reflections.

Once again, the background should be eliminated only when the next step involves the removal of the Kα_2 components or other unwanted peaks from the pattern to perform a peak search using algorithms that do not account for its presence (e.g., as in the DMSNT[16]). Other methods that also yield both peak positions and their intensities with a better precision (e.g., profile fitting) do not require Kα_2 stripping, and therefore, background subtraction in these cases should be avoided. The background is interpolated using various analytical functions, and their coefficients are refined simultaneously with other profile parameters, including structural parameters during Rietveld refinement. Hence, the background is accounted in a more flexible way, and experimental data are not subjected to an irreversible modification.[17]

The following functions are commonly used in background (b_i) interpolation,[18] where i varies from 1 to n, and n is the total number of data points measured in a whole powder diffraction pattern or in the region included in the processing:

- Polynomial function, which approximates the background as

$$b_i = \sum_{m=0 \text{ or } -1}^{N} B_m (2\theta_i)^m \tag{13.1}$$

- Here and below, B_m, are background parameters that can be refined and N is the order of the polynomial. The summation in Eq. (13.1) is usually carried out

---

[16] DMSNT—Data Management Software from Scintag Inc. (1999), later rebranded as WinXRD—Data Collection and Analysis Package from Thermo ARL. It is no longer supported. However, it is mentioned here because it was used to prepare some illustrations and examples in this text.

[17] Indeed, it is possible and always recommended to save a backup copy of the unmodified experimental data file. If for any reason, the background should be reevaluated, it is easy to do so, provided a copy of the original data file exists.

[18] Many other functions can be used to approximate the background. All of them should be continuous functions of Bragg angle in the processed range.

beginning from $m = 0$. However, to account for the often increased background at low angles, an additional (hyperbolic) factor with $m = -1$ may also be included.

- Chebyshev polynomial (type I or type II, either shifted or not). It is represented as a function of an argument x_i, which is defined as

$$x_i = \frac{2(2\theta_i - 2\theta_{\min})}{2\theta_{\max} - 2\theta_{\min}} - 1 \tag{13.2}$$

- where $2\theta_{\min}$ and $2\theta_{\max}$ are the minimum and maximum Bragg angles in the powder diffraction pattern. The background is calculated as:

$$b_i = \sum_{m=0}^{N} B_m t_m(x_i) \tag{13.3}$$

- and the Chebyshev function, t_m, is defined such that

$$t_{n+1}(x) + 2x \cdot t_n(x) + t_{n-1}(x) = 0 \tag{13.4}$$

- where $t_0 = 1$ and $t_1 = x$. Values of the function t are calculated using tabulated coefficients.
- Fourier polynomial, in which the background is represented as the following sum of cosines:

$$b_i = B_1 + \sum_{m=2}^{N} B_m \cos(2\theta_{m-1}) \tag{13.5}$$

- Diffuse background function to account for a peculiar scattering from amorphous phases (e.g., see Fig. 13.2, specimen C) or from a noncrystalline sample holder:

$$b_i = B_1 + B_2 Q_i + \sum_{m=1}^{N-2} \frac{B_{2m+1} \sin(Q_i B_{2m+2})}{Q_i B_{2m+2}} \tag{13.6}$$

where $Q_i = 2\pi/d_i$ and the "d-spacing" is calculated for each point, $2\theta_i$, of the powder diffraction pattern.

In Eqs. (13.1–13.6), the variable N typically varies from 6 to 12 when the entire powder diffraction pattern is of concern. In some instances, when profile fitting is applied to short fragments of the powder diffraction pattern, the most suitable background function is that given by Eq. (13.1) with N set to 1 or 2, representing a linear or parabolic background.

13.2.2 Smoothing

Smoothing is a numerical conditioning procedure employed to suppress statistical noise, which is present in any powder diffraction pattern as a result of random intensity measurement errors (Eq. 12.8). It improves the visual appearance of the powder diffraction pattern. For example, smoothing can make quickly collected data (say in a 15 min experiment) look similar to a pattern collected in a longer (e.g., in an overnight) experiment and may help with certain automatic procedures, such as background subtraction, $K\alpha_2$ stripping, and unbiased peak search.

Numerical conditioning, however, does not improve data quality. Moreover, it causes broadening of Bragg peaks and loss of resolution (e.g., see Figs. 12.22 and 12.23, which illustrate broadening caused by the varying receiving slit) and may result in the disappearance of weak peaks when overdone. On the contrary, increasing experiment time improves the pattern since it reduces statistical spreads.[19] An example of the original and smoothed patterns is shown in Fig. 13.5.

The most typical smoothing approach is often called box-car smoothing. It involves averaging intensities of current and neighboring data points using different

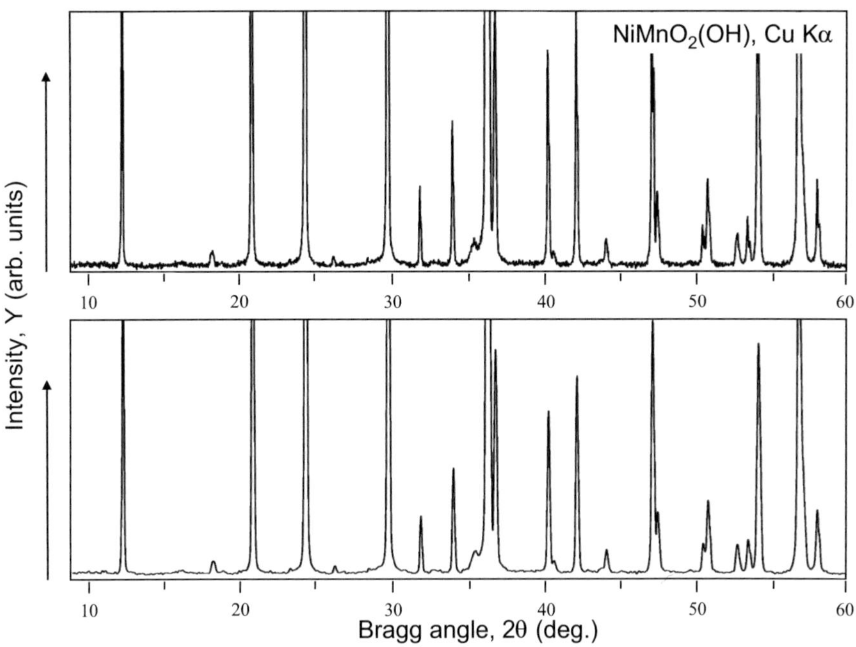

Fig. 13.5 The powder diffraction pattern from Fig. 13.3 with subtracted background prior to (*top*) and after (*bottom*) smoothing by using an analogue of Eq. (13.7) with seven sequential data points

[19] As described in Sect. 12.3.1, a twofold reduction in relative statistical errors requires a fourfold increase in data collection time.

weights. The weight is the largest for the point being smoothed, and it decreases rapidly for points located farther away. For example, when five points are employed, the weights (w_i) can be set at 1 for the point in the middle (Y_0), 0.5 for the nearest neighbors ($Y_{\pm 1}$), and 0.25 for the next nearest neighbors ($Y_{\pm 2}$).

Thus, the smoothed intensity of each point is the weighted sum of the intensities of five sequential data points divided by the sum of weights:[20]

$$Y_0^{smoothed} = \frac{w_2 Y_{-2} + w_1 Y_{-1} + w_0 Y_0 + w_1 Y_1 + w_2 Y_2}{2w_2 + 2w_1 + w_0} \tag{13.7}$$

where w_0, w_1, and w_2 are 1, 0.5, and 0.25, respectively. In this example, the weight of the next data point decreases by 2 as it moves away from the central point, though different dependencies can be applied. For instance, weights w_0, w_1, and w_2 could follow a linear progression of 1, 2/3, and 1/3, respectively. More complex, nonlinear weighting schemes, such as a Gaussian distribution of weights, may also be used. The number of points and smoothing coefficients (weights) can vary depending on the specific smoothing approach. Generally, the smoother pattern is obtained when more points are employed, or when the weights change less drastically from one point to the next.

Another commonly used smoothing approach is based on the fast Fourier transformation (FFT) algorithm. The number of Fourier coefficients in the original pattern is equal to the total number of the observed data points. In this case, the reverse Fourier transformation results in the original pattern. Each Fourier coefficient corresponds to a signal of specific frequency observed in the original pattern: the higher-order coefficients represent the higher-frequency signals. Thus, when high-order coefficients are set to zero or lowered, the reverse FFT produces a pattern similar to the original but with removed or reduced high-frequency noise, or in other words, a smoother pattern. Setting more high-order coefficients at zero produces stronger smoothing. As a result, the removal of high-frequency noise visually improves the pattern, but at the same time, more and more fine details (weak or narrowly split peaks) are lost. The loss of weak features in a pattern is a common problem in any smoothing algorithm.

Similar to background removal, smoothing *should never be* applied to a powder diffraction pattern that may later be used for profile fitting or Rietveld refinement.[21] Although smoothing might improve certain figures of merit (e.g., R_p, R_{wp}, and χ^2, see Eqs. (15.18, 15.19, and 15.22) in Sect. 15.6.2), it is likely to significantly distort lattice parameters and almost certainly alter all intensity-sensitive structural parameters, including coordinates, displacement, and population parameters. The only reliable and justifiable way to improve the true quality of the full-profile fit is to

[20] Except for the first (the last) few points in the pattern, where smoothing is either truncated to include only the points after (or before) Y_0, or the point remains "unconditioned."

[21] Unlike background removal, where the subtraction of a constant background may be permissible (see the Footnote 15 on page 386), this statement has no exceptions.

conduct a more accurate powder diffraction experiment through careful sample preparation and/or longer counting times.

13.2.3 *Kα_2 Stripping*

The presence of dual wavelengths in conventional X-ray sources, or in other words the presence of the Kα_2 component in both the incident and diffracted beams, complicates powder diffraction patterns by adding a second set of reflections from every reciprocal lattice point. They are located at slightly different Bragg angles when compared with those of the main (Kα_1) component. This decreases resolution and increases overlapping of Bragg peaks, both of which have an adverse effect on an unbiased peak search.

Since every Kα_1/Kα_2 double peak is caused by scattering from a single reciprocal lattice point, the *d*-spacing remains constant and the scattered intensity is proportional to the intensities of the two components in the characteristic spectrum. Using Braggs' law, the following equation reflects the relationship between the positions of the diffraction peaks in the doublet:

$$\sin\theta_1/\lambda_{K\alpha_1} = \sin\theta_2/\lambda_{K\alpha_2} \tag{13.8}$$

Further, the integrated intensities of the two peaks are related as:

$$I_{K\alpha_1} : I_{K\alpha_2} = 2 : 1 \tag{13.9}$$

Assuming that the peaks in the doublet have identical shapes, which is reasonable because the separation between the two is usually small, Eqs. (13.8 and 13.9) may be applied to every point of the observed peak shape, but not to the background. The Kα_2 stripping usually starts from the lowest Bragg angle point of the first observed peak, which should not be in the range of any other diffraction maximum, and moves toward the last point in the pattern. An example of Kα_2 stripping is shown in Fig. 13.6.

It is easy to see that this simple approach is far from ideal and the removal of the Kα_2 contributions is far from perfect. The inaccuracies occur because it is difficult to eliminate the background precisely. A higher quality pattern usually results in a better Kα_2 stripping, as illustrated in Fig. 13.6c where the same pattern was smoothed before removing the background and Kα_2 components. Visible peak broadening due to the smoothing treatment is noteworthy.

One of the unfortunate results of Kα_2 elimination is the distortion of the high-angle slopes of all Bragg reflections. Therefore, Kα_2 stripping is a valid step in the preparation of the powder pattern for the following automatic peak search, but it *should never be* performed as a part of data conditioning of the powder diffraction pattern for fitting and/or Rietveld refinement.

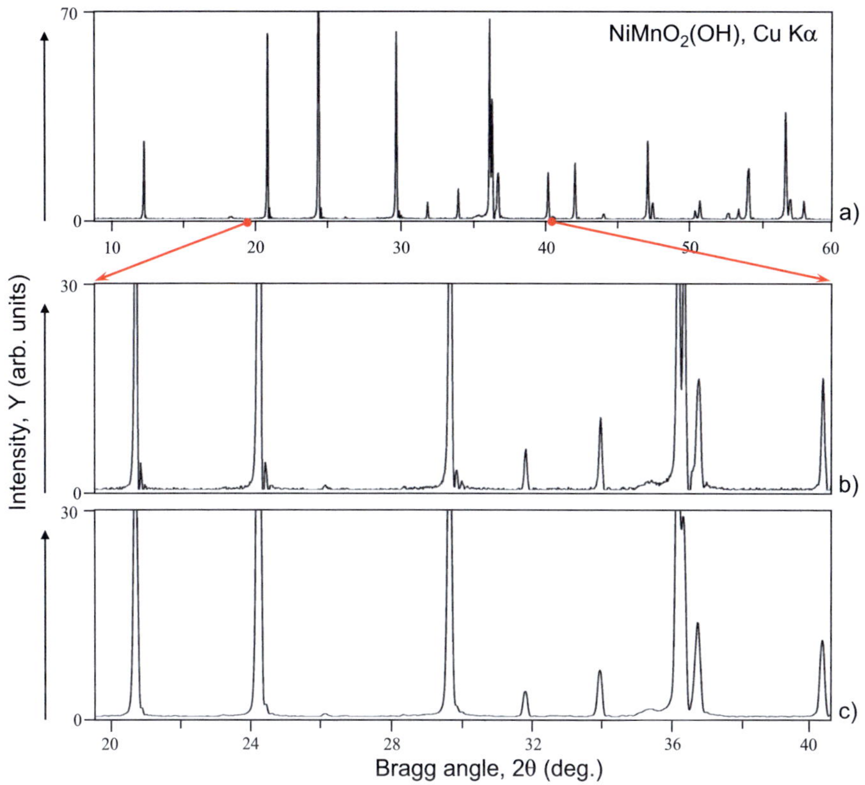

Fig. 13.6 An illustration of $K\alpha_2$ stripping (**a**). The expanded view of the range between ~20° and 40° 2θ highlighting inaccuracies on the high Bragg angle sides of the strongest peaks (**b**). The expanded view of the same region when the background subtraction and $K\alpha_2$ stripping were preceded by a 7 point-smoothing (**c**). Although inaccuracies on high Bragg angle sides are less visible after smoothing, the "improvement" is accompanied by the loss of resolution, which is easily seen from comparison of the two peaks at ~36° in (**b**) and (**c**)

13.2.4 Peak Search

A fast and reliable peak search or peak localization is essential for conducting both qualitative and quantitative phase as well as for performing a database search to match patterns. One of the most reliable (in terms of peak recognition), but far from the fastest techniques, is locating peaks manually, i.e., visually. This can be done in two ways: using the position of the peak maximum or the mid-point of the peak's full width at half maximum. Both cases require removing the $K\alpha_2$ contribution. If this is not done, the positions of both low-angle peaks (where $K\alpha_1$ and $K\alpha_2$ contributions essentially coincide) and high-angle peaks (where these components are nearly completely resolved) will be determined for the $K\alpha_1$ component. However, locations of peaks in the mid-angle range are determined somewhere between those of the $K\alpha_1$

and Kα_2 components. Ideally, they should correspond to the weighted average Kα wavelength (see Table 6.1), but in reality this is difficult to achieve, especially when peak tops are used as their positions. When Kα_2 components are stripped before locating peaks, this problem is avoided and the positions of all peaks correspond to the Kα_1 part of the characteristic spectrum.

An automatic peak search, being one-dimensional, represents the simplest case of the more general two- or three-dimensional image recognition problem. While image recognition is easily performed by the human eye and a brain, it becomes challenging to formalize and automate when random errors are present. As a result, many different approaches and methods have been developed. Two of the most commonly used methods in peak recognition, discussed here, are the second derivative method and the profile scaling technique.

The *second derivative method,* as is suggested from its name, involves calculating the derivatives, preceded by conditioning of the original patterns through a combination of background subtraction, Kα_2 stripping, and, if necessary, smoothing. This method is extremely sensitive to noise. Therefore, for fast-measured patterns with substantial random errors, smoothing is practically mandatory. The second derivative method consists of calculating first and then the second derivatives of $Y(2\theta)$ with respect to 2θ and utilizing them in the determination of peak positions. The derivatives can be easily computed numerically as:

$$\frac{\partial Y_i}{\partial 2\theta_i} = \frac{Y_{i+1} - Y_i}{s} \text{ and } \frac{\partial^2 Y_i}{(\partial 2\theta_i)^2} = \frac{Y_{i+2} - 2Y_{i+1} + Y_i}{s^2} \tag{13.10}$$

where Y_i, Y_{i+1}, and Y_{i+2} are the intensities of three consecutive data points and s is the data collection step. Instead of smoothing, it is possible to use a polynomial fit in the vicinity of every data point with the point in question located in the middle of a sequence. Once the coefficients of the polynomial are determined, both the first- and second-order derivatives are easily calculated analytically. For example, for a third-order polynomial

$$y = ax^3 + bx^2 + cx + d \tag{13.11}$$

the first and second derivatives, respectively, are

$$y' = 3ax^2 + 2bx + c \tag{13.12}$$

and

$$y'' = 6ax + 2b \tag{13.13}$$

where x is the Bragg angle. When the argument of the polynomial is chosen such that the point Y_i (where the derivatives are calculated) is at the origin of coordinates along the 2θ axis—i.e., when $x = 2\theta - 2\theta_i$—the corresponding derivatives simplify to c (Eq. 13.12) and $2b$ (Eq. 13.13).

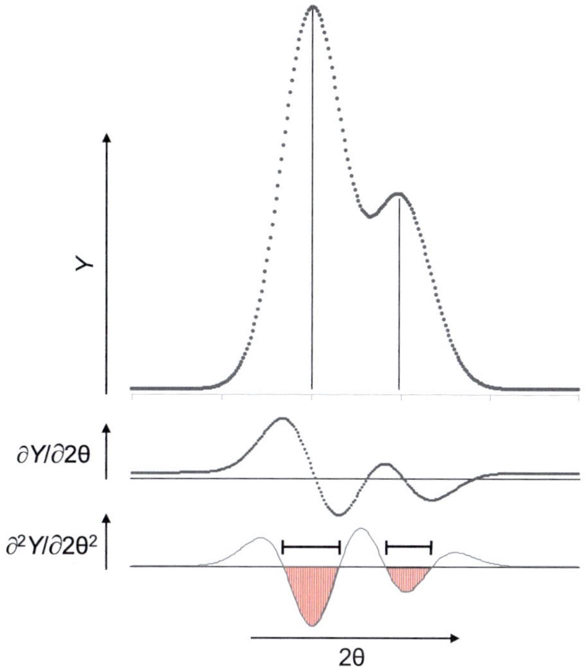

Fig. 13.7 Intensity distribution in two partially resolved Bragg peaks (*top*) and the corresponding first (*middle*) and second (*bottom*) derivatives. The second derivative forms series of sequential negative regions (*hatched*), which represent both the maximum of each Bragg peak (coinciding with the minimum of the corresponding negative region) and the estimate of peak width near its half maximum (the width of the associated negative sequence)

Some peak-search algorithms use the first derivative, which is reliable only for simple, well-resolved patterns. The second derivative method works better with complex data. An example in Fig. 13.7, top shows the profile representing two partially resolved Bragg peaks together with the first (middle) and second (bottom) derivatives. The first derivative is zero at the peak maximum and it changes sign from positive to negative when the Bragg angle increases. The second derivative reveals each peak more reliably as a sequence of negative values of the function, which are hatched in Fig. 13.7 for clarity.

To improve the detection, negative sequences in the second derivative are usually fitted to a parabolic function, thus resulting in a better precision of Bragg peak positions. The width of a Bragg peak can be estimated as the range of the associated negative region, since the second derivative changes its sign at each inflection point.

The example from Fig. 13.7 is simulated and therefore unaffected by noise. Thus, the ranges where the second derivative becomes negative can be detected with confidence. When processing real data, false peaks are often found, especially in the background regions. To avoid finding an excessive number of false peaks, an automatic peak search is usually coupled with several limiting parameters, which should be established empirically for a given quality and complexity of the data. For example, the minimum observed intensity above which the search is conducted excludes incompletely removed background; the minimum intensity above which a detected maximum would be considered as a peak excludes weak peaks; the minimum number of sequential negative values of the second derivative, which

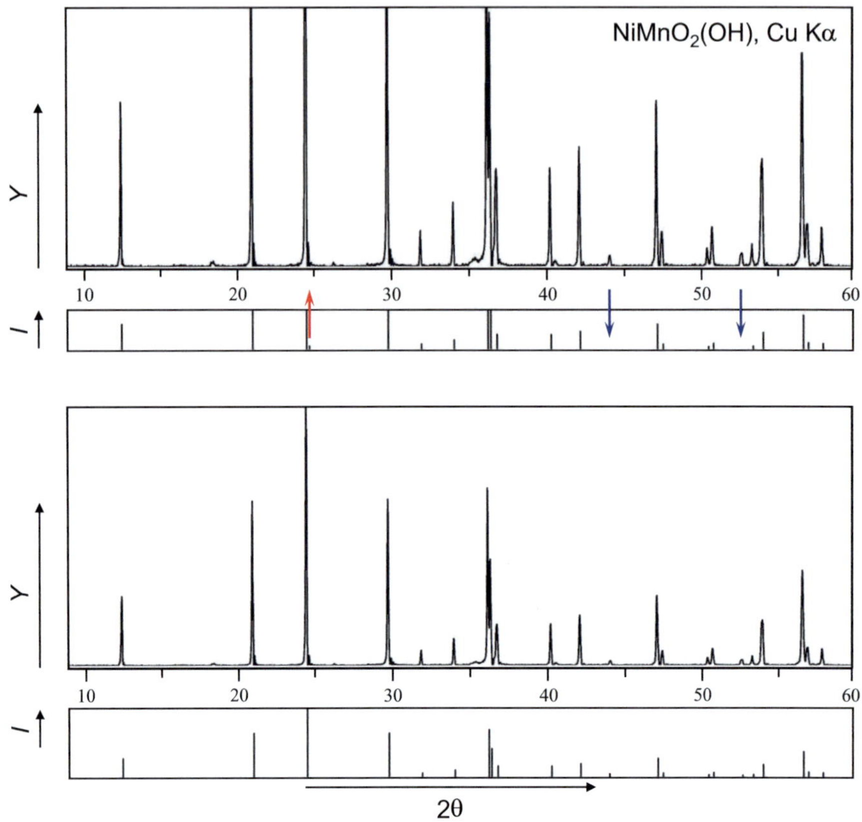

Fig. 13.8 Automatic peak search conducted using a second derivative method (*top*) and manually corrected reduced pattern (*bottom*). The red upward arrow placed on the digitized pattern shows a false peak (which was eliminated manually) and the *blue downward arrows* show the missed peaks (which were added manually)

would be considered as the manifestation of a peak, excludes noise. The results of an automatic peak search can be further improved by adding real and/or removing false peaks manually. An example of such a peak search is shown in Fig. 13.8.

The *profile scaling* peak-search algorithm employs a realistic analytical peak shape, for example, Pearson-VII or pseudo-Voigt functions (see Sect. 8.5.1). This approach does not require Kα_2 stripping or smoothing. Background subtraction may still be necessary if the fitting algorithm does not account for its presence. The simplest method uses the chosen peak-shape function with preset parameters, which may be adjusted manually. An improved approach is to use a well-resolved and strong Bragg peak to determine the peak-shape parameters, which are better suited to the actual data. The resulting normalized analytical shape is then moved along the diffraction pattern and its intensity (which is simply a multiplier, i.e., scale factor) is calculated by means of a linear least-squares technique (see Sect. 14.12.1) to produce the best fit. Regions, which meet certain criteria, are stored as observed peaks.

If necessary, an automatic peak search may be repeated using the difference between the observed data and the sum of profiles of all detected peaks. As a result, weak and/or poorly resolved Bragg reflections, missed in the earlier search may be found. This simple profile scaling method yields relatively accurate peak positions and integrated intensities and, if anticipated by the algorithm and realized in a computer code, their FWHMs. Its use is growing proportionally to the increasing computer speed. Since the search remains automatic, it may still require the adjustment of several empirical parameters to exclude the excessive appearance of false peaks or to improve the detection of weak Bragg reflections.

13.2.5 Profile Fitting

Profile fitting is the most accurate, although the slowest and the most painstaking procedure resulting in observed peak positions, full widths at half maximum, and integrated intensities of individual Bragg reflections. It is based on minimization of the difference between observed and calculated profiles using a nonlinear least-squares technique (see Sect. 15.5). The calculated profile is represented as a sum of scaled profiles of all individual Bragg reflections detected in the whole pattern or in any part of the pattern, plus an appropriate background function (see Sect. 13.2.1). Individual peak profiles are described by one of the common peak-shape functions, typically pseudo-Voigt or Pearson-VII (see Sect. 8.5.1). For conventional neutron diffraction data, a pure Gaussian function is typically employed.[22] Generally, three types of parameters can be optimized during the least-squares fit:

- *Peak positions* (2θ): which are normally refined for the $K\alpha_1$ components. If present, the locations of the $K\alpha_2$ constituents are determined by Eq. (13.8).
- *Peak-shape parameters*: These include full width at half maximum (H), asymmetry (α), and exponent (β) for Pearson-VII or mixing parameter (η) for pseudo-Voigt functions. All peak-shape parameters are typically refined for $K\alpha_1$ reflections, with the corresponding $K\alpha_2$ components assumed to have identical H, α, β (or η) values to $K\alpha_1$. In some applications, peak-shape parameters may be fixed at commonly observed values or adjusted manually.
- *Integrated intensity* (I): This serves as a scaling factor for each individual peak shape, see Eq. (8.21). Typically, the integrated intensity of the $K\alpha_1$ reflection is a free parameter and the intensity of the $K\alpha_2$ part (if present) is constrained as given in Eq. (13.9).

Overall, up to five parameters per diffraction peak can be refined: 2θ, I, H, α, and β (or η). In order to proceed with profile fitting using nonlinear least-squares

[22] It is worth noting that if the available software does not include a separate Gauss peak-shape function, it can easily be obtained from the pseudo-Voigt function by setting the mixing parameter, η, to 1.

refinement, all parameters should be assigned reasonable initial values. This is usually achieved in the following way:

- Approximate peak positions can be obtained through visual localization, from the automatic peak search, or calculated from known unit cell dimensions.
- Approximate peak-shape parameters can be preset to some practical or default values. They may be visually estimated from the pattern (the easiest is the full width at half maximum) and/or determined from a single, well-resolved strong Bragg peak.
- Approximate integrated intensity is easily established automatically: when only "scale factors" of individual peak shapes are of concern, a linear least-squares technique can be employed to find them relatively precisely (see Sect. 14.12.1 for a description of the method and Eq. (8.21), which indicates that the governing equations are indeed linear with respect to I).

All initial parameters can be approximate but they should be sufficiently precise to ensure that the nonlinear least-squares minimization converges. Approximate peak positions are among the most important, and they should fall within the range of each peak, or better yet, in the range of their full widths at half maximum. Usually this relatively vague localization of Bragg reflections is not a problem, even when peak tops are chosen to represent initial peak positions. However, when processing clusters of Bragg reflections that significantly overlap with each other, the approximate peak locations should be as precise as possible to ensure the stability of the least-squares minimization. In such cases, peak-shape parameters—and sometimes even the position of Bragg peaks—could be kept constant during the initial cycles of the least-square refinement and gradually released (included in the optimization) only after the refinement has stabilized and all other fitted parameters have been optimized.

Depending on the quality of the pattern, profile fitting can be conducted in several different ways. They differ in how peak positions and peak-shape parameters are handled, while assuming that integrated intensities are always refined independently for each peak, and that a single set of parameters describes a background within the processed range:

- All possible variables (positions and shapes) are refined independently for each peak or with some constraints. For example, the asymmetry parameter is usually a common variable for all peaks, while the full width at half maximum or even all peak-shape function parameters may be common for all peaks, especially if a relatively narrow range of Bragg angles is processed. When justified by the quality of data, an independent fit of all or most parameters produces the best results.

  However, a major problem with this approach, where all parameters are free and unconstrained, arises when clusters of reflections include both strong and weak Bragg peaks. In such cases, the peak-shape and other parameters for weak Bragg peaks may become unreasonable or even take on physically nonsensical values, such as negative intensities. Furthermore, when several strong reflections heavily overlap—typically when the difference in peak positions is only a small fraction of the full width at half maximum—their positions and especially integrated intensities strongly correlate. As a result, a nonlinear least-squares minimization may become unstable.

- Positions of Bragg peaks are refined independently, but the peak-shape function parameters are treated as corresponding functions of Bragg angle, and as above asymmetry variable is kept the same for all peaks. (see Sect. 8.5.1, Eqs. (8.22–8.25), and the following explanations). A major benefit of this approach is a more stable refinement of both the positions and intensities of weak Bragg peaks when they are randomly intermixed with strong reflections. However, a significant drawback is its inability to correctly determine peak-shape parameters when only weak peaks are present in the region being processed or when a few strong peaks are grouped together, which prevent the stable determination of relevant nonlinear dependencies over a broad range of Bragg angles.
- Peak locations are defined by refining the lattice parameters, while peak positions are calculated using Eqs. (8.2–8.7) (see Sect. 8.4.1). Peak-shape parameters are handled as described in item 2 above and rarely as in item 1. This approach is possible only when unit cell dimensions are at least approximately known. Therefore, it transitions from preliminary data processing to full-pattern decomposition using methods like Pawley or Le Bail, which are discussed later (Sect. 15.4). This refinement is often used to obtain accurate lattice parameters without employing other structural details. A major benefit of this approach is the relatively precise integrated intensities, which are used for solving the crystal structure from first principles (see Sect. 10.2). However, a significant drawback is that any full-pattern decomposition approach requires knowledge of the lattice parameters and symmetry, making it unsuitable for an unbiased determination of both the positions and integrated intensities of Bragg reflections.

Examples of profile fitting shown here were obtained using the DMSNT application. It employs two peak-shape functions: the Pearson-VII for symmetric peaks and the split Pearson-VII to treat the asymmetric peaks. All peak-shape parameters can be refined independently; all or any of them can be fixed. There is no mechanism to constrain peak-shape parameters, for example, to make some or all of them common for several peaks or to treat them as corresponding functions of 2θ. Therefore, in many cases when substantial peak overlapping is observed, and/or when data are of relatively low quality and resolution, profile fitting becomes unstable and does not converge. Moreover, background must be subtracted prior to profile fitting, as its refinement is not implemented.

The following examples were obtained using powder diffraction data collected from a polycrystalline sample of the orthorhombic polymorph of $NiMnO_2(OH)$. Data collection was carried out by using Cu Kα radiation on a Scintag XDS2000 powder diffractometer with a step $\Delta 2\theta = 0.02°$. The background was subtracted by manually specifying background points (see Fig. 13.4c) and Bragg reflections were located using an automatic peak search. No $K\alpha_2$ stripping/smoothing of the data had been performed.

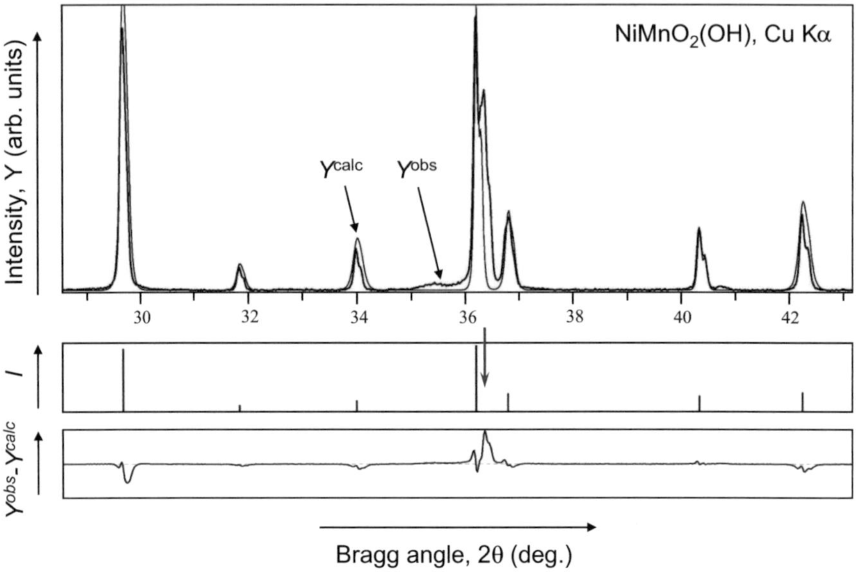

Fig. 13.9 Observed (*thick line*) and calculated (*thin line*) intensity profiles in a fragment of the powder diffraction pattern of $NiMnO_2(OH)$. The position of the missing peak is indicated by a *blue downward facing arrow*. Symmetrical Pearson-VII function with default peak-shape parameters was used in this example

Figure 13.9 illustrates initial observed and calculated profiles between 29 and 43° 2θ obtained without refinement of peak-shape parameters; integrated intensities were determined from linear least-square fitting. A symmetrical Pearson-VII peak-shape function, accounting for both $K\alpha_1$ and $K\alpha_2$ components, was employed. This and the following figures describing profile fitting consist of three parts: the box on top illustrates the observed and calculated patterns; the middle histogram shows positions and heights of the accounted Bragg peaks; and the bottom chart displays the difference between the observed and calculated profiles. The bottom box is of fixed height, and the difference profile is scaled to this height to clarify the details.

As indicated in Fig. 13.9, one strong peak in the middle (around 36.3°) had been overlooked during the automatic search. Its absence is easily detected from the analysis of the difference plot. The peak was included into the next step and the result is shown in Fig. 13.10.

Profile fitting was performed using both unit weights and weights based on statistical spreads of intensity data.[23] The resulting plots are shown in Figs. 13.11

[23]Weighting in both linear and nonlinear least squares is described in Sects. 14.12.1 and 15.5, respectively. When unit weights are employed, each data point contributes to the least squares solution equally. When weights are based on statistical spreads, this usually means that each data point is included into the least-squares minimization with the weight inversely proportional to the square of the corresponding statistical error. Thus, $w_i = 1/Y_i$ (also see Eq. (12.8)). Another weighting scheme commonly used in profile fitting is $w_i = 1/(Y_i + \sqrt{Y_i})$.

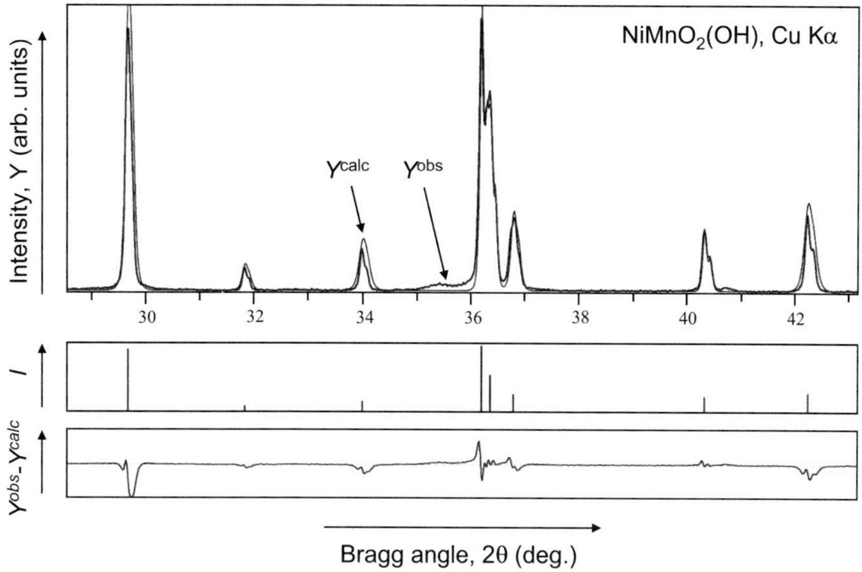

Fig. 13.10 Observed and calculated intensity in a fragment of the powder diffraction pattern of $NiMnO_2(OH)$ after adding the missing peak (compare with Fig. 13.9)

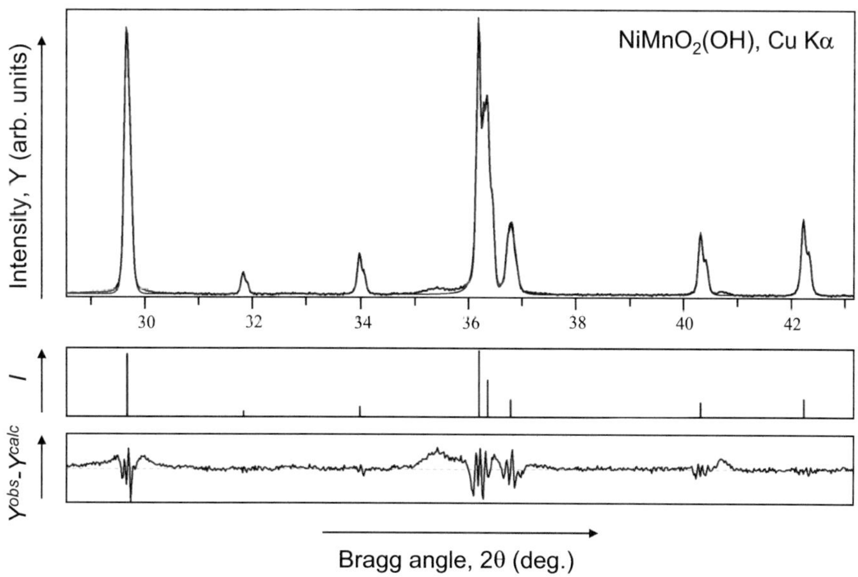

Fig. 13.11 Observed and calculated intensity in a fragment of the powder diffraction pattern of $NiMnO_2(OH)$ after least-squares refinement using unit weights (compare with Fig. 13.9)

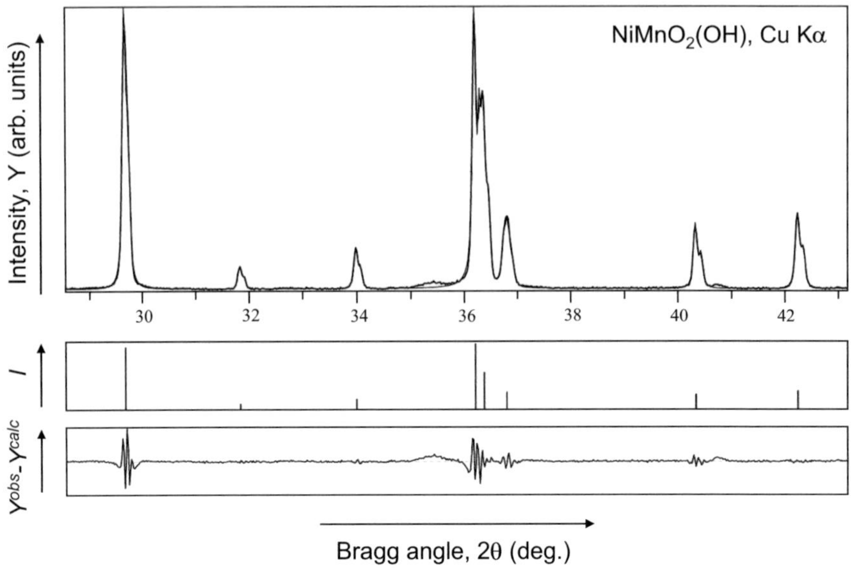

Fig. 13.12 Observed and calculated intensity in a fragment of the powder diffraction pattern of $NiMnO_2(OH)$ after least-squares refinement using weights inversely proportional to Y_i^{obs}

and 13.12, respectively. Unit weights result in a good fit only at peak tops. When the contributions of both low- and high-intensity data points have been equalized by introducing an appropriate weighting scheme, both peak tops and bases are represented equally well. It is easy to see by comparing Figs. 13.11 and 13.12, that weights based on intensity errors make the least-squares fit more reliable (remember that the scales in these two difference plots are different and they have been scaled to fit the entire chart for better visibility). This weighting scheme was retained through the end of the profile fitting.

The difference plots in Figs. 13.11 and 13.12 point to the presence of two broad peaks near 35° and 41° 2θ. The overall improvement after these peaks were included in the fit is shown in Fig. 13.13. We note that absolute differences between the observed and calculated profiles in the vicinities of strong reflections are usually greater when compared to those in the background and weak peaks regions. However, relative variances ($\Delta Y_i/Y_i$) do not differ substantially.

The $Y_i^{obs} - Y_i^{calc}$ under peaks changes sign several times, and they are distributed nearly evenly both above and below the zero-difference line. This behavior indicates that the discrepancies between Y_i^{obs} and Y_i^{calc} are due to random intensity errors. If this is the case, profile fitting is likely correct, and the discrepancies can be reduced (i.e., the quality of fit can be improved), if desired, by employing better quality data, for example, collected in a longer experiment. Distinct nonrandomness in the distribution in $Y_i^{obs} - Y_i^{calc}$ under the peaks points to one or several problems with profile fitting, which may be due to the wrong peak-shape function, excessive asymmetry when axial divergence was too high, a small amount of second phase

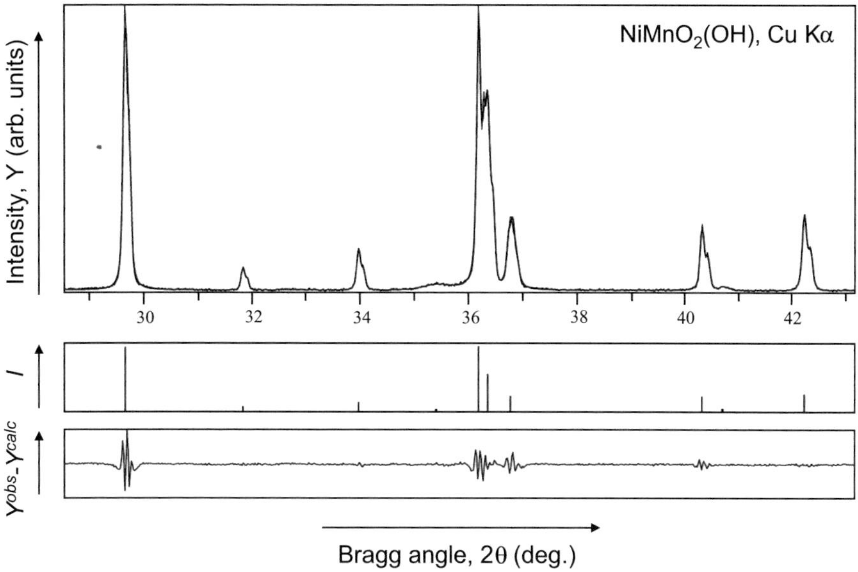

Fig. 13.13 Observed and calculated intensity in a fragment of the powder diffraction pattern of $NiMnO_2(OH)$ after adding two broad peaks at 2θ about 35.4 and 40.8°

if the sample was inhomogeneous or, perhaps, peak splitting that manifests pseudosymmetry. Then, the measured Bragg reflections' profiles should be visually examined and analyzed to identify the reasons and properly account for the observed systematic discrepancies.

Examples of fitting a whole profile simultaneously are shown in Figs. 13.14 and 13.15. The former depicts the observed and calculated intensities after initial fitting has been performed using a Pearson-VII function, proper weights ($w_i = 1/Y_i$), and peak positions found during an automatic peak search. The difference plot clearly reveals six weak peaks that were overlooked. The latter figure illustrates the result after all distinguishable peaks have been included in the fit.

To illustrate small differences that may occur in profile fitting, consider the example shown in Fig. 13.16. Here, the fit was performed using the same experimental data but with a different algorithm implemented in the WinCSD[24] software. This algorithm uses a different peak-shape function (pseudo-Voigt) and it also allows refinement of a whole pattern or any fraction of the pattern. The full width at half maximum can be constrained to a single parameter common for all peaks within the range, or it may be refined for each peak individually. The asymmetry (α)

[24] L.G. Akselrud, P.Yu. Zavalii, Yu.N. Grin, V.K. Pecharsky, B. Baumgartner, E. Wolfel, "Use of the CSD program package for structure determination from powder data", Mater. Sci. Forum **133–136**, 335 (1993). https://doi.org/10.4028/www.scientific.net/MSF.133-136.335; L. Akselrud, Yu. Grin, WinCSD: software package for crystallographic calculations (Version 4), J. Appl. Crystallogr. **47**, 803 (2014). https://doi.org/10.1107/S1600576714001058

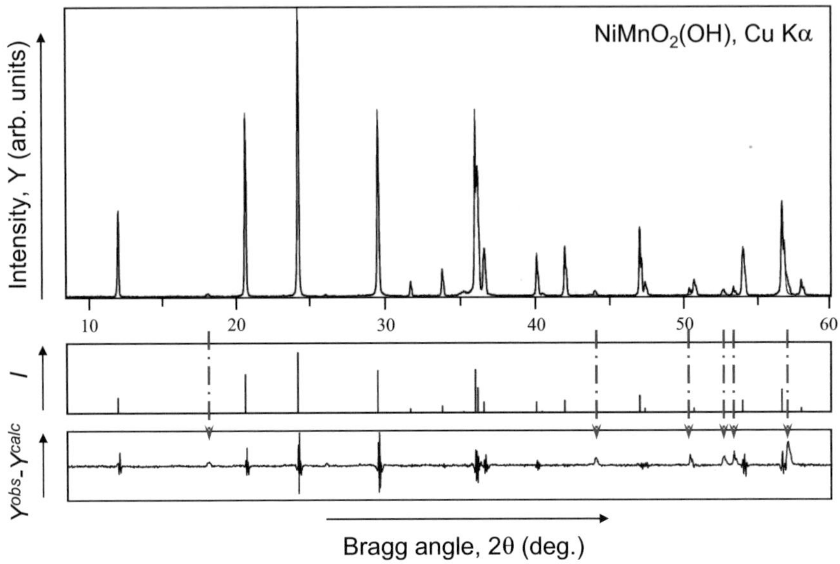

Fig. 13.14 Observed and calculated intensity in the powder diffraction pattern of $NiMnO_2(OH)$ after fitting using a Pearson-VII function. Downward facing dash-dotted arrows indicate the positions of six weak Bragg peaks not included in the fit

and mixing parameter (η) are always common for all peaks within the range included into the fit. The background is refined using a polynomial function. In this particular example, full widths at half maximum were refined separately for each individual peak, while α and η were common for all peaks.

The values of the observed Bragg angles, full widths at half maximum, and integrated intensities obtained using both the DMSNT and WinCSD programs are assembled in Table 13.1. The differences in Bragg angles and intensities of sharp peaks are statistically insignificant. The average FWHM varies from ~0.08° at the beginning to ~0.14° at the end of the pattern. Thus, the broad peak at $2\theta \cong 35.4°$ likely belongs to an impurity phase.[25] Full widths at half maximum of several weak Bragg reflections are noticeably inflated, which often happens when weak peaks are fitted using individual FWHMs. As seen in Fig. 13.17, the general trend in the two distributions of FWHM as functions of Bragg angle is normal. However, the deviations from dependencies discussed in Sect. 8.5.1 are substantial, especially for weak peaks. The spread in FWHMs can be reduced by increasing the quality of the data (for example, by improving counting statistics) and/or by improving the stability of nonlinear least squares using a choice of FWHM constraints.

[25] Weak Bragg reflection at $2\theta \cong 18.1°$ also belongs to an impurity, present in this powder sample. Its full width at half maximum determined using DMSNT is nearly identical to those of neighboring peaks, while WinCSD indicates a much broader peak shape.

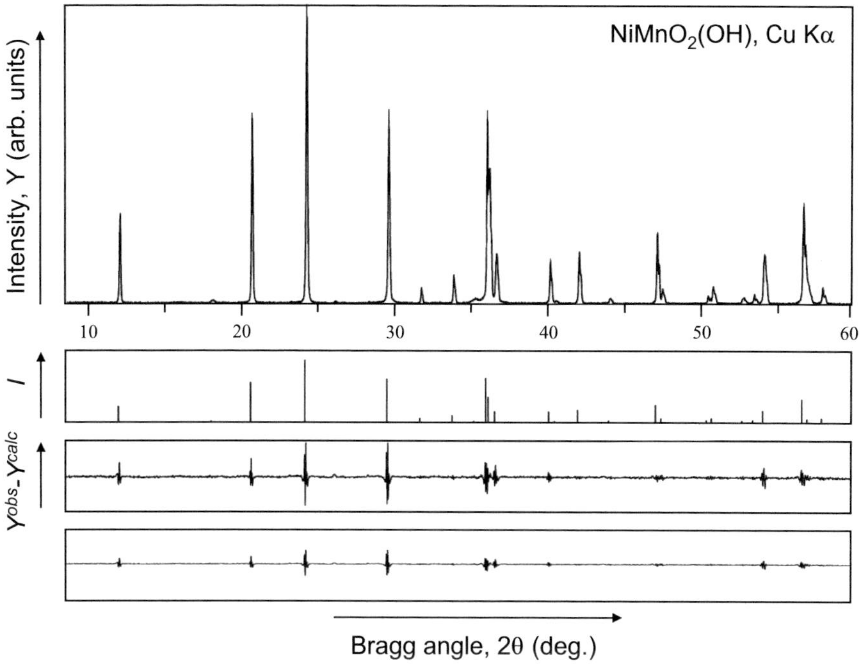

Fig. 13.15 Observed and calculated intensity in the powder diffraction pattern of $NiMnO_2(OH)$ after the completion of profile fitting using the DMSNT algorithm. A symmetrical Pearson-VII function was employed and all present Bragg peaks were included in the fit. The box at the bottom shows the difference between the observed and calculated intensities using the same scale as on the plot of both Y^{obs} and Y^{calc}

The most significant differences in the algorithms employed in both DMSNT and WinCSD are listed in Table 13.2. Overall, DMSNT results in slightly narrower peaks than WinCSD, while the difference in peak positions is only in thousandths of a degree. The largest observed difference is 0.009° for the first peak (disregarding the impurity peaks at $2\theta \cong 18.1$ and ~35.4°), most likely due to a different treatment of asymmetry.[26] The stability of the refinement is better in WinCSD because fewer parameters per peak vary independently.

The examples provided here are intended to illustrate the general approach to profile fitting and the analysis of both the digital and graphical results. The analysis of graphical data after profile fitting is nearly identical to that used during Rietveld refinement and full-pattern decomposition, which is discussed in a greater detail starting in Chap. 15. Since DMSNT and WinCSD are no longer available or

[26] Profile fitting using DMSNT has been performed employing a symmetrical Pearson-VII function. Thus, asymmetry, which has the greatest effect on low Bragg angle peaks, was not accounted for. In contrast, WinCSD employs a simple yet effective asymmetry model (as long as the asymmetry is not severe), where full widths are different on both sides of the peak maximum, see Sect. 8.5.2.

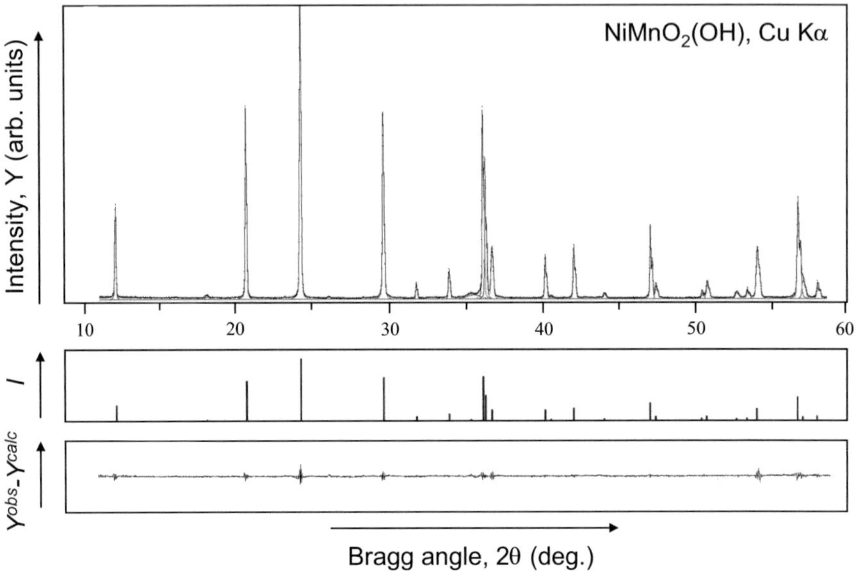

Fig. 13.16 Observed and calculated intensity in the powder diffraction pattern of $NiMnO_2(OH)$ after the completion of profile fitting employing the WinCSD algorithm. Pseudo-Voigt function was employed and all present Bragg peaks were included in the fit. The box at the bottom shows the difference between the observed and calculated intensities using the scale identical to that on the plot of both Y^{obs} and Y^{calc}

supported, we encourage readers to explore other tools that best suits their needs. In addition to the commercial software mentioned earlier (see Footnotes 4–8 on page 378), profile fitting can be performed using standalone freeware software such as GSAS/EXPGUI[27] and GSAS-II,[28] Rietica,[29] FullProf,[30] FOX,[31]

[27] A.C. Larson and R.B. Von Dreele, GSAS: General structure analysis system. LAUR 86–748 (2004); B. H. Toby, EXPGUI, a graphical user interface for GSAS, J. Appl. Cryst. 34, 210–213 (2001), https://doi.org/10.1107/S0021889801002242; available at https://subversion.xray.aps.anl.gov/trac/EXPGUI/wiki/GSASIntro

[28] B.H. Toby and R.B. Von Dreele, GSAS-II: the genesis of a modern open-source all purpose crystallography software package, **46**, 544 (2013). https://doi.org/10.1107/S0021889813003531; available for download from https://advancedphotonsource.github.io/GSAS-II-tutorials/

[29] LHPM-Rietica (authors B.A. Hunter and C.J. Howard) may be downloaded from http://www.rietica.org/

[30] https://www.ill.eu/sites/fullprof/

[31] V. Favre-Nicolin, R. Ĉerný, FOX, "free objects for crystallography": a modular approach to ab initio structure determination from powder diffraction, J. Appl. Cryst. **35**, 734 (2002). https://doi.org/10.1107/S0021889802015236; R. Černý, V. Favre-Nicolin, J. Rohlíček, and M. Hušák, FOX, Current State and Possibilities, Crystals, **7**, 322 (2017), https://doi.org/10.3390/cryst7100322; can be download from https://objcryst-fox.readthedocs.io/en/latest/index.html; can be downloaded from http://jana.fzu.cz/

Table 13.1 Results of profile fitting of the powder diffraction pattern of $NiMnO_2(OH)$ using DMSNT and WinCSD software as shown in Figs. 13.15 and 13.16, respectively

DMNST				WINCSD		
2θ (°)	*Y*(max)[a]	FWHM (°)	Area[b]	2θ (°)	FWHM (°)	Area[b]
11.998	235	0.084	226	12.007	0.087	236
18.098[c]	**10**	**0.089**	**15**	**18.095**[c]	**0.221**	**16**
20.678	641	0.066	544	20.683	0.082	534
24.219	1000	0.080	1000	24.222	0.097	1000
29.624	678	0.074	656	29.625	0.095	646
31.789	53	0.083	46	31.792	0.075	45
33.937	106	0.072	103	33.939	0.089	96
35.395[c]	**11**	**0.708**	**118**	**35.533**[c]	**1.070**	**120**
36.144	672	0.079	687	36.144	0.088	626
36.312	397	0.086	359	36.312	0.091	419
36.726	160	0.126	246	36.728	0.138	237
40.257	156	0.087	159	40.259	0.092	157
40.637	8	0.142	15	40.641	0.111	9
42.159	192	0.088	210	42.159	0.098	203
44.148	18	0.137	36	44.157	0.134	25
47.212	270	0.083	277	47.214	0.092	268
47.570	52	0.100	72	47.566	0.122	64
50.567	28	0.095	31	50.568	0.097	29
50.893	62	0.119	87	50.893	0.128	86
52.834	20	0.174	36	52.840	0.192	43
53.557	34	0.092	46	53.555	0.117	40
54.213	176	0.155	292	54.213	0.162	312
56.864	367	0.113	513	56.864	0.127	503
…	…	…	…	…	…	…

[a] Peak heights normalized to 1000
[b] Area represents integrated intensity; normalized to 1000
[c] Impurity peaks are shown in **bold**

JANA,[32] and others. Regardless of the specific software, the core principles and concepts discussed in this chapter apply to most programs designed for both preliminary processing and the further analysis of powder diffraction patterns.

[32] V. Petříček, L. Palatinus, J. Plášil, and M. Dušek, Jana2020—a new version of the crystallographic computing system Jana, Z. Kristallogr, **238**, 271 (2023), https://doi.org/10.1515/zkri-2023-0005

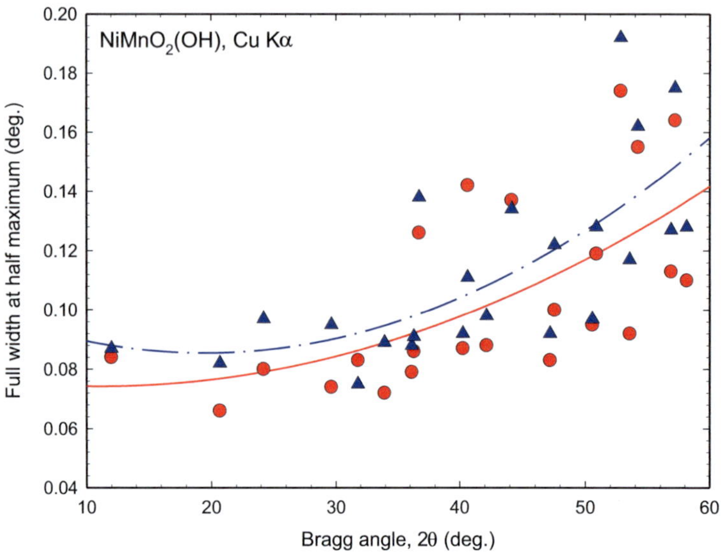

Fig. 13.17 The distribution of full widths at half maximum as a function of Bragg angle obtained using DMSNT (*red circles, solid line*) and WinCSD (*blue triangles, dash-dotted line*) algorithms. The lines represent parabolic fits of the two sets of data to illustrate the trend

Table 13.2 Major differences in profile fitting algorithms realized in DMSNT and WinCSD

Property	DMSNT	WinCSD
Peak-shape function	Pearson-VII or split Pearson-VII	Pseudo-Voigt
Parameters per peak	4 (Pearson-VII) or 6 (split Pearson-VII)	3
Common parameters	0	5
Background	Must be subtracted	Refined as a second-order polynomial
FWHM	Individual	Individual or common for all peaks
Exponent (β)	Individual	N/A
Mixing parameter (η)	N/A	Common for all peaks
Asymmetry	Individual using split Pearson-VII	Common for all peaks

13.3 Phase Identification and Quantitative Analysis

Each powder diffraction pattern is characterized by a unique distribution of both Bragg peak positions and intensities. Peak positions are defined by the unit cell dimensions, while reflection intensities are established by the distribution of atoms within the unit cell of each crystalline phase present in the sample (see Table 8.2). As a result, every individual crystalline compound has its own "fingerprint" enabling the use of powder diffraction data in phase identification.[33]

A digitized representation of powder data is compact and particularly useful for comparison with other patterns, provided an appropriate database is available. Each entry in such a database typically includes not only the digitized pattern but also symmetry information and other valuable information such as phase name, chemical composition, references, basic physical and chemical properties, and sometimes crystal structure. Powder diffraction databases are extensively used for both simple identification of compounds (qualitative analysis) and the quantitative determination of crystalline phases in a mixture (quantitative analysis).

13.3.1 Crystallographic Databases

Phase identification using powder diffraction data requires comparing several key features[34] of its digitized pattern with those of known compounds or phases. This is typically done by searching a powder diffraction database for records that match experimentally measured and digitized pattern. Therefore, access to a powder diffraction database or a relevant subset, along with an appropriate search-match tool, is essential.

The most comprehensive and widely used powder diffraction database is the Powder Diffraction File™ (PDF), maintained and continuously updated by the International Centre for Diffraction Data (ICDD®). The PDF is a commercial database, and information about both the ICDD and the Powder Diffraction File is available online.[35] This database is quite unique in that it contains either or both the experimentally measured and calculated digitized powder patterns for hundreds of thousands of compounds, including minerals, metals and alloys, inorganic materials, organic compounds, and pharmaceuticals. The PDF is available as a whole or in

---

[33] The diffraction pattern from a single crystal is also unique but due to complexity of a three-dimensional distribution of intensities, phase recognition based on the pattern is much more difficult to formalize, and instead, the unit cell dimensions and symmetry are commonly used as search parameters. Powder data are one-dimensional, and they can be converted into digitized patterns, which are in a way unique barcodes enabling automated pattern recognition.

[34] Most often search-match procedure is accomplished by using positions of the strongest Bragg reflections with largest d-spacing, as these are least affected by variations in data collection and processing parameters and are the most unique representatives of the specific diffraction pattern. A method known as the "Hanawalt search," in which the initial match is based on the three strongest Bragg peaks present in a powder diffraction pattern [J.D. Hanawalt, H.W. Rinn, and L.K. Frevel, Chemical analysis by X-ray diffraction, Ind. Eng. Chem., Anal. Ed. **10**, 457 (1938), https://doi.org/10.1021/ac50125a001], remains in use today. For more details, see also R. Jenkins and R.L. Snyder, Introduction to X-ray powder diffractometry, Wiley, NY (1996). https://doi.org/10.1002/9781118520994

[35] https://www.icdd.com

subsets. Historically, each record in the database was called a "card,"[36] but the terms "entry" or "record" are now commonly used too.

In recent decades, the database has been expanded to include full-profile experimental data in addition to the digitized patterns. It also includes atomic coordinates for structures refined using the Rietveld method. Beyond X-ray and fixed-wavelength neutron diffraction data, the database now also hosts time-of-flight neutron data and simulated electron diffraction patterns. Additionally, searchable property subfiles have been added, making the PDF a valuable resource for both data mining and materials exploration. One of the most valuable features of the PDF's development is the inclusion of data from other databases, particularly single-crystal data from ICSD and CSD. This data is used to calculate digitized powder patterns on-the-fly, which can then be used in the search-match procedure for phase identification.

An example of a PDF record is shown in Fig. 13.18, with the major fields or blocks numbered as following:

1. PDF card number: Displayed as either two or three numbers separated by dashes, depending on whether the old or new notation scheme is used.[37]
2. Quality: A brief description (historical) of the data quality, data source, or processing type, such as: Star (*)—the highest quality; Low Precision; Indexed; Calculated; Rietveld; or Blank—if unspecified.
3. General information: Includes the chemical, moiety, or structural formula and names, as well as mineral or other names, if any.
4. Experimental conditions: Information such as radiation, wavelength, and other experimental details.
5. Crystallographic data: Includes the crystal system and space group, unit cell dimensions, and other related data such as calculated and measured gravimetric density, F_N ($N \leq 30$) indexing figure of merit (see Sect. 14.4.1), and unit cell volume in Å^3.

[36] Early versions of the Powder Diffraction File (also known as the JCPDS file) were distributed on index cards. The first edition of the file dates back to the 1941 release containing 4000 cards describing powder diffraction patterns of ~1300 compounds. It was compiled by the Joint Committee on Chemical Analysis by X-ray Powder Diffraction Methods and published by the American Society for Testing and Materials (ASTM). In 1969, the Joint Committee on Powder Diffraction Standards (JCPDS) was registered as a Pennsylvania nonprofit corporation and the current name (International Centre for Diffraction Data) was adopted in 1978. See W. Wong-Ng, H.F. McMurdie, C.R. Hubbard, and A.D. Mighell, JCPDS-ICDD research associateship (cooperative program with NBS/NIST), J. Res. Natl. Inst. Stand. Technol. **106**, 1013 (2001). https://doi.org/10.6028%2Fjres.106.052

[37] The PDF card (record, or entry) number consists of two parts: set number (48 in this example) and sequential number of the entry in the set (1152 for this particular card). Recently, with the integration of several databases into the ICDD's files, an additional two-digit prefix has been added. This prefix is coded as following: 00 stands for a native ICDD record, 01 refers to ICSD data, 02 to CSD, 03 to NIST, and 04 to Pauling File. The new, full reference for this card, therefore, becomes 01-048-1152.

(1) 48-1152 Quality: Indexed (2)

(3)
Li0.6 V1.67 O3.67 ! H2 O
Lithium Vanadium Oxide Hydrate

(4)
Rad:CuKa1 Lambda:1.54056 Filter: d sp:Diffractometer
Cutoff: Int:Diffractometer I/Icor:
Ref:Whittingham, M., SUNY at Binghamton, MaterialsResearch Center, NY, USA.Chyrayil, T., Zavalij, P., Whittingham, M., (1

(5)
Sys:Tetragonal S.G.:I4/mmm
a:3.7047±0.0003 b: c:15.804±0.002
α: β: γ: Z:2 mp
Ref2
Dx:2.53 Dm:2.541 SS/FOM: F30=46.5(0.0161,40) Volume[CD]:216.91

(6)
εα: ηωβ: εγ: Sign: 2V:
Ref3

Color:

(7)
Prepared by hydrothermal treatment of tetramethylammonium hydroxide, vanadium pentoxide and \Li O H\ acidified to pH 2-5 for 3 days at 200 C. Pattern taken at 23(1) C.

32 reflections in pattern.

(8)

2θ	Int.	h k l	2θ	Int.	h k l	2θ	Int.	h k l	2θ	Int.	h k l
11.2026	100	0 0 2	50.5721	8	0 2 2	72.0262	4	2 2 0	83.7228	1	0 1 13
22.4967	19	0 0 4	54.6668	3	0 2 4	73.1843	2	2 2 2	84.1343	1	0 3 5
24.6618	9	0 1 1	55.7443	2	1 2 1	76.5173	1	2 2 4			
29.4652	50	0 1 3	58.0669	3	0 1 9	77.4598	1	0 3 1			
33.9955	1	0 0 6	58.3367	13	1 2 3	79.4091	2	1 2 9			
34.2095	14	1 1 0	58.3367	13	0 0 10	79.6864	4	0 3 3			
36.0710	1	1 1 2	58.4543	4	1 1 8	79.6864	4	0 2 10			
37.3772	4	0 1 5	63.3383	3	1 2 5	81.7407	2	1 1 12			
47.1058	19	0 1 7	69.4008	10	1 1 10	82.1813	2	1 3 0			
49.1443	16	0 2 0	70.4377	7	1 2 7	83.3159	1	1 3 2			

(9)

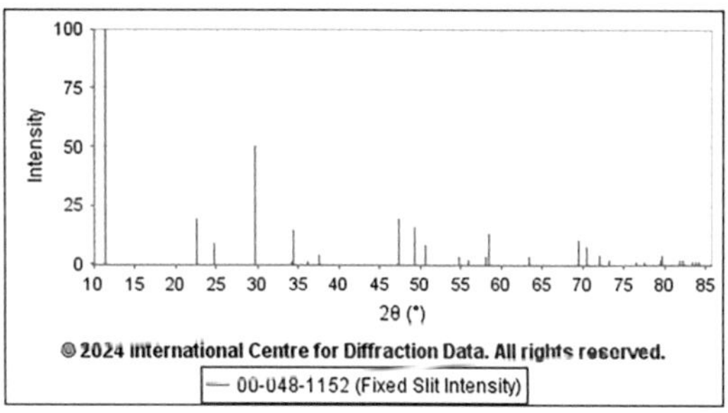

Fig. 13.18 An example of a record extracted from the ICDD Powder Diffraction File (The output shown here was obtained using LookPDF routine from the DMSNT applications. Other programs may display PDF cards in different format)

6. Properties and references: Relevant properties and their corresponding references, if any.
7. Comments: Includes the source and preparation of the compound, as well as preparation conditions like temperature and pressure.
8. Digitized pattern table: A table of diffraction peaks with *d*-spacing or 2θ angle, intensity normalized to 100 or 1000, and, if known, Miller indices (*hkl*).

Digitized pattern histogram: A histogram showing intensity versus Bragg angle (2θ). The ICDD's PDF is well-suited for identifying digitized powder patterns, and many manufacturers of powder diffractometers offer optional software for searching this database. However, the PDF is not a complete database—something nearly impossible to achieve. The information included in the PDF is primarily collected from published powder data and from records produced upon ICDD request or through the Grant-in-Aid program.[38] At the time of writing this book, the ICDD database exists in three formats:

- PDF-2: A classic format database containing both experimental and calculated powder patterns. It allows for search-match using digitized powder patterns and also provides limited search capabilities in other fields.
- PDF-4: A cost-effective relational database distributed in several subsets (see Table 13.3) and provides searchable access to all data fields.
- PDF-5: A comprehensive database with over 1.1 million entries, containing both powder and single-crystal data, designed for rapid phase identification and automated quantitative analyses.

In addition to its vast number of included entries and comprehensive quality control, the usefulness of the Powder Diffraction File is established by its ability to perform searches based strictly on the digitized patterns—without prior knowledge of unit cell dimensions or other crystallographic and chemical information. Similar searches can also be conducted using several different existing databases, such as the Pauling File and Mineralogy Database, and possibly a few others (see Table 13.3); however, these are not as comprehensive as the PDF. For example, the Pauling File is underdeveloped concerning multinary compounds, while the Mineralogy Database is dedicated to naturally occurring and synthetic minerals. More detailed information (as of 2002) about a variety of crystallographic databases can be found in a special joint issue of *Acta Crystallographica*, Sections B and D (see also references 6–13 in Sect. 13.4).[39]

When experimental data remain unidentified after a digitized pattern-based search-match, different databases should be checked before concluding that a material is new. Continuing searches, however, usually require unit cell dimensions, so the powder pattern should be indexed before proceeding. Various databases are dedicated to different classes of compounds and contain different types of information, as shown in Table 13.3.

[38] ICDD makes limited funds available to researchers interested in processing and submitting new experimental patterns for incorporation into the Powder Diffraction File. More information about the ICCD's Grant-in-Aid program can be found at https://www.icdd.com

[39] Acta Crystallographica is an international journal published by the International Union of Crystallography in five sections: Section A (Foundations of Crystallography); Section B (Structural Science); Section C (Crystal Structure Communications); Section D (Biological Crystallography); and Section E (Structure Reports Online). Special joint issue: Acta Cryst. **B58**, 317–422 (2002) and Acta Cryst. **D58**, 879–920 (2002), https://journals.iucr.org/b/issues/2002/03/01/

Table 13.3 Selected computer searchable crystallographic databases

Database	Content/compounds	No. of entries[a]
ICDD[b]—Powder Diffraction File	PDF-2, experimental and calculated patterns	359,400+ 252,600+ with I/Ic
	PDF-4/Axiom	110,800+ 80,700+ w/ xyz 79,500+ w/ I/Ic
	PDF-4/Minerals	53,400+ 43,900+ w/ xyz 41,300+ w/ I/Ic
	PDF-5+	1,104,100+ 626,140+ w/ xyz 997,300+ w/ I/Ic
LPF[c]—Pauling File	Inorganic ordered solids. Contains structural, diffraction, phase diagrams, and physical property data	82,000 crystal structures 55,000 material property 15,000 phase diagrams
ICSD[d]—Inorganic Crystal Structure Data	Inorganic crystal structures, with atomic coordinates, 1913 to date	299,045+ 10,670 structure types
CSD[e]—Cambridge Structural Database	Crystal structures of organic and metal–organic compounds	1.29 million
PDB[f]—Protein Data Bank	Structures of proteins and nucleic acids	224,201
IZA[g]—Zeolite database	All zeolite structure types: drawings, framework, and simulated patterns	256 framework types
Mineralogy Database[h]	Mineral descriptions, structures and properties, diffraction data (3 lines)	4714
COD[i]—Crystallography Open Database	Open-access crystal structures of organic, inorganic, minerals	517,253

[a] xyz—entries with atomic coordinates; I/Ic—entries with known corundum number and therefore are suitable to quantitative analysis

[b]°Release 2025. The International Centre for Diffraction Data (https://www.icdd.com/)

[c]°The multinary edition of database developed in cooperation between JST (Japan Science and Technology Corporation, Tokyo, Japan; https://crystdb.nims.go.jp/) and Pauling File (Material Phases Data System, Vitznau, Switzerland; https://paulingfile.com/)

[d]°Release 2024-1. The ICSD is produced by FIZ (Fachinformationszentrum) Karlsruhe, Germany (https://www.fiz-karlsruhe.de/)

[e]°Release Jun. 2024. Cambridge Crystallographic Data Centre (CCDC) (https://www.ccdc.cam.ac.uk/)

[f]°Release Apr. 2024. PDB is maintained by the Research Collaboratory for Structural Bioinformatics (https://www.rcsb.org/)

[g]°Revised Jan. 2024. IZA (International Zeolite Association) zeolite database is maintained by IZA structure commission. Available on-line at https://www.iza-structure.org/databases/

[h]°Update Sep. 2012. Mineralogy Database is available on-line at https://webmineral.com/

[i]°As of Aug. 2024. COD (Crystallography Open Database) https://www.crystallography.net/cod/

For example, two comprehensive databases, ICSD and CSD, contain crystallographic data and structural information about inorganic and organic and metal–organic compounds, respectively. The NIST database, however, encompasses all types of compounds but provides only crystal data with references. Other databases[40] are dedicated to specific classes of materials, such as metals and alloys, proteins and macromolecules, minerals, or zeolites. Search-match utilities are usually provided with databases, or they may be obtained separately.

Given a large variety and differences in the contents of existing databases, a material can be identified from its powder diffraction pattern by first, searching an appropriate powder diffraction database/file using a digitized powder diffraction pattern. If a search was successful, the identification may be considered complete after matching the entire digitized pattern, not just the few key features included in the search. If a search was unsuccessful, the pattern should be indexed, and unit cell dimensions should be determined (see Chap. 14). When both symmetry and unit cell are known, all relevant databases should be searched.[41] In the case of a suspected match, the powder diffraction pattern should be modeled from the known crystal structure of a material found in a database. Both the observed and computed patterns should be compared as a whole to ensure proper identification.

Since the ICDD Powder Diffraction File now includes not only experimental powder data but also powder patterns calculated from single-crystal data, conducting searches and finding matches is much easier and requires fewer steps than previously discussed. This greatly improves automation, simplifies phase identification process, and considerably expands the applicability of the powder method for qualitative phase analysis.

13.3.2 Phase Identification

Qualitative and quantitative phase analyses are essentially two sides of the same coin, as they answer two fundamental questions: "What?" and "How much?" respectively, regarding the crystalline phases present in a powder. Although phase identification—i.e., qualitative analysis—might seem straightforward (it involves comparing the positions and intensities of observed Bragg reflections with those stored in a database), the problem is far from trivial. The complexity in finding the right pattern arises from unavoidable experimental errors present in all patterns, both the analyzed ones and those located in the database, as well as from ambiguities intrinsic to image comparison. Thus, phase identification may be performed visually, automatically, or as is often the case, through a combination of both.

The manual approach is simple and feasible when a small subset of database entries is singled out as potential matches with the observed powder pattern. It is almost always employed as a final step to confirm the automatic search results and to select the best match among several feasible patterns. A manual search can be

[40] List of databases related to crystallography can be found at https://www.iucr.org/resources/data/databases

[41] It makes little sense to search organic or metal–organic structures database if the material from which powder diffraction data were collected is inorganic and vice versa.

conducted visually by comparing raw or digitized experimental patterns with those found in a database or digitally by comparing lists of a few strong, usually low-angle reflections. Purely manual searches are justified when there are a few unidentified reflections in a pattern—a situation when an automatic search is ineffective and typically fails.

Manual phase identification can be performed using searchable or alphabetical PDF indexes available from the ICDD. These indexes include detailed instructions on the required information and how to conduct the search. A searchable index is divided into groups (classes) according to the *d*-spacing of the second strongest line, and phases within the same group are sorted according to the *d*-spacing of the strongest line. An alphabetical index is typically used when exact or approximate elemental composition of the phase is known.[42]

An automatic search-match can be done much faster and, most importantly, by using multiple Bragg reflections to search through the enormous arrays of data contained in a typical database. The algorithms employed for automatic searches vary widely depending on the software used; however, parameters that are critical in any search typically include the following:

- The number of Bragg reflections that should match in their positions (*d*-spacings, *Q*-values, or Bragg angles), and sometimes in their relative intensities.
- The number of strongest reflections from a database record included in the comparison.
- Window (or tolerance): the acceptable difference in positions between the observed and database peaks. As long as the deviation remains within this tolerance (i.e., within the window), peaks are considered matching. The window may be specified as a range in 2θ, *d*-spacing, or other common measures of peak positions.

Automatic searches may often generate a massive number of matches, from which the correct solution, if any, must be selected manually—more precisely, visually—by the user. To aid in visual selection, matching patterns are usually sorted by a numerical figure of merit that considers the average difference in peak positions, the number of matching reflections, and optionally, matching relative intensities. A search-match utility can easily generate a list of several hundred potentially matching patterns, and the user can display histograms extracted from the Powder Diffraction File one-by-one alongside with the observed powder pattern, as shown in Fig. 13.19.

Depending on both the complexity and quality of the powder pattern, the number of found "matches" can become overwhelming. For example, if several hundred matching patterns are found, search parameters or restrictions should be adjusted to narrow the range within which the patterns are considered as matching. However,

[42] In the pre-personal computer era, the search-match procedure was performed manually using the Hanawalt method (see Footnote 34 on page 409). The original handwritten search index is on display at the ICDD headquarters in Newtown Square, PA.

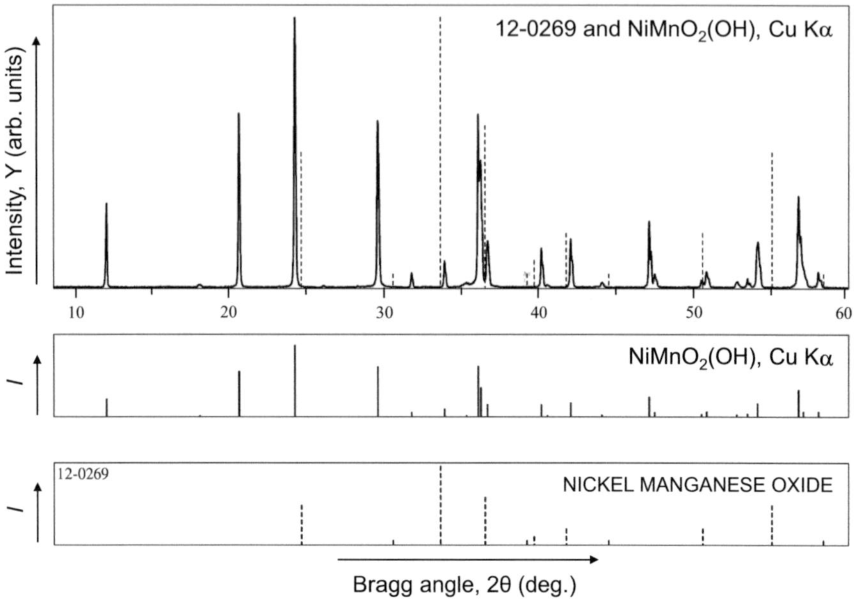

Fig. 13.19 Illustration of an incorrect match: Bragg peaks observed in the experimental diffraction pattern of $NiMnO_2(OH)$, shown on top and in the middle (*solid lines*), do not match those present in the nickel manganese oxide, ICDD PDF card No. 12-0269 shown at the bottom and overlapped with the plot on top (*dashed lines*)

adjusting parameters and applying stricter constraints should not be done at the very beginning, as the correct solution may be easily overlooked. A common practice is to start with the default search-match parameters and, if the search is unsuccessful, relax the restriction. Conversely, if the matching list is overwhelming, the restrictions can be tightened.

A powerful way to narrow the search field is to impose restrictions on the elemental composition. For example, in DMSNT, the list of chemical elements can be specified and used in combination with the following search options:

- "*Inclusive OR*" limits the search to phases containing at least one of the listed elements, along with any other elements not included in the list.
- "*Inclusive AND*" considers only phases containing all listed elements, plus any chemical element not included in the list.
- "*Exclusive OR*" checks phases containing any combination of the listed elements, but no other chemical elements are allowed.
- "*Exclusive AND*" seeks only patterns from phases containing all listed elements and nothing else.

The last option is the most restrictive, while the first alternative is the most relaxed. For example, when V and O are included in the list, then "*Exclusive AND*" limits the search only to vanadium oxides, while compounds containing other elements (e.g., vanadium hydroxides, hydrates, etc.) will not be considered and analyzed. Conversely, "*Inclusive AND*" searches among all compounds containing both V and O in combination with any other chemical elements. The latter option may be useful, for example, when intercalates of vanadium oxides are suspected or studied.

Another approach is used in the EVA[43] search-match algorithm, which allows users to specify constraints for any of the fields incorporated in the database, such as subfiles or a class of the compound (e.g., inorganic, mineral), experiment conditions, chemical composition, and more. The most powerful feature, and therefore most frequently used, is imposing constraints on chemical composition, which are set by marking selected chemical elements as "Must match," "May match," "At least one shall match," and "Exclude from match." Another powerful option in EVA is that search can be conducted not only using peak positions but also by analyzing the entire raw experimental pattern without any preliminary treatment. The latter speeds up the search-match process enormously, while the former remains valuable, especially when identifying minor phases represented by only a few weaker peaks.

When visually comparing potentially matching patterns, the following important issues should always be considered:

- When a database record contains a few strong reflections, all should be present in the analyzed experimental pattern. If even one of the strong peaks is missing in the analyzed pattern or has very low intensity, the match is likely incorrect—unless there is an extremely strong preferred orientation or another legitimate reason for the discrepancy.
- Relative intensities should be analyzed carefully, as significant discrepancies between experimental data and database entries may arise due to factors such as different wavelengths, diffractometer geometry, sample shape, or the presence and extent of preferred orientation. Preferred orientation is an important factor, and in many cases, it is unavoidable. Additionally, texture may be substantially different in different experiments, such as between the one being analyzed and the one in the database. Therefore, the following rule should be applied when comparing intensities: a strong reflection in the database record should correspond to a strong peak in the analyzed pattern, and a weak reflection in the database record should correspond to a low-intensity peak in the analyzed pattern.

Even though all automatically found patterns are ranked according to certain matching criteria, it is always recommended to visually analyze at least several solutions—preferably all that appear reasonable.

Once again, we consider the experimental data used as an example throughout this chapter (Fig. 13.3). The data were digitized through background subtraction, $K\alpha_2$ stripping, and smoothing, followed by automatic peak detection. The PDF search-match was restricted to phases containing Ni, Mn, and O, using the "Inclusive

[43] Bruker DIFFRAC.EVA (2020). https://www.bruker.com/en/products-and-solutions/diffractometers-and-x-ray-microscopes/x-ray-diffractometers/diffrac-suite-software/diffrac-eva.html

AND" option. In total, approximately 20 matching patterns were identified. One of the suspected matches, $NiMnO_3$ (card No. 12-0269), is depicted in Fig. 13.19. Although this pattern is not the best match according to the calculated figure of merit, we use it to illustrate the difference between a good match (Fig. 13.20) and a poor match (Fig. 13.19).

As shown in Fig. 13.19, most Bragg reflections only approximately coincide (within a few tenths of a degree). Several strong peaks present in the database record have no match in the measured pattern, for example, the reflection at $2\theta \approx 55°$. The strongest peak at ~33.7° matches only approximately, but its intensity is much greater than what was observed experimentally. Therefore, this "match" may be easily dismissed, especially since better matches with much higher figures of merit were found.

Two PDF records coincide with the experimental pattern much better than all others, both in peak positions and intensities. These records are compared with the observed profile in Fig. 13.20, while the corresponding PDF cards are shown in Fig. 13.21 for $NiMnO_2(OH)$ (card No. 43-0318) and Fig. 13.22 for $NiMnO_3$ (card No. 49-1170). These two compounds could be isostructural, as their patterns are nearly identical. However, the first record, 43-0318, is unindexed, which raises doubts about the quality of its digitized pattern (see the upper right corner in Fig. 13.21).

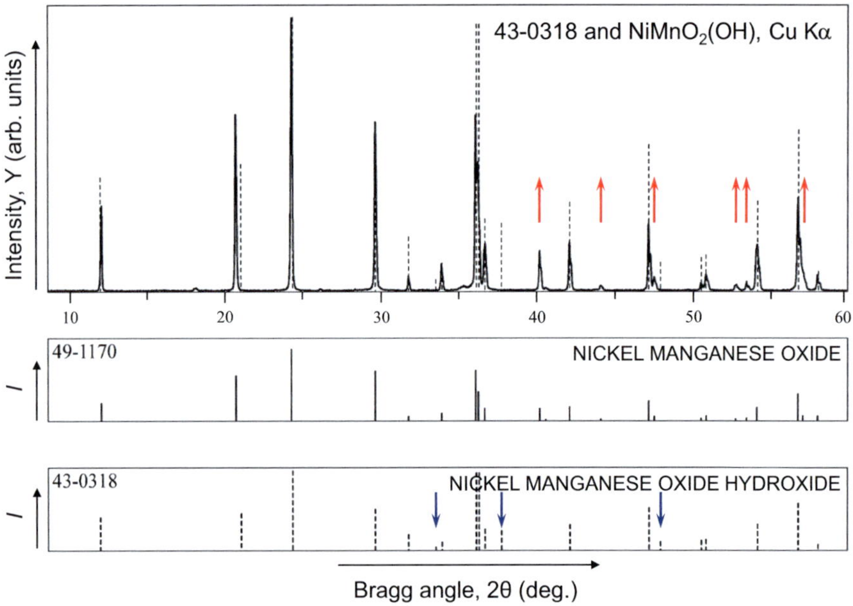

Fig. 13.20 Experimental powder diffraction pattern of $NiMnO_2(OH)$ (*top*) compared with digitized PDF records 49-1170 (*middle, solid lines*) and 43-0318 (*bottom, dashed lines*). Downward *blue arrows* indicate peaks present in the 43-0318 record but absent in the measured pattern. Upward *red arrows* on the experimental pattern indicate observed Bragg peaks that are missing in the 43-0318 record for nickel manganese oxide hydroxide

43-0318 Quality: Doubtful quality

Ni Mn O2 (O H)
Nickel Manganese Oxide Hydroxide

Rad:FeKa	Lambda:1.9373	Filter:	d sp:Diffractometer
Cutoff:	Int:Diffractometer	I/Icor:	
Ref:Yamamoto, N., 33 48, (1986)			

Sys:		S.G.:		
a:	b:	c:		
α:	β:	γ:	Z:	mp
Ref2				
Dx:	Dm:	SS/FOM:	Volume[CD]:0	
εα:	ηωβ:	εγ:	Sign:	2V:
Ref3				

Color:

Prepared by hydrothermal treatment at 200-320 C and 100 MPa of an equimolar mixture of \Ni (O H)2\ and $GG-MnOOH in a 1N \N H4 O H\ solution. Chemical analysis (wt.%): Ni 36.0, Mn 34.0, H 0.58, O 29.4. O assigned because unindexed.

21 reflections in pattern.

2θ	Int.	h k l	2θ	Int.	h k l	2θ	Int.	h k l	2θ	Int.	h k l
11.9241	41.3		37.8012	24.5		63.4445	12				
21.0048	46.2		42.1740	32.1							
24.3059	99.9		47.2279	53.3							
29.6253	51.6		47.9692	10.3							
31.7969	19.6		50.5841	12							
33.5754	4		50.8863	13							
33.9689	10		54.2674	32.6							
36.1914	97.3		56.8980	58.7							
36.3582	97.3		58.1954	7							
36.7434	26.1		61.6166	18.5							

Fig. 13.21 Example of the unindexed PDF card (also see Fig. 13.20, *bottom*). In this case, all observed reflections are identified by their Bragg angles and relative intensities but without the Miller indices. As a result, the "Doubtful quality" mark has been assigned to this record. This usually indicates the need for an independent verification before relying on the listed digitized pattern for positive identification of a polycrystalline material. Note that the original experimental data were collected using Fe Kα radiation (see field No. 3). Bragg angles, however, are listed for Cu Kα radiation, and these were recalculated by the search-match utility using the Bragg's equation

Careful analysis of Fig. 13.20 shows that six observed peaks (indicated by red arrows) are not present in record 43-0318, and three weak to medium intensity peaks (blue arrows) in the record have no match in the experimental pattern. The second record, 49-1170, is almost a perfect match, but a hydrogen atom is missing in its chemical formula, as was later determined from neutron diffraction data.[44]

Achieving success in qualitative analysis by employing any search-match utility becomes increasingly challenging as the complexity of the powder diffraction pattern grows, particularly when the material is a mixture of several phases. Positive phase identification can be easier after removing peaks corresponding to known

[44] R. Chen, P.Y. Zavalij, M.S. Whittingham, J.E. Greedan, N.P. Raju, and M. Bieringer, The hydrothermal synthesis of the new manganese and vanadium oxides, $NiMnO_3H$, MAV_3O_7 and $MA_{0.75}V_4O_{10}$·$0.67H_2O$ (MA = CH_3NH_3), J. Mater. Chem. **9**, 93 (1999). https://doi.org/10.1039/A805312J

49-1170 Quality: Quality Data

Ni Mn O3
Nickel Manganese Oxide

Rad:CuKa1	Lambda:1.54056	Filter:	d sp:Diffractometer
Cutoff:	Int:Diffractometer	I/Icor:	

Ref:Whittingham, S., SUNY at Binghamton, MaterialsResearch Center, NY, USA., (1997)

Sys:Orthorhombic		S.G.:A21am		
a:2.8609±0.0001	b:14.6482±0.0005	c:5.2703±0.0002		
α:	β:	γ:	Z:4	mp
Ref2				
Dx:4.861	Dm:4.861	SS/FOM: F30=103.1(0.0081,36)	Volume[CD]:220.86	

εα:	ηωβ:	εγ:	Sign:	2V:
Ref3				

Color:

Prepared by hydrothermal treatment of tetramethylammonium permanganate, nickel acetate and lithium carbonate for 2 days at 200 C. Pattern taken at 23(1) C.

55 reflections in pattern.

2θ	Int.	h k l	2θ	Int.	h k l	2θ	Int.	h k l	2θ	Int.	h k l
12.0679	25	0 2 0	47.6194	6	1 5 1	66.2340	1	0 10 1	80.4338	2	1 1 4
20.7417	55	0 2 1	50.6172	3	1 3 2	68.6303	14	1 9 1	80.7356	16	0 12 1
24.2722	100	0 4 0	50.9411	9	0 6 2	69.1962	2	2 2 1	80.7356	16	2 4 2
29.6756	65	0 4 1	52.8898	3	0 8 1	70.6260	5	2 4 0	81.7659	3	1 11 1
31.8434	5	1 1 0	53.6083	3	0 2 3	71.5591	12	0 0 4	83.0093	10	1 3 4
33.9955	10	0 0 2	54.2640	27	1 7 0	72.8763	3	0 2 4	83.2543	6	0 6 4
36.1914	62	0 2 2	56.9133	50	1 5 2	73.2173	4	2 4 1	88.1190	12	1 5 4
36.3735	43	1 3 0	57.2185	12	1 7 1	74.8306	3	0 8 3	88.1190	12	0 12 2
36.7901	23	0 6 0	58.1954	8	0 4 3	75.5715	1	2 0 2	88.6939	6	1 9 3
40.3213	16	1 3 1	61.6257	44	0 8 2	76.3434	3	1 9 2	89.1644	11	1 11 2
40.7014	1	0 6 1	62.2876	6	1 1 3	76.8677	11	2 2 2	93.0153	3	2 4 3
42.2153	22	0 4 2	65.1455	21	2 0 0	77.2366	3	2 6 0	95.6870	10	1 7 4
44.2102	2	1 5 0	65.3921	2	0 6 3	78.4823	4	1 7 3	96.0050	11	2 8 2
47.2644	28	1 1 2	65.6043	1	1 7 2	79.2909	4	1 11 0			

Fig. 13.22 Indexed PDF card (also see Fig. 13.20, *middle*) (Space-group symmetry is listed as $A2_1am$, which is $Cmc2_1$ in a standard setting). Every observed Bragg reflection has been indexed, and the corresponding F_{30} figure of merit (see Sect. 14.4.1) is excellent. Based on these and other established criteria, the quality mark is "Quality Data," indicating that the included digitized pattern can generally be trusted for positive phase identification

phases from the search list and then continuing the database search. However, matching all possible records with the entire diffraction pattern can also be effective. The first approach increases the chances of detecting and identifying minor phases, while the second method helps avoid overlooking suitable records due to nearly complete peak overlaps.

Regardless of the chosen approach, specifying the elemental composition of the material is always helpful, as it imposes essential constraints and limits the number of feasible solutions for visual analysis.

An example of successful phase identification in a multiple-phase sample is shown in Fig. 13.23. The search was conducted using restricting chemical composition to all inorganic compounds containing silicon and oxygen. No single match was adequate to interpret all strong observed Bragg peaks. However, two records—lithium silicate Li_2SiO_3 and quartz SiO_2—combined cover all strong peaks and

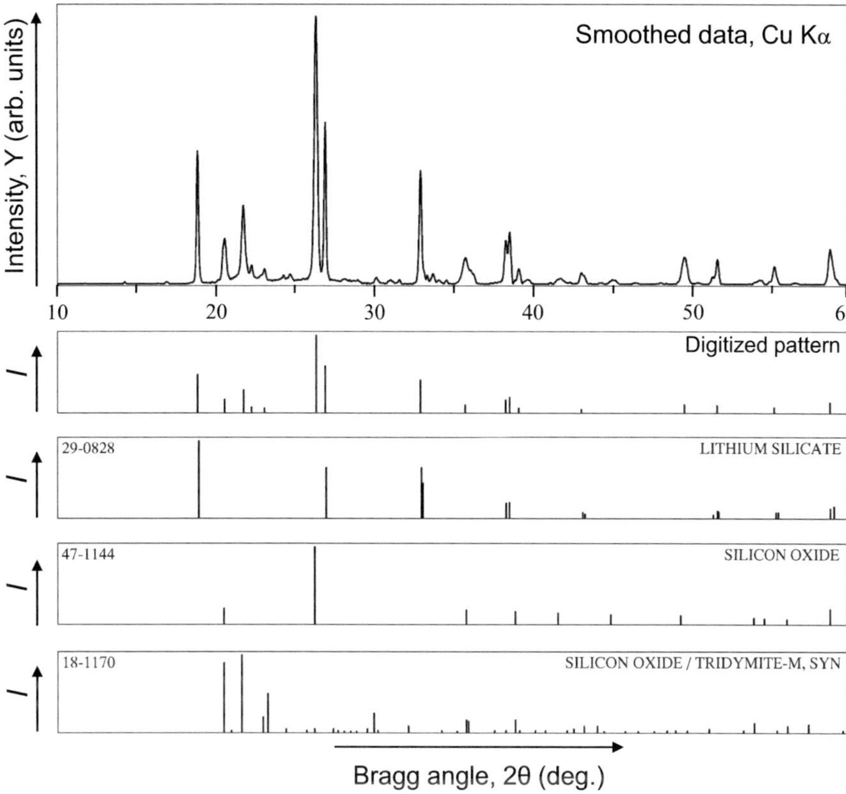

Fig. 13.23 The results of a qualitative analysis of a multiple-phase sample. Three crystalline phases are clearly identifiable: lithium silicate—Li_2SiO_3, silicon oxide—SiO_2 (quartz), and a different polymorph of silicon oxide—tridymite. A low-quality diffraction pattern collected during a fast experiment was employed in this example. The data shown on top were smoothed, the background was subtracted, and the Kα_2 components were stripped before the digitized pattern (shown below the smoothed profile) was obtained using an automatic peak search. Note that many weak Bragg reflections were missed in the peak search

majority of the medium-intensity ones. Remaining Bragg peaks have lower intensity (less than one-third of the maximum) and correspond to tridymite (SiO_2), which is present in a lower concentration than both Li_2SiO_3 and quartz. A few weak reflections in this pattern remain unidentified, likely due to the low quality of data (the experimental pattern shown in Fig. 13.23 has been smoothed) or a small amount of an impurity phase that does not contain silicon or oxygen.

Overall, the phase identification in a multiple-phase material consisting of more than two phases is difficult and often has no reasonable solution in a "blind" search, especially when none of the phases has been positively identified prior to the search using a different experimental technique. Further, chances for success decrease proportionally to the increased complexity of the measured powder diffraction

Table 13.4 Characteristics of selected standard reference materials (SRM) that can be used in quantitative analysis based on powder diffraction data

SRM	Formula	Purity (wt%)	Linear absorption μ (cm^{-1}) for Cu Kα	Mass absorption μ/ρ (cm^2/g)	Size (μm)	Corundum number, I/I_c
660a	LaB_6		1163.0	247.0	15 μm sieve	
640c	Si		139.3	59.8	4.9 μm	
676a	Al_2O_3	99.0(11)	121.1	30.4	Sub-μm	
674b	ZnO	95.3(6)	267.9	47.2	0.201(3)	4.95(1)
	TiO_2	89.5(6)	540.2	127.2	0.282(10)	3.44(1)
	Cr_2O_3	95.9(6)	468.1	179.1	0.380(14)	1.97(2)
	CeO_2	91.4(6)	2240.0	310.6	0.381(5)	12.36(9)

pattern, unless the number of possible components with different crystal structures in the mixture is limited to just a few.

13.3.3 Quantitative Analysis

Quantitative phase analysis is used to determine the concentration of various phases in a mixture after the identity of each phase has been established. This task can be quite complex, as several critical requirements and conditions must be met to achieve satisfactory accuracy.

Proper alignment and, especially, calibration of the diffractometer are very important. Calibration should be performed by examining one or several mixtures made from carefully prepared and well-characterized materials. Generally, any of the many available standard reference materials (SRM) may be used,[45] including a specially developed standard for quantitative analysis. The latter is the SRM-674b, a standard for powder diffraction intensity, which is a mixture of four stable oxides with different absorption coefficients: ZnO (wurtzite structure), TiO_2 (rutile structure), Cr_2O_3 (corundum structure), and CeO_2 (fluorite structure). A broad range of absorption coefficients is needed to match as close as possible absorption of the sample in order to minimize microabsorption effects that are discussed latter in this section on page 430. The main characteristics of this and other standards are listed in Table 13.4.

In addition to instrumental factors, specimen preparation and properties introduce several key features that may have a detrimental influence on the accuracy of quantitative phase analysis. Sample-related factors cannot be avoided completely, but their effects should be minimized as much as possible and accounted for in all

[45] E.g., see: Standard reference materials catalog 1992–1993. NIST, Special publication 260, NIST, Gaithersburg, MD, U.S.A. (1992), p.122. Up to date information about availability and pricing of X-ray diffraction standards for powder diffraction may be found online at https://www.nist.gov/srm

calculations. The main problems in quantitative analysis, the nature and form of the employed sample, are as follows:

- Preferred orientation: This can substantially affect the relative intensities of various groups of Bragg reflections (see Sect. 8.6.6). It should be minimized during sample preparation of both the investigated sample and the standard, if a standard is used.
- Absorption: Absorption varies for phases with different chemical compositions and gravimetric densities (see Sect. 8.6.5). It should always be accounted for.

Several methods of quantitative analysis have been developed and extensively tested. These methods can be grouped into broad categories, with the most commonly used approaches are described here.

1. *The absorption-diffraction method* employs a standard intensity (I^0_{hkl}) from a pure phase and the intensity of the same Bragg peak (I_{hkl}) observed in the mixture. The phase concentration in a mixture can be calculated by using Klug's equation:

$$X_a = \frac{\left(I_{a,hkl}/I^0_{a,hkl}\right)\left(\mu/\rho\right)_b}{\left(\mu/\rho\right)_a - \left(I_{a,hkl}/I^0_{a,hkl}\right)\left[\left(\mu/\rho\right)_a - \left(\mu/\rho\right)_b\right]} \tag{13.14}$$

   where X_a is the mass fraction of phase a in the mixture, $I_{a,\ hkl}$ and $I^0_{a,hkl}$ are intensities of the selected Bragg reflection, *hkl*, for phase a in the mixture and in the pure state, respectively, and $(\mu/\rho)_{a,\ b}$ are the mass absorption coefficients for phases a and b, respectively.

   Equation (13.14) makes use of the fact that the scattered intensity is proportional to the amount of a particular phase, e.g., see Eqs. (7.5–7.7), with a correction to account for different absorption of X-rays by two components in the mixture. Since the ratio of intensities from a pure phase and a mixture is used, diffraction patterns for both the pure material and the analyzed mixture must be measured under identical instrumental settings, as well as with identical sample characteristics such as preparation, shape, amount, packing density, surface roughness, and so on. Klug's equation becomes a simple intensity ratio when two phases have identical absorption coefficients, i.e., when $\mu/\rho_a = \mu/\rho_b$. We note that the composition of the second phase (or a mixture of all other phases) should be known in order to determine its mass absorption coefficient. Otherwise, mass absorption should be determined experimentally. Ignoring absorption effects can drastically reduce the accuracy of quantitative analysis.
2. *Method of standard additions* or spiking method consists of adding known amounts of pure component a to a mixture containing X_a and X_b of a and b phases. It requires the preparation of several samples and measurement of several diffraction patterns containing different yet known additions (Y_a) of phase a. Other phases in the mixture are not analyzed but at least one of them

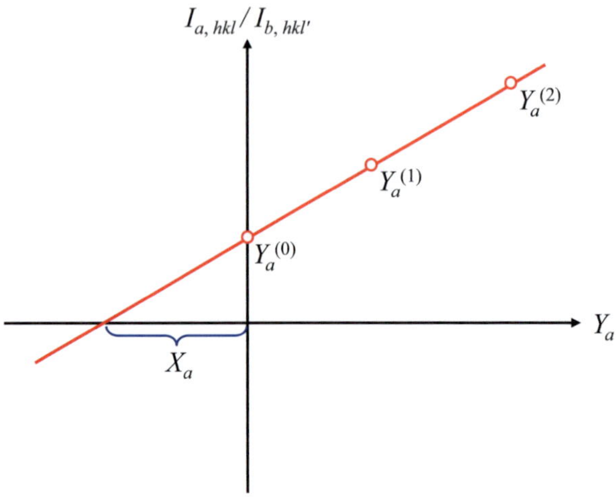

Fig. 13.24 Illustration of the standard addition method for quantitative analysis. The plot of the $I_{a,hkl}/I_{b,hkl'}$ intensity ratio as a function of the known amount (Y_a) of added phase a. The point marked $Y_a^{(0)}$ corresponds to the original two-phase mixture. The unknown amount of phase a in the original sample is X_a

(b) should have a reference Bragg peak (hkl)′, which does not overlap any reflection from phase a. The intensity ratio for this method is given by:

$$\frac{I_{a,(hkl)}}{I_{b,(hkl)'}} = K' \frac{X_a + Y_a}{X_b} \tag{13.15}$$

Assuming a constant mass of phase b (K'/X_b), Eq. (13.15) simplifies into:

$$\frac{I_{a,(hkl)}}{I_{b,(hkl)'}} = K(X_a + Y_a) \tag{13.16}$$

where K is the slope of the plot of $I_{a,hkl}/I_{b,hkl'}$ versus Y_a, established during measurements of mixtures with known additions of phase a, $I_{a,\,hkl}$ is the intensity of the selected peak for phase a, and $I_{b,hkl'}$ is the intensity of the selected peak for phase b. The unknown amount of phase a, X_a, is determined from the intercept of the calibration line with the Y_a axis, as shown in Fig. 13.24. The major advantage of this technique is that it enables quantitative analysis in the presence of unknown phase(s) without the need to know (or measure) absorption coefficients.

3. *Internal standard method* is likely the most commonly used approach in a quantitative phase analysis. It is based on the following relationship:

$$\frac{I_{a,(hkl)}}{I_{b,(hkl)'}} = K \frac{X_a}{Y_b} \tag{13.17}$$

where K is the slope of the plot of $I_{a,hkl}/I_{b,hkl'}$ versus X_a/Y_b. In Eq. (13.17), X_a is the unknown amount of the analyzed phase a, and Y_b is the known amount of the added standard phase b that is different from that present in the sample. In the same way as in the standard addition method, several measurements are needed to determine the slope K individually for each analyzed phase. The calibration line of Eq. (13.17) is then used to determine the content of phase a by measuring the $I_{a,hkl}/I_{b,hkl'}$ intensity ratio for a mixture of the analyzed sample with the known amount of the added internal standard Y_b.

4. *The reference intensity ratio method* is based on the experimentally established intensity ratio between the strongest Bragg peaks in the examined phase and in a standard reference material. The most typical reference material is corundum, and the corresponding peak is (113). The reference intensity ratio (k) is quoted for a 50:50 (wt.%) mixture of the material with corundum, and it is known as the "corundum number." The latter is commonly accepted and listed for many compounds in the ICDD's Powder Diffraction File. Even though this method is simple and relatively quick, careful account and/or experimental minimization of preferred orientation effects are necessary to obtain reliable quantitative results.
5. *Full-pattern decomposition* using Le Bail's or Pawley's techniques (see Sect. 15.4) produces intensities of individual Bragg peaks. Thus, multiple reflections from each phase can be used to compute intensity ratios required in methods described in items 1–4 listed here, which increases the accuracy of the analysis. The use of multiple Bragg peaks in evaluating an average intensity ratio to some extent diminishes the detrimental influence of preferred orientation, as long as it remains small to moderate. This method, however, requires lattice parameters and therefore is applicable to indexed patterns only. The phase composition is determined using any of the first four methods listed here by using intensities of several strong or all Bragg peaks instead of a single reflection. An interesting approach was proposed by Toraya.[46] In the first step of a two-step procedure, the individual patterns of single component phases are decomposed separately with scale factors fixed at unity. In the second step, the experimental pattern of a mixture is fitted using individual intensities obtained for pure phases. During this step, the intensities obtained in the first step are kept fixed, but the scale factors are refined. These scale factors (s_i) are proportional to the fraction of the ith phase in the mixture[47] and are used to calculate the weight fraction (w_i) of the ith phase as shown in Eq. (13.18), which also includes absorption correction.[48]

---

[46] H. Toraya, Applications of whole-powder-pattern fitting technique in materials characterization, Adv. X-Ray Anal. **37**, 37 (1994). https://doi.org/10.1154/S0376030800015494

[47] We note that the definition of the scale factor in the full pattern decomposition is different from that in the Rietveld refinement, where it is proportional to the number of the unit cells per volume as discussed in the following section.

[48] H. Toraya, S. Tsusaka, Quantitative phase analysis using the whole-powder-pattern decomposition method. I. Solution from knowledge of chemical compositions, J. Appl. Cryst. **28**, 392 (1995). https://doi.org/10.1107/S0021889894014986

$$w_i = s_i \prod_{j \neq i}^{N} \mu_j / \sum_{j=1}^{N} \left(s_j \prod_{k \neq j}^{N} \mu_k \right) \tag{13.18}$$

where μ_j or μ_k represent the linear absorption coefficients of the corresponding phase in the mixture of N phases.

Both the accuracy and limits of detection in a quantitative analysis are dependent on the method used, the quality of the experimental data, and other factors. A lower limit of detection is usually accepted as the concentration equivalent of two standard deviations of the observed background level. For example, if the average background is 100 counts and the maximum observed Bragg peak has a peak intensity of 1100 counts, then two standard deviations of the background are: $2 \times \sqrt{100} = 20$. Thus, the detection limit for this phase is estimated at $2/(1100 - 100) \times 100\ \% = 2\%$. The accuracy of the quantitative phase analysis is difficult to estimate rigorously. It varies considerably and is often claimed to be between 1% and 5%. The full-pattern decomposition (if lattice dimensions are known), combined with any of the methods described in this section that are based on the intensity of an individual peak, is probably the most accurate approach in quantifying phase composition. However, when crystal structures of all phases are known, the Rietveld method can be employed. It uses standard scale factors (Eq. 8.41) instead of observed intensities and, therefore, provides a more realistic estimate of the accuracy, because uncertainties in calculated concentrations can be easily estimated from standard uncertainties in the corresponding scale factors.

13.3.4 Phase Contents from Rietveld Refinement

The quantitative analysis methods described in the previous section are based on the integrated intensities of individual Bragg peaks (one or several). These methods generally do not require prior knowledge of either the lattice parameters or the crystal structure of phases in question. However, employing full-pattern decomposition, which requires lattice parameters and symmetry, results in higher accuracy since intensities of many or all of the Bragg peaks can be utilized. Conversely, all of these methods require an internal or external standardization, as this is the most accurate way to uniformly scale the integrated intensities used in the analysis. This requirement can be avoided by applying the Rietveld technique, which is discussed in Sect. 15.7, with numerous examples beginning in Chap. 16. A practical example illustrating quantitative phase analysis using Rietveld refinement can be found in Sect. 16.5.

The Rietveld method is one of the fastest and, arguably, the most reliable tool for quantifying phase contents, particularly because it can account for weak to moderate preferred orientation. This method uses intensities normalized to the scattering by a single unit cell (see Sects. 7.1.3, 8.6, and 9.1). As a result, neither internal nor external standards are necessary. However, this convenience comes at the cost of requiring knowledge of the atomic structure of each phase present in the mixture.

The main principle of this method is that the intensities calculated from the crystallographic data, normalized to the content of a single unit cell of each phase, are scaled to match the corresponding observed intensities in the same diffraction pattern using a phase-specific scale factor K (see Eq. 8.41, as well as Eqs. 15.29 and 15.30). Therefore, the scale factors of the individual phases are representative of the total number of the unit cells of each phase present in the irradiated volume of the sample.[49] The latter directly follows from Eqs. (7.6 and 7.7) after recalling that the proportionality coefficient, C, is a constant for any given powder diffraction experiment. Thus, the scale factors can be easily converted into weight, volume, or molar fractions of the respective phase. The weight fraction, w', of a particular phase can be calculated from the scale factor (K) as

$$w' \approx K \cdot ZMV \tag{13.19}$$

where Z is the number of formula units in the unit cell, M is the molecular mass of the formula unit, and V is the unit cell volume of the phase in question.

Clearly, individual weight fractions determined from Eq. (13.19) must be normalized so that the total is unity,[50] which may be done as follows:

$$w_i = w_i' / \sum_j w_j' = (KZMV)_i / \sum_j (KZMV)_j \tag{13.20}$$

The volume (v) fraction can be expressed as:

$$v_i = \left(KV^2\right)_i / \sum_j \left(KV^2\right)_j \tag{13.21}$$

Similar equation can be derived to determine the molar fraction of a phase from its scale factor s in the multiple-phase mixture, but we leave this as an exercise for the reader.

[49] R.J. Hill, C.J. Howard, Quantitative phase analysis from neutron powder diffraction data using the Rietveld method. J. Appl. Cryst. **20**, 467 (1987). https://doi.org/10.1107/S0021889887086199

[50] When some of the minor phases remain unidentified, or their structures are unknown, or a substantial amount of an amorphous phase is present, the Rietveld method can still be used but Eq. (13.20) is no longer valid. If this is the case, then the weight fractions should be normalized using an internal or an external standard as detailed in Sect. 13.3.5.

13.3.5 Determination of Amorphous Content or Degree of Crystallinity

More often than not, polycrystalline samples contain amorphous or low-crystallinity component(s), or phase(s) in both crystalline and amorphous form(s). This may make the Rietveld-based quantitative analysis inapplicable or at best may reduce its precision. While the Rietveld method may still result in reliable weight ratios among the crystalline components in the sample, in many applications it is important to know the absolute weight fractions that include the content of the amorphous part of the specimen. For example, knowing the ratio of amorphous and crystalline forms in pharmaceutical compounds, polymers, and cements is critical in assessing their usefulness.[51] This ratio is often referred to as the amorphous content or the degree of crystallinity. Regardless of whether they are crystalline or amorphous, the fractional contents of all components of the analyzed powder sample must add up to unity.

If a sample contains only a single phase in both amorphous and crystalline forms, there is a relatively straightforward method of their quantification, which is different from those discussed earlier. It consists of measuring the total intensity of all sharp Bragg (crystalline) peaks and all broad amorphous halos. The resultant total intensity ratio represents the ratio between the crystalline and amorphous phases. This method is not as precise as the Rietveld method, mainly due to inaccuracies in the amorphous part of the scattered intensity, since the presence of broad halos (humps) makes background determination and subtraction extremely difficult. Furthermore, determining the background is often biased. However, in some cases, no other method can be employed. This is particularly true in the analysis of many polymers, where the crystalline phase cannot be prepared in the pure form to use as a standard, nor its structure can be refined due to the absence of a three-dimensionally periodic atomic structure.

When the atomic structure of all crystalline phases is known, the amorphous content, or the content of a minor phase with unknown structure, can be determined using the Rietveld refinement results. This, however, implies that a standard should be used as the sum of weight fractions of all crystalline phases is no longer unity. Either an internal or external standard or both can be used as described next.[52]

External standard for determining absolute phase composition using Rietveld method is not equivalent to an external standard used in the absorption–diffraction

[51] P. Bergese, I. Colombo, D. Gervasoni, L.E. Depero, Assessment of the X-ray diffraction-absorption method for quantitative analysis of largely amorphous pharmaceutical composites, J. Appl. Cryst. **36**, 74 (2003). https://doi.org/10.1107/S002188980201926X; P.S. Whitfield, L.D. Mitchell, Quantitative Rietveld analysis of the amorphous content in cements and clinkers, J. Mater. Sci. **38**, 4415 (2003). https://doi.org/10.1023/A:1026363906432; L.D. Mitchell, J.C. Margeson, and P.S. Whitfield, Quantitative Rietveld analysis of hydrated cementitious systems, Powder Diffraction, **21**, 111 (2006). https://doi.org/10.1154/1.2204056

[52] P.M. Suherman, A. van Riessen, B. O'Connor, D. Li, D. Bolton, H. Fairhurst, Determination of amorphous phase levels in Portland cement clinker. Powder Diffraction, **17**, 178 (2002). https://doi.org/10.1154/1.1471518

method (see page 423). It may be—and often is—different from the phases present in the sample to avoid the difficulties associated with preparing a specific phase in a pure form. Moreover, it avoids mixing a standard with the specimen. The diffraction patterns of both the standard and the specimen are measured separately, but under the same conditions. The scale factors (K) obtained from the Rietveld refinement of both the standard and the analyzed specimen are used to calculate already normalized weight fractions (w) of the crystalline phases as follows:

$$w_i = \frac{K_i(ZMV)_i}{K_S(ZMV)_S} w_S^c \frac{(\mu/\rho)_m}{(\mu/\rho)_S} \tag{13.22}$$

where subscripts i and s refer to the phase i in the specimen and the standard Z, M, and V have the same meaning as in Eq. (13.19), μ/ρ_m and μ/ρ_s are mass absorption coefficients for the measured sample and the standard, and w_S^c is weight percent of the crystalline phase in the standard; $w_S^c = 1$, if no amorphous component is present in the standard.

The weight fraction of the amorphous component (w_A) is determined from the difference:

$$w_A = 1 - \sum_{i=1}^{N} w_i \tag{13.23}$$

The accuracy of this method depends critically on the accuracy of the absorption coefficients. The latter can be calculated quite easily (see Eq. 8.49), but their actual values may be far from reality due to possible differences in packing densities of the specimen and the standard. Thus, care must be taken to avoid different packing, which is rarely easy. Another reliable method is to determine the ratio of the absorption coefficients by experimentally measuring Compton scattering,[53] which however requires conducting an additional X-ray emission experiment for both the specimen and the standard.

Internal standard for absolute phase composition[54] using Rietveld method can similarly be—and often is—different from phases present in the sample. This differs from how the internal standard is used in the method of standard additions or spiking (see page 423). Here, a known amount of the standard material is mixed with the specimen. The resulting sample containing w_S fraction of the standard is analyzed in the same way as was described earlier for an external standard. The experimental diffraction pattern is measured only once using a sample mixed with the standard. The weight fractions (w_i^r) obtained from the Rietveld refinement are normalized to

[53] S. Pratapa, B. H. O'Connor, and I-M. Low, Use of Compton scattering measurements for attenuation corrections in Rietveld phase analysis with an external standard. Powder Diffraction **13**, 166 (1998). https://doi.org/10.1017/S0885715600010046

[54] D.L. Bish, S.A. Howard, Quantitative phase analysis using the Rietveld method. J. Appl. Cryst. **21**, 86 (1988). https://doi.org/10.1107/S0021889887009415

match the known content of the crystalline standard phase w'_S[55] and then recalculated to the weight fractions in the original sample (w_i). The combined expression is as follows:

$$w_i = w_i^r \frac{w'_S}{w_S^r} \tag{13.24}$$

Next, the content (weight fraction) of the amorphous portion in the original sample is complementary to the total content of the crystalline phases as in Eq. (13.23). However, when the content(s) of the crystalline phase(s) is not of concern, the amorphous content (w_A) is expressed using only the weight fraction of the standard:

$$w_A = \frac{1 - w'_S/w_S^r}{1 - w_S} \tag{13.25}$$

Thus, quantitative analysis using the Rietveld refinement is straightforward, when the crystal structure of all major phases is known. In practice, this is quite easy since the weight fractions are automatically calculated by the majority of the Rietveld refinement programs. Analysis is also quite accurate when a sample being examined meets all the criteria of an ideal powder specimen, see Sect. 12.1. However, the ideal specimen for quantitative analysis has additional requirements with respect to the packing density and the size of the particles. The former is important and should be uniform, when the external standard is used, to account for different absorption by the sample and the standard, while the latter is most important when the internal standard is employed. The particle size must be small enough for each of them to be transparent for the X-ray wavelength used. This is the consequence of the so-called *microabsorption effect* first outlined by Brindley[56] and later applied to Rietveld analysis by Taylor and others.[57] Brindley showed that microabsorption correction factor τ for phase i can be expressed as:

$$\tau_i = \frac{1}{V_i}\int_0^{V_i} \exp[-(\mu_i - \mu_S)l]dv \tag{13.26}$$

[55] $w'_S = w_S \cdot w_S^c$, if the standard material contains only w_S^c fraction of crystalline phase.

[56] G.W. Brindley, The effect of grain or particle size on X-ray reflection from mixed powders and alloys, considered in relation to the quantitative determination of crystalline substances by X-ray methods, Phil. Mag. **36**, 347 (1945). https://doi.org/10.1080/14786444508520918

[57] J.C. Taylor, Computer programs for standardless quantitative analysis of minerals using the full powder diffraction profile, Powd. Diff. **6**, 2 (1991). https://doi.org/10.1017/S0885715600016778; R.S. Winburn, S.L. Lerach, B.R. Jarabek, M.A. Wisdom, D.G. Grier, G.J. McCarthy, Quantitative XRD analysis of coal combustion by-products by the Rietveld method. Testing with standard mixtures, Adv. X-ray Anal. **42**, 389 (2000).

where V_i is the volume of particle i, v is the fraction of volume in correct orientation for diffraction, l is the length of the beam path in the particle, and μ_I and μ_S are the linear absorption coefficients for the ith phase and the sample, respectively.

This complex integral equation (13.26) was solved numerically and tabulated as a function of $(\mu_i - \mu_S)D$, where D is the particle size. The corrected weight fractions, w^c, may then be calculated by re-normalizing the fractions obtained from the Rietveld refinement as follows:

$$w_i^c = \frac{w_i}{\tau_i} / \sum_{j=1}^{N} \frac{w_j}{\tau_j} \tag{13.27}$$

The main problem in this approach is in obtaining the average particle (not grain) size l for each phase, which should be measured experimentally, for example, using scanning electron microscope or laser scattering. If the accurate particle size is known, the resulting weight fractions are quite precise. However, wrong particle size can yield results that are less accurate compared to those obtained without using any particle size correction. It is noteworthy that many methods based on the integrated intensity discussed in Sect. 13.3.3 are also affected by the microabsorption effects. The Rietveld refinement from the fully crystalline samples can also be affected by microabsorption when phases with different particle size and absorption coefficients are present. This, however, may not be obvious from the output files. Conversely, accuracy in the determination of amorphous content when microabsorption effects are strong and/or improperly accounted for may be clearly seen as a meaningless negative amorphous content. Unfortunately, there is no clear indication of an error when the content of amorphous phase is overestimated.

The significance of effects related to absorption and microabsorption in influencing the accuracy of the quantitative analysis and difficulties in choosing proper corrections (packing density and particle size) makes prevention or reduction of these effects an important issue. In general, the following rules should be followed as closely as possible:[58]

- Both the sample and the standard (if any) should be prepared very carefully in accordance with general rules for powder sample preparation (Sect. 12.1).
- When external standard is used, the uniform packing density should be assured, or actual absorption should be measured.
- The particle size[59] for all components in the mixture should be minimized or optimized. The calculated value of μD (where D is the particle size) should correspond to the "fine" and "medium" categories as defined by Brindley. Keep

[58] The authors would like to express their appreciation to the Rietveld users mailing list (rietveld_l@ill.fr) and, especially, to Reinhard Kleeberg for useful discussions and help with the formulation of these rules.

[59] The particle size and absorption coefficients along with other characteristics for commonly available standard reference materials are listed in Table 13.4.

in mind that when grinding some materials, particles can aggregate, which may increase the microabsorption effect.
- The standard, especially an internal one, should be selected so that its absorption (see Footnote 59) closely matches that of the sample.
- Selection of shorter wavelengths (e.g., Mo versus Cu radiation) is highly beneficial as it significantly reduces absorption. Ideally, neutron diffraction can be used, as absorption in this case is much smaller.

13.4 Additional Reading

1. L.S. Zevin and G. Kimmel, Quantitative X-ray diffractometry, Springer, New York (1995)
2. R. Jenkins and R.L. Snyder, Introduction to X-ray powder diffractometry, Wiley, New York (1996)
3. R.J. Hill, Data collection strategies: fitting the experiment to the need, in: R.A. Young, Ed., The Rietveld method. IUCr monographs on crystallography, 5, Oxford University Press, Oxford (1993)
4. H.P. Klug, and L.E. Alexander, X-ray diffraction procedures for polycrystalline and amorphous materials, Second Edition. Wiley, New York (1974)
5. B.D. Cullity S.R. Stock, Elements of X-Ray Diffraction, Third Edition, Pearson Education Limited (2013)
6. J. Faber and T. Fawcett, The Powder Diffraction File: present and future, Acta Cryst. **B58**, 325 (2002). https://doi.org/10.1107/S0108768102003312
7. S.N. Kabekkodu, J. Faber, and T. Fawcett, New Powder Diffraction File (PDF-4) in relational database format: advantages and data-mining capabilities, Acta Cryst. **B58**, 333 (2002). https://doi.org/10.1107/S0108768102002458
8. M.A. Van Hove, K. Hermann and P.R. Watson, The NIST Surface Structure Database—SSD version 4, Acta Cryst. **B58**, 338 (2002). https://doi.org/10.1107/S0108768102002434
9. P.S. White, J.R. Rodgers, and Y. Le Page, CRYSTMET: a database of the structures and powder patterns of metals and intermetallics, Acta Cryst. **B58**, 343 (2002). https://doi.org/10.1107/S0108768102002902
10. A. Belsky, M. Hellenbrandt, V.L. Karen, and P. Luksch, New developments in the Inorganic Crystal Structure Database (ICSD): accessibility in support of materials research and design, Acta Cryst. **B58**, 364 (2002). https://doi.org/10.1107/S0108768102006948
11. F.H. Allen, The Cambridge Structural Database: a quarter of a million crystal structures and rising, Acta Cryst. **B58**, 380 (2002). https://doi.org/10.1107/S0108768102003890
12. H.M. Berman, T. Battistuz, T.N. Bhat, W.F. Bluhm, P.E. Bourne, K. Burkhardt, Z. Feng, G.L. Gilliland, L. Iype, S. Jain, P. Fagan, J. Marvin, D. Padilla, V. Ravichandran, B. Schneider, N. Thanki, H. Weissig, J.D. Westbrook, and C. Zardecki, The Protein Data Bank, Acta Cryst. **D58**, 899 (2002). https://doi.org/10.1107/S0907444902003451

13. H.M. Berman, J.D. Westbrook, Z. Feng, L. Iype, B. Schneider, and C. Zardecki, The Nucleic Acid Database, Acta Cryst. **D58**, 889 (2002). https://doi.org/10.1107/S0907444902003487
14. N.V.Y. Scarlett, I.C. Madsen, L.M.D. Cranswick, T. Lwin, E. Groleau, G. Stephenson, M. Aylmore, N. Agron-Olshina. Outcomes of the International Union of Crystallography Commission on Powder Diffraction Round Robin on Quantitative Phase Analysis: samples 2, 3, 4, synthetic bauxite, natural granodiorite and pharmaceuticals. J. Appl. Cryst. **35**, 383 (2002). https://doi.org/10.1107/S0021889802008798
15. D.L. Bish, S.A. Howard. Quantitative phase analysis using the Rietveld method. J. Appl. Cryst. **21**, 86 (1988). https://doi.org/10.1107/S0021889887009415

13.5 Problems

1. Consider three powder diffraction patterns, which are shown in Figs. 13.25, 13.26, and 13.27. For each pattern, select all applicable processing steps and explain your reasoning, assuming the goal is to produce digitized (reduced) powder patterns for phase identification.

   (a) Smooth the data (Yes/No/Probably)
   (b) Eliminate background (Yes/No/Probably)
   (c) Strip Kα_2 contributions (Yes/No/Probably)

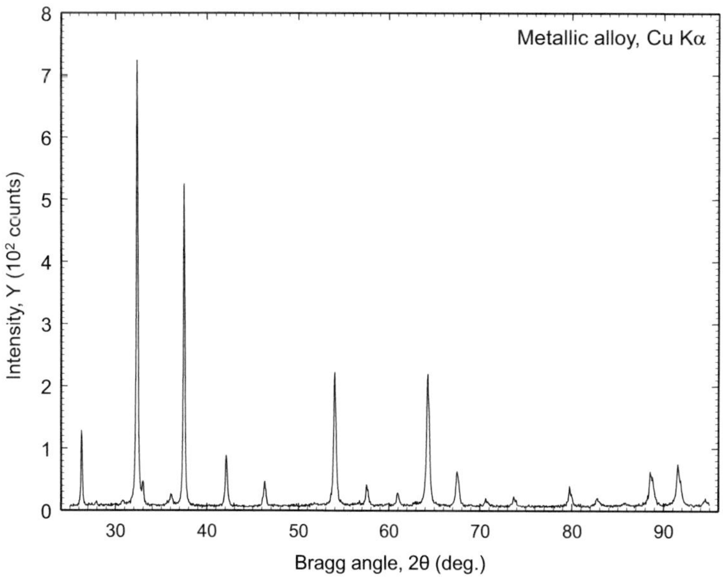

Fig. 13.25 Powder diffraction pattern collected using a conventional X-ray source

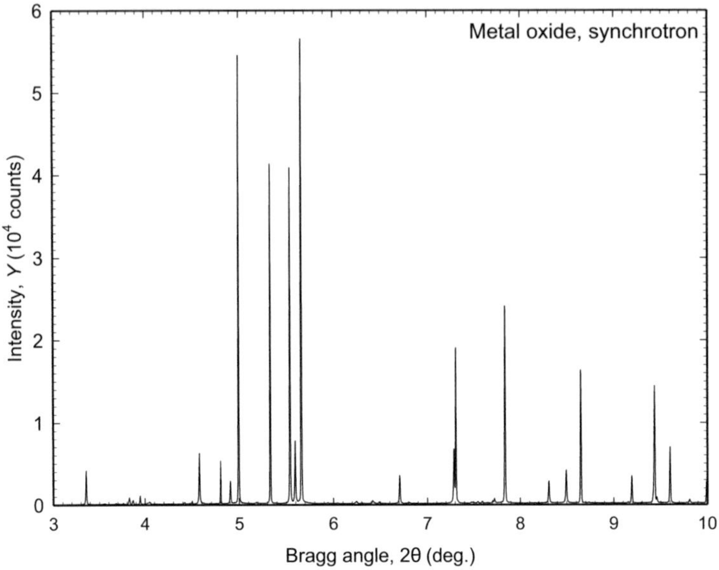

Fig. 13.26 Powder diffraction pattern collected using a synchrotron X-ray source. (Data courtesy of Dr. M.J. Kramer)

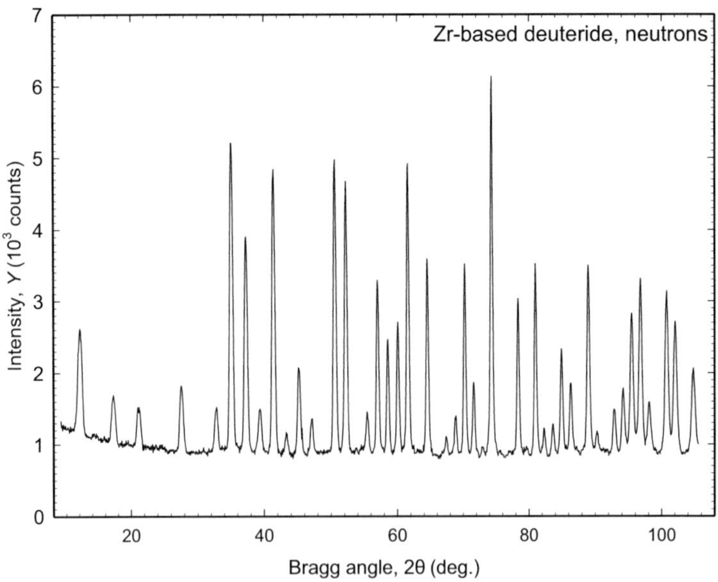

Fig. 13.27 Powder diffraction pattern collected using a reactor-based neutron source. (Data courtesy of Dr. W.B. Yelon)

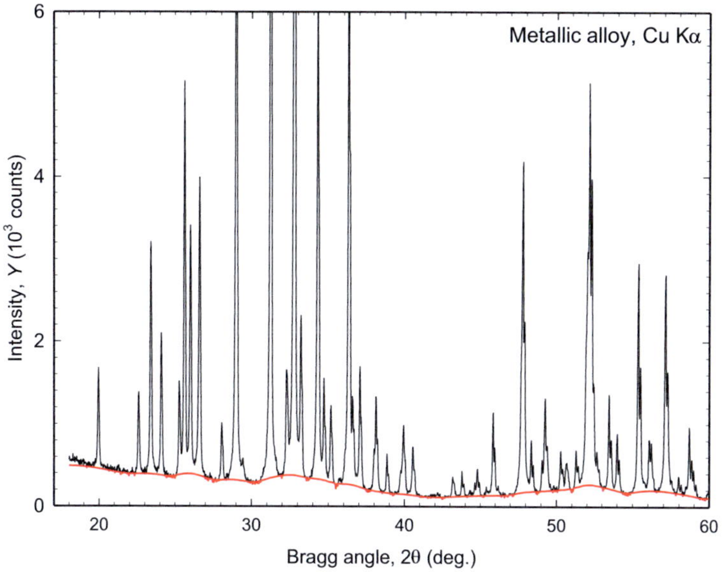

Fig. 13.28 Example of the automatically determined background (*thick red line* at the bottom of the plot shown using the scale identical to the experimental data)

2. A powder diffraction pattern collected from a metallic alloy was processed into two digitized patterns. In the first case, the background was eliminated as illustrated in Fig. 13.28, and in the second case, as shown in Fig. 13.29. In both cases, preliminary processing was continued as follows: $K\alpha_2$ components were stripped, and Bragg peak positions and intensities were determined using an automatic peak search. Compare the reliability of the digitized patterns obtained and explain your reasoning.
3. Consider the powder diffraction pattern shown in Fig. 13.30 and answering Yes/No/Maybe: Is this pattern suitable for phase identification? Is the material suitable for crystal structure determination using powder diffraction? Explain your reasoning. What other conclusions (if any) can be drawn from a visual analysis of this pattern?
4. Diffraction data[60] (Table 13.5) were collected from a white ceramic plate. Using the Mineral Database[61] and 3 strongest of the 15 observed peaks identify the material.

[60] In Problems 4–8, the data were collected on a powder diffractometer with Bragg–Brentano geometry using Cu Kα radiation. Errors in d-spacing should not exceed 0.02 Å for $d > 3$ Å; otherwise, they should be less than 0.01 Å.

[61] This is a freely accessible database available at https://www.webmineral.com/

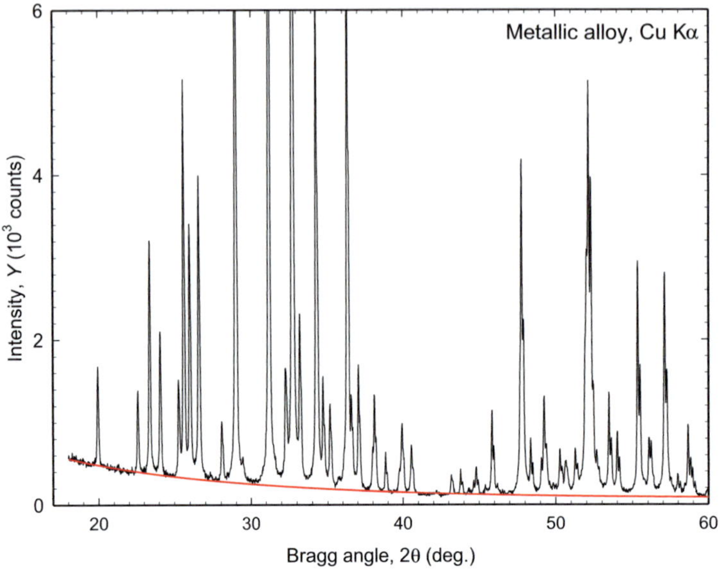

Fig. 13.29 Example of the background represented by a polynomial (*thick red line* at the bottom of the plot shown using the scale identical to the experimental data)

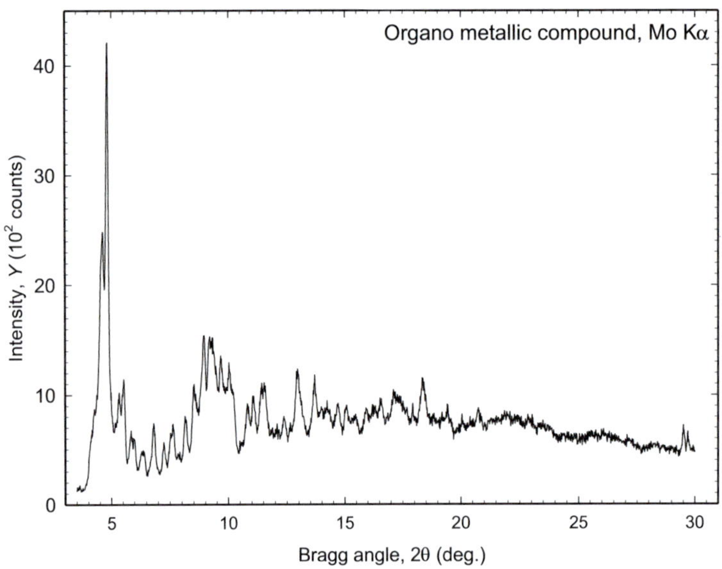

Fig. 13.30 Powder diffraction pattern collected from an organometallic compound on a Rigaku TTRAX powder diffractometer using Mo Kα radiation. The data were collected in a continuous scanning mode: scan rate was 5 deg/min, sampling step 0.01°

Table 13.5 Digitized pattern representing data collected from a white ceramic plate

I/I_0	2θ (°)	*d* (Å)	I/I_0	2θ (°)	*d* (Å)	I/I_0	2θ (°)	*d* (Å)
57	25.556	3.4827	99	57.465	1.6023	9	77.202	1.2346
88	35.125	2.5528	8	61.268	1.5117	7	80.648	1.1903
39	37.747	2.3812	38	66.481	1.4052	8	88.949	1.0995
100	43.324	2.0867	57	68.165	1.3745	10	91.139	1.0787
49	52.514	1.7412	17	76.834	1.2396	20	95.203	1.0431

Table 13.6 Digitized pattern representing data collected from a light-blue colored powder

I/I_0	2θ (°)	*d* (Å)	I/I_0	2θ (°)	*d* (Å)	I/I_0	2θ (°)	*d* (Å)
100	12.778	6.9223	4	41.846	2.1570	6	53.388	1.7147
36	25.724	3.4603	10	43.508	2.0784	3	57.955	1.5900
10	33.514	2.6717	3	49.088	1.8543	2	58.613	1.5737
11	36.438	2.4637	3	51.238	1.7815	2	62.381	1.4873
6	39.753	2.2656	3	52.856	1.7307	3	62.764	1.4792

Table 13.7 Digitized pattern representing data collected from a powder containing fluorine. The powder is stable between room temperature and ~500 °C

I/I_0	2θ (°)	*d* (Å)	I/I_0	2θ (°)	*d* (Å)	I/I_0	2θ (°)	*d* (Å)
41	25.850	3.4438	68	33.080	2.7057	37	49.555	1.8380
14	28.123	3.1704	26	34.120	2.6256	18	50.743	1.7977
17	29.081	3.0680	26	40.017	2.2512	17	51.554	1.7713
100	31.905	2.8026	29	46.858	1.9372	13	52.288	1.7482
43	32.235	2.7747	14	48.259	1.8842	19	53.154	1.7217

Table 13.8 Digitized pattern representing data collected from a manganese-containing powder

I/I_0	2θ (°)	*d* (Å)	I/I_0	2θ (°)	*d* (Å)	I/I_0	2θ (°)	*d* (Å)
100	28.630	3.1153	66	56.560	1.6258	18	72.249	1.3066
52	37.296	2.4090	16	59.275	1.5577	7	86.457	1.1246
13	40.947	2.2022	7	64.736	1.4388	9	93.578	1.0569
15	42.744	2.1137	13	67.136	1.3931	20	100.604	1.0011
5	46.026	1.9703	15	72.125	1.3085	5	102.886	0.9850

5. Diffraction data (Table 13.6) were collected from a light-blue colored powder. Using the Mineral Database and 3 strongest of the 15 observed peaks identify the material.
6. Diffraction data (Table 13.7) were collected from a powder containing fluorine. Using the Mineral Database and 3 strongest of the 15 observed peaks identify the material. Additional information about the powder: no weight loss has been detected during a thermogravimetric experiment carried out between ~25 and ~500 °C.
7. Diffraction data (Table 13.8) were collected from a powder containing manganese. Using the Mineral Database and 3 strongest of the 15 observed peaks identify the material.

Table 13.9 Digitized X-ray diffraction pattern representing data collected from a two-phase powder containing Al, Mn, and O

I/I_0	2θ (°)	*d* (Å)	I/I_0	2θ (°)	*d* (Å)	I/I_0	2θ (°)	*d* (Å)
33	25.460	3.4956	13	40.947	2.2022	66	56.560	1.6258
100	28.630	3.1153	15	42.744	2.1137	58	57.379	1.6045
52	35.031	2.5594	59	43.233	2.0909	16	59.275	1.5577
52	37.296	2.4090	5	46.026	1.9703			
23	37.654	2.3869	29	52.426	1.7439			

8. Diffraction data (Table 13.9) were collected from a two-phase powder containing Mn, Al, and O. Results of mass spectroscopic analysis with respect to all known chemical elements show that there are no other elements present in concentration exceeding 100 parts per million by weight. Using the Mineral Database and 6 strongest of the 13 observed peaks, identify both compounds that are present in the mixture.

Chapter 14
Determination and Refinement of the Unit Cell

As we established in Chap. 1, crystal lattices, which represent the periodic three-dimensional crystal structures of materials, are constructed by translating an identical elementary parallelepiped—the unit cell of a lattice—in three dimensions. Even when a crystal structure is aperiodic, it may still be represented by a three-dimensional unit cell in a lattice that occupies a superspace with more than three dimensions. In this case, conventional translations are perturbed by one or more modulation functions with different periodicity, as was discussed in Chap. 5.

Since the unit cell remains unchanged throughout the infinite lattice, the crystal structure of a material may be considered solved when both the shape and the content of the unit cell, including the spatial distribution of atoms within it, have been established. The determination of any crystal structure inevitably begins with identifying the shape, symmetry, and dimensions of the unit cell—that is, the lengths of the three unit cell edges (a, b, and c) and the values of the three angles (α, β, and γ) between the pairs of the corresponding unit cell vectors (see Fig. 1.4 for reference).

14.1 The Indexing Problem

In powder diffraction, the first step in solving the crystal structure is to determine the true unit cell, a task made challenging by the fact that the experimental data are a one-dimensional projection of the three-dimensional reciprocal lattice, recorded as a function of a single independent variable: the Bragg angle. As a result, directional information is lost, and only the lengths of the reciprocal lattice vectors can be measured in a powder diffraction experiment. This contrasts with single-crystal scattering, where both the length and direction of each vector in reciprocal space are preserved, provided the intensity of a corresponding Bragg peak exceeds the background and is measurable.

The loss of directions is illustrated in Fig. 14.1 for a two-dimensional case. As shown, when different vectors from the two-dimensional reciprocal lattice are

P. Zavalij, V. Pecharsky, *Fundamentals of Powder Diffraction and Structural Characterization of Materials*, https://doi.org/10.1007/978-3-031-91504-8_14

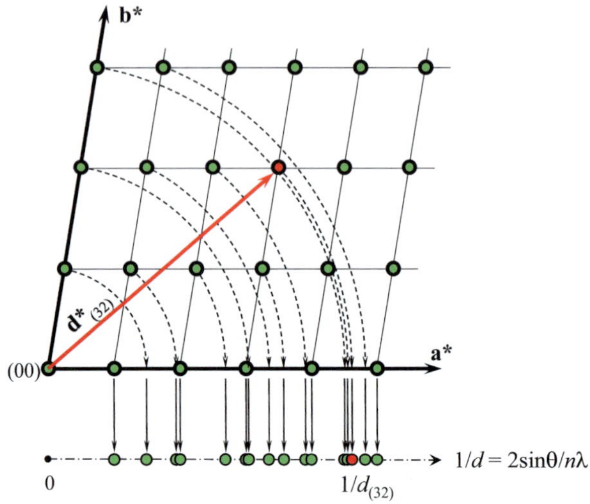

Fig. 14.1 The illustration of a two-dimensional reciprocal lattice (*top*) and its one-dimensional projection on the 1/*d* axis (*bottom*). The scales in the two parts of the drawing are identical because $1/d = d^*$. The 1/*d* axis in the figure is shifted downward from the origin of the reciprocal lattice for clarity. The reciprocal lattice point (32) is shown as a *red circle* both in the lattice and in its projection together with the corresponding reciprocal vector $\mathbf{d}^*_{(32)}$

projected onto the 1/*d* axis (which can be chosen arbitrarily as long as it intersects the origin of the reciprocal lattice), they all lose their "sense" of direction and are distinguishable only by their lengths. It is worth reminding one's self that the distribution of points along the 1/*d* axis in Fig. 14.1 determines the Bragg angles at which scattered intensity maxima can be observed in a powder diffraction experiment, as directly follows from the Braggs' law: $d^* = 1/d = 2 \sin \theta/n\lambda$.

Regardless of the nature of the diffraction experiment, finding the unit cell in a conforming lattice is a matter of selecting the smallest parallelepiped in reciprocal space, which completely describes the array of the experimentally registered Bragg peaks. Obviously, the selection of both the lattice and the unit cell should be consistent with crystallographic conventions (see Sect. 2.11), which impose certain constraints on the relationships between unit cell symmetry and dimensions.

Since each point in reciprocal space represents a series of crystallographic planes, the description of diffraction data by means of any lattice may therefore be reduced to assigning triplets of Miller indices to every observed Bragg peak based on the selected unit cell. Recalling the definitions of both the direct and reciprocal lattices (Eqs. 1.1 and 1.7, respectively) and considering Fig. 14.1 (or Fig. 8.10, which illustrates a three-dimensional case), the assignment of indices in a periodic lattice is based on Eq. (14.1).[1] The latter establishes relationships between the unit vectors

[1]Conventional lattices may be perturbed by functions with different periodicity, e.g. by sinusoidal or saw-tooth-like modulations, see Chap. 5. In the simplest case (one-dimensional modulation), Eq. (14.1) becomes $\mathbf{d}^*_{hkl} = h\mathbf{a}^* + k\mathbf{b}^* + l\mathbf{c}^* + m\mathbf{q}$ assuming that the perturbation function is periodic and has the modulation vector $\mathbf{q}$. In a case of three-dimensional modulation, a total of six indices (*h*, *k*, *l*, *m*, *n*, and *p*) are required to identify every point observed in reciprocal super-space: $\mathbf{d}^*_{hkl} = h\mathbf{a}^* + k\mathbf{b}^* + l\mathbf{c}^* + m\mathbf{q}_1 + m\mathbf{q}_2 + m\mathbf{q}_3$, where $\mathbf{q}_1$, $\mathbf{q}_2$, and $\mathbf{q}_3$ are the modulation vectors of the corresponding perturbation functions. If this is the case, vectors $\mathbf{q}_1$, $\mathbf{q}_2$, and $\mathbf{q}_3$ should be established in addition to **a***, **b***, and **c*** before assignment of indices can be performed. Since

$\mathbf{a}^*$, $\mathbf{b}^*$, and $\mathbf{c}^*$ and the corresponding reciprocal vectors $\mathbf{d}^*_{hkl}$ in terms of triplets of integers h, k, and l.

$$\mathbf{d}^*_{hkl} = h\mathbf{a}^* + k\mathbf{b}^* + l\mathbf{c}^* \tag{14.1}$$

This process is commonly known as indexing of diffraction patterns, and in three dimensions it usually has a unique and easy solution when both the lengths and directions of reciprocal vectors, $\mathbf{d}^*_{hkl}$, are available. On the contrary, when only the lengths, d^*_{hkl}, of the vectors in the reciprocal space are known, the task may become extremely complicated, especially if there is no additional information about the crystal structure other than the array of numbers representing the observed $1/d_{hkl}\left[\equiv d^*_{hkl}\right]$ values.

The difficulty and reliability of indexing are closely related to the absolute accuracy of the array of d^*_{hkl} values, i.e., to the absolute accuracy with which positions of Bragg reflections have been determined. For example, when the accuracy is nearly ideal, e.g., when the calculated rather than the experimental Bragg angles are used to determine d^*_{hkl}, indexing is usually straightforward and consistent even for large, low-symmetry unit cells. However, the presence of both random and systematic errors combined with multiple and sometimes severe overlaps of Bragg reflections—often observed in complex crystal structures characterized by low symmetry and large unit cell volume—reduces the accuracy of peak positions and, therefore, decreases the chances of successful indexing (or confidence in it) from first principles.[2]

As established earlier (see Sect. 8.4), the interplanar distances, d, are related to both the unit cell dimensions and Miller indices of the families of crystallographic planes by means of a well-defined function, which in general form can be written as follows:

$$d_{hkl} = f(h, k, l, a, b, c, \alpha, \beta, \gamma) \tag{14.2}$$

Peak positions, θ_{hkl}, are measurable from a powder diffraction experiment, for example, using any of the approaches discussed in Chap. 13. The observed d-spacing for any given combination (hkl) is then determined using the Braggs' equation, in which λ is the wavelength of the X-ray radiation used to collect the data and $n = 1$, meaning that only the first order reflections are included into the consideration:

---

even the three-dimensional diffraction pattern of a modulated structure is a projection of four- to six-dimensional superspace, indexing of single crystal diffraction data is quite complex. It is rarely, if ever, successful from first principles when only powder diffraction data are available.

[2]Similar to structure solution from first principles, the ab initio indexing implies that no prior knowledge about the symmetry or approximate unit cell dimensions of the crystal lattice exists. Indexing from first principles, therefore, usually means that Miller indices are assigned based strictly on the relationships between the observed Bragg angles.

$$d_{hkl}^{obs} = \frac{\lambda}{2\sin\Theta_{hkl}^{obs}} \tag{14.3}$$

By combining Eqs. (14.2) and (14.3), the observed positions of Bragg peaks may be used to calculate the corresponding unit cell dimensions, but first, the triplets of integer indices, *h*, *k*, and *l*, should be assigned to all observed diffraction maxima, or in other words, all observed Bragg peaks should be indexed in agreement with Eqs. (14.1)–(14.3). The algorithm of the indexing process in powder diffraction is usually dependent on whether or not the shape and dimensions of the unit cell are known at least approximately.

14.2 Known Versus Unknown Unit Cell Dimensions

Indexing of powder diffraction data when unit cell dimensions are known with certain accuracy includes:

1. Generating a list of all possible combinations of symmetrically independent *hkl* triplets, which can be observed within the studied range of Bragg angles.
2. Calculating interplanar distances using the generated list of *hkl*, the best estimate of the unit cell dimensions (a, b, c, α, β, and γ), and the appropriately simplified form of Eq. (14.2) that are given in Eqs. (8.2)–(8.7).
3. Assigning *hkl* triplets to the observed Bragg peaks by matching d^{obs} and d_{hkl}^{calc} (or θ^{obs} and $\theta_{hkl}^{\text{calc}}$) based on the minimum difference between the pairs of values.

Refining the unit cell dimensions using θ^{obs} or d^{obs} coupled with the assigned *hkl* triplets, i.e., using $\theta_{hkl}^{\text{obs}}$ or d_{hkl}^{obs}, respectively.

Although the indexing process may take several iterations, each resulting in a more accurate assignment of indices and in a better approximation of the unit cell, finding the best solution is usually trivial.

When both the symmetry of the lattice and unit cell dimensions are unknown, the ab initio indexing of powder diffraction data often becomes a trial-and-error process and finding the correct unit cell may be a challenge. This occurs because the assignment of *hkl* triplets to each observed Bragg peak is done without prior knowledge of the unit cell parameters (a total of six in the most general case). Clearly, this task is equivalent to restoring the directions of all observed reciprocal vectors based only on their lengths, so to say one needs to restore a three-dimensional image from a single one-dimensional projection. Referring to Fig. 14.1, it is nearly as easy to obtain the lower part of the figure from its upper part, as it is difficult to reconstruct the latter if only the former is known.

The difficulty of the ab initio indexing may be further illustrated using a noncrystallographic geometrical example by considering a cone, a cylinder, and a sphere, all of which have identical radii. Under certain conditions, it is possible that their projections are reduced to identical circles in two dimensions and then to

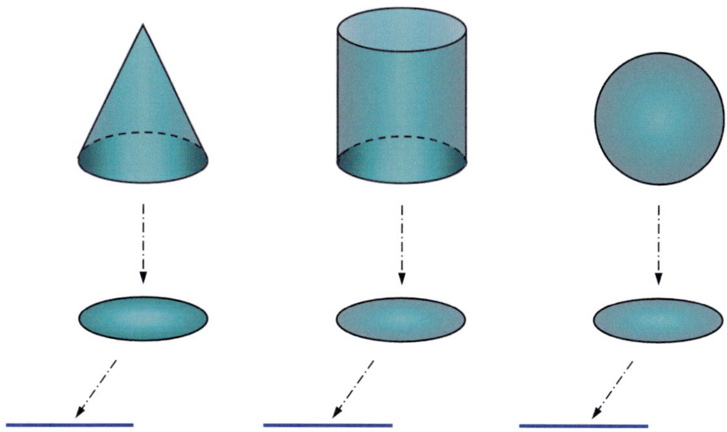

Fig. 14.2 The illustration of three indistinguishable one- and two-dimensional projections obtained from three different three-dimensional objects. The projection directions are shown by *dash-dotted arrows*

indistinguishable lines in one dimension,[3] as shown in Fig. 14.2. Assuming that there is no additional information about these objects and their projections, it is impossible to restore the correct shape of the object in three dimensions based on a single one-dimensional or even two-dimensional projection.

The problem of indexing powder diffraction data is intricate but not as hopeless as it may appear from Fig. 14.2 due to the presence of governing laws, Eqs. (14.1)–(14.3). They define a set of rules for the reconstruction of the reciprocal lattice, in which a vector of a given length may only realize a limited number of orientations. The ab initio indexing is, therefore, possible because vector directions should be such that their ends form a three-dimensional lattice. A two-dimensional example is found in Fig. 14.3, where the reciprocal lattice is identical to that shown in Fig. 14.1 and is depicted with both the positive and negative directions of the two basis lattice vectors **a*** and **b***.

In the absence of numerous overlapping Bragg peaks, which is the same as the absence of numerous independent reciprocal lattice vectors with identical lengths, the solution of the ab initio indexing problem is relatively straightforward. This is, however, the case only in the highest symmetry crystal system, i.e., cubic. As the symmetry of the lattice lowers, and especially when the unit cell volume of the direct lattice increases, multiple vectors with equal lengths but different directions will appear in the reciprocal lattice, e.g., $\mathbf{d}^*_{(32)}$ and $\mathbf{d}^*_{(4\bar{2})}$, where $d^*_{(32)} = d^*_{(4\bar{2})}$ in Fig. 14.3. Furthermore, numerous vectors will have nearly identical lengths but different directions, for example, $\mathbf{d}^*_{(03)}$, $\mathbf{d}^*_{(41)}$, $\mathbf{d}^*_{(2\bar{3})}$, and $\mathbf{d}^*_{(1\bar{3})}$ in addition to $\mathbf{d}^*_{(32)}$ and $\mathbf{d}^*_{(4\bar{2})}$.

[3] Strictly speaking, the shape of each object could be recognized from two-dimensional shadows (projections) if the objects are semitransparent. Recognition becomes impractical from one-dimensional projections.

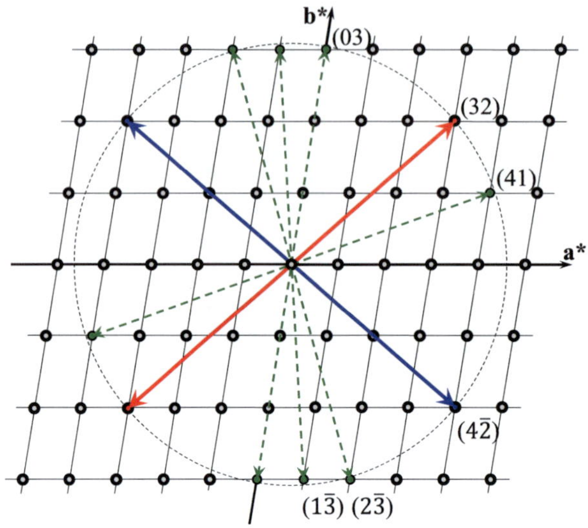

Fig. 14.3 The illustration of a reciprocal lattice showing both the length and orientation of the reciprocal vector $\mathbf{d}^*_{(32)}$. A second vector in this lattice with identical length but different orientation is $\mathbf{d}^*_{(4\bar{2})}$ and the identity of their lengths is coincidental, i.e., not mandated by lattice symmetry. In addition, there are several vectors [$\mathbf{d}^*_{(03)}$, $\mathbf{d}^*_{(41)}$, $\mathbf{d}^*_{(2\bar{3})}$, and $\mathbf{d}^*_{(1\bar{3})}$] with their lengths nearly identical to those of $\mathbf{d}^*_{(32)}$ and $\mathbf{d}^*_{(4\bar{2})}$. Note that since this two-dimensional reciprocal lattice has a twofold symmetry axis perpendicular to the plane of the projection and intersecting the origin of coordinates, each vector has its symmetrical equivalent in the opposite direction (the indices of any pair of the two symmetrically equivalent vectors have the same values but opposite signs)

Since both the resolution of the instrument and the accuracy of Bragg angle measurements are finite, the proper indexing of complex powder diffraction pattern(s) may be difficult or nearly impossible from first principles.

We conclude this section with a simple notion: it is impossible to solve the crystal structure of a material using an incorrect unit cell. Thus, proper indexing of the experimental powder diffraction pattern is of utmost importance, and below we shall consider various strategies leading to the solution of the indexing problem and finding of the most precise unit cell dimensions.

14.3 Indexing: Known Unit Cell

As mentioned earlier, indexing or assignment of *hkl* triplets using known unit cell dimensions is usually a trivial task. It may be completely formalized and therefore handled by a computer program automatically or nearly automatically with minimal intervention by the user. However, it is often true that the unit cell dimensions are

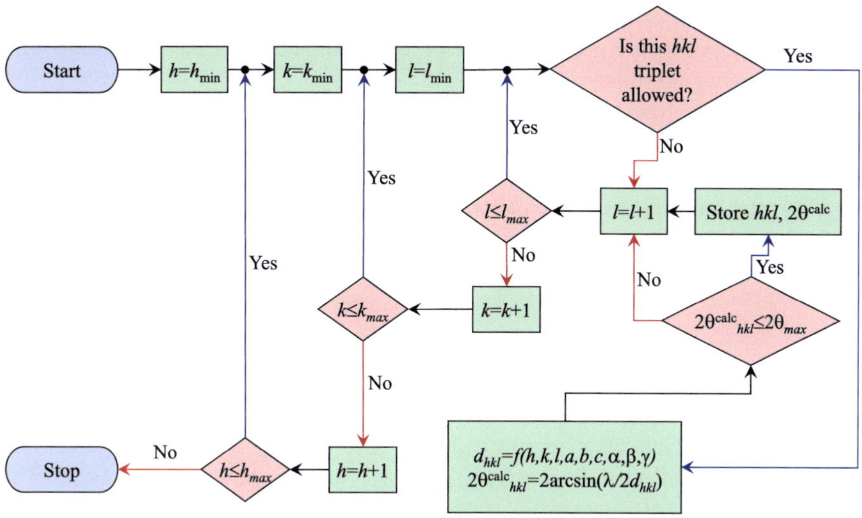

Fig. 14.4 The flowchart illustrating a generic algorithm designed to generate all allowed combinations of Miller indices and compute Bragg angles from known unit cell dimensions. Only the values with $2\theta_{hkl}^{calc}$ not exceeding the maximum observed 2θ are retained for further use. Both minimum and maximum *h*, *k*, and *l* values are calculated from the maximum observed 2θ

known only approximately or they are simply guessed. If this is the case, indexing is usually performed in several iterations as described by the following algorithm:

1. Expected Bragg peak positions, $2\theta_{hkl}^{calc}$, shall be computed for all possible combinations of Miller indices using the best available estimate of the unit cell dimensions coupled with the appropriate form of Eq. (14.2), suitably simplified to reflect the symmetry of the reciprocal lattice. Only those combinations of *h*, *k*, and *l* that are allowed by symmetry of the crystal lattice should be included in the calculations. The proper computational process is illustrated by a flowchart in Fig. 14.4. When available, Bragg peak positions coupled with known *hkl* triplets taken from a literature reference or from a database (e.g., the ICDD's Powder Diffraction File[4]) can be used instead of a computed list of $2\theta_{hkl}^{calc}$.
2. Assignment of indices should always start from a peak observed at the lowest Bragg angle and proceed toward the higher Bragg angles. The low Bragg angle peaks are usually well-resolved and are located far apart from one another (e.g., see Fig. 14.1), their calculated positions are least affected by inaccuracies in an initial approximation of the unit cell, and therefore, it is relatively easy to decide which triplet of indices corresponds to which observed peak. The indexing should only continue as long as there is no ambiguity in the assignment of indices.
3. Using all Bragg peaks which have been indexed and the associated observed Bragg angles, more accurate unit cell dimensions and, if applicable, systematic

[4] See Sect. 13.3 and https://www.icdd.com/ for more information about the Powder Diffraction File.

experimental errors, for example, sample displacement, sample transparency, or zero shift, which are described in Sect. 8.4.2, should be refined by means of a least-squares technique (see Sect. 14.12).

4. Using the improved unit cell dimensions obtained in Step 3, the process is repeated from Step 1 until all observed Bragg peaks have been indexed.[5] The indexing is considered complete when index assignments and the refined lattice parameters remain unchanged after the last iteration in comparison with the same from the previous cycle.

14.3.1 High-Symmetry Indexing Example

We illustrate the indexing approach described above by using the experimental diffraction data collected from a $LaNi_{4.85}Sn_{0.15}$ sample shown in Fig. 14.5.[6] The observed positions of Bragg peaks were determined using a profile fitting procedure and these are listed in Table 14.1 together with their relative integrated intensities (I/I_0) and full widths at half maximum (FWHM). The least-squares standard deviations in the observed Bragg angles did not exceed 0.003°.[7]

The crystal structure of the powder is hexagonal: space group is P6/mmm and the lattice parameters are $a = 5.04228(6)$, $c = 4.01170(5)$ Å, as determined from Rietveld refinement using diffraction data with $2\theta_{max} = 120°$. To better illustrate the indexing process, we will use a spreadsheet in the calculation of Bragg angles using both the approximate and refined unit cell dimensions, rather than relying on specialized software. Assume that we only know lattice parameters of this material to within ~0.10 Å, and that the best available approximation to begin with is $a = 4.95$ and $c = 4.10$ Å.

The simplified form of Eq. (14.2) in the hexagonal crystal system is

$$\frac{1}{d^2} = \frac{4}{3} \times \frac{h^2 + hk + k^2}{a^2} + \frac{l^2}{c^2} \tag{14.4}$$

By combining Eqs. (14.4) and (14.3), we obtain the following equation relating Bragg angles, lattice parameters, and Miller indices in the hexagonal system:

[5] When a powder diffraction pattern contains a few weak impurity reflections, indexing of all Bragg peaks may be impossible. Generally, all unindexed peaks should be explained by identifying impurity phase(s) or by shape anomalies when compared to the main phase.

[6] Same data are available online in the ASCII data file *Ch14Ex01_CuKa.xy*.

[7] The content of Table 14.1 can be found online in the file *Ch14Ex01_CuKa.pks*.

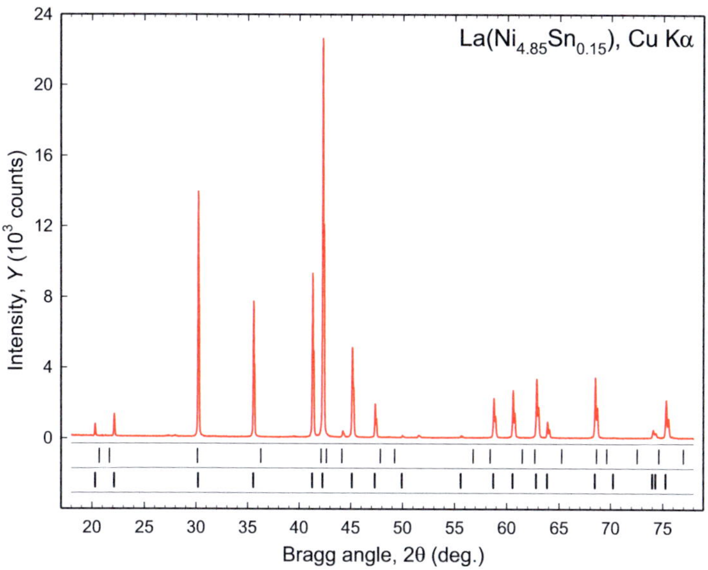

Fig. 14.5 A fragment of the diffraction pattern collected from a $LaNi_{4.85}Sn_{0.15}$ powder on a Rigaku TTRAX rotating anode powder diffractometer using Cu $K\alpha_{1,2}$ radiation. The data were collected in a step-scan mode with a step 0.02° of 2θ and counting time 4 s. As explained later (see 14.5 and Table 14.3, respectively), the two sets of vertical bars indicate locations of Bragg peaks calculated using the first (the upper set of bars) and the second (the lower set of bars) approximations of the unit cell dimensions

$$2\theta_{hkl} = 2\arcsin\left(\lambda\sqrt{\frac{1}{3}\times\frac{h^2+hk+k^2}{a^2}+\frac{1}{4}\times\frac{l^2}{c^2}}\right) \quad (14.5)$$

To ensure that no possible combination of indices have been missed (see Fig. 14.4), it is best to create a spreadsheet, in which the columns are labeled with values of l varying from $l_{min} = -l_{max}$ to l_{max}. The value of l_{max} is determined from Eq. (14.5) by substituting $2\theta_{max}$ for 2θ and letting $h = k = 0$. The rows are labeled with the values of h and k. In general, both h and k should vary between their respective minimum and maximum values determined in the same way as l_{min} and l_{max}.[8] In the hexagonal crystal system, however, there is no need to include negative values of indices due to limitations imposed by symmetry of the reciprocal lattice, and the smallest values of all three indices are set at 0. Further, reciprocal lattice points that are different from one another by a permutation of h and k are symmetrically equivalent in this hexagonal lattice. As a result, we have an additional restriction, i.e., $h \le k$ (or $h \ge k$) that limits the possible combinations of indices. No other

[8] A more precise definition of $|h_{max}|$, $|k_{max}|$, and $|l_{max}|$ can be given in terms of the maximum reciprocal lattice vector length, which is $\left|\mathbf{d}^*_{hkl}\right|_{max} \le 2/\lambda$.

Table 14.1 Relative integrated intensities (I/I_0), Bragg angles, and full widths at half maximum (FWHM) of Bragg peaks observed in the $LaNi_{4.85}Sn_{0.15}$ powder diffraction pattern collected using Cu Kα radiation in the range 18° ≤ 2θ ≤ 83° (see Fig. 14.5)

I/I_0	2θ (°)[a]	FWHM (°)
20	20.288	0.070
43	22.105	0.077
513	30.198	0.076
305	35.548	0.078
394	41.285	0.082
1000	42.272	0.085
18	44.211	0.099
274	45.130	0.106
89	47.332	0.091
4	49.965	0.109
10	51.517	0.140
6	55.621	0.094
136	58.742	0.117
149	60.584	0.112
210	62.824	0.123
54	63.866	0.115
213	68.485	0.124
35	74.050	0.140
7	74.353	0.140
153	75.298	0.140
4	78.958	0.175
9	79.629	0.175
123	81.357	0.175
55	81.632	0.175

[a] Bragg angles are listed for the location of the $K\alpha_1$ component in the doublet, λ = 1.540593 Å

limitations are imposed on the allowed combinations of indices in the case of $LaNi_{4.85}Sn_{0.15}$ because the space-group symmetry of the material is P6/mmm, which according to Table 9.8 has no forbidden reflections.

The spreadsheet used to calculate the values in Table 14.2 can be found online.[9] The calculated Bragg peak positions from Table 14.2 are plotted as the upper set of vertical bars in Fig. 14.5. It is easy to recognize that only four lowest Bragg angle peaks can be decidedly indexed using the first approximation of the unit cell dimensions. The corresponding assignment of the triplets of indices is shown in Table 14.3, which is the result of combining the experimental data listed in Table 14.1 with the calculated Bragg peak positions listed in Table 14.2.

Based on the results of the first round of indexing, the least-squares refinement of the lattice parameters yields the following values: $a = 5.047(1)$, c = 4.017(2) Å. The

[9]The file name is Ch14Ex01_BraggAngles.xls.

Table 14.2 Bragg angles[a] calculated using the first approximation of the unit cell dimensions for $LaNi_{4.85}Sn_{0.15}$: $a = 4.95$, $c = 4.10$ Å. Cells with the dashes correspond to the combinations of indices, which cannot be observed in the range $0° < 2\theta \leq 83°$

	l				
hk	0	1	2	3	4
00	–	21.658	44.142	68.615	–
01	20.703	30.137	49.229	72.539	–
02	42.124	47.848	62.657	–	–
03	65.240	69.622	82.159	–	–
11	36.267	42.636	58.407	80.159	–
12	56.772	61.486	74.598	–	–
22	76.992	81.113	–	–	–
13	80.764	–	–	–	–

[a] Bragg angles are listed for the location of the $K\alpha_1$ component in the doublet, $\lambda = 1.540593$ Å

Table 14.3 Index assignments in the powder diffraction pattern of for $LaNi_{4.85}Sn_{0.15}$ after the first iteration assuming $a = 4.95$, $c = 4.10$ Å

I/I_0	*hkl*	$2\theta^{calc}$	$2\theta^{obs}$	$2\theta^{obs}–2\theta^{calc}$	$FWHM^{obs}$
20	010	20.703	20.288	−0.415	0.070
43	001	21.658	22.105	0.447	0.077
513	011	30.137	30.198	0.061	0.076
305	110	36.267	35.548	−0.719	0.078

Table 14.4 Bragg angles[a] calculated using the second approximation of the unit cell dimensions for $LaNi_{4.85}Sn_{0.15}$ obtained after a least-squares refinement of lattice parameters employing the data from Table 14.3: $a = 5.047$, $c = 4.017$ Å. Cells with the dashes correspond to the combinations of indices, which cannot be observed in the range $0° < 2\theta \leq 83°$

	l				
hk	0	1	2	3	4
00	–	21.658	44.142	68.615	–
01	20.703	30.137	49.229	72.539	–
02	42.124	47.848	62.657	–	
03	65.240	69.622	82.159	–	–
11	36.267	42.636	58.407	80.159	–
12	56.772	61.486	74.598	–	–
22	76.992	81.113	–	–	–
13	80.764	–	–	–	–

[a] Bragg angles are listed for the location of the $K\alpha_1$ component in the doublet, $\lambda = 1.540593$ Å

recalculated Bragg angles are shown in Table 14.4 and they are plotted in Fig. 14.5 as the lower set of vertical bars. It is clear that nearly all observed Bragg peaks may now be unambiguously indexed as shown in Table 14.5, where all differences between the observed and calculated 2 are smaller than a fraction of the peaks full width at half maximum.

A least-squares refinement of the unit cell dimensions of $LaNi_{4.85}Sn_{0.15}$ based on the observed Bragg angles and indices listed in Table 14.5 yields the following

Table 14.5 Assignment of indices after the unit cell dimensions of $LaNi_{4.85}Sn_{0.15}$ have been refined using four lowest Bragg angle peaks: $a = 5.047$, $c = 4.017$ Å

I/I_0	*hkl*	$2\theta^{calc}$	$2\theta^{obs}$	$2\theta^{obs}–2\theta^{calc}$	$FWHM^{obs}$
20	010	20.301	20.288	−0.013	0.070
43	001	22.111	22.105	−0.006	0.077
513	011	30.193	30.198	0.005	0.076
305	110	35.546	35.548	0.002	0.078
394	020	41.277	41.285	0.008	0.082
1000	111	42.260	42.272	0.012	0.085
18[a]	–	–	44.211	–	0.099
274	002	45.104	45.130	0.026	0.106
89	021	47.314	47.332	0.018	0.091
4	012	49.931	49.965	0.034	0.109
10[b]	–	–	51.517	–	0.140
6	120	55.586	55.621	0.035	0.094
136	112	58.703	58.742	0.039	0.117
149	121	60.552	60.584	0.032	0.112
210	022	62.783	62.824	0.041	0.123
54	030	63.836	63.866	0.030	0.115
213	031	68.445	68.485	0.040	0.124
–	003	70.238	–	–	–
35	013	73.979	74.050	0.071	0.140
7	122	74.276	74.353	0.077	0.140
153	220	75.251	75.298	0.047	0.140
4	130	78.903	78.958	0.055	0.175
9	221	79.570	79.629	0.059	0.175
123	113	81.271	81.357	0.086	0.175
55	032	81.561	81.632	0.071	0.175

[a,b] The 111[a] and 002[b] Bragg reflections of an impurity phase (a solid solution of Sn in Ni)

unit cell: $a = 5.0421(1)$, $c = 4.0118(1)$ Å. The lattice parameters have been refined together with a zero-shift correction, which was determined to be 0.032°. The final indexing of this diffraction pattern is shown in Table 14.6, where the observed Bragg angles have been corrected for the zero-shift error by adding 0.032° to each observed 2θ.

There are two weak Bragg peaks in this diffraction pattern, which could not be indexed since they do not belong to the hexagonal crystal lattice with the established unit cell. As we will see later (Chap. 16), these two peaks manifest the presence of a small amount of an impurity phase—a solid solution of Sn in Ni, which has a face-centered cubic crystal structure with the space-group symmetry and $a = 3.543$ Å. Only one theoretically possible Bragg peak of the major hexagonal phase (003) is unobserved in this powder diffraction pattern because its intensity is below the limits of detection.

Table 14.6 Final assignment of indices for the powder diffraction pattern of $LaNi_{4.85}Sn_{0.15}$. The refined unit cell dimensions are: $a = 5.0421(1)$, $c = 4.0118(1)$ Å; zero shift is 0.032°

I/I_0	hkl	$2\theta^{calc}$	$2\theta^{obs\,a}$	$2\theta^{obs}-2\theta^{calc}$	$FWHM^{obs}$
20	010	20.321	20.320	−0.001	0.070
43	001	22.140	22.137	−0.003	0.077
513	011	30.228	30.230	0.002	0.076
305	110	35.582	35.580	−0.002	0.078
394	020	41.319	41.317	−0.002	0.082
1000	111	42.307	42.304	−0.003	0.085
18[b]	–	–	44.243	–	0.099
274	002	45.165	45.162	−0.003	0.106
89	021	47.366	47.364	−0.002	0.091
4	012	49.997	49.997	0.000	0.109
10[c]	–	–	51.549	–	0.140
6	120	55.644	55.653	0.009	0.094
136	112	58.779	58.774	−0.005	0.117
149	121	60.620	60.616	−0.004	0.112
210	022	62.864	62.856	−0.008	0.123
54	030	63.906	63.898	−0.008	0.115
213	031	68.524	68.517	−0.007	0.124
–	003	70.343	–	–	–
35	013	74.088	74.082	−0.006	0.140
7	122	74.372	74.385	0.013	0.140
153	220	75.337	75.330	−0.007	0.140
4	130	78.995	78.990	−0.005	0.175
9	221	79.665	79.661	−0.004	0.175
123	113	81.392	81.389	−0.003	0.175
55	032	81.668	81.664	−0.004	0.175

[a] The observed Bragg angles are listed as $2\theta^{meas.} + 0.032°$, where $2\theta^{meas.}$ is the as-measured Bragg angle, to account for the determined zero-shift error

[b,c] The 111[b] and 002[c] Bragg reflections of an impurity phase (a solid solution of Sn in Ni)

The column containing full widths at half maximum of the observed Bragg peaks is a useful tool in deciding whether the final differences between the observed and calculated Bragg angles are satisfactory or not—their absolute values should be lower than a small fraction of the corresponding FWHM. Furthermore, the observed FWHMs can be used in a computerized indexing procedure. Since the resolution of a well-aligned laboratory powder diffractometer is typically high enough to distinguish a pair of overlapping Bragg peaks with comparable intensities when their positions differ only by about ½–¼ of their FWHMs, a fraction of the experimentally determined FWHM may be used instead of a random tolerance parameter during the indexing. When the difference $|2\theta^{obs}| - |2\theta^{calc}|$ is less than the tolerance, the index triplet is assigned to the peak; otherwise, the assignment is not performed.

It is important to note that in the example considered in this section, the initial approximation of the unit cell dimensions was quite inaccurate. Nonetheless, the indexing was straightforward due to the small unit cell and high symmetry. As a result, only a limited number of reflections were possible within the range of measured Bragg angles, and most neighboring Bragg peaks were clearly resolved in the diffraction pattern.

Rather inaccurate lattice parameters can result from a comparison with known structures, serving as a basis for the initial guess. Considerable differences between the real and guessed unit cell dimensions can make indexing challenging, especially in case of large unit cells or low-symmetry crystal structures. In many real cases, the best-possible accuracy in the initial unit cell dimensions is critical for completing the indexing task in a reasonable time, that is, in a reasonable number of iterations. The whole pattern can rarely be indexed using the initial and imprecise approximation of lattice parameters due to inaccuracies in both the unit cell dimensions and in the measured peak positions, especially when systematic errors in the measured Bragg angles (e.g., zero-shift, sample-displacement, and/or transparency effects) are present.

14.3.2 Other Crystal Systems

Indexing of powder diffraction data in crystal systems other than hexagonal, when unit cell dimensions are approximately known, follows the procedure described in Sect. 14.3.1, except that the proper form of Eq. (14.2) should be used in Eqs. (14.4) and (14.5). In low-symmetry crystal systems, such as monoclinic and triclinic, one or two indices, respectively, must include negative values. These indices should vary from $-i_{max}$ to $+i_{max}$, where $i = h$, k, or l, for a complete generation of the list of possible *hkl*. Referring to the example of the two-dimensional reciprocal lattice in Fig. 14.3, it is easy to see that $d^*_{32} \neq d^*_{\bar{3}2}$, and therefore, the list of possible Bragg angles must include a set of reciprocal points with index h varying from $-h_{max}$ to h_{max} and k varying from 0 to k_{max} for completeness. This covers the upper half of the circle drawn in the reciprocal lattice. We note that all symmetrically independent combinations will also be generated when h varies from 0 to h_{max} but k varies from $-k_{max}$ to k_{max}, which corresponds to a semicircle on the right of Fig. 14.3.

The minimum and maximum values of Miller indices in three dimensions are fully determined by the symmetrically independent fraction of the reciprocal lattice

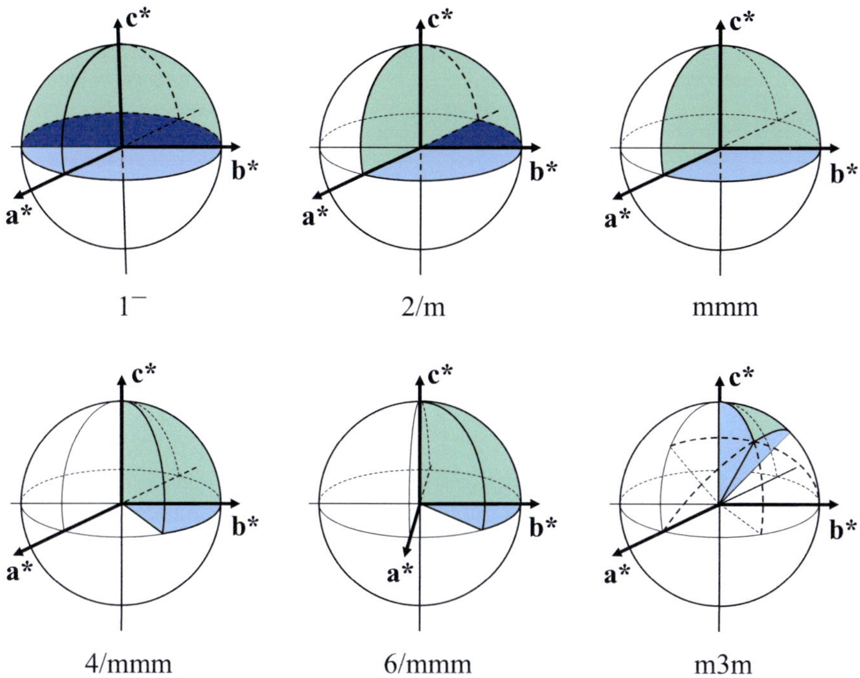

Fig. 14.6 Schematic representations in the reciprocal space of the sphere volume fractions ($r = 1/\lambda$) where the list of *hkl* triplets should be generated in each of the six "powder" Laue classes. This ensures that all symmetrically independent points in the reciprocal lattice are included in the calculation of Bragg angles using the proper form of Eq. (14.2). The monoclinic crystal system is shown in the alternative setting, with the unique twofold axis parallel to **c*** instead of the standard setting where the twofold axis is parallel to **b***

as shown in Fig. 14.6 for the six distinguishable "powder" Laue classes. The same conditions are also listed in Table 14.7.[10]

Instead of using a spreadsheet to generate a list of possible hkl values and calculate Bragg angles from known or approximate unit cell dimensions, it is possible to use one of several computer programs available for download from the

[10] Both Table 14.7 and Fig. 14.6 account for the differences among "powder" Laue classes, which are distinguishable at this stage, and are suitable for indexing of powder diffraction patterns. For example, in Laue classes 6/m and 4/m ("powder" Laue classes 6/mmm and 4/mmm, respectively), the intensities of *hkl* and *khl* reflections are different, although the corresponding Bragg angles remain identical. Consult the International Tables for Crystallography, Vol. A for proper intensity relationships in other Laue classes.

Table 14.7 Symmetrically independent combinations of indices in six "powder" Laue classes

"Powder" Laue class	Range of indices and limiting conditions[a]	Independent fraction of sphere volume[b]
Triclinic, $\bar{1}$	$-h\ldots+h$; $-k\ldots k$; $0\ldots+l$	½
Monoclinic, 2/m	$-h\ldots+h$; $0\ldots k$; $0\ldots+l$	¼
Orthorhombic, mmm	$0\ldots+h$; $0\ldots+k$; $0\ldots+l$	⅛
Tetragonal, 4/mmm	$0\ldots+h$; $0\ldots+k$; $0\ldots+l$; $h \le k$	$^1/_{16}$
Hexagonal (=trigonal), 6/mmm	$0\ldots+h$; $0\ldots+k$; $0\ldots+l$; $h \le k$	$^1/_{24}$
Cubic, m3m	$0\ldots+h$; $0\ldots+k$; $0\ldots+l$; $h \le k$; $k \le l$	$^1/_{48}$

[a] The range and limiting conditions match the illustrations in Fig. 14.6

[b] Any fraction of the sphere that is symmetrically equivalent to the shaded area in Fig. 14.6 is acceptable

International Union of Crystallography[11] or the International Tables for Crystallography websites. Nearly all of these programs are simple to use and require minimal data input. This input typically includes symmetry, provided as a space-group symbol or crystal system name, unit cell dimensions, wavelength, and maximum Bragg angle to limit the amount of output data.

Further, nearly every available crystallographic software package includes a procedure for generating a complete list of possible hkl reflections, along with the corresponding interplanar distances and Bragg angles calculated from the known unit cell dimensions.

Another option is to use structure drawing software, such as the freeware programs VESTA[12] or Mercury.[13] These easy-to-use, intuitive programs require input of the unit cell parameters, space-group symmetry, and crystal structure parameters (atomic coordinates). They can generate not only a list of reflections but also a calculated powder diffraction profile. The latter can only be produced if atomic coordinates are available; otherwise, a dummy atom must be added for the software to function.

[11] https://www.iucr.org/resources/other-directories/software, https://journals.iucr.org/j/software/, and https://www.iucr.org/resources/commissions/computing/software-museum

[12] K. Momma and F. Izumi, VESTA 3 for three-dimensional visualization of crystal, volumetric and morphology data, J. Appl. Crystallogr. **44**, 1272 (2011), https://doi.org/10.1107/S0021889811038970

[13] C. F. Macrae, I. Sovago, S. J. Cottrell, P. T. A. Galek, P. McCabe, E. Pidcock, M. Platings, G. P. Shields, J. S. Stevens, M. Towler and P. A. Wood, Mercury 4.0: from visualization to analysis, design and prediction, J. Appl. Cryst. **53**, 226 (2020), https://doi.org/10.1107/S1600576719014092

14.4 Reliability of Indexing

Regardless of the tools used to index the powder pattern, the most reliable solution should result in the minimal discrepancies in the series of simultaneous equations (Eq. 14.5 or its equivalent for a different crystal system). These equations are constructed with the observed 2θ on the left side and the assigned index triplets and refined unit cell dimensions on the right side of each equation. The combined discrepancy can be easily established algebraically, for example, by summing the squared differences

$$\varepsilon = \sum_{i=1}^{N} \left(2\theta_{h_i k_i l_i}^{\mathrm{obs}} - 2\theta_{h_i k_i l_i}^{\mathrm{calc}} \right)^2 \tag{14.6}$$

However, achieving the minimum ε is usually not enough to assume that the correct unit cell has been found. In Eq. (14.6), N represents the number of the observed Bragg peaks. Obviously, when the unit cell dimensions are increased (e.g., see Eq. 14.5) or symmetry is lowered (see Table 14.7), this results in an increased number of possible combinations of indices in the same range of Bragg angles. Ultimately, infinitesimal ε can be reached when the total number of possible combinations of *hkl* triplets, and therefore, unique *d* values, approaches infinity but N remains constant.

Due to this ambiguity, several additional criteria should be considered before accepting an indexing result—and thus a final unit cell—especially during the ab initio indexing. The related norms to consider are as follows (assuming all other factors are equal):

- The preference generally is given to the unit cell with the highest symmetry because high symmetry translates into a small number of symmetrically independent reciprocal lattice points (see Fig. 14.6) and, therefore, the lowest number of possible Bragg reflections. For example, when the following three unit cells (Table 14.8) result in the successful indexing, the best of the three is represented by a tetragonal symmetry.
- The unit cell with the smallest volume usually represents the best solution because it leaves the smallest number of possible index triplets unassigned to at least one of the observed Bragg peaks. In the example shown in Table 14.9, the third row is likely the best solution.
- The preference shall be given to a solution which results in the smallest number of possible *hkl* triplets in the examined range of Bragg angles. In other words, if among several solutions, which are otherwise equivalent, one has a centered

Table 14.8 Example of three unit cells with nearly identical volumes but different symmetry

Crystal system	*a* (Å)	*b* (Å)	*c* (Å)	β (°)	Bravais lattice
Monoclinic	7.128	9.253	7.127	89.98	P
Orthorhombic	9.253	7.127	7.129	90	P
Tetragonal (best)	7.127	7.127	9.255	90	P

Table 14.9 Example of three unit cells with identical symmetry but different volumes

Crystal system	*a* (Å)	*b* (Å)	*c* (Å)	*V* (Å^3)	Bravais lattice
Orthorhombic	7.128	9.253	5.613	370.21	P
Orthorhombic	9.253	7.128	11.226	740.42	P
Orthorhombic (best)	4.627	5.613	7.128	185.12	P

Table 14.10 Example of two unit cells with identical symmetry and volumes but different Bravais lattices

Crystal system	*a* (Å)	*b* (Å)	*c* (Å)	*V* (Å^3)	Bravais lattice
Tetragonal (best)	7.127	7.127	9.255	470.28	I
Tetragonal	7.127	7.127	9.255	470.28	P

lattice and all the others are primitive, the centered lattice is typically the best (see Table 14.10).

- Finally, the resulting ε (Eq. 14.6) or a similar algebraic measure should be at its minimum for the most preferred indexing solution.

We note that the benchmarks listed here must be applied all together. For example, it is nearly always possible to choose the highest symmetry crystal system (i.e., cubic) and a large unit cell to dubiously assign index triplets to all observed Bragg peaks, leaving many possible unassigned, and obtain acceptable ε. This happens because the density of points in the reciprocal lattice is proportional to the volume of the unit cell in the direct space. As shown in Fig. 14.7, all Bragg peaks observed in the powder diffraction pattern of $LaNi_{4.85}Sn_{0.15}$ can be indexed in a primitive cubic unit cell with $a \cong 24.74$ Å. Generally, random and large unit cells result in a massive number of unobserved Bragg peaks and more often than not such indexing is fictitious and the unit cell is incorrect.[14]

It is worth noting, however, that when the true crystal structure is derived from a minor distortion of a small unit cell—such as in the formation of a so-called superstructure, or when the real structure is perturbed by a long-period, small-amplitude modulation function—many Bragg peaks, often the majority, may have extremely low intensity and become undetectable. These cases require special attention during the indexing, and detailed consideration of related subjects exceeds the scope of this book.

[14] In real structures, unit cell volumes can vary by several orders of magnitude. For example, among metallic materials, the unit cell volume can be as small as ~20 Å^3 for a pure metal like iron (Fe), which has two atoms in a body-centered cubic unit cell. Conversely, it can be as large as ~20,000 Å^3 for some complex intermetallic compounds, such as β-Mg_2Al_3 [S. Samson, The crystal structure of the phase β Mg_2Al_3, Acta Cryst. **19**, 401 (1965) https://doi.org/10.1107/S0365110X65005133] and $Tb_{117}Fe_{52}Ge_{112}$ [V.K. Pecharsky, O.I. Bodak, V.K. Belsky, P.K. Starodub, I.R. Mokra, and E.I. Gladyshevsky, Crystal structure of $Tb_{117}Fe_{52}Ge_{112}$, Sov. Phys. Crystallogr. **32**, 194 (1987); https://www.crystallography.net/cod//1541110.html], both of which contain over 1000 atoms in the face-centered cubic unit cells. While most crystalline materials have unit cell volume from several hundred to several thousand cubic Angströms. However, in proteins, unit cell volume typically ranges from ~10^5 to ~10^7 Å^3.

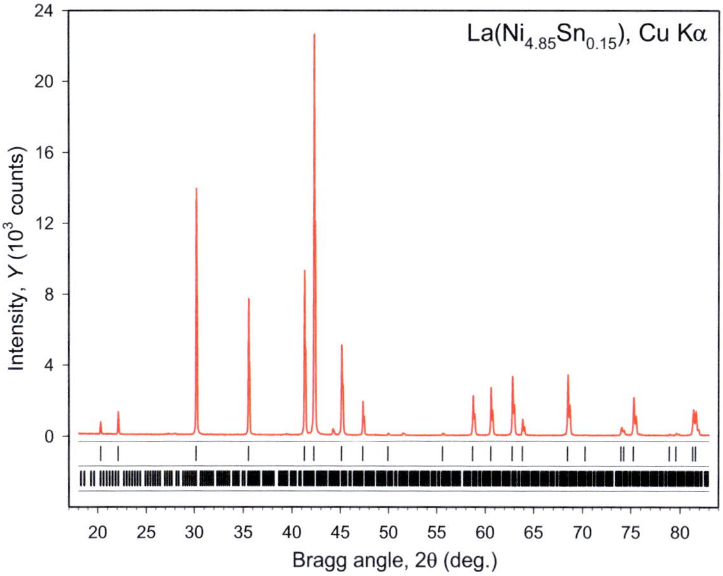

Fig. 14.7 The illustration of an incorrect indexing of the powder diffraction pattern of $LaNi_{4.85}Sn_{0.15}$. The *upper set of vertical bars* corresponds to the locations of Bragg peaks calculated using the correct hexagonal lattice with $a = 5.0421$, $c = 4.0118$ Å. The *lower set of bars* represents dubious indexing using a large primitive cubic unit cell with $a = 24.74$ Å. The latter unit cell results in an overwhelming number of possible reflections, only a few percent of which match the observed Bragg reflections

The quality and reliability of indexing are routinely characterized by means of various numerical figures of merit (FOMs). These figures become especially important when indexing is performed using a computer program. Numerous FOMs have been introduced and used with variable success in the powder diffraction data indexing process. We consider two of the most frequently used figures of merit. Both have been adopted by the International Centre for Diffraction Data to characterize the quality of indexing of patterns included into the Powder Diffraction File.

14.4.1 The F_N Figure of Merit

The so-called F_N figure of merit has been introduced by Smith and Snyder[15]: It is defined as

[15] G.S. Smith and R.L. Snyder, F_N: a criterion for rating powder diffraction patterns and evaluating the reliability of powder pattern indexing, J. Appl. Cryst. **12**, 60 (1979). https://doi.org/10.1107/S002188987901178X

$$F_N = \frac{N}{N_{\text{poss}}} \times \frac{1}{\overline{|\Delta 2\theta|}} = \frac{N^2}{N_{\text{poss}} \sum_{i=1}^{N} \left| 2\theta_i^{\text{obs}} - 2\theta_i^{\text{calc}} \right|} \tag{14.7}$$

where

- N is the number of the observed Bragg peaks,
- N_{poss} is the number of independent Bragg reflections possible within 2θ range of the observed diffraction peaks used in indexing,
- $\overline{|\Delta 2\theta|} = \frac{1}{N}\sum_{i=1}^{N} |\Delta 2\theta_i| = \frac{1}{N}\sum_{i=1}^{N} \left| 2\theta_i^{\text{obs}} - 2\theta_i^{\text{calc}} \right|$ is the average absolute difference between the observed and calculated $2\theta_i$.

In other words, N_{poss} is the number of symmetrically independent points in the reciprocal lattice, limited by a sphere with a diameter of d^*_{max} as determined by Eq. (14.3) using maximum observed Bragg angle, θ_{max}. Additional restrictions are imposed on N_{poss} in high-symmetry crystal systems: when reciprocal lattice points are not related by symmetry but when they have identical reciprocal vector lengths due to specific unit cell shape (e.g., *h*05 and *h*34 in the cubic, or 05 *l* and 34 *l* in the tetragonal crystal systems, see Eqs. 8.2 and 8.3), then only one is included into calculating N_{poss}. On the contrary, when the identity of the reciprocal lattice vectors lengths is coincidental, as shown for $d_{(32)} = d_{(4\bar{2})}$ in Fig. 14.3, both are included in the calculated N_{poss}. The F_N figure of merit is usually reported in the form

$$F_N = \text{Value}\left(\overline{|\Delta 2\theta|}, N_{\text{poss}}\right) \tag{14.8}$$

For example, in the indexed powder diffraction pattern of $LaNi_{4.85}Sn_{0.15}$ shown in Table 14.6 after the unit cell dimensions have been refined, the resulting F_N, excluding the two impurity peaks, is $F_{22} = 208.4(0.005, 23)$. The same figure of merit, calculated using the indexing results and unrefined unit cell dimensions (Table 14.5), is $F_{22} = 26.1(0.037, 23)$. The best FOM, $F_{22} = 208.4(0.005, 23)$, is interpreted as follows: the figure of merit for the 22 observed Bragg peaks is 208.4, the average absolute 2θ difference is 0.005°, and the total number of possible symmetrically inequivalent Bragg peaks in this range of Bragg angles is 23. The best indexing result usually has the highest F_N. Although it is difficult to establish strict guidelines on the values of F_N corresponding to a reasonable indexing, the F_N usually should be greater than 10; the lowest average 2θ difference should be lower than 0.02°, and the number of possible Bragg peaks, N_{poss}, should be the same or only slightly larger than the number of the observed peaks, N.

14.4.2 The M_{20} (M_N) Figure of Merit

Another frequently used figure of merit, M_{20}, has been introduced by de Wolff[16] and is defined as follows:

$$M_{20} = \frac{1}{N_{\text{poss}}} \times \frac{Q_{20}}{2|\overline{\Delta Q}|} = \frac{10Q_{20}}{N_{\text{poss}} \sum_{i=1}^{20} \left|Q_i^{obs} - Q_i^{\text{calc}}\right|} \quad (14.9)$$

where

- N_{poss} has the same meaning as in the F_N, except it is computed for $\theta_{\max} = \theta_{20}$, i.e., it is the number of symmetrically independent points in the reciprocal lattice up to the 20th observed Bragg peak,
- $Q = d^{*2} = 1/d^2$ represents the square of the length of the reciprocal vector,
- Q_{20} is the corresponding Q-value for the 20th observed Bragg peak,
- $|\overline{\Delta Q}| = \frac{1}{20}\sum_{i=1}^{20}|\Delta Q_i| = \frac{1}{20}\sum_{i=1}^{20}\left|Q_{i=1}^{\text{obs}} - Q_{i=1}^{\text{calc}}\right|$ is the average absolute difference between the observed and calculated Q_{hkl} for the first 20 Bragg peaks.

The M_{20} FOM is always calculated for the first 20 observed reflections unless the total number of the observed Bragg peaks is less than 20. When $N_{\text{obs}} > 20$, the value of M_N can be reported in addition to M_{20}. In these cases, Eq. (14.9) is modified as follows:

$$M_N = \frac{1}{N_{\text{poss}}} \times \frac{Q_N}{2|\overline{\Delta Q}|} = \frac{NQ_N}{2N_{\text{poss}}\sum_{i=1}^{N}\left|Q_i^{\text{obs}} - Q_i^{\text{calc}}\right|} \quad (14.10)$$

Once again referring to the two examples of indexing (Tables 14.5 and 14.6) the respective M_{20} values are 39.1 and 278.1. Similar to F_N, the more reliable indexing yields the higher M_{20}. However, it is even more difficult to specify an "acceptable" range of the M_{20} values. Compared to F_N, the M_{20} FOM is strongly dependent on both the complexity of the pattern and the unit cell volume. Therefore, one should look for a solution (or solutions) with the M_{20} figure of merit, which are distinctly larger than the others. If all results have similarly low M_{20} FOMs (e.g., all between 5 and 6), it is highly likely that none of them is the correct solution to the indexing problem. It is important to note that F_N and M_{20} figures of merit are useful for comparing different indexing solutions for the same pattern, but not for different patterns.

[16] P.M. de Wolff, A simplified criterion for the reliability of a powder pattern indexing, J. Appl. Cryst. **1**, 108 (1968). https://doi.org/10.1107/S002188986800508X

14.5 Introduction to Ab Initio Indexing

The complexity of indexing powder diffraction data without prior knowledge of either the symmetry or the dimensions of the unit cell (i.e., assigning indices from first principles based strictly on the relationships between the measured lengths of the reciprocal lattice vectors) is inversely proportional to the symmetry of the lattice—the higher the symmetry, the simpler the indexing process. Although actual indexing is now almost always performed using one or several freely available or commercial computer programs,[17] we believe that it is important to consider essential mathematical background, which was successfully used for many years to solve indexing problems "manually."

We begin with the cubic crystal system, where the assignment of indices is straightforward, and then explore the theory behind ab initio indexing in crystal systems with tetragonal and hexagonal symmetry.[18] As with any experimental work, experience is paramount, and we hope that the contents of this section may help the reader to achieve accurate solutions of real-life indexing tasks.

Accurate indexing from first principles rests on the four cornerstones:

1. The availability of Bragg peaks observed at the *lowest possible Bragg angles*. These peaks are critically important because they have the simplest indices (usually $-2 \leq h \leq 2$, $-2 \leq k \leq 2$, and $-2 \leq l \leq 2$), which considerably limits the possibilities of locating the corresponding vectors in the reciprocal lattice and therefore simplifies the whole process of restoring reciprocal lattice vector directions from their lengths.
2. *A small number of extinct*[19] *Bragg peaks*, especially at low Bragg angles. For example, if due to a variety of reasons, all observed Bragg peaks considered during the indexing have one of the indices divisible by two, then the resulting unit cell dimension is half of the true value.
3. *The high absolute accuracy,* including the absence of systematic errors affecting the measured Bragg angles. This requirement is obvious as only the lengths, but not the directions, of the reciprocal vectors are measurable in powder diffraction. Even a small systematic error (e.g., sample-displacement and/or zero-shift errors) may considerably affect the outcome of the indexing because it usually has the greatest impact on the lowest Bragg angle peaks, which are critical for successful

[17] Comprehensive collections of various crystallographic software products can be found on the International Union of Crystallography websites: https://www.iucr.org/resources/other-directories/software, https://www.iucr.org/resources/commissions/computing/software-museum, and https://journals.iucr.org/j/software/

[18] An excellent description of the ab initio indexing in all crystal systems can be found in H. Lipson and H. Steeple, Interpretation of X-ray powder diffraction patterns, Macmillan, London; St. Martin's Press, New York (1970).

[19] Unless extinctions are systematic, see Sects. 9.3 and 9.4.

indexing (see item 1 in this list, as well as Eq. 14.30 and Fig. 14.19 *later in this book*). In any case, systematic experimental errors should be minimized by proper alignment of the instrument. If necessary, additional experiment may be performed with a well-characterized internal standard added to the studied powder, thus enabling one to eliminate systematic errors from the data before attempting ab initio assignment of indices.

4. *The absence of impurity Bragg peaks* in the array of experimental data. If this requirement is not met, and one or more impurity peaks are included in the indexing attempt, it is nearly always true that the reciprocal vectors from the impurity phase do not fit the reciprocal lattice of the major phase. If a solution is found, the resulting unit cell is highly likely to be incorrect, as it describes vectors from both the major phase and the impurity phase(s) in the same arbitrary reciprocal lattice. Certain ab initio indexing procedures, for example, those that are incorporated into TREOR and ITO (see Sects. 14.10.1 and 14.10.3), automatically skip some Bragg peaks that do not fit in a trial lattice. This feature may help in indexing patterns containing some impurities, but it may also result in the incorrect, usually overly simplified solution when Bragg peaks from a major phase are accidentally excluded.

14.6 Cubic Crystal System

Cubic symmetry of a material is always a good guess when a powder diffraction pattern contains just a few peaks, or when there are sequences of observed Bragg peaks that appear nearly equally spaced (see Fig. 8.7), especially at low Bragg angles. Clearly, some experience in working with powder diffraction data is needed to make this judgment from a visual analysis of $Y(\theta)$. However, the ab initio indexing algorithm is so simple in the case of cubic symmetry, that nearly always it is justifiable to attempt cubic indexing to either confirm or dismiss this option.

When the crystal system is cubic, the general form of Eq. (14.2) is simplified to

$$\frac{1}{d_{hkl}^2} = \frac{h^2 + k^2 + l^2}{a^2} \tag{14.11}$$

or in reciprocal space

$$\frac{1}{d_{hkl}^2} = {d_{hkl}^*}^2 = \left(h^2 + k^2 + l^2\right) a^{*2} \tag{14.12}$$

Since h, k, and l are integers, the sums of their squares ($h^2 + k^2 + l^2$) are also integer numbers. Thus, Eq. (14.12) can be written as follows:

$$Q_{hkl} = A_{hkl} a^{*2} \tag{14.13}$$

where $Q_{hkl} = d^{*2}_{hkl} = 1/d^2_{hkl}$ and $A_{hkl} = (h^2 + k^2 + l^2)$. The former is established from the experiment for each Bragg peak using Eq. (14.3) and the latter is a positive integer, which depends on the values of h, k, and l in the index triplet. Certain whole numbers, for example, 7, 15, and others, are given by

$$A = i^2(8j - 1) \tag{14.14}$$

where $i = 1, 2, 3, \ldots$ and $j = 1, 2, 3, \ldots$ cannot be represented as a sum of squares of any three integers and therefore are forbidden in Eq. (14.13).

Assume that we have a set of experimental data where the observed Bragg angles have been converted into an array of Q-values. Then, if the crystal lattice is cubic, the following system of simultaneous equations can be written to associate each Bragg peak with a certain combination of hkl triplets:

$$\begin{array}{lcl} Q_{h_1k_1l_1} = A_{h_1k_1l_1} a^{*2} & & Q_{h_1k_1l_1}/a^{*2} = A_{h_1k_1l_1} \\ Q_{h_2k_2l_2} = A_{h_2k_2l_2} a^{*2} & \Rightarrow & Q_{h_2k_2l_2}/a^{*2} = A_{h_2k_2l_2} \\ \ldots & & \ldots \\ Q_{h_Nk_Nl_N} = A_{h_Nk_Nl_N} a^{*2} & & Q_{h_Nk_Nl_N}/a^{*2} = A_{h_Nk_Nl_N} \end{array} \tag{14.15}$$

As follows from the second form of Eq. (14.15), the observed array of Q-values should have a common divisor (a^{*2}), which results in the array of integers or nearly integers, considering the finite accuracy of the measured Bragg angles.

The indexing algorithm to test for cubic symmetry is shown in the form of a flowchart in Fig. 14.8. After the array of Bragg angles have been converted to Q-values, the next step is to normalize it and find the integers A^1, A^2, …, A^N. The simplest way to do so is to divide all Q-values by the smallest number present in the array, which is $Q_{h_1k_1l_1} = Q^1$. If the resulting array of A^1, A^2, …, A^N contains nearly whole numbers, the lattice parameter is calculated and the corresponding values of the $h_1k_1l_1$ triplets are determined based on the values of A^i after verification that no forbidden integers (e.g., 7, 15, etc., see Eq. 14.14) are present in the array A.

If the first normalization step results in clearly noninteger values in the array A, indexing still may be completed when the obtained A^1, A^2, …, A^N are multiplied by 2, 3, 4, etc. If the crystal system is truly cubic, a simple visual analysis of the array A after the first normalization usually enables one to determine the needed integer multiplier easily. For example, when decimal fractions of all A-values are close to 0 and 0.5, multiplying every number in the array A by 2 will result in all integers. Similarly, when the fractions are ~0, ~0.33, and ~0.66, the multiplier is 3, and so on. When the algorithm shown in Fig. 14.8 is realized as a computer program, then visual analysis of data in A is usually impractical and the value of n is determined

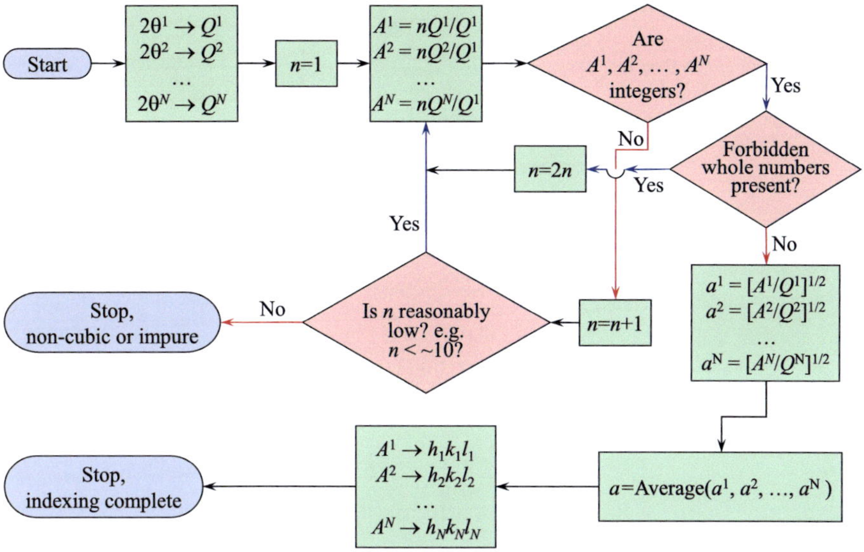

Fig. 14.8 The flowchart illustrating the algorithm of the ab initio indexing when cubic symmetry is suspected. It is assumed that the array of Q-values is sorted in the ascending order from Q^1 to Q^N

automatically, based on somewhat arbitrary tolerances that establish which value is taken as a whole number and which is not.

In some cases, the resulting array A may contain forbidden integer(s). Considering Eq. (14.14) and Table 14.11, two of the first 20 positive integers, 7 and 15, cannot be represented as sums of squares of any three integers. Their presence signals that the crystal lattice is body-centered. In a body-centered lattice, the simplest sequence of integers, which can be obtained using the indexing algorithm based on finding a common divisor, is shown in the corresponding column in Table 14.11 in parentheses (1, 2, …, **7**, 8, 9, …). The correct A-values are then found by doubling every integer, which results in the combination of indices where $h + k + l$ is always even

14.6.1 Primitive Cubic Unit Cell: LaB_6

We will now apply the indexing algorithm described here to experimental powder diffraction data collected from two different cubic materials. The first one is the standard reference material LaB_6, which was available from the National Institute of Standards and Technology (NIST) under the catalogue number SRM-660.[20] The

[20] SRM-660 has been replaced by SRM-660c. Consult https://www.nist.gov/srm for details.

Table 14.11 The first 20 positive integers and the corresponding *hkl* triplets in the primitive (P), body-centered (I), and face-centered (F) cubic lattices

	Centering					
	P		I		F	
Integer	*hkl*	A	*hkl*	A[a]	*hkl*	A
1	001	1				
2	011	2	011	2 (1)		
3	111	3			111	3
4	002	4	002	4 (2)	002	4
5	012	5				
6	112	6	112	6 (3)		
7	–		–		–	
8	022	8	022	8 (4)	022	8
9	003, 122	9				
10	013	10	013	10 (5)		
11	113	11			113	11
12	222	12	222	12 (6)	222	12
13	023	13				
14	123	14	123	14 (**7**)		
15	–		–		–	
16	004	16	004	16 (8)	004	16
17	014, 223	17				
18	114, 033	18	114, 033	18 (9)		
19	133	19			133	19
20	024	20	024	20 (10)	024	20

[a] The numbers in parentheses indicate the simplest sequence of integers, which describes the relationships between the sums of h^2, k^2, and l^2 in the body-centered cubic lattice. The presence of forbidden integers (e.g., 7, shown in **bold**) enables one to differentiate between the primitive and body-centered cubic lattices during the ab initio indexing

experimental powder diffraction pattern is shown in Fig. 14.9 and the certified lattice parameter of the material is $a = 4.15695(6)$ Å. The observed Bragg peak positions were determined using a profile fitting procedure and the least-squares standard deviations in the observed Bragg angles did not exceed 0.001° of 2θ (Table 14.12).

Table 14.12 shows how the indexing has been performed. First, Q^{obs} are calculated for each $2\theta^{obs}$:

$$Q^{obs} = \frac{1}{d^2} = \frac{4\sin^2\theta^{obs}}{\lambda^2} \tag{14.16}$$

Second, the values from this column are normalized by dividing them by the smallest observed Q, which is 0.05740. This yields the column of data marked as Q^{obs}/Q^{min}.

Analysis of this column indicates that it contains nearly integer numbers from 1 to 22, and that the deviations from the whole increase systematically with the increasing Bragg angle. The latter is generally observed due to the low accuracy of

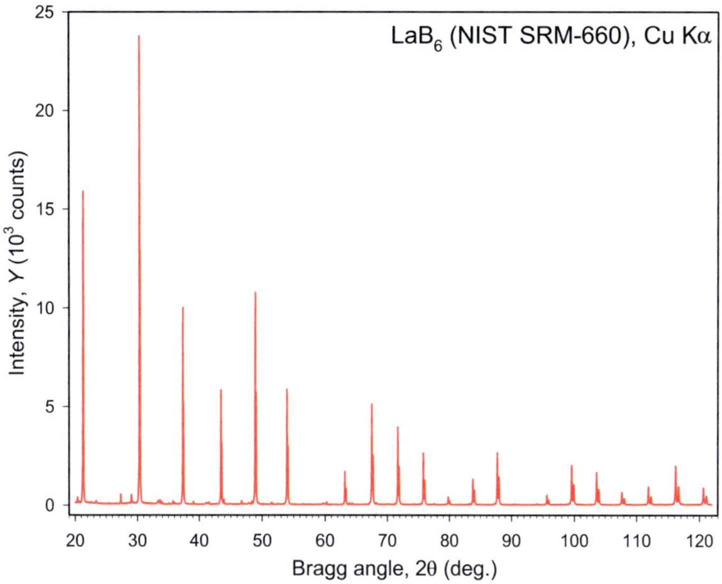

Fig. 14.9 The X-ray powder diffraction pattern collected from SRM-660 LaB_6 powder on a Scintag XDS2000 powder diffractometer using Cu Kα radiation. The data were collected in a step-scan mode with a step 0.02° of 2θ (The ASCII data file with diffraction data is available online, file name *Ch14Ex02_CuKa.xy*). Weak peaks visible at low Bragg angles (~20.5°, 27°, 29°, 34.5°, …, 60.2°) belong to an unidentified impurity

the first Q-value (the divisor), but it also may be due to the presence of a systematic error in the observed Bragg angles. Proceeding according to the algorithm depicted in Fig. 14.8, we find the corresponding integers in column A_{hkl} by rounding the values in the previous column and establish the respective *hkl* triplets for each observed Bragg peak.

The last column in Table 14.12 contains the values of the lattice parameter calculated from individual Bragg peaks by means of

$$a = \sqrt{A^i / Q^i} \tag{14.17}$$

When one uses the indexing results shown in Table 14.12 to perform a least-squares refinement of both the lattice parameter and sample displacement or zero shift, the resulting values are as follows: $a = 4.1574(1)$ Å and the zero shift is 0.078°. The corresponding F_N figure of merit is $F_{20} = 384.6(0.003, 20)$. The difference between the obtained lattice parameter and that which is considered a standard value ($a = 4.15695$ Å, see p. 413) is acceptable considering the absence of special procedures adopted by NIST in certifying the lattice parameter of the standard.

Table 14.12 Example of the ab initio indexing in the cubic crystal system using all observed Bragg peaks present in the pattern shown in Fig. 14.9[a]

I/I_0	$2\theta^{obs}$	$FWHM^{obs}$	Q^{obs}	Q^{obs}/Q^{min}	A_{hkl}	*hkl*	*a*, Å
632	21.270[b]	0.065	0.05740	1.000	1	001	4.1739
1000	30.304	0.063	0.11514	2.006	2	011	4.1677
429	37.357	0.062	0.17286	3.011	3	111	4.1660
250	43.423	0.062	0.23064	4.018	4	002	4.1645
485	48.874	0.064	0.28843	5.025	5	012	4.1636
276	53.905	0.066	0.34623	6.032	6	112	4.1629
84	63.134	0.070	0.46186	8.046	8	022	4.1619
269	67.461	0.071	0.51966	9.053	9	003, 122	4.1616
204	71.661	0.070	0.57753	10.061	10	013	4.1611
141	75.758	0.074	0.63535	11.069	11	113	4.1609
26	79.788	0.081	0.69327	12.078	12	222	4.1605
80	83.761	0.079	0.75109	13.085	13	023	4.1603
173	87.703	0.084	0.80889	14.092	14	123	4.1602
31	95.576	0.079	0.92454	16.107	16	004	4.1600
146	99.547	0.092	0.98242	17.115	17	014, 223	4.1598
122	103.560	0.095	1.04024	18.122	18	114, 033	4.1598
53	107.649	0.109	1.09815	19.131	19	133	4.1596
71	111.836	0.106	1.15609	20.141	20	024	4.1593
172	116.145	0.111	1.21398	21.149	21	124	4.1591
87	120.625	0.127	1.27193	22.159	22	233	4.1589
						$a_{Average} = 4.1621$	
					Standard deviation = 0.0037		

[a] The ASCII data file with diffraction data is available online, file name *Ch14Ex02_CuKa.xy*

[b] Bragg angles are listed for the location of the $K\alpha_1$ component in the doublet, $\lambda = 1.540593$ Å

14.6.2 Body-Centered Cubic Unit Cell: $U_3Ni_6Si_2$

The second example is shown in Fig. 14.10 and Table 14.13. The observed Bragg peak positions were determined using a profile fitting procedure and the least-squares standard deviations in the observed Bragg angles did not exceed 0.003° of 2θ. Considering the results shown in Table 14.13, the column labeled Q^{obs}/Q^{min} contains nearly whole numbers, but unlike the identical column in Table 14.12, here one finds forbidden integer values: ~7, ~15, and ~23 (all three are shown in bold in Table 14.13).

Thus, the column labeled A_{hkl} is obtained by rounding the values from the previous column and multiplying them by 2. The corresponding *hkl* triplets confirm a body-centered lattice ($h + k + l = 2n$), and the last column contains the values of the lattice parameter calculated from the individual Bragg peaks using Eq. (14.17).

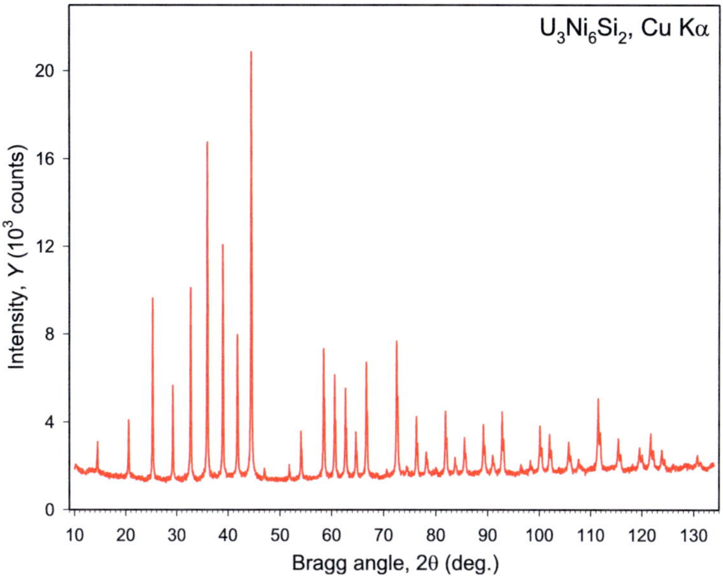

Fig. 14.10 The X-ray powder diffraction pattern of $U_3Ni_6Si_2$ collected on an HZG-4a powder diffractometer using filtered Cu Kα radiation. The data were collected in a step-scan mode with a step 0.02° of 2θ and counting time 25 s (The ASCII data file with the diffraction data is available online, file name *Ch14Ex03_CuKa.xy*.). When compared, for example, with Figs. 14.5 and 14.9, the increased background is noteworthy, which occurs as a result of its incomplete elimination when using a β-filter. Data courtesy of Dr. L.G. Akselrud

Even though the numbers in the column Q^{obs}/Q^{min} still show a systematic increase in the deviations from whole numbers (the reason is the low absolute accuracy of the divisor, $Q^{min} = 0.02682$), the lattice parameters listed in the last column do not reveal the presence of any kind of a systematic error in the observed Bragg angles. Using the average lattice parameter, $a = 8.6443$ Å, the indexing result listed in Table 14.13 yields $F_{20} = 194.7(0.005, 22)$: two of the first 22 Bragg peaks are not observed in this diffraction pattern (233 and 062) and one whole number (28) is forbidden. A least-squares refinement yields the unit cell dimension nearly identical to the average value shown here and results in a negligible zero-shift correction. We leave the least-squares refinement as an exercise to the reader.

14.7 Tetragonal and Hexagonal Crystal Systems

The quadratic form of Eq. (14.2) in the hexagonal crystal system is found in Eq. (14.4) and its analogue in the tetragonal crystal system is

Table 14.13 Example of the ab initio indexing in the cubic crystal system using first 20 Bragg peaks present in the pattern shown in Fig. 14.10[a]

I/I_0	$2\theta^{obs}$	FWHM^{obs}	Q^{obs}	Q^{obs}/Q^{min}	A_{hkl}	*hkl*	*a*, Å
58	14.493[b]	0.165	0.02682	1.000	2	011	8.6363
115	20.530	0.138	0.05352	1.996	4	002	8.6452
338	25.216	0.120	0.08030	2.995	6	112	8.6441
178	29.196	0.122	0.10706	3.992	8	022	8.6445
427	32.733	0.138	0.13382	4.990	10	013	8.6446
696	35.961	0.117	0.16060	5.989	12	222	8.6441
462	38.953	0.121	0.18736	**6.987**	14	123	8.6443
328	41.761	0.145	0.21410	7.984	16	004	8.6448
1000	44.425	0.134	0.24086	8.982	18	033, 114	8.6448
29	46.954	0.133	0.26747	9.975	20	024	8.6472
29	51.754	0.133	0.32102	11.972	24	224	8.6465
124	54.044	0.133	0.34788	12.973	26	134, 015	8.6451
348	58.425	0.147	0.40143	**14.970**	30	125	8.6448
271	60.544	0.143	0.42828	15.972	32	044	8.6439
232	62.613	0.141	0.45504	16.970	34	334, 035	8.6440
125	64.638	0.161	0.48172	17.965	36	244, 006	8.6448
321	66.641	0.155	0.50855	18.965	38	235, 116	8.6442
15	70.542	0.173	0.56196	20.957	42	145	8.6451
442	72.464	0.173	0.58876	21.957	44	226	8.6448
26	74.375	0.173	0.61570	**22.961**	46	136	8.6436
						$a_{\text{Average}} = 8.6443$	
					Standard deviation = 0.0021		

[a] Excel spreadsheet file Ch14Ex03_BodyCenteredCubic.xls is available online
[b] Bragg angles are listed for the location of the Kα_1 component in the doublet, $\lambda = 1.540593$ Å

$$\frac{1}{d^2} = \frac{h^2 + k^2}{a^2} + \frac{l^2}{c^2} \tag{14.18}$$

Following the approach illustrated using Eqs. (14.11)–(14.13) and maintaining similar notations, Eqs. (14.4) and (14.18) can be written as

$$Q_{hkl} = A_{hk}^{t,h} a^{*2} + C_l c^{*2} \tag{14.19}$$

where $Q_{hkl} = d_{hkl}^{*2} = \frac{1}{d_{hkl}^2}, A_{hk}^t = (h^2 + k^2), A_{hk}^h = (h^2 + hk + k^2)$, and $C_l = l^2$, and noting that in the hexagonal crystal system $a^* = 2/(a\sqrt{3})$. Hence, positions of Bragg peaks found in the diffraction patterns of materials that belong to either the tetragonal or hexagonal crystal systems can be represented by the following series of simultaneous equations, where A_{hk}^t and A_{hk}^h are substituted by A_{hk}:

$$\begin{aligned} Q_{h_1k_1l_1} &= A_{h_1k_1}a^{*2} + C_{l_1}c^{*2} \\ Q_{h_2k_2l_2} &= A_{h_2k_2}a^{*2} + C_{l_2}c^{*2} \\ &\ldots \\ Q_{h_Nk_Nl_N} &= A_{h_Nk_N}a^{*2} + C_{l_N}c^{*2} \end{aligned} \tag{14.20}$$

The solution of Eq. (14.20) could be found after calculating all possible differences between the observed pairs of Q_{hkl}. This leads to the following series of equations:

$$\begin{aligned} & Q_{h_2k_2l_2} - Q_{h_1k_1l_1} = (A_{h_2k_2} - A_{h_1k_1})a^{*2} + (C_{l_2} - C_{l_1})c^{*2} \\ & \ldots \\ & Q_{h_Nk_Nl_N} - Q_{h_1k_1l_1} = (A_{h_Nk_N} - A_{h_1k_1})a^{*2} + (C_{l_N} - C_{l_1})c^{*2} \\ & Q_{h_3k_3l_3} - Q_{h_2k_2l_2} = (A_{h_3k_3} - A_{h_2k_2})a^{*2} + (C_{l_3} - C_{l_2})c^{*2} \\ & \ldots \\ & Q_{h_Nk_Nl_N} - Q_{h_{N-1}k_{N-1}l_{N-1}} = (A_{h_Nk_N} - A_{h_{N-1}k_{N-1}})a^{*2} + (C_{l_N} - C_{l_{N-1}})c^{*2} \end{aligned} \tag{14.21}$$

As follows from Eq. (14.21), when two Bragg peaks have the same value of l, for example, h_ik_i0 and h_jk_j0, or h_ik_i1 and h_jk_j1, or h_ik_i2 and h_jk_j2, and so on, the resulting difference is only a function of A_{hk} and a^*:

$$Q_{h_ik_il_i} - Q_{h_jk_jl_j} = \left(A_{h_ik_i} - A_{h_jk_j}\right)a^{*2} \tag{14.22}$$

Similarly, when h and k are identical, but l is different, for example, $01l_i$ and $01l_j$, or $11l_i$ and $11l_j$, or $12l_i$ and $12l_j$, and so on, some of the equations in 14.21 are transformed into:

$$Q_{h_ik_il_i} - Q_{h_jk_jl_j} = \left(C_{l_i} - C_{l_j}\right)c^{*2} \tag{14.23}$$

Solving the indexing problem becomes a matter of identifying the differences that result in whole numbers when divided by a common divisor (a^{*2} and c^{*2}, respectively). The expected whole numbers are shown in Tables 14.14, 14.15, and 14.16 for several small h, k, and l. It only makes sense to consider these small values because successful indexing is critically dependent on the availability of low Bragg angle peaks, which usually have small values of indices.

The resulting whole numbers, expected as multipliers for the differences in Q-values given by Eq. (14.23) (Table 14.14), are dissimilar from those expected for Eq. (14.22) (Tables 14.15 and 14.16), and this property may be employed to distinguish between c^{*2} and a^{*2}.

Table 14.14 Possible values of integer multipliers for the differences defined in Eq. (14.23) for both tetragonal and hexagonal crystal systems

	l^2_i				
l^2_j	0	1	4	9	16
0	–	1	4	9	16
1		–	3	8	15
4			–	5	12
9				–	7
16					–

Table 14.15 Possible values of integer multipliers for the differences defined in Eq. (14.22) for the tetragonal crystal system

	(h_ik_i)				
(h_jk_j)	(00); $A_{hk} = 0$	(01); $A_{hk} = 1$	(11); $A_{hk} = 2$	(02); $A_{hk} = 4$	(12); $A_{hk} = 5$
(00); $A_{hk} = 0$	–	1	2	3	4
(01); $A_{hk} = 1$		–	1	3	4
(11); $A_{hk} = 2$			–	2	3
(02); $A_{hk} = 4$				–	1
(12); $A_{hk} = 5$					–

Table 14.16 Possible values of integer multipliers for the differences defined in Eq. (14.22) for the hexagonal crystal system

	(h_ik_i)				
(h_jk_j)	(00); $A_{hk} = 0$	(01); $A_{hk} = 1$	(11); $A_{hk} = 3$	(02); $A_{hk} = 4$	(12); $A_{hk} = 7$
(00); $A_{hk} = 0$	–	1	3	4	7
(01); $A_{hk} = 1$		–	2	3	6
(11); $A_{hk} = 3$			–	1	4
(02); $A_{hk} = 4$				–	3
(12); $A_{hk} = 7$					–

Given the background considered here, the indexing of experimental powder diffraction data assuming tetragonal or hexagonal symmetry may be carried out using the algorithm illustrated in Fig. 14.11. First, all possible differences between the observed Q-values are computed for several low Bragg angle peaks. Second, the array of the obtained differences is analyzed to find the most frequently occurring small values. Third, the found values are tested with respect to whether the indexing of all observed Bragg peaks is possible assuming that one of the quantities is a^{*2} and another is c^{*2}.

If the indexing is successful, both the crystal system and lattice parameters are determined. If not, the small values found can be tested after they have been divided by a whole number, usually between 2 and 4. Again, if indexing is possible, the problem is solved. However, if indexing is not possible, one or both of the small values that have been identified as suspected as a^{*2} and c^{*2} should be discarded, and the search for a suitable pair of a^{*2} and c^{*2} continues. When all potential candidates

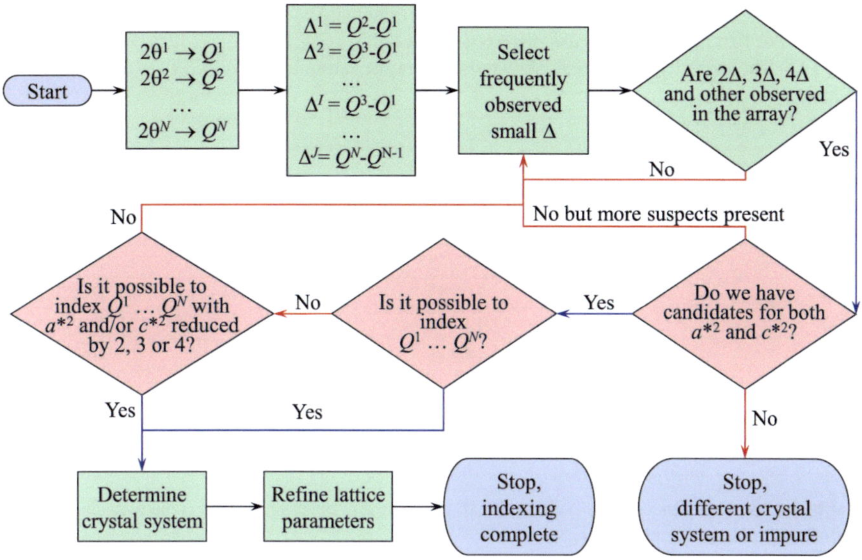

Fig. 14.11 The flowchart illustrating the algorithm of the ab initio indexing assuming tetragonal or hexagonal symmetry of the reciprocal lattice

have been tested and no solution has been found, then likely the assumption of tetragonal or hexagonal symmetry is wrong, or the powder diffraction data contain impurity Bragg peaks at low angles, which makes reasonable indexing impossible.

14.7.1 *Indexing Example: $LaNi_{4.85}Sn_{0.15}$*

We illustrate the use of the technique described here by indexing the powder diffraction pattern of $LaNi_{4.85}Sn_{0.15}$ (Fig. 14.5 and Table 14.1). Note that in the process we make no assumptions about the exact symmetry of the material except that we suspect that it may be either tetragonal or hexagonal. In our indexing attempt, we will limit all calculations to the first seven observed Bragg peaks excluding the weak impurity peak observed at $2\theta = 44.211°$ as shown in Table 14.17. The observed Bragg angles as determined directly from a profile fitting procedure are employed without correcting for any kind of a systematic error.[21]

[21] Although the presence of a systematic error may, in general, hinder an indexing attempt, neglecting a small zero-shift error in this case (see Table 14.6) is forgivable because only a small region of Bragg angles (from 20° to 45° 2θ) is employed. When larger arrays of data are included in an ab initio indexing, they should be corrected for all known systematic errors, if any, for example, by employing an internal standard.

Table 14.17 Relative integrated intensities (I/I_0), Bragg angles, and full widths at half maximum (FWHM) of first eight Bragg peaks observed in the $LaNi_{4.85}Sn_{0.15}$ powder diffraction pattern, see Fig. 14.5. The impurity Bragg peak shown using a strike-through font has been excluded from the indexing attempt

I/I_0	2θ (°)[a]	FWHM (°)
20	20.288	0.070
43	22.105	0.077
513	30.198	0.076
305	35.548	0.078
394	41.285	0.082
1000	42.272	0.085
~~18~~	~~44.211~~	~~0.099~~
274	45.130	0.106

[a] Bragg angles are listed for the location of the $K\alpha_1$ component in the doublet, $\lambda = 1.540593$ Å

Table 14.18 Differences in Q-values for the first seven observed Bragg peaks in the powder diffraction pattern of $LaNi_{4.85}Sn_{0.15}$

	Qj						
Qi	0.0523	0.0619	0.1144	0.1570	0.2095	0.2191	0.2482
0.0523	–	0.0097	0.0621	0.1048	0.1572	0.1668	0.1959
0.0619		–	0.0524	0.0951	0.1475	0.1572	0.1862
0.1144			–	0.0427	0.0951	0.1048	0.1338
0.1570				–	0.0524	0.0621	0.0911
0.2095					–	0.0097	0.0387
0.2191						–	0.0290
0.2482							–

The values of Bragg angles listed in Table 14.17 have been converted into the Q-values using Eq. (14.3). Possible differences (Eq. 14.21) have been calculated and listed in Table 14.18. They are sorted and analyzed as shown in Table 14.19. The first column in this table includes the Q-values of the seven observed Bragg peaks (in **bold**) in addition to all calculated differences.

Considering Eqs. (14.22) and (14.23), Tables 14.14, 14.15, and 14.16, and Fig. 14.11, it is quite clear that all computed differences should be analyzed for the occurrence of nearly identical small values, which potentially may correspond to the differences resulting in a^{*2} and c^{*2} or their whole multiples. Once the repetitive numbers are found, the next step is to find out whether the array of differences and observed Q-values contains whole multiples or whole fractions of the found quantities. The tested whole numbers should be correlated with Tables 14.14, 14.15, and 14.16.

The initial candidate in the first column of Table 14.19 is 0.0097, which occurs twice. The average and the results of its multiplication by 2, 3, and 4—the simplest whole numbers that are present in Tables 14.14, 14.15, and 14.16—are shown in

Table 14.19 Illustration of the indexing of the first seven Bragg peaks observed in the powder diffraction pattern of $LaNi_{4.85}Sn_{0.15}$ using Eqs. (14.20)–(14.23)

Q_{obs}, Diff.	Mean[a]	×2[a]	×3[a]	×4[a]	a^{*2}, c^{*2}	Eq. (14.19)	Q_{calc}	*hkl*
0.0097								
0.0097	0.0097	0.0193	0.0290	0.0387	?			
0.0290								
0.0387								
0.0427								
0.0523								
0.0524	0.0524	*0.1047*	0.1571	0.2095	a^{*2}	**1a^{*2}**	**0.0524**	**010**
0.0524								
0.0619								
0.0621	0.0620	0.1241	0.1861	0.2481	c^{*2}	**1c^{*2}**	**0.0620**	**001**
0.0621								
0.0911								
0.0951								
0.0951								
0.1048								
0.1048								
0.1144						**1a^{*2}+ 1c^{*2}**	**0.1144**	**011**
0.1338								
0.1475								
0.1570						**3a^{*2}**	**0.1571**	**110**
0.1572								
0.1572								
0.1668								
0.1862								
0.1959								
0.2095						**4a^{*2}**	**0.2095**	**020**
0.2191						**3a^{*2}+ 1c^{*2}**	**0.2191**	**111**
0.2482						**4c^{*2}**	**0.2481**	**002**

[a] All values in the table are listed with four digits after the decimal point but the actual computations were performed with a better accuracy

columns 2–5. If the computed average corresponds to either a^{*2} or c^{*2}, then these products should also be often observed in the array combining both the differences and observed Q-values. As seen from Table 14.19, this is not the case since only one occurrence of triple and quadruple multiples of the suspected value is found, and 0.0097 as a candidate for a^{*2} or c^{*2} is dismissed as unsuitable. Furthermore, this value appears too small because it results in $d \cong 10$ Å, which is too large considering the simplicity of the powder diffraction pattern (Fig. 14.5).

The next significant value is 0.0524, which appears three times as the differences between pairs of first seven observed Bragg peaks. Not only does this value occur more frequently than any other smaller quantity found in the table, but when multiplied by two, it yields 0.1047, a value that appears twice in the array—these nearly identical numbers are shown in *italics*. Additionally, multiplying 0.0524 by three results in 0.1571, which occurs three more times (all are underlined). There is also one instance where 4 × 0.0524 equals 0.2095, which is double underlined. Hence, this value seems to be an excellent candidate for one of the reciprocal lattice parameters. After consulting Tables 14.14, 14.15, and 14.16, it is clear that 2 × c^{*2} is not expected to be seen but 2 × a^{*2} should be observed quite frequently in both the tetragonal and hexagonal crystal systems.

Proceeding in a similar fashion with the triple occurrence of the next small value (the average is 0.0620), we find that the array of differences and observed *Q*-values has no occurrences of 2 × 0.0620 = 0.1241, but both 3 × 0.0620 = 0.1861 and 4 × 0.0620 = 0.2481 have one occurrence in the table. Hence, as follows from Table 14.14, the value of 0.0620 is an outstanding candidate for c^{*2}.

The next step is to verify whether the found candidates for a^{*2} and c^{*2} (both are shaded in Table 14.19) result in the complete indexing of the existing seven Bragg peaks. By using Eq. (14.19), all observed *Q*-values are nearly equal to the sums of Aa^{*2} and Cc^{*2}, where *A* and *C* are whole numbers, which are listed in the corresponding column in Table 14.19 in bold. Strictly speaking, the whole diffraction pattern should be indexed following the same approach, but we leave this to the reader as an exercise.

At this point (or after the whole pattern has been indexed), the analysis of the observed values of *A* enables one to establish whether we deal with the tetragonal or hexagonal crystal systems. As seen in Table 14.19, the whole multipliers of a^{*2} are 1 and 3, and 3 is only possible in the hexagonal crystal system for $h = 1, k = 1$. After a simple calculation using the average values of a^{*2} and c^{*2} listed in Table 14.19, we find approximate values of *a* and *c* as 5.046 and 4.015 Å, respectively. A least-squares refinement of the lattice parameters using the entire array of indexed Bragg peaks obviously yields the same lattice parameters as were established before (see Table 14.6).

It is worth noting that the considered example is relatively easy because first, the lattice is primitive and second, there were no extinct Bragg reflections among the first seven observed diffraction maxima. If some Bragg peaks are missing, then the list of the generated differences (column 1 in Table 14.19) becomes incomplete and the task of identifying the quantities corresponding to a^{*2} and c^{*2} becomes considerably more complex. This is especially true when the material has a Bravais lattice other than primitive since additional lattice translations cause multiple systematic absences in the list of possible *hkl*. In any case, the final indexing should always be checked by calculating one or more figures of merit after refinement of lattice parameters, which for this pattern has been done earlier (see Sect. 14.4).

14.8 Automatic Ab Initio Indexing Algorithms

The complexity of solving the indexing problem increases rapidly as the lattice symmetry decreases. For instance, in the orthorhombic crystal system, the dimensions of the reciprocal unit cell, which affect the governing reciprocal lattice equation (Eq. 14.1), depend on three unknown parameters (a^*, b^*, and c^*), while in the monoclinic and triclinic crystal systems the number of unknown parameters becomes four (a^*, b^*, c^*, and β^*) and six (a^*, b^*, c^*, α^*, β^*, and γ^*), respectively. As a result, manual indexing of low-symmetry powder patterns becomes extremely difficult and time-consuming. It is important to emphasize that the indexing problem in powder diffraction arises not from the number of unknown parameters or low symmetry, but from the challenge of recovering the Miller indices, or in other words, the three-dimensional nature of the diffraction data, which is lost during the experiment due to the projection of the reciprocal lattice from three-dimensional vector space onto one-dimensional scalar axis 2θ. Therefore, automatic indexing using various algorithms and software becomes irreplaceable. The use of computers also speeds up high-symmetry cases and, at the end, allows one to perform a comprehensive search for indexing solution in all crystal systems.

The solution of the indexing problem can be performed using several different computational algorithms, which have been realized in a variety of automatic indexing programs.[22] All of them use one of the two fundamentally different approaches to the ab initio indexing by manipulating either direct-space parameters (i.e., unit cell dimensions) or reciprocal space parameters (i.e., reflection indices) as free variables in order to describe all or almost all observed diffraction peaks using a reasonable crystal lattice. In addition to this principal difference, different types of data are used when searching for a unit cell and/or evaluating indexing solutions. Thus, reciprocal space techniques employ Bragg angles, while direct-space methods often use either as-measured or somehow modified profiles. The latter is computationally more intense than the former, and it was made possible by recent manifold increases in the speed of computers. One of the major benefits of the direct-space approaches is in the avoidance of peak searches and profile fitting, which may become biased when the complexity of a powder pattern increases.

A brief description of rapidly evolving direct-space methods is given in Sect. 14.8.1, including a list of currently available software. Reciprocal space methods, which form the basis of three commonly used indexing programs (TREOR, ITO, and DICVOL), are discussed in Sect. 14.8.2. These three applications—TREOR, ITO, and DICVOL—have been cited an order of magnitude more often than any other

[22](a) J. Bergmann, A. Le Bail, R. Shirley, V. Zlokazov, Renewed interest in powder diffraction data indexing. Z. Kristallogr. **219**, 783 (2004). https://doi.org/10.1524/zkri.219.12.783.55862;
(b) P.-E.Werner, Autoindexing. in: Structure determination from powder diffraction data. IUCr monographs on crystallography 13. W. I. F. David, K. Shankland, L.B. McCusker, and Ch. Baerlocher, Eds., Oxford University Press, Oxford, New York (2002).

indexing software,[23] and therefore, they are considered in more detail in Sect. 14.10 and used in practical indexing examples in Sect. 14.11.

14.8.1 Indexing in Direct Space

Direct-space indexing employs either the *grid-search method*, when unit cell dimensions vary with a certain increment within certain limits, or an alternative random *Monte Carlo search* for unit cell parameters. Both methods share a single goal: to achieve the best description of the measured powder diffraction pattern.

In the *grid-search method*, an attempt to index the entire diffraction pattern is made after every incremental step in lattice parameters. The increments depend on both the accuracy and complexity of the diffraction pattern, but ~0.01 Å for the unit cell edges (a, b, and c) and ~0.1° for the angles (α, β, and γ) should be sufficient in most cases. The maximum length of the unit cell edges can be estimated from the d-spacing of the first Bragg peak (d_{max}) observed in the diffraction pattern. In the majority of low-symmetry cases (triclinic through orthorhombic crystal systems), the maximum size of the unit cell edge should not exceed $2d_{max}$, while in the high-symmetry cases it should be set at 4 (tetragonal/cubic) to 6 (hexagonal/trigonal) d_{max}. When indexing superlattices, in which many possible reflections may be missing, higher limits on the maximum unit cell dimensions may be required.

The grid search is the simplest but also the slowest indexing method. Obviously, each crystal system should be tested separately, as the number of free variables has a critical influence on the computation time. For example, a total of 4×10^6 different unit cells must be checked, assuming a tetragonal or hexagonal crystal system with unit cell dimensions in the range between 2.01 and 22 Å using 0.01 Å increment ($2000 \times 2000 = 4{,}000{,}000$). In a triclinic crystal system, with unit cell edges between 2.01 and 12 Å and angles between 90.1° and 120°, a total of 4.2×10^{14} combinations should be tested even when using larger increments (0.02 Å and 0.2°, respectively). Assuming that 1000,000 unit cells can be tested in 1 s,[24] an unrestricted and exhaustive search in the tetragonal or hexagonal case will take ~4 s, but one will have to wait nearly 13 years and 4½ months to test all possible combinations and see the answer in a triclinic crystal system.

Restrictions dictated by crystallography may and are indeed used to reduce the computation time significantly. For example, the maximum expected unit cell volume can be evaluated from the density of diffraction peaks observed in a certain range of Bragg angles. Further, the following additional restrictions can be imposed: in the monoclinic crystal system $a \leq c$ and in the orthorhombic and triclinic crystal

[23] According to reference 22(a) on page 475.

[24] This assumption is unrealistic using even the most powerful quad processor PC available in early 2008. A more rational estimate is between ~10^3 and ~10^4 unit cells per second for a well-optimized computer code.

systems $a \leq b \leq c$, because in these cases the solution is invariant to a permutation of unit cell edges. Modern high-speed computers can handle the problem only in high-symmetry cases, because even with all applicable restrictions, the grid-search indexing in a triclinic crystal system may take from a few days to over a year, dependent on the size of the unit cell.

Later, direct-space indexing has been expanded by introducing *Monte Carlo search* of the unit cell, which is much faster than the grid-search technique. In the Monte Carlo approach, the unit cell is generated randomly and then compared with the experimental pattern. Random search may also be time-consuming, even more so than a simple grid search. However, it becomes much faster when optimized so that the next tested cell is not entirely random but instead represents a random deviation from the best-known cell at that point in time. This is usually achieved by using genetic, simulated annealing or other global optimization algorithms. Essentially the same techniques are also applicable to solving crystal structures in direct space as discussed in Sect. 15.1.3.

Many advanced methods and approaches for direct-space indexing have been developed and implemented in various applications, such as EFLECH,[25] McMaille[26], SVD-Index,[27] X-Cell,[28] LSI (Least-Squares Indexing) and LP-Search,[29] the differential evolution algorithm (FOX),[30] a genetic algorithm search method (GAIN),[31] and others.[32]

---

[25] EFLECH/INDEX—an automatic peak-hunting (EFLECH) and random cell parameters search (INDEX) based on covariance matrix from EFLECH. J. Bergmann, R. Kleeberg, EFLECH/INDEX—a program for peak search/fit and indexing, IUCr CPD Newsletter, 5 (1999); https://www.iucr.org/__data/assets/pdf_file/0016/21634/cpd21.pdf

[26] A. Le Bail, Monte Carlo indexing with McMaille, Powder Diffraction, **19**, 249 (2004); https://doi.org/10.1154/1.1763152

[27] SVD-Index—solving linear equations with Monte Carlo search. A.A. Coelho, Indexing of powder diffraction patterns by iterative use of singular value decomposition, J. Appl. Cryst. **36**, 86 (2003); https://doi.org/10.1107/S0021889802019878

[28] X-Cell—a new implementation of successive dichotomy approach in Materials Studio. M.A. Neumann, X-Cell: a novel indexing algorithm for routine tasks and difficult cases, J. Appl. Crystallogr, **36**, 356 (2003); https://doi.org/10.1107/S0021889802023348

[29] A.A. Coelho, A. Kern, Advances in indexing of powder diffraction patterns: Iterative use of least squares and Monte-Carlo based whole powder pattern decomposition. The Third Pharmaceutical Powder X-ray Symposium (PXRD-3). (2004); https://www.icdd.com/assets/ppxrd/03/reginfo/Alan_Coelho.pdf

[30] R. Černý, V. Favre-Nicolin, J. Rohlíček, M. Hušák, Z. Matěj, R. Kužel, Expanding FOX: Auto-indexing, grid computing, profile fitting. p. 16 in: CPD Newsletter "Real-Space and Hybrid Methods for Structure Solution from Powders", Issue 35, (2007); https://www.iucr.org/__data/assets/pdf_file/0013/21622/CPD_35_total.pdf

[31] International Tables for Crystallography Vol. **H**, Ch. 3.4.4.3.2, 277 (2019); https://onlinelibrary.wiley.com/iucr/itc/Ha/ch3o4v0001/

[32] See review of indexing program by R. Shirley, Overview of powder-indexing program algorithms (history and strengths and weaknesses), IUCr Computing Commission Newsletter **2**, 48 (2003); https://www.iucr.org/__data/assets/pdf_file/0004/6394/iucrcompcomm_jun2003.pdf; International Tables for Crystallography Vol. **H**, Ch. 3.4.4. Software packages for indexing and examples of their use, 270 (2019); https://onlinelibrary.wiley.com/iucr/itc/Ha/ch3o4v0001/

As an example, a brief description of two techniques for Monte Carlo indexing is provided here. In the *genetic algorithm*, a trial set of lattice parameters is generated randomly, and this population undergoes a series of processes that mimic natural evolution, such as mating, mutation, and natural selection. The quality of generated solutions, or in other words, the fitness of individual members of the population, is evaluated from weighted profile residual, R_{wp} (see Eq. 15.19 in Sect. 15.6.2 for definition), followed by a full-profile fitting. As the population grows, the $R_{wp}(a,b,c,\alpha,\beta,\gamma)$ hyper-surface is explored for the presence of a global minimum, which corresponds to a correct indexing solution.[33] This approach is implemented in the indexing application GAIN, which employs full-pattern indexing so there is no need for initial determination of peak positions. Impurity phases in reasonably small concentrations present little to no problems in arriving to a satisfactory solution. Conversely, the use of full-profile fitting even in a simplified form slows down the indexing, which, however, may or may not be critical, especially considering the high speed of modern multi-core computers.

In the Monte Carlo approach, incorporated in McMaille application, the full profile is also used but in a simplified way. Here, positions of Bragg peaks, usually found automatically, are used to generate a surrogate pattern, which is then employed for comparison with patterns calculated from the generated unit cells. Profile residual, R_p (see Eq. 15.18 in Sect. 15.6.2), is employed as a test criterion. When a match is below some preselected R_p level and the required fraction of Bragg peaks is indexed, the unit cell dimensions are further adjusted in small steps, once again employing the Monte Carlo algorithm because least-square refinement is less efficient. These unit cell adjustments usually substantially decrease the R_p. The process resembles the simulated annealing approach, which also is quite effective in other applications. In order to overcome a potential from converging to a false minimum, all solutions with R_p lower than some preset value (R_{max}) are saved, and only when a solution with R_p falling below the acceptance level (R_{min}) is found such unit cell is considered correct. This approach is much faster than the previous one partly because of simple peak-shape functions (Gauss or a box function) and the use of much broader peaks in the surrogate pattern compared to those present in the experimental pattern. In addition to making the indexing practically insensitive to small systematic errors (e.g., due to specimen displacement, zero-shift, or transparency effects, see Sect. 8.4.2), this method senses correct or potentially correct solutions even when a trial random unit cell is relatively far from the optimum. All of this translates into a shorter computation time.

A few of special features incorporated in the McMaille application are worth noting. One is the ability to simultaneously attempt indexing of two phases in a mixture. Another is a fully automated (black box) regime that may be greatly

[33] B.M. Kariuki, S.A. Belmonte, M.I. McMahon, R.L. Johnston, K.D.M. Harris, R.J. Nelmes, A new approach for indexing powder diffraction data based on whole-profile fitting and global optimization using a genetic algorithm, J. Synchrotron Rad. **6**, 87 (1999); https://doi.org/10.1107/S0909049598017932

appreciated by either the beginners or occasional users. McMaille also incorporates the grid-search indexing technique, which can be performed with much larger increments due to a possibility of further optimizing promising unit cells as described here.

In direct-space indexing, grid search is known as the "cannot-miss-the-cell" technique (if only one can wait long enough), while the random search finds a solution much faster, though the chances to miss a correct unit cell are still small. This holds true assuming that proper indexing parameters have been chosen (which comes with experience) and excluding special cases, such as a superstructure or an accidental relationship between two or more of the unit cell parameters causing a severe overlap of unrelated reflections. The latter scenario is a recipe for failure, regardless of the indexing methods used, if they are applied without critical analysis of the results. Every such case requires special handling.

14.8.2 Indexing in Reciprocal Space

The reciprocal space approach is more effective than the grid search, especially in low-symmetry cases. This approach uses only several low Bragg angle peaks that are chosen as a basis set, and then an exhaustive permutation-based assignment of various combinations of *hkl* triplets to each peak from the basis set is carried out. Index permutation algorithms are more complex to implement than the grid-search algorithm but they are many orders of magnitude faster.[34] This occurs because the indices of low Bragg angle peaks are three relatively small integers. Today, index permutation is the most common technique used in various indexing computer programs.

The reciprocal space indexing can be implemented in several different ways. Two of them are *trial-and-error* and *zone-search* methods. The first one is more efficient in high-symmetry crystal systems (from cubic to orthorhombic) but becomes slow for low-symmetry crystal systems (especially triclinic), while the second method

[34] Consider a triclinic crystal system, where a minimum of six independent Bragg reflections are required to determine the unit cell. Assuming that the maximum value of each of the three indices is 1 and recalling that two of them should vary from -1 to 1 (see Table 14.7), a total number of possible combinations for one Bragg reflection is $3 \times 3 \times 2 - 1 = 17$ [the set (000) cannot be observed and is excluded from the consideration]. In an exhaustive search without imposing any limitations, a total of ~$17^6 \cong 2.4{\times}10^7$ combinations among all six reflections resulted. This represents about 7 orders (!) of magnitude reduction in the computation time when compared to the unrestricted exhaustive grid search in direct space. The same example also highlights the critical role of the lowest Bragg angle reflections in finding a solution of the indexing problem: when the maximum index to be considered is increased to 2, the total number of combinations to be tested in an unrestricted exhaustive grid search rises to ~$1.6{\times}10^{11}$, and it becomes ~$2.1{\times}10^{14}$ when the maximum value of index increases to 3.

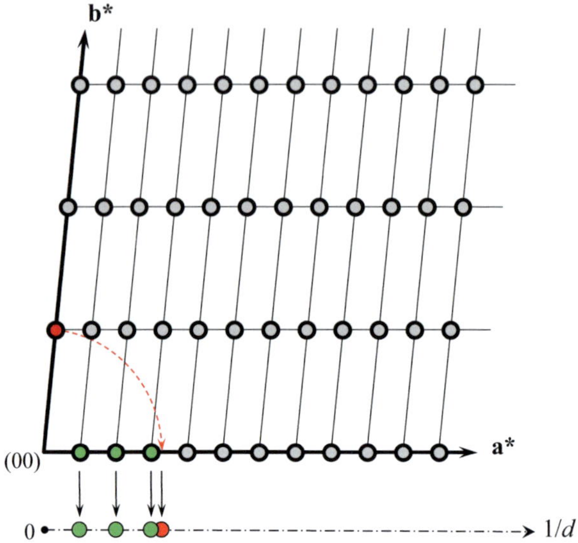

Fig. 14.12 The illustration of the two-dimensional lattice with one long (**b***) and one short (**a***) reciprocal lattice vectors. If the three lowest Bragg angle peaks (*filled circles*) are selected as a basis set for indexing, all of them are collinear and only depend on a^*. The remaining two lattice parameters (b^* and γ^*) cannot be determined from this basis set

works quite effectively and is fast with low symmetries (from triclinic to orthorhombic).[35]

The *trial-and-error* method is based on assigning indices to a minimum required number of low Bragg angle peaks—the so-called basis set. The minimum number of peaks in the basis set is equal to the number of the individual unit cell parameters, which varies from 1 to 6, depending on the crystal system. The values of indices in the triplets vary between certain preselected minimum and maximum values of h, k, and l. Each permutation is followed by the determination of a trial unit cell and by an attempt to index ~10 to 30 consecutive reflections at progressively higher Bragg angles in the resulting unit cell. Successful solutions, those producing the fully or almost fully indexed diffraction pattern, are stored along with the computed figure (s) of merit for further analysis and automatic or manual selection of the best indexing solution.

The success of this approach critically depends on the selection of the basis set, which generally should contain more Bragg peaks than the number of the unknown unit cell dimensions. Potential caveats include but are not limited to the following: the selected basis set contains reciprocal vectors which are collinear or coplanar (see Fig. 14.12); there is one or more impurity peak(s) in the basis set or in the list of Bragg peaks included into the consideration during the indexing attempt and inadequate selection of the minimum and maximum h, k, and l values for index permutations.

[35] We note that the power of modern multi-core computers makes both methods equally fast. Hence, the only important matter that remains is the ability to arrive to a solution with notably better figures of merit than false solutions. From this point of view, two approaches are somewhat different when working with high or low symmetry data.

The exhaustive permutation technique may be improved by eliminating collinear reciprocal lattice vectors from the basis set, which can be done by analysis of the relationships between the observed Q-values. As follows from Eq. (14.1), when two different $1/d = \sqrt{Q}$ are related to one another by a whole multiplier, the two are likely collinear, and only the smallest is usually retained in the basis set.

Needless to say, different crystal systems should be tested from the highest to the lowest symmetry until a satisfactory solution is found. More often than not, automatic trial-and-error indexing yields multiple solutions, generally with different figures of merit, and the final decision is still up to the researcher.

Zone-search method begins with searching for one-dimensional and two-dimensional zones. It then builds a three-dimensional zone using common rows in the two-dimensional lattices, which is used to attempt indexing all observed Bragg reflections. This automatic indexing technique is more sophisticated when compared to a trial-and-error approach, but it is still based on Eq. (14.1). First, the analysis of numerical relationships between the observed Q-values is made to identify zones that are invariant with respect to two indices, for example, $h00$, $h10$, $0k0$, and so on. Second, these are combined to identify zones which are invariant with respect to one index, and finally, zones where all three indices in the triplets vary independently are found and analyzed. To a certain extent, zone searching resembles the algorithm described earlier for the manual indexing in the tetragonal and hexagonal crystal systems; for example, zones that are invariant to both k and l should satisfy Eq. (14.24).

$$\boldsymbol{d}^*_{(hkl)} = h\boldsymbol{a}^* + \text{const} \tag{14.24}$$

The zone-search indexing method does not require an assumption about the crystal system, and therefore, it results in a primitive lattice in most cases. When the lattice is confirmed by the subsequent indexing of all observed Bragg peaks, it shall be converted to one of the 14 standard Bravais lattices. The latter is achieved in a process known as the reduction of the unit cell.

14.9 Unit Cell Reduction Algorithms

Low-symmetry unit cells can be selected in a variety of ways regardless of which method was used for indexing. For example, in the case of an orthorhombic crystal system, all three vectors **a**, **b**, and **c** can be permuted; in a monoclinic crystal system, **a**, **c**, and **d** = ±(**a** + **c**) can be switched in a setting with $\beta \neq 90°$, while in a triclinic case, lattice vectors may be selected in a number of different ways. In order to compare and analyze different indexing solutions, the lattice must be reduced to a certain unique, preferably standard form. It is usually achieved by applying the following rules (which sometimes are already incorporated in the indexing process itself by imposing specific restrictions on the assigned indices):

In the orthorhombic crystal system, the unit cell dimensions should be such that $a \leq b \leq c$.

- In the monoclinic crystal system, $a \leq c$, assuming a standard setting with b as the unique axis.
- In the triclinic crystal system, the reduction becomes more complicated due to possible multiple choices of the basis vectors in the lattice.

The first two reduction rules are normally employed only during the indexing. They usually do not produce a standard choice of the unit cell since at this stage the space-group symmetry and often even the lattice type are not involved. For example, in the orthorhombic space-group symmetry Pnma (a standard setting), the condition $a \leq b \leq c$ is not necessarily obeyed.

There are two broadly accepted methods of transforming the unit cell. One of them was introduced by Delaunay[36] and then applied to a transformation of a randomly selected unit cell by Ito.[37] This technique is known as the Delaunay-Ito method. In order to achieve complete standardization, a different method should be employed. This approach, originally introduced by Niggli,[38] results in the so-called Niggli-reduced unit cell.

14.9.1 Delaunay-Ito Transformation

The Delaunay-Ito transformation is performed on three lattice vectors $\mathbf{v}_1$, $\mathbf{v}_2$, and $\mathbf{v}_3$, and a fourth vector, $\mathbf{v}_4 = -\mathbf{v}_1 - \mathbf{v}_2 - \mathbf{v}_3$ (an inverted body diagonal of a parallelepiped) as shown in Fig. 14.13 (left). In this figure, four vectors $\mathbf{v_i}$ are associated with the corners of a tetrahedron, while six scalar products, $s_{ij} = \mathbf{v}_i \cdot \mathbf{v}_j = |\mathbf{v}_i||\mathbf{v}_j| \cos \alpha_{ij}$ (for i $\neq$ j), are associated with the edges of the same tetrahedron. The unit cell is considered reduced when angles α_{ij} between each pair of vectors $\mathbf{v}_i$ are greater than or equal to 90°, and therefore, all scalar products s_{ij} are negative or zero.

The transformation is carried out by changing the sign of any scalar product that is greater than zero, simultaneously with modifying other relevant parameters of the tetrahedron as shown in Fig. 14.13 (right). This procedure is equivalent to adding vectors and is repeated until all scalar products, s_{ij}, become negative or zero.

From the four possible triplets of resulting vectors ($\mathbf{v}_1$, $\mathbf{v}_2$, $\mathbf{v}_3$ to $\mathbf{v}_2$, $\mathbf{v}_3$, $\mathbf{v}_4$), the one that has shortest vectors is selected because the angles among these vectors are

[36] B. Delaunay, Neue Darstellung der geometrischen Kristallographie, Z. Kristallogr. **84**, 109 (1933). https://doi.org/10.1524/zkri.1933.84.1.109

[37] T. Ito, A general powder X-ray photography, Nature **164**, 755 (1949). https://doi.org/10.1038/164755a0

[38] P. Niggli, Krystallographische und strukturtheoretische Grundbegriffe. Band 7, 1. Teil, Handbuch der Experimentalphysik, Akademische Verlagsgesellschaft, 108 (1928). https://www.amazon.com/Krystallographische-strukturtheoretische-Grundbegriffe-P-Nigglie/dp/B004KZU3SU

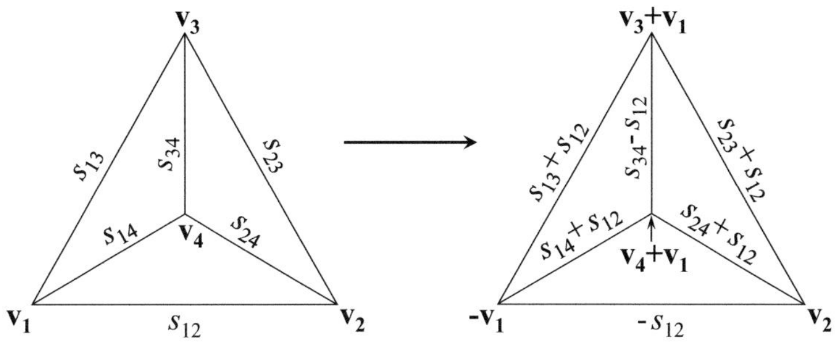

Fig. 14.13 The schematic of Delaunay-Ito transformation. *Left*—the four unit cell vectors ($\mathbf{v}_1$, $\mathbf{v}_2$, $\mathbf{v}_3$, and $\mathbf{v}_4 = -\mathbf{v}_1 - \mathbf{v}_2 - \mathbf{v}_3$) are associated with the corners of the tetrahedron, while the six scalar products between the corresponding pairs of the vectors (s_{12}–s_{34}) are linked to the edges of the tetrahedron. Assuming that $s_{12} > 0$, the transformation is carried out as shown on the *right*: the sign of s_{12} is changed; its value is subtracted from that on the opposite edge and added to those on all adjacent edges; the direction of $\mathbf{v}_1$ (or $\mathbf{v}_2$) is reversed and the new vectors $\mathbf{v}'_3$ and $\mathbf{v}'_4$ are determined as $\mathbf{v}_3 + \mathbf{v}_1$ and $\mathbf{v}_4 + \mathbf{v}_1$ (or $\mathbf{v}_3 + \mathbf{v}_2$ and $\mathbf{v}_4 + \mathbf{v}_2$, respectively)

closest to 90°. The Delaunay–Ito transformed primitive cells can be classified into 24 types according to the relationships between unit cell vectors and their scalar products. They are easily converted into one of the 14 Bravais lattices. For example, if $\mathbf{v}_1 = \mathbf{v}_2$ and $s_{13} = s_{23} = 0$ $\alpha_{13} = \alpha_{23} = 90°$, the standard unit cell is orthorhombic C-centered, with lattice vectors, $\mathbf{v}_i^{\text{ortho}}$, calculated as follows:

$$\begin{aligned} v_1^{\text{ortho}} &= v_1 + v_2 \\ v_2^{\text{ortho}} &= -v_1 + v_2 \\ v_3^{\text{ortho}} &= v_3 \end{aligned} \tag{14.25}$$

The unit cell transformation using Delaunay–Ito method can be easily automated as is done in the ITO indexing computer code, which is discussed in Sect. 14.10.3. The Delaunay–Ito transformed unit cell, however, may not be the one with the shortest possible vectors, although the latter is conventionally defined as a standard unit cell.

14.9.2 Niggli Reduction

The Niggli approach defines the reduced unit cell in terms of the shortest possible vectors.[39] In other words:

[39] P.M. de Wolff, B. Gruber, Niggli lattice characters: Definition and graphical representation, Acta Cryst. **A47**, 29 (1991). https://doi.org/10.1107/S0108767390009485

$$\nu_1 + \nu_2 + \nu_3 = \text{minimum} \tag{14.26}$$

Unfortunately, this simple condition cannot be evaluated directly, and in order to find a properly reduced unit cell, it is be replaced by a total of five simultaneous inequalities[40]:

$$\begin{aligned} &1)\ \nu_1 \le \nu_2 \le \nu_3 \\ &2)\ |s_{12}| \le \nu_3/2 \\ &3)\ |s_{23}| \le \nu_2/2 \\ &4)\ |s_{13}| \le \nu_1/2 \\ &5)\ |s_{23} \pm s_{13}| \le (\nu_1 + \nu_2 \pm 2s_{12}) \end{aligned} \tag{14.27}$$

All angles between lattice vectors are then obtuse or are acute. The unit cell defined by Eq. (14.27) is unique, but the reduction algorithm is not as simple as in the Delaunay–Ito method.[41] It is worth mentioning than if all angles are obtuse, the Niggli unit cell coincides with the Delaunay–Ito unit cell. When all angles are acute, the Delaunay–Ito unit cell contains only two of the shortest vectors, and the third vector is one of the diagonals in the Niggli unit cell. In a similar way as Delaunay–Ito unit cells, the Niggli cells form 44 classes (called characters), and each class is related to one of the 14 Bravais lattices.

Regardless of which indexing method was employed, the resulting unit cell (especially when it is triclinic) shall be reduced using either Delaunay–Ito or Niggli method in order to enable the comparison of different solutions and to facilitate database and literature searches. Further, the relationships between reduced unit cell parameters must be used to properly determine the Bravais lattice. The Niggli-reduced cell is considered standard and therefore is preferable.

It is worth noting that not all automatic indexing computer codes perform unit cell reduction and even when they do, the reduced unit cell may not be standard. Therefore, it is advisable to always check/reduce the obtained unit cell using available software.

14.10 Automatic Ab Initio Indexing: Software Programs

A variety of first-principles indexing methods has been implemented in numerous computer programs, each with various modifications. Most of these indexing programs,[42] along with other crystallographic software, were collected and hosted by

---

[40] L. Zuo, J. Muller, M.-J. Phillippe, and G. Esling, A refined algorithm for the reduced-cell determination, Acta Cryst. **A51**, 943 (1995). https://doi.org/10.1107/S0108767395006672

[41] I. Křivý, B. Gruber, A unified algorithm for determining the reduced (Niggli) cell, Acta Cryst. **A32**, 297 (1976). https://doi.org/10.1107/S0567739476000636

[42] Some of which are discussed in Sect. 14.8.1, see Footnotes 25–32 on page 477.

the Collaborative Computational Project Number 14 (CCP14), which is no longer supported and has effectively ceased to exist. The existing copy of CCP14[43] website contains only selected links and references. Nonetheless, these implementations and developments have not been lost. The most advanced and widely used indexing algorithms, as well as unit cell refinement and full-profile fitting methods, along with newly developed ones, have been incorporated into popular software packages, such as the standalone freeware GSAS/GSAS-II, LHPM-Rietica, FullProf, FOX, JANA,[44] EXPO,[45] or commercial like JADE[46] (ICDD/MDI). These algorithms are also integrated into software packages from diffractometer manufacturers, including Diffrac.Eva and Topas (Bruker), HighScore (Malvern Panalytical), SmartLab II (Rigaku), among others.[47] A comprehensive overview on indexing methods and programs, compiled by Robin Shirley, can be found at IUCr Computing Commission Newsletter.[48] The most recent analysis of the indexing methods is available in Sect. 3.4[49] of the International Tables for Crystallography, Vol. H (Powder Diffraction), which also discusses software packages for indexing and examples of their use in Sect. 3.4.4.

In this section, we review several commonly used standalone automatic indexing programs: TREOR, DICVOL, and ITO, which represent different indexing methods. Our aim is to familiarize novices with ab initio indexing in general, as well as the capabilities and limitations of these programs, while also providing practical indexing examples. The following four sections are not intended to replace the manuals of these indexing utilities, as each program comes with a detailed manual explaining the required format for input data and the meaning of the resulting output.

All three computer codes (TREOR, DICVOL, and ITO) have been extensively tested by hundreds of researchers and undergone multiple revisions by both the original developers and experienced crystallographers. As a result, they are highly reliable, provided adequate high-quality powder diffraction data are used. As previously mentioned, the following are the keys to successful indexing from first principles: *precision*, *precision*, and *precision* of experimental Bragg angles.

[43] http://ccp14.cryst.bbk.ac.uk

[44] See Footnote 32 on page 107, Chap. 13.

[45] A. Altomare, C. Cuocci, C. Giacovazzo, A. Moliterni, R. Rizzi, N. Corriero, and A. Falcicchio, EXPO2013: a kit of tools for phasing crystal structures from powder data, J. Appl. Cryst. **46**, 1231 (2013); https://doi.org/10.1107/S0021889813013113; https://www.ba.ic.cnr.it/softwareic/expo/

[46] https://www.icdd.com/mdi-jade/

[47] See Footnotes 4–7 on page 378, Chap. 13.

[48] R. Shirley, Overview of powder-indexing program algorithms (history and strengths and weaknesses), IUCr Computing Commission Newsletter **2**, 48 (2003), https://www.iucr.org/__data/assets/pdf_file/0004/6394/iucrcompcomm_jun2003.pdf

[49] A. Altomare, C. Cuocci, A. Moliternia, and R. Rizzia, Ch. 3.4. Indexing a powder diffraction pattern, in: the International Tables for Crystallography, Vol. **H**, 270–28 (2019), https://onlinelibrary.wiley.com/iucr/itc/Ha/ch3o4v0001/

14.10.1 TREOR

TREOR[50] (including its updated versions, TREOR90[51] and N-TREOR09[52]) is a semi-exhaustive trial-and-error indexing program that operates by permutating indices within a selected basis set of the lowest Bragg angle peaks. The algorithm is described as semi-exhaustive because certain limitations on the possible permutations of indices are incorporated to improve speed. Another modification of the TREOR indexing method further increases speed by adapting the algorithm for use with parallel computing and graphics processing units (GPUs).[53] TREOR90 includes an analysis of the dominant axial zones (i.e., *h*00, 0*k*0, and 00*l*). In the case of a monoclinic crystal system, the so-called short-axis test and (020)-finding algorithm are employed to increase chances of indexing when unit cell edges have unusually different lengths. This program works with any crystal system, but it is most effective for those with high and medium symmetries, such as cubic, hexagonal, tetragonal, and orthorhombic systems. Triclinic indexing uses lower index limits and is extremely time-consuming. The basis set consists of up to six reflections, with the peaks suspected to represent collinear reciprocal lattice vectors automatically excluded. The default combinations of peaks in the basis sets, maximum values of indices, and sums of indices are listed in Table 14.20.

The limiting values of indices may be freely modified by the user as described in TREOR's manual, while the sequence of peaks in the basis set may only be changed to a certain extent as described in the footnotes to Table 14.20. In the majority of cases, program defaults work quite well and any changes in them should be attempted only when no acceptable solution has been found.

Although no strict limits have been set, usually no more than 25 lowest Bragg angle peaks total should be used in the indexing attempt. TREOR90 may successfully index powder diffraction patterns containing a few impurity peaks by automatically skipping up to three Bragg reflections, which do not fit into a found unit cell. If some peaks were skipped, one should always remember the very important

---

[50] P.-E. Werner, L. Eriksson and M. Westdahl, A semi-exhaustive trial-and-error powder indexing program for all symmetries, J. Appl. Cryst. **18**, 367 (1985), https://doi.org/10.1107/S0021889885010512

[51] A. Altomare, C. Giacovazzo, A. Guagliardi, A.G.G. Moliterni, R. Rizzi, P.-E. Werner, New techniques for indexing: N-TREOR in EXPO. J. Appl. Cryst. **33**, 1180 (2000), https://doi.org/10.1107/S0021889800006427; TREO90 is also included in Crystal Impact (https://www.crystalimpact.com/match/help/idh_indexing_treor.htm) and FullProf (https://www.ill.eu/sites/fullprof/php/programsdccc.html?pagina=Treor90) software packages.

[52] A. Altomare, G. Campi, C. Cuocci, L. Eriksson, C. Giacovazzo, A. Moliterni, R. Rizzi and P.-E. Werner, Advances in powder diffraction pattern indexing: N-TREOR09, J. Appl. Cryst. 42, 768 (2009), https://doi.org/10.1107/S0021889809025503

[53] I. Šimeček, J. Rohlíček, T. Zahradnický, and D. Langr, A new parallel and GPU version of a TREOR-based algorithm for indexing powder diffraction data, J. Appl. Cryst. **48**, 166 (2015), https://doi.org/10.1107/S1600576714026466

Table 14.20 Order of peaks in basis sets and limitations imposed on the values of indices for the basis-set peaks used in TREOR

Crystal system	Basis set (in terms of peak number)	max h,k,l[a]	max $h+k+l$[a]
Cubic	(1), (2)	4,4,4	6
Tetragonal, hexagonal	(1,2), (1,3), (2,3)	4,4,4 4,4,4	4 4
Orthorhombic	(1,2,3), (1,2,4), (1,2,5), (1,3,4), (2,3,4), (1,2,6) or (N,1,2), (N,1,3), (N,2,3)[b]	2,2,2 2,2,2 2,2,2	3 4 4
Monoclinic	(1,2,3,4), (1,2,3,5), (1,2,4,5) or (N,1,2,3) (N,1,2,4), (N,1,3,4)[b] and additionally (1,3,4,5), (1,2,3,6), (2,3,4,5), (1,2,3,7)[c]	2,2,2 2,2,2 2,2,2 2,2,2	2 3 3 4
Triclinic	(1,2,3,4,5,6), (1,2,3,4,5,7), (1,2,3,4,5,8), (1,2,3,4,6,7), (1,2,3,4,6,8), (1,2,3,5,6,8), (1,2,3,5,6,7), (1,2,3,5,7,8), (1,2,3,4,5,9)[d]	1,1,1 1,1,1 1,1,1 1,1,1 1,1,1 1,1,1	1 2 2 2 3 3

[a] Absolute value of h, k, and l is assumed. Max $h + k + l$ values are for each basis set reflection, respectively

[b] N is defined as a nonzero value in the keyword SELECT, e.g., in the orthorhombic crystal system SELECT = 5 results in the following basis sets (1,2,5), (1,3,5), and (2,3,5)

[c] Additional sets that are added to the basis set when MONOSET keyword value is greater than 3, 4, 5, or 6, respectively

[d] The first observed peak is always (100), and the second observed peak always has positive indices

statement (capitalized in the manual): do not accept unindexed peaks unless you are able to explain them.

The key to a successful indexing is not a complete absence of impurity peaks (a few may be present), but it is the accuracy with which peak positions have been determined and the absence of significant systematic errors. Yet another important piece of advice, given in the manual, should always be followed: do not waste computer time on bad data. Since the cost of computer time continuously lowers, but the cost of labor continuously rises, this statement could be rephrased: do not waste *your* time on bad data. The latter is indeed applicable to any type of data analysis.

Several other items should be considered while using TREOR. Indexing starts from the cubic crystal system and may proceed through all crystal systems down to triclinic. By default, however, the indexing stops at the orthorhombic symmetry. Therefore, in order to check monoclinic solutions, the value of a keyword MONO should be set to the maximum desired monoclinic angle, for example, MONO = 130 will result in searching for solutions with angles β varying from 90° to 130°. Indexing in a triclinic unit cell can be attempted by including the instruction TRIC = 1. Another way of testing all crystal systems is to specify a negative value of the maximum unit cell volume using keyword VOL, for example, VOL = −2000. TREOR reduces the triclinic unit cell only for the best solution, and therefore, numerous nonreduced variations of the same unit cell may be produced.

When an acceptable solution is found, the program terminates, and therefore, indexing in lower symmetry unit cells will not be performed. A solution is considered acceptable if the minimum figure of merit, M_N, where N is the smallest of 20 and the number or present peaks, is reached or exceeded. It is advisable to check lower symmetry indexing in order to avoid situations when accidental relationships between the real unit cell parameters result in successful high-symmetry indexing. In majority of the cases, the incorrect high-symmetry unit cell has a large volume that cannot be explained by the increased symmetry of the lattice. This can be done by disabling any of the high-symmetry crystal systems: for example, cubic indexing can be suppressed by using the keyword KS = 0, and tetragonal and hexagonal indexing are not attempted when THS = 0 (consult the manual for more keywords).

An alternative way of continuing the indexing in low-symmetry lattices is to increase the minimum figure of merit, M_N, by adding or changing the keyword MERIT. If this keyword is missing (default), the minimum M_N is set to 6, which is quite low, given the accuracy of both the modern diffractometers and data processing algorithms (e.g., see the corresponding figures of merit described in Sect. 14.4.2). Any solution with M_N greater than 10 is considered to be a true unit cell, and the program execution terminates. In order to continue the search for other possible solutions, the value of MERIT should be increased, e.g., by setting MERIT = 50 or higher.

By default, TREOR adjusts the data for possible systematic errors in the first seven peaks by using higher-order peaks. Sometimes this procedure may be not quite right, and when no solution has been found, it is recommended to suppress this correction by adding the instruction IDIV = 0.

The successful indexing result (if any)—the one with M_N greater than a set minimum—is stored in a condensed output file. All intermediate solutions with M_N less than the specified minimum but greater than 6 and no more than 3 unindexed lines (unless different limits are specified) are stored in a general output file. Therefore, using unrealistically high MERIT = 1000 and VOL = −2000 would result in checking all crystal systems and unit cells with volumes smaller than 2000 Å^3.

14.10.2 DICVOL

DICVOL[54] method is an exhaustive trial-and-error indexing program with variation of parameters by successive dichotomy and partitioning of the unit cell volume. This program works with all crystal systems; however, low-symmetry monoclinic and

[54] J.W. Visser, A fully automatic program for finding the unit cell from powder data, J. Appl. Cryst. 2, 89 (1969). https://doi.org/10.1107/S0021889869006649

especially triclinic indexing may take some time. The indexing strategy is based on searching for a solution from high to low symmetry using partitioning of the unit cell volume in 400 Å^3 increments, except for a triclinic case, when the maximum volume is estimated from the density of diffraction peaks in the pattern.

Similar to TREOR, any crystal system can be included or excluded from the indexing process, and increasing a minimum figure of merit, M_N (the default is 5) can be used to generate low-symmetry results. Input data may also be used to set maximum lengths of the unit cell edges, monoclinic angle, and unit cell volume. Errors in peak positions can be provided for each Bragg reflection individually, or they may be set to the same value for all observed peaks. The default is 0.03° of 2θ, which is somewhat excessive for good-quality diffractometer data, so the value of 0.01° or 0.02°, depending on the resolution, is recommended. For more details, consult the manual available along with the program. As a rule, indexing should be performed in three steps by selecting the following crystal systems: cubic through orthorhombic; monoclinic; and finally triclinic, if needed.

Originally, DICVOL as well as its 1991 version, DICVOL91,[55] was not designed to account for possible impurity peaks or to correct for systematic errors. However, later releases, such as DICVOL04,[56] DICVOL06, and DICVOL14,[57] include an option to specify a maximum number of unindexed Bragg peaks to ignore and can estimate and correct for the presence of zero-shift error. Note that being one of several popular indexing programs, DICVOL91 is also included in other software packages, such as FullProf.[58]

No formal limits on the amount of input Bragg peaks have been established in this program, but it is recommended that 20 or more lowest Bragg angle peaks are used for a reliable indexing. It is worth noting that if the true unit cell belongs to one of the high-symmetry crystal systems (e.g., tetragonal), an attempt to index the diffraction pattern in lower symmetry crystal systems (e.g., orthorhombic, monoclinic, and triclinic) will usually result in the solution with $a \cong b$ (orthorhombic), $a \cong b$ and $\beta \cong 90°$ (monoclinic), and/or $a \cong b$ and $\alpha \cong \beta \cong \gamma \cong 90°$ (triclinic). Independent unit cell reduction (see Sect. 14.9) should be employed to test any resulting low-symmetry unit cell in order to compare it with the literature or database records.

---

[55] A. Boutlif and D. Louër, Indexing of powder diffraction patterns for low symmetry lattices by the successive dichotomy method, J. Appl. Cryst. 24, 987 (1991), https://doi.org/10.1107/S0021889891006441

[56] A. Boultif and D. Louër, Powder pattern indexing with the successive dichotomy method, J. Appl. Cryst. 37, 724 (2004). https://doi.org/10.1107/S0021889804014876

[57] D. Louër and A. Boultif, Some further considerations in powder diffraction pattern indexing with the dichotomy method, Powder Diffraction, **29** (S7) (2014), https://doi.org/10.1017/S0885715614000906

[58] https://www.ill.eu/sites/fullprof/php/programs64ba.html?pagina=Dicvol

14.10.3 ITO

The ITO[59] algorithm utilizes a zone-search indexing method combined with the Delaunay–Ito technique (see Sect. 14.9.1) to transform the most probable unit cell. The most commonly used versions of codes are ITO13 and ITO15, either as standalone programs or as a part of software package, such as WinCSD or Fullprof.[60] The program searches for an indexing solution using the following algorithm:

Identifies and reduces potential zones by consistently using the 20 lowest Bragg peaks, testing and refining them with a least-square method.

- Builds lattices by identifying zones with a common row and calculating the angles between them.
- Reduces the resultant unit cell using the Delaunay–Ito method and then transforms it into one of the 14 Bravais lattices if the lattice is not primitive.
- Finally, attempts to index all available peaks, and if the indexing is successful, calculates the figure(s) of merit.

The program works quite effectively with low-symmetry crystal systems: triclinic, monoclinic, and orthorhombic. The resulting centered lattice in the monoclinic and orthorhombic symmetry (if centering has been found) is an important argument in accepting such a solution, since it is less likely to be a random unit cell when compared with any primitive lattice. Lattice centering analysis is not performed in higher symmetry cases; nevertheless, suggestions are provided at the end of the output and should be carefully examined.

The program requires at least 20 lowest Bragg angle peaks to find a reasonable solution and will not work with fewer peaks. A total of 30–35 consecutive Bragg reflections are recommended. The maximum number of peaks is 40. Peaks in excess of 20 are indexed using only the best found solution. Similar to TREOR, ITO may successfully index powder diffraction patterns, which contain several impurity peaks. The maximum number of skipped unindexed peaks is one of the user-specified parameters. ITO can check for potential zero-shift errors (by default no checking is performed) and allows a zero-shift correction to be applied to all peaks before indexing. In addition, in the final least-squares refinement of the unit cell dimensions, the zero-shift error in Bragg angles can be determined (by default).

The program works in a straightforward fashion, and only a few parameters can be varied in the case of failure. These are: several tolerances, the number of peaks for zone search, and the zero-shift correction. As a last resort, low-intensity (potential impurity peaks) and/or high-angle peaks (if they were included in the unsuccessful run) may be removed from the list one-by-one. The default values for tolerances for two- and three-dimensional zone search are set to 3.0 and 4.5, respectively, in the

[59] J.W. Visser, A fully automatic program for finding the unit cell from powder data, J. Appl. Cryst. **2**, 89 (1969). https://doi.org/10.1107/S0021889869006649

[60] https://www.ill.eu/sites/fullprof/php/programsa0be.html?pagina=Ito15

units of Q, which are defined in this program as $10^4/d^2$. However, it is recommended to set them as low as 2.0 and 2.5 when the accuracy in peak positions is at or better than 0.01° of 2θ. The final advice, given by the author (J.W. Visser) should always be remembered: "Finding the unit cell depends for 95% on the quality of the input data. A random error of 0.03° 2θ can usually be tolerated, but a systematic zero point error of 0.02° is probably disastrous."

14.10.4 Selecting a Solution

TREOR and ITO programs have an option to input intensity for each observed diffraction peak, but they are included only for informative purposes and are never analyzed during the indexing. Nevertheless, it is often helpful to see the intensity next to indices when analyzing the results, especially for those Bragg peaks, which do not fit into the proposed lattice.

All three programs TREOR, DICVOL, and ITO allow optional input of the information about the measured gravimetric density and formula weight in order to estimate the number of formula units expected in the found unit cell (see Sect. 15.2). The latter should be an integer number compatible with the unit cell symmetry. For example, in a primitive monoclinic lattice, it should typically be a multiple of 2 or 4. The agreement between the number of formula units in the unit cell and lattice symmetry may be used as an additional stipulation when selecting the most probable solution.

Automatic indexing programs usually generate multiple solutions, and the following criteria should be considered when deciding which one is the best—in other words, which solution likely represents the true unit cell:

1. The correct solution should result in a high figure of merit (though not necessarily the highest) and most, if not all, diffraction peaks should be indexed. Unindexed Bragg reflections must be explained, for example, by low intensity, by noticeably different peak widths, or they should be identified as corresponding to an impurity phase.
2. As is stated in the DICVOL manual: "… a solution from the first 20 (or N) lines … must index the complete powder diffraction pattern … ." The best way to evaluate the quality of the indexing is to perform lattice parameters refinement and to calculate a figure of merit using all measured and indexed Bragg peaks. Alternatively or in addition to this evaluation, full-profile fitting using LeBail or Pawley methods could be applied, which would likely be the next step if the goal is the precise determination of lattice parameters, solving the crystal structure, or simply refining the crystal structure.
3. Every or almost every observed peak must correspond to a calculated reflection after accounting for systematic absences, and vice versa, at least in a low Bragg angle region of the pattern where all peaks are well-resolved.

4. The presence of distinct systematic absences and an unambiguous determination of the diffraction group make a particular indexing solution especially probable. Exceptionally encouraging are centered lattices as well as space groups such as $P2_1/c$, Pbca, $I4_1/amd$, $R\bar{3}c$, and many others, which contain multiple glide planes and/or screw axes, distinguishable from the list of unobserved (extinct or forbidden) Bragg reflections.
5. When the unit cell is found, it can be used to search multiple databases in addition to the ICDD Powder Diffraction File. A positive search result usually confirms indexing, while a negative outcome does not necessarily mean that the indexing is wrong—a particular crystal structure may be novel and/or not yet included in a database.
6. Finally, in the case of a new material, the correctness of the *ab initio* indexing is generally ensured by solving and refining the crystal structure,[61] which should make both chemical and physical sense—that is, it should exhibit correct bond distances and angles, reasonable coordination polyhedra, appropriate oxidation states, and so on. The correctness of indexing may also be proven by other means, for example, by selected area electron diffraction from an individual grain or micro crystal, high-resolution electron microscopy, and similarity with known structures.

All three of the briefly described computer programs may only work consistently when the quality of input data is very high. This implies that the accuracy of peak positions is outstanding, normally at least 0.02° or better, which is usually achieved by maintaining a well-aligned goniometer and by using a proper profile fitting procedure. It is important to ensure that all possible systematic errors are eliminated beforehand, since they can lead to unpredictable results and/or failure to find a solution. When experimental data are affected by a small systematic error, reducing the number of Bragg peaks in the initial indexing attempt may be helpful.

Each program is distributed with ready-to-use files containing test/demo example (s). We encourage the readers to familiarize themselves with examples after reading the manuals before trying to perform ab initio indexing of a new diffraction pattern.

Input files from Example 4 (*Ch14Ex04*) for DICVOL, ITO, and TREOR can be used as templates for user data. Users need to only replace the 2θ Bragg angles, and optionally the TREOR intensity, with their own data. Some refinement setup parameters—such as selecting crystal systems for indexing or adjusting the accuracy of Bragg angles—may require modification based on the program manual.

Alternatively, readers can consider newer indexing programs that are now commonly incorporated into software packages, such as GSAS-II, TOPAS, HighScore, and others. Many of these programs employ automatic peak-search routines, eliminating the need to manually identify peaks. Some also use algorithms that process the entire diffraction pattern without relying on individual peaks. These methods simplify the indexing process, although they may take longer to find a solution.

---

[61] In some cases, an experienced researcher can be confident in the indexing without solving the crystal structure, if the pattern is relatively simple, the peaks are sharp and well-resolved, and all previous criteria are satisfied.

14.11 Ab Initio Indexing Examples

We present four practical indexing examples. The first example details the indexing of a relatively simple, high-symmetry hexagonal pattern, with all input data files available as an electronic supplement from Springer. The remaining examples are illustrated without proving all input data files, as we expect readers to use existing data and independently reproduce the indexing results. The second and third examples cover indexing of low-symmetry patterns: a complex monoclinic and a simpler triclinic structure, respectively. The fourth example highlights the challenges that can arise from accidental relationships between unit cell dimensions.

14.11.1 *Hexagonal Indexing:* $LaNi_{4.85}Sn_{0.15}$

Experimental data from the $LaNi_{4.85}Sn_{0.15}$ sample are especially useful for this illustration because as established earlier, this diffraction pattern has been successfully indexed manually. We also know that the data are affected by a small systematic error, which can be eliminated by introducing a zero-shift correction of 0.032°, and that there are two low-intensity Bragg peaks, which belong to an impurity phase (see Sects. 14.3 and 14.7).

14.11.1.1 Indexing Using TREOR90

The input data file[62] for the TREOR90 program contains the following information, where the comments after the exclamation symbols are not included in the data file:

Sample: $LaNi_{4.85}Sn_{0.15}$	! Title line
20.288	! Bragg angle of the first observed peak
22.105	! Bragg angle of the second observed peak
...	! Bragg angles in ascending order
81.632	! Bragg angle of the last observed peak
CHOICE=3,	! Blank line terminating experimental data
WAVE=1.540593,	! Instruction specifying input as 2θ angles
END*	! Instruction specifying the wavelength
0.000	! End of instructions
	! The value of the zero shift, if known

In our first indexing attempt, we use all observed Bragg peaks, including the two peaks from an impurity phase at $2\theta = 44.211°$ and $51.517°$ (see Fig. 14.5), without correcting experimental data for the known zero-shift error. Automatic indexing results in the following solution: $a = 18.537$, $c = 4.3681$ Å in the hexagonal crystal

[62] Data file *Ch14Ex04_Treor90-All.dat* is available online.

system with F_{24} = 16(0.008, 205) and M_{20} = 21. One weak Bragg peak at $2\theta = 78.958°$ (relative intensity is 0.4%) remains unindexed. This solution is likely incorrect, which is easily concluded from the F_N figure of merit indicating that only ~12% of the possible Bragg peaks have been observed.

After removing the two impurity Bragg peaks,[63] the indexing result is $a = 6.179$, $c = 10.097$ Å in the tetragonal crystal system with F_{22} = 12(0.019, 98) and $M_{20} = 16$. Both figures of merit are lower than in the first solution, and this is also an incorrect indexing result.

Even though it appears that this program cannot find an indexing solution in this relatively simple case, we should note that so far the program was used only with default settings, including the minimum acceptable figure of merit. The default value is set at $M_{20} = 10$, and once the solution with M_{20} exceeding 10 is found, no further attempt to test lattices with lower (i.e., next in the line: cubic, tetragonal, hexagonal, orthorhombic, monoclinic, and triclinic) symmetry is made. This is the reason of the incorrect solution when all but the two impurity peaks were included in the clean data file.

A simple modification of the default figure of merit by including the instruction MERIT = 40, which is nearly double the best M_{20} observed so far[64] yields a correct solution: hexagonal lattice with $a = 5.0443$, $c = 4.0135$ Å and the corresponding figures of merit are as follows: F_{22} = 126 (0.008, 23) and $M_{20} = 189$. The lattice parameters and figures of merit are slightly different from what has been established earlier (see Table 14.6 and associated text), which is the result of the incomplete refinement of the lattice parameters carried out by the program after the solution has been found.

It is interesting to point out that the same modification (adding the instruction MERIT = 40 to the original data file) results in a different solution in the hexagonal crystal system: $a = 8.7371$, $c = 4.0135$ Å with the corresponding figures of merit: F_{24} = 59 (0.008, 53) and $M_{20} = 77$. One of the two impurity peaks was left unindexed, but the second impurity peak accidentally fits into this enlarged unit cell. A simple analysis of Miller indices of all indexed Bragg peaks, which can be found in the output file produced by the program, indicates that in all cases except one, the sums $h^2 + hk + k^2$ are divisible by three. Considering Eq. (14.4), the reduction of the a lattice parameter by a factor of $\sqrt{3}$ produces the correct unit cell with $a = 5.0444$ Å. The only peak with the sum $h^2 + hk + k^2 \neq 3n$ (where n is an integer) is the impurity peak observed at $2\theta = 51.517°$.

14.11.1.2 Indexing Using ITO13

The input data file[65] for the ITO13 program contains the following information (the comments after the exclamation symbols are not included in the data file):

[63] Data file *Ch14Ex04_Treor90-Clean.dat* is available online.

[64] Data file *Ch14Ex04_Treor90-HighMerit.dat* is available online.

[65] Data file *Ch14Ex04_ITO13-All.dat* is available online.

Sample: $LaNi_{4.85}Sn_{0.15}$	! Title line
20.288	! Parameter line (may be completely blank)
22.105	! Second parameter line (may be blank)
...	! Bragg angle of the first observed peak
81.632	! Bragg angle of the second observed peak
0	! Bragg angles in ascending order
END	! Bragg angle of the last observed peak
	! Zero or blank line terminating this set of data
	! End of data file

The complete description of how to modify various default values by using different fields in the two parameter lines in the file is found in the documentation to the program. Even if all parameters are left at their defaults, the two blank lines must precede the first Bragg peak. The program can process multiple sets of data placed in the same file, and these sets are separated by a single blank line or by a line containing 0 in the first position. The line containing the word END indicates the end of the input data file. The ITO13 program finds the correct unit cell: hexagonal crystal system, $a = 5.0423$, $c = 4.0119$ Å. They are equal to those found before (Table 14.6) within one standard deviation. The program used the default value of the wavelength ($\lambda = 1.5406$ Å), which can be modified by inserting the corresponding value in positions 21–30 of the first parameter line.

14.11.1.3 Indexing Using DICVOL91

The input data file[66] for the DICVOL91 program contains the following information (the comments after the exclamation symbols are not included in the data file) using DICVOL:

Sample: $LaNi_{4.85}Sn_{0.15}$	! Title line
24 2 1 1 1 1 0 0	! Parameter line 1
0 0 0 0 0 0 0	! Parameter line 2
1.540593 0 0 0	! Parameter line 3
0 0 0	! Parameter line 4
20.288	! Bragg angle of the first observed peak
22.105	! Bragg angle of the second observed peak
...	! Bragg angles in ascending order
81.632	! Bragg angle of the last observed peak
	! Blank line terminating input data

The first parameter line contains the following information: number of consecutive Bragg peaks to use counting from the first (24), type of input data ($2 = 2\theta$ in degrees), then six single digit flags specifying which crystal systems should be examined in the following order: cubic, tetragonal, hexagonal, orthorhombic, monoclinic, and triclinic. When the switch is set to 1, the crystal system is tested, and when

[66] Data file *Ch14Ex04_Dicvol91-All.dat* is available online.

it is set to 0, no solution in this crystal system is attempted. The first parameter line in the example data file is set to search for a solution in all crystal systems except monoclinic and triclinic.

The second parameter line is used to specify the maximum possible values of *a*, *b*, and *c* (first three quantities in the line with the defaults set at 20 Å), then minimum and maximum unit cell volume (next two values with the defaults 0 and 1500 $Å^3$), and finally, minimum and maximum β if the monoclinic crystal system is to be tested (the last two values, defaults are set at 90° and 125°, respectively). The second parameter line in our example selects all corresponding parameters as their default values.

The third parameter line contains the following information: the first number is the wavelength of the used radiation in Angströms and the second, third, and fourth quantities represent the molecular weight of the formula unit in atomic mass units (a.m.u.), the measured density in g/cm^3, and the error in the measured density, respectively. The defaults are set at zero meaning that none of these characteristics of the material is known.

The last (fourth) parameter line should contain three numbers (with zeros representing the selection of default values) and their meaning is as follows. The first value indicates how the errors in the experimental data are handled. By default, the measurement errors are assumed constant value of 0.03° 2θ. The second value specifies the minimum acceptable M_N figure of merit, where *N* is the number of used Bragg peaks, which is identical to the first number in the first parameter line, and its default is 5. The last number in this line is the value of the zero shift to be added to the observed experimental data.

The attempt to find an indexing solution using all available data results in the best figures of merit, $F_{24} = 10(0.010, 248)$ and $M_{24} = 13$ for a large orthorhombic unit cell with $a = 17.475$, $b = 7.569$, and $c = 4.013$ Å. The failure of indexing using the array of Bragg peaks containing two impurity data points was expected, because this program cannot leave any of the data points unindexed. It is worth noting, however, that there is one correct unit cell dimension in this incorrect solution—the value of the *c* lattice parameter has been found accurately. Further, this solution results in nearly all Bragg peaks having index *h* divisible by 4 and index *k* divisible by 3. This outcome is a very good indication that input data do contain one or more impurity peaks and that the solution is false.

To obtain a proper solution, it is necessary to eliminate the impurity Bragg peaks from the indexing process. Although it is easy to do in this case since we know which peaks do not belong to the major component, in real-life situations it may not be as simple unless the structure of the impurity phase(s) is known. Provided the impurity phase is a minority phase in the specimen, the best way to eliminate impurity peaks is to proceed with the gradual elimination of the weakest Bragg peaks from the indexing attempt.

We begin by first eliminating all peaks with intensity less than 1% of the strongest Bragg peak. This produces the truncated array of data[67] and a still incorrect but

[67] Data file *Ch14Ex04_Dicvol91–99%.dat* is available online.

different solution in the orthorhombic crystal system (note that the first impurity peak has intensity ~1.8% of the strongest). A second iteration is to eliminate all Bragg peaks with intensity lower than 2% of the strongest.[68] This leaves only 16 Bragg peaks and leads to two solutions with very high figures of merit. These are as follows:

1. Hexagonal crystal system with $a = 5.0448$, $c = 4.0137$ Å, $F_{16} = 74(0.009, 23)$, and $M_{16} = 159$.
2. Orthorhombic crystal system with $a = 4.3687$, $b = 4.0137$, $c = 2.5230$ Å, $F_{16} = 77(0.008, 28)$, and $M_{16} = 167$.

Given the rules described earlier (see Sect. 14.4) and the insignificant difference in the figures of merit, the preference should be given to the highest symmetry crystal system, which is the solution No.1, even if we did not know the correct unit cell of this material. Furthermore, the value of the lattice parameter c in the orthorhombic solution is too small considering the size of one of the components in the $LaNi_{4.85}Sn_{0.15}$ compound: $r_{La} = 1.87$°A, which restricts the shortest unit cell dimension to $2 \times r_{La} = 3.74$°A or higher.

The indexing process in this case should be finalized by using the found lattice parameters to assign indices to the weak Bragg reflections that have been eliminated from the indexing and by refinement of the lattice parameters employing all available experimental data. This can be easily done following the procedure described earlier in Sect. 14.3.

The two solutions (hexagonal and orthorhombic) are actually similar, since any hexagonal lattice can also be described in the lower symmetry orthorhombic base-centered lattice, as shown in Fig. 14.14. Obviously, the orthorhombic solution, found by DICVOL91, represents the primitive orthorhombic unit cell with one-fourth the volume of the conforming base-centered orthorhombic unit cell with the following unit cell dimensions:

$$a_{\mathrm{O}} = \frac{1}{2} a_{\mathrm{H}} \sqrt{3}; \quad b_{\mathrm{O}} = c_{\mathrm{H}}, \quad \text{and} \quad c_{\mathrm{O}} = \frac{1}{2} b_{\mathrm{H}}$$

14.11.2 *Monoclinic Indexing: $(CH_3NH_3)_2Mo_7O_{22}$*

In this example, we consider the ab initio indexing of the diffraction pattern collected from dimethylammonium heptamolybdate $(CH_3NH_3)_2Mo_7O_{22}$ powder. Profile fitting of the powder diffraction data (Fig. 14.15) resulted in 30 peaks observed below 30° of 2θ (see Table 14.21[69]). The indexing has been conducted using

[68] Data file *Ch14Ex04_Dicvol91–98%.dat* is available online.

[69] Same data are available online in the file *Ch14Ex05_CuKa.pks*.

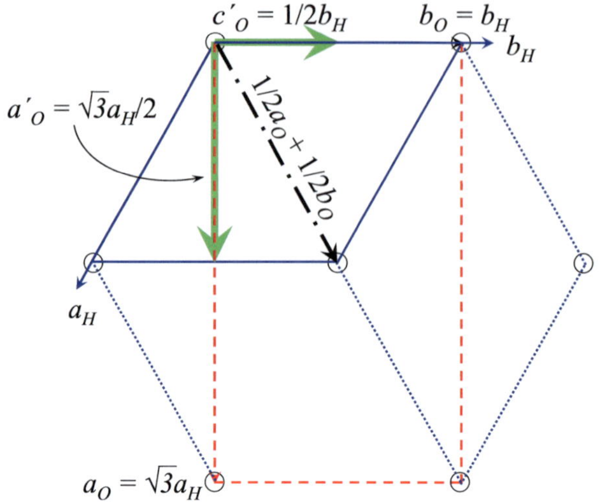

Fig. 14.14 The relationship between the lattice parameters of the hexagonal unit cell (*blue solid and dotted lines*) and the related orthorhombic unit cell (*red dashed lines*). The unit cell parameter perpendicular to the plane of the projection is identical in both crystal systems. The smaller orthorhombic unit cell found using the DICVOL91 indexing program is indicated by the *thick green vectors* (a'_O and c'_O). *Open circles* show lattice points and the *black dash-dotted vect*or illustrates the C-translation in the conforming orthorhombic lattice

TREOR, DICVOL, and ITO. Attempts to use TREOR produce no reasonable solution in high-symmetry crystal systems, including orthorhombic. Because of the complexity of the pattern, the maximum unit cell volume was increased from a default (1500) to 2500 Å^3 by adding the instruction VOL = 2500.

Keeping the maximum unit cell volume at 2500 Å^3, indexing attempts using TREOR were extended to include the monoclinic crystal system by adding the instruction MONO = 130.[70] The program finds several solutions, and the one listed in the first row in Table 14.22 appears reasonable. Despite the relatively low figure of merit ($M_{20} = 9$), it is promising because all Bragg peaks are indexed in the base-centered lattice.

To confirm the feasibility of this monoclinic unit cell, indexing was attempted using both DICVOL and ITO. The latter gives no acceptable solution, while the former results in three unit cells. The best of the three is listed in the second row in Table 14.22. All Bragg peaks can be indexed, but this time in a body-centered lattice. The two remaining solutions correspond to base-centered unit cells with all peaks indexed. They have slightly lower figures of merit but the same unit cell volume and parameter *b*. As an exercise, use experimental data from Table 14.21 and run DICVOL to see all unit cells.

The unit cell edges, *a* and *c*, and angle β in all three solutions are related to one another as shown in Fig. 14.16. Here, **a, c**, and angle β correspond to those in the first

[70] See data file *Ch14Ex05_Treor90.dat* online.

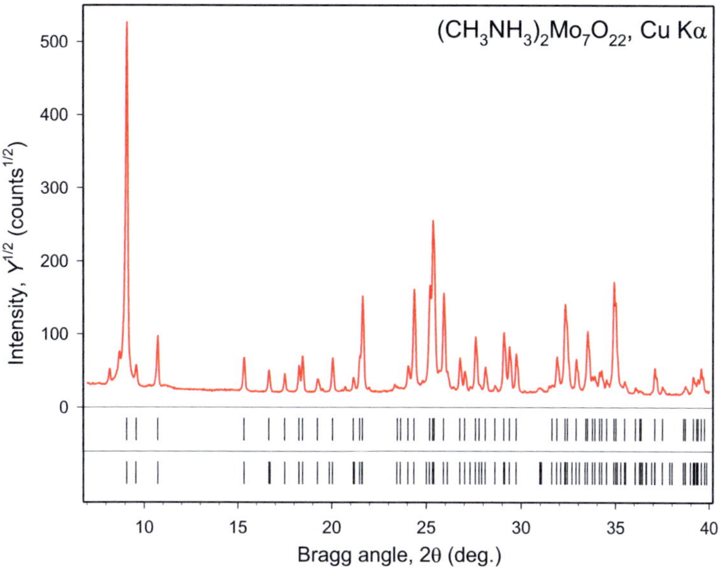

Fig. 14.15 Powder diffraction pattern of $(CH_3NH_3)_2Mo_7O_{22}$ collected on a Scintag XDS2000 diffractometer using Cu Kα radiation in a step-scan mode with $\Delta 2\theta = 0.02°$ and counting time 30 s. The square root of intensity is plotted as a function of Bragg angle for clarity. The two sets of *vertical bars* illustrate the following: *top*—positions of the observed Bragg peaks, *bottom*—the same calculated in the space group C2/c for the solution No. 1 from Table 14.22. The two weak Bragg peaks preceding the strongest peak at $2\theta \cong 9°$ have been excluded from the indexing because their presence yields unreasonably large unit cells (These peaks probably belong to an impurity phase. It is quite possible because the $(CH_3NH_3)_2Mo_7O_{22}$ powder was prepared hydrothermally (see Chap. 22), which sometimes results in the presence of several metastable phases in the material)

row in Table 14.22. The inverse of the body diagonal of the parallelogram **ac**, **d** = −**a**–**c**, can be used as the third choice of the unit cell edge giving a total of three possible selections of the unit cell: **a** and **c**, **a** and **d**, and **d** and **c**. Since the first pair produces a base-centered lattice, then the second combination remains base-centered, but the third unit cell should be body-centered, which is not standard in the monoclinic crystal system. The transformation to a standard setting (**a'**, **b'**, and **c'**) is shown in Table 14.22 in column 6 with the resulting unit cell dimensions shown in columns 7 and 8.

As can be seen from Fig. 14.15, the calculated positions of Bragg peaks match the observed diffraction pattern quite well. The lack of reflections $h0l$ with $l = 2n + 1$ in addition to base-centered systematic absences clearly points to space groups C2/c or Cc. Because of relatively low figures of merit, indexing with different sample shift corrections was conducted and sample shift of −0.1 mm results in the improved figures of merit as shown in Table 14.22 in Rows 3–5 for TREOR, DICVOL, and ITO, respectively. Interestingly, TREOR finds a unit cell with good figure of merit and half the volume but leaves some reflections unindexed, which is a disadvantage

Table 14.21 Relative integrated intensities (I/I_0), Bragg angles (listed for the location of the Kα_1 component in the doublet, $\lambda = 1.540593$ Å), and full widths at half maximum (FWHM) of Bragg peaks observed in the range $7 \leq 2\theta \leq 30°$ in the $(CH_3NH_3)_2Mo_7O_{22}$ powder diffraction pattern collected using Cu Kα radiation (see Fig. 14.15)

I/I_0	2θ (°)[a]	FWHM (°)
1000	9.077	0.089
10	9.574	0.089
34	10.716	0.085
17	15.328	0.084
8	16.653	0.084
6	17.496	0.084
10	18.264	0.084
18	18.449	0.084
4	19.240	0.084
17	20.030	0.084
4	21.130	0.081
15	21.466	0.081
89	21.614	0.081
1	23.466	0.081
1	23.641	0.081
10	24.033	0.077
109	24.343	0.077
110	25.162	0.080
257	25.321	0.080
68	25.394	0.080
98	25.896	0.080
16	26.758	0.081
8	27.016	0.081
38	27.581	0.081
1	27.771	0.081
11	28.093	0.081
1	28.600	0.081
45	29.069	0.081
29	29.368	0.081
21	29.725	0.081

[a] Bragg angles are listed for the location of the Kα_1 component in the doublet, $\lambda = 1.540593$ Å

of the algorithm that allows skipping Bragg peaks. DICVOL and ITO find the unit cell identical to that established previously, but with a better fit. In the case of ITO, however, this solution was not the best according to figure of merit: the best solution had merit $M_{20} = 39.4$, body-centered lattice, and was similar to that found by TREOR, with half the volume but with only 17 out of 20 lowest angle Bragg peaks indexed.

Other attempts, including eliminating some weak and suspicious Bragg peaks, did not result in a better indexing solution. Thus, this unit cell was considered as true after it was additionally confirmed by refinement of lattice parameters using all Bragg peaks observed up to $2\theta = 60°$. The final confirmation of the indexing

Table 14.22 Indexing solutions describing powder diffraction data of $(CH_3NH_3)_2Mo_7O_{22}$

Solution number	*a*, *b*, *c*	β, *V*	M_N	Lattice centering	**a'**,**b'**, **c'**	*a'*,*b'*,*c'*	β'	Source
1	20.624	109.93	9_{20}	C				TREOR
	5.525							
	19.592	2098.7						
2	23.105	127.11	9.9_{30}	I	**−a**	23.105	122.95	DICVOL
	5.524				**b**	5.524		
	20.626	2099.3			**a + c**	19.601		
3	20.589	109.94	40_{20}[a]	C	**a + 2c**	23.060	122.93	TREOR[b]
	5.514				**b**	5.514		
	9.7790	1044.9			**−2c**	19.561		
4	23.060	127.12	20.5_{30}	I	**−a**	23.060	122.93	DICVOL[b]
	5.514				**b**	5.514		
	20.590	2087.7			**a + c**	19.561		
5[c]	20.564	109.92	10.6_{20}	C	**a + c**	23.043	122.97	ITO[b]
					b	5.512		
	5.512				**−c**	19.546		
	19.546	2082.9						

[a] Four peaks unindexed; found volume is half that from other solutions
[b] 2θ corrected for a sample shift error, δ, as $2\theta^{corr} = 2\theta^{obs}$—180/π·2δ·cosθ/R (δ = −0.1 mm; goniometer radius *R* = 250 mm)
[c] Second best solution according to the figure of merit. The best FOM is 39.4 with half the unit cell volume and only 17 of 20 reflections indexed; also see solution No. 3

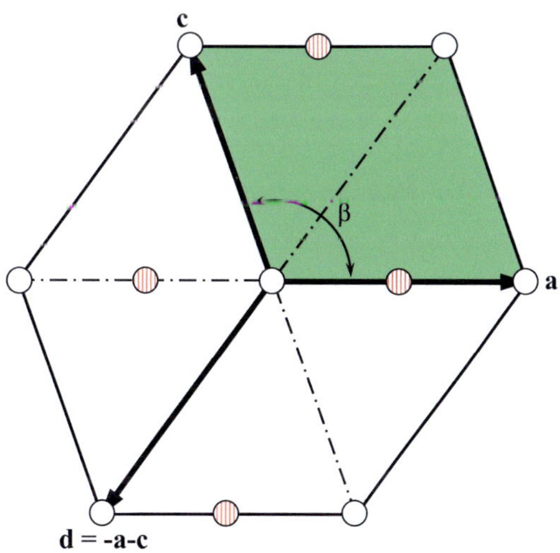

Fig. 14.16 Alternative axes selection in the monoclinic crystal system. *Open and hatched points* represent lattice points. The *open points* are located in the plane, while the *hatched points* are raised by ½ of the full translation in the direction perpendicular to the plane of the projection. Unit cells based on the vectors **a** and **c** or **a** and **d** correspond to a base-centered (C) lattice, while the unit cell based on the vectors **c** and **d** corresponds to a body-centered lattice

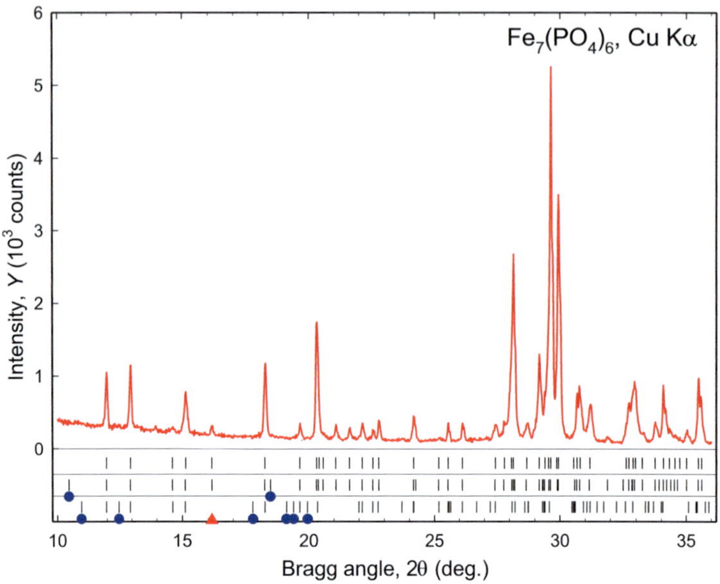

Fig. 14.17 Powder diffraction pattern of $Fe_7(PO_4)_6$ collected on a Scintag XDS2000 diffractometer using Cu Kα radiation in a step-scan mode with $\Delta 2\theta = 0.02°$ and counting time 30 s. The three sets of *vertical bars* illustrate the following: *top*—positions of the observed Bragg peaks, *middle*—positions of Bragg peaks calculated using ITO solution No. 1 (correct), and *bottom*—the same calculated using ITO solution No. 2 (incorrect); both solutions are listed in Table 14.24. *Blue circles* indicate unobserved reflections, and *red triangle* marks the only observed reflection below $2\theta = 20°$, which remained unindexed in second solution

solution was obtained after the crystal structure was determined and refined as discussed in Chap. 22.

14.11.3 *Triclinic Indexing: $Fe_7(PO_4)_6$*

Indexing in a triclinic crystal system generally should be attempted if no solution or only highly questionable solutions exist in higher symmetry crystal systems. In this example, we use diffraction data collected from an iron phosphate powder (Fig. 14.17). A total of 34 individual Bragg peaks observed below 32° 2θ (Table 14.23)[71] were refined using profile fitting and included in the indexing process.

Indexing this powder diffraction pattern using TREOR and DICVOL and assuming any symmetry higher than triclinic produces no solution. The ITO algorithm also fails using the data listed in Table 14.23, and therefore, the observed Bragg angles

[71] Same data are available online in the data file *Ch14Ex06_CuKa.pks*.

Table 14.23 Relative integrated intensities (I/I_0), Bragg angles, and full widths at half maximum (FWHM) of Bragg peaks observed in the range $10 \leq 2\theta \leq 32°$ in the $Fe_7(PO_4)_6$ powder diffraction pattern collected using Cu Kα radiation (see Fig. 14.17)

I/I_0	2θ (°)[a]	FWHM (°)
133	11.977	0.082
168	12.946	0.087
14	14.619	0.125
151	15.132	0.137
25	16.149	0.082
212	18.277	0.088
35	19.652	0.088
343	20.306	0.088
26	20.402	0.088
10	20.569	0.088
37	21.071	0.088
31	21.622	0.088
46	22.129	0.088
27	22.547	0.088
55	22.781	0.088
77	24.166	0.075
8	25.168	0.075
44	25.545	0.075
52	26.119	0.075
52	27.435	0.078
47	27.798	0.078
177	28.084	0.078
467	28.156	0.078
59	28.686	0.078
284	29.174	0.078
128	29.408	0.078
269	29.546	0.078
1000	29.636	0.078
87	29.874	0.078
757	29.938	0.078
12	30.543	0.078
136	30.673	0.078
138	30.796	0.078
142	31.182	0.078

[a] Bragg angles are listed for the location of the $K\alpha_1$ component in the doublet, $\lambda = 1.540593$ Å

were corrected by different sample-displacement errors followed by repetitive indexing attempts. When the observed Bragg angles were modified by sample displacement $\delta = -0.15$ mm using the goniometer radius 250 mm, a good indexing solution has been obtained. It is shown in the first row of Table 14.24.[72] The

[72] And is also found online in the data file Ch14Ex06_ITO-Indexed.out.

Table 14.24 The indexing of $Fe_7(PO_4)_6$ powder data obtained using ITO15, DICVOL91, and TREOR90, shown in columns (*a*,*b*,*c*) and (α,β,γ) followed by the reduced unit cell parameters in (*a'*,*b'*,*c'*) and (α',β',γ')

No.	*a*, *b*, *c*	α, β, γ	V	M_N	**a',b',c'**	*a'*, *b'*, *c'*	α', β', γ'	Source
1	7.9791 9.5752 6.4736	111.045 101.620 67.321	424.77	56.8_{20}	**−c** **−a** **b + c**	6.475 7.977 9.434	104.68 108.78 101.59	ITO
2	8.1990 8.9216 7.1689	99.955 100.503 111.595	462.61	40.1_{20}	**c** **a** **b**	7.169 8.199 8.922	111.60 99.96 100.50	ITO[a]
3	9.8147 9.5710 6.4750	111.056 79.219 131.423	424.51	30.8_{34}	**−c** **−a** **b + c**	6.474 7.979 9.438	104.66 108.76 101.62	DICVOL
4	6.4662 9.4427 9.8219	123.000 79.305 71.281	424.51	36_{20}	**a** **−a + b + c** **−b**	6.466 7.979 9.443	104.65108.72101.66	TREOR[b] (defaults)
5	6.4780 7.9764 9.5697	67.307 68.916 78.386	424.54	56_{20}	**−a** **b** **a−c**	6.478 7.976 9.431	104.67108.77101.61	TREOR (MERIT = 50)
6	6.4902 7.9634 9.4521	104.330 109.028 101.642	425.65	–	**a** **b** **c**	6.4902 7.9634 9.4521	104.33 109.03 101.64	Single crystal data

[a] The second best solution according to M_N (indicated in red) is an example of the incorrect indexing: only 14 out of 20 lowest Bragg angle peaks are indexed
[b] Despite an acceptable FOM, seven peaks remain unindexed and the unit cell is correct

presence of a considerable systematic error, which may occur due to an improperly positioned specimen and/or due to intrinsic reasons (e.g., a weakly absorbing sample), explains why the first indexing attempt failed, even though the algorithm incorporated into the ITO program foresees this kind of an error. It is worth mentioning that sample-displacement or zero-shift error can be estimated by comparing positions of several lowest Bragg peaks with the corresponding second-order peaks.[73]

The complete list of solutions[74] obtained in this indexing attempt contains five different unit cells with the following M_{20} FOMs: 57, 40, 12, 5, and 4. The first two are markedly higher than the others (also see the first two rows in Table 14.24). The first solution indexes all 34 Bragg peaks, has the highest figure of merit, and has the smallest unit cell. Additionally, there are only 23 possible reflections within the Bragg angle range corresponding to the first 20 observed peaks (see Table 14.25).

The second solution appears much worse: despite the high figure of merit and only slightly higher volume, it leaves 6 out of the first 20 reflections unindexed even though there are a total of 28 calculated Bragg peaks in this range of angles. Other solutions have much lower figures of merit, higher volumes, and leave too many unindexed peaks.

Some of the unreliable solutions can be eliminated by decreasing the tolerances in 2θ for zone searches. However, when choosing a tolerance that is too low, the true solution may be easily missed. To avoid this, the tolerance should be reduced gradually, and only when excessive amounts of different and unreliable solutions emerge. In our example, the tolerance that was set by default between 3 and 4 appears to be a good choice.

Table 14.24 also contains the results of triclinic indexing using the DICVOL and TREOR algorithms. With all crystal systems tested, DICVOL results in a single solution, shown in the third row. All reflections were indexed, and the figures of merit are quite high: $M_{34} = 30.8$ and $F_{34} = 104.5(0.0081, 40)$. Employing TREOR with all parameters set to their default value (except instruction TRIC = 1), three solutions where found. Two of them with unit cell volume of ~800 Å^3 have very low figures of merit and leave many peaks unindexed. The third one has the same cell volume (~424 Å^3) as the earlier solutions and is listed in Row 4. This solution has 6 out of 34 Bragg reflections unindexed, which is due to the low accuracy of the determined unit cell. To check other possible solutions, the keyword MERIT = 50 was inserted in the input data file. This modification results in more than 20 different solutions with a unit cell volume of ~424 Å^3. The best solution has all 34 reflections indexed, high figures of merit, and it is listed in Row 5.

As is obvious from Columns 2 and 3 in Table 14.24, different indexing algorithms result in different choices of the unit cell for the same lattice, and therefore,

[73] The *d*-spacing of the first- and second-order Bragg peaks should be related as 2:1 in the absence of sample-displacement and/or zero-shift errors. If this ratio is different, the related systematic error can be computed from the Braggs' law (see Eqs. 8.12–8.17 on pp. 162–163).

[74] See the data file Ch14Ex06_ITO–Indexed.out online.

Table 14.25 The best solution from indexing the powder diffraction pattern of $Fe_7(PO_4)_6$ using Ito method: $a = 7.9791$ Å, $b = 9.5752$ Å, $c = 6.4736$ Å, $\alpha = 111.045°$, $\beta = 101.620°$, and $\gamma = 67.321°$

$2\theta_{obs}$	√Intensity[a]	$2\theta_{corr}$[b]	$2\theta_{calc}$	$2\theta_{corr}-2\theta_{calc}$	*h*	*k*	*l*
–			10.539		0	1	0
11.977	36	12.035	12.050	−0.015	1	0	0
12.946	40	13.005	13.009	−0.004	1	1	0
14.619	11	14.677	14.686	−0.009	0	0	1
15.132	38	15.191	15.182	0.009	0	1	1
16.149	15	16.207	16.219	−0.012	1	1	1
18.277	45	18.335	18.339	−0.004	1	0	1
–			18.577		1	1	0
19.652	18	19.710	19.708	0.002	1	0	1
20.306	58	20.364	20.365	−0.001	1	2	1
20.402	16	20.460	20.445	0.015	1	2	0
20.569	10	20.627	20.632	−0.005	0	1	1
21.071	19	21.128	21.159	−0.031	0	2	0
21.622	17	21.679	21.693	−0.014	0	2	1
22.129	21	22.186	22.183	0.003	1	1	1
22.547	16	22.604	22.605	−0.001	1	1	1
22.781	23	22.839	22.851	−0.012	2	1	0
24.166	27	24.223	24.226	−0.003	2	0	0
–			24.317		2	1	1
25.168	9	25.225	25.240	−0.015	1	1	1
25.545	20	25.602	25.604	−0.002	2	2	1
26.119	22	26.176	26.179	−0.003	2	2	0
27.435	22	27.491	27.492	−0.001	2	0	1
27.798	21	27.854	27.854	0.000	1	2	0
28.084	41	28.141	28.140	0.001	0	1	2
28.156	67	28.213	28.224	−0.011	1	3	1
–			28.262		1	1	2
28.686	24	28.743	28.736	0.007	1	2	1
29.174	52	29.230	29.225	0.005	1	2	2
–			29.366		2	0	1
–			29.397		1	2	1
29.408	35	29.465	29.459	0.006	0	2	1
29.546	51	29.603	29.611	−0.008	0	0	2
29.636	100	29.692	29.697	−0.005	2	1	0
29.874	29	29.930	29.948	−0.018	2	1	1
29.938	86	29.994	30.003	−0.009	1	3	0
30.543	10	30.600	30.628	−0.028	0	2	2
30.673	36	30.729	30.731	−0.002	0	3	1

[a] Normalized to 100
[b] Corrected for a sample shift $\delta = -0.15$ mm: $2\theta_{corr} = 2\theta_{obs}$—$180/\pi \cdot 2\delta \cdot \cos\theta/R$ and $R = 250$ mm

unit cell reduction is important to compare the results, especially in triclinic symmetry. The reduced unit cell dimensions are listed in Table 14.24 in Columns 6–8. Obviously, all of them are represented by the same unit cell, except the incorrect solution shown in Row 2. The triclinic unit cell was confirmed by a single-crystal diffraction experiment, as shown in Row 6.

14.11.4 *Pseudo-Hexagonal Indexing: Li[B(C_2O_4)$_2$]*

This example illustrates the indexing of lithium *bis*(oxalato)borate, Li[B(C_2O_4)$_2$],[75] which exhibits an accidental hexagonal-like relationship between two of the unit cell edges in the orthorhombic structure. The challenge in this and similar cases is not the indexing itself, but the selection of the correct indexing solution, as both the true and pseudo-symmetric unit cells are often found in the list of suggested solutions. Difficulties arise from accidental relationships between some or all of the unit cell parameters, which result in nearly identical Bragg angles for numerous symmetrically unrelated reflections. As a result, these reflections overlap completely and cannot be deconvoluted or resolved. For example, an orthorhombic lattice with two identical or nearly identical unit cell edges *a* and *b* will result in indistinguishable *d*-spacings and Bragg angles for any pair of *hkl* and *khl* reflections. This accidental relationship between the unit cell dimensions is the same as that dictated by symmetry in the tetragonal crystal system, and therefore, Eq. (8.5) degenerates into Eq. (8.3). Obviously, such a pseudo-tetragonal pattern can be indexed in the tetragonal crystal system with higher (and often substantially higher) figure of merit (Eqs. 14.7 and 14.9), simply because in the orthorhombic crystal system the number of possible reflections, N_{poss}, is almost twice the number of the observed peaks due to significant overlap of unrelated peaks with identical Bragg angles. Therefore, full crystal structure determination is typically necessary to confirm the correctness of the chosen indexing solution.[76]

The powder diffraction pattern shown in Fig. 14.18 was collected from very fine, highly hydroscopic white powder of Li[B(C_2O_4)$_2$]. The pattern has two distinct features. First, a noticeable peak broadening with full widths at half maximum ranging from 0.15° at low angles to 0.30° 2θ at high angles, which is due to a small grain size since this compound is formed by decomposition of solvates. The second feature is a rapid reduction of the peak intensity with increasing Bragg angle, so that there are only a few broad peaks at 2θ > 45° (not shown). This is expected because this compound is made of only light chemical elements, with oxygen being

[75] P.Y. Zavalij, S. Yang, M.S. Whittingham, Structures of potassium, sodium and lithium bis (oxalato)borate salts from powder diffraction data. Acta Cryst. **B59**, 753 (2003). https://doi.org/10.1107/s0108768103022602

[76] As noted above, correct unit cell may be also confirmed using other methods, such as selected area electron diffraction or high-resolution electron microscopy.

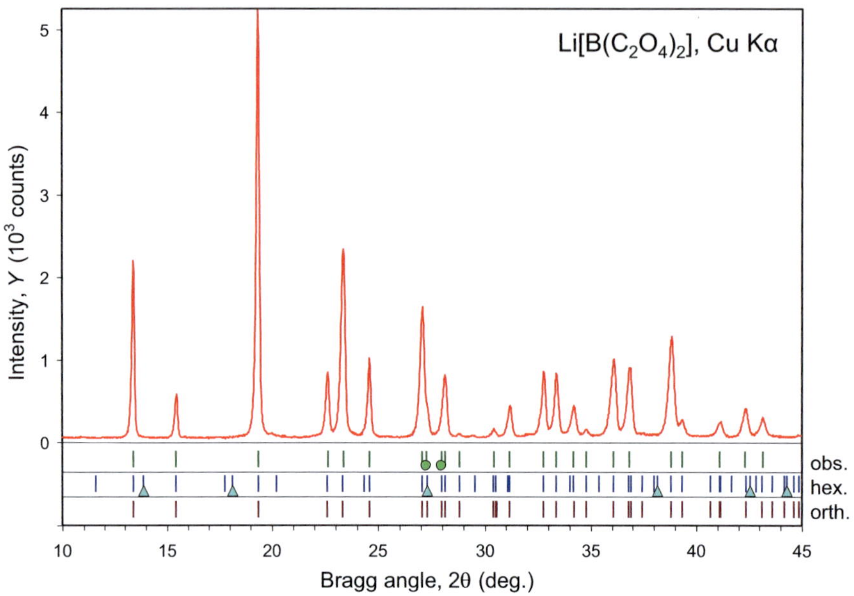

Fig. 14.18 Powder diffraction pattern of $Li[B(C_2O_4)_2]$ collected on a Scintag XDS2000 diffractometer using Cu Kα radiation in a step-scan mode with $\Delta 2\theta = 0.02°$ and counting time 2 s (The ASCII data file with diffraction data is available online, file name *Ch14Ex07_CuKa.xy*.). The three sets of vertical bars illustrate positions of the observed Bragg peaks (*top*), positions of the Bragg peaks calculated for an incorrect hexagonal unit cell (*middle*), and the same, calculated for a correct orthorhombic unit cell (*bottom*) with all systematic absences for space group Pnma excluded. All relevant numerical values are listed in Table 14.26. The *blue triangles* indicate Bragg reflections that are extinct (not allowed by symmetry) in space group $P6_3/mmc$, and the *green circles* mark two Bragg peaks excluded from the indexing due to heavy overlap and low accuracy

the heaviest. Both these features make the unit cell indexing and the crystal structure determination even more challenging.

Thus, profile fitting was performed only in the region $10° \le 2\theta \le 45°$, which is shown in Fig. 14.18. This resulted in a total of 24 peaks that are listed in Table 14.26.[77] Two of them at $2\theta = 27.255°$ and 27.965° (shown in bold in Table 14.26 and marked with the green circles in Fig. 14.18) were excluded from indexing because their intensities are relatively low, while they are located near the bases of much stronger peaks. Ab initio indexing of 22 well-resolved peaks using TREOR90, ITO15, and DICVOL91 resulted in a variety of solutions, two of which deserve special attention since both assign indices to every of the 22 peaks and have distinctly higher figure of merits than the others. These are the hexagonal and orthorhombic unit cells listed in Table 14.27. Their volume ratio is 2:1 ratio. The edges of both unit cells are related as follows:

[77] Only peaks below 40° are listed in the table. All data are available online in the data file *Ch14Ex07_CuKa.pks*.

Table 14.26 Relative integrated intensities (I/I_0), Bragg angles (2θ listed for the location of Cu $K\alpha_1$ component), and full widths at half maximum (FWHM) of Bragg peaks observed in the range $10° \leq 2\theta \leq 40°$ in the $Li[B(C_2O_4)_2]$ powder diffraction pattern shown in Fig. 14.18

			Hexagonal indexing[a] in $P6_3/mmc$			Orthorhombic indexing[a] in Pnma		
I/I_0	$2\theta_{obs}$,°	FWHM,°	*hkl*	$2\theta_{calc}$,°	$\Delta 2\theta$,°	*hkl*	$2\theta_{calc}$,°	$\Delta 2\theta$,°
–			010	6.662				
–			110	11.589				
350	13.399	0.141	020	13.397	0.002	011, 002	13.397	0.002
90	15.417	0.141	011	15.408	0.008	101	15.408	0.008
–			120	17.770				
1000	19.352	0.150	021	19.343	0.009	111, 102	19.343	0.009
–			030	20.180				
171	22.629	0.200	121	22.626	0.002	112	22.626	0.002
580	23.360	0.200	220	23.351	0.009	020, 013	23.351	0.009
–			130	24.320				
212	24.578	0.200	031	24.584	−0.006	103	24.584	−0.006
386	27.035	0.200	040	27.035	0.000	022, 004	27.036	−0.001
65	**27.255**	0.200	Unindexed*			113	27.274	−0.019
20	**27.965**	0.200	002	27.978	−0.013	200	27.978	−0.013
184	28.129	0.200	131	28.117	0.012	121	28.117	0.012
7	28.808	0.200	012	28.803	0.005	201	28.803	0.005
–			230	29.518				
27	30.404	0.219	112	30.391	0.013	210	30.391	0.013
–			041	30.520		122, 104	30.520	
–			140	31.073				
105	31.162	0.219	022	31.157	0.005	211, 202	31.158	0.004
206	32.750	0.219	231	32.763	−0.013	114	32.764	−0.014
209	33.366	0.219	122	33.362	0.004	212	33.362	0.004
–			050	33.991				
111	34.179	0.219	141	34.185	0.007	123	34.186	−0.008
23	34.763	0.219	032	34.763	0.000	203	34.763	0.000
–			330	35.370				
256	36.064	0.219	240	36.042	0.022	031, 024, 015	36.042	0.022
262	36.828	0.219	222	36.776	0.052	220, 213	36.776	**0.052**
–			051	36.885	−0.057	105	36.886	**−0.058**
–			132	37.426		221	37.427	
–			150	37.997				
336	38.814	0.219	241	38.804	0.010	131, 124, 115	38.804	0.010
52	39.321	0.219	042	39.323	−0.002	222, 204	39.323	−0.002

[a] For both hexagonal and orthorhombic unit cells, Bragg angles were calculated using parameters from the TREOR indexing solutions listed in Table 14.27

Table 14.27 The solution from indexing the powder diffraction pattern of $Li[B(C_2O_4)_2]$ using 22 Bragg peaks listed in Table 14.26

Unit cell	*a*, Å	*b*, Å	*c*, Å	*V*, Å^3	M_{20}	F_N	Application
Hexagonal	15.194		6.363	1272.2	36	64(0.0087, 45)	TREOR
a	15.217	26.320	6.364	2548.7	22		ITO
Orthorhombic	6.362	7.598	13.168	636.5	33	51(0.0072, 59)	TREOR
b	6.360	7.593	13.167	635.8	35		ITO
	6.361	7.595	13.177	636.6	39	60(0.0063, 59)	DICVOL

[a] This is an orthorhombic C-centered lattice with $a = b\sqrt{3}$ which is alternative representation of the hexagonal lattice

[b] ITO automatically recommends converting this solution into a hexagonal unit cell with $a = 15.20$ Å and $c = 6.36$ Å

$$a_{\mathrm{orth}} = c_{\mathrm{hex}};\quad b_{\mathrm{orth}} \cong \frac{1}{2} a_{\mathrm{hex}};\quad c_{\mathrm{orth}} \cong \sqrt{3} b_{\mathrm{orth}} \cong \frac{\sqrt{3}}{2} a_{\mathrm{hex}} \tag{14.28}$$

Considering all the information available at this point, it can be concluded that both reciprocal lattices are topologically similar, and their one-dimensional projections are identical, which poses a problem of choice. As follows from Table 14.27, TREOR indexing yields higher FOMs for the hexagonal unit cell, while the M_{20} FOM from ITO is better for the orthorhombic solution, even though the ITO application suggests converting the found orthorhombic unit cell to the hexagonal cell. As a matter of fact when all 24 peaks are used in the indexing, ITO also results in a noticeably better FOM for a hexagonal solution, which shows how important is accuracy in the peak positions.

Naturally, the hexagonal unit cell was chosen as the best solution mainly because it has a higher symmetry and it results in about one possible reflection for every observed Bragg peak, while the orthorhombic unit cell leads to two or three possible reflections for every observed peak. We also note here that the first two selection criteria discussed in Sect. 14.10.4 do not weigh heavily toward any of the two obtained solutions: first, the difference in the figures of merit is inconclusive because they depend on the application, and second, both solutions yield reasonable systematic absences that correspond to commonly occurring space groups. These are $P6_3/mmc$ (or $P6_3mc$, $P\bar{3}1c$, and so on) in the hexagonal case and Pnma or $Pna2_1$ in the orthorhombic case (see Tables 9.6, 9.7, and 9.8).

Nevertheless, selecting the hexagonal unit cell was only an educated guess, which was later proven to be wrong. All attempts to solve the crystal structure using various structure solution methods and software in all possible hexagonal and trigonal space groups, including the lowest symmetry P3, failed. In addition, several attempts to obtain single crystals large enough for unit cell determination were made. However, recrystallization from a variety of solvents always led to solvated crystals, which turn into the same fine white powder after the loss of a solvent.[78] During attempts to solve the crystal structure, all models led to a high intensity of the first Bragg peak that is possible in the hexagonal crystal system (010, see Table 14.26), which in reality is too low to be observed, which raised doubts in the correctness of the hexagonal unit cell.

Conversely, solving crystal structure in the orthorhombic crystal system, and therefore, confirming the accuracy of the orthorhombic indexing solution, was only a matter of choosing the right space-group symmetry. It is worth noting that only real space methods were successful in solving the orthorhombic structure because of a significant overlap of Bragg reflections and rapidly decreasing intensities at high Bragg angles. For more details on selection of space-group symmetry and structure determination, we refer the reader to the original paper, see Footnote 75 on page 507.

14.12 Precise Lattice Parameters and Linear Least Squares

After the powder diffraction pattern has been successfully indexed, the next step is to establish the unit cell dimensions with the highest possible precision. By combining Eqs. (14.2) and (14.3), one can see that the errors in the lattice parameters only depend on the errors in the measured Bragg angles assuming that Miller indices and λ are known exactly:

$$f(h, k, l, a, b, c, \alpha, \beta, \gamma) = {}^{\lambda}\!/_{2\sin\theta_{(hkl)}} \tag{14.29}$$

By differentiating the right-hand side of Eq. (14.29), we find that the absolute error in the interplanar distance, σd, is a function of both the error in Bragg angle, $\sigma\theta$, and the Bragg angle itself:

$$\sigma d = \frac{\lambda \cos\theta}{2\sin^2\theta}\sigma\theta \tag{14.30}$$

The dependencies of σd on Bragg angle for several constant $\sigma\theta$ are shown in Fig. 14.19, illustrating that to achieve high precision in lattice parameters, powder diffraction data must be collected at the highest possible Bragg angles.

Ideally, the Bragg angles close to $\theta = 90°$ ($2\theta = 180°$) should be available to claim the precision of the unit cell dimensions equivalent to that of the precision of the used wavelength. Unfortunately, measurements at $2\theta = 180°$ are impossible, and in most commercial powder diffractometers, the highest reachable Bragg angle is

[78] Unrelated to the subject of this book but an interesting fact is that in this manner crystal structures of five different solvates were determined. See P.Y. Zavalij, S. Yang, and M.S. Whittingham, Structural chemistry of new lithium bis(oxalato)borate solvates. Acta Cryst. **B60**, 716 (2004). https://doi.org/10.1107/S0108768104022153

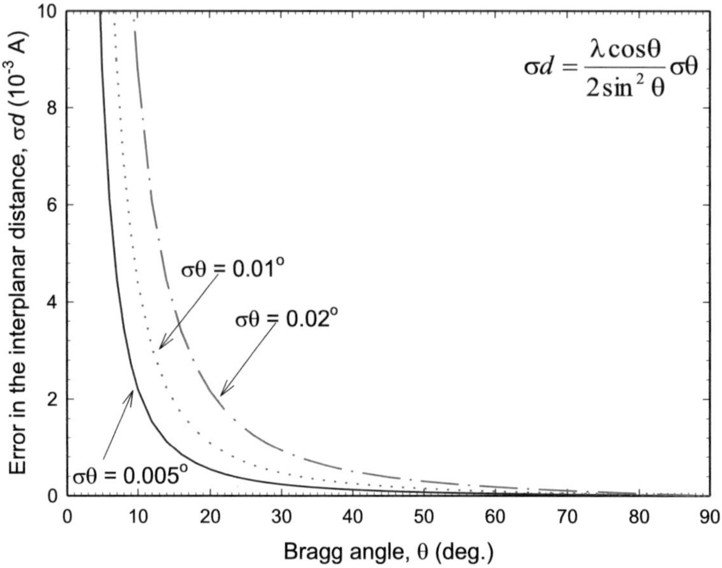

Fig. 14.19 The dependence of the absolute error in the interplanar distance, σd, on Bragg angle for several different constant errors in the measured Bragg angle when using Cu Kα_1 radiation, $\lambda = 1.540593$ Å

limited by $2\theta = 140°–160°$. Thus, in nearly every instance, all available data are used to determine lattice parameters using least-squares refinement. The use of the least-squares technique is fully justifiable as it also allows one to refine the most critical corrections to the observed Bragg angles that arise from the presence of systematic errors in the experimental data due to sample displacement or zero shift.

14.12.1 Linear Least Squares

Assume that we need to find a solution of the system of n simultaneous linear equations with m unknown parameters. In general form, this system of equations can be represented as

$$\begin{aligned} a_{11}x_1 + a_{12}x_2 + \ldots + a_{1m}x_m &= y_1 \\ a_{21}x_1 + a_{22}x_2 + \ldots + a_{2m}x_m &= y_2 \\ \ldots \\ a_{n1}x_1 + a_{n2}x_2 + \ldots + a_{nm}x_m &= y_n \end{aligned} \quad (14.31)$$

In a matrix and vector notation, Eq. (14.31) becomes

$$\boldsymbol{Ax} = \boldsymbol{y} \tag{14.32}$$

where

$$\mathbf{A} = \begin{pmatrix} a_{11} & a_{12} & \dots & a_{1m} \\ a_{21} & a_{22} & \dots & a_{2m} \\ \dots & \dots & \dots & \dots \\ a_{n1} & a_{n2} & \dots & a_{nm} \end{pmatrix}; \quad \mathbf{x} = \begin{pmatrix} x_1 \\ x_2 \\ \dots \\ x_m \end{pmatrix}; \quad \mathbf{y} = \begin{pmatrix} y_1 \\ y_2 \\ \dots \\ y_n \end{pmatrix} \tag{14.33}$$

When $n < m$, Eq. (14.31) has an infinite number of solutions with respect to a set of n unknowns, and each particular solution depends on certain assumptions which were made about the values of the remaining m—n parameters. When $n = m$, Eq. (14.31) has one exact solution, which exists only when det(**A**) ≠ 0. Conversely, when $n > m$, the solution of Eq. (14.31) can be obtained in two fundamentally different ways.

First, it is possible to randomly select m equations and find m parameters, if any, that exactly satisfy each of the m selected equations. This solution, however, will likely be far from the best for the remaining n–m equations (e.g., see the varying lattice parameter determined in this way in Table 14.12).

Second, it is possible to find vector **x**, which would be the best solution for all n existing equations using the least-squares technique. The least-squares solution of Eq. (14.31) is obtained by rearranging it into the following form:

$$\begin{aligned} &a_{11}x_1 + a_{12}x_2 + \dots + a_{1m}x_m - y_1 = \varepsilon_1 \\ &a_{21}x_1 + a_{22}x_2 + \dots + a_{2m}x_m - y_2 = \varepsilon_2 \\ &\dots \\ &a_{n1}x_1 + a_{n2}x_2 + \dots + a_{nm}x_m - y_n = \varepsilon_n \end{aligned} \tag{14.34}$$

and then finding the solution that best satisfies all equations is equivalent to finding the minimum of the following function:

$$\Phi(x_1, x_2, \dots x_m) = \sum\nolimits_{i=1}^{n} \varepsilon_i^2 \tag{14.35}$$

The best solution of Eq. (14.34) (and Eq. 14.31) is found by calculating partial derivatives of the function defined in Eq. (14.35) with respect to x_1, x_2, …x_m and equating each to zero, which are the conditions of the minimum of $\Phi(x_1, x_2, \dots x_m)$:

$$\begin{aligned} &\frac{\partial \Phi(x_1, x_2, \dots, x_m)}{\partial x_1} = 0 \\ &\frac{\partial \Phi(x_1, x_2, \dots, x_m)}{\partial x_2} = 0 \\ &\dots \\ &\frac{\partial \Phi(x_1, x_2, \dots, x_m)}{\partial x_m} = 0 \end{aligned} \tag{14.36}$$

Equation (14.36) has a single solution, if any, since we replaced the system containing n equations and m unknowns (Eqs. 14.31 and 14.34) with the system in which $n = m$. After calculating partial derivatives, regrouping the coefficients, and rearranging each of the equations, we get the following equivalent of Eq. (14.36):

$$
\begin{aligned}
& x_1 \sum_{i=1}^{n} a_{i1}^2 + x_2 \sum_{i=1}^{n} a_{i1}a_{i2} + \ldots + x_m \sum_{i=1}^{n} a_{i1}a_{im} = \sum_{i=1}^{n} a_{i1}y_i \\
& x_1 \sum_{i=1}^{n} a_{i2}a_{i1} + x_2 \sum_{i=1}^{n} a_{i2}^2 + \ldots + x_m \sum_{i=1}^{n} a_{i2}a_{im} = \sum_{i=1}^{n} a_{i2}y_i \\
& \ldots \\
& x_1 \sum_{i=1}^{n} a_{im}a_{i1} + x_2 \sum_{i=1}^{n} a_{m}a_{i2} + \ldots + x_m \sum_{i=1}^{n} a_{im}^2 = \sum_{i=1}^{n} a_{im}y_i
\end{aligned}
\tag{14.37}
$$

The coefficients on the left-hand side of Eq. (14.37) are product of the transpose of matrix **A** (Eq. 14.32) and the matrix **A** itself, while the coefficients on the right-hand side of Eq. (14.37) are the product of the transpose of matrix **A** and vector **y** (Eq. 14.32). Thus, letting $\mathbf{A}^\mathbf{T}$ be the transpose of matrix **A**

$$
\mathbf{A}^\mathbf{T} = \begin{pmatrix} a_{11} & a_{21} & \ldots & a_{n1} \\ a_{12} & a_{22} & \ldots & a_{n2} \\ \ldots & \ldots & \ldots & \ldots \\ a_{1m} & a_{2m} & \ldots & a_{nm} \end{pmatrix} \tag{14.38}
$$

Equation (14.37) can be written in a matrix and vector notation as:

$$
\left(\mathbf{A}^\mathbf{T}\mathbf{A}\right)\mathbf{x} = \left(\mathbf{A}^\mathbf{T}\mathbf{y}\right) \tag{14.39}
$$

Its solution (and thus the least-squares solution of Eq. 14.31) is found from Eq. (14.40), where $(\mathbf{A}^\mathbf{T}\mathbf{A})^{-1}$ is the inverse of the square matrix, which is the product of $\mathbf{A}^\mathbf{T}$ and **A**.

$$
\mathbf{x} = \left(\mathbf{A}^\mathbf{T}\mathbf{A}\right)^{-1}\left(\mathbf{A}^\mathbf{T}\mathbf{y}\right) \tag{14.40}
$$

Equations (14.36), (14.37), and (14.39) are identical and they represent different notations of the so-called normal equations in the least-squares method. Two noteworthy properties of the elements of the normal equation matrix $(\mathbf{A}^\mathbf{T}\mathbf{A})$ or, which is the same, the coefficients near the unknowns in normal equations (Eq. 14.37) are:

- The elements located on the main diagonal of the matrix are always positive.
- The matrix is symmetrical with respect to the main diagonal.

14.12.2 Precise Lattice Parameters from Linear Least Squares

We now consider the most general form of Eq. (14.29), i.e., that relating the unit cell dimensions, Miller indices, wavelength, and Bragg angles in the triclinic crystal system:

$$\frac{1}{V^2}\begin{bmatrix} h^2b^2c^2\sin^2\gamma + k^2a^2c^2\sin^2\beta + l^2a^2b^2\sin^2\alpha + \\ +2hkabc^2(\cos\alpha\cos\beta - \cos\gamma) + 2hlab^2c(\cos\alpha\cos\gamma - \cos\beta) + \\ +2kla^2bc(\cos\beta\cos\gamma - \cos\alpha) \end{bmatrix}$$
$$= 4\sin^2\theta/\lambda^2 \tag{14.41}$$

Here, V is the volume of the unit cell, which is given as

$$V = abc\sqrt{1 - \cos^2\alpha - \cos^2\beta - \cos^2\gamma + 2\cos\alpha\cos\beta\cos\gamma} \tag{14.42}$$

By comparing Eq. (14.41) with Eq. (14.31), it is easy to see that the linear least squares technique is not directly applicable in this case since Eq. (14.41) is nonlinear with respect to the unknowns (a, b, c, α, β and γ). However, when Eq. (14.41) is rewritten in reciprocal space

$$\begin{pmatrix} h^2a^{*2} + k^2b^{*2} + l^2c^{*2} + 2hka^*b^*\cos\gamma^* + 2hla^*c^*\cos\beta^* + \\ +2klb^*c^*\cos\alpha^* \end{pmatrix}$$
$$= 4\sin^2\theta/\lambda^2 \tag{14.43}$$

it becomes linear with respect to a different set of the unknowns:

$$\left(S_{11}h^2 + S_{22}k^2 + S_{33}l^2 + 2S_{12}hk + 2S_{13}hl + 2S_{23}kl\right) = 4\sin^2\theta/\lambda^2$$

or

$$\mathbf{h}^\mathbf{T}\mathbf{S}\mathbf{h} = (h\ \ k\ \ l)\begin{pmatrix} S_{11} & S_{12} & S_{13} \\ S_{12} & S_{22} & S_{23} \\ S_{13} & S_{23} & S_{33} \end{pmatrix}\begin{pmatrix} h \\ k \\ l \end{pmatrix} = 4\sin^2\theta/\lambda^2 \tag{14.44}$$

where the new parameters, S_{ij}, are defined as follows:

$$
\begin{aligned}
S_{11} &= a^{*2} = \frac{b^2c^2\sin^2\gamma}{V^2} \\
S_{22} &= b^{*2} = \frac{a^2c^2\sin^2\beta}{V^2} \\
S_{33} &= c^{*2} = \frac{a^2b^2\sin^2\alpha}{V^2} \\
S_{12} &= a^*b^*\cos\gamma^* = \frac{abc^2(\cos\alpha\cos\beta - \cos\gamma)}{V^2} \\
S_{13} &= a^*c^*\cos\beta^* = \frac{ab^2c(\cos\alpha\cos\gamma - \cos\beta)}{V^2} \\
S_{23} &= b^*c^*\cos\alpha^* = \frac{a^2bc(\cos\beta\cos\gamma - \cos\alpha)}{V^2}
\end{aligned}
\tag{14.45}
$$

By comparing Eq. (14.44) with Eqs. (14.31)–(14.33), it is easy to see that matrix **A** and vector **y** are constructed as follows:

$$
\boldsymbol{A} = \begin{pmatrix} h_1^2 & k_1^2 & l_1^2 & 2h_1k_1 & 2h_1l_1 & 2k_1l_1 \\ h_2^2 & k_2^2 & l_2^2 & 2h_2k_2 & 2h_2l_2 & 2k_2l_2 \\ \cdots & \cdots & \cdots & \cdots & \cdots & \cdots \\ h_n^2 & k_n^2 & l_n^2 & 2h_nk_n & 2h_nl_n & 2k_nl_n \end{pmatrix}
\tag{14.46}
$$

$$
\boldsymbol{y} = \begin{pmatrix} 4\sin^2\theta_1/\lambda^2 \\ 4\sin^2\theta_2/\lambda^2 \\ \cdots \\ 4\sin^2\theta_n/\lambda^2 \end{pmatrix}
\tag{14.47}
$$

and that the least-squares solution according to Eq. (14.40) results in a vector

$$
\boldsymbol{x} = \begin{pmatrix} S_{11} \\ S_{22} \\ S_{33} \\ S_{12} \\ S_{13} \\ S_{23} \end{pmatrix}
\tag{14.48}
$$

from which the unit cell dimensions of the direct lattice are calculated using Eq. (14.45).

So far, we considered the application of a liner least-squares technique in the case when no systematic error has been present in the observed powder diffraction data. However, as we already know, in many cases the measured Bragg angles are affected by a systematic sample-displacement or zero-shift error. The first systematic error affects each data point differently and considering Eq. 12.4 (Sect. 12.1.5), when a sample-displacement error, s, is present in the data, Eq. (14.44) becomes

$$
\begin{aligned}
&\left(S_{11}h^2 + S_{22}k^2 + S_{33}l^2 + 2S_{12}hk + 2S_{13}hl + 2S_{23}kl\right) = \\
&= \frac{4\sin^2\left(\theta + \frac{s}{R}\cos\theta\right)}{\lambda^2}
\end{aligned}
\tag{14.49}
$$

In Eq. (14.49), *R* is the radius of the goniometer. The second systematic error—zero shift—adds a small constant value, $\delta\theta_0$, to each observed Bragg angle, which results in the following equivalent of Eq. (14.44):

$$\left(S_{11}h^2 + S_{22}k^2 + S_{33}l^2 + 2S_{12}hk + 2S_{13}hl + 2S_{23}kl\right) = \frac{4\sin^2(\theta + \delta\theta_0)}{\lambda^2} \quad (14.50)$$

Both Eqs. (14.49) and (14.50) introduce one additional parameter in the least-squares refinement, which is sample displacement divided by the goniometer radius, *s*/*R*, or zero shift, $\delta\theta_0$, respectively. Regardless of the fact that the contribution from both parameters is nonlinear, the linear least-squares technique can still be applied after the following simplifications.

From trigonometry, we know that:

$$\begin{aligned} \sin(\alpha + \beta) &= \sin\alpha\cos\beta + \cos\alpha\sin\beta \\ \sin 2\alpha &= 2\sin\alpha\cos\alpha \end{aligned} \quad (14.51)$$

Hence,

$$\sin^2(\theta + x) = \sin^2\theta\cos^2 x + \cos^2\theta\sin^2 x + \frac{1}{2}\sin 2\theta\sin 2x \quad (14.52)$$

Recall that *x*, which represents an error in the Bragg angle, is usually quite small, then $\cos^2 x \cong 1$, $\sin^2 x \cong 0$, and $\sin 2x \cong 2x$. Whence, Eq. (14.52) is simplified to

$$\sin^2(\theta + x) \cong \sin^2\theta + x\sin 2\theta \quad (14.53)$$

By substituting the result obtained in Eq. (14.53) into Eqs. (14.49) and (14.50), they are transformed into

$$\begin{aligned} &\left(S_{11}h^2 + S_{22}k^2 + S_{33}l^2 + 2S_{12}hk + 2S_{13}hl + 2S_{23}kl\right) - \\ &-\frac{s}{R}\frac{4}{\lambda^2}\cos\theta\sin 2\theta = \frac{4\sin^2\theta}{\lambda^2} \end{aligned} \quad (14.54)$$

and

$$\begin{aligned} &\left(S_{11}h^2 + S_{22}k^2 + S_{33}l^2 + 2S_{12}hk + 2S_{13}hl + 2S_{23}kl\right) - \\ &-\delta\theta_0\frac{4}{\lambda^2}\sin 2\theta = \frac{4\sin^2\theta}{\lambda^2} \end{aligned} \quad (14.55)$$

Matrix **A** is then modified to

$$\mathbf{A} = \begin{pmatrix} h_1^2 & k_1^2 & l_1^2 & 2h_1k_1 & 2h_1l_1 & 2k_1l_1 & \frac{4\cos\theta_1 \sin 2\theta_1}{\lambda^2} \\ h_2^2 & k_2^2 & l_2^2 & 2h_2k_2 & 2h_2l_2 & 2k_2l_2 & \frac{4\cos\theta_2 \sin 2\theta_2}{\lambda^2} \\ \cdots & \cdots & \cdots & \cdots & \cdots & \cdots & \\ h_n^2 & k_n^2 & l_n^2 & 2h_nk_n & 2h_nl_n & 2k_nl_n & \frac{4\cos\theta_n \sin 2\theta_n}{\lambda^2} \end{pmatrix} \tag{14.56}$$

or

$$\mathbf{A} = \begin{pmatrix} h_1^2 & k_1^2 & l_1^2 & 2h_1k_1 & 2h_1l_1 & 2k_1l_1 & \frac{4\sin 2\theta_1}{\lambda^2} \\ h_2^2 & k_2^2 & l_2^2 & 2h_2k_2 & 2h_2l_2 & 2k_2l_2 & \frac{4\sin 2\theta_2}{\lambda^2} \\ \cdots & \cdots & \cdots & \cdots & \cdots & \cdots & \\ h_n^2 & k_n^2 & l_n^2 & 2h_nk_n & 2h_nl_n & 2k_nl_n & \frac{4\sin 2\theta_n}{\lambda^2} \end{pmatrix} \tag{14.57}$$

to account for the sample-displacement and zero-shift errors, respectively, and the least-squares solution produces the following vectors:

$$\mathbf{x} = \begin{pmatrix} S_{11} \\ S_{22} \\ S_{33} \\ S_{12} \\ S_{13} \\ S_{23} \\ s/R \end{pmatrix} \tag{14.58}$$

or

$$\mathbf{x} = \begin{pmatrix} S_{11} \\ S_{22} \\ S_{33} \\ S_{12} \\ S_{13} \\ S_{23} \\ \delta\theta_0 \end{pmatrix} \tag{14.59}$$

When the symmetry of the material is higher than triclinic, Eqs. (14.54) and (14.55) contain fewer unknowns and the least-squares procedure is simplified.

The least-squares technique described here assumes that each data point (i.e., Bragg peak) is measured with the same experimental error and, therefore, equally contributes to the resulting solution that represents the refined unit cell dimensions and/or correction for a systematic error, if any. When a realistic estimate of individual errors in Bragg angles is available, it is possible to adjust the contributions from the individual Bragg peaks to reflect higher or lower precision of Bragg angles. This is realized by introducing individual weights into the calculation of the normal equations.

Thus, each row of matrix **A**, each element of vector **y** (see Eq. 14.33), and each column of the transpose matrix $\mathbf{A}^{\mathrm{T}}$ (see Eq. 14.38) are changed by the multiplier that is inversely proportional to the square root of the experimental error in the corresponding experimental data point. Alternatively, the weighted least-squares solution may be expressed as follows:

$$\mathbf{x} = \left(\mathbf{A}^{\mathbf{T}}\mathbf{W}\mathbf{A}\right)^{-1}\left(\mathbf{A}^{\mathbf{T}}\mathbf{W}\mathbf{y}\right) \tag{14.60}$$

where **W** is the square matrix representing individual weights:

$$\mathbf{W} = \begin{pmatrix} w_1^2 & 0 & \ldots & 0 \\ 0 & w_2^2 & \ldots & 0 \\ \ldots & \ldots & \ldots & \ldots \\ 0 & 0 & \ldots & w_n^2 \end{pmatrix} \tag{14.61}$$

The standard uncertainties (or standard deviations) for each parameter determined according to the least-squares method are calculated from

$$\sigma(x_j) = \sqrt{\frac{\left(\mathbf{A}^{\mathbf{T}}\mathbf{W}\mathbf{A}\right)_{jj}^{-1}\sum_{k=1}^{n} w_k\left(Q_k^{obs} - Q_k^{calc}\right)^2}{n-m}}, \quad j = 1, \ldots, m \tag{14.62}$$

where

- n is the number of equations (Bragg peaks),
- m is the number of unknown parameters (from 2 to 7, assuming that sample-displacement or zero-shift error is always refined),
- $\left(\mathbf{A}^{\mathbf{T}}\mathbf{W}\mathbf{A}\right)_{jj}^{-1}$ is *the* corresponding diagonal element of the inverse normal equation matrix,
- w_k is the *corresponding* weight, or unity if no weight is applied,
- $(Q_k^{obs} - Q_k^{calc})$ is the difference between the observed and calculated $1/d^2$.

Next, we illustrate the least-squares refinement of the lattice parameter for the LaB_6 compound, which was fully indexed earlier (see Table 14.12). Least-squares refinement of lattice parameter (Eq. 14.40), assuming unit weights and using all 20 available Bragg peaks, results in $a = 4.1599(3)$ Å. The differences between the observed and calculated 2θ values are shown in Fig. 14.20, which clearly reveals a systematic dependence of Δ2θ on the Bragg angle. Such behavior is typically indicative of a systematic error, likely due to zero shift or sample displacement, or a combination of both.

The fact that the differences between the observed and calculated Bragg angles change sign (Fig. 14.20) is intrinsic to a least-squares technique, which simply minimizes the function defined in Eq. (14.35). As a result, the refined lattice parameter of LaB_6, $a = 4.1599(3)$ Å, is far from its standard value of $a = 4.15695(6)$ Å. When the sample-displacement error has been refined together with the lattice parameter, this yields $a = 4.1583(1)$ Å and $s/R = 0.00632$. Conversely, when zero-shift error is refined instead of sample displacement, the resultant unit cell dimension becomes $a = 4.1574(1)$ Å and zero shift $\delta\theta_0 = 0.078°$. The corresponding

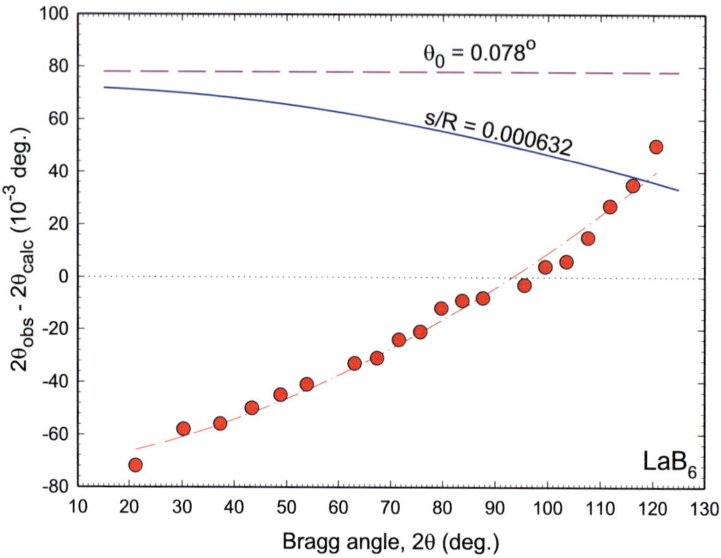

Fig. 14.20 The differences between the observed and calculated Bragg angles after least-squares refinement of the lattice parameter of LaB_6 without accounting for the presence of any kind of systematic error (*open circles*) using $a = 4.1599(3)$ Å. The *dash-dotted line* drawn through the data points is a guide for the eye. The *solid line* represents corrections of the observed Bragg angles using the refined in the next step sample-displacement error ($s/R = 0.00632$) and the *dashed line* represents a similar correction by using the determined zero-shift error ($\delta\theta_0 = 0.078°$)

corrections of the observed Bragg angles are shown in Fig. 14.20 and the respective sets of differences between the observed and calculated Bragg angles are depicted in Fig. 14.21.

It is easy to recognize that the effect of a systematic error has been removed from the experimental data in each case since in Fig. 14.21 the differences are distributed nearly randomly around zero and they are much smaller when compared to those in Fig. 14.20. The F_{20} figures of merit shown in Fig. 14.21 are only different by a few percent.

However, if one compares the values of the lattice parameter obtained when a different kind of a systematic error was assumed and accounted for in the data, the difference between the two is statistically significant (4.1583 vs. 4.1574 Å for sample-displacement and zero-shift effects, respectively). This is expected given the different contribution from different errors, as seen in Fig. 14.20. Usually, both effects are present in experimental data. However, refining both contributions simultaneously is not feasible due to the strong correlation between the sample-displacement and zero-shift parameters, as both curves follow the same trend in the measured 2θ range between 10° and 130° (see Fig. 14.22).

Considering the resultant unit cell dimensions, it appears that the zero-shift error has the largest influence on the discussed experimental data, since the refined lattice parameter (a = 4.1574 Å) is the closest match with the standard value of $a = 4.15695$ Å.

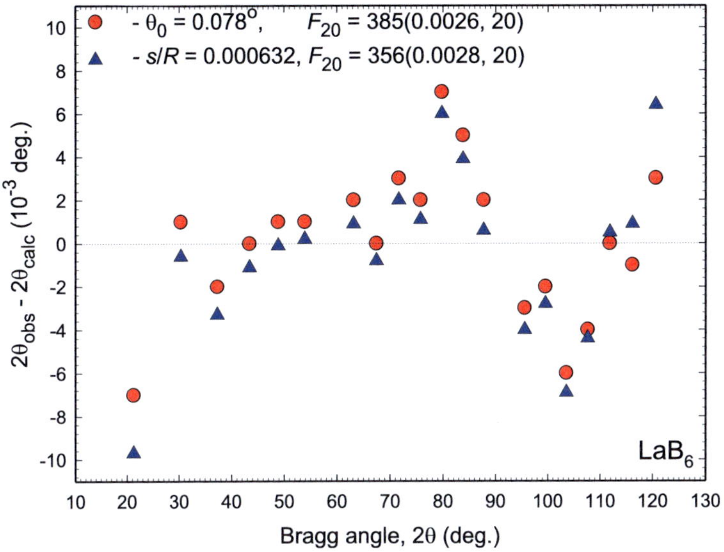

Fig. 14.21 The differences between the observed and calculated Bragg angles after the least-squares refinement of the lattice parameter of LaB_6 simultaneously with the zero-shift error (*red circles*) or simultaneously with the sample-displacement error (*blue triangles*)

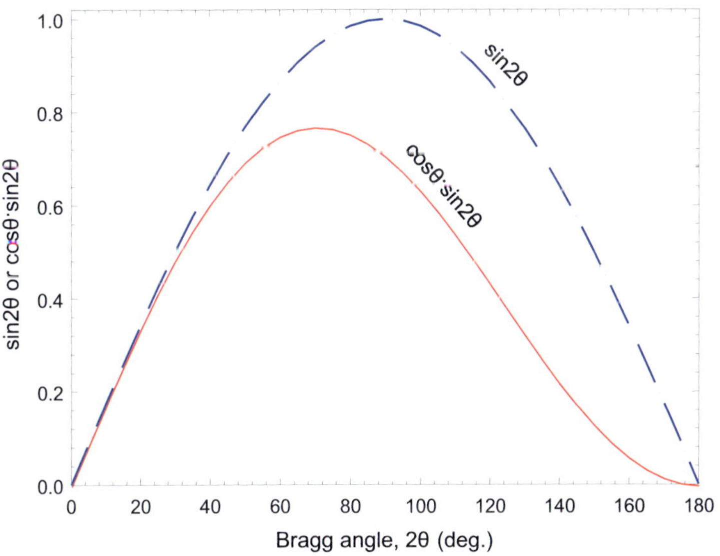

Fig. 14.22 The behavior of $\cos\theta \sin 2\theta$ (sample displacement, Eq. 14.56) and $\sin 2\theta$ (zero shift, Eq. 14.57) as functions of Bragg angle. The two functions show similar behavior, especially at low Bragg angles, and therefore, the two parameters strongly correlate with one another when both are simultaneously included in the least-squares refinement

Lattice parameters, determined by any of the three indexing programs considered in this chapter are usually refined using the least-squares method. On the one hand, their accuracy is quite satisfactory for using unit cell dimensions in database searches or as starting values in full-pattern fitting, whether through the Le Bail or Pawley technique (Sect. 15.4) or in Rietveld refinement (Sect. 15.7).

On the other hand, lattice parameters refined during the indexing are not quite precise because of missed or improperly handled systematic errors. Moreover, only a limited subset of Bragg reflections, usually at low angles, and therefore, most affected by various systematic errors, is used in ab initio indexing. When precise unit cell dimensions are required, their accuracy after automatic indexing can be improved by including additional Bragg reflections into the least-squares procedure, without the need for full-profile methods. The entire pattern should be indexed as described in Sect. 14.3, and all resolved Bragg peaks should be included in the least-squares minimization.

It is noteworthy that the full-profile fitting methods result in the highest precision of lattice parameters achievable for a given dataset, as there is no need for biased resolution of the overlapped peaks. Additionally, these methods include corrections for more systematic effects, not just for zero shift and sample displacement, but also for absorption effect, beam spill, sample porosity, just to name a few.

Considering the pattern of $(CH_3NH_3)_2Mo_7O_{22}$, shown in Fig. 14.15, its complexity arises from the relatively large unit cell and monoclinic symmetry coupled with the typical resolution of a conventional laboratory diffractometer. As a result, a substantial Bragg peak overlap is observed, especially at high angles. The first 30 resolved peaks below $2\theta = 30°$ were indexed as shown in Sect. 14.11.2. The remaining 57 resolvable Bragg peaks between 30 and 51° 2θ were indexed manually, using a solution with the highest figure of merit from Table 14.22. A least-squares refinement employing all 87 Bragg peaks resulted in the following lattice parameters:

$$a = 23.0875(9)\ \text{Å}, b = 5.5191(5)\ \text{Å}, c = 19.5789(9)\ \text{Å}, \beta = 122.924(3)\,°$$

The differences between the observed and calculated Bragg angles are shown in Fig. 14.23 as blue squares. They clearly indicate the presence of a systematic error. Moreover, several peaks with larger differences are far away from the gradually changing $2\theta_{obs}$—$2\theta_{calc}$ behavior. These peaks are marked with red circles in the figure, and all of them are weak and/or are heavily overlapped with strong neighboring Bragg reflections. A second least-squares refinement of the lattice parameters together with sample-displacement optimization after excluding the five marked reflections produces.

$$a = 23.0641(8)\ \text{Å}, b = 5.5131(2)\ \text{Å}, c = 19.5601(6)\ \text{Å}, \beta = 122.930(1)\,°.$$

and sample-displacement parameter, $s/R = -0.00042(1)$. Much lower discrepancies between the observed and calculated 2θ are noteworthy. Standard uncertainties of

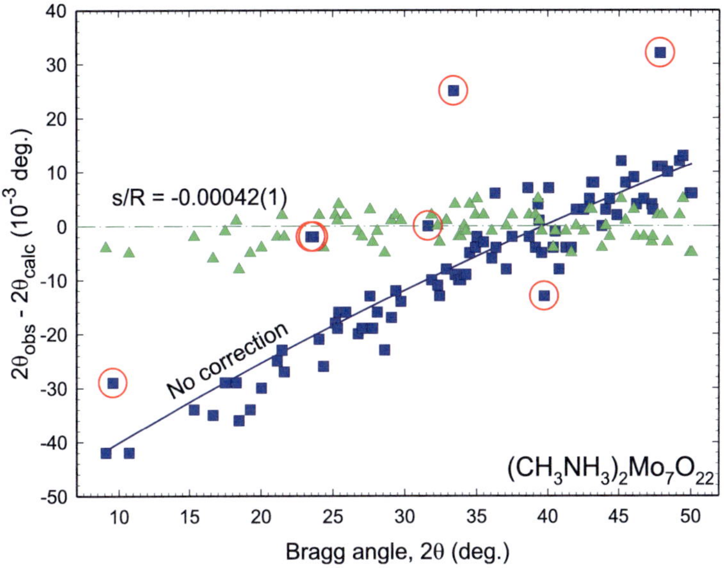

Fig. 14.23 Differences between the observed and calculated Bragg angles after the least-squares refinement of lattice parameters of the monoclinic $(CH_3NH_3)_2Mo_7O_{22}$. The *blue squares* and *green triangles* correspond to refinements without and with the sample-displacement correction, respectively. The data points enclosed inside red circles were excluded from the second refinement due to the low intensity, strong overlap and, therefore, low accuracy of the corresponding observed Bragg angles

lattice parameters are also reduced when the sample displacement was accounted for. Adding more Bragg reflections, those that are located above $2\theta = 51°$, produces even lower standard deviations, but the resultant unit cell dimensions become biased by subjective assignments of Miller indices due to severe overlapping, which causes low accuracy in the observed peak positions. Ambiguities occur despite narrow peaks, which have full widths at half maximum under ~0.1° of 2θ.[79]

Note that the unit cell dimensions obtained from the full-pattern fitting using Rietveld refinement of the tile compound (see in Table 22.4) are a = 23.0707(3), b = 5.51522(7), c = 19.5669(2) Å, β = 122.930(1)°, with standard uncertainties three times smaller than those from fitting individual peaks. The lattice dimensions obtained here differ from Rietveld data by as much as 10 s.u., which is not unusual, considering that the Rietveld refinement used twice the angular range and a much larger array of the reflections.

[79]The list of 87 resolved Bragg peaks is found in the ASCII data file *Ch14Ex08_CuKa.pks* online.

14.13 Concluding Remarks

The reliability and precision of the established unit cell dimensions are not only functions of the quality of the collected experimental data but also depend on the presence of measurable diffraction peaks within certain Bragg angle ranges. As illustrated in Fig. 14.24, the availability of low Bragg angle reflections is crucial for the successful assignment of indices, while high Bragg angle peaks are necessary to determine lattice parameters with maximum precision. In practice, the entire range of Bragg angles is used in routine least-squares refinements of the unit cell dimensions, particularly when the data are affected by small systematic instrumental and specimen-related errors.

Hence, when only lattice parameters are of concern and given the choice of available X-ray energies, medium to long wavelengths—such as those from Cu, Fe, Co, or Cr anodes—should be employed when using conventional X-ray sources. The range of wavelengths from ~1.5 to ~2.3 Å ensures that both low and high Bragg

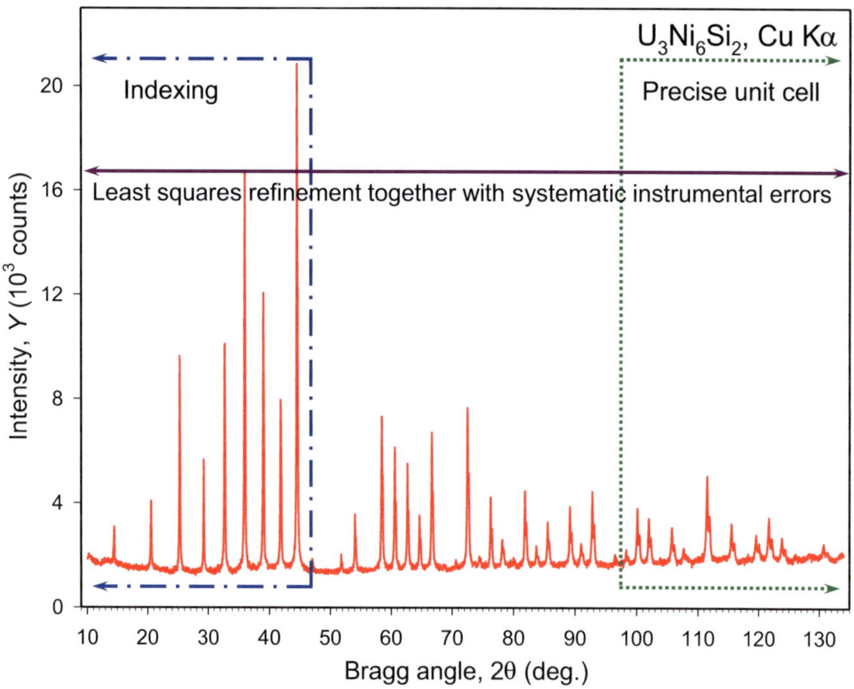

Fig. 14.24 The X-ray powder diffraction pattern of $U_3Ni_6Si_2$ (see also Fig. 14.10) illustrates regions critical for successful indexing and precise determination of unit cell dimensions. The boundaries of both the low and high Bragg angle regions are diffuse and can vary from one pattern to another

angle reflections are well-resolved and measured with sufficient accuracy, leading to the most precise unit cell dimensions.

14.14 Additional Reading

1. P.-E. Werner, Autoindexing, in: Structure determination from powder diffraction data. IUCr monographs on crystallography 13. W. I. F. David, K. Shankland, L.B. McCusker, and Ch. Baerlocher, Eds., Oxford University Press, Oxford (2002).
2. H. Lipson and H. Steeple, Interpretation of X-ray powder diffraction patterns, MacMillan, London (1970).
3. P.-E. Werner, L. Eriksson and M. Westdahl, TREOR, a semi-exhaustive trial-and-error powder indexing program for all symmetries, J. Appl. Cryst. **18**, 367 (1985). https://doi.org/10.1107/S0021889885010512
4. A. Boutlif and D. Louër, Indexing of powder diffraction patterns for low symmetry lattices by the successive dichotomy method, J. Appl. Cryst. **24**, 987 (1991). https://doi.org/10.1107/S0021889804014876
5. J.W. Visser, A fully automatic program for finding the unit cell from powder data, J. Appl. Cryst. **2**, 89 (1969). https://doi.org/10.1107/S0021889869006649
6. E. Prince and P.T. Boggs, Chap. 8.1. Least squares, in: International Tables for Crystallography, Vol. C, Mathematical, physical and chemical tables, First online edition, E. Prince, Ed. (2006). https://it.iucr.org/C/ and references therein.
7. J. Bergmann, A. Le Bail, R. Shirley, V. Zlokazov. Renewed interest in powder diffraction data indexing. Z. Kristallogr. **219**, 783 (2004). https://doi.org/10.1524/zkri.219.12.783.55862
8. R. Shirley, Overview of powder-indexing program algorithms (history and strengths and weaknesses), IUCr Computing Commission Newsletter, **2**, 48 (2003). https://www.iucr.org/__data/assets/pdf_file/0004/6394/iucrcompcomm_jun2003.pdf
9. A. Altomare, G. Campi, C. Cuocci, L. Eriksson, C. Giacovazzo, A. Moliterni, R. Rizzi and P.-E. Werner, Advances in powder diffraction pattern indexing: N-TREOR09, J. Appl. Cryst. 42, 768 (2009). https://doi.org/10.1107/S0021889809025503
10. A. Altomare, C. Cuocci, A. Moliternia, and R. Rizzi, Chapter 3.4. Indexing a powder diffraction pattern, in International Tables for Crystallography. Vol. H. Ch. 3.4, pp. 270 (2019). https://onlinelibrary.wiley.com/iucr/itc/Ha/ch3o4v0001/references/
11. IUCr Crystallographic software list. https://www.iucr.org/resources/other-directories/software
12. Crystallographic source code museum created by Armel LeBail and currently hosted by IUCr. https://www.iucr.org/resources/commissions/computing/software-museum

14.15 Problems

1. Consider the X-ray powder diffraction pattern of $LaNi_{11.6}Ge_{1.4}$ shown in Fig. 14.25 and found in the file *Ch14Pr01_CuKa.xy* online. A total of 27 individual Bragg peaks are measurable up to $2\theta = 91°$. They are listed in Table 14.28 and also found in the file *Ch14Pr01_CuKa.pks* online. Peak positions, intensities, and full widths at half maximum have been determined using a profile fitting procedure. The observed Bragg angles are listed for the $K\alpha_1$ component in the doublet, $\lambda = 1.540593$ Å.

   (a) Using a spreadsheet, perform indexing of the powder diffraction pattern assuming cubic crystal system.
   (b) Analyze the combinations of Miller indices of the observed reflections, and determine possible space groups of the material.
   (c) Using the average value of the lattice parameter, calculate both F_N and M_{20} figures of merit for your indexing result. Is the determined unit cell realistic?
   (d) Perform least-squares refinement of the unit cell parameter using Eq. (14.40) and a spreadsheet (do not use any freely or commercially available software). Note that the normal equation matrix for the case of a single parameter is a single number.

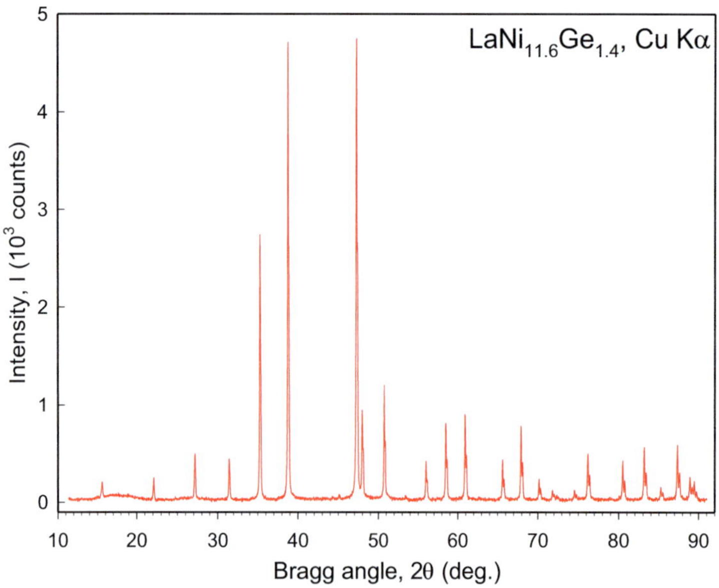

Fig. 14.25 The X-ray powder diffraction pattern of $LaNi_{11.6}Ge_{1.4}$ collected on an HZG-4a powder diffractometer using Cu Kα radiation. Numerical data are available in the file *Ch14Pr01_CuKa.xy* online. (Data courtesy of Dr. L.G. Akselrud)

Table 14.28 Relative integrated intensities (I/I_0), Bragg angles, and full widths at half maximum (FWHM) of Bragg peaks observed in the $LaNi_{11.6}Ge_{1.4}$ powder diffraction pattern collected using Cu Kα radiation in the range 11° ≤ 2θ ≤ 91° (see Fig. 14.25). Numerical data are available in the file *Ch14Pr01_CuKa.pks*

I/I_0	2θ (°)[a]	FWHM (°)
33	15.561	0.165
31	22.099	0.128
87	27.131	0.122
80	31.423	0.116
487	35.255	0.104
847	38.745	0.100
1000	47.232	0.114
191	47.945	0.114
228	50.716	0.105
86	55.950	0.116
172	58.455	0.117
194	60.890	0.118
90	65.596	0.113
172	67.882	0.118
47	70.133	0.119
25	71.796	0.130
11	72.343	0.118
23	74.542	0.119
127	76.172	0.132
4	76.691	0.132
3	78.887	0.132
97	80.467	0.128
147	83.114	0.141
32	85.233	0.138
158	87.336	0.151
65	88.909	0.157
45	89.434	0.137

[a] Bragg angles are listed for the location of the Cu $K\alpha_1$ component in the doublet, λ = 1.540593 Å

2. Consider the X-ray powder diffraction pattern of $CeRhGe_3$ shown in Fig. 14.26 and found in the file *Ch14Pr02_MoKa.xy* online. A total of 55 individual Bragg peaks are measurable up to 2θ = 46.5° and these are listed in Table 14.29 and also found in the file *Ch14Pr02_MoKa.pks* online. Peak positions, intensities, and full widths at half maximum have been determined using a profile fitting procedure. The observed Bragg angles are listed for the $K\alpha_1$ component in the doublet, λ = 0.709317 Å.

   (a) Using TREOR, ITO, and DICVOL, perform indexing of the powder diffraction pattern. Make sure that you obtain a solution in each of the three programs.
   (b) Analyze the combinations of Miller indices of the observed Bragg reflections and determine possible space groups describing symmetry of the material.

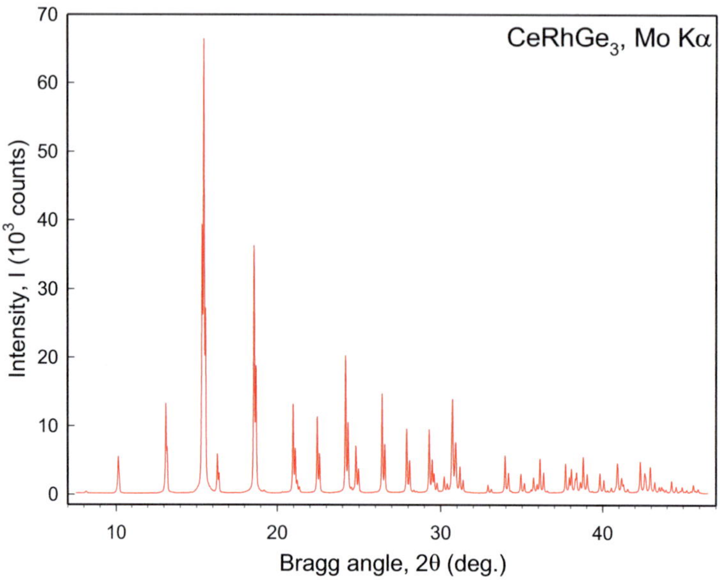

Fig. 14.26 The X-ray powder diffraction pattern of $CeRhGe_3$ collected on a Rigaku TTRAX rotating anode powder diffractometer using Mo Kα radiation. Numerical data are available in the file *Ch14Pr02_MoKa.xy* online

(c) Perform least-squares refinement of the unit cell dimensions using all available data without refining any kind of a systematic error.[80] Analyze the differences between the observed and calculated 2θ and decide whether the refinement of a zero-shift or sample-displacement error is warranted. If it is, refine lattice parameters together with a zero-shift or sample-displacement error.

3. Consider the X-ray powder diffraction pattern of $SrSi_2$ shown in Fig. 14.27 and found in the file *Ch14Pr03_MoKa.xy* online. A total of 20 individual Bragg peaks are measurable up to $2\theta = 33°$. They are listed in Table 14.30 and also found in the file *Ch14Pr03_MoKa.pks* online. Peak positions, intensities, and full widths at half maximum have been determined using a profile fitting procedure. The observed Bragg angles are listed for the $K\alpha_1$ component in the doublet, $\lambda = 0.709317$ Å.

[80]If you do not have one, refinement software may be downloaded from IUCr crystallographic software list (https://www.iucr.org/resources/other-directories/software) or crystallographic source code museum created by Armel LeBail and hosted by IUCr (https://www.iucr.org/resources/commissions/computing/software-museum)

Table 14.29 Relative integrated intensities (I/I_0), Bragg angles, and full widths at half maximum (FWHM) of Bragg peaks observed in the $CeRhGe_3$ powder diffraction pattern collected using Mo Kα radiation in the range 7.5° ≤ 2θ ≤ 46.5° (see Fig. 14.26). Numerical data are available in the file *Ch14Pr02_MoKa.pks*

I/I_0	2θ (°)[a]	FWHM (°)
5	8.100	0.091
110	10.091	0.073
270	13.086	0.068
789	15.303	0.070
1000	15.413	0.064
115	16.245	0.065
774	18.555	0.069
4	20.283	0.065
269	20.928	0.065
32	21.172	0.065
227	22.399	0.064
418	24.143	0.065
13	24.483	0.085
157	24.766	0.073
303	26.362	0.066
206	27.866	0.071
5	28.296	0.058
200	29.249	0.067
60	29.547	0.075
54	30.167	0.075
87	30.625	0.069
301	30.682	0.074
18	30.905	0.070
76	31.129	0.068
24	32.844	0.070
120	33.905	0.068
9	34.057	0.090
66	34.872	0.072
9	35.507	0.071
49	35.654	0.071
112	36.055	0.070
10	36.288	0.078
92	37.633	0.067
81	37.997	0.073
63	38.316	0.070
129	38.737	0.074
4	39.060	0.061
64	39.776	0.070
10	40.005	0.061
14	40.483	0.071

(continued)

Table 14.29 (continued)

I/I_0	2θ (°)[a]	FWHM (°)
81	40.848	0.078
41	40.888	0.071
23	41.236	0.072
1	41.379	0.074
113	42.270	0.074
47	42.599	0.082
77	42.893	0.075
20	43.446	0.076
22	43.606	0.081
43	44.232	0.074
6	44.758	0.072
20	44.883	0.073
5	45.413	0.080
31	45.573	0.078
4	45.721	0.087

[a] Bragg angles are listed for the location of the Mo $K\alpha_1$ component in the doublet, $\lambda = 0.709317$ Å

(a) Conduct ab initio indexing of the powder diffraction pattern manually, i.e., using a spreadsheet rather than any kind of crystallographic software. Compute the F_N and M_{20} figures of merit and discuss both the probability of the determined unit cell and the accuracy of the observed Bragg angles.
(b) Now, perform the indexing of the same pattern using TREOR, ITO, and DICVOL. Make sure that you obtain a solution in each of the three programs.
(c) Analyze the combinations of Miller indices of the observed reflections and determine possible space groups, which characterize the symmetry of the material.

4. Consider the X-ray powder diffraction pattern of $Li_2Sn(OH)_6$ shown in Fig. 14.28 and found in the file *Ch14Pr04_CuKa.xy* online. A total of 38 individual Bragg peaks are measurable up to $2\theta = 51°$ and these are listed in Table 14.31. The values in Table 14.31 have been corrected for the sample-displacement error. All 88 (uncorrected) peaks observed below $2\theta = 71°$ can be found in the file *Ch14Pr04_CuKa.pks* online. Peak positions (listed for the $K\alpha_1$ component in the doublet, $\lambda = 1.540593$ Å), intensities, and full widths at half maximum have been determined using a profile fitting procedure.

   (a) Using data from Table 14.31 and TREOR, ITO, and DICVOL, perform indexing of the powder diffraction pattern. Try to obtain solution in each of the three programs.
   (b) Analyze the observed Miller indices and determine possible space groups of the material.

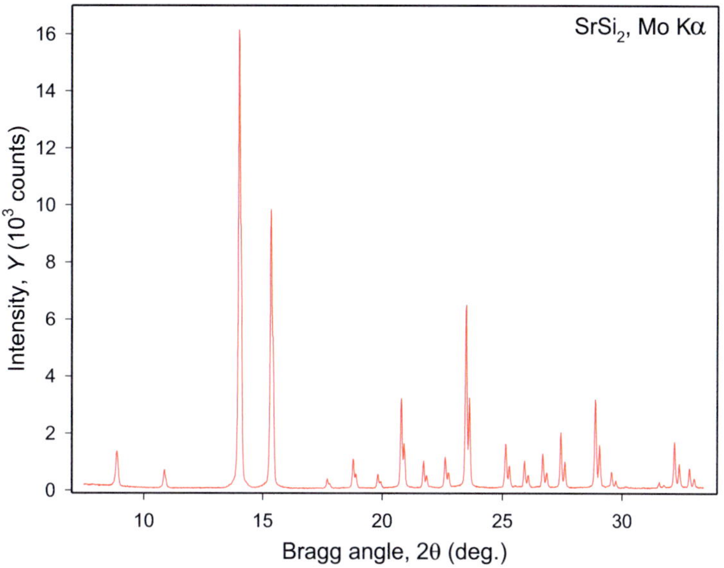

Fig. 14.27 The X-ray powder diffraction pattern of $SrSi_2$ collected on a Rigaku TTRAX rotating anode powder diffractometer using Mo Kα radiation. Numerical data are available in the file *Ch14Pr03_MoKa.xy* online

Table 14.30 Relative integrated intensities (I/I_0), Bragg angles, and full widths at half maximum (FWHM) of Bragg peaks observed in the powder diffraction pattern of $SrSi_2$ collected using Mo Kα radiation in the range $8° \leq 2\theta \leq 33°$ (see Fig. 14.27). Numerical data are available in the file *Ch14Pr03_MoKa.pks*

I/I_0	2θ (°)[a]	FWHM (°)
80	8.872	0.091
38	10.851	0.080
1000	13.999	0.080
626	15.337	0.081
20	17.713	0.081
64	18.794	0.080
29	19.815	0.079
203	20.791	0.081
54	21.724	0.076
396	23.484	0.079
101	25.128	0.084
57	25.912	0.080
72	26.672	0.080
114	27.418	0.076
200	28.848	0.081
34	29.540	0.079
12	31.533	0.081
104	32.170	0.077

[a] Bragg angles are listed for the location of the Mo $K\alpha_1$ component in the doublet, $\lambda = 0.709317$ Å

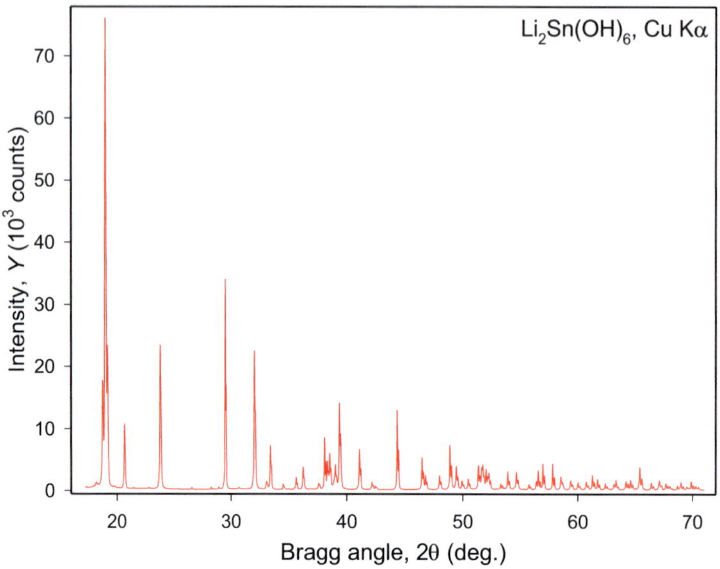

Fig. 14.28 The X-ray powder diffraction pattern of $Li_2Sn(OH)_6$ collected on a Scintag XDS2000 powder diffractometer using Cu Kα radiation. Numerical data are available in the file *Ch14Pr04_CuKa.xy* online

(c) Perform the least-squares refinement of the unit cell dimensions using all available data (file *Ch14Pr04_CuKa.pks* online) without refining any kind of a systematic error and then refine lattice parameters together with a zero-shift or a sample-displacement error.

5. Consider the X-ray powder diffraction pattern of $tea_2Mo_6O_{19}$ (*tea* is tetraethylammonium, $[N(C_2H_5)_4]^+$), which is shown in Fig. 14.29 and found in the file *Ch14Pr05_CuKa.xy* online. A total of 44 individual Bragg peaks are measurable up to $2\theta = 49.5°$ and these are listed in Table 14.32 and also found in the file *Ch14Pr05_CuKa.pks* online. Peak positions, intensities, and full widths at half maximum have been determined using a profile fitting procedure. The observed Bragg angles are listed for the Cu $K\alpha_1$ component in the doublet, $\lambda = 1.540593$ Å.

(a) Using TREOR, ITO, and DICVOL, perform ab initio indexing of the powder diffraction pattern. Try to obtain solution in each of the three programs. Figure out a systematic error using different orders of the same reflections. *Indexing hint*: the unit cell is relatively large; therefore, increase maximum volume if no acceptable solution is found, and use only 20 lowest Bragg angle peaks.

(b) Analyze the combinations of Miller indices of the observed Bragg reflections and determine possible space groups describing symmetry of the material.

Table 14.31 Relative integrated intensities (I/I_0), Bragg angles, and full widths at half maximum (FWHM) of Bragg peaks observed in the powder diffraction pattern of $Li_2Sn(OH)_6$ collected using Cu Kα radiation in the range $18° \leq 2\theta \leq 51°$ (see Fig. 14.28). Numerical data are available in the file *Ch14Pr04_CuKa.pks*

I/I_0	2θ (°)[a]	FWHM (°)
197	18.802	0.096
1000	19.041	0.108
288	19.213	0.107
150	20.740	0.090
337	23.916	0.083
1	25.361	0.083
344	29.541	0.054
334	32.114	0.074
18	33.143	0.074
103	33.478	0.074
12	34.576	0.074
1	34.980	0.074
28	35.689	0.074
55	36.275	0.074
16	37.620	0.069
105	38.141	0.069
63	38.375	0.069
83	38.589	0.069
9	38.845	0.069
18	38.990	0.069
47	39.070	0.069
193	39.428	0.069
29	39.503	0.069
96	41.137	0.075
1	41.721	0.075
19	42.211	0.075
3	42.565	0.075
1	43.545	0.075
7	42.508	0.075
158	44.405	0.063
79	46.520	0.076
33	46.825	0.076
36	48.045	0.076
106	48.948	0.076
3	49.132	0.076
64	49.489	0.076
23	49.981	0.076
26	50.517	0.076

[a] Bragg angles are listed for the location of the Cu $K\alpha_1$ component in the doublet, $\lambda = 1.540593$ Å. Peak positions are already corrected for the sample shift –0.15 mm, assuming goniometer radius of 250 mm

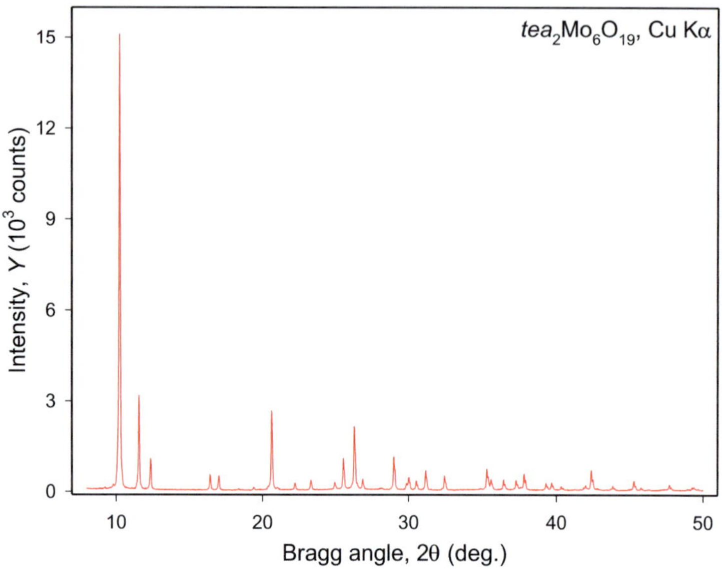

Fig. 14.29 The X-ray powder diffraction pattern of $tea_2Mo_6O_{19}$ collected on a Scintag XDS2000 powder diffractometer using Cu Kα radiation. Numerical data are available in the file *Ch14Pr05_CuKa.xy* online

(c) Perform the least-squares refinement of the unit cell dimensions using all available data without refining any kind of a systematic error. Analyze the differences between the observed and calculated 2θ and decide whether the refinement of zero-shift or sample-displacement error is warranted. If it is, refine lattice parameters together with a zero-shift or a sample-displacement error.

6. Consider the X-ray powder diffraction pattern of MnV_2O_5 shown in Fig. 14.30 and found in the file *Ch14Pr06_CuKa.xy* online. A total of 53 individual Bragg peaks are measurable up to $2\theta = 62°$ and these are listed in Table 14.33 and also found in the file *Ch14Pr06_CuKa.pks* online. Peak positions, intensities, and full widths at half maximum have been determined using a profile fitting procedure. The observed Bragg angles are listed for the $K\alpha_1$ component in the doublet, $\lambda = 1.540593$ Å.

   (a) Using TREOR, ITO, and DICVOL, perform ab initio indexing of the powder diffraction pattern. Try to obtain solution in each of the three programs.
   (b) Analyze the combinations of Miller indices of all observed Bragg reflections and determine possible space groups describing symmetry of the material.
   (c) Perform least-squares refinement of the unit cell dimensions using all available data without refining any kind of a systematic error. Analyze the differences between the observed and calculated 2θ and decide whether the

Table 14.32 Relative integrated intensities (I/I_0), Bragg angles, and full widths at half maximum (FWHM) of Bragg peaks observed in the *tea*$_2$Mo$_6$O$_{19}$ powder diffraction pattern collected using Cu Kα radiation in the range $10° \leq 2\theta \leq 49.5°$ (see Fig. 14.29). Numerical data are available in the file *Ch14Pr05_CuKa.pks*

I/I_0	2θ (°)[a]	FWHM (°)
1000	10.241	0.085
182	11.573	0.070
65	12.376	0.073
30	16.408	0.076
30	16.986	0.076
4	19.393	0.050
163	20.615	0.074
14	22.217	0.074
20	23.297	0.074
17	24.933	0.074
73	25.523	0.074
155	26.286	0.074
6	26.479	0.074
22	26.851	0.074
3	28.138	0.074
78	28.991	0.074
13	29.842	0.074
28	30.004	0.074
20	30.508	0.074
45	31.153	0.074
32	32.427	0.074
2	34.830	0.075
52	35.282	0.075
6	35.425	0.075
24	35.567	0.075
24	36.419	0.075
24	37.259	0.075
4	37.531	0.075
41	37.798	0.075
16	39.282	0.071
19	39.667	0.071
8	40.316	0.071
5	41.842	0.082
9	41.972	0.082
55	42.355	0.082
4	42.747	0.082
3	43.698	0.082
9	43.824	0.082
24	45.242	0.081
6	45.725	0.081

(continued)

Table 14.32 (continued)

I/I_0	2θ (°)[a]	FWHM (°)
15	47.671	0.081
3	48.911	0.081
7	49.244	0.081
5	49.357	0.081

[a] Bragg angles are listed for the location of the Cu Kα_1 component in the doublet, $\lambda = 1.540593$ Å

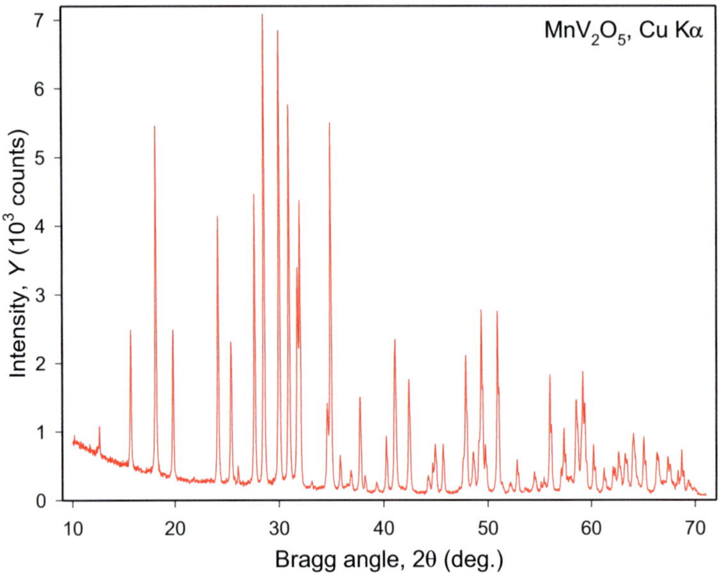

Fig. 14.30 The X-ray powder diffraction pattern of MnV_2O_5 collected on a Scintag XDS2000 powder diffractometer using Cu Kα radiation. Numerical data are available in the file *Ch14Pr06_CuKa.xy* online

data are affected by a zero-shift or a sample-displacement error. If a systematic error is substantial, refine lattice parameters together with a zero-shift or a sample-displacement error.

Table 14.33 Relative integrated intensities (I/I_0), Bragg angles, and full widths at half maximum (FWHM) of Bragg peaks observed in the MnV_2O_5 powder diffraction pattern collected using Cu Kα radiation in the range 12° ≤ 2θ ≤ 62° (see Fig. 14.30). Numerical data are available in the file *Ch14Pr06_CuKa.pks*

I/I_0	2θ (°)[a]	FWHM (°)
36	12.595	0.062
261	15.710	0.124
681	18.152	0.113
281	19.801	0.109
519	24.091	0.113
307	25.367	0.113
31	26.049	0.113
592	27.630	0.113
1000	28.520	0.113
988	29.972	0.113
829	30.912	0.113
443	31.715	0.113
578	31.935	0.113
172	34.542	0.111
786	34.856	0.111
78	35.778	0.111
51	36.788	0.111
229	37.673	0.111
34	38.157	0.111
27	39.268	0.152
136	40.215	0.152
456	41.035	0.152
314	42.378	0.152
40	44.182	0.126
63	44.630	0.126
123	44.877	0.126
123	45.631	0.126
82	47.584	0.123
363	47.828	0.123
107	48.546	0.123
125	49.100	0.123
464	49.335	0.123
106	49.720	0.123
484	50.890	0.123
10	51.336	0.123
28	52.091	0.123
88	52.774	0.123
13	53.593	0.123
51	54.474	0.123
23	55.132	0.123

(continued)

Table 14.33 (continued)

I/I_0	2θ (°)[a]	FWHM (°)
29	55.407	0.123
320	55.983	0.123
63	57.063	0.131
182	57.339	0.131
35	57.698	0.131
41	57.965	0.131
325	58.526	0.131
116	59.021	0.131
296	59.156	0.131
124	59.367	0.131
148	60.186	0.131
78	61.212	0.131
15	61.610	0.131

[a] Bragg angles are listed for the location of the Cu $K\alpha_1$ component in the doublet, $\lambda = 1.540593$ Å

Chapter 15
Solving Crystal Structure from Powder Diffraction Data

Assuming that both the crystal system—i.e., the "powder" Laue class—and lattice parameters of a material have been established, the next step is to solve its crystal structure to determine the distribution of atoms within the unit cell. This problem is generally far from trivial, and while many structure solutions in powder diffraction are unique, they share commonalities. All involve linking reciprocal and direct spaces, i.e., a powder pattern and the distribution of electron or nuclear density in a crystal lattice. Although traveling the structure determination process in powder diffraction is far from a smooth ride, like cruising down an interstate in the latest model Tesla, knowing where to enter, how to proceed, and where and when to exit is equally vital. Familiarity with crystallographic features of related materials, along with the basic chemical and physical properties of the material in question, such as probable oxidation and coordination states, expected connectivity of atoms, and shortest interatomic distances, is essential.

In this chapter, we explore practical applications of kinematical diffraction theory to solving crystal structures from powder diffraction data. When considering several rational examples in reciprocal space, we implicitly assume that the crystal structure of each sample is unknown and must be solved based solely on the information obtained from a powder diffraction experiment using just a few other basic properties of a polycrystalline material. The solution of a number of crystal structures in direct space is based on the earlier known structural data and supported by the results of powder diffraction analysis, such as unit cell dimensions and symmetry.

15.1 Ab Initio Methods of Structure Solution

The process of solving the crystal structure (Fig. 15.1) typically begins by analyzing systematic absences to determine the space-group symmetry of a material or at least to identify all probable space groups if the corresponding diffraction class includes more than one (see Sects. 9.3 and 9.4). The next step involves determining the

P. Zavalij, V. Pecharsky, *Fundamentals of Powder Diffraction and Structural Characterization of Materials*, https://doi.org/10.1007/978-3-031-91504-8_15

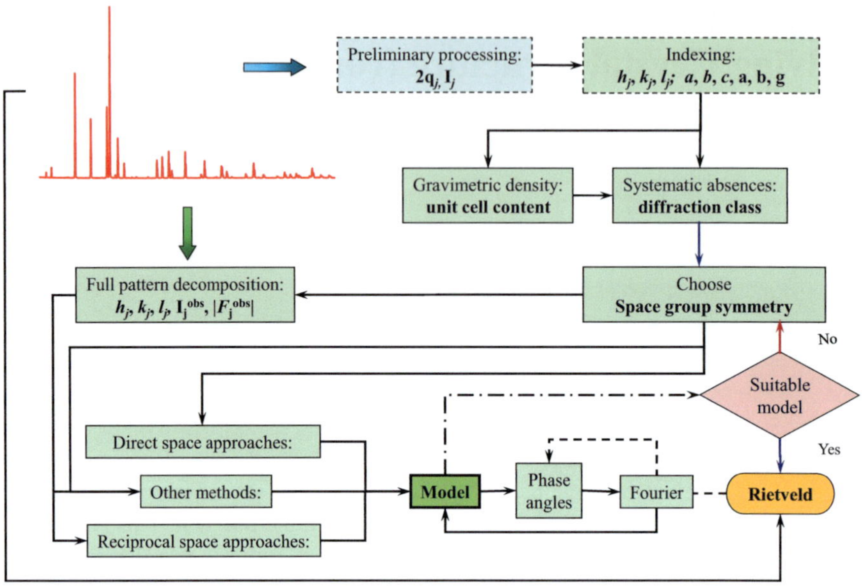

Fig. 15.1 The flowchart illustrating crystal structure determination from powder diffraction data. Preliminary processing and indexing are described in Chaps. 13 and 14, respectively, and have been assumed accomplished at an earlier stage

content of the unit cell, i.e., establishing or estimating how many atoms of each type, molecules, or groups of atoms are present in one unit cell. In some cases, this simple information can help narrow the space-group selection, particularly when the multiplicities of site positions are high (e.g., in high-symmetry groups) and a site with the lowest multiplicity places inside the unit cell more atoms of a given type than necessary.

Once the content of the unit cell has been established, a model of the crystal structure can be built using either direct or reciprocal space techniques, or a combination of both, known as dual-space methods. Direct-space approaches do not require immediate use of the observed integrated intensities, whereas reciprocal space methods rely on them.

15.1.1 Conventional Reciprocal Space Methods

In any of the reciprocal space method that relies exclusively on observed structure factors, the powder diffraction pattern must be deconvoluted, and the integrated intensities of all, or as many as possible, individual Bragg reflections should be determined with maximum precision. Only then can Patterson or direct phase-angle determination techniques be employed to construct a partial or complete structural model. The theoretical background supporting these two methods is reviewed in Sect. 10.2.

The Patterson technique is often called the heavy-atom method because it is most effective when a few atoms in the unit cell are significantly heavier than the others and, therefore, have much stronger scattering X-ray scattering ability. Squared structure amplitudes are used as coefficients in the Fourier transformation, and the resulting Patterson map provides a distribution of the interatomic vectors within the unit cell. A structure model is then obtained in direct space by analyzing the Patterson function also known as function of interatomic vectors.

Direct phase determination method is based on generating probable phases that were lost (cannot be measured) during the diffraction experiment. Observed structure amplitudes which have been normalized are employed together with computed phases in the Fourier transformation, yielding the so-called E-maps, which contain direct images of the electron density distribution, where maxima correspond to positions of atoms in the unit cell.

Until recently, Patterson and direct methods were the most commonly used "classical" approaches for solving structures. Direct phase determination methods have been particularly successful in solving structures from single-crystal data. However, their application in powder diffraction has been gradually increasing as the quality of powder data improves, better deconvolution techniques are developed, and more precise individual structure factors become available. Nowadays, however, these methods are being increasingly replaced by dual-space charge flipping methods and direct-space modeling.

15.1.2 Conventional Direct-Space Modeling

As an example of a direct-space approach, consider a material, which is an intermetallic compound. Many intermetallics form series of closely related structures, the so-called isostructural compounds, where the coordinates of atoms in the unit cell remain nearly identical and only the distribution of different kinds of atoms varies among available crystallographic sites. If this is the case, the crystal structure can be solved by comparing it with known structure types, by searching for matching symmetry, lattice centering, unit cell contents, chemical similarity of the components, and analogous relationships among the unit cell dimensions. Another example is an organic compound with the well-known configuration of its molecule. In some cases, it may be relatively easy to model the packing of the identical molecules in the unit cell, and perhaps, optimize their positions, orientations, and, if required, conformations by using energy minimization or other principles.

There are many ways to build a model of the crystal structure without initially relying on the intensities of individual Bragg reflections, which are often obscured in powder diffraction due to partial or complete overlap. Most direct-space approaches are, in essence, trial-and-error methods and typically include some or all of the following components:

- Obtaining a structural model from the analysis of potentially isostructural compounds with partially or completely different chemical composition but identical or similar stoichiometry. The search for isostructural compounds may be conducted using a digitized powder pattern, unit cell dimensions, stoichiometry, and/or other suitable parameters.[1,2]
- Obtaining a partial structural model from known similar or closely related compounds, for example, those with an identical framework or layers. Similar structures are usually found from close relationships among all or some unit cell dimensions. For example, layered intercalates with the same type of the host layer should have two similar unit cell dimensions.[3]
- When structures are simple or building blocks are known, geometrical modeling of a trial structure or several possible structures can be performed (e.g., modeling of zeolites from polyhedra or more complex building blocks). The resulting model may then be optimized or directly tested by using powder diffraction data.[4]
- Direct-space modeling is a particularly powerful tool for intermetallic and related structures, many of which are derived from close packing of incompressible spheres. When positions of large atoms in the unit cell are known, the smaller atoms will likely occupy voids of sufficient size.

Trial-and-error techniques clearly require some, and often extensive, chemical, crystallographic, and physical knowledge about a specific class of materials, as well as access to a structural database and some experience in structural analysis.

15.1.3 Unconventional Direct, Reciprocal, and Dual-Space Methods

No matter how advanced, every numerical data processing technique has intrinsic limitations, particularly when the data complexity increases or when some required information is missing, as in single-crystal diffraction from macromolecules and proteins, or in powder diffraction from conventional materials with large unit cell volume or low symmetry. Poor crystallinity and an excess of weakly scattering elements in macromolecular compounds, on the one hand, and significant overlap of

---

[1] Five examples of using the ICSD database in solving crystal structure by structural analogy are discussed in: J.A. Kaduk, Use of the Inorganic Crystal Structure Database as a problem solving tool, Acta Cryst. **B58**, 370 (2002), https://doi.org/10.1107/S0108768102003476

[2] An example of deriving the crystal structure of $Mn_7(OH)_3(VO_4)_4$ from the isostructural $Zn_7(OH)_3(SO_4)(VO_4)_3$ found by comparing unit cell dimensions while looking for identical symmetry is given in Chap. 23.

[3] Chapter 22 describes details of solving the crystal structure of a layered intercalated compound—methylammonium heptamolybdate—by analogy with a thallium compound.

[4] An example of solving the crystal structure of the monoclinic $Gd_5Si_2Ge_2$ based on the known crystal structure of the orthorhombic Gd_5Ge_4 is found in Chap. 19

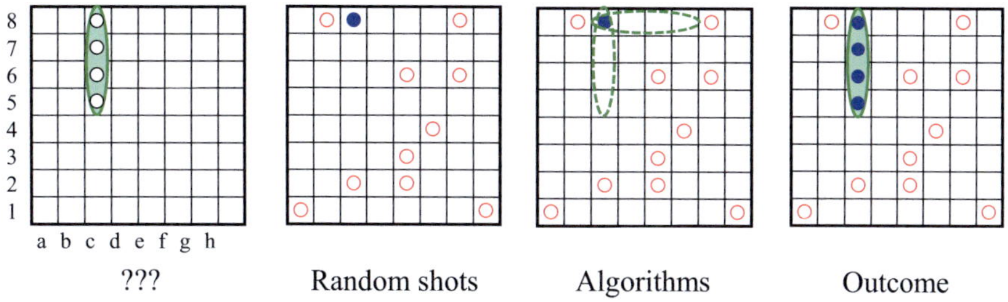

Fig. 15.2 The Monte Carlo method can be illustrated as a game of battleship. Assume that the unknown location of a *four-dot* battleship on the left represents a structure to be found. First, a player makes some random shots (*empty red circles*) until the ship is hit once, shown as a *filled blue circle*. Next, the player applies algorithms (i.e., a battleship is in the *vertical* or *horizontal* direction). Finally, based on the outcome of the random sampling and the algorithm, the player determines the location of the other player's ship, thus solving the structure (The figure and the caption are adopted with modifications from https://en.wikipedia.org/wiki/Monte_Carlo_method)

Bragg reflections, on the other, affect both the quality and quantity of the available diffraction data and therefore cause problems in both cases that are similar in many ways. As a result, the conventional phase-angle determination methods may become and often turn out to be ineffective.

Therefore, novel techniques potentially applicable to solving crystal structures are under continuous testing and development. A collective monograph by David, Shankland, McCusker, and Baerlocher[5] on the structure determination from powder diffraction data provides an excellent discussion of the problem and introduces different approaches that may be used in the structure solution. Topical updates regarding these fast-developing methods can be found in latest reviews by David and Shankland,[6] and Černý and Favre-Nicolin.[7] In this chapter, unconventional structure solution methods are discussed only briefly to the extent needed to understand the material covered. These are: the genetic algorithm, maximum entropy, maximum likelihood, and simulated annealing (or parallel tempering) methods.

The direct space methods based on global optimization[8] generate the initial model randomly using Monte Carlo approach (see Fig. 15.2) or sampling based on the Boltzmann distribution, which is also known as the reverse Monte Carlo method. Optimization that follows is conducted to satisfy the so-called cost function (one or

[5] Structure determination from powder diffraction data. IUCr monographs on crystallography 13. W. I. F. David, K. Shankland, L.B. McCusker, and Ch. Baerlocher, Eds., Oxford University Press, Oxford (2002).

[6] W. I. F. David and K. Shankland. Structure determination from powder diffraction data. Acta Cryst. **A64**, 52 (2008). https://doi.org/10.1107/S0108767307064252

[7] R. Černý and V. Favre-Nicolin, Direct space methods of structure determination from powder diffraction: principles, guidelines and perspectives. Z. Kristallogr. **222**, 105 (2007). https://doi.org/10.1524/zkri.2007.222.3-4.105

[8] More details about the global optimization method and its various algorithms can be found at https://www.mathworks.com/help/gads/what-is-global-optimization.html

several), which in this case is a powder diffraction pattern, either a whole profile or extracted integrated intensities, or simply the difference between the observed and calculated patterns expressed as R_p or R_B (see Eqs. 15.18 and 15.20, respectively, in Sect. 15.6.2 for the definition of both parameters). Other cost functions, such as potential energy, atomic coordination including bond valence sum,[9] Van der Waals distances, and others can be used in addition to the diffraction data. As a matter of fact, the direct-space optimization can be performed even without the diffraction data, which allows predicting the crystal structure by using optimal geometry and/or minimum energy as cost functions, while varying the unit cell parameters and the atomic coordinates.[10] Moreover, these methods can be and are used for ab initio structure determination of low-crystallinity materials, for example, nanoclusters, using a pair-distribution function.[11]

Genetic algorithm is the direct-space optimization method based on the evolution principle, in which only the members that fit best into the environment survive. The improved subsequent generation is obtained by considering the current state of a complex system and events that are equivalent to mating, mutation, and natural selection. In powder diffraction, the fit is defined as profile residual, R_P (see Sect. 15.6.2). The system, represented by a crystal structure, is split into fragments; each "survives" or "dies" depending on how it affects the fit (the cost function). In addition to structure solution,[12] this method can be and has been applied to unit cell determination from powder data (also see Sect. 14.8.1).[13] Another example is the application of the genetic algorithm method to solving protein structures by deconvoluting a Patterson function.[14]

Maximum entropy method is a powerful numerical technique, which is based on Bayesian estimation theory and is often applied to derive the most probable values of missing data. For example, rolling a dice with six faces numbered 1–6 gives equal probability to see each face up, $p_i = 1/6$, assuming a completely random distribution.

---

[9] I.D. Brown, D. Altermatt, Bond-valence parameters obtained from a systematic analysis of the inorganic crystal structure database. Acta Cryst. **B41**, 244 (1985). https://doi.org/10.1107/S0108768185002063

[10] A. Le Bail, Inorganic structure prediction with GRINSP. J. Appl. Cryst. **38**, 389 (2005). https://doi.org/10.1107/S0021889805002384

[11] P. Juhás, D.M. Cherba, P.M. Duxbury, W.F. Punch, S.J.L. Billinge, Ab initio determination of solid-state nanostructure. Nature (London) **440**, 655 (2006).

[12] K.D.M. Harris, R.L. Johnston and B.M. Kariuki, The genetic algorithm: Foundations and applications in structure solution from powder diffraction data, Acta Cryst. **A54**, 632 (1998). https://doi.org/10.1107/S0108767398003389; K. Shankland, B. David, and T. Csoka, Crystal structure determination from powder diffraction data by the application of a genetic algorithm, Z. Kristallogr. **212**, 550 (1997). https://doi.org/10.1107/S0108767307064252

[13] B.M. Kariuki, S.A. Belmonte, M.I. McMahon, R.L. Johnston, K.D.M. Harris, and R.J. Nelmes, A new approach for indexing powder diffraction data based on whole-profile fitting and global optimization using a genetic algorithm, J. Synchrotron Rad. **6**, 87 (1999). https://doi.org/10.1107/S0909049598017932

[14] G. Chang and M. Lewis, Using genetic algorithms for solving heavy-atom sites, Acta Cryst. **D50**, 667 (1994). https://doi.org/10.1107/s0907444994000727

In this case, the average value of multiple observations is (1 + 2 + 3 + 4 + 5 + 6)/6 = 3.5. The problem is in finding the probability of events (faces up) when the distribution is not random and only the average is known, for example, when the average is 2.5. The maximum entropy method results in the highest unbiased unique probability of events (e.g., faces up in the case of a dice) using the entropy function, $\sum p_i \log p_i$. This method, similar to the genetic algorithm approach, finds applications in various fields of scientific analysis, especially in image processing and reconstruction. In powder diffraction, the maximum entropy technique is successfully employed both to restore lost phase angles[15] and to determine the relative intensities of overlapped reflections.[16] Crystal structure determination using the maximum entropy approach solves the phase problem and, as a result, is considered a reciprocal space method.

Maximum likelihood method, similar to maximum entropy, works in reciprocal space and results in finding a raw model that has the best chance to be improved by applying small steps to achieve full agreement between the observed and calculated structure amplitudes,[17] rather than the final structure. Maximum likelihood and maximum entropy techniques are often combined.[18]

Various energy minimization methods that operate in direct space can be employed in order to optimize positions, orientations, and conformations of molecules or structural fragments. The challenge lies in the presence of multiple local minima along the path to the global energy minimum, which requires the initial structure model to be close to the global minimum. This makes the search for the lowest energy exceedingly slow. Different approaches and optimization functions may be employed to speed up the process. Typically, the potential energy of a system is used in combination with some other criteria.

The **simulated annealing** method is used to explore local minima quickly by generating multiple trial models using Monte Carlo or grid-search methods. Simulated annealing resembles the physical process of annealing by virtually heating a sample to a certain temperature, and it includes several control parameters defining the search of the global minimum that are analogous to the real annealing process. They are the initial temperature, the rate of its decrease, and the magnitude of random atomic jumps. The initial temperature should be set high, somewhere near

---

[15] C.J. Gilmore, Maximum entropy and Bayesian statistics in crystallography: a review of practical applications, Acta Cryst. **A52**, 561 (1996). https://doi.org/10.1107/S0108767396001560

[16] W. Dong and C.J. Gilmore, The ab initio solution of structures from powder diffraction data: the use of maximum entropy and likelihood to determine the relative amplitudes of overlapped reflections using the pseudophase concept, Acta Cryst. **A54**, 438 (1998). https://doi.org/10.1107/S0108767398000063

[17] A.J. Markvardsen, W.I.F. David, and K. Shankland, A maximum-likelihood method for global-optimization-based structure determination from powder diffraction data, Acta Cryst. **A58**, 316 (2002). https://doi.org/10.1107/S010876730200510X

[18] G. Bricogne, A multisolution method of phase determination by combined maximization of entropy and likelihood. III. Extension to powder diffraction data, Acta Cryst. **A47**, 803 (1991). https://doi.org/10.1107/S010876739100541X

the melting point; if it is too low, no changes occur, or they occur very slowly and when it is too high, the structure may "melt" and become amorphous.[19]

A modification of the simulated annealing method known as **parallel tempering** optimizes or follows several configurations in parallel. It is faster when applied to complex structures and does not require the use of usual trial parameters such as the "initial temperature" and the "annealing rate." In general, parallel tempering method avoids being trapped in the false local minima in the parameters space.[20] Simulated annealing methods can also be used for solving structure in reciprocal space by annealing reflection phases[21] as well as for unit cell determination.

Recently developed **dual-space methods** appear to be working in both reciprocal and direct space. Starting reflection phases are generated using either random sampling or Patterson function and are used in the following Fourier transformation to compute the electron density distribution. The latter is modified using certain rules and used to calculate new reflection phases, which optionally may also be modified. This process is recycled until a certain set of criteria is met. The obtained structure model is then analyzed and stored, or if no improvement is detected, a new set of starting phases is generated. We note that this algorithm mimics a typical pathway of completing the structure determination. However, the latter is performed in an interactive fashion and modifies electron density by improving the structural model by adding missing or removing incorrectly placed atoms, while the dual-space methods are automatic and modify the electron density itself, not the atomic structure. Sheldrick[22] introduced the sphere of influence algorithm to predict potential atomic positions. The electron density in such positions is changed to zero if it is negative or sharpened if it is positive. This method was developed for determination of macromolecular structures using single-crystal data. It is insensitive to noise resulting from the rapid truncation of the high Bragg angle data because of their low intensity. It is also insensitive to the presence of disordered solvent molecules and therefore has been successfully applied to solve large crystal structures and/or when using low-resolution data, as well as to structure determination from powder diffraction data.

---

[19] A.A. Coelho, Whole-profile structure solution from powder diffraction data using simulated annealing, J. Appl. Cryst. **33**, 899 (2000) https://doi.org/10.1107/S002188980000248X

[20] V. Favre-Nicolin, R. Ĉerný, FOX, "free objects for crystallography": a modular approach to ab initio structure determination from powder diffraction. J. Appl. Cryst. **35**, 734 (2002) https://doi.org/10.1107/S0021889802015236

[21] G. M. Sheldrick, Phase Annealing in SHELX-90: Direct methods for larger structures. Acta Cryst. **A46**, 467 (1990), https://doi.org/10.1107/S0108767390000277

[22] G.M. Sheldrick, Macromolecular phasing with SHELXE. Z. Kristallogr. **217**, 644 (2002), https://doi.org/10.1524/zkri.217.12.644.20662; G.M. Sheldrick. A short history of SHELX. Acta Cryst. **A64**, 112 (2008), https://doi.org/10.1107/S0108767307043930

Another dual-space method, the **charge flipping algorithm**,[23] modifies the electron density below a certain threshold by changing its sign (flipping the charge). The charge flipping algorithm has several benefits: it is truly ab initio; electron density is represented by a grid rather than by atoms. Hence, no information about atom types, composition, charge, etc. is required. This includes symmetry, which is set to the lowest P1, and actual symmetry is determined later based on the relationship between the reflection phases. In addition to periodic structures, this method can be used to solve commensurately modulated and quasicrystalline structures.[24] It has been successfully employed in structure determinations from powder diffraction data.[25] For example, the crystal structure of a new polymorph of iodic acid, δ-HIO_3, was determined using the charge flipping method with SHELXT[26] software, based on the integrated intensities obtained from a Le Bail full-pattern decomposition.[27]

Practically all nontraditional methods for solving crystal structures have been initially developed for either powder or single-crystal diffraction data to manage intrinsic incompleteness or poor quality that cannot be improved experimentally. Despite a variety of structure solution approaches, traditional direct phase determination methods, being ab initio methods, are commonly and successfully used when powder diffraction data are adequate.[28] Patterson methods also work quite well but they also require the presence of a heavy atom (or atoms) in addition to adequate data quality and, perhaps, more extensive crystallographic expertise. Nontraditional methods undergo rapid developments in both theory and implementations, and therefore, they are becoming common and are successfully and more often applied to structure determinations from powder diffraction data. They are irreplaceable, when dealing with low-quality and/or poorly resolved powder diffraction patterns,

[23] G. Oszlányi, A. Süto, The charge flipping algorithm. Acta Cryst. **A64**, 123 (2008), https://doi.org/10.1107/s0108767307046028; L. Palatinus, The charge flipping algorithm in crystallography, Acta Cryst. **B69**, 1 (2013). https://doi.org/10.1107/S2052519212051366

[24] L. Palatinus, G. Chapuis, SUPERFLIP—a computer program for the solution of crystal structures by charge flipping in arbitrary dimensions, J. Appl. Cryst. **40**, 786 (2007). https://doi.org/10.1107/S0021889807029238

[25] A.A. Coelho, A charge-flipping algorithm incorporating the tangent formula for solving difficult structures, Acta Cryst. **A63**, 400 (2007), https://doi.org/10.1107/S0108767307036112; A.A. Coelho, TOPAS-Academic V7 (2020), http://www.topas-academic.net/

[26] G.M. Sheldrick, SHELXT—Integrated space-group and crystal-structure determination, Acta Cryst. A71, 3 (2015). https://doi.org/10.1107/S2053273314026370

[27] T. Wu, P.Y. Zavalij, and M.R. Zachariah, Crystal structure of a new polymorph of iodic acid, δ-HIO_3, from powder diffraction, Powder Diffraction, **32**, 261 (2017), https://doi.org/10.1017/S0885715617000859

[28] C. Giacovazzo, Direct methods and powder data: state of the art and perspectives, Acta Cryst. **A52**, 331 (1996), https://doi.org/10.1107/S0108767395013651

substantial peak overlap due to low symmetry or pseudosymmetry,[29] large and complex structures including aperiodic and quasicrystalline solids, low-crystallinity materials, and nanomaterials. A recent demonstration that even a crystal structure of a protein can be solved from powder data is worth noting.[30] The crystal structure of a small protein, consisting of 67 residues with 554 protein atoms and 101 water molecules within a unit cell that has volume of 64,879 $Å^3$ was determined by molecular replacement—the technique common in protein crystallography. The model was further refined using maximum likelihood refinement.[31] Further, these methods, being less demanding to the data quality and largely automated, are increasingly applied to routine periodic structures as well.

15.1.4 Validation and Completion of the Model

Regardless of how the model was created, it must be validated and/or completed by computing the electron density distribution, or, if neutron diffraction data are available, the nuclear density distribution[32] within the unit cell to confirm the placement of known atoms and locate missing atoms, if any (see Sects. 10.1 and 10.2). At this stage of the structure solution process, the observed structure factors—and thus the intensities of individual Bragg peaks—are required. Fourier map calculations may be repeated several times using observed structure amplitudes and calculated phase angles, incorporating additional atoms found from previous distributions of electron (nuclear) densities, and adjusting the coordinates of the existing atoms to align with the corresponding peaks on the Fourier maps.

A crystal structure solution does not end with the development of a plausible model. Once the model is fully constructed, multiple structural and profile parameters should be refined[33] to achieve the best-possible agreement between the observed and calculated powder diffraction patterns—in other words, between the

[29] An example of a large structure solved using simulated annealing methods in the presence of pseudo-symmetry in: I.R. Evans, J.A.K. Howard, J.S.O. Evans, α-$Bi_2Sn_2O_7$—a 176 atom crystal structure from powder diffraction data, J. Mater. Chem. **13**, 2098 (2003), https://doi.org/10.1039/B305211G

[30] R. B. Von Dreele, P. W. Stephens, G. D. Smith and R. H. Blessing, The first protein crystal structure determined from high-resolution X-ray powder diffraction data: a variant of T3R3 human insulin-zinc complex produced by grinding, Acta Cryst. **D56**, 1549 (2000), https://doi.org/10.1107/S0907444900013901

[31] I. Margiolaki, J.P. Wright, M. Wilmanns, A.N. Fitch, N. Pinotsis, Second SH3 domain of ponsin solved from powder diffraction, J. Am. Chem. Soc. **129**, 11,865 (2007), https://doi.org/10.1021/ja073846c

[32] Nuclear density determined from neutron diffraction describes distribution of the scattering power rather than the distribution of the nuclei in the unit cell.

[33] Especially when powder data are complex, it may be necessary to use Rietveld technique to re-determine individual intensities before all structural details can be established using Fourier or differential Fourier maps.

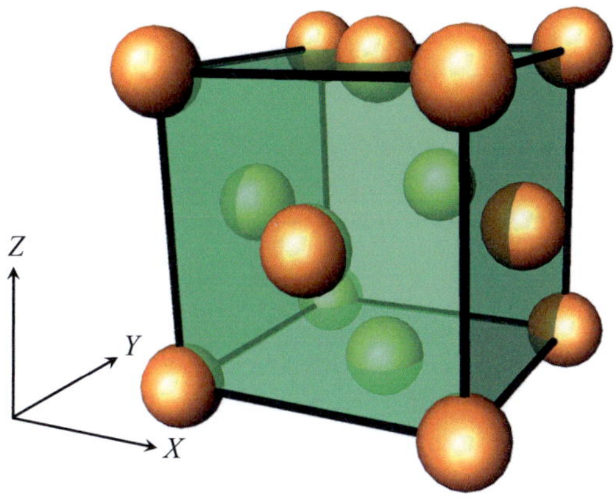

Fig. 15.3 One unit cell in the crystal structure of elemental copper illustrating that a point located in a corner contributes ⅛ of an atom and a point located on a face contributes ½ of an atom to the content of the unit cell. Similarly, a point located on an edge would contribute ¼ of an atom. Thus, the overall content of this unit cell is 8 × ⅛ + 6 × ½ = 4 Cu atoms

crystal structure and its observed reciprocal space image. Apparently, the crystal structure must be both chemically (e.g., connectivity, oxidation states, charge balance, valence, coordination, etc.) and physically (e.g., interatomic distances, valence and torsion angles, coordination polyhedra, etc.) reasonable. Rietveld refinement, discussed in this chapter, is a crucial step in both the validation and completion of the model.

15.2 The Content of the Unit Cell

Consider Fig. 15.3, which illustrates the crystal structure of elemental copper. If the lattice parameters are known, so is the volume of its unit cell. Furthermore, if the total number of atoms in the unit cell is known, the gravimetric density of a material can be easily calculated by dividing the mass of all atoms located in one unit cell by its volume.

The inverse calculation is also possible: the content of the unit cell may be determined from the known chemical composition, lattice parameters, and density of a material. Assume that the total mass of all atoms located in one unit cell is m. Also, assume that the unit cell volume is V. The latter is known from diffraction analysis as soon as lattice parameters have been established (see Eq. 14.42). Thus, provided the gravimetric density (ρ) of the crystalline material has been measured, the mass of one unit cell can be easily calculated:

$$m = \rho V \tag{15.1}$$

It is usually the case that the chemical composition of a material is known as the stoichiometry of its molecule, or in general, as its formula unit. Molecular mass or formula unit mass, M, is given as

$$M = \sum_{j}^{n} x_j A_j \tag{15.2}$$

where n is the number of different atom types in a molecule or in a formula unit, x_j is the total number of atoms of type j, and A_j is the molar mass of atoms of type j. The number of molecules or formula units, Z, in the unit cell is, therefore, found from

$$Z = \frac{m}{M} = \frac{\rho V}{M} \tag{15.3}$$

Equation 15.3 may be transformed into a more useful form, where the mass of the formula unit is given in a.m.u., the unit cell volume is in Å^3, and the density is in g/cm^3:

$$Z = \frac{\rho V / 10^{24}}{M / 6.023 \times 10^{23}} = 0.6023 \frac{\rho V}{M} \tag{15.4}$$

For example, the experimentally measured density of copper is 8.92 g/cm^3, the unit cell dimension of its cubic unit cell is a = 3.615 Å, and the molecular mass of a formula unit is 63.55 a.m.u., which is the molar mass of copper, one atom per formula unit. Thus, Eq. (15.4) results in Z = 3.99 ≅ 4 atoms per unit cell. The same equation may be used to calculate the density of a material when its crystal structure has been established. It is worth noting that the computed value of the material's gravimetric density is known as the X-ray density, and it is usually slightly higher than the measured density because real materials always have some defects and porosity that are not accounted in Eq. (15.4).

This method of calculating the contents of the unit cell requires experimentally measured density,[34] which is not always available. Nonetheless, even when the gravimetric density cannot be measured, Eq. (15.4) still can be used, especially when the chemical composition of the material is known or when working with molecular compounds. In these cases, Z can be estimated from restrictions imposed by symmetry and from approximately known range of densities for a specific class of compounds. For example, Eq. (15.4) is easily rearranged as

[34]The two methods, most commonly used to measure gravimetric density, are pycnometric and flotation. In a pycnometric technique, the volume of the material is determined from the volume of a fluid displaced by the known mass of a material. In a flotation approach, a small particle, which can be a single crystal, is placed in a low-density fluid, where it sinks. A high-density fluid is then slowly added until neutral buoyancy of the particle is reached. The gravimetric density of the particle is then determined from the density of the mixture of two fluids, provided their amounts are known. Obviously, the two fluids should form an ideal solution, i.e., volume effects of mixing should be negligible.

$$\rho = 1.66\frac{ZM}{V} = 1.66\frac{nzM}{V} \tag{15.5}$$

where n is the minimum number of molecules per unit cell, which is usually the multiplicity of a general site position,[35] and z is a variable integer, which corresponds to the number of molecules with mass M in the asymmetric (1/nth) part of the unit cell. For instance, in the space group $P2_1/m$, n = 4, and by varying z, the acceptable value of density ρ can be achieved.

Consider tetraphenylphosphonium (TPP) tetravanadate, which crystallizes in the monoclinic crystal system in the space group $P2_1/c$.[36] Its unit cell volume is $V = 6806.6(6)$ Å^3. The molecular weight of $[(C_6H_5)_4P]_2V_4O_{11}$ is 1058.55 a.m.u. The density range for this class of compounds (not the easiest but certainly a good example) may be estimated between 1.2 and 1.6 g/cm^3 given the presence of both the metal-oxide core and large organic molecule. The multiplicity of the general site position in this space group is 4. The V_4O_{11} unit, however, can be centrosymmetric because it has four vanadium and ten oxygen atoms, plus an additional oxygen atom that can be located in the center of inversion, and there are two TPP molecules in the formula unit. Therefore, a minimum multiplier for the number of molecules per unit cell (n) should be set to $4/2 = 2$. Now, let's consider density while varying z. For z = 1 and z = 2, the densities of 0.516 and 1.033 g/cm^3 are unreasonably low. For z = 4, the density is 2.065 g/cm^3, which is too high. Therefore, z = 3 is the only feasible option, resulting in a density of 1.549 g/cm^3. This value falls within the expected range, giving a total of Z = nz = 6 formula units in the unit cell. Given that the multiplicity of the general site is 4, there are 1.5 formula units in the asymmetric part of the unit cell. Consequently, there should be 3 TPP ions in the general position, along with 6 V_4O_{11} formula units (24 vanadium atoms and 66 oxygen atoms) in the unit cell. This implies there are 6 vanadium atoms and 16 oxygen atoms in the general position with multiplicity 4 and 1 oxygen atom in special position with multiplicity 2.[37] This is consistent with the symmetry as in $P2_1/c$ space group, which has 4 special positions located at the inversion centers with a multiplicity of 2.[38]

[35] In general, n in Eq. (15.5) can be the multiplicity of a general or any special position or their sum. The latter makes it complicated but still keeps the solution integer with respect to z. Note that symmetry of any special position and the molecule should agree; remember, some special positions, i.e., those at the center of inversion where all three coordinates are fixed, can be occupied by only one atom.

[36] S. Sharma, A. Ramanan, P.Y. Zavalij, M.S. Whittingham, The first example of a novel one-dimensional cyclic tetrameric metavanadate: $[PPh_4]_2V_4O_{11}$, CrystEngComm, **4**, 601 (2002). https://doi.org/10.1039/b209979a

[37] Alternatively, it can be represented as 1 V_4O_{11} unit in a general position (n = 4) and 1 in a special position (n = 2), when considering V_4O_{11} building blocks instead of individual atoms.

[38] See Wyckoff Positions of Space Groups in Space-group symmetry at Bilbao Crystallographic Server (https://www.cryst.ehu.es/) or International Tables for Crystallography, Vol. A, Space-group symmetry, M. I. Aroyo, Ed., Second online edition, (2016). https://it.iucr.org/Ac/

Table 15.1 Pearson's symbols used to designate 14 types of Bravais lattices

Crystal system	Bravais lattice	First two parts of Pearson's symbol
Cubic	Primitive, P	*c*P
	Body-centered, I	*c*I
	Face-centered, F	cF
Tetragonal	Primitive, P	*t*P
	Body-centered, I	*t*I
Hexagonal/trigonal	Primitive, P	*h*P
	Rhombohedral, R	*h*R
Orthorhombic	Primitive, P	*o*P
	Base-centered, C	*o*C
	Body-centered, I	*o*I
	Face-centered, F	*o*F
Monoclinic	Primitive, P	*m*P
	Base-centered, C	*m*C
Triclinic (anorthic)	Primitive, P	*a*P

15.3 Pearson's Classification

As noted in Sect. 15.1, when the material of interest is an intermetallic alloy, the solution of its crystal structure may be simplified because intermetallics often form series of isostructural compounds. Stoichiometries of the majority of intermetallic phases are not restricted by "normal" valence and oxidation states of atoms and ions. Therefore, crystal structures of metallic alloy phases are conveniently coded using the classification suggested by W.B. Pearson,[39] where each type of the crystal structure is assigned a specific code (symbol), which is constructed from three components as follows:

- The first position in a structure type symbol is occupied by a small letter designating the crystal system of the material: *c* for *c*ubic, *t* for *t*etragonal, *h* for *h*exagonal, trigonal, and rhombohedral, *o* for *o*rthorhombic, *m* for *m*onoclinic, and *a* for triclinic (*a*northic).
- The second position in the symbol is occupied by a standard notation of Bravais lattice. Thus, the first two elements in the Pearson's symbol are letters and they classify all available alloy structures according to 14 Bravais lattices, as shown in Table 15.1.
- The third (and last) position in Pearson's symbols is occupied by the total number of atoms located in one unit cell of the compound.

[39] W.B. Pearson, Handbook of lattice spacings and structures of metals, Vol. 2, Pergamon Press, New York (1967); W.B. Pearson, The crystal chemistry and physics of metals and alloys, Wiley-Interscience, New York (1972).

For example, considering the crystal structure of copper, which has cubic face-centered lattice (Fig. 15.3) and a total of four atoms in the unit cell, its Pearson's symbol is *c*F4. On the other hand, if the material has Pearson's symbol *o*I32, this means that its crystal structure is orthorhombic, and one body-centered unit cell contains a total of 32 atoms.

Pearson's classification is insensitive to both chemical compositions and stoichiometries of metallic alloys. It is quite useful because all known intermetallic crystal structures are grouped according to their structural symbols, which are quite simple. Thus, once the symmetry and the content of the unit cell of a new alloy phase have been established, it only makes sense to search for potentially isostructural compounds among those that have identical Pearson's symbols.

Tens of thousands of intermetallic phases have been systematized and classified using Pearson's symbols. They are listed in a source commonly known as Pearson's Handbook.[40] The handbook also provides detailed information about the coordinates of atoms in unit cells of all known structure types of metals, alloys, and related phases, which makes it a valuable tool in the structure solution of metallic materials.

15.4 Finding Structure Factors from Powder Diffraction Data

When the ab initio solution of a crystal structure is attempted from powder diffraction data using reciprocal space methods, eventually integrated intensities and structure factors of individual Bragg reflections are required, see Fig. 15.1. A simple numerical integration (Eq. 8.40) is rarely applicable; intensities are almost always determined after decomposition (or deconvolution) of partially overlapped Bragg reflections. Sometimes, this decomposition is performed peak-by-peak or group-by-group, as was described in Chap. 13. However, more commonly, individual observed structure factors are determined by using the so-called full-pattern decomposition techniques.

In addition to determining individual observed structure factors, full-pattern deconvolution serves several supplementary functions:

- First, to verify the correctness of the indexing solution or helps select one among several possible indexing solutions. This can be easily established by visually comparing the observed and calculated patterns to ensure that every observed peak has a corresponding calculated Bragg reflection in addition to a simple comparison of profile residuals R_p or R_{wp} (see Eqs. 15.18 and 15.19 in Sect. 15.6.2).

[40] Pearson's handbook crystallographic data for intermetallic phases. Desk ed./P. Villars, ASM International (1997). https://openlibrary.org/books/OL3732017M/Pearson's_handbook

- Second, it allows for precise determination of the unit cell dimensions without the need for semimanual profile fitting (see Chap. 13), which may be biased when there is a significant peak overlap.
- Third, it is used to obtain reasonable peak-shape parameters, including background function, which can be further employed in the structure solution methods that utilize the full pattern as a cost function, e.g., the same as one used in simulated annealing using FOX software, see Chap. 25.
- Fourth, it helps estimate the best figures of merit (see Sect. 15.6.2), achievable in a Rietveld refinement using the existing set of diffraction data, as well as the structure-only related residual R_B (or R_F^2), see Eq. (15.20) in Sect. 15.6.2.

The two related full-pattern decomposition methods in common use today were suggested by Pawley[41] and by Le Bail et al.[42] Pawley's approach is based on Eq. (8.21) and when $K\alpha_1/K\alpha_2$ doublets are present, full-pattern decomposition is performed by solving the following system of equations using a least-squares minimization:

$$\begin{aligned} Y_1 &= b_1 + \sum_{k=1}^{m} I_k[y_k(x_k) + 0.5y_k(x_k + \Delta x_k)] \\ Y_2 &= b_2 + \sum_{k=1}^{m} I_k[y_k(x_k) + 0.5y_k(x_k + \Delta x_k)] \\ &\dots \\ Y_n &= b_n + \sum_{k=1}^{m} I_k[y_k(x_k) + 0.5y_k(x_k + \Delta x_k)] \end{aligned} \tag{15.6}$$

The notations used in Eq. (15.6) are identical to Eq. (8.21). Individual integrated intensities (I_k) are treated as free least-squares parameters. Peak-shape function parameters are represented as described in Sect. 8.5.1, and Bragg peak positions, which affect the values of x_k, are established by the unit cell dimensions, see Sect. 8.4. The background, b_i, where $1 \leq i \leq n$, and n is the total number of measured data points, is modeled by one of the several functions described in Chap. 13 (see Eqs. 13.1–13.6).

When peak-shape functions and their parameters, including Bragg reflection positions, are known precisely and the background is modeled by a polynomial function with j coefficients, the solution of Eq. (15.6) is trivial because all equations are linear with respect to the unknowns (B_j, see Eq. 13.1, and I_k). It facilitates the use of the linear least-squares algorithm described in Sect. 14.12.1. In practice, it is nearly always necessary to refine both peak-shape and lattice parameters in addition

[41] G.S. Pawley, Unit-cell refinement from powder diffraction scans, J. Appl. Cryst. **14**, 357 (1981). https://doi.org/10.1107/S0021889881009618

[42] A. Le Bail, H. Duroy, and J.L. Fourquet, Ab initio structure determination of $LiSbWO_6$ by X-ray powder diffraction, Mat. Res. Bull. **23**, 447 (1988). https://doi.org/10.1016/0025-5408(88)90019-0; This method is commonly known as Le Bail extraction.

to B_j and I_k to achieve a better precision of the resultant integrated intensities. Thus, a nonlinear least-squares minimization technique (see Sect. 15.5) is usually employed during full-pattern decomposition using Eq. (15.6).

Pawley's method works best when the complexity of a powder diffraction pattern is relatively low, with a few hundred or so Bragg peaks in total. However, when the number of independent Bragg peaks (m) exceeds several thousand, the size of the least-squares normal equation matrix increases proportionally to m^2. As a result, Pawley's algorithm may become unstable, particularly when there are multiple low-intensity Bragg reflections and/or severe peak overlap. This instability is quite common when the total number of measurable Bragg peaks exceeds approximately 1000 in the scanned range. In the case of complex patterns, the algorithm often struggles to keep all integrated intensities, I_k, nonnegative, as indicated by Eq. (15.6). This is due to the strong correlation in least-squares refinement between the integrated intensities of Bragg reflections that are severely overlapped (located too close to each other). To address this issue, the refined integrated intensity can be constrained to remain positive.

Le Bail's approach is also based on Eq. (15.6) and it differs from the Pawley's method in that the individual intensities remain unaltered during each least-squares cycle.[43] They are extracted from a total observed intensity of the pattern between the least-squares cycles after subtraction of the background. The extraction is performed using a decomposition approach that has been employed in the Rietveld method since its development,[44] in which intensity observed in every point of the powder pattern is divided among different reflections in proportion to their calculated intensities:

$$y_{k,i}^{obs} = p_{k,i}\left(Y_i^{obs} - b_i\right), \text{where } p_{k,i} = \frac{y_{k,i}^{calc}}{\sum_{k=1}^{m} y_{k,i}^{calc}} \tag{15.7}$$

Thus, by removing integrated intensities from the least-squares refinement, the Le Bail method effectively resolves the issue of negative intensities.

In Eq. (15.7), $y_{k,i}^{obs}$ is the pseudo-observed intensity of the kth reflection in the ith point, $p_{k,\,i}$ is the fractional contribution of the kth reflection to the ith point, Y_i^{obs} is the observed total intensity in the ith point, and $y_{k,i}^{calc}$ is the calculated intensity of the kth reflection in the ith point. The main difference between Rietveld and Le Bail decompositions is in the calculated intensities. The former technique uses intensities computed from the model of the crystal structure, while the latter approach uses intensity obtained from the previous cycle during the decomposition. Initially, all

[43] A. Le Bail, Whole powder pattern decomposition methods and applications: A retrospection, Powder Diffraction, **20**(4), 316 (2005), https://doi.org/10.1154/1.2135315

[44] H.M. Rietveld, Line profiles of neutron powder-diffraction peaks for structure refinement, Acta Cryst. **22**, 151 (1967). https://doi.org/10.1107/S0365110X67000234; H.M. Rietveld, A profile refinement method for nuclear and magnetic structures, J. Appl. Cryst. **2**, 65 (1969). https://doi.org/10.1107/S0021889869006558

"calculated" intensity values in the Le Bail's method are set to arbitrary identical quantities, typically unity.

It is worth noting that in the Le Bail's decomposition, the number of free least-squares variables becomes independent of the number of Bragg reflections and only background, peak-shape, and lattice parameters are refined during each least-squares cycle. A small inconvenience of Le Bail's approach is that the unit cell dimensions should be known with a greater precision than in Pawley's method. It also takes more least-squares refinement cycles to complete, but the fit converges in much more complex cases. The disadvantage of the approach suggested by Le Bail is the separate handling of intensities for $K\alpha_1$ and $K\alpha_2$ peaks during the decomposition according to Eq. (15.7) and correction for the proper intensity ratio after the decomposition, which increases the time required for the completion of the fitting process. As far as suitable software is of concern, it is essential to check the manual and find out how the presence of the $K\alpha_1/K\alpha_2$ doublet is handled because some computer codes do not address this issue and the resulting intensities of $K\alpha_1$ and $K\alpha_2$ components may become unreasonable.

A variety of software, such as GSAS-II and TOPAS, allows users to deconvolute powder diffraction pattern to determine lattice and peak-shape function parameters, as well as individual intensities and, consequently, the observed structure factors of all possible Bragg reflections for further use in crystal structure determination.

15.5 Nonlinear Least Squares

Both full-pattern decomposition and Rietveld refinement rely on the nonlinear least-squares minimization of the differences between the observed and calculated profiles. Therefore, a brief overview of the nonlinear least-squares method is provided here.

Assume that we are seeking the best solution to a system of n simultaneous equations with m unknown parameters ($n \gg m$), where each equation is a nonlinear function of the unknowns $x_1, x_2, \ldots, x_m$. In a general form, this system of equations can be represented as

$$
\begin{aligned}
&f_1(x_1, x_2, \ldots, x_m) = y_1 \\
&f_2(x_1, x_2, \ldots, x_m) = y_2 \\
&\ldots \\
&f_n(x_1, x_2, \ldots, x_m) = y_n
\end{aligned}
\tag{15.8}
$$

Obviously, a linear least-squares algorithm described in Sect. 14.12.1 is not directly applicable to find the best solution of Eq. (15.8). In some instances, it may be possible to convert each equation in Eq. (15.8) into a linear form by appropriate substitutions of variables and thus reduce the problem to a linear case. In general, the least-squares solution of Eq. (15.8) is obtained by expanding the left-hand side of every equation using Taylor series and truncating the expansion after the first partial derivatives of the respective functions.[45] Hence, Eq. (15.8) may be converted into:

$$\begin{aligned}
&\frac{\partial f_1\left(x_1^0,\ldots,x_m^0\right)}{\partial x_1}\Delta x_1+\ldots+\frac{\partial f_1\left(x_1^0,\ldots,x_m^0\right)}{\partial x_m}\Delta x_m\cong y_1-f_1\left(x_1^0,\ldots,x_m^0\right)\\
&\frac{\partial f_2\left(x_1^0,\ldots,x_m^0\right)}{\partial x_1}\Delta x_1+\ldots+\frac{\partial f_2\left(x_1^0,\ldots,x_m^0\right)}{\partial x_m}\Delta x_m\cong y_2-f_2\left(x_1^0,\ldots,x_m^0\right)\\
&\ldots\\
&\frac{\partial f_n\left(x_1^0,\ldots,x_m^0\right)}{\partial x_1}\Delta x_1+\ldots+\frac{\partial f_n\left(x_1^0,\ldots,x_m^0\right)}{\partial x_m}\Delta x_m\cong y_n-f_n\left(x_1^0,\ldots,x_m^0\right)
\end{aligned}\tag{15.9}$$

as long as the corresponding derivatives exist and are finite. Equation (15.9) is linear with respect to Δx_1, Δx_2, …, Δx_n and its solution is obtained by applying conventional linear least-squares technique as

$$\Delta\mathbf{x}=\left(\mathbf{A}^{\mathbf{T}}\mathbf{W}\mathbf{A}\right)^{-1}\left(\mathbf{A}^{\mathbf{T}}\mathbf{W}\mathbf{y}\right)\tag{15.10}$$

The notations in Eq. (15.10) are as follows:

$$\Delta\mathbf{x}=\begin{pmatrix}x-x_1^0\\x-x_2^0\\\ldots\\x-x_m^0\end{pmatrix}=\begin{pmatrix}\Delta x_1\\\Delta x_2\\\ldots\\\Delta x_m\end{pmatrix}\tag{15.11}$$

$$\mathbf{y}=\begin{pmatrix}y_1-f_1\left(x_1^0,\ldots,x_m^0\right)\\y_2-f_2\left(x_1^0,\ldots,x_m^0\right)\\\ldots\\y_n-f_n\left(x_1^0,\ldots,x_m^0\right)\end{pmatrix}\tag{15.12}$$

[45] Taylor series is an expansion of a real function about a point. In the case of a function of one variable, $f(x)$, the expansion about a point $x = x^0$ is given as

$$f(x)=f\left(x^0\right)+\Delta x f'\left(x^0\right)+\frac{\Delta x^2}{2!}f''\left(x^0\right)+\ldots+\frac{\Delta x^n}{n!}f^n\left(x^0\right)+R_n$$

where $\Delta x = (x - x^0)$ and R_n is a remainder. When $f(x_1, x_2, \ldots x_k)$ is a function of k variables, the expansion about a point $(x_1^0, x_2^0, \ldots x_k^0)$ is obtained by substituting each derivative in the equation above with the sum of partial derivatives multiplied, respectively, by Δx_1, Δx_2, …, Δx_k taken to the appropriate power and divided by the corresponding $n!$. Brook Taylor (1685–1731) was an English mathematician best known for representing a function as an infinite sum of terms—Taylor series. See https://en.wikipedia.org/wiki/Brook_Taylor for a brief biography.

$$\mathbf{W} = \begin{pmatrix} w_1^2 & 0 & \dots & 0 \\ 0 & w_2^2 & \dots & 0 \\ \dots & \dots & \dots & \dots \\ 0 & 0 & \dots & w_n^2 \end{pmatrix} \tag{15.13}$$

where $\mathbf{W}$ is a square matrix representing individual weights (w_i) for each of the available n data points, and $\mathbf{A}^\mathbf{T}$ is the transpose of $\mathbf{A}$.

The refined parameters are computed by using both the set of the original $x_1^0, x_2^0, \dots x_m^0$, which represents the initial approximation of the unknowns, and the vector $\Delta\mathbf{x}$, which has been obtained from least squares (Eq. 15.10), as:

$$\mathbf{x} = \mathbf{x}^0 + \Delta\mathbf{x} = \begin{pmatrix} x_1^0 + \Delta x_1 \\ x_2^0 + \Delta x_2 \\ \dots \\ x_m^0 + \Delta x_m \end{pmatrix} \tag{15.14}$$

The standard uncertainty[46] for each refined parameter according to the least-squares method is calculated from

$$\sigma(x_j) = \sqrt{\frac{(\mathbf{A}^\mathbf{T}\mathbf{WA})_{jj}^{-1} \sum_{i=1}^{n} w_i (y_i)^2}{n - \mathrm{m}}}, \quad j = 1, \dots, m \tag{15.15}$$

where

- n is the number of equations in Eq. (15.9),
- m is the number of unknown parameters in Eq. (15.9),
- $(\mathbf{A}^\mathbf{T}\mathbf{WA})_{jj}^{-1}$ is the corresponding diagonal element of the inverse normal equation matrix,
- w_i is the corresponding weight,
- y_i is the corresponding element of the vector $\mathbf{y}$.

The major differences between the nonlinear least-squares technique and the linear least-squares method, described in Sect. 14.12.1, are as follows:

- The substitution of the original Eq. (15.8) with Eq. (15.9) requires the knowledge of initial (i.e., approximate) values of parameters to be refined, which are represented by the set $x_1^0, x_2^0, \dots x_m^0$. That is why the crystal structure must be solved before it can be refined.
- The least-squares solution (Eq. 15.10) results in the shifts (vector $\Delta\mathbf{x}$, Eq. 15.11), which shall be added to the corresponding initial parameters, as shown in Eq. (15.14).

[46]Recently, the term "standard uncertainty" (s.u.) is becoming more common than the "standard deviation" (s.d.).

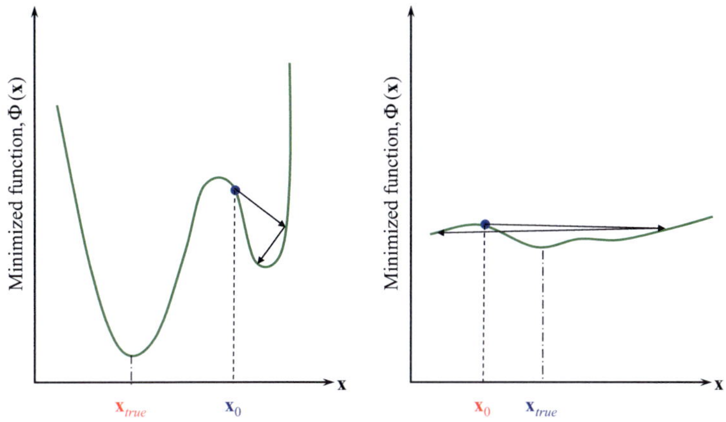

Fig. 15.4 Two examples when the nonlinear least-squares technique may fail in finding the best solution of Eq. (15.9): *left*—the initial approximation ($\mathbf{x}_0$) is located near a false minimum; *right*—the minimum is poorly defined. The *arrows* represent the possible outcomes of two least-squares cycles. In the case on the left, the minimization ends in a false minimum. In the case on the right, the obtained shifts have correct signs but wrong magnitudes and instead of converging (i.e., instead of all $|\Delta x_i|$ becoming smaller), their absolute values continue to increase. True solutions (i.e., global minima) are marked as $\mathbf{x}_{true}$

- Because Eq. (15.9) is not exact, usually more than 1 cycle of a least-squares refinement is necessary to achieve a full convergence: during the second and following least-squares cycles, the new set of parameters as obtained in the previous step from Eq. (15.14) is used as the initial approximation. Thus, nonlinear least-squares refinement is an iterative process, where the result of the next iteration depends on the result obtained during the prior iteration.
- Because of the iterative nature of nonlinear least squares, convergence may be difficult to achieve, especially when the initial approximation is far from correct (Fig. 15.4, left) or when the minimized function (see Eqs. 14.34, 14.35 and 15.9) is poorly defined. The latter often occurs when certain least-squares parameters correlate.[47] Instead of converging, nonlinear least squares may diverge and become unstable, as illustrated in Fig. 15.4 (right).

Therefore, various numerical conditioning techniques should be employed to improve both the convergence and stability of the method. Their detailed consideration exceeds the scope of this book, except for two commonly used approaches that are briefly mentioned below, and the reader is referred to a large amount of special literature covering this subject.

[47] Considering, for example, Eqs. (8.41 and 9.3), when the phase scale (K) is refined together with the population parameters of all atoms (g^j), a complete correlation results: an increase of the phase scale by a factor k ++ is completely offset by the reduction of all population parameters by the factor $\sqrt{k}$.

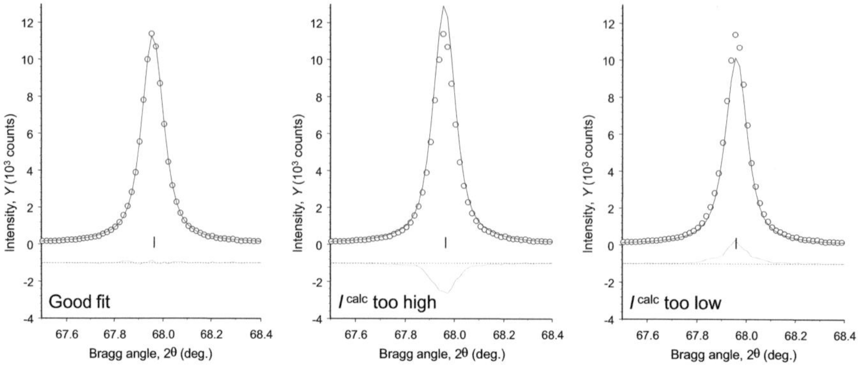

Fig. 15.5 The observed (*circles*), calculated (*solid lines*), and difference (*solid lines* at the *bottom*) profiles. The *dotted line* at the *bottom* corresponds to $Y_i^{obs} - Y_i^{calc} = 0$. The *short vertical lines* correspond to the calculated position of the Bragg peak. The *panel on the left* shows a good fit. The *panel in the middle* shows the result when the calculated integrated intensity is overestimated, while the *panel on the right* is for the underestimated integrated intensity

The simplest way is to proceed with smaller steps as obvious from Fig. 15.5, and a simple way of doing this is to modify the shifts Δx_i before applying them (Eq. 15.14) by the so-called damping factors $m_i{\cdot}\Delta x_i$, where $0 < m_i \leq 1$. The damping factors can be different for different parameters or group of parameters. This reduces the risk of jumping out of the minimum. Unfortunately, this applies to both the global and false minima: while damping factors can help stabilize the refinement, they may also trap the refinement in the false minimum.

Another approach is to modify the normal equation matrix $\mathbf{A^TWA}$ or, sometimes, its inverse as is done in a commonly used Marquardt damping (compare with Eq. 15.10):[48]

$$\Delta \mathbf{x} = \left(\mathbf{A^TWA} + \lambda \mathbf{D}\right)^{-1} \left(\mathbf{A^TWy}\right) \tag{15.16}$$

where $\boldsymbol{D}$ is usually either a diagonal of a normal equation matrix or the unity matrix, and λ is a damping factor. In practice, λ is set large at the beginning of the refinement to ensure stability and convergence and is later decreased by a user or automatically as the algorithm begins to converge. Ideally, but not necessarily, damping factor should be zero (or unity in GSAS, which uses 1 + λ as the input damping factor) during the final stages of refinement, which is the same as not using the damping. The advantage of this damping approach is that, if set correctly, only parameters that correlate are affected significantly.

[48] D. Marquardt, An algorithm for least squares estimation of nonlinear parameters: SIAM, J. Appl. Math. **11**, 431 (1963). https://doi.org/10.1137/0111030

The inverse of the normal equation matrix, $(\mathbf{A}^{\mathrm{T}}\mathbf{W}\mathbf{A})^{-1}$, may be used to evaluate the correlation coefficients (ρ_{ij}) among the pairs of free least-squares variables (x_i and x_j):

$$\rho_{ij} = \left(\mathbf{A}^{\mathrm{T}}\mathbf{W}\mathbf{A}\right)_{ij}^{-1} / \sqrt{\left(\mathbf{A}^{\mathrm{T}}\mathbf{W}\mathbf{A}\right)_{ii}^{-1}\left(\mathbf{A}^{\mathrm{T}}\mathbf{W}\mathbf{A}\right)_{jj}^{-1}} \tag{15.17}$$

The correlation coefficients vary from 0 to 1 (absolute) and when they are in the range from −0.5 to 0.5, the associated parameters show little to no correlation. When $|\rho_{ij}|$ is unity, the corresponding variables (x_i and x_j) are fully (100%) correlated and one of them should be excluded from the refinement (kept fixed). It is useful to check the matrix of the correlation coefficients during the refinement, especially when the nonlinear least-squares process appears unstable, which is usually detected as a continuous worsening of one or more numerical figures of merit (see Sect. 15.6.2) in addition to erratic changes of the values of some free variables included into the refinement. The analysis of correlation coefficients may help in identifying the nonobvious and serious problems and show which parameters strongly correlate, i.e., depend on each other, and therefore, decide which of them cannot be successfully refined together or provide a reason to invoke damping of correlations that are not severe.

15.6 Quality of Profile Fitting

Since both Pawley and Le Bail full-pattern decompositions are based on finding a least-squares solution of Eq. (15.6), the problem may be considered solved, and the pattern sufficiently deconvoluted when the best-possible fit between the calculated intensities, Y_i^{calc}, and the observed intensities, Y_i^{obs}, is achieved. In fact, the left-hand sides of each equation in Eq. (1.6) represent Y_i^{obs}, while the right-hand sides represent Y_i^{calc}.

It is, therefore, of utmost importance to have certain assessment tools in order to make a decision that the fitting process has converged to a true minimum, especially considering the iterative nature of the nonlinear least-squares algorithm. Two approaches are in common use today. The first one is visual examination of the difference $Y_i^{obs} - Y_i^{calc}$, which ideally must be zero or nearly zero for any $1 \le i \le n$, where n is the total number of the data points in the measured profile. Visual analysis requires some experience to quickly determine the most critical source(s) of potential mismatch between Y_i^{obs} and Y_i^{calc} and make adjustments to a selection of free least-squares variables. The second approach utilizes a set of numerical figures of merit that quantify the quality of the least-squares fit, which can also be used to estimate the reliability of both the fit and the extracted integrated intensities and observed structure factors. Unlike visual examination of data, numerical figures of merit provide no clues about potential source(s) of problems with the least-squares fit.

Therefore, visual examination of the difference plot should be used alongside numerical figures of merit to evaluate the coherence of the fit.

15.6.1 Visual Assessment of the Quality of Profile Fitting

As noted by McCasker et al.,[49] any issue in approximating the Bragg peak profile results in a characteristic difference in the $(Y_i^{obs} - Y_i^{calc})$ profile. Therefore, visual examination of the difference usually provides important clues about which additional parameters must be adjusted (refined). All examples found in this section have been obtained using X-ray powder diffraction data[50] collected from a sample of $HoIn_3$ compound. It crystallizes in cubic space group $Pm\overline{3}m$ with the unit cell dimension of ~4.57 Å. After the powder was prepared by grinding in an agate mortar with a pestle, it was screened through a 35 μm sieve and heat-treated at 600 ° C for 1 h to anneal strains. Experimental data have been collected using monochromatic Cu $K\alpha_1$ radiation on a Malvern Panalytical X'Pert Pro diffractometer equipped with Johansson monochromator. In every case, peak-shape function was pseudo-Voigt (Eq. 8.24) with asymmetry correction in Finger, Cox, and Jephcoat approximation (see Footnote 22 on page 199, Chap. 8).

Figure 15.5 illustrates a good fit of a single Bragg peak together with the other two fits in which computed integrated intensity has been either overestimated or underestimated. When intensity is too high, the difference has a characteristic minimum centered at the calculated position of the Bragg peak. A maximum is found at the calculated location of the Bragg peak when the computed integrated intensity is too low.

A different situation is illustrated in Fig. 15.6, where the discrepancies are due to problems in determining the location of the Bragg peak. When the calculated profile is displaced into the high Bragg angle region, the difference is a combination of an asymmetric peak followed by an asymmetric valley. When the displacement is toward the low Bragg angle, the profile of the difference is reversed, so that a valley is followed by a maximum. These types of discrepancies result from either an incorrect determination of the unit cell dimensions or from an improper treatment of sample-displacement or zero-shift errors.

Figure 15.7 illustrates the case when poor fits are related to wrong full widths at half maximum. When FWHM is too large, the difference profile exhibits a peak enclosed between two shallow asymmetric minima. When the full width at half maximum is too small, a sharp minimum is centered between two shallow asymmetric humps. The FWHM-related discrepancies are common when anisotropic peak broadening is present.

---

[49]L.B. McCusker, R.B. Von Dreele, D.E. Cox, D. Louër, and P. Scardi, Rietveld refinement guidelines, J. Appl. Cryst. **32**, 36 (1999). https://doi.org/10.1107/S0021889898009856

[50]Experimental data used in Figs. 15.5, 15.6, 15.7, and 15.8 are courtesy of Dr. Ya. Mudryk.

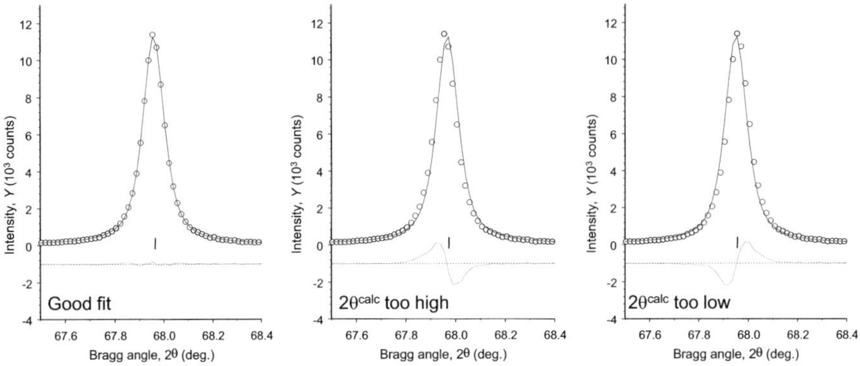

Fig. 15.6 The *panel on the left* shows a good fit. The *panel in the middle* shows the result when the calculated Bragg angle is overestimated, while the *panel on the right* is for the underestimated Bragg angle. See Fig. 15.5 for the explanation of notations

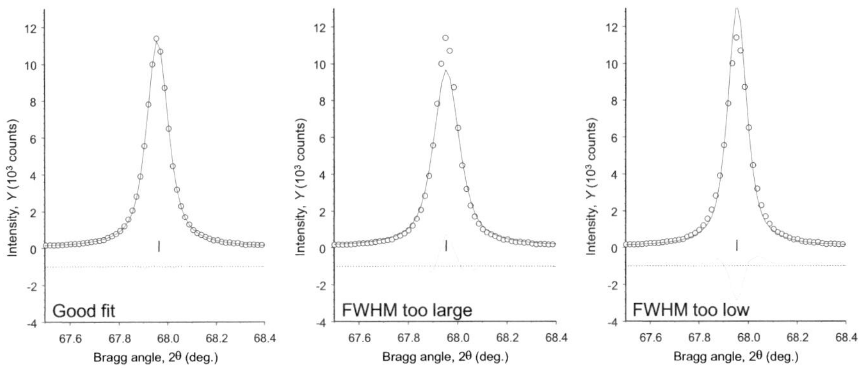

Fig. 15.7 The *panel on the left* shows a good fit. The *panel in the middle* shows the result when the calculated FWHM is overestimated, while the *panel on the right* is for the underestimated FWHM. See Fig. 15.5 for the explanation of notations

Finally, problems related to improperly accounted asymmetry are illustrated in Fig. 15.8. Here, when asymmetry has been overestimated, a shallow minimum in the difference profile is followed by a reasonable fit of the top of the peak, and then a shallow maximum, and when asymmetry has been underestimated, the locations of the minimum and the maximum are reversed.

We note that cases illustrated in Figs. 15.5, 15.6, 15.7, and 15.8 are due to problems caused by a discrepancy of one kind. In real least-squares fits of powder diffraction profiles, it is common to have a combination of reasons for a less-than-ideal fits. Wrong FWHM may be in addition to incorrect intensity, or wrong integrated intensity may be complicated by problems related to an incorrect approximation of peak asymmetry. Furthermore, when $K\alpha_1/K\alpha_2$ doublets are used to record powder diffraction patterns, difference profiles illustrated in Figs. 15.5, 15.6, 15.7, and 15.8 will be more complex. All the aforementioned issues complicate visual

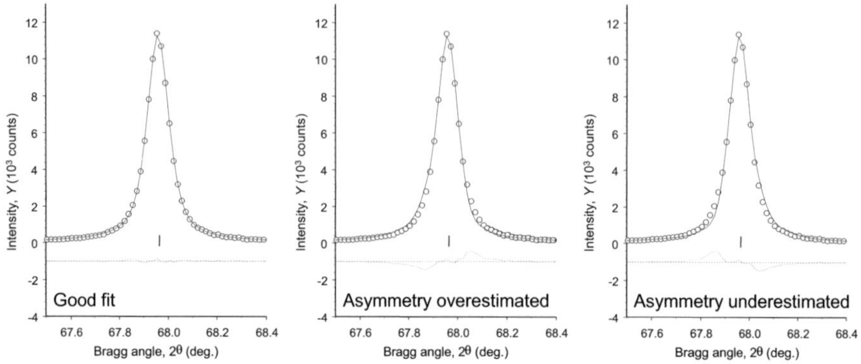

Fig. 15.8 The *panel on the left* shows a good fit. The *panel in the middle* shows the result when peak asymmetry is overestimated, while the *panel on the right* is for the underestimated asymmetry. See Fig. 15.5 for the explanation of notations

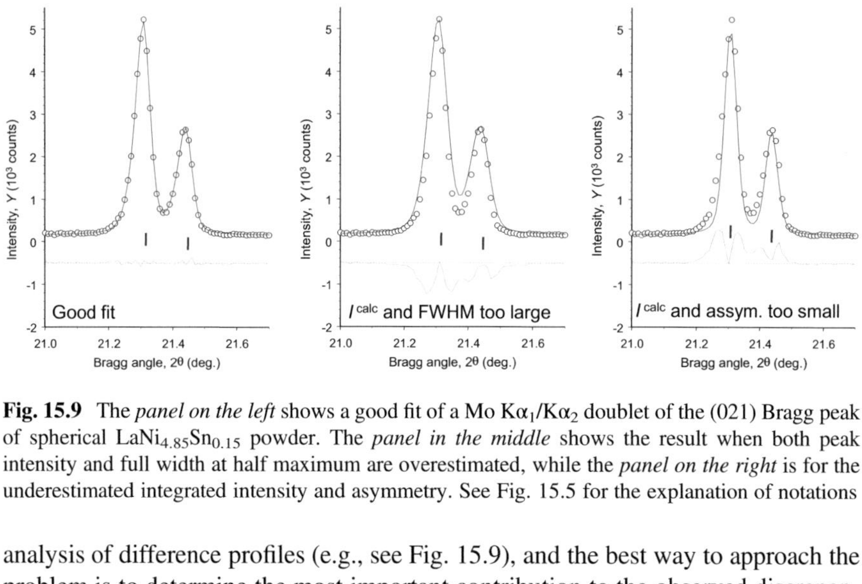

Fig. 15.9 The *panel on the left* shows a good fit of a Mo $K\alpha_1/K\alpha_2$ doublet of the (021) Bragg peak of spherical $LaNi_{4.85}Sn_{0.15}$ powder. The *panel in the middle* shows the result when both peak intensity and full width at half maximum are overestimated, while the *panel on the right* is for the underestimated integrated intensity and asymmetry. See Fig. 15.5 for the explanation of notations

analysis of difference profiles (e.g., see Fig. 15.9), and the best way to approach the problem is to determine the most important contribution to the observed discrepancies, and therefore, adjust the choice of most critical fitting parameters in the refinement process as early as possible.

15.6.2 Figures of Merit

The following figures of merit are customarily used to characterize both the full-pattern decomposition and Rietveld refinement quality.

The profile residual (or reliability) factor, R_p:

$$R_p = \frac{\sum_{i=1}^{n} \left| Y_i^{obs} - Y_i^{calc} \right|}{\sum_{i=1}^{n} Y_i^{obs}} \times 100\% \tag{15.18}$$

The weighted profile residual, R_{wp}:

$$R_{wp} = \left[\frac{\sum_{i=1}^{n} w_i \left(Y_i^{obs} - Y_i^{calc} \right)^2}{\sum_{i=1}^{n} w_i \left(Y_i^{obs} \right)^2} \right]^{1/2} \times 100\% \tag{15.19}$$

The Bragg residual, R_B (this figure of merit is quite important in Rietveld refinement but has no use during full-pattern decomposition because only observed Bragg intensities are meaningful in both Pawley and Le Bail methods):

$$R_B = \frac{\sum_{j=1}^{m} \left| I_j^{obs} - I_j^{calc} \right|}{\sum_{j=1}^{m} I_j^{obs}} \times 100\% \tag{15.20}$$

The expected profile residual, R_{exp}:

$$R_{exp} = \left[\frac{n - p}{\sum_{i=1}^{n} w_i \left(Y_i^{obs} \right)^2} \right]^{1/2} \times 100\% \tag{15.21}$$

The goodness of fit, χ^2 (sometimes referred as chi-squared):

$$\chi^2 = \frac{\sum_{i=1}^{n} w_i \left(Y_i^{obs} - Y_i^{calc} \right)^2}{n - p} = \left[\frac{R_{wp}}{R_{exp}} \right]^2 \tag{15.22}$$

In Eqs. (15.18–15.22), the following notations have been used:

- n is the total number of points measured in the powder diffraction pattern.
- Y_i^{obs} is the observed intensity of the ith data point.
- Y_i^{calc} is the calculated intensity of the ith data point.
- w_i is the weight of the ith data point, which is usually taken as $w_i = 1/\sigma_i = 1/Y_i^{obs}$ (see Eq. 12.8 in Sect. 12.3.1 for the definition of spread, σ_i).
- m is the number of independent Bragg reflections.
- I_j^{obs} is the "observed" integrated intensity of the jth Bragg peak, which has been calculated after Y_i^{obs} have been partitioned according to the calculated intensities of the contributing Bragg peaks (see Eq. 15.7).
- I_j^{calc} is the calculated integrated intensity of the jth Bragg peak.
- p is the number of free least-squares parameters.

All figures of merit except one (Eq. 15.20) include a contribution from the background. This raises an interesting point about the "reliability of profile reliability factors" (pardon the tautology) when the correctness of the structural model is of concern, which is further discussed in Chap. 16 onward.

The importance of taking the background out of the equation can easily be illustrated when the background is high, i.e., the same order of magnitude or higher than the intensity of Bragg peaks. This is the case when, for example, an iron-rich or a gadolinium rich (see Fig. 11.25) compound is measured using Cu Kα radiation without diffracted beam monochromatization, or a hydrogen (not deuterium) containing sample is studied by neutron diffraction (see Fig. 20.7). It is easy to see that denominators in Eqs. (15.18, 15.19, and 15.21) (but not in Eqs. 15.20 and 15.22) are proportional to the total count of the scattered intensity and therefore are highly affected by the level of the background. However, the numerators contain the differences of the observed and calculated intensities, and therefore, they are independent of the background provided the fit of the background is reasonable. Because of this, strong backgrounds lead to low residuals and even extremely poor fits may result in excellent profile residuals. The residuals become more realistic when the observed scattered intensity in the denominator is reduced by the contribution from the background. Thus, the profile residual with the background contribution subtracted[51] is given as follows:

$$R_{\mathrm{pb}} = \frac{\sum_{i=1}^{n}\left|Y_i^{obs} - Y_i^{calc}\right| \cdot \frac{\left|Y_i^{obs} - Y_i^{back}\right|}{Y_i^{obs}}}{\sum_{i=1}^{n}\left|Y_i^{obs} - Y_i^{back}\right|} \times 100\% \tag{15.23}$$

where Y^{back} refers to the background contribution to the profile and the corresponding weighted profile residual as:

$$R_{\mathrm{wpb}} = \left[\frac{\sum_{i=1}^{n} w_i\left(\left(Y_i^{obs} - Y_i^{calc}\right) \cdot \left(Y_i^{obs} - Y_i^{back}\right)/Y_i^{obs}\right)^2}{\sum_{i=1}^{n} w_i\left(Y_i^{obs} - Y_i^{back}\right)^2}\right]^{1/2} \times 100\% \tag{15.24}$$

A simple analysis of Eqs. (15.18–15.20 and 15.22) indicates that a better fit results in lower values of all residuals. Unfortunately, there are no exact thresholds for R_p, R_{wp}, and/or R_B, below which a fit is acceptable, good, or excellent. To a certain degree, the "absolute quality" of the result is established by the relationship between R_{wp} and R_{exp}, that is, by the value of χ^2. The expected residual (Eq. 15.21) characterizes the quality of experimental data because a larger denominator means a better counting statistic (see Sect. 12.3.1). The last figure of merit—the goodness of fit—should therefore approach unity when R_{wp} approaches R_{exp}. In practice, this is

[51] A.C. Larson and R.B. Von Dreele, General Structure Analysis System (GSAS), Los Alamos National Laboratory Report LAUR 86–748 (2004). https://subversion.xray.aps.anl.gov/trac/EXPGUI/wiki/GSASIntro

rarely achieved even with high-quality powder diffraction data when R_{exp} are quite low, which is usually due to the inadequacy of peak-shape and background functions.

Another useful figure of merit is the Durbin–Watson d-statistic:

$$d = \frac{\sum_{i=2}^{n}\left(\Delta Y_i/\sigma_i - \Delta Y_{i-1}/\sigma_{i-1}\right)^2}{\sum_{i=1}^{n}\left(\Delta Y_i/\sigma_i\right)^2} \quad (15.25)$$

where $\Delta Y_i = Y_i^{obs} - Y_i^{calc}$ and σ_i is the corresponding statistical error (see Eq. 12.8). According to Hill and Flack,[52] the weighted Durbin–Watson statistic may be used to provide a sensitive measure of refinement progress and to quantify serial correlation between adjacent least-squares residuals in a full-pattern decomposition or Rietveld refinement based on step-scan experimental data. It remains discriminating when other residuals fail, for example, when comparing the results obtained at different step sizes. Ideally, the d-parameter (Eq. 15.25) should be close to 2, thus indicating the absence of serial correlation, although in practice it often deviates considerably from this ideal value. In general, if correlation is insignificant, d is closer to 2 than another parameter, Q, which is defined as:

$$Q = 2\left[\frac{n-1}{n-p} - \frac{3.0902}{\sqrt{n+2}}\right] \quad (15.26)$$

where n and p are the number of experimental observations and free least-squares parameters, respectively. When $d < Q < 2$, serial correlation is positive and sequential observations have the same sign, while in the case of $d > 4 - Q > 2$, serial correlation is negative and sequential observations have opposite signs.[53]

15.7 The Rietveld Method

The determination of a crystal structure may be considered complete only when multi-pattern variables and the model's crystallographic parameters have been fully refined against the observed powder diffraction data. Importantly, the refined model should remain reasonable from both physical and chemical perspectives. The

[52] R.J. Hill and H.D. Flack, The use of the Durbin–Watson d-statistic in Rietveld analysis, J. Appl. Cryst. **20**, 356 (1987). https://doi.org/10.1107/S0021889887086485

[53] Ideally, experimental data (intensity) should be randomly distributed both above and below the calculated intensity profile. If there are multiple sequences with all observed points above or all below the calculated intensity values, it is said that serial correlation occurs. In other words, the d-statistic reflects correlation between adjacent least-squares residuals, and it can be used as an indicator that refined parameters are unbiased.

refinement technique most commonly employed today is based on the idea proposed by Rietveld in the mid-1960s.[54,55] The essence of Rietveld's approach is that experimental powder diffraction data are utilized without extraction of the individual integrated intensities or structure factors. Instead, all structural and instrumental parameters are refined by fitting a calculated profile to the observed data—in other words, by minimizing the difference between the observed and calculated profiles.

To a certain extent, the Rietveld method (also known as the full-pattern or the full-profile refinement) is similar to the full-pattern decomposition using Pawley and/or Le Bail algorithms, except that the values of the integrated intensities are no longer treated as free least-squares variables (Pawley) or determined iteratively after each refinement cycle (Le Bail).[56] They are included into all calculations as functions of relevant geometrical, specimen, and structural parameters (see Sect. 8.6 and Chap. 9).

Full-profile refinement is computationally intense and employs the nonlinear least-squares method (Sect. 15.5), which requires a reasonable initial approximation of many free variables. These usually include peak-shape parameters, unit cell dimensions, and coordinates of all atoms in the model of the crystal structure. Other unknowns (e.g., constant background, scale factor, overall atomic displacement parameter, and to greater extend peak-shape parameters, etc.) may be simply guessed at the beginning and then effectively refined as the least-squares fit converges to a global minimum. When either Le Bail's or Pawley's fitting is employed for full-pattern decomposition prior to Rietveld refinement, it is logical to use suitably determined relevant parameters—such as background, peak-shape, zero shift, sample displacement, and, of course, unit cell dimensions—as the initial approximation. If full-pattern fitting was not performed, it is advisable to do so, especially in complex cases (e.g., multiple phases or large unit cells) to obtain the best-possible starting value for the nonstructural parameters.

The successful practical use of the Rietveld method, while directly related to the quality of powder diffraction data (the higher the quality, the better the outcome), largely depends on the user's experience and ability to properly select a sequence in which various groups of parameters are refined. Despite the method's relatively long history, nearly everyone familiar with it has their own set of "unique" strategies for

[54] Hugo M. Rietveld (b. 1932). Dutch physicist who between 1964 and 1966 demonstrated that accurate determination of crystal and magnetic structures is possible using neutron diffraction data from powders. His approach was later extended to X-rays. At present, crystal structures of hundreds, if not thousands of polycrystalline materials, are studied and refined every year using the Rietveld method. In recognition of the "Rietveld method," the Royal Swedish Academy of Sciences awarded the author, Dr. Hugo M. Rietveld, the Aminoff Prize in 1995.

[55] The following two papers are considered seminal: H.M. Rietveld, Line profiles of neutron powder-diffraction peaks for structure refinement, Acta Cryst. **22**, 151 (1967), https://doi.org/10.1107/S0365110X67000234; H.M. Rietveld, A profile refinement method for nuclear and magnetic structures, J. Appl. Cryst. **2**, 65 (1969), https://doi.org/10.1107/S0021889869006558

[56] We introduce this analogy for clarity, even though both Pawley's and Le Bail's techniques were developed following Rietveld's work.

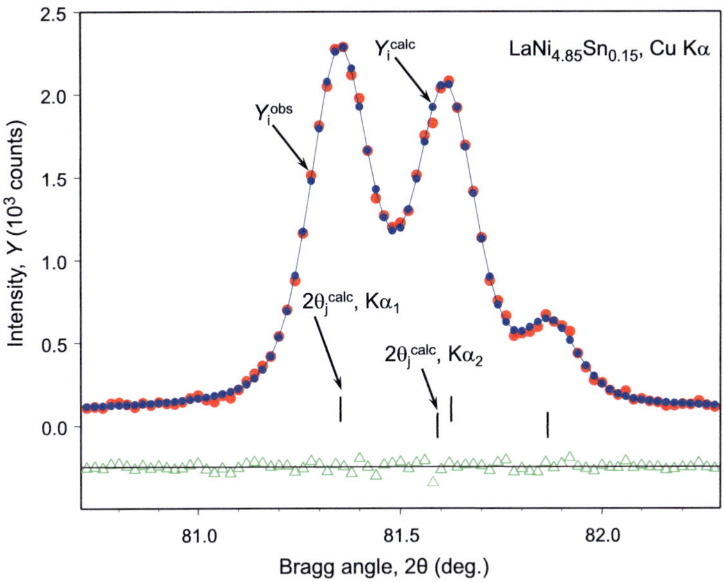

Fig. 15.10 A fragment of a powder diffraction pattern of $LaNi_{4.85}Sn_{0.15}$ collected on a Rigaku TTRAX rotating anode powder diffractometer using Cu Kα radiation with a step $\Delta 2\theta = 0.02°$. The observed scattered intensity (Y_i^{obs}) is represented by *red circles*; calculated intensity (Y_i^{calc}) is shown by *blue circles* connected with *thin solid line*. The differences between the observed and calculated intensities are depicted by *green triangles*. A *thick solid line* across the triangles corresponds to $Y_i^{obs} - Y_i^{calc} = 0$. The differences between the observed and calculated intensities are typically plotted on the same scale as Y_i^{obs} and Y_i^{calc}, with a constant vertical displacement for clarity. *Vertical tick marks* indicate the calculated positions of Bragg peaks

making the refinement stable, complete, and successful. Therefore, in the remainder of this chapter, we introduce the basic theory of Rietveld's approach. A series of hands-on examples demonstrating the solution and refinement of crystal structures with varying degrees of complexity is presented in the following chapters of this book. Each example is supplemented by actual experimental data available online, providing readers with numerous opportunities to follow our reasoning and develop and test their own strategies for successfully determining crystal structures from powder diffraction data.

Consider Fig. 15.10, which shows both the observed and calculated powder diffraction patterns of $LaNi_{4.85}Sn_{0.15}$ over a narrow range of Bragg angles. Assuming that:

- We have an adequate structural model that makes both physical and chemical sense.
- The model yields correct (i.e., close to experimentally observed) integrated intensities.
- We have a suitable peak-shape and background functions, the fully refined crystal structure of a material should result in a calculated powder diffraction pattern that

closely resembles the collected data. In other words, the difference between the measured and calculated powder diffraction profiles should approach zero. This fundamental concept, applied to the entire powder diffraction pattern, is the foundation of the Rietveld method.

The development of the Rietveld method, particularly the work that demonstrated its applicability to conventional X-ray powder diffraction data,[57] initiated a remarkable era that continues today. More and more complex crystal structures are now routinely solved and fully refined using the basic experimental technique of powder diffraction. Although the method's utility is somewhat limited by the one-dimensional nature of the data and the instrumental resolution, its power is astonishing. A simple property—the observed scattered intensity as a function of Bragg angle—combined with the computed powder diffraction profile provides sufficient evidence of the correctness of a crystal structure model.

No other technique, including the more sophisticated single-crystal diffraction method, matches the visual elegance of the Rietveld method. When applied properly, however, the Rietveld method competes in accuracy and often surpasses single-crystal diffraction, especially when the material is only available as a fine-grained powder, untextured thin film, or in other states unsuitable for single-crystal studies. Furthermore, with specimen size differing by several orders of magnitude—micrograms for single-crystal diffraction versus hundreds of milligrams for powder diffraction—the latter provides a much better representation of the material's bulk structure.

It is important to remember, however, that the Rietveld method requires an existing crystal structure model and does not provide guidance on how to create one from first principles. Therefore, the Rietveld technique is primarily a powerful refinement and optimization tool. It can also be used to uncover subtle structural details that may have been missed during a partial or complete ab initio structure solution, as demonstrated in examples found in the following chapters of this book.

Finally, the continuing development of Rietveld refinement methods, driven by the growing demand for precise structural information from complex structures, extends well beyond improving the accuracy and quality of routine structure determinations. The technique is becoming increasingly valuable in the following applications:

- The structures of complex organic and metal-organic compounds can be reliably determined using rigid-body description.

[57] J. Malmros and J.O. Thomas, Least-squares structure refinement based on profile analysis of powder film intensity data measured on an automatic microdensitometer, J. Appl. Cryst. **10**, 7 (1977), https://doi.org/10.1107/S0021889877012680; R.A. Young, P.E. Mackie, and R.B. Von Dreele, Application of the pattern-fitting structure-refinement method of X-ray powder diffractometer patterns, J. Appl. Cryst. **10**, 262 (1977), https://doi.org/10.1107/S0021889877013466; C.P Khattak and D.E. Cox, Profile analysis of X-ray powder diffractometer data: structural refinement of $La_{0.75}Sr_{0.25}CrO_3$. J. Appl. Cryst. **10**, 405 (1977). https://doi.org/10.1107/S0021889877013855

- The structures of small- and medium-sized proteins and other macromolecules can be refined using stereochemical restraints.
- Charge density distribution can be studied from powder diffraction data due to the greatly improved accuracy of powder diffraction experiments and profile simulations.[58,59]
- The structures of low-crystallinity and nano-crystalline materials can be refined using pair-distribution function.
- A wealth of noncrystallographic structural information can be extracted from Rietveld refinement, including reliable phase composition (such as amorphous content), grain size, microstrain information, texture analysis, and others.

15.7.1 Fundamentals of the Rietveld Method

During refinement using the Rietveld method, a system of equations is solved through nonlinear least-squares minimization:[60]

$$\begin{aligned} Y_1^{calc} &= kY_1^{obs} \\ Y_2^{calc} &= kY_2^{obs} \\ &\dots \\ Y_n^{calc} &= kY_n^{obs} \end{aligned} \tag{15.27}$$

Here, Y_i^{obs} is the observed, and Y_i^{calc} is the calculated intensity at point i of the powder diffraction pattern. The parameter k represents the pattern scale factor, which is typically set to $k = 1$ since scattered intensity is measured on a relative scale, and k is absorbed by the phase scale factor (see, for example, Eqs. 15.29 and 15.30). Lastly, n is the total number of the measured data points. Hence, a digital powder diffraction pattern, where the scattered intensity at each point is measured with high

[58] K. Tolborg, M.R.V. Jørgensen, S. Christensen, H. Kasai, J. Becker, P. Walter, A.-C. Dippel, J. Als-Nielsen, and B. B. Iversen, Accurate charge densities from powder X-ray diffraction—a new version of the Aarhus vacuum imaging-plate diffractometer, Acta Cryst. B73, 521 (2017), https://doi.org/10.1107/S2052520617006357

[59] K. Tolborg, M.R V. Jørgensen, S. Christensen, H. Kasai, J. Becker, P. Walter, A.-C. Dippel, J. Als-Nielsen, and B. B. Iversen, Accurate charge densities from powder X-ray diffraction—a new version of the Aarhus vacuum imaging-plate diffractometer, Acta Cryst. **B73**, 521 (2017), https://doi.org/10.1107/S2052520617006357

[60] In this section, we are concerned with an experiment, which consists of a single pattern. The Rietveld technique may also be used to conduct refinement of the crystal structure employing multiple patterns collected from the same material. For example, conventional X-ray data collected using different wavelengths, conventional and synchrotron X-rays, conventional or synchrotron X-rays, and neutron source may be used simultaneously in a combined Rietveld refinement. The fundamentals of the combined Rietveld refinement are briefly considered in Sect. 16.5.

accuracy, is essential for the successful implementation of this technique. In the Rietveld method, the minimized function Φ is given by

$$\Phi = \sum_{i=1}^{n} w_i \left(Y_i^{obs} - Y_i^{calc} \right)^2 \tag{15.28}$$

where w_i is the weight assigned to the *i*th data point and *k* (Eq. 15.27) is unity. The summation in Eq. (15.28) is carried over all measured data points, *n*.[61]

Considering Eq. (8.21) and taking into account the one-dimensionality of powder data, which introduces multiple Bragg reflection overlaps, the expanded form of Eq. (15.28) in the simplest case, when the powder is a single-phase crystalline material, becomes

$$\Phi = \sum_{i=1}^{n} w_i \left(Y_i^{obs} - \left[b_i + K \sum_{j=1}^{m} I_j y_j (x_j) \right] \right)^2 \tag{15.29}$$

for a single wavelength experiment or

$$\Phi = \sum_{i=1}^{n} w_i \left(Y_i^{obs} - \left[b_i + K \sum_{j=1}^{m} I_j \{ y_j (x_j) + 0.5 y_j (x_j + \Delta x_j) \} \right] \right)^2 \tag{15.30}$$

for dual wavelength ($K\alpha_1$ + $K\alpha_2$) data. In Eqs. (15.29 and 15.30), b_i is the background at the *i*th data point, *K* is the phase scale factor, *m* is the number of Bragg reflections contributing to the intensity of the *i*th data point, I_j is the integrated intensity of the *j*th Bragg reflection, $y_j(x_j)$ is the peak-shape function, and Δx_j is the difference in positions of $K\alpha_1$ and $K\alpha_2$ components in the doublet, and $x_j = 2\theta_j^{calc} - 2\theta_i$.

A simple analysis of these two equations indicates that the experimental minimization of the background, which generally contains little or no useful structural information, is of utmost importance for a successful outcome of a full-profile-based refinement. When the background is low, so that $b_i \ll K \sum I_j y_j(x_j)$, the functions given in both Eqs. (15.29 and 15.30) are defined by contributions from the integrated

[61] In a typical experiment, *n* varies from ~10^3 to ~10^4 data points. In powder diffraction, the value of *n* is determined only by the scanned Bragg angle range and by the data collection step, $\Delta 2\theta$: $n = (2\theta_{max} - 2\theta_{min})/\Delta 2\theta + 1$. It is unrelated to both the complexity of the crystal structure and the number of Bragg reflections, *M*, which can be observed between $2\theta_{min}$ and $2\theta_{max}$. Obviously, *M* is proportional to the number of symmetrically independent points in a "visible" fraction of the reciprocal lattice. A simple estimate based on the volumes of the Ewald's sphere and the unit cell of a reciprocal lattice results in $M \propto V/\lambda^3$, where *V* is the volume of the primitive unit cell in direct space and λ is the wavelength. Thus, as unit cell volume increases, the number of possible Bragg reflections (*M*) also increases, but the number of observations (*n*) in powder diffraction remains constant, provided experimental conditions (λ, $2\theta_{max}$, $2\theta_{min}$, and $\Delta 2\theta$) remain constant.

intensities and peak-shape parameters. Conversely, when the background is high, i.e., when $b_i \cong K \sum I_j y_j(x_j)$ and, especially when $b_i \gg K \sum I_j y_j(x_j)$, the function, which is minimized during a least-squares refinement, becomes nearly entirely dependent on the adequacy of the background and not the integrated intensities and peak shapes. Therefore, in general, a structural model cannot be satisfactorily refined using data collected in the presence of a large background.

In the absence of a background and assuming that the measured intensity is only affected by statistical errors (see Eqs. 12.8 and 12.9), the weight can be given as

$$w_i = \left[Y_i^{obs}\right]^{-1} \tag{15.31}$$

In practice, the weight is usually calculated without subtracting the background, which yet again emphasizes the importance to have the latter at its practical minimum.

When a powder diffraction pattern is collected from a material, which is a mixture of several (p) phases, the contribution from every crystalline phase is accounted for by modifying Eqs. (15.29 and 15.30) as follows:

$$\Phi = \sum_{i=1}^{n} w_i \left(Y_i^{obs} - \left[b_i + \sum_{l=1}^{p} K_l \sum_{j=1}^{m} I_{l,j} y_{l,j}(x_{l,j}) \right] \right)^2 \tag{15.32}$$

$$\varPhi = \sum_{i=1}^{n} w_i \left(Y_i^{obs} - \left[b_i + \sum_{l=1}^{p} K_l \sum_{j=1}^{m} I_{l,j} \{ y_{l,j}(x_{l,j}) + 0.5 y_{l,j}(x_{l,j} + \Delta x_{l,j}) \} \right] \right)^2 \tag{15.33}$$

Considering Eqs. (15.32 and 15.33), it is clear that each additional crystalline phase adds multiple Bragg peaks plus a new scale factor along with a set of corresponding peak-shape and structural parameters into the nonlinear least squares. Even though mathematically they are easily accounted for, the finite accuracy of measurements and the limited resolution of even the most advanced powder diffractometer usually result in lowering the quality and stability of the Rietveld refinement in the case of multiple-phase samples. Thus, when the precision of structural parameters is of concern, it is best to work with single-phase materials, where Eqs. (15.29 and 15.30) are applicable. Conversely, since individual scale factors may be independently determined, Rietveld refinement of multiple-phase powder diffraction patterns offers an opportunity for a quantitative analysis of a mixture or a multiple-phase crystalline material.[62]

[62] In this text, we are not specifically concerned with quantitative phase analyses of multiple phase mixtures, except for a single example considered in Chap. 16. Interested reader is referred to an excellent overview given by R.J. Hill, Data collection strategies: fitting the experiment to the need, in: R.A. Young, Ed., The Rietveld method, Oxford University Press, Oxford, New York (1993).

15.7.2 Classes of Rietveld Refinement Parameters

Analytical expressions for the background (Eqs. 13.1–13.6), integrated intensity (Eqs. 8.41–8.47, 8.51–8.65, and 9.1–9.22), and peak shape (Eqs. 8.22–8.39) have been considered earlier. The minimum of the corresponding function, defined by one of the relevant formulae (Eqs. 15.29, 15.30, 15.32, or 15.33) can be found using a nonlinear least-squares technique (see Eqs. 15.8–15.14). Thus, the following groups of independent least-squares parameters may be and are usually refined using the Rietveld method:

(a) 1–12 parameters representing the background, although there may be as many as 36.
(b) Sample-displacement, sample transparency, or zero-shift corrections.
(c) Peak-shape function parameters, which usually include full width at half maximum (FWHM) as a function of 2θ, asymmetry, and other relevant variables, which depend on the type of a function selected to represent a peak shape. In a multi-phase diffraction pattern, these parameters may be constrained to be identical or refined independently for each identified phase, except typically for the asymmetry, which is common to all phases. Asymmetry is mainly a function of the axial and in-plane divergence of the X-ray beam (see Fig. 8.15). Refining these parameters independently is warranted if supported by the quality of the data and significant differences in the physical states of the phases in the specimen.
(d) Unit cell dimensions, usually from 1 to 6 independent parameters for each crystalline phase present in the specimen.
(e) Preferred orientation, and if necessary, absorption, porosity, and extinction parameters, which usually are independent for each phase.
(f) Scale factors, one for each phase (K_l) in all multiple sets of powder diffraction data. However, the ratio of scale factors for the same phases across different datasets may be constrained to be identical, particularly when the same sample is measured using difference wavelengths.
(g) Positional parameters of all independent atoms in the model of the crystal structure of each crystalline phase, usually from 0 to 3 per atom.
(h) Population parameters, if certain site positions are occupied partially or by different types of atoms simultaneously, usually one per site.[63]
(i) Atomic displacement parameters, which may be treated as an overall displacement parameter (one for each phase or a group of atoms) or individual atomic displacement parameters, with the number of independent variables between one (isotropic approximation) and six (anisotropic approximation) per site.

The least-squares parameters listed in items (a) through (d) are the same in Pawley and Le Bail full-pattern decomposition and in Rietveld refinement. Other variables,

[63] When more than two types of atoms occupy the same site, more than one variable per site may be adjusted. However, these cases are usually extremely difficult to refine sensibly.

Table 15.2 A suggested parameter turn-on sequence in a single-phase, single-pattern Rietveld refinement using constant wavelength X-ray or neutron data (adopted from Young)

Parameter or group of parameters	Linear in Eqs. (15.29–15.33)	Stable	Comment	Sequence
Phase scale	Yes	Yes	[a]	1
Specimen displacement	No	Yes	[b]	1
Linear background	Yes	Yes	[c]	2
Lattice parameters	No	Yes	[d]	2
More background	Generally no	Fairly	[c]	2 or 3
W	No	Poorly	[e]	3 or 5
Coordinates of atoms	No	Fairly	[f]	3
Preferred orientation	No	Fairly	[f]	4, if at all
Population and isotropic displacements	No	Varies	Correlated	5
U, *V*, other profile parameters	No	No	[e]	Last
Anisotropic atomic displacement parameters	No	Varies		Last, if at all
Zero shift	No	Yes	[b]	1, 5, if at all

[a] When the scale factor is far off, or when the model of the crystal structure is wrong or too far from reality, the refined scale factor may become incorrect
[b] The specimen displacement parameter typically varies from sample to sample and often accounts for effects such as sample transparency and similar factors. With a properly aligned goniometer, the zero-shift error should be negligible
[c] When the background is significant, at least its constant component should be estimated and included from the start. In a polynomial approximation, the background parameters are linear. However, using more than necessary can lead to severe correlations between them
[d] When one or more lattice parameters are incorrect, one or more calculated Bragg peaks may lock onto an incorrect observed peak, leading to a false minimum
[e] *U*, *V*, and *W* tend to be highly correlated, and various combinations of their significantly different values can produce very similar peak widths
[f] When coordinates of all atoms are refined, the plot of the observed and calculated diffraction data should be used to determine whether preferred orientation should be included. Coordinates may strongly correlate in the presence of pseudosymmetry

specifically those listed in items (e) through (i), are unique to the Rietveld method. While it is nearly impossible to prescribe a universal, rigid order for including various groups of physically distinct parameters in a Rietveld refinement, the most common turn-on sequence for activating parameters—based on their importance and potential for least-squares instability, as suggested by Young[64]—is illustrated in Table 15.2.

The turn-on sequence described in Table 15.2 may be and often is altered depending on various factors, including data quality, the accuracy of the initial structural model, and knowledge of the instrumental contributions to profile

[64] R.A. Young, Introduction to the Rietveld method, in: R.A. Young, Ed., The Rietveld method, Oxford University Press, Oxford, New York (1993).

parameters. It is important to recognize that it is rarely, if ever, possible to refine all relevant variables simultaneously from the start, due to the complexity of the problem and the risk of an unstable least-squares process caused by strong, often severe, correlation between refined parameters. As Young noted, a systematic, one-by-one turn-on sequence is almost always the most effective approach to identify which parameter is causing the trouble when the refinement does proceed as expected.

15.7.3 Restraints, Constraints, and Rigid Bodies

We already mentioned that all chemical and physical information must be considered in order to succeed and/or speed up the process of solving the crystal structure from powder diffraction data. The same is true for Rietveld refinements, especially when diffraction data are of low resolution and/or the structures are complex. Only relatively simple structures can be refined using good data without any or with minimal restrictions. Instead of employing numerical tools in order to stabilize the refinement and obtain a sensible result (e.g., using damping, see Sect. 15.5) that are not related to the chemical knowledge, the expected structural features can be used to control the stability of the refinement by enhancing the powder diffraction data with crystallographic and chemical information. This is especially important and often needed when dealing with structural complexity exacerbated by limited diffraction data. The result is the increased ratio of the available to the resultant information, in other words, the greater number of observations per refined parameter.

Obviously, structural and chemical restrictions biasing the least-squares refinement must be used carefully because if incorrect information is fed in, there is no chance for a correct outcome. We note that this approach is extensively used in single-crystal refinements, not only when dealing with complex macromolecular or disordered structures but also in routine structures when handling hydrogen atoms in groups with well-known configurations. In both single-crystal and powder structure refinements, the chemical and structural information is introduced using the mechanism of the so-called *restraints*, *constraints*, and *rigid bodies*, which are described here to the extent necessary to familiarize the reader with the meaning and possible applications.

In common use, the words "restraint" and "constraint" are nearly synonymous, but in crystallography, they have distinct meanings and are used for different purposes. Specifically, restraints and constraints in structure refinement are different since they describe different kinds of restrictions imposed on the structural parameters. Thus, *restraints*[65] mean approximate limitations or relationships between

[65] "Those who restrain desire, do so because theirs is weak enough to be restrained." William Blake.

parameters and therefore are sometimes called *soft restraints* or *soft constraints*. In contrast, *constraints*[66] impose *strict limitations* or *exact relationships* between specific parameters. In practice, constraints reduce the number of free variables, while restraints effectively increase the number of observations (see Eq. 15.34), thereby stabilizing the nonlinear least-squares refinement.

15.7.3.1 Restraints

Restraints are defined by a user based on known chemical and structural information. Each type of restraints is included in the refinement as a set of observations (a histogram in GSAS), in addition to the main set (histogram), i.e., the intensities (Y) of the experimental profile. The most comprehensive application of restraints is implemented in GSAS as described in detail in the corresponding manual.[67] Several common types of restraints are shown in Eq. (15.34) that represents a combined minimization function.

$$\begin{aligned}\Phi = &\sum_{i=1}^{N^Y} w_i^Y \left(Y_i^{obs} - Y_i^{calc}\right)^2 + f^c \sum_{i=1}^{N^c} w_i^c \left(c_i^{obs} - c_i^{calc}\right)^2 \\ &+ f^\delta \sum_{i=1}^{N^\delta} w_i^\delta \left(\delta_i^{exp} - \delta_i^{calc}\right)^2 + f^\alpha \sum_{i=1}^{N^\alpha} w_i^\alpha \left(\alpha_i^{exp} - \alpha_i^{calc}\right)^2 \\ &+ f^p \sum_{i=1}^{N^p} w_i^p \left(-p_i^{calc}\right)^2 + f^P \sum_{i=1}^{N^P} w_i^P \left(-P_i^{calc}\right)^2 \\ &+ f^z \sum_{i=1}^{N^z} w_i^z \left(-z_i^{calc}\right)^2 + \ldots\end{aligned} \tag{15.34}$$

where Y are profile intensities; f is weight factor; w is weight of each restraint. The restrained type of parameters are as follows: c—chemical composition or charge balance, δ—distances, α—valence angles, p—displacement from a plane, P—positive Pawley extracted integrated intensity, and z—positive pole figure.

The **weight factors** (f^j, $j = c, \delta, \ldots$) set for each type of the restraint are needed to scale the restraints with one another and with the experimental profile. In other words, they define how important restraints are relatively to the pattern, which always has unity weight. Usually, f^j are set large at the beginning and then are gradually decreased as the refinement converges.

The **weights** (w^j, $j = c, \delta, \ldots$) are defined as $1/\sigma^2$ with σ being expected or desired deviation from the average value of the restrained parameter (r^{calc}), so that r is kept within the $r^{calc} - \sigma \leq r \leq r^{calc} + \sigma$ range. In reality, the refined parameters falling or not within the desired range depend on the value of the weight factor for a particular restraint and also on the correctness of the selected restrained model. Therefore, all

[66] "Software constraints are only confining if you use them for what they're intended to be used for." David Byrne.

[67] A.C. Larson and R.B. Von Dreele, General Structure Analysis System (GSAS), Los Alamos National Laboratory Report LAUR 86-748 (2004). The GSAS manual is available at https://subversion.xray.aps.anl.gov/EXPGUI/gsas/all/GSAS%20Manual.pdf

deviations from the expected r^{calc} values must be thoroughly analyzed after the refinement is completed.

Restraints on the composition (c) are typically imposed when a chemical element is statistically disordered in two or more sites, as often happens in naturally occurring minerals. For a particular element, the chemical composition is calculated as a sum over all n involved sites:

$$c^{calc} = \sum_{i=1}^{n} s_i m_i g_i \tag{15.35}$$

where g is the fractional site occupation, m is the site multiplicity, and s is the multiplier. When s is set to unity, the content of a particular chemical element is restrained to match the known chemical composition. Compositional restraint(s) can also be imposed using valence or ionic charge to balance the electroneutrality of the unit cell.[68] In this case, s is set to charge (valence) of the chemical element occupying a particular site.

Restraints on interatomic distances (δ) and bond angles (α) are applied most often compared to other common types of restraints since these are readily available and their expected values are usually known quite well, for example, for simple inorganic ions and organic molecules and groups. Examples of imposing the distance and angle restraints for a tetrahedral tetramethylammonium ion can be found in Sect. 21.3.2, and restrained refinement of acetaminophen is described in Sect. 25.3.

When a molecule or a functional group is flat, the root mean square (RMS) displacements of atoms from the plane (p) can be restrained to be within $\pm\sigma$ from zero. RMS, which is also known as the quadratic mean, is calculated as:

$$x_{RMS} = \sqrt{\frac{1}{n}\sum_{i=1}^{n} x_i^2} \tag{15.36}$$

Restraints can also be applied to profile and other parameters . Thus, integrated intensities extracted using Pawley method (P) can be restrained to ensure that they remain positive, as is done in GSAS when negative intensities that result from the least-squares Pawley extraction (see Sect. 15.4) are restrained to be zero, while positive intensities remain unaffected. Yet another restraint (z) is designed to keep pole figure positive (since a negative value is unphysical) by restricting coefficients of the spherical harmonics in the preferred orientation correction (see Eq. 8.62 and the remainder of Sect. 8.6.6), or in other corrections that are described by spherical harmonics.

More restraints are introduced as more complex structures and diffraction patterns are analyzed. For example, torsion angle pseudopotentials (single and coupled),

---

[68] See the online tutorial "Setting charge balance restraint in GSAS," which is available at http://ccp14.cryst.bbk.ac.uk/solution/gsas/charge_balance_restraint.html

chiral volumes, Van der Waals distances, and hydrogen bond lengths[69] restraints were used together with all of the restraints mentioned here in the first work describing refinement of a macromolecular structure using Rietveld method.[70] In complex protein structures, the number of introduced restraints can be extremely large, significantly increasing the total number of observations. In a recent report, combined Rietveld and stereochemical refinement of hen egg white lysozyme protein complex was performed using 5750 measured data points and a total of 5391 restraints, nearly doubling the number of experimental observations. This was enough to fit as many as 3279 parameters.[71]

15.7.3.2 Constraints

Constraints are defined as exact relationships between least-squares parameters. There are several types of constraints: those imposed by symmetry, user-defined constraints, and the rigid-body constraints.

Symmetry constraints are dictated by symmetry relationships and are automatically imposed by most crystallographic applications, often without being noticed. Thus, lattice symmetry establishes specific relationships between the unit cell dimensions as shown in Table 2.10. Point-group symmetry of a site in a special position (see Sect. 3.6) establishes certain relationships between the coordinates of an atom occupying the site as was illustrated in Fig. 3.8. Site symmetries higher than 1 also impose constraints on the anisotropic atomic displacement parameters. For example, $U_{22} = U_{11}$ for an atom located on a fourfold axis; $U_{22} = U_{11}$ and $U_{12} = ½\ U_{11}$ for an atom located on a three- or sixfold axis. The relationships between the atomic coordinates can easily be obtained from symmetry operations of the corresponding special site, and these are listed in the International Tables for Crystallography, but constraints applicable to atomic displacement parameters are not as straightforward. However, numerous structure refinement applications (both single crystal and powder) automatically deduce and impose these constraints, and therefore, the algorithm will not be discussed here.

Lattice symmetry also imposes constraints on other parameters, not related to the crystal structure. For example, coefficients of spherical harmonics used to model complex preferred orientations[72] (Sect. 8.6.6) and peak broadening due to

[69] We note that for Van der Waals contacts and hydrogen bonds only minimum distances are restrained.

[70] R.B. Von Dreele, Combined Rietveld and stereochemical restraint refinement of a protein crystal structure, J. Appl. Cryst. **32**, 1084 (1999), https://doi.org/10.1107/S002188989901064X

[71] R.B. Von Dreele, Binding of N-acetylglucosamine oligosaccharides to hen egg-white lysozyme: a powder diffraction study. Acta Cryst. **D61**, 22 (2005), https://doi.org/10.1107/S0907444904025715

[72] Note that constraints on spherical harmonics coefficients imposed by symmetry are different from restraints that are meant to ensure physically meaningful pole figure (Eq. 15.34).

anisotropic stress (Eq. 8.35 in Sect. 8.5.1). In both cases, applications such as GSAS automatically apply required constraints.

User-defined constraints are often employed to control site occupancies when disorder is present in one or more sites (see Eq. 9.4 in Sect. 9.1.1). For example, if three atoms (A, B, and C) occupy the same site, and assuming full total occupancy of the site, the following relationship (or constraint) applies: $g_A + g_B + g_C = 1$. Thus, only two of the fractional occupancies should be refined, for instance, g_A and g_B can be treated as free least-squares variables, while $g_C = 1 - g_A - g_B$. Other restrictions may also be imposed if corresponding information is available. For example, if the contents of A and B are known to be the same, only one occupation factor is independent (e.g., g_A). In this case, $g_B = g_A$ and $g_C = 1 - 2g_A$.

Constraining site occupancies is only one of the many possible kinds of user-imposed constraints. Profile and other parameters often require constraining. For example, when working with a multi-phase sample, it is reasonable to constraint sample displacement (or other corrections on the peak positions) to be the same for all phases. The same is true for peak asymmetry since a major contribution in this case comes from goniometer optics (see Fig. 12.21). Often, during the initial stages of multi-phase refinements, it may be assumed that peak broadening is the same for all phases, especially if this appears to be the case visually, and appropriate constraints are set. These constraints, however, should be removed during the final stages of refinement.

Finally, a researcher can introduce additional constraints (and restraints), but only if there are compelling reasons to do so. This is only feasible if the software being used allows setting constraints for parameters believed to require them.

15.7.3.3 Rigid Bodies

Rigid bodies are molecules or groups that are treated as a whole (rigidly), described using the Cartesian coordinate system,[73] or, in some cases, using Euler angles.[74] The most recent developments in the rigid-body algorithms in the Rietveld refinement have been reviewed by Dinnebier,[75] and many applications of the technique have been incorporated in GSAS and GSAS-II by Toby and Von Dreele.[76] Here, we only give a brief description in order to help the reader with a general understanding of the rigid-body approach and its possible applications.

---

[73]Cartesian coordinate system uses three mutually perpendicular coordinate axes with equal scaling, e.g. 1 Å or 1 nm.

[74]The Euler angles were developed by Leonhard Euler to describe the orientation of a rigid body. For more details, see https://en.wikipedia.org/wiki/Euler_angles

[75]R.E. Dinnebier, Rigid bodies in powder diffraction. A practical guide. Powder Diffraction, **14**, 84 (1999), https://doi.org/10.1017/S0885715600010356

[76]B.H. Toby and R.B. Von Dreele, GSAS-II: the genesis of a modern open-source all purpose crystallography software package, 46, 544 (2013), https://doi.org/10.1107/S0021889813003531

Transformations of the crystallographic coordinates (**x**) into the Cartesian coordinates (**X**) and back are performed as:

$$\mathbf{X} = \mathbf{H}(\mathbf{x} - \mathbf{t}) \text{ and } \mathbf{x} = \mathbf{H}^{-1}\mathbf{X} + \mathbf{t} \tag{15.37}$$

where

$$\mathbf{H} = \begin{pmatrix} a & b\cos\gamma & c\cos\beta \\ 0 & b\sin\alpha^*\sin\gamma & 0 \\ 0 & -b\cos\alpha^*\sin\gamma & c\sin\beta \end{pmatrix} \tag{15.38}$$

and **t** is the center of the rigid body consisting of n atoms defined as

$$\mathbf{t} = \frac{1}{n}\sum_{i=1}^{n}\mathbf{x}_i \tag{15.39}$$

The location of the center and orientation of the rigid body in the Cartesian coordinate system is fully defined by six parameters: three coordinates of the center describing the displacement of the rigid body from the origin and three angles describing the orientation of the rigid body. Thus, instead of $3n$ coordinates in general, the number of parameters is reduced to no more than six and both the location and orientation of the rigid body can easily be adjusted. If the rigid body is located on a finite symmetry element, the number of parameters is reduced. For example, a molecule located on a mirror plane can only move in the plane (two positional parameters), and it can only rotate around a normal to the plane (one orientational parameter); a molecule located on a twofold axis has one positional (along the axis) and one rotational (around the axis) parameter.

Thermal motions of a rigid body are described using the so-called **TLS** matrices:

$$\mathbf{T} = \begin{pmatrix} T_{11} & T_{12} & T_{13} \\ T_{12} & T_{22} & T_{23} \\ T_{13} & T_{23} & T_{33} \end{pmatrix}, \mathbf{L} = \begin{pmatrix} L_{11} & L_{12} & L_{13} \\ L_{12} & L_{22} & L_{23} \\ L_{13} & L_{23} & L_{33} \end{pmatrix}, \text{ and } \mathbf{S} = \begin{pmatrix} S_{11} & S_{12} & S_{13} \\ S_{21} & S_{22} & S_{23} \\ S_{31} & S_{32} & S_{33} \end{pmatrix} \tag{15.40}$$

where the **T** is translation, **L** is libration, and **S** is a screw matrix that describes thermal motions. Matrices **T** and **L** are symmetric ($M_{ij} = M_{ji}$), and matrix **S** is not. The screw matrix defines mixing of the translational and librational motions and is often 0, e.g., for a centrosymmetric rigid body. The maximum number of the **TLS** parameters is 20 (6 + 6 + 9 − 1); one parameter is subtracted since only two diagonal elements of **S** are independent and they are usually refined as $S_{AA} = S_{22} - S_{11}$ and

$S_{BB}=S_{33} - S_{22}$. When the center of a rigid body is located in a special position, the site symmetry imposes additional constraints[77] on the allowed rigid-body motions. As a result, the number of independent parameters (which are the elements of the **TLS** matrices) is reduced, sometimes significantly.

The tensor **U** (see Eq. 9.10) representing thermal motions of the *i*th atom $(1 \leq i \leq n)$ is obtained from the **TLS** parameters as follows:

$$\mathbf{U}_i = \mathbf{T} + \mathbf{A}_i\mathbf{S} + \mathbf{S}^T\mathbf{A}_i^T + \mathbf{A}_i\mathbf{L}\mathbf{A}_i^T \tag{15.41}$$

where matrix $\mathbf{A}_i$ is constructed from Cartesian coordinates of an atom:

$$\mathbf{A}_i = \begin{pmatrix} 0 & Z_i & -Y_i \\ -Z_i & 0 & X_i \\ Y_i & -X_i & 0 \end{pmatrix} \tag{15.42}$$

The anisotropic atomic displacement parameters U_{ij} (see Eq. 9.9) can be obtained from the tensor **U** by using the so-called **g** matrix, which is the same as that used to calculate the *d*-spacings, $(1/d)^2 = \mathbf{h}^T\mathbf{g}\mathbf{h}$, also see Eqs. (8.10, 14.44, and 14.45):

$$U_{ij} = \frac{(\mathbf{gUg})_{ij}}{\sqrt{g_{ii}g_{jj}}}, \text{ where } \mathbf{g} = \begin{pmatrix} a^{*2} & a^*b^*\cos\gamma^* & a^*c^*\cos\beta^* \\ a^*b^*\cos\gamma^* & b^{*2} & b^*c^*\cos\alpha^* \\ a^*c^*\cos\beta^* & b^*c^*\cos\alpha^* & c^{*2} \end{pmatrix} \tag{15.43}$$

The **TLS** approach is far from trivial, particularly regarding the symmetry-imposed relationships between matrix elements. Careful attention is required to ensure a stable and accurate refinement; however, it is important not to overdo it.[78] Nevertheless, the mechanism of rigid bodies offers many advantages. These are: elimination of meaningless changes in the geometry; a substantial reduction of the number of free variables, which can be determined with a higher accuracy; a better convergence to a correct structure when compared to the unconstrained refinement; inclusion of hydrogen atoms at early stages; and reliable anisotropic atomic displacement parameters, which is usually difficult to expect from powder data. The following rules proposed by Dinnebier should be followed:

- If the center of the rigid body is also the center of gravity, the components of the **S** matrix can normally be set to zero.
- Refining only the diagonal components of the **T** matrix constrained to be identical and keeping all elements of **L** and **S** equal to zero is the same as refining an overall isotropic temperature factor for the rigid body.

[77] V. Shomaker, K.N. Trueblood, On the rigid body motion of molecules in crystal, Acta Cryst. **B24**, 63 (1986). https://doi.org/10.1107/S0567740868001718

[78] "If one sticks too rigidly to one's principles, one would hardly see anybody." Agatha Christie.

- Refining all elements of the **T** matrix, with **L** = **S** = 0, is the same as refining an overall anisotropic temperature factor for the rigid body.
- For planar molecules, it is often sufficient to refine only the components of the **T** matrix and the diagonal elements of the **L** matrix.
- If the rigid body is located on a finite symmetry element, some elements of the **TLS** matrices must be set to zero or constrained accordingly.

An example on the rigid-body refinement of a complex 16 atoms ligand in a metal-organic compound can be found in recent paper by Nielson et al.[79] In addition, a variety of tutorials and practical guides on how to use the rigid-body refinement can be found on the GSAS web page.[80]

15.7.4 Figures of Merit and Quality of Rietveld Refinement

Similar to both Le Bail's and Pawley's full-pattern decompositions, the quality of the refinement using the Rietveld method is quantified by the corresponding figures of merit: profile residual, R_p, weighted profile residual, R_{wp}, Bragg residual, R_B, expected residual, R_{exp}, and goodness of fit, χ^2 (see Eqs. 15.18–15.22). The Durbin–Watson *d*-statistic (Eqs. 15.25 and 15.26) may be used to quantify a serial correlation between adjacent least-squares residuals in a Rietveld refinement based on step-scan experimental data. As noted earlier, all but one (R_B) residuals depend on both the profile and structural parameters. The Bragg residual becomes especially significant during Rietveld refinement because it is the only figure of merit, which is nearly exclusively dependent on structural parameters and therefore primarily characterizes the accuracy of the model of the crystal structure.[81]

Regardless of the importance of various numerical figures of merit used to measure the quality of the Rietveld refinement, none of the residuals is a substitute for the plots of the observed and calculated powder diffraction patterns supplemented by the difference, $\Delta Y_i = Y_i^{obs} - Y_i^{calc}$, plotted on the same scale (see Fig. 15.5, 15.6, 15.7 and 15.8). A standard in the modern representation of the

[79] R.B. Nielsen, K.O. Kongshaug, H. Fjellvåg, Delamination, synthesis, crystal structure and thermal properties of the layered metal-organic compound $Zn(C_{12}H_{14}O_4)$. J. Mat. Chem. **18**, 1002 (2008). https://doi.org/10.1039/B712479A

[80] https://subversion.xray.aps.anl.gov/trac/EXPGUI/wiki/GSASIntro

[81] Bragg residual, Eq. (15.20), as calculated in Rietveld refinement, uses true calculated integrated intensities but the "observed" integrated intensities are never actually measured experimentally. They are simply calculated by prorating the observed experimental profile proportionally to the contributions from multiple overlapped calculated reflection profiles after the background has been subtracted (see Eq. 15.7), followed by the numerical integration (see Eq. 8.40). In this regard, Bragg residual also depends on the profile parameters, although this dependence is far less critical when compared to R_p, R_{wp}, and χ^2. In some references, R_B, which is based on the square roots of the integrated intensities, is used as an equivalent of R_F. The latter employs the absolute values of the observed and calculated structure factors: $R_F = \Sigma(|K|\mathbf{F_i}^{obs}| - |\mathbf{F_i}^{calc}||)/\Sigma(K|\mathbf{F_i}^{obs}|)$.

refinement results also requires the inclusion of tick marks to indicate the calculated positions of Bragg peaks (e.g., see Fig. 12.16). For dual-wavelength data, showing the positions of Bragg reflections only for $K\alpha_1$ might be sufficient.

The need for a graphical representation of the results is especially important because both R_p and R_{wp} absorb the contribution from the background. In extreme cases, when the background is high, it is possible that the corresponding numerical figures of merit appear to be excellent due to extremely large denominators in Eqs. (15.18 and 15.19), but neither the model nor the fit of Bragg peaks makes much sense. When the background is unusually high, both the observed and calculated powder diffraction patterns should be plotted after a constant component of the background has been subtracted to enable easy examination of a potential inadequacy of the selected peak-shape function and/or other unusual discrepancies not visible on top of a vast background. Truly, the numbers may be biased, but the figure can be trusted when it comes to assessing the quality of Rietveld fits.

15.7.5 Common Problems and How to Deal with Them

Every powder diffraction pattern presents its own problems when subjected to Rietveld refinement, but some of them are more common than the others. As a result, certain guidelines have been developed under the auspices of the Commission on Powder Diffraction of the International Union of Crystallography.[82] In this section, we follow the suggestions outlined by L.B. McCusker et al. (see Footnote 82), while also encouraging readers to consult this excellent paper. Aside from the obvious, i.e., a variety of errors related to data files used in Rietveld refinement, McCusker, Von Dreele, Cox, Louër, and Scardi provide a brief description of common problems. In the following subsections, we support this description with numerous practical examples found in Chap. 16 and beyond.

15.7.5.1 Poor Fit of the Background

Most commonly, problems with the background occur when the primary, rather than diffracted beam has been conditioned by a crystal monochromator (e.g., see Fig. 11.25). Complicated backgrounds are common when a sample has been enclosed in any kind of a controlled environment (see Sect. 11.4) due to additional scattering on windows and other parts of sample attachment. In some cases, it may be useful to eliminate complex background contribution from the surroundings by measuring it without the sample, and in others, changing the background function or increasing the number of parameters used to fit the background will address the

[82] L.B. McCusker, R.B. Von Dreele, D.E. Cox, D. Louër, and P. Scardi, Rietveld refinement guidelines, J. Appl. Cryst. **32**, 36 (1999). https://doi.org/10.1107/S0021889898009856

problem. It may be necessary to do both to achieve a sufficient approximation of the background in some powder patterns. Needless to say, measuring the background without a sample must be done using configuration (sample holder, optics, and power settings) that is identical to those employed during the actual measurement of the powder diffraction pattern.

Complexity of the background may also be due to factors intrinsic to a sample, such as poor crystallinity, extreme disorder, or nanocrystallinity. When deviations from the conventional, weakly Bragg angle dependent behavior are minor, adding a few free variables to fit the background usually helps. Otherwise, the best approach is to use tools developed for total scattering analysis[83] rather than Rietveld refinement.

15.7.5.2 Poor Fit of Peak Shapes

This usually occurs when one or more critical peak-shape parameters have been overlooked or became unphysical during previous refinement steps. The latter often happens when too many of the peak-shape-related parameters are set as free variables in the initial stages of refinement. Common solutions include

- Checking the difference profile (see Figs. 15.5, 15.6, 15.7, 15.8 and 15.9) to identify the factor responsible for discrepancies in the majority of strong, well-resolved Bragg peaks. Once the source of the problem is identified, the incorrect parameter should be reset and/or the overlooked parameter included in the refinement.
- It is possible that the selected peak-shape function does not accurately describe this particular experiment. The best way to find the right function is to examine a well-characterized standard available from NIST,[84] such as LaB_6 (SRM-660a) or Si powder (SRM-640g[85]). Once the function that fits peak shapes of the standard is found, it should be used to fit profiles of all other samples measured using the same diffractometer and the same goniometer optics.
- Asymmetry correction is commonly overlooked. Make sure that an appropriate model (see Sect. 8.5.2) is selected and relevant parameters are included in the refinement.
- Sometimes, peak widths exhibit strong anisotropy due to anisotropic strain, extreme particle shape anisotropy, or stacking faults. In such cases, it may be

[83] T. Egami and S.J.L. Billinge, Underneath the Bragg peaks. Pergamon Materials Series (Pergamon, Amsterdam, 2003).

[84] SRM is Standard Reference Material. Consult https://nist.gov/srm for details.

[85] J.P. Cline, M.H. Mendenhall, D. Black, A. Henins, J.B. Prothero, Certification of Standard Reference Material® 640 g: Line Position and Line Shape Standard for Powder Diffraction (Silicon Powder), NIST Special Publication 260 (SP) NIST SP 260-245, (2024), https://doi.org/10.6028/NIST.SP.260-245

necessary to use appropriate corrections to the conventional, smoothly varying FWHM as a function of Bragg angle (e.g., see Eq. 8.34).

15.7.5.3 Mismatch Between Calculated and Observed Peak Positions

Most commonly, this problem occurs when experimental data have been affected by errors due to zero-shift, sample-displacement, or sample transparency. Setting either the zero-shift or sample-displacement parameter as a free variable will help. Sometimes, a problem may be less obvious. One example is when a model used to approximate peak asymmetry is inadequate. As a result, Rietveld refinement of the unit cell dimensions may not converge properly when asymmetry is strong. Another example is when there is a problem with the goniometer, slipping through a data point, or getting stuck at the same position during data collection, which remains undetected by the data collection software. If these mechanical errors are reproducible, measuring a standard will help to pinpoint the cause. Yet another possible reason is the presence of a subtle structural distortion or incorrectly assumed lattice symmetry (such as forcing cubic symmetry upon a structure which is tetragonal or rhombohedral). A careful analysis of mismatches, anisotropy of peak shapes, and/or signs of peak splitting is required to find out the reality.

15.7.5.4 The Tails of Strong Peaks Are Cut Off Prematurely

This problem often occurs when Bragg peaks are more Lorentzian than they are Gaussian (see Fig. 8.12). Then, Bragg intensity spreads far away to the left and to the right from the peak center, and if the calculated peak-shape function is truncated prematurely, distinct steps are observed upon a close examination of the calculated profile (see Fig. 16.6). The solution is to increase the range over which the peak profile is calculated.

15.7.5.5 The Relative Intensities Do Not Match

An obvious reason may be an incorrect or incomplete model of the crystal structure. A solution is to either find the correct model or to complete it by positioning missing atoms in the unit cell. Assuming that the crystal structure model is accurate and complete, various other factors may cause intensity mismatch.

One of the reasons could be preferred orientation. Typically, Bragg peaks with specific combinations of indices are too strong, while others are too weak. For example, if observed intensities of all Bragg peaks (00*l*) are higher than the corresponding calculated intensities but the relationships between observed and calculated intensities are opposite for all resolved (*h*00), (0*k*0), and (*hk*0) reflections, most likely the sample is textured along the [001] direction. Refining preferred orientation using any of the models discussed above (see Sect. 8.6.6) may solve the problem but only when the texture is moderate. If this does not help, the experiment has to be redone with special attention to minimizing texture.

Another potential problem with intensity mismatch is when coarse powder was used to prepare the specimen (see Figs. 12.13–12.15), and/or when the sample was not spinning. It is usually recognized by seeing a few observed intensities that are too high but none is too low. This situation cannot be corrected analytically, so the experiment must be repeated with a much finer powder while spinning the specimen.

15.7.5.6 Presence of Weak Unindexed Bragg Peaks

These may be either due to crystalline impurities, or they may indicate formation of a superstructure (doubling, tripling, etc. of one or more unit cell dimensions). The best way to identify the source is to repeat synthesis of the same material. If the unindexed peaks remain where they were, plus their relative intensities remain similar when compared to intensities of the main (indexed) phase, chances are that these unindexed peaks signal the formation of a superstructure. When a different batch of the same material results in a different set of unindexed peaks, or some of the peaks have disappeared or changed relative intensity, these are likely impurity peaks. An attempt should be made to index them and to assign them to a particular crystalline solid.

15.7.5.7 Problems with Converging Refinement

If the structural model is complete, most likely some or all initial approximations are too far away from their true values, or there are correlations between several parameters (see Fig. 15.4) . Problem causing variables may be identified by monitoring their shifts, standard deviations, and correlations during several consecutive refinement cycles. Opposite signs with gradually increasing amplitudes of the shifts and/or unusually large standard deviations are good indicators of the problematic parameter. Damping should be applied as needed. It may be necessary to refine parameters in groups rather than altogether. For example, population parameters are usually correlated with atomic displacement parameters, and these may need to be refined separately. In some cases, constraints and restrains must be added to avoid least-squares instabilities. Care must always be taken to ensure that the data support the number and types of parameters being refined ($n \gg m$ in Eq. 15.8).

15.7.6 Termination of Rietveld Refinement

The nonlinear least-squares technique finds a set of increments that are added to a set of free variables representing an initial approximation. Parameters, obtained in this way, are carried over into the next refinement cycle as a more precise initial approximation. In some cases, it may take a few refinement cycles to achieve the best fit—in other words, to minimize the corresponding function—while in many instances, the number of required least-squares steps may be quite large. Especially in Rietveld refinement, where various groups of parameters have a different and often unrelated physical origin, the ability to detect the completion of the minimization—or, in other words, the complete convergence of the least squares—is essential.

The critical variables to watch during the refinement are indeed a set of standard figures of merit (FOMs), such as R_p, R_{wp}, R_B, and χ^2. When the Rietveld algorithm is stable, they should gradually decrease and then level off showing minimal fluctuations about certain minimum values, which are experiment- and structure-specific. In some cases, FOMs may begin to rise. More often than not, their steady or erratic increase indicates the undesired divergence in the nonlinear least squares, typically due to strong or severe correlations between two or more variables. If this condition is detected, the refinement should be stopped without saving the results and the array of free variables reanalyzed to introduce proper constraints.

Even if the refinement is stable, it should be terminated at a certain point. Because of the finite accuracy of both the data and computations, it is unrealistic to wait until increments of all least-squares parameters become zero. The latter usually never happens anyway, due to simplifications introduced in the nonlinear least-squares algorithm (see Eq. 15.9). In addition to the stabilization of all figures of merit near their respective minima, another important factor that should be considered is the relationships between the increments and their corresponding standard deviations after each least-squares cycle. It is generally accepted that least-squares refinement can be considered converged when the absolute values of all increments become smaller than the estimated standard deviations of the corresponding free variables,[86] while ensuring the model remains rational.

15.8 Concluding Remarks

In the remainder of this book, we present multiple practical examples of structure solutions. In most cases, individual intensities and structure factors are extracted by using Le Bail's algorithm for full-pattern decomposition. This technique is preferred over Pawley's approach because Le Bail's algorithm is generally more stable and has been incorporated into several freely available software programs that are coupled with Patterson and Fourier calculations. These are: LHPM-Rietica,[87] GSAS and

[86] Usually lower than a small (at least 1/10th) fraction of the standard deviation.

[87] LHPM-Rietica (authors B.A. Hunter and C.J. Howard) may be downloaded from http://www.rietica.org/

GSAS-II,[88] and FOX.[89] Other well-developed and tested computer codes, such as EXPO[90]—an integrated package for full-pattern decomposition and crystal structure solution by direct methods—are also available.

In some of the more complex examples, however, we will employ manual profile fitting. The latter approach is less "user-friendly" in terms of automation.

By solving crystal structures of different classes of materials, we illustrate just a few of the possible approaches to ab initio structure solution from powder diffraction data. It is important to note that identical parameters to those in our examples can only be expected if the same software is used for full-profile refinement, as small but detectable differences in Rietveld method implementation can vary by developers.[91] Furthermore, even when the same version of a program is used to process the same experimental data, small deviations may occur due to subjective decisions, such as when to terminate the refinement. In this case, however, the differences should be within a few standard uncertainties. Unless otherwise noted, the processing of individual structure factor data, direct phase determination, and Patterson-, Fourier- and E-map calculations presented in the remaining chapters were performed using WinCSD application.[92]

---

[88] GSAS (authors A.C. Larson and R.B. Von Dreele) may be downloaded from https://subversion.xray.aps.anl.gov/trac/EXPGUI/wiki/GSASIntro and GSAS-II (authors B.H. Toby and R.B. Von Dreele) from https://advancedphotonsource.github.io/GSAS-II-tutorials/; B.H. Toby and R.B. Von Dreele, GSAS-II: the genesis of a modern open-source all purpose crystallography software package, 46, 544 (2013). https://doi.org/10.1107/S0021889813003531

[89] R. Černý, V. Favre-Nicolin, J. Rohlíček, M. Hušák, Z. Matěj, R. Kužel, Expanding FOX: Auto-indexing, grid computing, profile fitting. p. 16 in: CPD Newsletter "Real-Space and Hybrid Methods for Structure Solution from Powders", Issue 35, (2007). https://www.iucr.org/__data/assets/pdf_file/0013/21622/CPD_35_total.pdf

[90] A. Altomare, C. Cuocci, C. Giacovazzo, A. Moliterni, R. Rizzi, N. Corriero, and A. Falcicchio, EXPO2013: a kit of tools for phasing crystal structures from powder data, J. Appl. Cryst. **46**, 1231 (2013), https://doi.org/10.1107/S0021889813013113; A. Altomare, F. Capitelli, N. Corriero, C. Cuocci1, A. Falcicchio, A. Moliterni, and R. Rizzi, The Rietveld Refinement in the EXPO Software: A Powerful Tool at the End of the Elaborate Crystal Structure Solution Pathway, Crystals, 8(5), 203 (2018), https://doi.org/10.3390/cryst8050203

[91] There is a variety of software applications that can be used to process "pseudo-single crystal" experimental diffraction data represented in a form of individual structure factors or their squares (the IUCr Crystallographic Software List: https://www.iucr.org/resources/other-directories/software). The most commonly used software products are SHELXS, SHELXD, SHELXT, and SHELXL (author G.M. Sheldrick), which are distributed free for academic use (consult SHELX home page at https://shelx.uni-goettingen.de/ for details). See details in: G.M. Sheldrick, Crystal structure refinement with SHELXL, Acta Cryst. **C71**, 3 (2015), https://doi.org/10.1107/S2053229614024218; and G.M. Sheldrick, SHELXT - Integrated space-group and crystal-structure determination, Acta Cryst. **A71** 3 (2015), https://doi.org/10.1107/S2053273314026370

[92] WinCSD is available from https://wincsd.eu: L.G. Akselrud, P.Yu. Zavalii, Yu.N. Grin, V.K. Pecharsky, B. Baumgartner, and E. Wolfel, "Use of the CSD program package for structure determination from powder data", Mater. Sci. Forum **133–136**, 335 (1993). https://doi.org/10.4028/www.scientific.net/MSF.133-136.335; L. Akselrud, Yu. Grin, WinCSD: software package for crystallographic calculations (Version 4), J. Appl. Crystallogr. **47**, 803 (2014), https://doi.org/10.1107/S1600576714001058

15.9 Additional Reading

1. J.A. Kaduk, S.J.L. Billinge, R.E. Dinnebier, N. Henderson, I. Madsen, R. Černý, M. Leoni, L. Lutterotti, S. Thakral, and D. Chateigner, Powder diffraction, Nat. Rev. Methods Primers, **1**, 77 (2021), https://doi.org/10.1038/s43586-021-00074-7
2. R.E. Dinnebier, A. Leinweber, and J.S.O. Evans, Rietveld Refinement. Practical Powder Diffraction Pattern Analysis using TOPAS, Walter de Gruyter GmbH, Berlin/Boston (2019), https://doi.org/10.1515/9783110461381-201
3. R.E. Dinnebier and S.J.L. Billinge, Powder Diffraction. Theory and Practice, RSC Publishing (2008), https://doi.org/10.1039/9781847558237
4. W. I. F. David, K. Shankland, L.B. McCusker, and Ch. Baerlocher, Structure determination from powder diffraction data. IUCr monographs on crystallography, Oxford University Press, Oxford (2002), https://doi.org/10.1107/S0108767306047349
5. Commission on Powder Diffraction Newsletter "Real-Space and Hybrid Methods for Structure Solution from Powders", Issue 35 (2007) at https://www.iucr.org/__data/assets/pdf_file/0013/21622/CPD_35_total.pdf
6. K.D.M. Harris, Circumventing a challenging aspect of crystal structure determination from powder diffraction data, Acta Cryst. **B78**, 96 (2022), https://doi.org/10.1107/S2052520622003717
7. A.A. Coelho, TOPAS and TOPAS-Academic: an optimization program integrating computer algebra and crystallographic objects written in C++, J. Appl. Cryst. **51**, 210 (2018), https://doi.org/10.1107/S1600576718000183
8. D. Louër, Advances in powder diffraction analysis, Acta Cryst. **A54**, 922 (1998). https://doi.org/10.1107/S0108767398007363
9. D.M. Poojary and A. Clearfield, Application of X-ray powder diffraction techniques to the solution of unknown crystal structures, Acc. Chem. Res. **30**, 414 (1997). https://doi.org/10.1021/ar960143j
10. K.D.M. Harris and M. Tremayne, Crystal structure determination from powder diffraction data, Chem. Mater. **8**, 2554 (1996). https://doi.org/10.1021/cm960218d
11. C. Giacovazzo, Direct methods and powder data: State of the art and perspectives, Acta Cryst. **A52**, 331 (1996). https://doi.org/10.1107/S0108767395013651
12. J.A. Kaduk, Use of the Inorganic Crystal Structure Database as a problem solving tool, Acta Cryst. **B58**, 370 (2002). https://doi.org/10.1107/S0108768102003476
13. A. Le Bail, SDPD—Structure Determination from Powder Diffraction—Database of bibliography and methods, http://cristal.org/
14. V. Favre-Nicolin and R. Černý, FOX, "Free objects for crystallography": a modular approach to ab initio structure determination from powder diffraction, J. Appl. Cryst. **35**, 734 (2002). https://doi.org/10.1107/S0021889802015236

15. L.B. McCusker, R.B. Von Dreele, D.E. Cox, D. Louër, and P. Scardi, Rietveld refinement guidelines, J. Appl. Cryst. **32**, 36 (1999). https://doi.org/10.1107/S0021889898009856
16. W. I. F. David, K. Shankland. Structure determination from powder diffraction data. Acta Cryst. **A64**, 52 (2008). https://doi.org/10.1107/S0108767307064252
17. R. Černý, V. Favre-Nicolin, Direct space methods of structure determination from powder diffraction: principles, guidelines and perspectives. Z. Kristallogr. **222**, 105 (2007). https://doi.org/10.1524/zkri.2007.222.3-4.105
18. G.M. Sheldrick, A short history of SHELX. Acta Cryst. **A64**, 112 (2008). https://doi.org/10.1107/S0108767307043930
19. R.E. Dinnebier, Rigid bodies in powder diffraction. A practical guide, Powder Diffraction **14**, 84 (1999). https://doi.org/10.1017/S0885715600010356
20. https://www.aps.anl.gov/Education/Powder-Diffraction-Educational-Materials
21. A.C. Larson and R.B. Von Dreele, General Structure Analysis System (GSAS), Los Alamos National Laboratory Report LAUR 86–748 (2004); B. H. Toby, EXPGUI, a graphical user interface for GSAS, J. Appl. Cryst. 34, 210–213 (2001). https://doi.org/10.1107/S0021889801002242
22. GSAS tutorial—Alumina example: https://www3.aps.anl.gov/X-ray-Science-Division/Powder_Diffraction_Crystallography/tutorial3/index.html
23. J. Cui, Q. Huang, and B. H. Toby, Magnetic structure refinement with neutron powder diffraction data using GSAS: A tutorial, Powder Diffraction, **21**, 71 (2006). https://doi.org/10.1154/1.2179805
24. B.H. Toby and R.B. Von Dreele, GSAS-II: the genesis of a modern open-source all purpose crystallography software package, J. Appl. Cryst. **46**, 544 (2013). https://doi.org/10.1107/S0021889813003531
25. B.H. Toby and R.B. Von Dreele, What's new in GSAS-II, Powder diffraction, **29** (S2) (2014), https://doi.org/10.1017/S0885715614000736
26. GSAS-II tutorials. Slides available at: https://subversion.xray.aps.anl.gov/pyGSAS/trunk/help/Tutorials.html; video (56 min) from author (Von Dreele): https://www.youtube.com/watch?v=Lq-0Omqcldw

Chapter 16
Crystal Structure of $LaNi_{4.85}Sn_{0.15}$: Using Cu and Mo Radiation Data

Consider a powder diffraction pattern of $LaNi_{4.85}Sn_{0.15}$, shown in Fig. 16.1.[1] From Chap. 14 (Sects. 14.3.1, 14.7.1, and 14.11.1), we already known that the alloy is hexagonal, with lattice parameters $a = 5.0421$, $c = 4.0118$ Å. Analysis of Table 14.6 indicates that there are no forbidden reflections, and from Tables 9.7 and 9.8, the possible space-group symmetries for this material are as follows: P6/mmm, $P\bar{6}m2$, $P\bar{6}2m$, P6mm, P622, P6/m, $P\bar{6}$, P6, $P\bar{3}1m$, P31m, P312, $P\bar{3}1m$, P3m1, P321, $P\bar{3}$, and P3.

The fact that a total of 16 (!) space groups are possible for this material could complicate the choice for the ab initio structure solution. However, its unit cell is relatively small. Moreover, since $LaNi_{4.85}Sn_{0.15}$ is an intermetallic compound, the highest symmetry space group (P6/mmm) is quite probable.

The measured gravimetric density of the alloy is 8.21 g/cm^3. After calculating the mass of its formula unit, $M = 441.4$ a.m.u., and the unit cell volume, $V = \sqrt{3}a^2c/2 = 88.327°A^3$, Eq. (15.4) results in Z = 0.99 ≅ 1. Thus, one unit cell of the compound contains 1 La, 4.85 Ni, and 0.15 Sn atoms. Because of the fractional amounts in the unit cell, the Ni and Sn atoms may be distributed statistically in their respective crystallographic sites (they have similar atomic radii: $r_{Ni} = 1.35$, $r_{Sn} = 1.45$ Å), but much larger La atoms ($r_{La} = 1.95$ Å) should occupy a separate crystallographic site. Thus, before proceeding with the full-pattern decomposition, it is advisable to check whether some of the space groups that are possible for this compound can be eliminated.[2] Analysis of the multiplicities of site positions

---

[1] Files *Ch16Ex01_CuKa.dat* and *Ch16Ex01_CuKa.xy*, which contain the same experimental data saved in two different formats, are available online at www.springer.com/FPDSCM3

[2] Generally, it is unnecessary to have the exact space-group symmetry information during the full-pattern decomposition. In fact, it is more practical to select the most symmetrical space group in the established diffraction group and avoid multiple and complete Bragg peak overlaps. For example, when both trigonal and hexagonal symmetry (primitive lattice) groups are possible, full-pattern decomposition is best performed in the hexagonal crystal system assuming Laue class 6/mmm. This eliminates Bragg doublets (e.g., *hkl* and *khl*, which are present in the Laue class 6/m) and

P. Zavalij, V. Pecharsky, *Fundamentals of Powder Diffraction and Structural Characterization of Materials*, https://doi.org/10.1007/978-3-031-91504-8_16

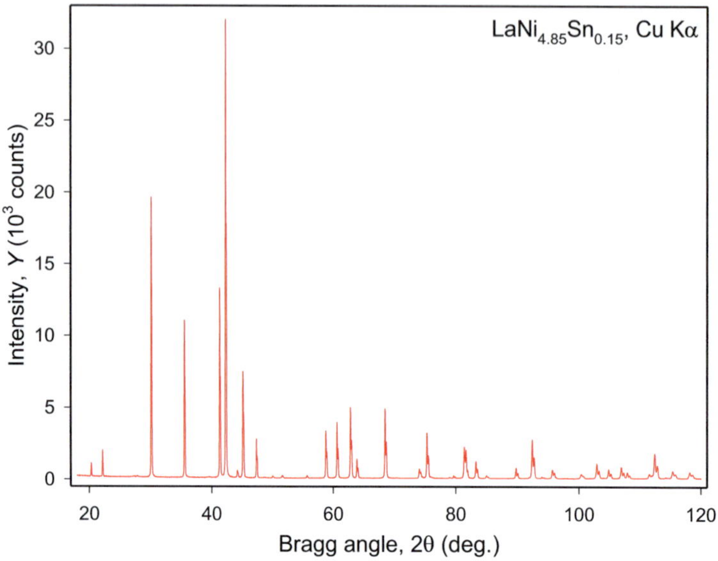

Fig. 16.1 The powder diffraction pattern collected from a sample of $LaNi_{4.85}Sn_{0.15}$ using Cu Kα radiation on a Rigaku TTRAX rotating anode diffractometer. The divergence slit was 0.75°, and the receiving slit was 0.03°. The experiment was carried out in a continuous scanning mode with a rate 0.5 deg/min and with a sampling step 0.02°. The powder used in this experiment was prepared by gas atomization from the melt, and therefore, particles were nearly spherical (see inset in Fig. 12.16)

of the 16 possible space groups indicates that none of the groups can be excluded outright because every space-group symmetry has one-, two-, and threefold sites (some of the trigonal space groups have only one- and threefold sites), which are sufficient to accommodate one La and a total of five Ni + Sn atoms.

In this example,[3] we use LHPM-Rietica. For readers who have no experience with the program, we suggest using both the manual, which is available for download with the software, and a web-based tutorial[4] in combination with the experimental data found online. Ideally, the results presented here should be reproduced nearly exactly, although small deviations may occur due to the nature of the nonlinear least squares.

---

quadruplets (e.g, hkl, $hk\bar{l}$, khl, and $k\bar{h}\bar{l}$, which should be considered in the Laue class $\bar{3}$), which are indistinguishable using both Le Bail and Pawley techniques. For structure solution, it may be necessary to add the missing combinations of indices to the array of structure amplitudes obtained in the highest symmetry space group.

[3] Due to its simplicity, this example is purely academic; however, it is an excellent sample to gain initial experience in deconvoluting powder diffraction patterns using Le Bail's technique. Therefore, it is considered in detail.

[4] Brett Hunter's LHPM-Rietica Rietveld tutorial is available at http://ccp14.cryst.bbk.ac.uk/tutorial/lhpm-rietica/index.html

Table 16.1 Figures of merit obtained at different stages during the full-pattern decomposition of the powder diffraction pattern of $LaNi_{4.85}Sn_{0.15}$ using Le Bail approach incorporated in LHPM-Rietica. Wavelengths used: $\lambda K\alpha_1 = 1.54059$ Å, $\lambda K\alpha_2 = 1.54441$ Å; $R_{exp} = 5.24\%$

Refined parameters	Illustration	R_p [a]	R_{wp} [a]	χ^2
Experimental data	Figure 16.1	–	–	–
Background (linear)	Figure 16.2	22.75	28.25	29.02
+ U, V, W, η_0	Figure 16.3	19.40	24.13	21.20
+ a, c, zero shift	Figure 16.4	7.75	12.23	5.44
+ Background (third order), asymmetry	Figure 16.5	5.61	9.16	3.06
All, plus broader base width	Figure 16.7	4.82	8.31	2.52
Second phase, all parameters	Figure 16.8	4.38	6.47	1.53

[a] Here and in all other similar tables, the residuals are listed in percent and χ^2 is dimensionless (see Eqs. 15.18–15.20 and 15.22)

16.1 Full-Pattern Decomposition

Full-pattern decomposition usually begins with the refinement of the background while keeping peak shape and instrumental parameters fixed at their default values and unit cell dimensions fixed at their best-known values. We begin with only two parameters representing the background as a straight line, which is sufficient as a first approximation (see Fig. 16.1). Peak shapes are represented using a pseudo-Voigt function.[5] The progression of the full-pattern decomposition is illustrated by the corresponding figures of merit, which are assembled in Table 16.1.

Initial refinement of only two background parameters serves an important function: it enables preliminary determination of the individual integrated intensities, which are initially set to unity (see Sect. 15.4), and the calculated diffraction pattern begins to resemble the observed diffraction data (Fig. 16.2). When this initial refinement step is completed and the refinement appears stable, it is possible to begin releasing other relevant parameters. The refinement has been performed in the order shown in Table 16.1; the results are illustrated in Figs. 16.3, 16.4, 16.5, 16.6, 16.7 and 16.8 and in Table 16.2.

The initial refinement with a linear background (Fig. 16.2) is far from an ideal fit. Although the unit cell dimensions are quite precise, we know that the experimental data are affected by a small systematic error in Bragg angles (see Table 14.6, which indicated a zero shift of 0.032° or an equivalent sample-displacement error). Further, peak-shape parameters have been set at their defaults, which may be unsuitable to describe the observed profile quite well. A relatively poor fit is reflected in high residuals as seen in Row 2 of Table 16.1. When major peak-shape (U, V, W, and η_0) and lattice parameters were refined along with zero shift, the fit improves considerably, as seen in Rows 3 and 4 in Table 16.1 and visualized in Figs. 16.3 and 16.4,

[5] Asymmetry was treated as in: C.J. Howard, The approximation of asymmetric neutron powder diffraction peaks by sums of Gaussians, J. Appl. Cryst. **15**, 615 (1982). https://doi.org/10.1107/S0021889882012783. See also Eq. (8.39).

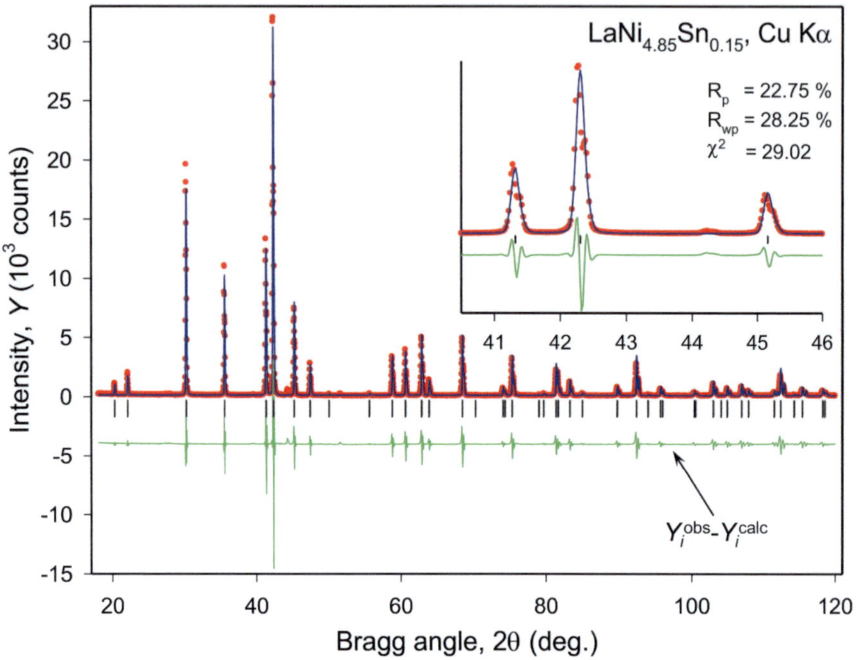

Fig. 16.2 The observed (*red circles*) and calculated (*blue lines*) powder patterns of $LaNi_{4.85}Sn_{0.15}$ after a linear background has been refined. Peak-shape parameters were kept at their default values and lattice parameters were fixed at $a = 5.0421$, $c = 4.0118$ Å. The vertical bars located just below the background level indicate calculated positions of Bragg peaks for λ $K\alpha_1$. Green curve in the bottom represents the difference $Y_i^{obs} - Y_i^{calc}$. The scales for the observed, calculated, and difference plots are identical. The inset highlights details in the vicinity of the strongest Bragg peak. Same notations are maintained in all similar figures below and throughout this book

respectively. For such a simple pattern, the order in which these parameters have been refined is practically irrelevant, but for the cases that are more complex it may be necessary to examine the differences, $Y_i^{obs} - Y_i^{calc}$ (see Sect. 15.6.1), and to decide which set of parameters is more likely to have the largest effect on the improved fit: they should be refined first.

In this example, lattice parameters and the zero-shift correction have a substantial impact on the quality of the fit and the weighted profile residual, R_{wp}, decreases nearly twofold (from ~24% to ~12%), while the refinement of peak-shape parameters decreases R_{wp} by only ~4%. Therefore, in this case, lattice parameters should have been refined first. However, it is not always obvious which parameters are more important and should be released at a particular stage of the least-squares refinement. Because of this, in complex cases, a trial-and-error approach is often employed.[6]

[6] An idiosyncrasy of the Le Bail's approach nearly always requires that early refinement includes the simplest suitable background function with all other relevant parameters kept fixed at their default or approximately known values. This ensures the proper determination of the individual integrated

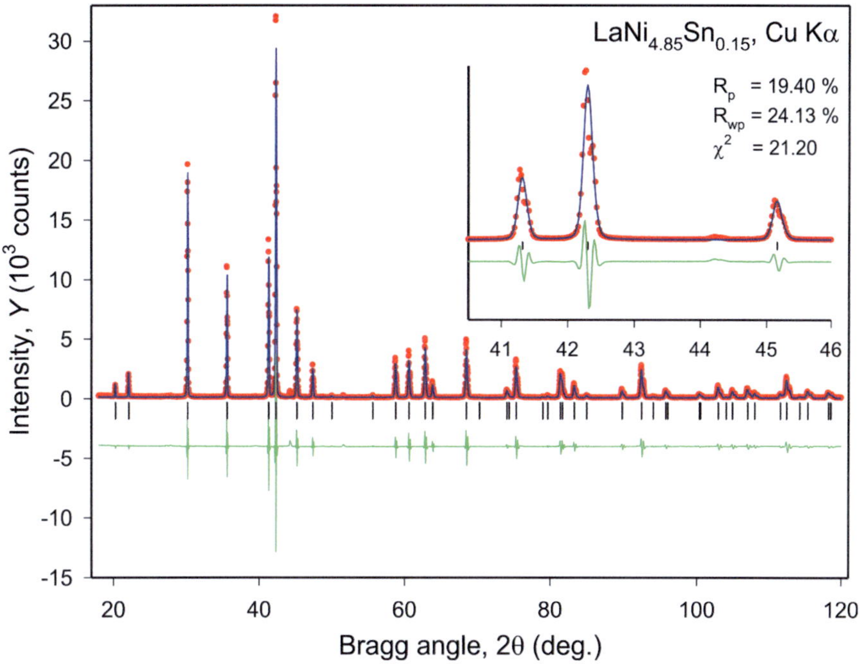

Fig. 16.3 The observed and calculated powder diffraction patterns of $LaNi_{4.85}Sn_{0.15}$ after peak-shape parameters, *U*, *V*, *W*, and η_0, were refined together with linear background

The result illustrated in Fig. 16.4 shows a satisfactory fit and the next groups of parameters included in the refinement were: a more complex background (third-order polynomial instead of a linear function) and peak asymmetry. The fit further improves and when the calculated and observed profiles match quite well (i.e., $Y_i^{obs} - Y_i^{calc}$ approach 0), as shown in Fig. 16.5, it is useful to examine the bases of the strongest Bragg peaks to establish whether the selected peak range is adequate or not. The peak width at the base was seven FWHMs so far (default) and, as seen in Fig. 16.6, it is too narrow. Therefore, from this point on, it was increased to 12 FWHMs, which further improves the fit (Fig. 16.7 and Row 6 in Table 16.1). The proper selection of peak widths at the base is best established visually, which should be gradually increased until "steps" indicated by arrows in Fig. 16.6 disappear. While it is possible to select an even larger peak width at the base, it is always better to choose a realistic minimum value because as the complexity of the pattern increases, the excessive widths of peaks may have a deleterious effect on the background function, especially at high Bragg angles.

The powder diffraction pattern is contaminated by a small amount of an impurity phase, which is a solid solution of Sn in Ni. Its structure is cubic, space group $Fm\bar{3}m$,

intensities, which are initially set to identical values, and the overall success of the full pattern decomposition.

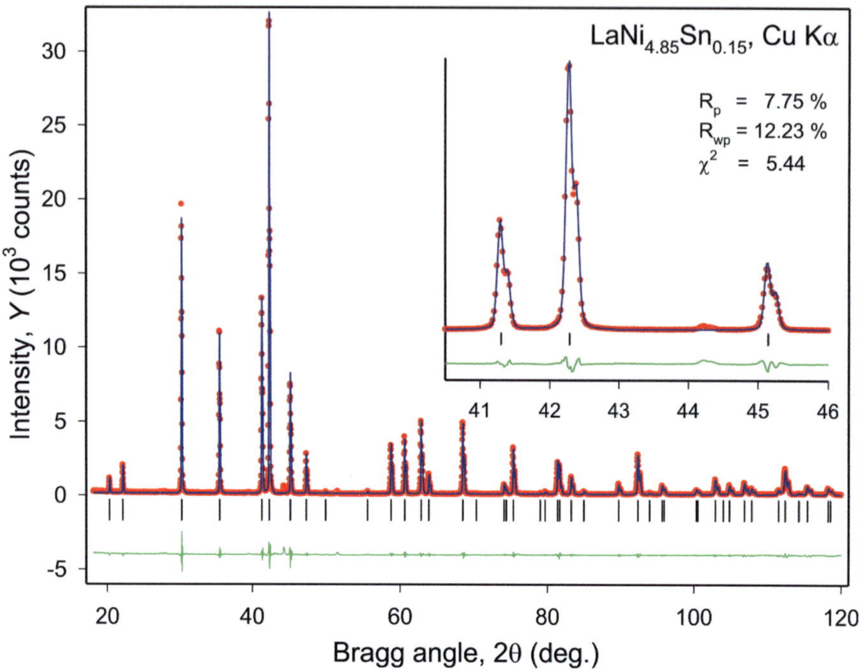

Fig. 16.4 The observed and calculated powder diffraction patterns of $LaNi_{4.85}Sn_{0.15}$ after zero-shift and lattice parameters were included in the refinement

and approximate lattice parameter is $a = 3.54$ Å. When its presence is accounted, the fit and all residuals further improve.[7] The corresponding parameters, as obtained after each refinement step, are assembled in Table 16.2. The inclusion of an impurity phase in the final step is not critical but is noticeable (see Table 16.1 and Fig. 16.8), because the amount of this phase is quite small, totaling ~2 vol. % according to a microstructural analysis, or ~2.5 wt. % according to the Rietveld refinement (see later in the text).

The results of the full-pattern decomposition are shown in Table 16.2 as a list of peak-shape and lattice parameters and also in Table 16.3 as a list of Miller indices with the corresponding individual $|\mathbf{F}^{obs}|^2$ and their standard deviations. The array of intensity data can now be processed by any suitable crystallographic software, and

[7] When the crystal structure of the impurity (or at least its lattice parameters) is unknown, the contaminated ranges of the powder diffraction pattern may be simply excluded from the least-squares refinement. In this example, the following ranges may be excluded to improve the fit: 43.6°–44.6°, 51°–52°, 75°–77°, and 97°–99° of 2θ. One impurity Bragg peak (at 2θ = 92.3°) nearly completely overlaps with the peak from the major phase in the sample. Hence, exclusion may not be a suitable alternative if there are multiple overlaps of Bragg peaks that belong to the main and impurity phases, especially when the amount of the latter is significant.

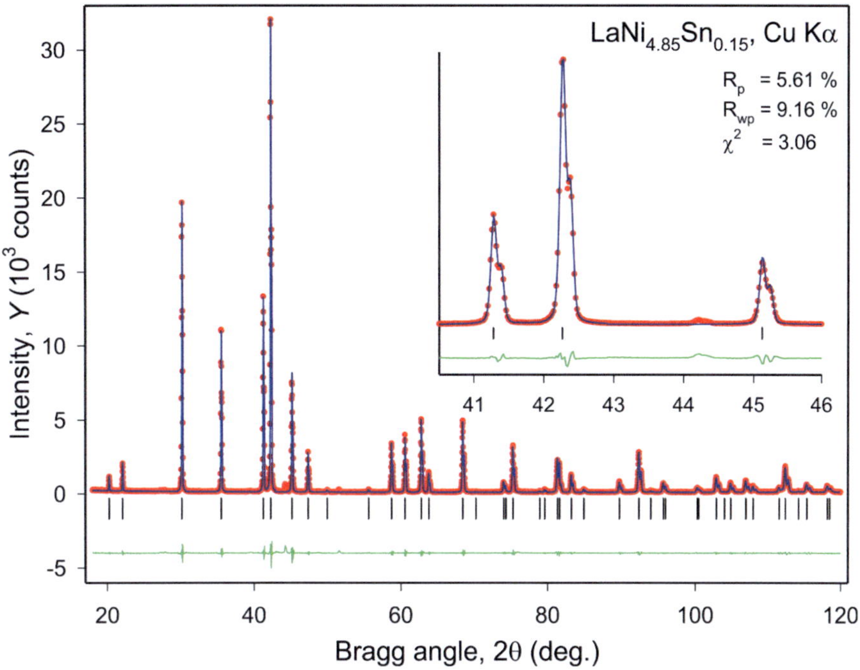

Fig. 16.5 The observed and calculated powder diffraction patterns of $LaNi_{4.85}Sn_{0.15}$ after the polynomial background and asymmetry were refined together with lattice, zero-shift, and peak-shape parameters (U, V, W, and η_0)

used to calculate a Patterson function, or in combination with direct phase-angle determination algorithm(s) to solve the crystal structure of $LaNi_{4.85}Sn_{0.15}$.

16.2 Solving the Crystal Structure

Strictly speaking, solving the crystal structure from first principles is not necessary in this case. Once the material's Pearson symbol is identified as hP6, the correct structure type of this alloy can be easily found by consulting Pearson's Handbook. Further, in such a small- and high-symmetry unit cell, the possibilities to place different atoms are quite limited due to geometrical constraints, and chances are quite high that this type of crystal structure is already known. Nonetheless, this example is useful to illustrate how a simple combination of available data about the chemical composition, unit cell contents, and symmetry provides required information, which may be used to locate all atoms in the unit cell.

As established earlier, a total of 16 space groups are possible for $LaNi_{4.85}Sn_{0.15}$, but the highest symmetry is quite likely because the material is an intermetallic compound. Hence, we first concentrate our attention on the space-group P6/mmm. If

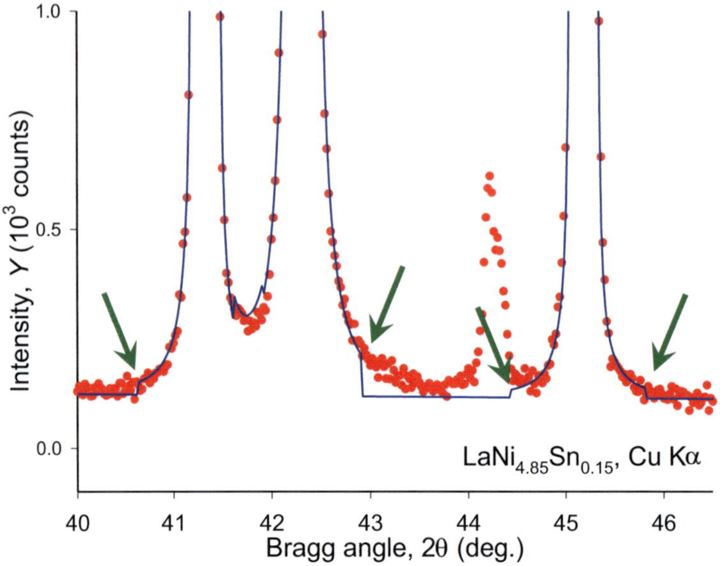

Fig. 16.6 Expanded view of the bases of the strongest Bragg peaks indicating that peak ranges should be increased. The additional Bragg peak at $2\theta \cong 44.2°$ is from Ni impurity

the structure is solved in this space-group symmetry, then there is no need to test lower symmetry groups. However, if the solution could not be found, the symmetry should be gradually lowered until the crystal structure is solved, see Fig. 15.1.

We already know that we must locate a total of one La and five Ni + Sn atoms in the unit cell. According to the International Tables for Crystallography,[8] only two sites in this space-group symmetry have multiplicity one and, therefore, are suitable to accommodate the La atom (Table 16.4). These are: 1(a) with coordinates of a point in 0,0,0 and 1(b) 0,0,½. Considering the fact that La is the strongest scattering atom in this crystal structure, it has the largest contribution to the phase angles of all reflections. Hence, Ni atoms should be easily located from a Fourier map.[9]

As is easy to verify by calculating interatomic distances, the La atoms can be accommodated in 1(a) or 1(b) sites, which differ only by a shift of the origin of coordinates, and La–La distances are identical regardless of where the La atom is placed. We place La in the 1(a) site. Calculation of phase angles always involves calculation of $\left|\mathbf{F}_{hkl}^{calc}\right|$, and therefore, it is also possible to compute the corresponding figure of merit, R_F, which is similar to the Bragg residual (Eq. 15.20) except that the

[8]International Tables for Crystallography, Vol. A, Space-group symmetry, M. I. Aroyo, Ed., Second online edition, (2016). https://it.iucr.org/A/

[9]The amount of Sn in the alloy is small, and assuming random distribution of Ni and Sn in the corresponding sites, the mixture is nearly pure nickel: 97 at.% Ni and 3 at.% Sn.

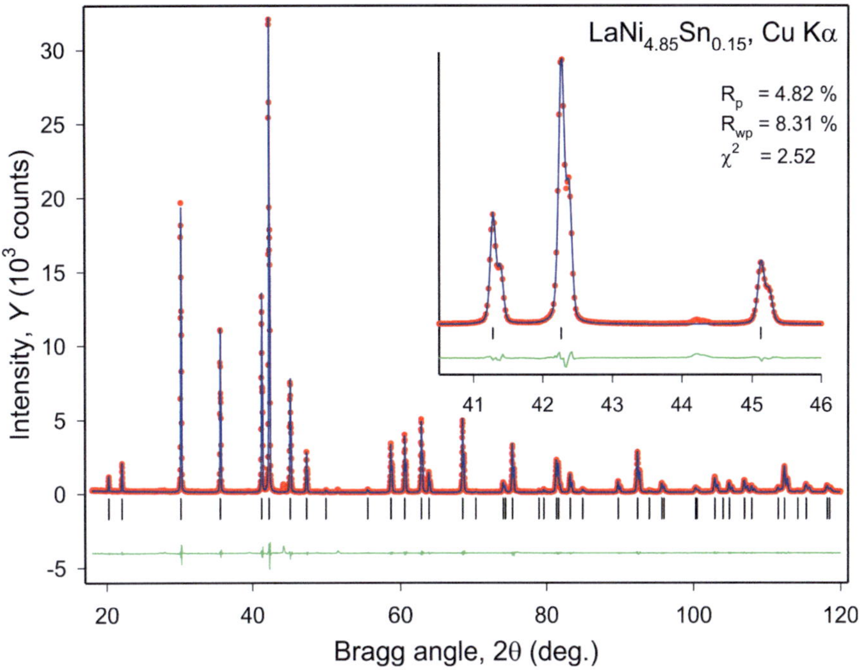

Fig. 16.7 The observed and calculated powder diffraction patterns of $LaNi_{4.85}Sn_{0.15}$ after all parameters have been refined with the increased peak base widths

integrated intensities are substituted with the absolute values of structure factors.[10] After a La atom is placed in the 1(a) site, the $R_F = 54.8\%$, which is quite good considering that five atoms are still missing from our model.

The coordinates and heights[11] of the electron density peaks after calculating a three-dimensional electron density distribution (Fig. 16.9) are listed in Table 16.5. The strongest peak (No. 1) confirms the placement of a La atom in the unit cell. It is easy to see that two additional and distinct peaks (No. 2 and 3) appeared on the Fourier map. Their heights are about half of the expected number of electrons in pure Ni, which is normal given the incomplete accuracy of reflection phases, which were calculated using only the La atom. Between Peaks No. 3 and 4, there is a sharp reduction in the heights of the electron density maxima (italicized in Table 16.5), and this feature usually indicates that no more atoms are located in the unit cell.

A simple calculation of the interatomic distances between the La atom in the 1 (a) site and all peaks listed in Table 16.5 shows that only the distances for Peaks 2 and 3 are a good match for La–Ni distances. The distance between Peaks 2 and

[10] This figure of merit is customarily used in single-crystal diffraction, where individual squared amplitudes $|\mathbf{F}_{hkl}^{obs}|^2$ are determined directly from the experiment.

[11] Peak heights are usually used instead of integrated peak volumes for simplicity.

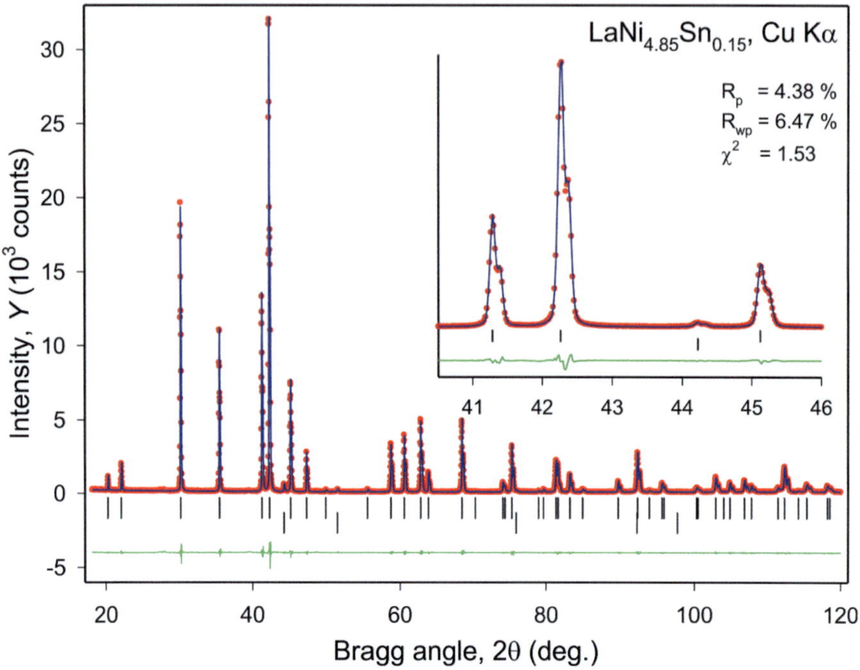

Fig. 16.8 The observed and calculated powder diffraction patterns of $LaNi_{4.85}Sn_{0.15}$ after a second phase (solid solution of Sn in Ni) was included in the refinement. Considering the low background (the peak-to-background ratio exceeds 200 for the strongest Bragg peak at $2\theta \cong 42.3°$), the final residuals are excellent

3 corresponds to the sum of radii of Ni atoms. Further, Peak No. 2 represents coordinates of the 2(c) sites, and Peak No. 3 corresponds to the 3(g) sites in space group P6/mmm. Thus, a total of six atoms have been placed in the unit cell of $LaNi_{4.85}Sn_{0.15}$, which is exactly as many as established from the gravimetric density of the alloy.

It is important to emphasize that La in the 1(a) site has been confirmed by the electron density calculation, although it is worth noting that in a heavy-atom approach it is often the case, that a single strongly scattering atom always appears on the Fourier map even when it has been placed incorrectly. However, if the location of the heavy atom is wrong, additional strong peaks in the electron density distribution will normally be incorrect, as can be easily established by the computation of the interatomic distances. Such a model of the crystal structure will be impossible to complete, as the missing atoms typically will not appear on the subsequent Fourier maps due to wrong phase angles.

After all three independent atoms (Peaks 1–3 in Table 16.5) have been included in computations assuming identical displacement parameters in an isotropic approximation ($B = 0.5$ Å^2), the resulting $R_F = 6.9\%$ without refinement. This value is excellent because (1) the powder diffraction pattern is relatively simple with

Table 16.2 Parameters obtained at different stages during the full-pattern decomposition of the powder diffraction pattern of $LaNi_{4.85}Sn_{0.15}$

Free parameters	Refined values
Initial (default) parameters	Background: $b_0 - b_4$ and b_{-1}, all zero $U = 0.0100$; $V = -0.0050$; $W = 0.0200$ $\eta_0 = 0.2000$; $\eta_1 = 0.0000$; $\eta_2 = 0.0000$ Asymmetry = 0.020; Anisotropy (001) = 0.0000 $a = 5.0421$ Å; $c = 4.0118$ Å; $\delta 2\theta = 0.000°$
Background (linear)	$b_0 = 146.58$; $b_1 = -0.448$
Background (linear), U, V, W, η_0	$b_0 = 154.24$; $b_1 = -0.699$ $U = 0.0164$; $V = 0.0599$; $W = -0.0023$; $\eta_0 = 0.2614$
Background (linear), U, V, W, η_0, a, c, zero shift	$b_0 = 157.95$; $b_1 = -0.698$ $U = 0.0403$; $V = -0.0081$; $W = 0.0056$; $\eta_0 = 0.4064$ $a = 5.0432$ Å; $c = 4.0126$ Å; $\delta 2\theta = 0.02°$
Background (third-order polynomial plus hyperbolic term, b_{-1}), U, V, W, η_0, asymmetry, a, c, zero shift	$b_0 = 41.921$; $b_1 = -0.112$; $b_2 = 4 \times 10^{-4}$; $b_3 = 3 \times 10^{-5}$; $b_{-1} = 3332.1$ $U = 0.0311$; $V = 0.0022$; $W = 0.0031$; $\eta_0 = 0.4152$ Asymmetry = −0.025 $a = 5.0419$ Å; $c = 4.0115$ Å; $\delta 2\theta = 0.049°$
Background (third order), U, V, W, η_0, η_1, η_2, asymmetry, anisotropy + 12 FWHM peak base, a, c, zero shift	$b_0 = 37.941$; $b_1 = -0.272$; $b_2 = 3 \times 10^{-3}$; $b_3 = 1 \times 10^{-5}$; $b_{-1} = 3470.6$ $U = 0.0283$; $V = -0.0046$; $W = 0.0058$ $\eta_0 = 0.1066$; $\eta_1 = 0.0086$; $\eta_2 = 0.0000$ Asymmetry = −0.016; Anisotropy (001) = 0.0165 $a = 5.0421$ Å; $c = 4.0116$ Å; $\delta 2\theta = 0.044°$
Same as above plus a second phase included. Due to a small amount of impurity, all of the peak-shape parameters of a second phase were constrained to be identical to those of the majority phase	$b_0 = 31.562$; $b_1 = -0.423$; $b_2 = 7 \times 10^{-3}$; $b_3 = 7 \times 10^{-6}$; $b_{-1} = 3654.2$ $U = 0.0281$; $V = -0.0042$; $W = 0.0056$ $\eta_0 = 0.1256$; $\eta_1 = 0.0080$; $\eta_2 = 0.0000$ Asymmetry = −0.016; Anisotropy (001) = 0.0165 $a = 5.0419$ Å; $c = 4.0114$ Å; $\delta 2\theta = 0.044°$ $a = 3.5416$ Å (second phase, $Ni_{1-x}Sn_x$)

minimum overlap, and (2) the powder particles used in the diffraction experiment were nearly ideal (spherical), thus preferred orientation effects were negligible. The following electron density distribution (Fig. 16.10 and Table 16.6) was obtained using the newly determined set of phase angles.

The major difference between the two Fourier maps shown in Figs. 16.9 and 16.10 and Tables 16.5 and 16.6 is that peak heights of the correctly placed atoms are

Table 16.3 The list of Bragg reflections with their corresponding observed structure factors squared, determined from Le Bail's full-pattern decomposition of the powder diffraction pattern of $LaNi_{4.85}Sn_{0.15}$. Only the data for the major phase are listed

h	*k*	*l*	$\|F^{obs}\|^2$	$\sigma\|F^{obs}\|^2$	*h*	*k*	*l*	$\|F^{obs}\|^2$	$\sigma\|F^{obs}\|^2$
0	1	0	85	1	0	3	2	1040	6
0	0	1	617	5	1	3	1	678	5
0	1	1	2012	6	0	2	3	255	4
1	1	0	3328	13	0	4	0	1846	17
0	2	0	5604	19	2	2	2	3862	19
1	1	1	7488	16	0	4	1	124	2
0	0	2	13,440	54	1	2	3	500	4
0	2	1	816	6	1	3	2	73	1
0	1	2	50	1	0	0	4	3436	36
1	2	0	75	2	2	3	0	87	1
1	1	2	1713	10	0	3	3	1842	13
1	2	1	1003	5	0	1	4	42	1
0	2	2	2979	14	2	3	1	548	5
0	3	0	1559	14	0	4	2	1509	12
0	3	1	3503	16	1	4	0	749	7
0	0	3	201	4	1	1	4	682	7
0	1	3	641	6	1	4	1	1787	9
1	2	2	60	1	2	2	3	129	2
2	2	0	5592	30	0	2	4	1291	10
1	3	0	78	2	1	3	3	502	4
2	2	1	166	3	2	3	2	91	1
1	1	3	2611	15					

Table 16.4 Low-multiplicity sites available in the space group P6/mmm

Site	Coordinates of symmetrically equivalent points			
1(a)	0,0,0			
1(b)	0,0,½			
2(c)	⅓,⅔,0	⅔,⅓,0		
2(d)	⅓,⅔,½	⅔,⅓,½		
2(e)	0, 0, *z*	0, 0, $\bar{z}$		
3(f)	½,0,0	0,½,0	½,½,0	
3(g)	½,0,½	0,½,½	½,½,½	
4(h)	⅓, ⅔, *z*	⅔, ⅓, *z*	⅓, ⅔, $\bar{z}$	⅔, ⅓, $\bar{z}$

much stronger than the heights of false peaks.[12] Further, the coordinates of false peaks vary, but the coordinates of true maxima remain the same. As is easy to verify by the calculation of distances, none of the peaks listed below Peak No. 3 in

[12]False peaks (e.g., Peak No. 4 in Table 16.5 which is easily recognizable in Fig. 16.9) appear on Fourier maps due to a variety of reasons: (1) the largest contribution comes from the truncation of the Fourier summation (Eq. 10.6) because only a limited amount of diffraction data is available (see Table 16.3); (2) the structure amplitudes are not exact, especially when powder diffraction data were used in combination with Le Bail's extraction; and (3) phase angles calculated using atomic parameters, which are not fully refined, are still imprecise because we used randomly assigned

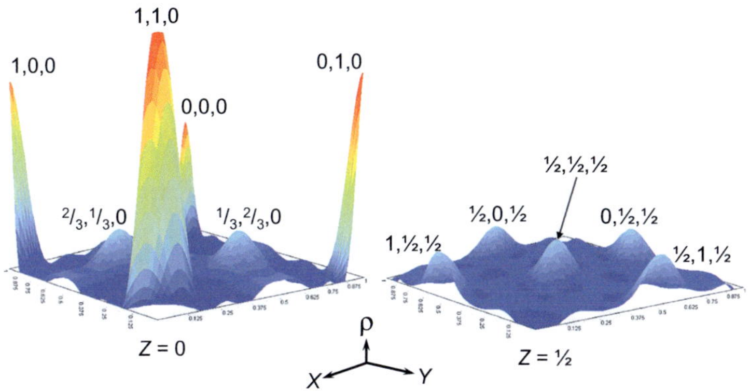

Fig. 16.9 The cross-sections of the three-dimensional Fourier map of $LaNi_{4.85}Sn_{0.15}$ at $Z = 0$ (*left*) and at $Z = ½$ (*right*) calculated using structure amplitudes listed in Table 16.3 and phase angles determined by the La atom placed in the 1(a) site. The triplets of numbers indicate the coordinates of the strongest peaks in the unit cell. The following groups of peaks are symmetrically equivalent to one another: 0,0,0; 1,0,0; 0,1,0; and 1,1,0 (all four correspond to the La atom in 1(a) and Peak No. 1 in Table 16.5); ⅓,⅔,0 and ⅔,⅓,0 (Peak No. 2 in Table 16.5); and ½,0,½; 0,½,½; ½,½,½; 1,½,½; and ½,1,½ (Peak No. 3 in Table 16.5)

Table 16.5 The three-dimensional electron density distribution in the symmetrically independent part of the unit cell of $LaNi_{4.85}Sn_{0.15}$ calculated using the observed structure factors determined from Le Bail's extraction (Table 16.3) and phase angles determined by the La atom placed in the 1 (a) site of the space group P6/mmm ($R_F = 54.8\%$)

Fourier map peak number	*x*	*y*	*z*	Peak height[a]
1	**0**	**0**	**0**	**58**
2	**0.6667**	**0.3333**	**0**	**14**
3	**0.5**	**0**	**0.5**	**14**
4	*0.5*	*0*	*0*	*4*
5	*0*	*0*	*0.288*	*3*
6	*0.462*	*0.124*	*0.240*	*2*
7	*0.540*	*0.249*	*0.512*	*2*
8	*0.211*	*0.075*	*0.516*	*1*
9	*0.231*	*0.200*	*0.151*	*1*

[a] Peak heights have been normalized to the number of electrons associated with the site occupied by La

Table 16.6 has a reasonable distance to the La and Ni atoms already located in the unit cell.

Considering the low R_F, the absence of new peaks on the second Fourier map, which may correspond to additional atoms, and the fact that the contents of the unit cell match that established from the gravimetric density of the material, we conclude that all atoms in the unit cell of $LaNi_{4.85}Sn_{0.15}$ have been located. It makes no sense

displacement parameters and assumed completely random distribution of Ni and Sn in two possible sites.

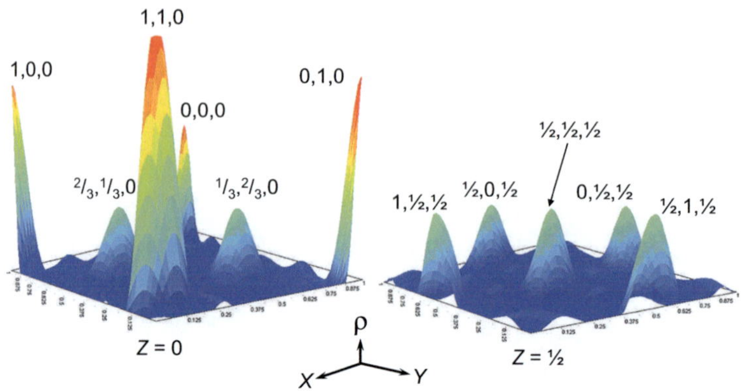

Fig. 16.10 The cross-sections of the three-dimensional Fourier map of $LaNi_{4.85}Sn_{0.15}$ at $Z = 0$ (*left*) and at $Z = ½$ (*right*) calculated using structure amplitudes listed in Table 16.3 and phase angles determined by the La atom placed in the 1(a) site and Ni atoms placed in the 2(c) and 3 (g) sites. The notations are identical to Fig. 16.9

Table 16.6 The three-dimensional electron density distribution in the symmetrically independent part of the unit cell of $LaNi_{4.85}Sn_{0.15}$ calculated using the observed structure factors determined from Le Bail's extraction (Table 16.3) and phase angles determined by the La atom placed in the 1 (a), and Ni atoms placed in the 2(c) and 3(g) sites of the space group P6/mmm ($R_F = 6.9\%$)

Fourier map peak number	*x*	*y*	*z*	Peak height
1	**0**	**0**	**0**	**70**
2	**0.5**	**0**	**0.5**	**27**
3	**0.6667**	**0.3333**	**0**	**25**
4	*0*	*0*	*0.304*	*3*
5	*0.141*	*0.067*	*0.5*	*2*

Table 16.7 Coordinates of atoms in the unit cell of $LaNi_{4.85}Sn_{0.15}$ as determined from powder diffraction data. All coordinates are fixed by symmetry of the corresponding sites and only population and displacement parameters can and should be refined using Rietveld technique

Atom	Site	*x*	*y*	*z*
La	1(a)	0	0	0
0.97Ni + 0.03Sn	2(c)	⅓	⅔	0
0.97Ni + 0.03Sn	3(g)	½	0	½

to proceed with the least-squares refinement of atomic parameters using structure factors determined from Le Bail's extraction, and the refinement of the crystal structure should be completed using the Rietveld technique (see Sect. 16.3). The coordinates and possible distribution of atoms are listed in Table 16.7.

It is easy to verify that if we would start with the La atom in the 1(b) site with coordinates 0,0,½, then the missing five Ni + Sn atoms will appear on a subsequent

Table 16.8 Coordinates of atoms (*x*, *y*, and *z*) and site occupancies (*n*) in the unit cell of $LaNi_{4.85}Sn_{0.15}$ according to the initial model of the crystal structure (compare with Table 16.7)

Atom	Site	*x*	*y*	*z*	*n*
La	1(a)	0	0	0	0.04167
0.97Ni1	2(c)	⅓	⅔	0	0.08083
0.03Sn1	2(c)	⅓	⅔	0	0.00250
0.97Ni2	3(g)	½	0	½	0.12125
0.03Sn2	3(g)	½	0	½	0.00375

Fourier map in 2(d)—⅓,⅔,½ and 3(f)—½,0,0 sites.[13] The two crystal structures are indeed identical to one another because they are only different by a translation vector (0,0,½); in other words, they correspond to a different choice of the unit cell origin.

According to Pearson's Handbook, this crystal structure is commonly known as the $CaCu_5$-type structure.[14] The parent $LaNi_5$ alloy is well-known for its excellent hydrogen storage properties, and many closely related alloys with the same crystal structure have been commercialized as electrode materials in rechargeable nickel–metal hydride batteries.

16.3 Rietveld Refinement Using Cu $K\alpha_{1,2}$ Radiation

To demonstrate the Rietveld refinement of this crystal structure, we begin with the profile and unit cell parameters determined from Le Bail's algorithm (Table 16.2) and the model of the crystal structure determined from sequential Fourier maps listed in Table 16.7. To account for the presumably statistical distribution of Ni and Sn atoms in the 2(c) and 3(g) sites in this crystal structure, the initial distribution of atoms in the unit cell has been assumed as listed in Table 16.8. The initial profile and structural parameters are available online in the input file for LHPM-Rietica.[15] We remind the reader that experimental diffraction data, collected on a Rigaku TTRAX rotating anode powder diffractometer using Cu Kα radiation in a continuous scan mode are also available online.[16]

The last column in Table 16.8 contains site occupancies by all atoms in the format required by LHPM-Rietica.[17] The occupancy of each site (n) is given as a product of the population parameter (g) and site multiplicity (m) divided by the multiplicity of the general site position (M):

[13] We leave verification of this statement to the reader as a self-exercise.

[14] W.Z. Haucke, Kristallstruktur von $CaZn_5$ and $CaCu_5$, Z. Anorg. Chem. **244**, 17 (1940). https://aflow.org/p/F7Y6/

[15] The file name is *Ch16Ex01a.inp*.

[16] The file name is *Ch16Ex01_CuKa.dat*.

[17] Some software products, e.g. GSAS, require specification of the population parameters as g, while the multiplicities of site positions are automatically included by the program.

Table 16.9 The progress of Rietveld refinement of the crystal structure of $LaNi_{4.85}Sn_{0.15}$ using powder diffraction data. Wavelengths used: $\lambda K\alpha_1 = 1.54059$ Å, $\lambda K\alpha_2 = 1.54441$ Å. In this and all other tables, the R_p, R_{wp}, and R_B are listed in percent; χ^2 is dimensionless

Refined parameters	R_p	R_{wp}	R_B	χ^2
Initial residuals: peak shape, background, zero shift and unit cell from Table 16.2, initial model from Table 16.8; see Fig. 16.11	19.75	25.11	21.11	22.91
Scale factor	11.26	14.47	9.35	7.62
Scale factor, peak-shape parameters, background (third-order polynomial), zero shift, unit cell; see Fig. 16.12	10.35	13.21	9.30	6.36
All of the above plus overall atomic displacement parameter	7.90	10.81	5.75	4.26
Individual atomic displacement parameters in isotropic approximation plus all peak-shape parameters, background, zero shift, unit cell, scale	6.77	9.95	3.98	3.61
Overall atomic displacement parameter but individual population parameters of 2(c) and 3(g) sites plus peak-shape parameters, background, zero shift, unit cell, scale	6.48	9.68	3.55	3.42
Individual atomic displacement parameters in isotropic approximation plus all peak-shape parameters, background, zero shift, unit cell, scale	6.44	9.68	3.58	3.42
Individual atomic displacement parameters in anisotropic approximation plus all peak-shape parameters, background, zero shift, unit cell, scale (i.e., all free variables)	6.26	9.57	3.10	3.34
Same as above plus a second phase (as Le Bail's extraction)—i.e., all free variables including those of an impurity phase; see Fig. 16.14	5.85	8.07	3.10	2.38

$$n = \frac{gm}{M} \tag{16.1}$$

The fractional population parameter, g, varies from 1 (fully occupied site) to 0 (completely unoccupied site). Thus, the fully occupied 1(a) site has occupancy by La: $n = 1/24 = 0.04167$; the 97% occupancy of 2(c) and 3(g) sites by Ni results in $n_{Ni1} = 0.97 \times 2/24 = 0.08083$ and $n_{Ni2} = 0.97 \times 3/24 = 0.12125$, respectively. The 3% occupancy of the same two sites by Sn yields $n_{Sn1} = 0.00250$ and $n_{Sn2} = 0.00375$, respectively. The overall atomic displacement parameter, $B = 0.5$ Å^2, has been assumed at the beginning of Rietveld refinement. The progression of the refinement using LHPM-Rietica is illustrated in Table 16.9 and in Figs. 16.11, 16.12, 16.13, and 16.14.

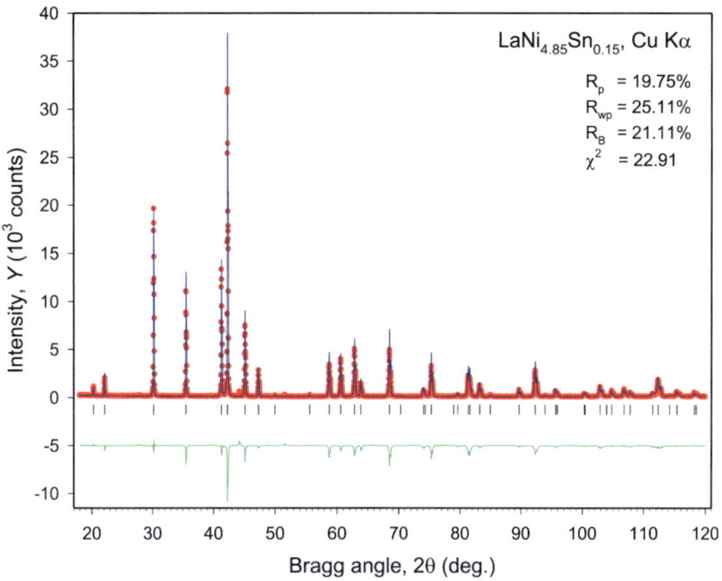

Fig. 16.11 The observed (*open circles*) and calculated (*solid line*) diffraction patterns of $LaNi_{4.85}Sn_{0.15}$. Scattered intensity was calculated using instrumental and lattice parameters determined during Le Bail's refinement and default scale factor ($K = 0.01$). The difference, $Y_i^{obs} - Y_i^{calc}$, is displaced by −5000 counts for clarity. *Vertical bars* in the lower part of the figure indicate calculated positions of the Kα$_1$ components of Bragg reflections in $LaNi_{4.85}Sn_{0.15}$

16.3.1 Scale Factor and Profile Parameters

Initial residuals (Row 1 in Table 16.9), calculated using profile parameters determined from the full-pattern decomposition employing Le Bail's algorithm and the default value of the scale factor ($K = 0.01$), are quite low.[18] The corresponding plot of the observed and calculated powder diffraction data is shown in Fig. 16.11, from which it is obvious that the intensities of nearly all observed Bragg peaks are lower than those of the calculated reflections. As follows from Eq. (15.30), this is indeed the effect of an overestimated scale factor. Thus, one of the critical parameters which should be determined at the beginning of every Rietveld refinement is the scale factor, K (also see Table 15.2). Several refinement cycles (all variables except K are fixed) are usually sufficient to reach convergence (Row 2, Table 16.9) since in this case the least-squares procedure, in fact, becomes linear.[19]

[18]This is usually not the case in the majority of Rietveld refinements because the default value of the scale factor selected in LHPM-Rietica ($K = 0.01$) is arbitrary. For example, see Table 16.11, where initial residuals are much higher due to the inadequacy of the default phase scale.

[19]One cycle is usually enough to determine K; additional cycles may be needed for a proper determination of the intensities of overlapped Bragg peaks.

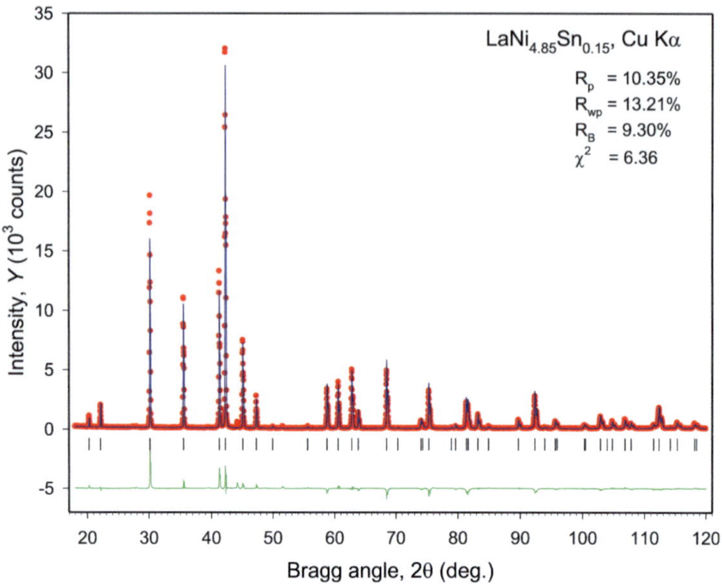

Fig. 16.12 The observed and calculated diffraction patterns of $LaNi_{4.85}Sn_{0.15}$. The scattered intensity was calculated using scale factor, instrumental, and lattice parameters determined during Rietveld refinement, and guessed overall atomic displacement parameter $B = 0.5$ Å^2. All notations are identical to Fig. 16.11

Fig. 16.13 One unit cell of the crystal structure of $LaNi_{4.85}Sn_{0.15}$ as determined from Rietveld refinement. The model reflects the different distribution of the Ni and Sn atoms between 2(c) and 3 (g) sites (dark- and light-gray, respectively). The displacement ellipsoids are shown at 99% probability

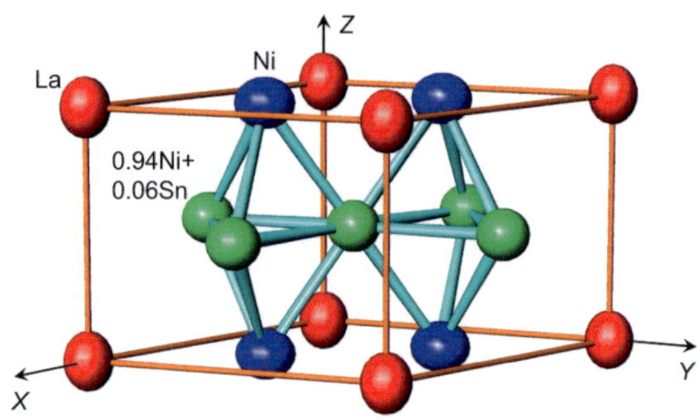

Under normal circumstances,[20] the scale factor is one of the few least-squares parameters, which is always kept as a free variable in a Rietveld refinement. Since in this example, all profile and lattice parameters have been refined using Le Bail's

[20] In rare cases, e.g., when all sites in the crystal structure are partially occupied, the scale factor may become strongly correlated with the population parameters. This requires special attention, and detailed consideration of all possibilities exceeds the scope of this text. One of the options is to use the known scale factor as a fixed parameter while refining population parameters of all crystallographic sites. This option is, however, seldom available because relative and not absolute intensities are customarily measured in a powder diffraction experiment.

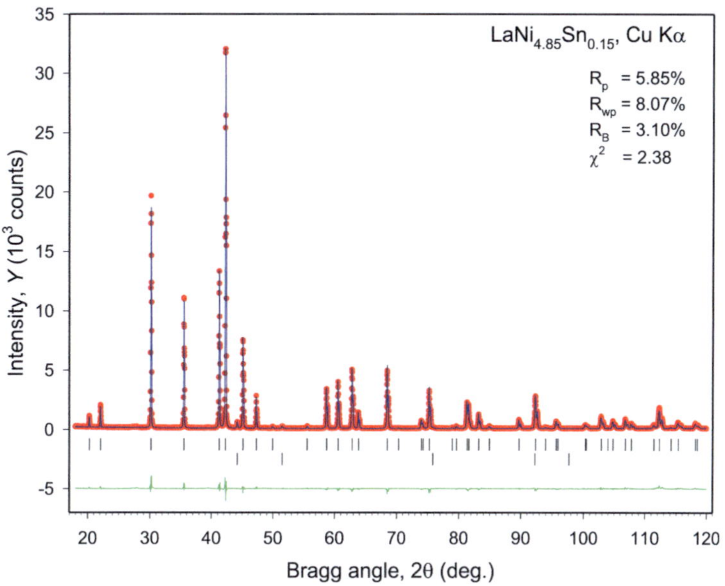

Fig. 16.14 The observed and calculated diffraction patterns of $LaNi_{4.85}Sn_{0.15}$ after the completion of Rietveld refinement. Notations are identical to Fig. 16.11, and the second set of vertical bars indicates the calculated positions of the $K\alpha_1$ components of Bragg peaks in the impurity phase (solid solution of Sn in Ni)

approach, the next reasonable step is to release all associated variables before proceeding with individual atomic parameters (the scale factor remains a least-squares free variable).[21] As expected, the resultant change in the figures of merit, listed in third row in Table 16.9, is small because the background, peak-shape, and lattice parameters are already quite accurate.

16.3.2 Overall Atomic Displacement Parameter

So far, the only "structural" parameter included in the refinement was the scale factor. This particular crystal structure has no free coordinates of individual atoms—all are fixed by the symmetry of the occupied special positions. Thus, given the relatively low values of all residuals and recalling that at the beginning we assumed an arbitrary value for the overall atomic displacement parameter ($B = 0.5$ Å^2), it is time to include it into the least-squares minimization. The inconsistency of the randomly chosen overall atomic displacement parameter is also easily distinguishable from the plot of the observed and calculated intensities, which is shown in

[21] In this structure, only individual displacement and population parameters can be refined.

Fig. 16.12. The intensities of the majority of Bragg peaks observed at low angles exceed calculated intensities, but at high angles ($2\theta > \sim65°$), the relationships become just the opposite. Given the effect of the atomic displacement parameters on the integrated intensity as a function of Bragg angle (see Eq. 9.5), it is easy to conclude that the actual overall B is considerably higher than 0.5 Å^2.

The refinement of the overall B results in a significant reduction of all residuals (fourth line in Table 16.9), which is also expected since now we have a realistic global estimate of thermal motions of atoms in the crystal lattice of $LaNi_{4.85}Sn_{0.15}$. The refined value of the overall atomic displacement parameter is $B = 1.32(2)$ Å^2, where the number in parenthesis indicates a standard deviation in the last significant digit.

16.3.3 Individual Parameters: Free and Constrained Variables

At this point, we may begin to refine atomic displacement parameters of the individual atoms. This is done by substituting individual B's (which were kept at 0) by the refined overall atomic displacement parameter. *The overall B after this substitution should be set to* 0. The distribution of atoms in the model of the crystal structure of $LaNi_{4.85}Sn_{0.15}$ is such that two of the sites, 2(c) and 3(g), see Table 16.8, are occupied by different types of atoms simultaneously in equal proportions. Generally, atomic displacement parameters of different atoms occupying the same crystallographic site should be constrained at identical values.[22]

In LHPM-Rietica and almost every other commonly available software product, any dependent parameter, $P_{dependent}$, may be constrained to any free least-squares parameter, P_{free}. The differentiation of the corresponding linear relationship between the two variables results in the numerical constraint, which is employed during the calculation of the normal equations matrix in the least squares and is applied to the shift obtained for the free parameter before adding it to the constrained variable:

$$P_{dependent} = f\left(P_{free}\right) \Rightarrow \Delta P_{dependent} = \frac{\partial f\left(P_{free}\right)}{\partial P_{free}} \Delta P_{free} \tag{16.2}$$

Considering Eq. (16.2) and assuming that $B_{Ni1} = B_{Sn1}$ and $B_{Ni2} = B_{Sn2}$, the proper constraints are $\Delta B_{Sn1} = 1 \times \Delta B_{Ni1}$ and $\Delta B_{Sn2} = 1 \times \Delta B_{Ni2}$. This implies that

[22]This is a reasonable simplification especially because it is nearly impossible to differentiate properly between the vibrational motions of different atoms located in the identical environment. Further, individual displacement parameters of atoms occupying the same crystallographic sites tend to strong correlation, thus causing severe instabilities of non-linear least squares. If a site has variable coordinate parameters, the latter should be constrained equal for all atoms occupying the site due to the same reasons.

individual isotropic atomic displacement parameters of Ni atoms are maintained as free least-squares variables and the same for Sn atoms are taken as constrained parameters.[23]

The resulting refinement further reduces the residuals (see Row 5 in Table 16.9) and yields the following distribution of the individual atomic displacement parameters:

$$\text{La in } 1(\text{a}), \quad B = 1.30(1)\ \text{Å}^2$$

$$0.97\text{Ni} + 0.03\text{Sn in } 2(\text{c}), \quad B = 1.81(2)\ \text{Å}^2$$

$$0.97\text{Ni} + 0.03\text{Sn in } 3(\text{g}), \quad B = 0.98(2)\ \text{Å}^2$$

It is obvious that atoms in the 2(c) site have much larger atomic displacement parameters than identical atoms in the 3(g) site. This situation is quite unusual for a simple intermetallic compound and likely indicates that our assumption about a statistical distribution of Ni and Sn in both crystallographic sites was incorrect. The enhanced isotropic atomic displacement parameter in the 2(c) site points to a lower scattering ability, while the reduced atomic displacement parameter in the 3(g) site points to a higher scattering factor when compared to the current distribution of atoms. Indeed, we may speculate that only 3(g) sites contain Sn atoms, which have greater scattering ability than Ni atoms. Another possibility is that the 2(c) sites are depleted in Sn, while some Sn atoms are still located there, and the 3(g) sites are enriched in Sn.

When the precision of X-ray diffraction data is high, which is the case here, it is possible to refine the population of different crystallographic sites to eliminate guesses and obtain a quantitative result. The best way to do so is to return to the overall displacement parameter and, instead of refining individual atomic displacement parameters, include the refinement of the individual population parameters in the corresponding sites.[24]

---

[23] In the hexagonal crystal system, a and b unit cell parameters are constrained by symmetry. ($a = b$). In LHPM-Rietica, lattice parameters are constrained automatically for all crystal systems. The input file, *Ch16Ex01b.inp*, with all other up-to-the-point parameters refined and properly constrained is found online (consult both the LHPM-Rietica manual and the on-line tutorial—Brett Hunter, LHPM-Rietica Rietveld accessible via http://ccp14.cryst.bbk.ac.uk/tutorial/lhpm-rietica/index.html)

[24] We note that since one of the sites, 1(a) seems to be fully occupied by La, the least-squares refinement of the population parameters of the two remaining sites, 2(c) and 3(g) may be carried out together with the scale factor. Only when the population of all sites is in question, special precautions should be taken to avoid severe correlation between the scale factor and all population parameters. When all sites are occupied partially, diffraction data alone normally do not provide an adequate answer because both K and g_j are simple multipliers, which affect the scattered intensity. Other experimental methods should be employed to establish and/or prove that defects exist on all lattice sites. One of these is measuring gravimetric density.

Assuming full occupancy of all sites, $g_{Ni1} = 1 - g_{Sn1}$ and $g_{Ni2} = 1 - g_{Sn2}$, the corresponding constraints (Eq. 16.2) should be set at $\Delta n_{Sn1} = -1 \times \Delta n_{Ni1}$ and $\Delta n_{Sn2} = -1 \times \Delta n_{Ni2}$.[25] This Rietveld refinement step results in the following occupancies of the two sites in question:

$$2(c) : n_{Ni1} = 0.085(1), n_{Sn1} = -0.002(1)$$

$$3(g) : n_{Ni1} = 0.118(1), n_{Sn1} = 0.007(1)$$

The negative occupancy by Sn of the 2(c) site has no physical sense, especially given that the absolute value of the occupancy is comparable with the standard deviation. Thus, this site appears to be pure Ni. Conversely, it is confirmed by the refinement that all Sn is segregated in the 3(g) sites. It is worth noting that when the chemical composition of the formula unit is calculated from the refined occupancies, the result is $LaNi_{4.83(2)}Sn_{0.17(2)}$, which matches the as-prepared chemical composition of the material within one standard deviation.[26]

Proceeding with the refinement of the individual atomic displacement parameters after removing Sn atoms from the 2(c) site and setting its occupancy to $n_{Ni1} = 0.08333$,[27] we find that the individual isotropic atomic displacement parameters of all atoms become much closer to one another. The resultant residuals are listed in Row 7 of Table 16.9.

16.3.4 Anisotropic Atomic Displacement Parameters

The only remaining degree of freedom in this crystal structure is to refine the displacement parameters of all atoms in the anisotropic approximation (the presence of preferred orientation is quite unlikely since the used powder was spherical and we leave it to the reader to verify its absence by trying to refine the texture using

---

[25] The file *Ch16Ex01c.inp*, in which all parameters are properly constrained, is found online.

[26] Refinement of the crystal structure is, therefore, a powerful chemical analysis technique. Unlike conventional chemical analysis, which only yields the bulk composition of the sample, powder diffraction facilitates accurate determination of the occupancies of different crystallographic sites by various chemical elements, or in other words, establishes precise chemical composition of the crystal at the atomic resolution. It should be noted that the results may be considered reliable only when the difference in the scattering by atoms in question is significant, in addition to a very high quality of experimental data. This is indeed the case here because normal scattering factors of Sn and Ni are related as ~1.8:1. We note that when using synchrotron radiation, it is also possible to tune the energy of the X-ray photons so that one of the elements scatters anomalously, e.g., see Eq. (9.15) and Fig. 9.5. Then, even elements that are neighbors in the periodic system, such as, for example, Cu and Zn that have nearly identical normal scattering factors of 29 and 30 electrons, respectively, become distinguishable because of large anomalous scattering components $\Delta f'$ and $\Delta f''$.

[27] The file *Ch16Ex01d.inp* is located online.

available experimental data). As noted in Sect. 9.1.2, special positions usually mandate certain relationships between the anisotropic atomic displacement parameters of the corresponding atoms. In the space group P6/mmm, the relevant constraints are as follows:

$$\begin{aligned} \text{La in } 1(a): &\quad \beta_{11}(\text{free}) = \beta_{22} = 2\beta_{12}; \beta_{33}\text{—free}; \beta_{13} = \beta_{23} = 0 \\ \text{Ni in } 2(c): &\quad \beta_{11}(\text{free}) = \beta_{22} = 2\beta_{12}; \beta_{33}\text{—free}; \beta_{13} = \beta_{23} = 0 \\ \text{Ni} + \text{Sn in } 3(g): &\quad \beta_{11}, \beta_{22}, \beta_{33}\text{—free}; \beta_{12} = 0.5\beta_{22}; \beta_{13} = \beta_{23} = 0 \end{aligned}$$

After the individual isotropic atomic displacement parameters were replaced by the properly constrained individual anisotropic displacement parameters (LHPM-Rietica uses β_{ij}, see Eq. 9.7), the refinement converges to the residuals listed in Row 8 of Table 16.9.[28]

16.3.5 *Multiple-Phase Refinement*[29]

The presence of a minor second phase impurity can be added either in the form of the actual structural model of Ni or as a Le Bail's phase, where only the unit cell and peak-shape parameters are considered. The latter option has been chosen since we are not interested in the crystal structure of this minor impurity, and it may be a difficult task, given its small contribution to the total scattered intensity.

The refinement with both crystalline phases contributing to the computation of the scattered intensity (Row 9, Table 16.9) converges rapidly and yields the residuals, which are only slightly higher than those obtained when no model of the crystal structure was present during full-pattern decomposition.

16.3.6 *Refinement Results*

The final parameters in the crystal structure of the $LaNi_{4.85}Sn_{0.15}$ as determined from the Rietveld refinement are listed in Table 16.10. The refined model of the crystal structure of the $LaNi_{4.85}Sn_{0.15}$ compound is shown in Fig. 16.13. All interatomic distances are normal, and the atomic displacement parameters of all atoms show little anisotropy, which is typical for many intermetallic compounds, structures of which are based on the close packing of spheres. The resultant observed and calculated diffraction patterns are plotted in Fig. 16.14 in a standard Rietveld format.

[28] The parameters of the fully refined structure can be found in the file *Ch16Ex01e.inp* online.

[29] The input file after the completion of the refinement (*Ch16Ex01f.inp*) is included online.

Table 16.10 Structural parameters of the major phase ($LaNi_{4.85}Sn_{0.15}$), fully refined by the Rietveld technique employing powder diffraction data collected using Cu Kα radiation (see Fig. 16.14). The refined unit cell dimensions are: $a = 5.04228(6)$, $c = 4.01170(5)$ Å

Atom	Site	x	y	z	g[a]	$10^4\beta_{11}$[b]	$10^4\beta_{22}$[b]	$10^4\beta_{33}$[b]	$10^4\beta_{12}$[b]
La	1(a)	0	0	0	1	156(2)	156(2)	258(3)	78(1)
Ni1	2(c)	⅓	⅔	0	1	227(3)	227(3)	183(5)	113(1)
Ni2	3(g)	½	0	½	0.944(2)	178(3)	174(4)	187(5)	87(2)
Sn	3(g)	½	0	½	0.056(2)	178(3)	174(4)	187(5)	87(2)

[a] The population parameters of the 3(g) site are listed as g, i.e., they represent fractional occupancies by Ni and Sn, refined assuming full overall occupancy of the site

[b] The following anisotropic displacement parameters are fixed by symmetry for all sites: $\beta_{13} = \beta_{23} = 0$

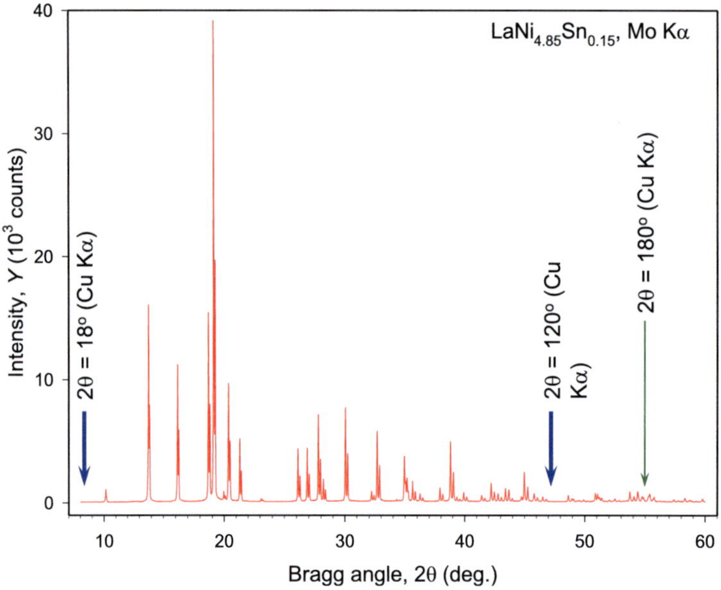

Fig. 16.15 The X-ray powder diffraction pattern of $LaNi_{4.85}Sn_{0.15}$ collected on a Rigaku TTRAX rotating anode powder diffractometer using Mo Kα radiation with a step $\Delta 2\theta = 0.01°$. The spherical $LaNi_{4.85}Sn_{0.15}$ powder was produced by high-pressure gas atomization. The two *short thick vertical arrows* show the range of $\sin\theta/\lambda$ examined using Cu Kα radiation (e.g., see Fig. 16.1), and the *long thin vertical arrow* indicates the fundamental upper limit of the reciprocal space ($2\theta = 180°$) accessible when using Cu Kα radiation

16.4 Rietveld Refinement Using Mo $K\alpha_{1,2}$ Radiation

The X-ray powder diffraction data were also collected from the same specimen of $LaNi_{4.85}Sn_{0.15}$ using Mo Kα radiation. The experimental data are shown in Fig. 16.15. One of the most obvious benefits of using higher energy photons is the greater volume of the reciprocal lattice, which can be examined experimentally. An important advantage of including the data measured at higher sinθ/λ into the Rietveld refinement is in obtaining more accurate values of the individual atomic displacement parameters together with the more precise site populations in this crystal structure. The better accuracy, when compared to Cu Kα radiation, becomes obvious from a simple analysis of Eqs. (9.3 and 9.5), which indicate that the effect of varying g is independent of the Bragg angle, while the varying B has the largest influence on the intensity, scattered at high sinθ/λ. Thus, by including intensity scattered at high sinθ/λ we reduce the correlation between the atomic displacement and population parameters.

Since we did not perform Le Bail's decomposition of this powder diffraction pattern, we illustrate the Rietveld refinement sequence in this case beginning from all profile parameters selected at their default values and the same starting model of the crystal structure as was used in the case of Cu Kα radiation data. The progression of the refinement is shown in Table 16.11.[30]

When peak-shape and background parameters are unknown, the Rietveld refinement strategy usually changes. At the beginning (first nine rows in Table 16.11), major attention has been paid to a sensible refinement of the peak shape, background, and lattice parameters along with the zero shift to achieve the best-possible agreement between the observed and calculated scattered intensity. Only then were the overall displacement and properly constrained individual population parameters included into the refinement. At this point, the examination of low Bragg angle peak shapes indicated that Howard's asymmetry approximation is not suitable, and asymmetry was changed to a more realistic Finger, Cox, and Jephcoat model, as indicated in the footnotes to Table 16.11.

Since the quality of this powder diffraction pattern is excellent, we can immediately switch to the refinement of the individual anisotropic parameters. Further, the presence of diffraction data at high sinθ/λ enables us to refine the individual displacement and population parameters in the 3(g) site simultaneously. Finally, the presence of an impurity was also accounted as the "Le Bail's" phase. The resultant observed and calculated powder diffraction patterns are shown in Fig. 16.16, and the final parameters of individual atoms are listed in Table 16.12.[31]

A comparison of the results presented in Table 16.12 with those shown in Table 16.10 indicates no discrepancies in the refined populations of Ni2 and Sn atoms, which simultaneously occupy the 3(g) site, although a small difference in the anisotropic displacement parameters of all atoms is evident. Regardless of this difference, as seen in Fig. 16.17, displacement ellipsoids have nearly identical orientations and shapes. The preference should be given to the results obtained using Mo Kα radiation because Bragg reflections at higher sinθ/λ were included into the determination of the individual anisotropic displacement parameters.

[30] The input file *Ch16Ex01g.inp* with initial variables and the experimental data file *Ch16Ex01_MoKa.dat* are available online.

[31] The fully refined input data file *Ch16Ex01h.inp* is located online.

Table 16.11 The progress of Rietveld refinement of the crystal structure of $LaNi_{4.85}Sn_{0.15}$ using powder diffraction data shown in Fig. 16.15. Wavelengths used: $\lambda K\alpha_1 = 0.70932$ Å, $\lambda K\alpha_2 = 0.71361$ Å

Refined parameters	R_p	R_{wp}	R_B	χ^2
Initial (all default, model from Table 16.8)[a]	617.8	808.3	608.7	3×10^4
Scale factor[a]	52.96	60.34	36.38	172.8
Scale factor plus linear background[a]	39.91	47.28	27.55	106.1
Scale, linear background plus *U*, *V*, *W*[a]	22.85	29.97	12.29	42.67
Scale, linear background, *U*, *V*, *W*, plus η_0[a]	20.44	26.51	10.13	33.39
Scale, linear background, *U*, *V*, *W*, η_0, *a*, and *c*[a]	14.93	19.59	9.21	18.24
Scale, linear background, *U*, *V*, *W*, η_0, *a*, *c*, plus asymmetry[a]	13.25	17.91	9.15	15.25
Scale, linear background, *U*, *V*, *W*, η_0, *a*, *c*, asymmetry, plus zero shift[a]	13.02	17.77	9.02	15.01
Scale, *U*, *V*, *W*, η_0, *a*, *c*, asymmetry, zero shift, background (third order)[a]	11.81	15.75	8.73	11.79
Scale, *U*, *V*, *W*, η_0, *a*, *c*, asymmetry, zero shift, background (third order), plus overall *B*[a]	9.13	12.96	4.33	7.99
Scale, *U*, *V*, *W*, η_0, *a*, *c*, asymmetry, zero shift, background (third order), overall *B*, plus individual population parameters[a]	8.47	12.18	2.76	7.06
Scale, *U*, *V*, *W*, η_0, *a*, *c*, zero shift, background (third order), overall *B*, individual population parameters. Asymmetry in FCJ approximation[b]	7.49	11.20	2.47	5.97
Scale, *U*, *V*, *W*, η_0, *a*, *c*, asymmetry, zero shift, background (third order), individual population parameters, plus individual anisotropic displacement parameters[b,c]	6.95	10.80	1.52	5.56
Scale, all peak-shape, lattice, zero shift, background (third order), individual population, and anisotropic parameters[b,c]	6.21	10.28	1.48	5.03
All as above plus a second phase (Le Bail's decomposition)[b,c]	5.71	8.41	1.33	3.38

[a] Pseudo-Voigt peak-shape function with Howard's asymmetry, see Eq. (8.39)
[b] Pseudo-Voigt peak-shape function. Asymmetry in Finger, Cox, and Jephcoat approximation, which better represents peak shapes measured on this powder diffractometer using Mo Kα radiation
[c] Population of the 2(c) site set to pure Ni, while the population of the 3(g) site remained free parameter

Conversely, the comparison of lattice parameters obtained using Cu Kα (Table 16.10) and Mo Kα (Table 16.12) data indicates small but statistically significant differences. Considering the fact that the longer wavelength experiment includes reflections at higher Bragg angles, the preference should be given to the unit cell dimensions obtained in the refinement based on Cu Kα radiation data.

Given the observed small variations in the Rietveld refinement results, it is much better to employ all available data while performing the combined least-squares fit of the model. This can be done using the majority of available Rietveld refinement programs and is illustrated in Sect. 16.5.

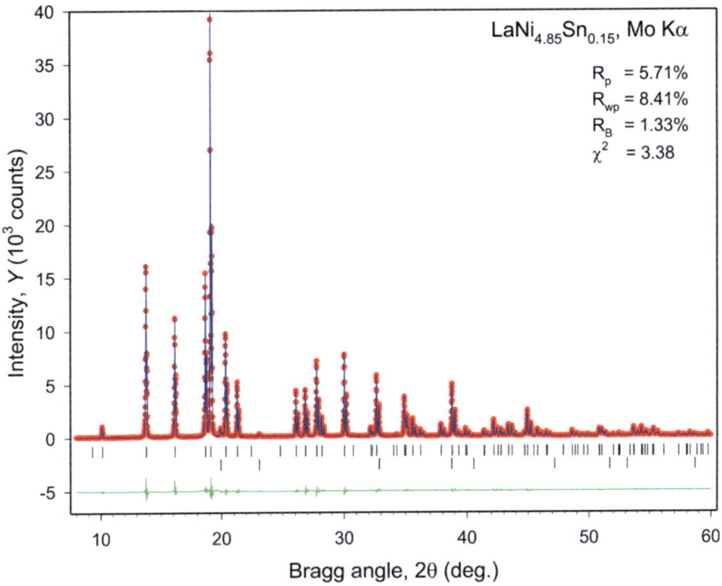

Fig. 16.16 The observed and calculated powder diffraction patterns of $LaNi_{4.85}Sn_{0.15}$ after the completion of Rietveld refinement. Notations are identical to Fig. 16.11. The second set of *vertical bars* indicates the calculated positions of the $K\alpha_1$ components of Bragg peaks in the impurity phase, which is a solid solution of Sn in Ni. The virtual absence of a Bragg peak at $2\theta \cong 9°$ and the presence of the same reflection at $2\theta \cong 20°$ when Cu Kα radiation was employed (see Fig. 16.14) are the result of differences in the anomalous scattering (see also Eqs. 9.15 and 9.22)

Table 16.12 Structural parameters of the major phase ($LaNi_{4.85}Sn_{0.15}$), fully refined by the Rietveld technique using powder diffraction data shown in Fig. 16.16. The refined unit cell dimensions are: a = 5.04479(3), c = 4.01303(3) Å. The following anisotropic displacement parameters are constrained by symmetry in all sites: $\beta_{13} = \beta_{23} = 0$

Atom	Site	x	y	z	g[a]	$10^4\beta_{11}$	$10^4\beta_{22}$	$10^4\beta_{33}$	$10^4\beta_{12}$
La	1(a)	0	0	0	1	135(2)	135(2)	252(4)	68(1)
Ni1	2(c)	⅓	⅔	0	1	194(2)	194(2)	130(4)	97(1)
Ni2	3(g)	½	0	½	0.944(2)	136(2)	104(3)	137(4)	52(2)
Sn	3(g)	½	0	½	0.056(2)	136(2)	104(3)	137(4)	52(2)

[a] The population parameters of the 3(g) site are listed as g, i.e., they represent fractional occupancies by Ni and Sn, refined assuming full overall occupancy of the site

16.5 Combined Refinement Using Different Sets of Diffraction Data

We will begin combined Rietveld refinement of the crystal structure of $LaNi_{4.85}Sn_{0.15}$ using profile parameters (background, zero shifts, and peak shape) determined for each set of diffraction data during independent refinements. The starting unit cell dimensions have been chosen as established from Cu Kα data, and

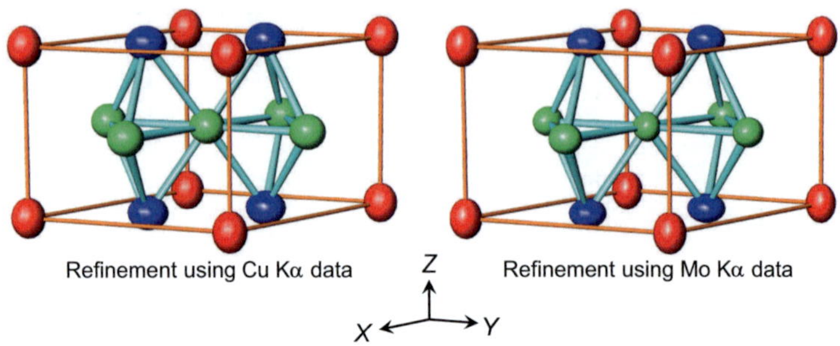

Fig. 16.17 Atomic displacement ellipsoids in the crystal structure of $LaNi_{4.85}Sn_{0.15}$, shown at 99% probability, as refined using Cu Kα (*left*) and Mo Kα (*right*) powder diffraction data

Table 16.13 The progress of the combined Rietveld refinement of the crystal structure of $LaNi_{4.85}Sn_{0.15}$ employing two sets of experimental data obtained using Cu Kα and Mo Kα radiations

	Cu Kα			Mo Kα		
Refined parameters	R_p	R_{wp}	R_B	R_p	R_{wp}	R_B
Initial	11.78	14.45	11.45	523.6	612.5	589.1
Scale factors[a]	6.97	10.08	4.40	22.44	27.02	8.12
Scale factors plus *a*, *c*, and zero shifts	8.00	11.38	4.60	7.55	11.39	1.40
As above plus peak shape, background	7.44	10.51	4.35	6.38	10.27	1.42
As above plus β_{ij} (all) and population parameter, *n*, in 3(g) site	7.29	10.39	4.13	6.43	10.30	1.53
All plus an impurity phase	6.87	9.02	4.13	5.99	8.62	1.40

[a] When multiple sets of data are used in a combined Rietveld refinement, the first set has a fixed scale, $k = 1$, but all other sets have their own scales in addition to a phase scale. Thus, in our case when we have two sets of diffraction data and one crystalline phase, two scale factors (K and $k_{Mo\ K\alpha}$) have been refined independently (see Eq. 16.3)

the starting population and individual atomic displacement parameters are taken from the Mo Kα result.[32] The progression of Rietveld refinement is illustrated in Table 16.13. The initial residuals are shown in the first row in Table 16.13, where as expected, the figures of merit for the second set of data (Mo Kα) are much worse than for the first (see Tables 16.9 and 16.11).

When more than one set of experimental diffraction data is employed in the combined Rietveld refinement, the minimized function (in the simplest case of Eq. 15.29) becomes

[32]The corresponding input data file *Ch16Ex01i.inp* is found online. For LHPM-Rietica, a single data file should contain both sets of powder diffraction data, and these can be found online in the file *Ch16Ex01_Cu&MoKa.dat*.

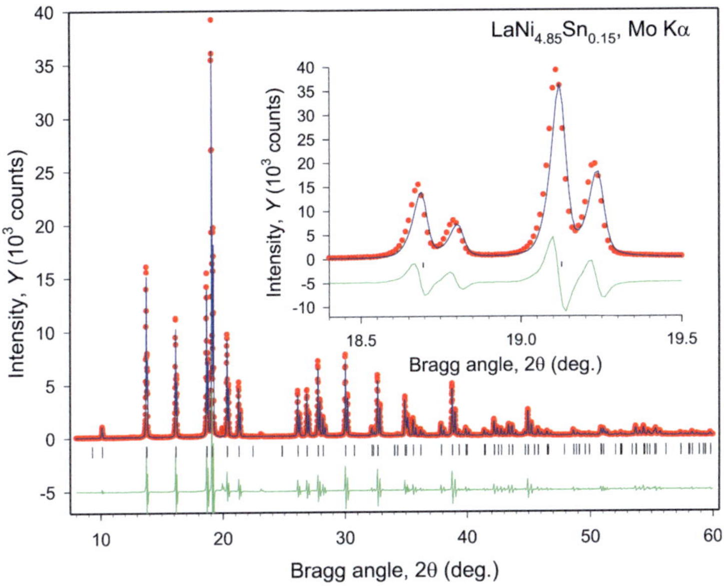

Fig. 16.18 The observed (Mo Kα radiation) and calculated powder diffraction patterns of $LaNi_{4.85}Sn_{0.15}$ after combined refinement of scale factors only. The inset shows the expanded view between 18.4° and 19.5° of 2θ to illustrate the inaccuracy of lattice parameters

$$\Phi = \sum_{s=1}^{h} \sum_{i=1}^{n_s} w_{s,i} \left[k_s \left(Y_{s,i}^{obs} - b_{s,i} \right) - K \sum_{j=1}^{m} I_j y_{s,j} \left(x_{s,j} \right) \right]^2 \qquad (16.3)$$

In Eq. (16.3), h is the number of different sets of powder diffraction data, n_s is the number of data points collected in the sth set, and k_s is the scale factor for the sth diffraction pattern, which appears because scattered intensity is measured on a relative scale. Other notations are identical to Eq. (15.29). Different scale factors, k_s and K, are simple multipliers. Hence, they strongly correlate and usually are not refined simultaneously. Constraining one of the scale factors (usually k_1, for the first diffraction dataset) at 1 enables successful refinement of the phase scale (K) and scale factors of all remaining sets of diffraction data (k_2, k_3, …, k_h). Eqs. (15.30, 15.32, and 15.33) are modified in the same way as Eq. (15.29) for a combined Rietveld refinement. Further, it is often the case that X-ray and neutron, or conventional X-ray and synchrotron data are used in combined refinements; therefore, the corresponding groupings of Eqs. (15.29, 15.30, 15.32, and 15.33), modified as shown in Eq. (16.3), are employed to express the minimized function.

Hence, the subsequent step is to refine both the phase scale factor and the scale factor of the Mo Kα experiment. As shown in row 2 of Table 16.13, these two variables have a tremendous effect on the resulting figures of merit. The residuals for Mo Kα data, however, remain far from the best, and a simple examination of the

Table 16.14 Structural parameters of the major phase ($LaNi_{4.85}Sn_{0.15}$), fully refined by the Rietveld technique using the combined powder diffraction data collected employing Cu Kα and Mo Kα radiations from the same specimen. The refined unit cell dimensions are: $a = 5.04430(3)$, $c = 4.01292(3)$ Å. The following restrictions are imposed by symmetry: $\beta_{13} = \beta_{23} = 0$

Atom	Site	*x*	*y*	*z*	*g*	$10^4\beta_{11}$	$10^4\beta_{22}$	$10^4\beta_{33}$	$10^4\beta_{12}$
La	1(a)	0	0	0	1	141(1)	141(1)	252(3)	71(1)
Ni1	2(c)	⅓	⅔	0	1	205(2)	205(2)	151(4)	103(1)
Ni2	3(g)	½	0	½	0.956(2)	142(2)	117(3)	139(3)	59(1)
Sn	3(g)	½	0	½	0.044(2)	142(2)	117(3)	139(3)	59(1)

Rietveld plot (Fig. 16.18) indicates that this is due to the inadequacy of the unit cell parameters determined from the Cu Kα experiment.

Thus, the next refinement step includes releasing unit cell dimensions and zero-shift parameters. The latter are usually different for different sets of diffraction data. As seen in Row 3 of Table 16.13, the overall fit becomes much better, though the residuals for Cu Kα data slightly increase. After including all possible variables into the refinement (Rows 4 and 5 in Table 16.13), the fit improves for both sets of experimental data.[33]

As expected, adding the contribution from the impurity phase (again as Le Bail's approximation) results in further reduction of the profile residuals, see Row 6 in Table 16.13. Structural parameters of the final model, as determined using the combined Rietveld refinement in the two-phase approximation are listed in Table 16.14.[34]

The resultant unit cell dimensions and structural parameters shown in Table 16.14 are closer to those obtained in an independent Rietveld refinement using Mo Kα data (see Table 16.12). Further, the population of the 3(g) site is slightly different from that obtained in both independent refinements. The resulting chemical composition of the major phase is $LaNi_{4.87(1)}Sn_{0.13(1)}$, which remains within two standard deviations from the alloy stoichiometry.

The Rietveld plots of both powder diffraction datasets are shown in Figs. 16.19 and 16.20. Visual analysis of both figures indicates a good fit, which was expected from the low residuals (Table 16.13). The model of the crystal structure (Table 16.14) appears to be complete and makes both physical (reasonable atomic displacement parameters) and chemical sense (no overlapping atoms, the 3(g) sites are occupied simultaneously by atoms that have close atomic volumes—Ni and Sn; the chemical composition of the major phase established from X-ray data is nearly identical to the known composition of the alloy). Therefore, the outcome of this crystal structure determination may be accepted as satisfactory.

The presence of an impurity phase in this specimen may also be accounted for by including its crystal structure into a "normal" Rietveld refinement process rather than

[33] The fully refined model of the crystal structure of $LaNi_{4.85}Sn_{0.15}$ without including the impurity phase is found in the file *Ch16Ex01j.inp* online.

[34] And found in the data file *Ch16Ex01k.inp* online.

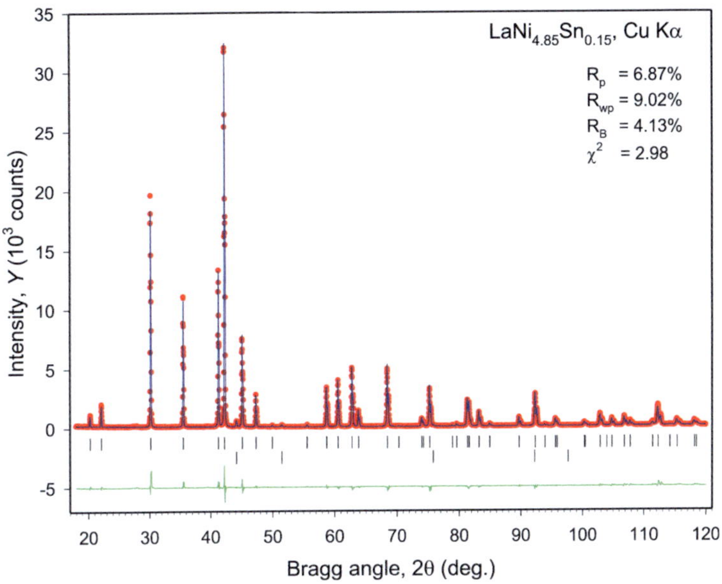

Fig. 16.19 The observed and calculated powder diffraction patterns of $LaNi_{4.85}Sn_{0.15}$ (Cu Kα radiation) after the completion of the combined Rietveld refinement

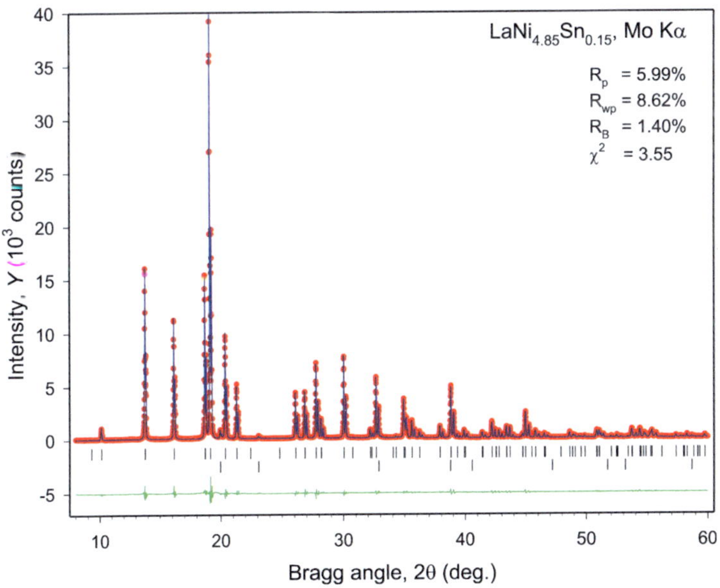

Fig. 16.20 The observed and calculated powder diffraction patterns of $LaNi_{4.85}Sn_{0.15}$ (Mo Kα radiation) after the completion of the combined Rietveld refinement

as a “Le Bail’s” phase.[35] It is unfeasible to refine the chemical composition of the Ni-based impurity because of its low concentration in the sample. Thus, only the scale factor, unit cell dimensions, and overall atomic displacement parameter have been refined for the impurity phase, while its chemical composition was assumed as $Ni_{4.85}Sn_{0.15}$. All residuals remain practically identical to those shown in the last row of Table 16.13.

As noted earlier (Sect. 13.3.4), Rietveld refinement offers an opportunity to determine the phase composition of the sample quantitatively. This information can be found in the general output file after Rietveld refinement; its extension is “out.” The corresponding concentrations are as follows: the major $LaNi_{4.85}Sn_{0.15}$ phase makes 97.5(4) and the impurity $Ni_{0.97}Sn_{0.03}$ phase accounts for 2.5(3) percent by weight of the alloy.

[35] For the sake of illustration, this has been done and the set of fully refined parameters can be found in the data file *Ch16Ex01m.inp*.

Chapter 17
Crystal Structure of $CeRhGe_3$: Low- Versus High-Symmetry Model

In this example, we will work with powder diffraction patterns of $CeRhGe_3$,[1] collected using Mo Kα radiation and thermal neutrons. We already discussed this compound in Problem 2, Chap. 14. The crystal system is tetragonal with $a = 4.3979(1)$ and $c = 10.0329(3)$ Å.[2]

Analysis of the possible systematic absences (see the solution of Problem 2 in Chap. 14) indicates a body-centered lattice with no additional forbidden reflections. Therefore, one of the following eight space groups is to be expected: I4/mmm, $I\bar{4}m2$, $I\bar{4}2m$, I4mm, I422, I4/m, $I\bar{4}$, or I4. The measured gravimetric density of the alloy is 7.79 g/cm^3. One unit cell ($V = 194.05$ Å^3) contains two formula units of $CeRhGe_3$ ($Z = 1.98 \approx 2$) or a total of two Ce, two Rh, and six Ge atoms. Similar to the previous example, none of the eight space groups can be excluded, as all contain two- and fourfold sites suitable to accommodate all three types of atoms in either an ordered or a disordered manner. Rh and Ge atoms have similar atomic volumes and may share same sites, while the much larger Ce atoms should occupy their own distinct sites.

17.1 Full-Pattern Decomposition

Full-pattern decomposition has been performed using LHPM-Rietica. Peak shapes have been represented using a pseudo-Voigt function. The progression of the full-pattern decomposition is shown by the corresponding figures of merit, which are

[1] V.K. Pecharsky and K.A. Gschneidner, Jr., unpublished. The alloy was prepared by arc-melting a stoichiometric mixture of pure components and then heat treated at 900 °C for 1 week. Metallographic examination indicated that the alloy was essentially a single-phase material.

[2] Files with experimental data (*Ch17Ex01_MoKa.dat*, *Ch17Ex01_MoKa.xy*, and *Ch17Ex01_Neut.dat*) are available online.

P. Zavalij, V. Pecharsky, *Fundamentals of Powder Diffraction and Structural Characterization of Materials*, https://doi.org/10.1007/978-3-031-91504-8_17

Table 17.1 Figures of merit obtained during Le Bail decomposition of the powder diffraction pattern of $CeRhGe_3$. Wavelengths used: $\lambda K\alpha_1 = 0.70932$ Å, $\lambda K\alpha_2 = 0.71361$ Å; $R_{exp} = 2.96\%$

Refined parameters	Illustration	R_p	R_{wp}	χ^2
Background (linear)	–	32.68	44.44	226.1
+ *a*, *c*	–	27.35	35.45	143.9
+ *U*, *V*, *W*, η_0, asymmetry	–	7.63	10.74	13.23
+ Background (fourth order), broader base	–	4.89	7.03	5.67
All, with Howard's[a] asymmetry	Figure 17.1	4.71	6.87	5.41
All, with FCJ[b] asymmetry	Figure 17.2	4.45	6.52	4.88

[a] C.J. Howard, The approximation of asymmetric neutron powder diffraction peaks by sums of Gaussians, J. Appl. Cryst. **15**, 615 (1982). https://doi.org/10.1107/S0021889882012783

[b] Asymmetry was treated in the two-parameter approximation given by L.W. Finger, D.E. Cox, and A.P. Jephcoat, A correction for powder diffraction peak asymmetry due to axial divergence, J. Appl. Cryst. **27**, 892 (1994). https://doi.org/10.1107/S0021889894004218

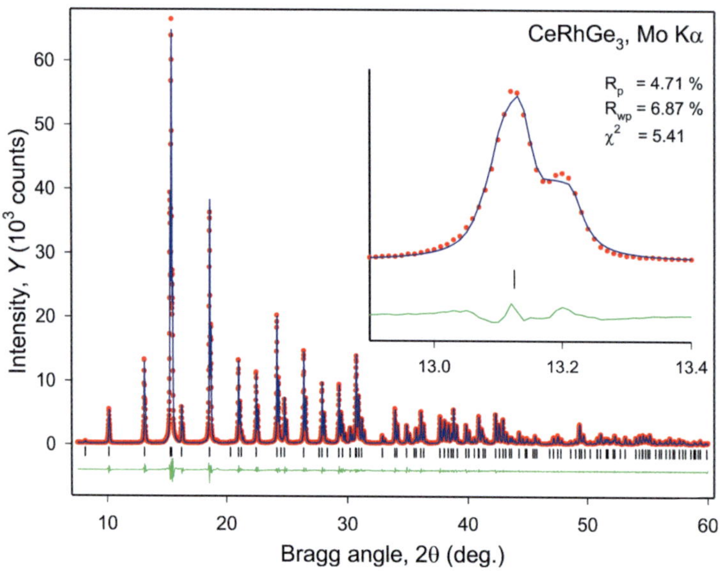

Fig. 17.1 The observed and calculated powder diffraction patterns of $CeRhGe_3$ after all parameters were refined in the same approximation as in Chap. 16. The data were collected from a ground sample of $CeRhGe_3$ using Mo Kα radiation on a Rigaku TTRAX rotating anode diffractometer. The divergence slit was 0.38°; the receiving slit was 0.03°. The experiment was carried out in a step-scan mode with a step 0.01° and counting time 4 s per step. The inset illustrates an inadequate asymmetry approximation

assembled in Table 17.1, and the results are illustrated in Figs. 17.1 and 17.2 and Table 17.2.

The use of Mo Kα radiation shifts all Bragg peaks to lower angles, and therefore, asymmetry is more severe than in the previous example, where Cu Kα radiation was used. As a result, the order in which parameters were refined was changed to avoid

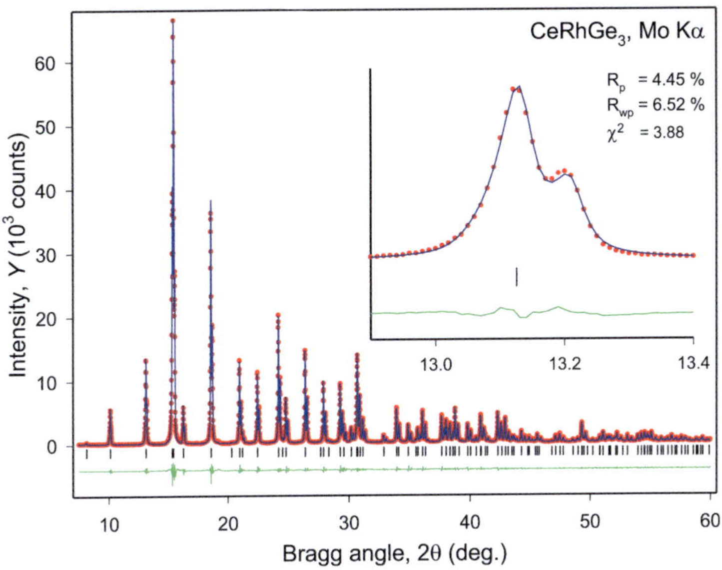

Fig. 17.2 The observed and calculated powder diffraction patterns of $CeRhGe_3$ after refinement of all parameters with an asymmetry correction in the Finger, Cox, and Jephcoat approximation. The inset illustrates an adequately treated asymmetry (compare with the inset in Fig. 17.1)

potential least-squares instability problems. When everything was refined in essentially the same approximation as in Chap. 16 (see Row 5 in Table 17.1), the resultant figures of merit are satisfactory, but a careful analysis of Fig. 17.1 indicates that the shapes of low-angle Bragg peaks are not approximated adequately.

An improvement (Fig. 17.2) was obtained by choosing a different asymmetry correction, suggested by Finger, Cox, and Jephcoat (FCJ), where the effects of axial divergence are treated by introducing a two-parameter asymmetry function. The residuals are lower (see Table 17.1), and this modification has an effect on the lattice parameters: the unit cell dimensions and zero shift refined using Howard's asymmetry approximation are $a = 4.3986$, $c = 10.0331$ Å, $\delta 2\theta = 0.044°$, while they become $a = 4.3981$, $c = 10.0320$ Å, $\delta 2\theta = 0.038°$ in the FCJ approximation. Given the fact that the least-squares standard deviations of lattice parameters are smaller than 10^{-4} Å, these differences are statistically significant and the result shown in Fig. 17.2 should be considered as a better outcome of the full-pattern decomposition.

17.2 Solving the Crystal Structure from X-Ray Data

Similar to the example in Chap. 16, $CeRhGe_3$ is an intermetallic compound, so it makes sense to assume the highest symmetry space group—I4/mmm—as a first attempt to solve the crystal structure. Cerium is the strongest scattering atom

Table 17.2 The list of Bragg reflections with their corresponding observed structure factors squared determined after Le Bail's full-pattern decomposition of the X-ray powder diffraction pattern of $CeRhGe_3$

h	*k*	*l*	$\|F^{obs}\|^2$	$\sigma\|F^{obs}\|^2$	*h*	*k*	*l*	$\|F^{obs}\|^2$	$\sigma\|F^{obs}\|^2$
0	0	2	13	0[a]	2	3	7	146	1
0	1	1	94	0	0	4	6	76	1
1	1	0	756	2	2	4	4	229	2
0	1	3	1435	2	0	2	10	39	0
1	1	2	1984	2	0	1	11	44	1
0	0	4	1054	4	0	3	9	273	2
0	2	0	4329	7	3	3	6	428	3
0	2	2	6	0	0	5	1	22	0
1	1	4	929	3	3	4	1	22	0
1	2	1	53	0	1	5	0	151	1
0	1	5	895	3	1	4	7	104	1
1	2	3	950	2	0	5	3	298	1
0	0	6	185	1	3	4	3	298	1
0	2	4	720	3	1	5	2	301	1
2	2	0	3276	8	2	4	6	54	0
2	2	2	2	0	2	2	10	0	0
1	1	6	1249	4	0	0	12	568	5
0	3	1	37	0	1	2	11	86	1
1	2	5	664	2	0	4	8	315	2
1	3	0	382	2	2	3	9	385	2
0	1	7	363	2	1	5	4	189	1
0	3	3	756	2	2	5	1	23	0
1	3	2	1006	2	1	3	10	170	1
0	2	6	141	1	1	1	12	272	2
2	2	4	558	3	0	5	5	226	1
0	0	8	806	6	3	4	5	226	1
1	3	4	520	2	3	3	8	53	1
2	3	1	19	0	2	5	3	271	2
0	3	5	593	3	0	2	12	612	4
1	1	8	78	1	4	4	0	1346	8
1	2	7	240	1	0	3	11	48	0
2	3	3	549	2	2	4	8	382	2
2	2	6	107	1	1	4	9	329	2
0	4	0	2012	9	4	4	2	7	0
0	2	8	871	4	1	5	6	390	2
0	1	9	696	3	0	1	13	50	0
0	4	2	343	2	2	5	5	193	1
1	3	6	724	3	3	5	0	149	1
1	4	1	33	0	0	5	7	107	1
2	3	5	407	2	3	4	7	107	1
3	3	0	502	3	3	5	2	309	2

(continued)

Table 17.2 (continued)

h	*k*	*l*	$\|F^{obs}\|^2$	$\sigma\|F^{obs}\|^2$	*h*	*k*	*l*	$\|F^{obs}\|^2$	$\sigma\|F^{obs}\|^2$
0	3	7	212	2	4	4	4	181	1
1	4	3	492	2	0	4	10	69	1
3	3	2	564	2	2	2	12	556	4
0	4	4	316	2	0	6	0	764	6
0	0	10	113	1	2	3	11	51	0
2	4	0	1530	6	0	6	2	45	0
2	2	8	541	3	3	5	4	162	1
1	2	9	383	2	1	6	1	0	0
2	4	2	25	0	3	3	10	153	1
3	3	4	277	2	1	2	13	27	0
1	1	10	306	2	0	0	14	9	0
1	4	5	317	2	1	3	12	129	1
1	3	8	46	0	1	5	8	101	1

[a] Some of the errors are listed as zeros because of automatic rounding to integers

Table 17.3 The three-dimensional distribution of the interatomic vectors (the Patterson function) in the symmetrically independent part of the unit cell of $CeRhGe_3$ calculated using the observed structure factors determined from Le Bail's extraction (Table 17.2)

Peak number	*u*	*v*	*w*	Peak height
1	**0**	**0**	**0**	**6000**
2	**0.5**	**0**	**0.253**	**1673**
3	**0**	**0**	**0.341**	**1666**
4	**0**	**0**	**0.5**	**1424**
5	**0**	**0**	**0.239**	**926**
6	**0.5**	**0**	**0.390**	**829**
7	*0.5*	*0.271*	*0.474*	*112*
8	*0.266*	*0*	*0.301*	*62*
9	*0.268*	*0*	*0.433*	*41*
10	*0.176*	*0.176*	*0.322*	*20*

(58 electrons), while Ge is the weakest (32 electrons). Considering the simplicity of the unit cell and the presence of only two heavy atoms per cell (Ce), the Patterson technique should be an adequate tool to solve this crystal structure.[3]

The file with the observed structure factors (Table 17.2) was processed accordingly, and the resultant distribution of the interatomic vectors in the asymmetric part of the unit cell in the space group I4/mmm is listed in Table 17.3. A cross-section of the three-dimensional distribution of the interatomic vectors in the unit cell of $CeRhGe_3$ at $V = 0$ is also illustrated in Sect. 10.2.1, Fig. 10.4, left. It only makes sense to consider vectors with reasonable lengths—those that are approximately equal or longer than the expected shortest interatomic distance. In the case of

[3] Patterson method has a high success rate when applied to intermetallic structures mainly because of the high resolution of the resultant three-dimensional distribution of interatomic vectors. The latter is due to notably greater minimal interatomic distances and lower atomic displacements when compared to other classes of compounds.

Table 17.4 Low-multiplicity sites available in the space group I4/mmm

Site	Coordinates of symmetrically equivalent points			
2 (a)	0,0,0;	½,½,½		
2 (b)	0,0,½;	½,½,0		
4 (c)	0,½,0;	½,0,0;	½,0,½;	0,½,½
4 (d)	0,½,¼;	½,0,¼;	½,0,¾;	0,½,¾
4 (e)	0,0,*z*;	0,0,−*z*;	½,½,½ + *z*;	½,½,½ − *z*
8(f)	¼,¼,¼; ¾,¾,¼;	¼,¾,¼; ¾,¼,¼;	¾,¾,¾; ¼,¼,¾;	¾,¼,¾; ¼,¾,¾
8 (g)	0,½,z; 0,½, $\bar{z}$;	½,0,z; ½,0, $\bar{z}$;	½,0,½ + z; ½,0,½ − z;	0,½,z; 0,½,½ − z
8 (h)	*x*,*x*,0; $\bar{x},\bar{x}$, 0;	*x*, $\bar{x}$, 0; $\bar{x}$, *x*, 0;	½ + *x*,½ + *x*,½; ½ − *x*,½ − *x*,½;	½ + *x*,½ − *x*,½; ½ − *x*,½ + *x*,½;
8(i)	*x*,0,0; $\bar{x}$, 0, 0;	0,*x*,0; 0, $\bar{x}$, 0;	½ + *x*,½,½; ½ − *x*,½,½;	½,½ + *x*,½; ½,½ − *x*,½
8(j)	*x*,½,0; $\bar{x}$, ½, 0;	½,*x*,0; ½, $\bar{x}$, 0;	½ + *x*,0,½; ½ − *x*,0,½;	0,½ + *x*,½; 0,½ − *x*,½

$CeRhGe_3$, they should exceed ~2.5 Å, and these are the six strongest peaks in the Patterson function, which are highlighted in bold, while the others are shown in italics in Table 17.3 ($r_{Ce} = 1.85$, $r_{Rh} = 1.35$, and $r_{Ge} = 1.25$ Å).

17.2.1 *Highest Symmetry Attempt*

After consulting the International Tables for Crystallography (see Table 17.4), there are two sites in the unit cell with I4/mmm symmetry that may accommodate two Ce atoms: 2(a) with the coordinates 0,0,0 and 2(b) with the coordinates 0,0,½. Both sites differ only by a (0,0,½) shift of the whole structure and produce no additional peaks on the Patterson map except 0,0,0 and its symmetrical equivalent ½,½,½. We note that since there are only two Rh atoms in the unit cell, they can be accommodated in the second twofold position. The corresponding Ce–Rh interatomic vector is 0,0,½ [$(0,0,0)_{Ce} - (0,0,½)_{Rh} = (0,0,½)_{Ce\text{–}Rh}$], and it is Peak No. 4 in Table 17.3. Thus, it appears that we have to test only one possible model: two Ce in 2(a) and two Rh in 2 (b) because switching the occupancy of these two sites is the result of shifting the entire structure by 0,0,½. The interatomic distance Ce–Rh (3.11 Å) is slightly shorter than the sum of the corresponding atomic radii, but it remains within acceptable limits.[4]

[4]When compared to sums of empirical atomic radii, 5–10% reduction of the interatomic distances is common in intermetallic phases.

Table 17.5 The three-dimensional electron density distribution in the symmetrically independent part of the unit cell of $CeRhGe_3$ calculated using the observed structure factors determined from Le Bail's extraction (Table 17.2) and phase angles determined by Ce atoms placed in the 2(a), and Rh atoms placed in the 2(b) sites of the space group I4/mmm ($R_F = 55.6\%$)

Peak number	*x*	*y*	*z*	Peak height
1	0	0	0	72
2	0.5	0	0.250	13
3	0	0	0.3424	11
4	0	0	0.434	11
5	0	0	0.243	7
6	0.5	0	0.372	5
7	0.5	0.25	0.472	2

Table 17.6 The three-dimensional electron density distribution in the symmetrically independent part of the unit cell of $CeRhGe_3$ calculated using the observed structure factors determined from Le Bail's extraction (Table 17.2) and phase angles determined by Ce atoms placed in the 2(a), and 0.25Rh + 0.75Ge atoms placed in the 4(d) sites of the space group I4/mmm ($R_F = 44.0\%$)

Peak number	*x*	*y*	*z*	Peak height
1	0	0	0	58
2	0.5	0	0.250	43
3	0	0	0.368	15
4	0.5	0	0	9
5	0.5	0	0.413	5

Phase angles were calculated using the model where Ce is in the 2(a) and Rh is in the 2(b) sites. The $R_F = 55.6\%$ is nearly identical to the earlier example, and the resultant electron density distribution is listed in Table 17.5. Analysis of Table 17.5 immediately indicates that the Rh atom, which was placed in the 2(b) site with coordinates 0,0,½, was not confirmed on the Fourier map because there is no peak with the coordinates 0,0,½ or ½,½,0, which are symmetrically equivalent due to the presence of the body-centered translation. Thus, our reasoning may have been flawed, and Rh and Ge could be distributed statistically between two fourfold sites or in a single eightfold site in this space-group symmetry instead of being ordered. The second strongest peak in the Fourier map has coordinates ½,0,¼, which corresponds to the 4(d) site in space group I4/mmm. After placing 25% Rh and 75% Ge in this site and removing Rh from the 2(b) site, the corresponding $R_F = 44.0\%$ and the subsequent three-dimensional Fourier map are shown in Table 17.6.[5]

Both atoms have been confirmed on the Fourier map and it appears that the next missing atom is located in the 4(e) site with coordinates 0,0,0.368. All interatomic

[5] Another possibility would be to place 2 Rh and 2 Ge (i.e., a 50:50 mixture of Rh and Ge in this site) assuming that the remaining fourfold site is fully occupied by Ge, and we leave this exercise to the reader.

Table 17.7 The three-dimensional electron density distribution in the symmetrically independent part of the unit cell of $CeRhGe_3$ calculated using the observed structure factors determined from Le Bail's extraction (Table 17.2) and phase angles determined by Ce atoms placed in the 2(a), and 0.25Rh + 0.75Ge atoms placed in the 4(d) and 4(c) [$z = 0.368$] sites of the space group I4/mmm ($R_F = 32.2\%$)

Peak number	x	y	z	Peak height
1	0	0	0	47
2	0	0	0.363	25
3	0.5	0	0.250	23
4	0	0	0.211	8
5	0	0	0.100	5

distances are normal and after including this atom into the computation of structure factors and phase angles, the corresponding $R_F = 32.2\%$. This value is quite high, but it is still worthwhile to calculate a third Fourier map, which is shown in Table 17.7.

All atoms present in the model have been confirmed (Peaks No. 1–3 in Table 17.7). The fourth peak, however, is only ~⅓ of Peak No. 3 and, therefore, it corresponds to more than ten electrons, because a statistical mixture of 25%Rh and 75%Ge has 35¼ electrons. Assuming that an atom or a fraction of atom is located in this position, it has prohibitively short distances (δ) with all other atoms, already located in the unit cell: $\delta_{1\text{–}4} = 2.12$ Å; $\delta_{2\text{–}4} = 1.57$ Å; and $\delta_{3\text{–}4} = 2.22$ Å. This result combined with a high R_F (compare $R_F = 32.2\%$ obtained in the last iteration with $R_F = 6.9\%$ when all atoms were found in the crystal structure of $LaNi_{4.85}Sn_{0.15}$) is usually a strong indicator that the symmetry of the material has been overestimated.

It is possible to use this model of the crystal structure and attempt Rietveld refinement (as will be illustrated later in this chapter), but we will proceed with testing other space groups from the list of eight possible groups (I4/mmm, $I\bar{4}m2$, $I\bar{4}2m$, I4mm, I422, I4/m, $I\bar{4}$, and I4). Analysis of space groups $I\bar{4}m2$ and $I\bar{4}2m$ indicates that available low-multiplicity sites are essentially identical to those of the space group I4/mmm. When these two groups are tested as described here, the resultant models are also quite suspicious, and we leave verification of this statement as a self-exercise for the reader.

17.2.2 Low-Symmetry Model

The next space group on the list is I4mm (Table 17.8). This group has no fixed origin along the *Z*-axis: one available twofold site 2(a) has coordinates 0,0,z, and the only available fourfold site has coordinates ½,0,z. We note that there is no reason to recalculate the Patterson function, because its symmetry remains I4/mmm. To ensure that we do not place any of the atoms incorrectly, we now position only two Ce atoms in the 2(a) site in this space group. Because the origin along the *Z*-axis can be chosen arbitrarily in this space group, it does not matter which *z*-coordinate is chosen

Table 17.8 The three-dimensional electron density distribution in the symmetrically independent part of the unit cell of $CeRhGe_3$ calculated using the observed structure factors determined from Le Bail's extraction (Table 17.2) and phase angles determined by the Ce atom placed in the 2(a) site of the space group I4mm with $z = 0.000$ ($R_F = 42.6\%$)

Peak number	x	y	z	Peak height
1	0	0	0	60
2	0	0	0.662	9
3	0.5	0	0.253	9
4	0	0	0.341	9
5	0	0	0.500	7
6	0	0	0.429	7

Table 17.9 The three-dimensional electron density distribution in the symmetrically independent part of the unit cell of $CeRhGe_3$ calculated using the observed structure factors determined from Le Bail's extraction (Table 17.2) and phase angles determined by the Ce and Rh atoms placed in the 2 (a) sites of the space group I4mm with $z = 0.000$ and 0.662, respectively ($R_F = 42.3\%$)

Peak number	x	y	z	Peak height
1	0	0	0	55
2	0	0	0.665	27
3	0	0	0.427	9
4	0.5	0	0.259	9
5	0	0	0.242	8
6	0	0	0.495	6

for Ce. After placing two Ce in 2(a) with z = 0.000, the R_F is 42.6%, and the resultant electron density distribution is shown in Table 17.9.

There is no sharp reduction of peak heights between any pair of peaks in Table 17.9 except after the first, and therefore, we proceed by adding only one twofold site for the next iteration. By choosing Peak No. 2 and assuming that it is the next strongest scattering atom, Rh, the distance between this peak and Ce atom in 2 (a) with $z = 0.000$ is normal. The residual did not change after Rh in 2(a) with z = 0.662 has been added to the model ($R_F = 42.3\%$). The subsequent Fourier map is shown in Table 17.9.

Proceeding slowly, we add the third peak as Ge and change the z-coordinate of Rh from 0.662 to 0.665 as had been determined from the latest electron density map (Table 17.9). All distances remain normal and the residual lowers to $R_F = 32.1\%$; the Fourier map calculated using phase angles determined by all three independent atoms is shown in Table 17.10.

The fourth peak is now about twice as high as the fifth, and all atoms already present in the unit cell have been confirmed. After changing z-coordinates of Rh and the first Ge atom to 0.666 and 0.429, respectively, and adding the coordinates of the fourth peak as Ge in 4(b) with $z = 0.262$, the $R_F = 9.7\%$, which is a much lower value when compared to that achieved in the space group I4/mmm. The Fourier map calculated using phase angles determined by all four independent atoms is shown in

Table 17.10 The three-dimensional electron density distribution in the symmetrically independent part of the unit cell of $CeRhGe_3$ calculated using the observed structure factors determined from Le Bail's extraction (Table 17.2) and phase angles determined by Ce in 2(a) with $z = 0.000$, Rh in 2 (a) with $z = 0.665$, and Ge in 2(a) with $z = 0.427$ in the space group I4mm ($R_F = 32.1\%$)

Peak number	*x*	*y*	*z*	Peak height
1	0	0	0	47
2	0	0	0.666	33
3	0	0	0.429	23
4	0.5	0	0.262	9
5	0.5	0	0.400	5
6	0	0	0.177	3

Table 17.11 The three-dimensional electron density distribution in the symmetrically independent part of the unit cell of $CeRhGe_3$ calculated using the observed structure factors determined from Le Bail's extraction (Table 17.2) and phase angles determined by Ce in 2(a) with $z = 0.000$, Rh in 2 (a) with $z = 0.666$, Ge in 2(a) with $z = 0.429$, and Ge in 4(b) with $z = 0.262$ in the space group I4mm ($R_F = 9.7\%$). See also Fig. 10.2, left, where this Fourier map is visualized, although with a different selection of the origin of coordinates

Peak number	*x*	*y*	*z*	Peak height
1	**0**	**0**	**0**	**42**
2	**0**	**0**	**0.665**	**32**
3	**0**	**0**	**0.428**	**22**
4	**0.5**	**0**	**0.265**	**20**
5	*0*	*0*	*0.173*	2
6	*0.25*	*0*	*0.700*	2

Table 17.11. As follows from this Fourier map, all four atoms have been confirmed, and in addition, there is a sharp reduction in peak heights after the fourth maximum (italicized) in Table 17.11. The latter is a clear indication that the structure solution is completed. Therefore, the model (Table 17.12), built in the noncentrosymmetric space group I4mm, may be considered a correct solution of the problem and all structural parameters of $CeRhGe_3$ should be finalized by Rietveld refinement (see Sect. 17.4).

Pearson symbol of this crystal structure is *t*I10, and after consulting Pearson's Handbook, it easy to find that it belongs to the $BaNiSn_3$-type structure.[6]

[6]W. Doerrscheidt and H. Schaefer, The structure of barium-platinum-tin ($BaPtSn_3$), barium-nickel-tin ($BaNiSn_3$) and strontium-nickel-tin ($SrNiSn_3$) and their relation to the thorium-chromium-silicon ($ThCr_2Si_2$) structure type, J. Less-Common Met. **58**, 209 (1978). https://doi.org/10.1016/j.jssc.2019.02.012

Table 17.12 Coordinates of atoms in the unit cell of $CeRhGe_3$ as determined from X-ray powder diffraction data in the space-group symmetry I4mm

Atom	Site	*x*	*y*	*z*
Ce	2(a)	0	0	0.000
Rh	2(a)	0	0	0.665
Ge1	2(a)	0	0	0.428
Ge2	4(b)	½	0	0.265

Table 17.13 Figures of merit obtained at different stages during the full-pattern decomposition of the powder diffraction pattern of $CeRhGe_3$ collected using thermal neutrons with the wavelength $\lambda = 1.494$ Å; $R_{exp} = 3.09\%$. The temperature of the experiment was 200 K

Refined parameters	Illustration	R_p	R_{wp}	χ^2
Background (linear)	–	27.11	34.38	124.0
+ *a*, *c*	–	20.57	27.18	77.52
+ *U*, *V*, *W*, η_0, asymmetry	–	3.87	5.33	2.99
All, background (fourth order).	Figure 17.3	3.54	4.78	2.40

17.3 Solving the Crystal Structure from Neutron Data[7]

We now consider a neutron powder diffraction pattern of $CeRhGe_3$. The full-pattern decomposition has been conducted using LHPM-Rietica, and peak shapes have been represented by means of the pseudo-Voigt function with asymmetry in Howard's approximation (see Footnote 1 on page 625). The progression of the Le Bail refinement is shown by the corresponding figures of merit, which are listed in Table 17.13. The obtained result is illustrated in Fig. 17.3, and the individual structure factors are listed in Table 17.14. A few additional intensity maxima appear between 35.2 and 39.1° 2θ due to instrumental contribution, and this region of the powder diffraction pattern was excluded from calculations.

Ignoring the dissimilarities in the values of both the scattered intensity and structure factors for X-rays (Fig. 17.2 and Table 17.2, respectively) and neutrons, which are due to the differences in the scattering factors,[8] the most significant difference between the X-ray and neutron diffraction profiles is in the widths of Bragg peaks and the mixing parameter of the pseudo-Voigt function as illustrated in Fig. 17.4. The Bragg peaks in the X-ray experiment are narrow, and they are well-described by a nearly pure Lorentzian. Conversely, Bragg peaks are much broader

[7] O. Zaharko, V.K. Pecharsky, and K.A. Gschneidner, Jr., unpublished. Neutron diffraction data were collected using the High Resolution Powder Diffractometer for Thermal Neutrons at the Paul Scherrer Institute, Switzerland (https://www.psi.ch/en/sinq/hrpt). The file with experimental data (*Ch17Ex01_Neut.dat*) is available online.

[8] Coherent neutron scattering lengths (b) are 4.84, 5.88, and 8.185 fm for Ce, Rh, and Ge, respectively. Since for these three elements, the neutron scattering ability accidentally increases with the decreasing atomic number, Ge is determined from neutron data with better accuracy than Ce. On the contrary, the X-ray scattering factors are proportional to the atomic number, and therefore, Ce is determined from X-ray diffraction with a better precision than Ge.

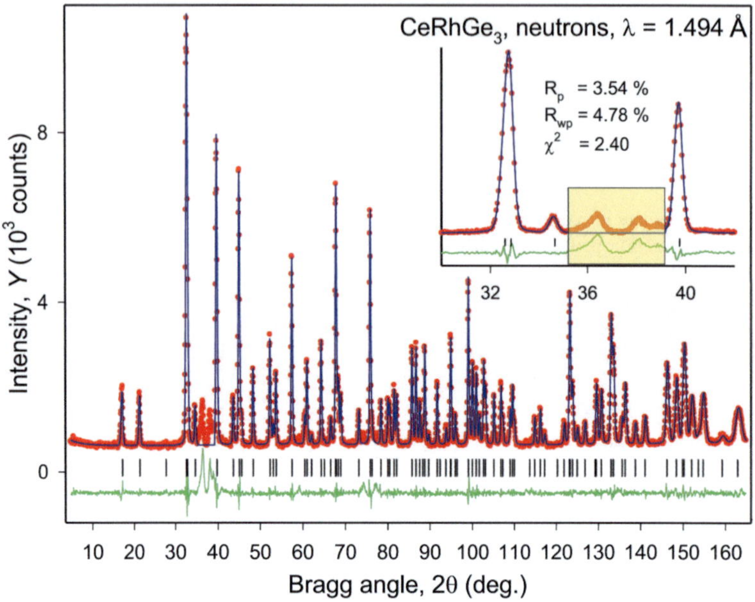

Fig. 17.3 The observed and calculated powder diffraction patterns of $CeRhGe_3$ after refinement of all parameters. The region $35.2 < 2\theta < 39.1°$ (*highlighted* in the inset) was excluded from the refinement because it contains three additional peaks due to instrumental contribution. The powder diffraction data were collected with a step 0.05° of 2θ

and they are closer to a pure Gaussian distribution in the neutron diffraction experiment.

Since we established that the correct space-group symmetry for this material is I4mm, we will not try any other space groups. The distribution of peaks in the Patterson function calculated using the observed structure factors obtained in the neutron powder diffraction experiment is shown in Table 17.15. The *U*0*W* cross-section of the three-dimensional Patterson function using the same powder diffraction data is illustrated in Sect. 10.2.1, Fig. 10.4, right. The height of an interatomic vector between two atoms of type *i* and *j* is proportional to the product of their scattering power, which in this case is proportional to the product of the corresponding coherent scattering lengths, b_ib_j. For example, Ce–Ce vectors should be proportional to b^2_{Ce} (~23 fm^2), while Ge–Ge vectors should be almost three times stronger: $b^2_{Ge} \cong 67$ fm^2.

Only the eight strongest independent peaks in Table 17.15 (highlighted in bold and separated by the italics from the rest) have meaningful lengths between ~2.4 and 5 Å. The differences in the heights of Patterson peaks between Tables 17.3 and 17.15 are expected because now Ge atoms are the strongest scattering species. Between the two sites (see Table 17.16) that may accommodate six Ge atoms, the 2(a) site results in the 0,0,0 vector, and the 4(b) site yields one additional ½,½,0 vector. Indeed, the second strongest vector found in Table 17.15 is identical to the latter

Table 17.14 The list of Bragg reflections and observed structure factors squared determined after Le Bail's full-pattern decomposition of the neutron powder diffraction data of $CeRhGe_3$

h	*k*	*l*	$\|F^{obs}\|^2$	$\sigma\|F^{obs}\|^2$	*h*	*k*	*l*	$\|F^{obs}\|^2$	$\sigma\|F^{obs}\|^2$
0	0	2	1667	10	2	4	0	14,985	71
0	1	1	537	3	2	2	8	7328	44
1	1	0	0	0	1	2	9	3329	21
0	1	3	2372	9	2	4	2	1299	10
1	1	2	8241	24	3	3	4	7445	44
0	0	4	3567	27	1	1	10	5300	34
0	2	0	17,951	65	1	4	5	2360	17
0	2	2	1564	11	1	3	8	2813	18
1	1	4	9641	36	2	3	7	1213	9
1	2	1	466	4	0	4	6	3318	25
0	1	5	2992	18	2	4	4	2834	18
1	2	3	2318	13	0	2	10	2225	17
0	0	6	3320	28	0	1	11	0	0
0	2	4	3301	21	0	3	9	2854	22
2	2	0	18,709	84	3	3	6	3603	27
2	2	2	1467	12	0	5	1	357	3
1	1	6	4597	27	3	4	1	357	3
0	3	1	512	5	1	5	0	0	0
1	2	5	3060	17	1	4	7	1123	9
1	3	0	210	2	0	5	3	1934	9
0	1	7	1584	13	3	4	3	1930	9
0	3	3	2250	10	1	5	2	5994	26
1	3	2	7326	28	2	4	6	3082	18
0	2	6	3616	23	2	2	10	1802	14
2	2	4	3038	21	0	0	12	10,617	78
0	0	8	9660	75	1	2	11	18	0
1	3	4	8333	34	0	4	8	6892	40
2	3	1	496	4	2	3	9	3171	19
0	3	5	3310	24	1	5	4	7187	30
1	1	8	3286	23	2	5	1	2825	13
1	2	7	1460	11	1	3	10	2458	11
2	3	3	2065	14	1	1	12	3134	20
2	2	6	3821	27	0	5	5	2415	13
0	4	0	15,318	89	3	4	5	2415	13
0	2	8	7911	45	3	3	8	2621	18
0	1	9	3449	25	2	5	3	1712	11
0	4	2	1217	10	0	2	12	11,012	47
1	3	6	4074	23	4	4	0	16,837	76
1	4	1	375	4	0	3	11	366	2
2	3	5	2611	18	2	4	8	6659	25
3	3	0	0	0	1	4	9	3265	15
0	3	7	1366	12	4	4	2	1146	6

(continued)

Table 17.14 (continued)

h	*k*	*l*	$\|F^{obs}\|^2$	$\sigma\|F^{obs}\|^2$	*h*	*k*	*l*	$\|F^{obs}\|^2$	$\sigma\|F^{obs}\|^2$
1	4	3	1916	11	1	5	6	3428	15
3	3	2	6705	37	0	1	13	931	5
0	4	4	2635	21	2	5	5	2513	10
0	0	10	1453	14					

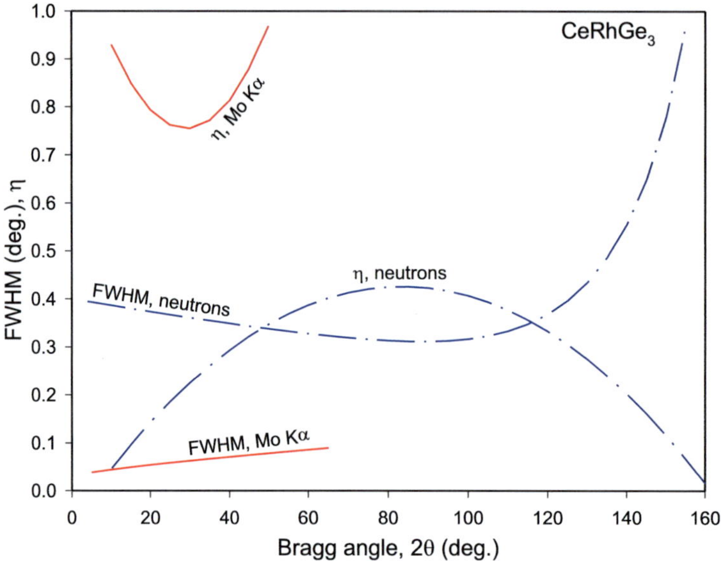

Fig. 17.4 Full widths at half maximum (FWHM) and mixing parameters (η) of the pseudo-Voigt function used to approximate peak shapes in the X-ray (Fig. 17.2) and neutron (Fig. 17.3) powder diffraction patterns collected from the same $CeRhGe_3$ powder

Table 17.15 The three-dimensional distribution of the interatomic vectors in the symmetrically independent part of the unit cell of $CeRhGe_3$ calculated using the observed structure factors determined from Le Bail's extraction employing neutron diffraction data (Table 17.14)

Patterson map peak number	*u*	*v*	*w*	Peak height
1	**0**	**0**	**0**	**6000**
2	**0**	**0**	**0.5**	**2675**
3	**0.5**	**0**	**0.355**	**1500**
4	**0.5**	**0**	**0.145**	**1500**
5	**0.5**	**0**	**0.250**	**1500**
6	**0**	**0**	**0.237**	**1124**
7	**0**	**0**	**0.346**	**687**
8	**0**	**0**	**0.404**	**489**
9	*0.125*	*0.125*	*0.088*	*266*
10	*0.218*	*0.218*	*0.457*	*260*

Table 17.16 Low-multiplicity sites available in the space group I4mm

Site	Coordinates of symmetrically equivalent points			
2(a)	$0,0,z$	$\frac{1}{2},\frac{1}{2},\frac{1}{2}+z$		
4(b)	$0,\frac{1}{2},z$	$\frac{1}{2},0,z$	$\frac{1}{2},0,\frac{1}{2}+z$	$0,\frac{1}{2},\frac{1}{2}+z$
8(c)	$x, x, z;\quad \bar{x},\bar{x},z$	$\bar{x},x,z;\quad x,\bar{x},z$	$\frac{1}{2}+x,\frac{1}{2}+x,\frac{1}{2}+z;\quad \frac{1}{2}-x,\frac{1}{2}-x,\frac{1}{2}+z$	$\frac{1}{2}-x,\frac{1}{2}+x,\frac{1}{2}+z;\quad \frac{1}{2}+x,\frac{1}{2}+x,\frac{1}{2}+z$
8(d)	$x, 0, z;\quad \bar{x},0,z$	$0, x, z;\quad 0,\bar{x},z$	$\frac{1}{2}+x,\frac{1}{2},\frac{1}{2}+z;\quad \frac{1}{2}-x,\frac{1}{2},\frac{1}{2}+z$	$\frac{1}{2},\frac{1}{2}+x,\frac{1}{2}+z;\quad \frac{1}{2},\frac{1}{2}-x,\frac{1}{2}+z$

Table 17.17 The three-dimensional nuclear density distribution in the symmetrically independent part of the unit cell of $CeRhGe_3$ calculated using the observed structure factors determined from Le Bail's extraction (Table 17.14) and phase angles determined by Ge in 4(b) with $z = 0.000$ and Ge in 2(a) with $z = 0.355$ in the space group I4mm ($R_F = 29.0\%$)

Fourier map peak number	x	y	z	Peak height
1	0	0	0.355	73
2	0.5	0	0	63
3	0	0	0.754	18
4	0	0	0.587	15
5	0	0	0.119	11
6	0	0	0.921	10

(0,0,½ + ½,½,½ = ½,½,0). Thus, the Patterson function points to a strongly scattering atom in the 4(b) site of the space group I4mm. Since the origin of coordinates here is not fixed along the *Z*-axis, we may choose any *z*-coordinate for the Ge atom in this site.

Assume that four Ge atoms are located in 4(b) with $z = 0.000$. Let us try to use the Patterson function (Table 17.15) and locate a second Ge, which should be in a twofold site because there is a total of six Ge atoms in the unit cell. The coordinates of a point in 2(a) are 0,0,z. Hence, the corresponding vector between the two independent Ge atoms should be ½,0,0 – 0,0,z = ½,0,–z ≡ ½,0,z due to a mirror plane perpendicular to *Z* at z = ½, which is present in the space group I4/mmm that describes Patterson symmetry. A second vector can be found from ½,0,0 + ½,½,½ – 0,0,z = 0,½,½ – z. Given the presence of a fourfold axis parallel to *Z*, this vector is identical to ½,0,½ – z. These are vectors No. 3 and 4 in Table 17.15 assuming $z = 0.355$ (or $z = 0.145$). Without other atoms in the model of the crystal structure, the two choices are equivalent, and any of the two *z*-coordinates may be selected to represent Ge atom in the 2(a) site. Thus, the two independent Ge atoms, according to the Patterson function, are as follows: 4Ge1 in 4(b): ½,0,0.000 and 2Ge2 in 2(a): 0,0,0.355. The corresponding residual $R_F = 29.0\%$. The coordinates of peaks found on the three-dimensional Fourier map calculated using phase angles determined from this partial model of the crystal structure are listed in Table 17.17.

The next strongest peak (No. 3) also belongs to the 2(a) site with the coordinate z = 0.754. Its shortest interatomic distances are $\delta_{3\text{–Ge1}} = 3.30$ and 3.36 Å and $\delta_{3\text{–Ge2}} = 3.27$ Å. All three correspond to Ce rather than to Rh, even though Ce is a light scattering atom when compared to Rh. Thus, after placing 2Ce in 2(a) with $z = 0.754$, the $R_F = 19.8\%$ and the subsequently calculated Fourier map is illustrated in Table 17.18.

So far three atoms have been confirmed, and the fourth strongest peak points to an atom that also belongs to the 2(a) site with $z = 0.580$. The only atom, which is missing from our model, is Rh. When 2 Rh atoms are placed in this site, the interatomic distance $\delta_{\text{Rh–Ce}} = 1.75$ Å, which is too short. The fourth peak is,

Table 17.18 The three-dimensional nuclear density distribution in the symmetrically independent part of the unit cell of $CeRhGe_3$ calculated using the observed structure factors determined from Le Bail's extraction (Table 17.15) and phase angles determined by Ge in 4(b) with $z = 0.000$, Ge in 2 (a) with $z = 0.355$, and Ce in 2(a) with $z = 0.754$ in space group I4mm ($R_F = 19.8\%$)

Fourier map peak number	*x*	*y*	*z*	Peak height
1	0	0	0.353	61
2	0.5	0	0	55
3	0	0	0.758	29
4	0	0	0.580	13
5	0	0	0.113	10
6	0.5	0	0.258	7

Table 17.19 The three-dimensional nuclear density distribution in the symmetrically independent part of the unit cell of $CeRhGe_3$ calculated using the observed structure factors determined from Le Bail's extraction (Table 17.15) and phase angles determined by Ge in 4(b) with $z = 0.000$, Ge in 2 (a) with $z = 0.355$, Ce in 2(a) with $z = 0.754$, and Rh in 2(a) with $z = 0.113$ in the space group I4mm ($R_F = 16.3\%$)

Fourier map peak number	*x*	*y*	*z*	Peak height
1	**0**	**0**	**0.349**	**48**
2	**0.5**	**0**	**0**	**47**
3	**0**	**0**	**0.108**	**33**
4	**0**	**0**	**0.759**	**24**
5	*0*	*0*	*0.233*	*6*
6	*0*	*0*	*0.000*	*3*

therefore, a false maximum and should be discarded.[9] After the next strongest peak (No. 5) is tested, all distances are normal. The $R_F = 16.3\%$ and the subsequent Fourier map is listed in Table 17.19, which now displays a sharp drop in peak heights after Peak No. 4. The cross-section of the nuclear density distribution in the *X*0*Z* plane calculated using the same powder diffraction data is shown in Fig. 10.2. It is worth noting that Fig. 10.2 represents a different origin of coordinates since its selection along the *Z*-axis is arbitrary in this space-group symmetry.

Since we did not change the coordinates of atoms to those refined from the previous Fourier map computations, we will do it now and assume the distribution of atoms in the unit cell of $CeRhGe_3$ as shown in Table 17.20. All atoms were assigned identical isotropic displacement parameters $B = 0.5$ Å^2. The resulting $R_F = 9.8\%$, which is quite close to that obtained using X-ray diffraction data (see the previous section). The coordinates of all atoms may be further refined by calculating another Fourier map. This is, however, unnecessary since the model of the crystal structure appears complete and all relevant structural parameters should

[9] False peaks may be stronger than the real peaks on the Fourier map, especially when a model is incomplete and/or accuracy of structure factors is low, which is the case here. One atom still missing from the model is Rh. It is the second strongest scattering atom and, therefore, phase angles are imprecise, thus resulting in a strong false peak.

Table 17.20 Coordinates of atoms in the unit cell of CeRhGe$_3$ as determined from neutron powder diffraction data in the space-group symmetry I4mm. $R_F = 9.8\%$

Atom	Site	x	y	z
Ge1	4(b)	0.5	0	0.000
Ge2	2(a)	0	0	0.349
Ce	2(a)	0	0	0.759
Rh	2(a)	0	0	0.108

be and are refined next using the Rietveld technique based on the available neutron powder diffraction data.

A comparison of Tables 17.13 and 17.20 indicates that the coordinates of the atoms are different in the two models. Nonetheless, the models are identical: the two crystallographic bases are related to one another by the center of inversion and by different origins of coordinates, which will be visualized later in Fig. 17.8.

17.4 Rietveld Refinement

The X-ray data were collected at room temperature, while the neutron scattering experiment ($\lambda = 1.494$ Å) was conducted at 200 K. Hence, combined Rietveld refinement is impossible because of the differences in the lattice and structural parameters of the alloy due to thermal expansion,[10] and we use the two sets of data independently.

17.4.1 X-Ray Data: Correct Low-Symmetry Model

Initial parameters for Rietveld refinement were assumed as follows: background, peak shape, unit cell dimensions, and zero shift as determined from Le Bail's full-pattern decomposition (see Sect. 17.1) and the model of the crystal structure from the ab initio solution in the space group I4mm (Table 17.13).

Refinement of all free variables except the coordinates and displacement parameters of individual atoms, beginning with the scale factor (Rows 1–3 in Table 17.21), results in low residuals, basically confirming the model of the crystal structure. When the coordinate parameters of atoms in the unit cell were included in the refinement (Row 4, Table 17.21), all residuals improve, especially R_B, which is

[10] Displacement parameters of atoms are also expected to be different as the temperature of the powder diffraction experiment varies. Further, it is also feasible that atomic positions may change due to generally anisotropic thermal expansion of crystal lattices. These considerations are in addition to the most obvious cause (different lattice parameters) preventing combined refinement using powder diffraction data collected at different temperatures. In general, material may also be polymorphic but this is not the case here, as was established in Sects. 17.2 and 17.3, also see Fig. 17.8.

Table 17.21 The progress of Rietveld refinement of the crystal structure of $CeRhGe_3$ using X-ray powder diffraction data. Wavelengths used: $\lambda K\alpha_1 = 0.70932$ Å, $\lambda K\alpha_2 = 0.71361$ Å

Refined parameters	R_p	R_{wp}	R_B	χ^2
Initial (profile from Le Bail, model from Table 17.13, overall $B = 0.5$ Å^2)	1468	1557	1526	3×10^5
Scale factor	8.87	11.46	7.05	15.03
Scale, all profile, *a* and *c*, overall displacement	8.25	10.91	6.45	13.66
All of the above plus coordinates of individual atoms[a]	5.88	8.04	3.12	7.43
All of the above plus individual isotropic displacement parameters	5.78	7.92	3.07	7.22
All of the above plus individual anisotropic displacement parameters[b]	5.75	7.86	3.00	7.11
All of the above plus extinction[c]	5.42	7.54	2.14	6.54

[a] In the space group I4mm, the origin of coordinates along the *Z*-axis is not fixed by symmetry. Therefore, the *z*-coordinate of one atom in the unit cell must be excluded from the least squares at all times to avoid severe correlation problems. We selected the *z*-coordinate of Ce atom ($z = 0.000$) as the fixed coordinate parameter

[b] All atoms in this crystal structure are located in special sites. This introduces certain relationships between individual anisotropic displacement parameters: $\beta_{22} = \beta_{11}$ for the atoms in 2(a) sites (Ce, Rh, and Ge1); $\beta_{12} = \beta_{13} = \beta_{23} = 0$ for all sites

[c] Extinction parameter 0.00001 was used as the initial approximation. Extinction (see Eqs. 8.41 and 8.64) may correlate with phase scale factor and, therefore, it may be necessary to keep the scale factor fixed during the initial refinement of extinction

lowered from 6.45% to 3.12%. Similar to the previous example from Chap. 16, the quality of the experimental data is quite high, and therefore, we easily refine individual isotropic and then individual anisotropic displacement parameters of all atoms.

Finally, as may be established by a trial-and-error approach, a small extinction correction further improves the agreement between the observed and calculated intensities, as shown in the last row of Table 17.21. Attempts to include preferred orientation assuming several possible texture axes (such as [100], [110], [001], and a few others) did not result in the improvement of the fit, thus indicating that in this experiment preferred orientation effects are nonexistent within the accuracy of the data.

The plot of the observed and calculated intensities is shown in Fig. 17.5, and the fully refined structural parameters are listed in Table 17.22. It is easy to verify by the calculation of the Fourier map that there are no additional atoms in the unit cell of this compound. All interatomic distances are normal. The crystal structure is shown in Fig. 17.6, from which it is easy to see that thermal ellipsoids of all atoms are reasonable.

When the final residuals obtained after the Rietveld refinement (Table 17.21) are compared with those obtained during Le Bail's full-pattern decomposition (Table 17.1), the differences are small, which serves as another confirmation of the correctness of the structural model in the space group I4mm. In this regard, it is useful to illustrate the Rietveld refinement of a different model of the same crystal structure (see Table 17.7), which was constructed earlier in the higher symmetry space group—I4/mmm. This model was discarded based on high R_F and also based on the presence of strong peaks on the Fourier map, which were too close to the atoms already located in the unit cell.

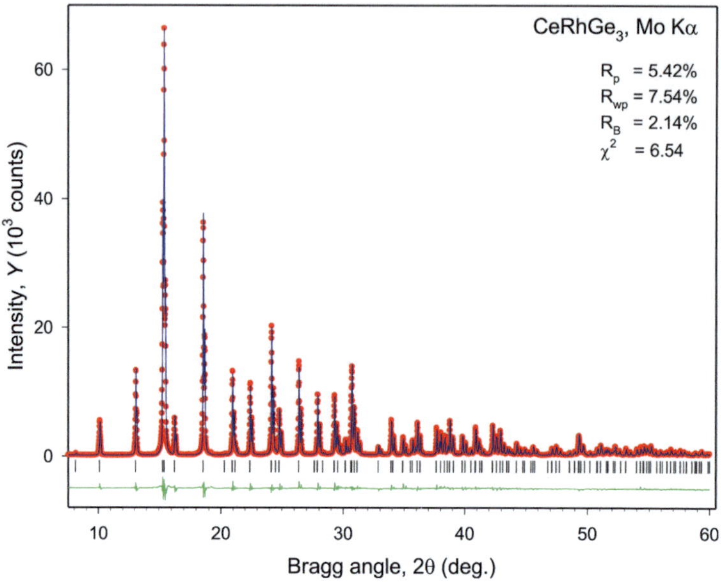

Fig. 17.5 The observed and calculated powder diffraction patterns of CeRhGe$_3$ after the completion of Rietveld refinement. The data were collected from a ground CeRhGe$_3$ powder dusted on a flat sample holder using a rotating anode Rigaku TTRAX powder diffractometer in a step-scan mode with a step $\Delta 2\theta = 0.01°$. All notations on the plot are identical to Fig. 16.2

Table 17.22 Structural parameters of CeRhGe$_3$, fully refined by the Rietveld technique employing powder diffraction data collected from a ground powder using Mo Kα radiation. The space group is I4mm. The unit cell dimensions are: $a = 4.39830(3)$, $c = 10.03259(8)$ Å. Some of the anisotropic displacement parameters are fixed by site symmetry: $\beta_{12} = \beta_{13} = \beta_{23} = 0$. All sites are fully occupied[a]

Atom	Site	x	y	z	$10^4 \times \beta_{11}$	$10^4 \times \beta_{22}$	$10^4 \times \beta_{33}$
Ce	2(a)	0	0	0.0000[b]	111(2)	111(2)	16(1)
Rh	2(a)	0	0	0.6589(1)	48(3)	48(3)	20(1)
Ge1	2(a)	0	0	0.4209(1)	70(4)	70(4)	24(1)
Ge2	4(b)	½	0	0.2615(1)	160(5)	58(5)	26(1)

[a] The refined model of the crystal structure can be found in the file *Ch17Ex01a.inp* online, and the experimental patterns are located in the data file *Ch17Ex01_MoKa.dat*
[b] This coordinate was fixed to determine the origin of coordinates

17.4.2 X-Ray Data: Wrong High-Symmetry Model

Beginning again with the background, peak-shape and unit cell parameters along with the zero shift determined during Le Bail's full-pattern decomposition, we attempt to perform Rietveld refinement of the crystal structure model shown in

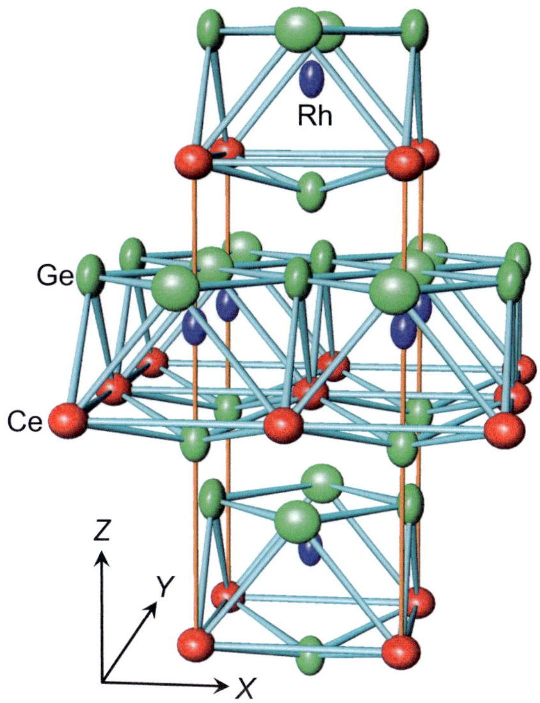

Fig. 17.6 The crystal structure of $CeRhGe_3$ as determined from Rietveld refinement in the space group I4mm (see Table 17.22 and Fig. 17.5). Displacement ellipsoids are shown at 99% probability. Also, see the Footnote 15 on page 647

Table 17.23.[11] The observed and calculated intensities after the full-profile least squares are plotted in Fig. 17.7.

Visual analysis of Fig. 17.7 immediately indicates that the agreement between the observed and calculated intensity is poor. Moreover, Rietveld refinement of this model of the crystal structure, during which the coordinates of atoms in the 4(e) site plus population of 4(d) and 4(e) sites were optimized together with the overall isotropic displacement parameter, results in the removal of Rh atoms from all sites: population of both sites by Rh becomes negative.[12]

Hence, this model has no chemical sense because Rh is indeed present in the material. Finally, all residuals (see Fig. 17.7) are much higher when compared to those obtained in both the earlier refinement (in the space group I4mm) and Le Bail's full-pattern decomposition (see Table 17.1). Thus, Rietveld refinement of the model in the space-group symmetry I4/mmm corroborates the conclusion made earlier about its inadequacy.

[11] The corresponding data file, *Ch17Ex01b.inp*, containing initial parameters for the Rietveld refinement using LHPM-Rietica is found online.

[12] See the data file *Ch17Ex01c.inp* online

Table 17.23 Coordinates of atoms in the unit cell of $CeRhGe_3$ as determined from X-ray powder diffraction data in the space-group symmetry I4/mmm (earlier discarded as wrong, see Table 17.7)

Atom	Site	*x*	*y*	*z*
Ce	2(a)	0	0	0
0.25Rh + 0.75Ge	4(d)	½	0	¼
0.25Rh + 0.75Ge	4(e)	0	0	0.363

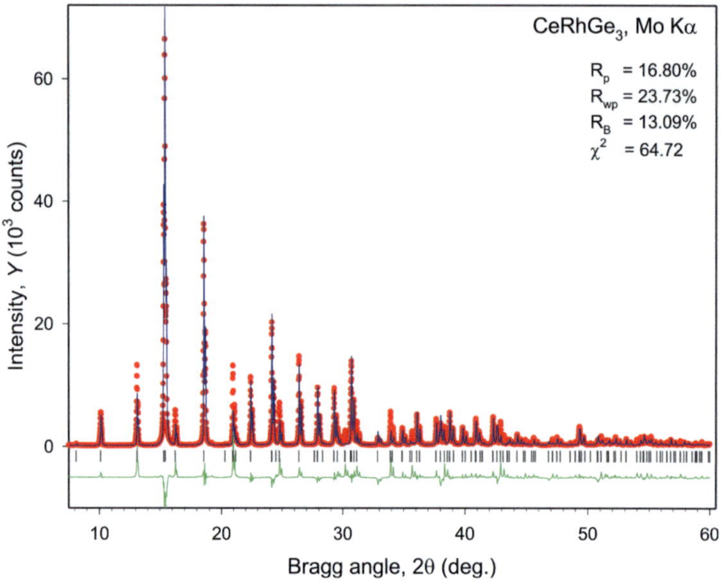

Fig. 17.7 The observed and calculated powder diffraction patterns of $CeRhGe_3$ after Rietveld refinement of the model in the space group I4/mmm. Compare with Fig. 17.5

17.4.3 Neutron Data

When we solved this crystal structure using neutron diffraction data, we found a model (Table 17.20) where the origin of coordinates was shifted with respect to that solved from X-ray diffraction (Tables 17.13 and 17.22). Here, we first use the coordinates of atoms determined from X-ray data[13] and then perform a refinement of the "original" model as established from a neutron diffraction experiment.[14]

Table 17.24 lists the residuals obtained at different stages of Rietveld refinement. After the determination of the scale factor, all residuals are quite low, and therefore, the completion of Rietveld refinement in this case presents no problems. The least-squares refinement of both models of $CeRhGe_3$ converges to identical residuals

[13]This fully refined crystal structure is found in the file *Ch17Ex02a.inp*.

[14]Data file *Ch17Ex02b.inp*. The measured neutron powder diffraction pattern is also found online in the data file *Ch17Ex01_Neut.dat*.

Table 17.24 The progress of Rietveld refinement of the crystal structure of $CeRhGe_3$ using neutron powder diffraction data[a] collected at T = 200 K. The wavelength used: $\lambda = 1.494$ Å

Refined parameters	R_p	R_{wp}	R_B	χ^2
Initial (profile parameters from Le Bail, model from Table 17.22 overall $B = 0.5$ Å^2)	38.44	49.21	78.08	253.7
Scale factor	4.29	5.62	2.56	3.31
Scale, all profile, *a* and *c*, overall displacement	4.03	5.30	2.08	2.96
Scale, all profile, *a* and *c*, overall displacement plus coordinates of individual atoms and individual isotropic displacement parameters[b]	3.97	5.24	1.89	2.91
All, individual anisotropic displacement parameters[c]	3.91	5.20	1.69	2.85
Fully refined "original" model from Table 17.20[d]	3.91	5.20	1.69	2.85

[a] The region $35.2 < 2\theta < 39.1°$ was excluded from the refinement, as previously done (see Fig. 17.3), because it contains three additional peaks due to instrumental contribution

[b] In the space group I4mm, the origin of coordinates along the *Z*-axis is not fixed. Therefore, the *z*-coordinate of one atom in the unit cell must be excluded from least squares at all time. We choose the *z*-coordinate of the Ce atom ($z = 0.000$) as a fixed coordinate parameter

[c] All atoms in this crystal structure are located in special site positions. This introduces certain relationships between anisotropic displacement parameters: $\beta_{22} = \beta_{11}$ for the atoms in 2(a) sites (Ce, Rh, and Ge1); $\beta_{12} = \beta_{13} = \beta_{23} = 0$ for all sites

[c] In this refinement, the *z*-coordinate of the Ge1 atom ($z = 0.000$) was chosen as a fixed coordinate parameter to define the origin of coordinates

(compare the two rows at the end of Table 17.24), thus confirming our earlier conclusion about their identity. The final parameters of the individual atoms are assembled in Table 17.25.

Both models of the crystal structure are shown in Fig. 17.8. It is easy to verify that thermal ellipsoids of individual atoms are identical in the two models, even though the β_{11} and β_{22} of Ge2 are switched, which is due to the different selection of the origin of coordinates. When compared to the results obtained at room temperature (Table 17.22 and Fig. 17.6), the individual displacement parameters of all atoms are reduced. This reduction is expected because thermal excitations in the crystal lattice are lowered when temperature decreases. Some differences in the anisotropy of Ge2 and Rh (compare Fig. 17.6 with Fig. 17.8) may be associated with both the reduction of temperature and with the different sensitivity of X-rays and neutrons: Ge has the lowest scattering ability for X-rays and it has the largest coherent scattering length for neutrons.[15]

[15] Anisotropic parameters obtained employing neutron data, where scattering occurs on nuclei, are more reliable. In the X-ray diffraction, the anisotropy reflects not only electron density deformations but also stronger and potentially unaccounted absorption or porosity effects. Overall, displacement anisotropy derived from powder diffraction data should be carefully analyzed, especially in the presence of preferred orientation.

Table 17.25 Structural parameters of $CeRhGe_3$ fully refined by Rietveld technique using neutron data. The space group is I4mm. The unit cell dimensions are: $a = 4.39180(4)$, $c = 10.0238(1)$ Å. Some of the anisotropic displacement parameters are fixed by symmetry: $\beta_{12} = \beta_{13} = \beta_{23} = 0$. All sites are fully occupied

Atom	Site	x	y	z	$10^4 \times \beta_{11}$	$10^4 \times \beta_{22}$	$10^4 \times \beta_{33}$
Model 1							
Ce	2(a)	0	0	0.0000	57(5)	57(5)	10(2)
Rh	2(a)	0	0	0.6598(3)	53(6)	53(6)	3(1)
Ge1	2(a)	0	0	0.4219(3)	43(4)	43(4)	12(1)
Ge2	4(b)	½	0	0.2630(3)	74(5)	41(4)	7(1)
Model 2							
Ce	2(a)	0	0	0.7630(3)	57(5)	57(5)	10(2)
Rh	2(a)	0	0	0.1033(2)	53(6)	53(6)	3(1)
Ge1	2(a)	0	0	0.3412(2)	43(4)	43(4)	12(1)
Ge2	4(b)	½	0	0.0000[a]	41(4)	74(5)	7(1)

[a] This coordinate was fixed to determine the origin of coordinates

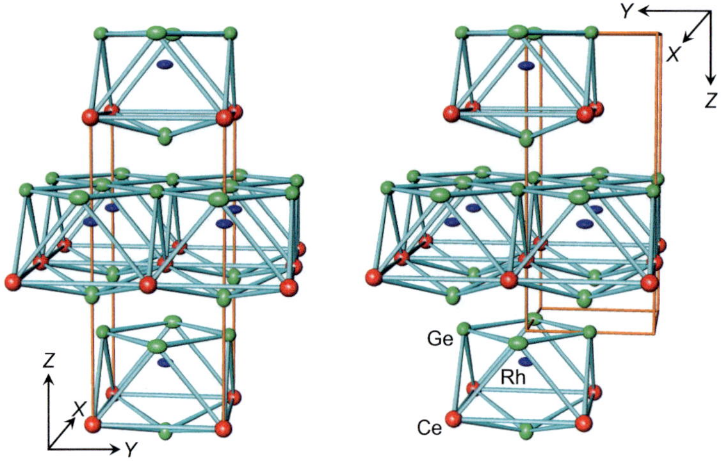

Fig. 17.8 The two models of the crystal structure of $CeRhGe_3$ refined using neutron powder diffraction data collected at T = 200 K. The thermal displacement ellipsoids are shown at 99% probability. The models are identical except for the inversion of the coordinate system and differently chosen origin of the unit cell

The Rietveld plot of the observed and calculated neutron diffraction patterns of $CeRhGe_3$ is shown in Fig. 17.9. Given excellent residuals and both physical and chemical rationale of the refined model, we conclude that the results presented in Table 17.25 are correct.

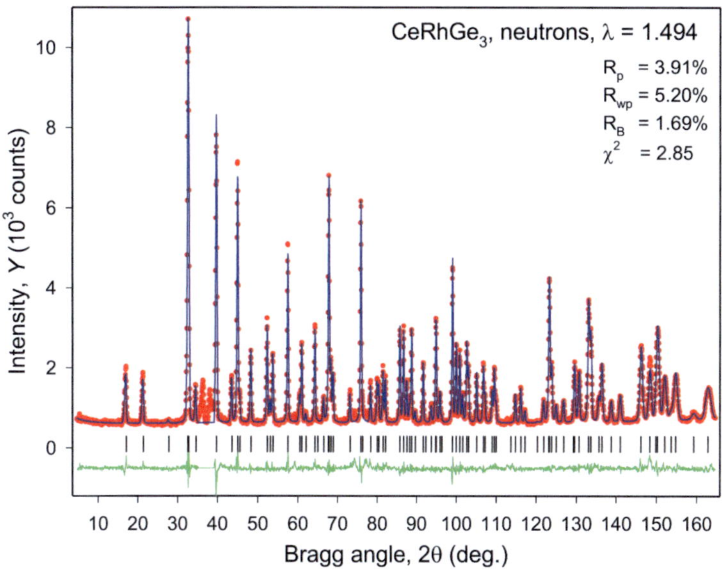

Fig. 17.9 The observed and calculated neutron powder diffraction patterns of CeRhGe$_3$ after the completion of Rietveld refinement. The region $35.2 < 2\theta < 39.1°$ was excluded from the refinement. (Data courtesy of Dr. O. Zaharko)

Chapter 18
Crystal Structure of Nd_5Si_4: Solved Using Direct Methods

The next example is also an intermetallic compound, Nd_5Si_4, which is the most complex among those considered so far. The powder diffraction pattern (Fig. 18.1)[1] has been indexed in a tetragonal crystal system with lattice parameters $a = 7.871$ and $c = 14.812$ Å. Analysis of the systematic absences shows that reflections $h00$ with $h \neq 2n$ and $00\,l$ with $l \neq 4n$ are extinct and points to two possible space groups: $P4_12_12$ or $P4_32_12$. The complexity of this example arises from a large primitive unit cell of the material, which results in over 440 Bragg reflections possible between 18 and 120° 2θ when using Cu Kα radiation. One of the major difficulties in the full-pattern decomposition here is in the uncertainty of the background at high Bragg angles, where multiple reflections heavily overlap. Therefore, the background baseline should be monitored at all times during the refinement.

18.1 Full-Pattern Decomposition

The progression of the Le Bail full-pattern decomposition is illustrated in Table 18.1, and the results are shown in Fig. 18.1. Bragg peaks were represented by the pseudo-Voigt function with Howard's asymmetry correction.[2]

The measured gravimetric density of the alloy is 5.96 g/cm^3 and one unit cell contains Z = 3.95 ≅ 4 formula units of Nd_5Si_4. The two possible space-group symmetries are enantiomorphous, and they cannot be distinguished using powder diffraction data. The results shown in Table 18.1 and Fig. 18.1 are obtained in the space group $P4_12_12$.

[1] Files with experimental data (*Ch18Ex01_CuKa.dat* and *Ch18Ex01_CuKa.xy*) are available online.

[2] Due to a large number of Bragg reflections, the list of individual structure factors squared is found online in the data file *Ch18Ex01_F2.dat*.

P. Zavalij, V. Pecharsky, *Fundamentals of Powder Diffraction and Structural Characterization of Materials*, https://doi.org/10.1007/978-3-031-91504-8_18

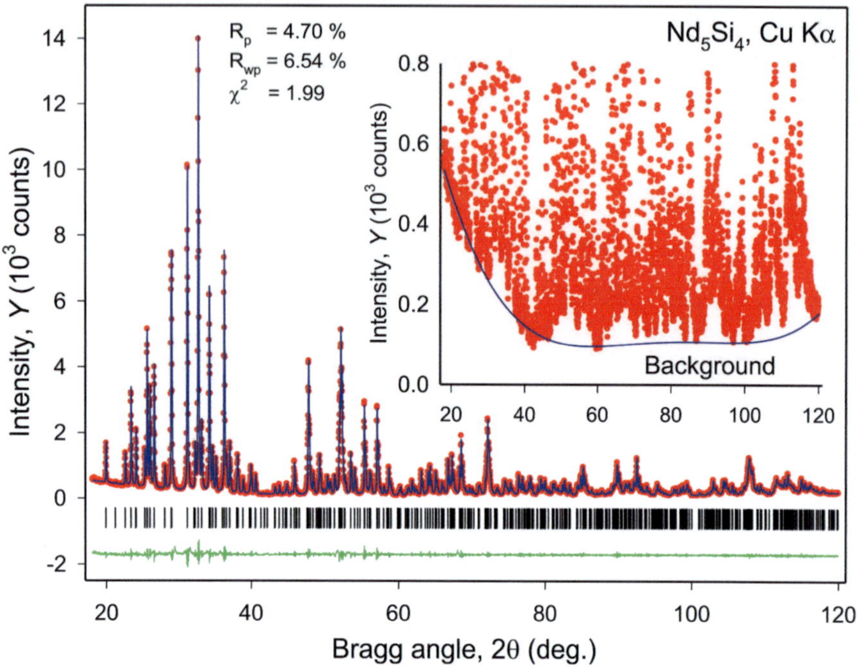

Fig. 18.1 The observed and calculated powder diffraction patterns of Nd_5Si_4 after refinement of all parameters. The powder diffraction data were collected from a ground sample of Nd_5Si_4 using Cu Kα radiation on a Rigaku TTRAX rotating anode diffractometer. The divergence slit was 0.75°, and the receiving slit was 0.03°. The experiment was carried out in a continuous scanning mode with a sampling step 0.02° and a scan rate of 0.5°/min. The inset shows the low-intensity region of the powder diffraction pattern with a properly determined background (*solid line*)

Table 18.1 Figures of merit obtained at different stages during the full-pattern decomposition of the Nd_5Si_4 powder diffraction pattern. Wavelengths used: $\lambda K\alpha_1 = 1.54059$ Å, $\lambda K\alpha_2 = 1.54441$ Å; $R_{exp} = 4.64\%$

Refined parameters	Illustration	R_p	R_{wp}	χ^2
Background (linear)	–	13.70	19.06	16.86
Background (fifth order)	–	10.42	14.15	9.29
+ a, c, U, V, W, η_0, asymmetry	–	5.14	7.03	2.30
All	Figure 18.1	4.70	6.54	1.99

18.2 Solving the Crystal Structure

The complexity of this crystal structure precludes easy interpretation of the Patterson function because there is a total of 20 Nd atoms in the unit cell. Therefore, we will try to solve the crystal structure of this alloy using direct methods. The space-group symmetry $P4_12_12$ contains only two possible sites, which are listed in Fig. 18.2.

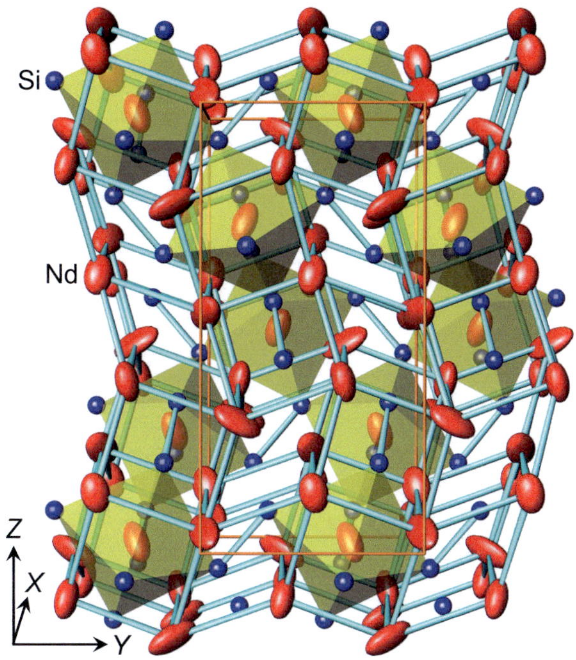

Fig. 18.2 The model of the crystal structure of Nd_5Si_4 shown with displacement ellipsoids of Nd atoms at 99% probability as determined in the process of the Rietveld refinement. Sizeable displacement anisotropy of Nd atoms may be indicative of the presence of unidentified experimental errors (also see the Footnote 15 on page 647)

Table 18.2 The three-dimensional E-map in the symmetrically independent part of the unit cell of Nd_5Si_4 calculated using the observed structure factors determined from Le Bail's extraction and directly determined phase angles in the space group $P4_12_12$

E-map peak number	*x*	*y*	*z*	Peak height
1	0.3698	0.0021	0.4543	36
2	0.1268	0.9925	0.8768	28
3	0.8118	0.1882	0.7500	19
4	0.2089	0.1725	0.6881	8
5	0.8661	0.9856	0.2168	7
6	0.1218	0.0046	0.0224	7
7	0.6805	0.1999	0.4311	7

The array of the individual structure factors determined from Le Bail's full-pattern decomposition was processed using WinCSD (see Footnote 24 on page 403, Chap. 13) and, according to a combined figure of merit, one of the possible solutions was notably better than the others. The subsequently calculated E-map is shown in Table 18.2 as a list of coordinates and heights of the electron density peaks.

The first two peaks in Table 18.2 correspond to atoms in the general site position 8(b) of the space group $P4_12_12$, and the third peak indicates an atom in the 4(a) site. Interatomic distances among peaks No. 1–3 vary from ~3.4 to ~3.7 Å, which match well the atomic radius of Nd (~1.85 Å). Thus, it appears that the direct phase-angle determination results in finding all Nd atoms, which are expected to be in one unit

Table 18.3 The three-dimensional Fourier map in the symmetrically independent part of the unit cell of Nd_5Si_4 calculated using the observed structure factors determined from Le Bail's extraction and phase angles determined by three Nd atoms: 8Nd1 in 8(b) with $x = 0.3698$, $y = 0.0021$, $z = 0.4543$; 8Nd2 in 8(b) with $x = 0.1268$, $y = 0.9925$, $z = 0.8768$; and 4Nd3 in 4(a) with $x = 0.3118$ in the space group $P4_12_12$ ($R_F = 31.1\%$)

Fourier map peak number	x	y	z	Peak height
1	0.3713	0.0053	0.4564	75
2	0.1276	0.9908	0.8780	72
3	0.8107	0.1893	0.7500	63
4	0.4238	0.1989	0.8100	13
5	0.1799	0.2036	0.3114	12
6	0.2322	0.0429	0.4980	10
7	0.1282	0.0793	0.1500	9
8	0.7451	0.0777	0.4221	9

Table 18.4 The three-dimensional Fourier map in the symmetrically independent part of the unit cell of Nd_5Si_4 calculated using the observed structure factors determined from Le Bail's extraction and phase angles determined by three Nd and two Si atoms: 8Nd1 in 8(b) with $x = 0.3698$, $y = 0.0021$, $z = 0.4543$; 8Nd2 in 8(b) with $x = 0.1268$, $y = 0.9925$, $z = 0.8768$; 4Nd3 in 4(a) with $x = 0.3118$; 8Si1 in 8(b) with $x = 0.4238$, $y = 0.1989$, $z = 0.8100$; and 8Si2 in 8(b) with $x = 0.1799$, $y = 0.2036$, $z = 0.3114$ in the space group $P4_12_12$ ($R_F = 27.3\%$)

Fourier map peak number	x	y	z	Peak height
1	0.3714	0.0057	0.4561	61
2	0.1280	0.9903	0.8779	59
3	0.8103	0.1897	0.7500	52
4	0.4220	0.2004	0.8095	17
5	0.1778	0.2070	0.3108	15
6	0.2371	0.0395	0.5003	7
7	0.8345	0.1294	0.3169	7
8	0.1239	0.0444	0.1529	7

cell of the compound (8 + 8 + 4 = 20). Recalling that Nd is the strongest scattering atom in this crystal structure, it makes little, if any, sense to further analyze the E-map: the locations of Si atoms should be easily revealed on a subsequent Fourier map (Table 18.3) after calculating reflection phases using the coordinates of the first three peaks as three independent Nd atoms. At this point, $R_F = 31.1\%$.

All three Nd atoms were confirmed on the Fourier map. Although there is no sharp reduction of peak heights among Peaks 4–8, calculation of the interatomic distances indicates that only Peaks 4 and 5 can be used as positions of silicon atoms. Both coordinate triplets correspond to the general site in the space group $P4_12_12$. Assuming that these are the missing Si atoms, the Nd–Si distances vary between ~2.9 and ~3.3 Å, and Si–Si distances are 2.50 Å ($r_{Si} = 1.10$ Å). The calculated $R_F = 27.3\%$ when all five atoms are placed as determined from the E-map (Nd) and the Fourier map (Si), and the coordinates of the electron density peaks on the subsequent Fourier map are shown in Table 18.4.

Table 18.5 Coordinates of atoms in the unit cell of Nd_5Si_4 as determined from X-ray powder diffraction data in the space-group symmetry $P4_12_12$ ($R_F = 27.1\%$)

Atom	Site	*x*	*y*	*z*
Nd1	8(b)	0.3714	0.0057	0.4561
Nd2	8(b)	0.1280	0.9903	0.8779
Nd3[a]	4(a)	0.3103	0.3103	0
Si1	8(b)	0.4220	0.2004	0.8095
Si2	8(b)	0.1778	0.2070	0.3108

[a] The coordinates of Nd3 were modified from 0.8103,0.1897,¾ (Table 18.4) to represent the triplet in a standard notation, i.e., *x*, *x*,0 instead of ½ + *x*,½-*x*,¾ with $x = 0.3103$ (see Table 18.6), by using the following transformation: $x-½,½-y, z-¾$

Table 18.6 Sites available in space-group symmetry $P4_12_12$

Site	Coordinates of symmetrically equivalent points			
4(a)	$x,x,0$	$x,\bar{x},½$	$½-x,½+x,¼$	$½+x,½-x,¾$
8(b)	x,y,z $y,x,\bar{z}$	$x,\bar{y},½+z$ $y,\bar{x},½-z$	$½-y,½+x,¼+z$ $½-x,½+y,¼-z$	$½+y,½-x,¾+z$ $½+x,½-y,¾-z$

The entire model of the crystal structure has been confirmed by the last Fourier map, which in addition has a twofold reduction in the heights of the observed electron density maxima after Peak No. 5 (double underlined). After the coordinates of all atoms have been changed as determined by the latest electron density distribution, the model of the crystal structure, which is shown in Table 18.5, results in $R_F = 27.1\%$.

The value of the residual is higher than any of the R_F's we have seen so far. This is associated with the multiple overlapping Bragg peaks at high angles, which becomes especially severe at $2\theta > \sim 70°$ (see Fig. 18.1). As a result, the full-pattern decomposition produces individual intensities and squared structure factors, which are affected by larger than usual errors. This is easy to verify by eliminating high Bragg angle reflections: when only reflections below 70° 2θ are included in the computation, the corresponding R_F becomes 20.3%. The Pearson symbol of this crystal structure is *t*P36, and after consulting Pearson's Handbook, it is easy to find that the crystal structure of Nd_5Si_4 belongs to the Zr_5Si_4-type.[3]

[3] H.U. Pfeifer and K. Schubert, Crystal structure of Zr_5Si_4, Z. Metallk. **57**, 884 (1966). https://materials.springer.com/isp/crystallographic/docs/sd_0261579

Table 18.7 The progress of Rietveld refinement of the crystal structure of Nd_5Si_4 using X-ray powder diffraction data. Wavelengths used: $\lambda K\alpha_1 = 1.54059$ Å, $\lambda K\alpha_2 = 1.54441$ Å

Refined parameters	R_p	R_{wp}	R_B	χ^2
Initial (profile parameters from Le Bail, model from Table 18.5, overall $B = 0.5$ Å^2)	1×10^5	1×10^5	1×10^4	6×10^6
Scale factor	10.17	13.09	7.64	7.95
Scale, all profile, *a* and *c*, overall *B*	9.46	12.23	6.57	6.97
Scale, all profile, *a* and *c*, overall *B* plus coordinates of individual atoms[a]	7.12	9.23	4.04	3.98
All of the above plus preferred orientation, [001]	7.04	9.18	3.89	3.93
All of the above, plus individual isotropic displacement parameters of Nd only (both Si atoms were constrained to have the same isotropic *B*)	7.01	9.12	3.68	3.89
All of the above, plus individual anisotropic displacement parameters of Nd,[b] both Si atoms were constrained to have the same isotropic *B*	6.92	9.01	3.52	3.80

[a] The coordinates of Nd3 in the 4(a) site are constrained by symmetry: $y = x$
[b] The anisotropic parameters of Nd3 in 4(a) site are constrained by symmetry: $\beta_{22} = \beta_{11}$; $\beta_{23} = -\beta_{13}$

18.3 Rietveld Refinement

The coordinates of individual atoms listed in Table 18.5 were used as a starting point together with the background, peak-shape, zero-shift, and lattice parameters determined from Le Bail's full-pattern decomposition.[4] The progress of the Rietveld refinement is illustrated in Table 18.7.

The initial model of the crystal structure results in acceptable residuals without refinement of coordinates and displacement parameters of individual atoms. When the coordinates of all atoms and the overall displacement parameter were included into the least squares, the residuals further improve (Row 4 in Table 18.7). The greatest improvement is observed in the Bragg residual, R_B, which is expected because this figure of merit is mostly affected by the adequacy of the structural model and it is least affected by the inaccuracies in profile parameters.

Unlike in any of the earlier examples, a small preferred orientation contribution is evident in this powder diffraction pattern after including a relevant parameter (Eq. 8.58) in the refinement, as shown in Row 5 in Table 18.7. An important issue to consider when refining preferred orientation is the direction of the texture axis. When the preferred orientation effects are strong, axis direction is usually easy to recognize from a simple analysis of the relationships between the observed and calculated intensities of groups of Bragg reflections with related indices. For example, if most or all Bragg reflections with indices (00 *l*) have observed intensities much stronger (or weaker) than calculated, this suggests that the preferred orientation axis is [001] (see Sects. 8.6.6 and 12.1.2). When the preferred orientation effects

[4] Data file *Ch18Ex01a.inp* is found online.

are small, the only feasible way to determine the direction of the preferred orientation axis in LHPM-Rietica is to refine the texture parameter with different texture axes beginning from its default value ($\tau = 1$, i.e., no preferred orientation is present). The axis, which results in the lowest residuals, is selected as the most probable. In this example, the following directions were tested as potential texture axes: [001], [010], [011], [111], and [112]. The best result was obtained for the [001] direction.

Refinement of the individual isotropic displacement parameters of all atoms yields a small negative *B* of Si1. It is unfeasible that Nd atoms are statistically mixed in the same sites with Si because their volumes are too different (~27 Å^3 for Nd versus ~7 Å^3 for Si). Given the density of the alloy, it is also impossible that all sites except this one are partially occupied. Therefore, the negative B_{Si1} is likely due to the fact that Si atoms have only a fraction of the scattering ability of Nd atoms, and individual displacement parameters of the former cannot be reliably determined from this experiment. Another possible reason is the nonideality of the selected peak-shape function or other small but unaccounted systematic errors. One of these is an unknown polarization constant of the employed monochromator (see Eq. 8.45). Another possibility is a more complex preferred orientation. As a result, the isotropic displacement parameters of two independent sites occupied by Si were constrained to be identical. In a way, the Si atoms were refined in an "overall isotropic" approximation.

The Rietveld refinement was finalized by optimizing the individual anisotropic displacement parameters of three independent Nd atoms. The refinement converges, and the resulting residuals are only slightly higher when compared to those obtained during Le Bail's full-pattern decomposition, thus confirming the model of the crystal structure. It is worth noting that when individual integrated intensities are re-determined at the end of Rietveld refinement in a usual way by prorating profile intensities proportionally to the calculated intensities of contributing Bragg peaks (Eq. 15.7), the resulting $R_F = 2.03\%$. This value is much smaller when compared to that calculated earlier in this chapter during the crystal structure solution when individual structure factors determined from the full-pattern decomposition were employed. In fact, such a low value of R_F would be considered an excellent result in a single-crystal diffraction experiment, especially because all 445 possible Bragg reflections were included into the calculation of R_F. The Fourier map calculated using the re-determined structure factors is clean, which means that there is a sharp reduction of peak intensity after the last Si atom, which indicates that there are no additional atoms in the unit cell of Nd_5Si_4.

The refined parameters of individual atoms are listed in Tables 18.8 and 18.9. The model of this crystal structure is shown in Fig. 18.2 together with the atomic displacement ellipsoids of Nd atoms.[5] The plot of the observed and calculated intensities is shown in Fig. 18.3. Low residuals, combined with the normal interatomic distances indicate that the solution of this crystal structure is adequate.

[5]The fully refined profile and structural data can be found online in the file *Ch18Ex01b.inp*.

Table 18.8 Coordinates of atoms in the crystal structure of Nd_5Si_4, fully refined by Rietveld technique using powder diffraction data collected from a ground powder employing Cu Kα radiation. The space group is $P4_12_12$. The unit cell dimensions are: a = 7.8714(1), c = 14.8117 (3) Å. All sites are fully occupied

Atom	Site	x	y	z
Nd1	8(b)	0.3687(2)	0.0104(2)	0.4530(1)
Nd2	8(b)	0.1292(2)	0.9844(2)	0.8742(1)
Nd3	4(a)	0.3115(2)	0.3115(2)	0
Si1	8(b)	0.4255(8)	0.2044(8)	0.8081(4)
Si2	8(b)	0.1632(8)	0.1911(8)	0.3079(3)

Table 18.9 Displacement parameters of atoms in the crystal structure of Nd_5Si_4

Atom	Site	$10^4 \times \beta_{11}$ or B	$10^4 \times \beta_{22}$	$10^4 \times \beta_{33}$	$10^4 \times \beta_{12}$	$10^4 \times \beta_{13}$	$10^4 \times \beta_{23}$
Nd1	8(b)	53(4)	24(4)	7(1)	−18(2)	−11(1)	1(1)
Nd2	8(b)	15(3)	21(3)	10(1)	−3(3)	3(1)	2(1)
Nd3[a]	4(a)	19(3)	19(3)	10(1)	−3(3)	5(1)	−5(1)
Si1[b]	8(b)	0.18(7)					
Si2[b]	8(b)	0.18(7)					

[a] In this site, the following relationships are imposed by symmetry: $\beta_{11} = \beta_{22}$ and $\beta_{23} = -\beta_{13}$
[b] Common isotropic displacement parameter, listed as B

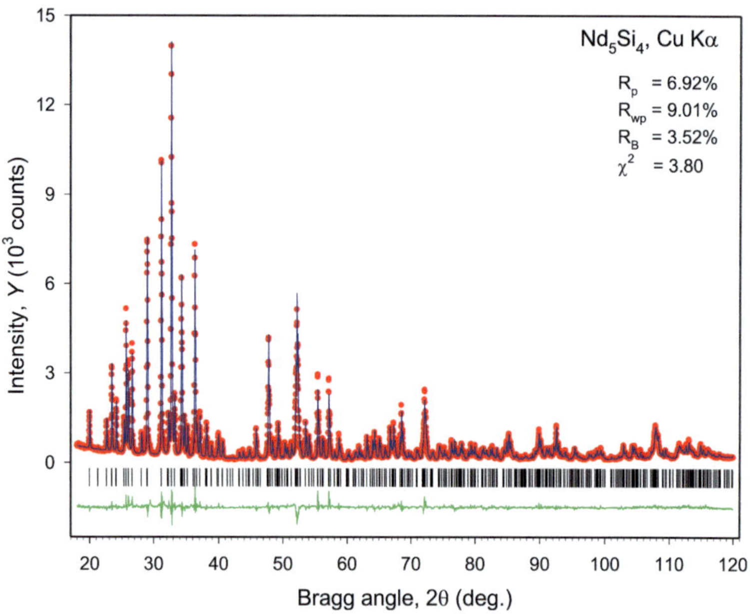

Fig. 18.3 The observed and calculated powder diffraction patterns of Nd_5Si_4 after the completion of Rietveld refinement. The data were collected from a ground Nd_5Si_4 powder dusted on a flat sample holder using a rotating anode Rigaku TTRAX powder diffractometer in a step-scan mode with a step $\Delta 2\theta = 0.02°$

Chapter 19
Empirical Methods for Solving Crystal Structures of Gd_5Ge_4, Gd_5Si_4, and $Gd_5Si_2Ge_2$

In addition to the reciprocal and direct-space techniques discussed in earlier sections, a wide range of approaches can be employed to build a crystal structure model in direct space. One of these methods, the geometrical method, was implicitly used in Chap. 16, where the location of a single La atom in the unit cell was determined through a simple analysis of the unit cell dimensions and from the availability of low-multiplicity sites in the space-group symmetry P6/mmm. A more complex case considered here is the solution of several crystal structures in the series of $Gd_5(Si_xGe_{1-x})_4$ alloys.[1] These examples illustrate the power of the powder diffraction method in detecting subtle details of atomic distribution in the unit cell, while also highlighting how structural information is crucial for establishing critical structure–property correlations.[2] It is worth noting that empirical techniques almost always require extensive literature searches to gather as much information as possible about the crystal structures of closely related materials.[3]

According to Smith, Tharp, and Johnson,[4] both the silicide and germanide of Gd at 5:4 stoichiometries belong to the same type of crystal structure as the

[1] V.K. Pecharsky and K.A. Gschneidner, Jr., Phase relationships and crystallography in the pseudobinary system Gd_5Si_4-Gd_5Ge_4, J. Alloys Comp. **260**, 98 (1997). https://doi.org/10.1016/S0925-8388(97)00143-6

[2] V.K. Pecharsky and K.A. Gschneidner, Jr., $Gd_5(Si_xGe_{1-x})_4$: An extremum material, Adv. Mater. **13**, 683 (2001). https://doi.org/10.1002/1521-4095(200105)13:9<683::AID-ADMA683>3.0.CO;2-O

[3] Examples considered in this section are also similar to the case of hydrated and anhydrous $FePO_4$ discussed later in Chap. 24. The major difference is the better crystallinity of the $Gd_5(Si_xGe_{1-x})_4$ materials and in the resulting higher quality of powder diffraction data, which facilitate straightforward Rietveld refinement of individual atomic and profile parameters without the need for preliminary quantum mechanical and/or geometrical optimizations.

[4] G.S. Smith, A.G. Tharp, and Q. Johnson, Rare earth—germanium and –silicon compounds at 5:4 and 5:3 compositions, Acta Cryst. **22**, 940 (1967). https://doi.org/10.1107/S0365110X67001902

P. Zavalij, V. Pecharsky, *Fundamentals of Powder Diffraction and Structural Characterization of Materials*, https://doi.org/10.1007/978-3-031-91504-8_19

Table 19.1 Coordinates of atoms in the unit cell of Sm_5Ge_4 as determined from single-crystal diffraction data in the space-group symmetry Pnma, after Smith et al., Acta Cryst. **22**, 269 (1967)

Atom	Site	*x*	*y*	*z*
Sm1	4(c)	0.2880	¼	0.0024
Sm2	8(d)	−0.0283	0.1004	0.1781
Sm3	8(d)	0.3795	0.8843	0.1612
Ge1	4(c)	0.1761	¼	0.3667
Ge2	4(c)	0.9132	¼	0.8885
Ge3	8(d)	0.2206	0.9551	0.4688

orthorhombic Sm_5Ge_4-type structure.[5] Furthermore, as reported by Holtzberg, Gambino, and McGuire,[6] extended solid solutions based on both binary compounds exist in the $Gd_5(Si_xGe_{1-x})_4$ system, along with the formation of an intermediate phase with an unknown crystal structure near the $Gd_5Si_2Ge_2$ stoichiometry. The atomic coordinates in the unit cell of Sm_5Ge_4 are listed in Table 19.1.

Powder diffraction patterns collected from three different samples, which belong to three different phase regions in the $Gd_5(Si_xGe_{1-x})_4$ system, are shown in Fig. 19.1. Both the similarities and differences are noteworthy: the patterns have distinct clusters of Bragg peaks in the regions $\sim 10 < 2\theta < \sim 18°$ and $\sim 21 < 2\theta < \sim 26°$; however, a conspicuous variation in peak intensities from one pattern to another is also observed.

This simple visual analysis of powder diffraction patterns typically indicates that there are detectable changes in the atomic structures of the materials in question, while the overall structural motif remains closely related. Based on this conclusion, and assuming that at least one of the materials belongs to the Sm_5Ge_4 type, it should be possible to determine details of atomic distribution in these three lattices, provided the diffraction data quality is sufficient.[7]

19.1 Crystal Structure of Gd_5Ge_4

As reported by Smith et al.[5], pure gadolinium germanide, Gd_5Ge_4, has the following unit cell dimensions: $a = 7.69$, $b = 14.75$, $c = 7.76$ Å. From the similarity of the electronic structure, chemical properties, and atomic volumes of Sm and Gd, the assumption about the identity of the crystal structures of Gd_5Ge_4 and Sm_5Ge_4 appears to be quite reasonable. Moreover, Si and Ge are electronic and

[5] G.S. Smith, Q. Johnson, and A.G. Tharp, Crystal structure of Sm_5Ge_4, Acta Cryst. **22**, 269 (1967). https://doi.org/10.1107/S0365110X67000453

[6] F. Holtzberg, R.J. Gambino, and T.R. McGuire, New ferromagnetic 5:4 compounds in the rare earth silicon and germanium systems, J. Phys. Chem. Solids **28**, 2283 (1967). https://doi.org/10.1016/0022-3697(67)90253-3

[7] Holtzberg et al., J. Phys. Chem. Solids **28**, 2283 (1967), https://doi.org/10.1016/0022-3697(67)90253-3, also noted the differences in the powder diffraction patterns of Gd_5Ge_4, $Gd_5Si_2Ge_2$, and Gd_5Si_4. However, considering the state of the art in X-ray powder diffraction analysis in 1960s, it was difficult, if at all possible, to establish the details of the distribution of atoms in these complex crystal structures.

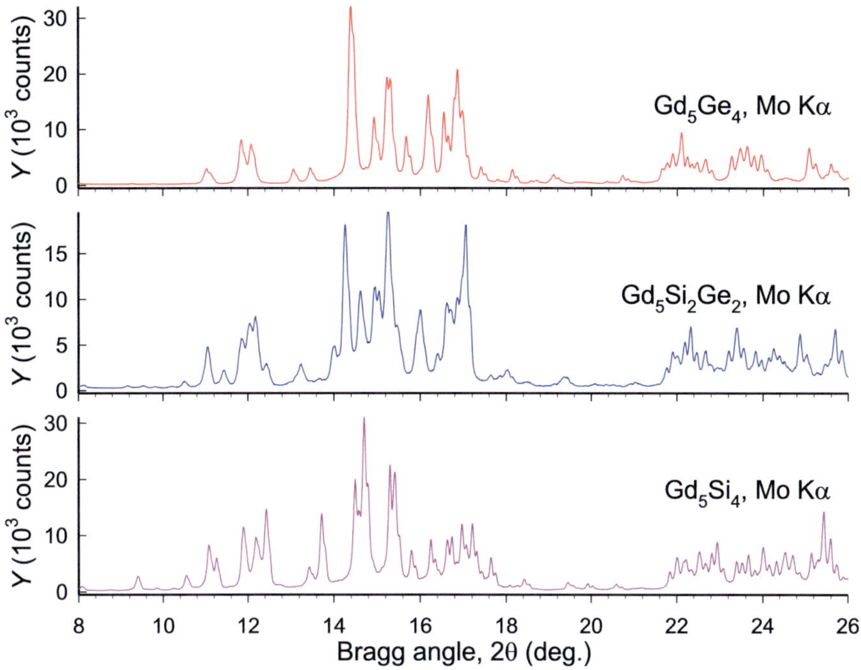

Fig. 19.1 Low Bragg angle regions of the powder diffraction patterns of Gd_5Ge_4, $Gd_5Si_2Ge_2$, and Gd_5Si_4. The three patterns are representative for materials in the Gd_5Ge_4-based solid solution, the intermediate intermetallic phase, and the Gd_5Si_4-based solid solution, respectively. The experimental data were collected on a Rigaku TTRAX rotating anode powder diffractometer using Mo Kα radiation from 8 to 50° 2θ in a step-scan mode with a step $\Delta 2\theta = 0.01°$ and counting time 15 sec/point. Distinct clusters of Bragg peaks observed in the regions ~10 < 2θ < ~18° and ~21 < 2θ < ~26° point to close relationships between the three crystal structures. Gradual redistribution of peak intensities from one pattern to another, however, cannot be solely associated with the expected change in lattice parameters due to the substitution of smaller Si for larger Ge atoms. Therefore, we conclude that three crystal structures are closely related, yet different

crystallographic twins, and it is likely that they may substitute for one another in an extended range of concentrations in metallic alloys. Nonetheless, assumptions like this usually require initial verification by computing the distribution of intensities in the powder diffraction pattern of the compound in question using proper or approximate unit cell dimensions coupled with the distribution of atoms taken from a prototype crystal structure. The result is illustrated in Fig. 19.2.

The unit cell dimensions for this sample were determined by employing Le Bail's method applied to the fragment of data from 8.5 to 27° 2θ ($a = 7.6984$, $b = 14.8278$, $c = 7.7849$ Å) along with the relevant peak-shape and background parameters, the results of which are illustrated in Fig. 19.3. All of them were used in the modeling of the calculated powder diffraction pattern using coordinates of atoms listed in Table 19.1 and assuming that Gd atoms occupy Sm-sites. Individual displacement parameters of all atoms were set at $B = 0.5$ Å^2.

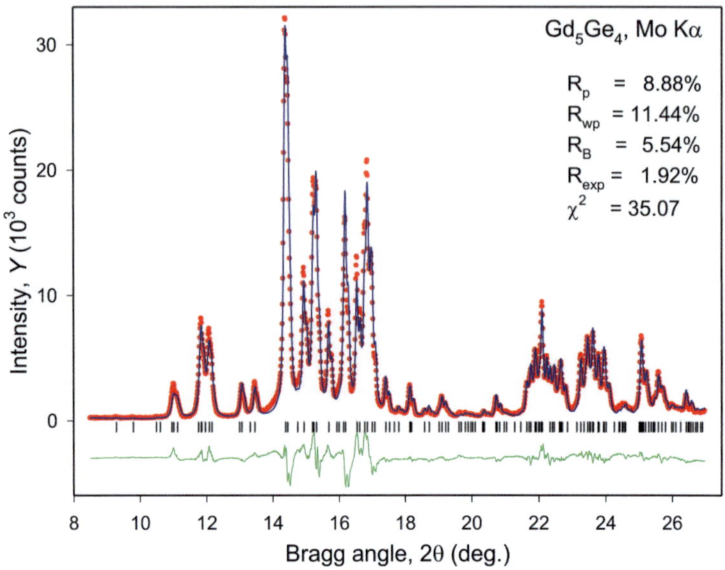

Fig. 19.2 The observed and calculated intensities in a fragment of a powder diffraction pattern of Gd_5Ge_4. The calculated intensity has been scaled to match the observed profile (The observed data are available in the data files *Ch19Ex01_MoKa.xy* and *Ch19Ex01_MoKa.dat* online)

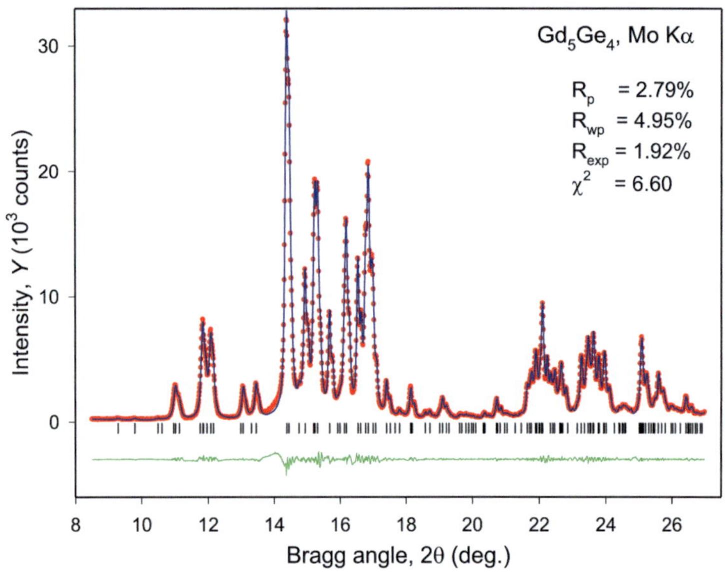

Fig. 19.3 The results of Le Bail's full-pattern decomposition of the powder diffraction pattern of Gd_5Ge_4. The discrepancies between the observed and calculated profiles are small and all residuals are low, indicating that the unit cell dimensions are accurately determined and that the chosen peak-shape function (pseudo-Voigt) is a good choice for this experiment

Even though some discrepancies between the observed and calculated profiles are obvious from Fig. 19.2, the overall good agreement and low residuals are solid indicators that the crystal structure of this material indeed belongs to the Sm_5Ge_4 type. All relevant parameters of individual atoms should be and are refined using the Rietveld method applied to the entire measured range of Bragg angles, as will be described in Sect. 19.4.1.

19.2 Crystal Structure of Gd_5Si_4

As reported by Smith et al. (see reference 4 on page 659), pure gadolinium silicide, Gd_5Si_4, has the following unit cell dimensions: $a = 7.45$, $b = 14.67$, $c = 7.73$ Å. The reduction of lattice parameters, when compared to germanide, is expected to give a smaller effective radius of Si in comparison with that of Ge.

Following the same approach—after determining the unit cell dimensions ($a = 7.4863$, $b = 14.7465$, $c = 7.7503$ Å), peak-shape, and background parameters using Le Bail's full-pattern decomposition (Fig. 19.4)—the observed and calculated powder diffraction profiles of the Gd_5Si_4 alloy are shown in Fig. 19.5. The coordinates of atoms listed in Table 19.1 were used in the computation of the scattered intensity assuming that Gd occupies Sm-sites and Si occupies the corresponding Ge

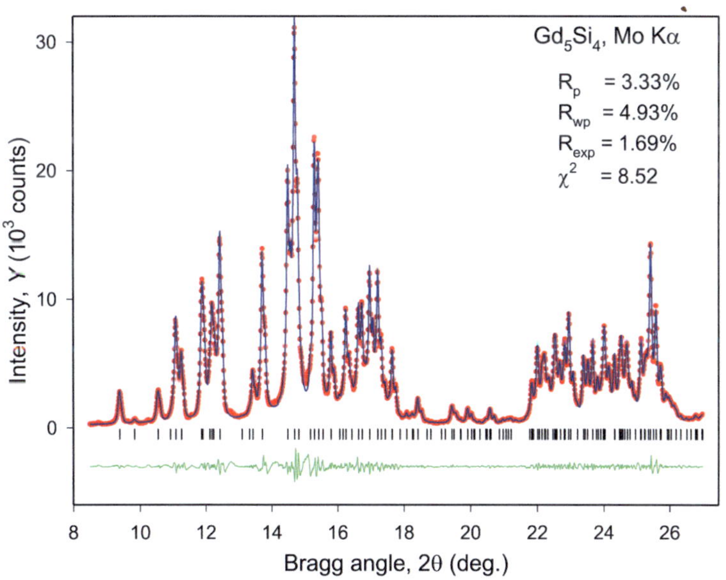

Fig. 19.4 The results of Le Bail's full-pattern decomposition of the powder diffraction pattern of Gd_5Si_4. The discrepancies between the observed and calculated profiles are small and all residuals are low, indicating that the unit cell dimensions are accurately determined and that the chosen peak-shape function (Pearson-VII) is a good choice for this experiment (The observed data are available in the data files *Ch19Ex02_MoKa.xy* and *Ch19Ex02_MoKa.dat* online)

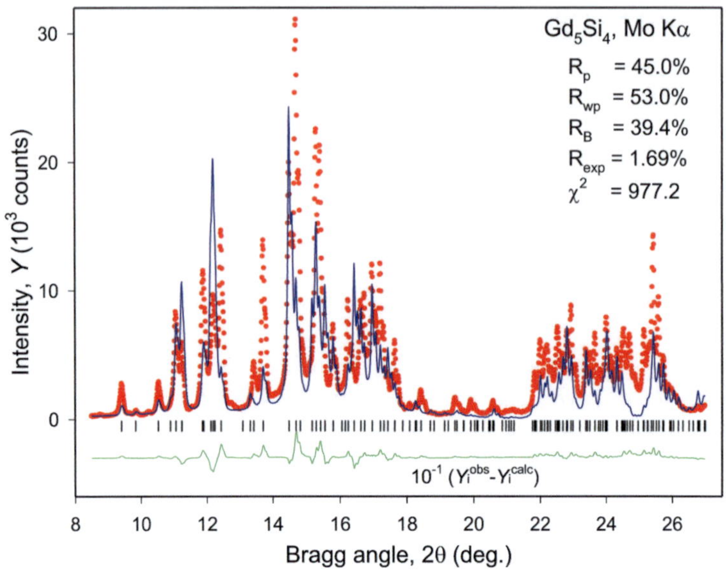

Fig. 19.5 The observed (*circles*) and calculated (*lines*) intensities in a fragment of powder diffraction pattern of Gd_5Si_4. The calculated intensity has been normalized to match the observed profile. Peak-shape, background, and lattice parameters employed to compute the calculated profile have been obtained by a full-pattern decomposition of the observed data using Le Bail's technique, as shown in Fig. 19.4. Note that the difference plot has been compressed tenfold for clarity

sites. Unlike in the case of Gd_5Ge_4 (Fig. 19.2), the match between the observed and calculated intensities in Fig. 19.5 is quite poor.

The Fourier map, calculated using this model of the crystal structure, confirms the locations of all Gd and two independent Si atoms, simultaneously indicating quite significant deviations from the coordinates listed in Table 19.1.[8] Thus, at this point, we may conclude that although the structure of the Gd_5Si_4 alloy is closely related to that of the Gd_5Ge_4 alloy, the distributions of atoms in their unit cells are not truly identical. It is feasible that by using Rietveld refinement, we can establish the necessary structural details. The structure completion process will be described in Sect. 19.4.2.

[8]We leave this as a self-exercise to the reader noting that several sequential Fourier maps, each preceded by the re-determination of the individual intensities, may be required to establish the coordinates of all atoms in the unit cell leading to an acceptable match between the observed and calculated intensities.

19.3 Crystal Structure of $Gd_5Si_2Ge_2$

Attempts to index the powder diffraction pattern of $Gd_5Si_2Ge_2$ in any unit cell in the space-group symmetry Pnma with dimensions intermediate between those of the Gd_5Ge_4 and Gd_5Si_4 alloys (see the pattern in the middle of Fig. 19.1) fail because not all peaks even at low Bragg angles can be indexed. Thus, we have to assume that the formation of the intermediate phase is accompanied by the change of the symmetry of the material. The ab initio indexing leads to a monoclinic unit cell with lattice parameters $a = 7.581$, $b = 14.809$, $c = 7.784$ Å, $\gamma = 93.19°$. The obtained unit cell dimensions are obviously related to those observed in the orthorhombic Gd_5Ge_4, see Sect. 19.1 and it is quite reasonable to assume that the intermediate intermetallic phase in the $Gd_5(Si_xGe_{1-x})_4$ system is a monoclinic distortion of the Sm_5Ge_4-type structure.

We begin with establishing the space-group symmetry of the material. In this case, a straightforward analysis of the systematic absences is difficult because of the overwhelming number of Bragg reflections, which are possible in the examined range of Bragg angles: a total of ~1500 reflections could be observed up to $2\theta = 50°$ when Mo Kα radiation is employed. However, considering a monoclinic distortion of the orthorhombic lattice, the analysis becomes relatively easy, as illustrated in Fig. 19.6. When the angle between the **a** and **b** basis vectors is no longer 90°, both the mirror plane m, which is perpendicular to **b**, and the glide plane n, perpendicular to **a**, are no longer possible. The glide plane, a, which is perpendicular to **c**, the screw axis 2_1 parallel to **c**, and the center of inversion remain unaffected by this distortion.

Thus, the likely resulting space-group symmetry is $P112_1/a$, which is a subgroup of Pnma. As also shown in Fig. 19.6, any atom located in a general site position of the space group Pnma (the multiplicity of this site is 8) breaks into two symmetrically independent atoms located in two general fourfold sites in the monoclinic symmetry (the open circles and the hatched circles in the right-hand part of the figure). Further, atoms located in special sites on mirror planes in the orthorhombic crystal system, where the value of the *y*-coordinate has been fixed at $y = ¼$ (see Table 19.1) are no longer special sites and their *y*-coordinates become free parameters.

Before proceeding with creating a model of the crystal structure of the monoclinically distorted $Gd_5Si_2Ge_2$ we will verify the correctness of both the lattice parameters and space-group symmetry ($P112_1/a$) by performing Le Bail's refinement of the experimental profile. The result shown in Fig. 19.7 indicates a high probability for this unit cell.

Considering Table 19.1 and Fig. 19.6, the approximate model of the crystal structure of $Gd_5Si_2Ge_2$ can be easily derived by splitting the coordinates of atoms located in 8(d) sites in the space-group symmetry Pnma into two 4(e) sites in the space group $P112_1/a$. The coordinates of atoms in the split positions are approximately related as *x,y,z* and *x,½-y,z*, where *x,y,z* are the coordinates of the symmetrically independent atoms in the 8(d) sites. The coordinate triplets of atoms located in 4(c) sites in the space group Pnma remain unchanged except for the fact that the

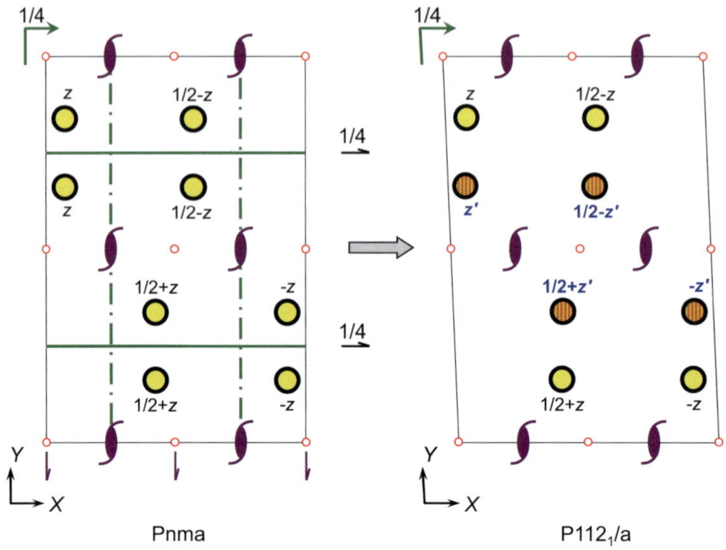

Fig. 19.6 The disappearance of mirror planes, m, and glide planes, n, during a monoclinic distortion of the orthorhombic unit cell corresponding to the space-group symmetry Pnma (*left*) when the angle between *a* and *b* unit cell edges deviates from 90° (*right*). The resulting space-group symmetry is $P112_1/a$: centers of inversion, twofold screw axes perpendicular to the *XY* plane, and glide planes, a, parallel to the *XY* plane remain unaffected by this distortion. The *open circles* on the left represent symmetrically equivalent atoms located in a general site position, 8(d), in the space group Pnma. The eightfold site splits into two independent 4(e) fourfold sites in the space-group symmetry $P112_1/a$, as indicated by both open and hatched circles on the right, for which $z' \cong z$

values of their *y*-coordinates are no longer fixed at $y = ¼$. The positions of all of the independent atoms obtained in this way, which represent the initial model of the $Gd_5Si_2Ge_2$ crystal structure, are listed in Table 19.2.

The feasibility of this model of the crystal structure has been verified by computing scattered intensity using all relevant parameters as determined from Le Bail's decomposition and coordinates of atoms from Table 19.2. It is easy to see from Fig. 19.8 that although the discrepancies are considerable, the calculated and observed profiles match better than in the case of Gd_5Si_4 (Fig. 19.5). Thus, we conclude that not only the model derived by a monoclinic distortion of the orthorhombic lattice is feasible, but also the structure of the $Gd_5Si_2Ge_2$ alloy is likely to be an intermediate between those of the gadolinium silicide and germanide. Obviously, this structure, as well as two others considered in this section, should be completed and fully refined using the Rietveld technique, which is described in Sect. 19.4.3. The relationships between these three crystal structures together with their influence on the magnetism of the alloys in the $Gd_5(Si_xGe_{1-x})_4$ system are also briefly mentioned in Sect. 19.5.

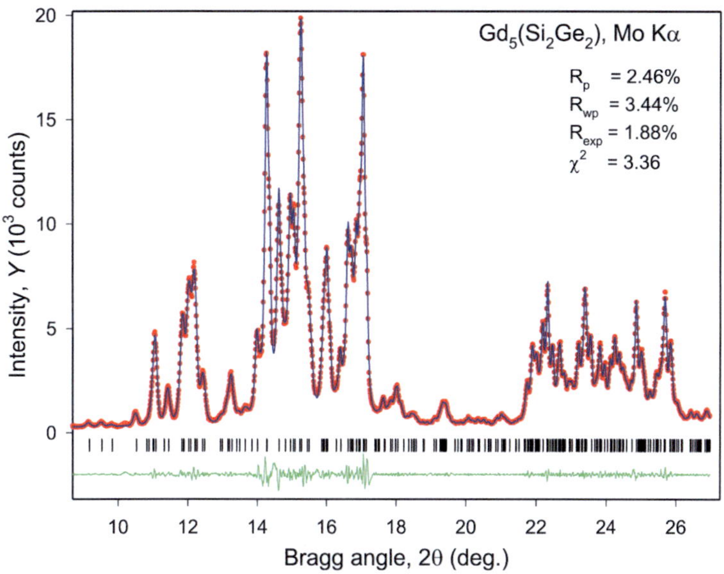

Fig. 19.7 The results of Le Bail's full-pattern decomposition of the powder diffraction pattern of $Gd_5Si_2Ge_2$. The discrepancies between the observed and calculated profiles are small and all residuals are low, indicating that the unit cell dimensions are accurately determined and that the chosen peak-shape function (pseudo-Voigt) is a good choice for this experiment (The observed data are available in the data files *Ch19Ex03_MoKa.xy* and *Ch19Ex03_MoKa.dat* online)

Table 19.2 Approximate coordinates of atoms in the unit cell of $Gd_5Si_2Ge_2$ as obtained by splitting general site positions from the space group Pnma into two sites in the space-group symmetry $P112_1/a$. As a first approximation, a completely random distribution of Si and Ge atoms in the corresponding sites is assumed

Atom	Site (Pnma)	Site ($P11/2_1a$)	*x*	*y*	*z*
Gd1	4(c)	4(c)	0.29	0.25	0.00
Gd2a	8(d)	4(c)	−0.03	0.10	0.18
Gd2b		4(e)	−0.03	0.40	0.18
Gd3a	8(d)	4(e)	0.38	0.88	0.16
Gd3b		4(e)	0.38	0.62	0.16
(Si,Ge)1	4(c)	4(e)	0.18	0.25	0.37
(Si,Ge)2	4(c)	4(e)	0.91	0.25	0.89
(Si,Ge)3a	8(d)	4(e)	0.22	0.96	0.47
(Si,Ge)3b		4(e)	0.22	0.54	0.47

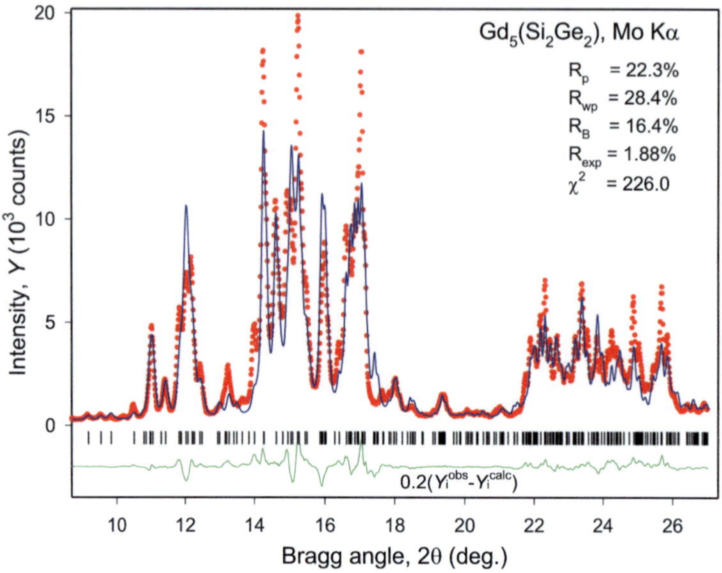

Fig. 19.8 The observed (*circles*) and calculated (*lines*) intensities in a fragment of powder diffraction pattern of $Gd_5Si_2Ge_2$. The calculated intensity has been normalized to match the observed profile. Peak-shape, background, and lattice parameters employed to compute the calculated profile have been obtained by a full-pattern decomposition of the observed data using Le Bail's technique, as shown in Fig. 19.7. Note, that the difference plot has been compressed fivefold for clarity (the differences are too large to be well seen)

19.4 Rietveld Refinement of Gd_5Ge_4, Gd_5Si_4, and $Gd_5Si_2Ge_2$

Considering the results obtained in Sects. 19.1–19.3, it appears that the three crystal structures are related to one another and that the closest to reality is the model of the germanide, Gd_5Ge_4 (Fig. 19.2). Therefore, we will begin with illustrating its refinement and then proceed with the second orthorhombic phase, Gd_5Si_4, even though it seems that the agreement between the observed and calculated diffraction data (Fig. 19.5) was the poorest for the binary silicide. Finally, we will establish the details of the crystal structure of $Gd_5Si_2Ge_2$, which appears to be a monoclinically distorted derivative of Gd_5Ge_4, as concluded earlier. In every case, we will employ all available experimental data, which were collected on a Rigaku TTRAX rotating anode powder diffractometer using Mo Kα radiation between 8.5 and 50° 2θ in a step-scan mode with a step $\Delta 2\theta = 0.01°$ and a counting time of 15 s/step. As in the previous chapters, here we will use LHPM-Rietica.

Table 19.3 The progress of Rietveld refinement of the crystal structure of Gd_5Ge_4 using X-ray powder diffraction data. Wavelengths used: $\lambda K\alpha_1 = 0.70932$ Å, $\lambda K\alpha_2 = 0.71361$ Å

Refined parameters	R_p	R_{wp}	R_B	χ^2
Initial (profile parameters from Le Bail, model from Table 19.1, overall $B = 0.6$ Å^2)	3×10^4	3×10^4	3×10^4	2×10^8
Scale factor	11.54	14.65	9.53	43.17
Scale, *U*, *V*, *W*, η, asymmetry, *a*, *b*, *c*, sample displacement, overall displacement parameter	7.36	9.57	5.14	18.45
All as above plus coordinates of all atoms and background	5.85	7.90	3.38	12.62
All, plus individual isotropic displacement parameters and preferred orientation along [001][a]	5.70	7.69	3.25	12.01

[a] The selection of the preferred orientation axis was based on the lowest residuals after attempting to refine texture in the March–Dollase approximation along the three major crystallographic axes

Table 19.4 Coordinates of atoms and individual isotropic displacement parameters in the crystal structure of Gd_5Ge_4. The space group is Pnma. The unit cell dimensions are: $a = 7.6997(3)$, $b = 14.8309(4)$, $c = 7.7861(3)$ Å. All sites are fully occupied

Atom	Site	*x*	*y*	*z*	*B* (Å^2)
Gd1	4(c)	0.2915(3)	¼	−0.0008(3)	0.84(4)
Gd2	8(d)	−0.0236(2)	0.1004(1)	0.1793(2)	1.48(3)
Gd3	8(d)	0.3782(2)	0.8839(1)	0.1623(2)	0.97(3)
Ge1	4(c)	0.1754(5)	¼	0.3653(6)	1.4(1)
Ge2	4(c)	0.9187(6)	¼	0.8905(6)	1.1(1)
Ge3	8(d)	0.2189(4)	0.9544(2)	0.4710(4)	1.5(1)

19.4.1 Gd_5Ge_4

The initial coordinates of atoms were taken from Table 19.1 along with the unit cell dimensions and all profile parameters determined from Le Bail's full-pattern decomposition.[9] The progress of Rietveld refinement is illustrated in Table 19.3.

Refinement quickly converges to low residuals, thus confirming the correctness of the model of the crystal structure of this compound. Refined individual parameters of all atoms are listed in Table 19.4. When individual displacement parameters are refined in an anisotropic approximation, residuals can be lowered further, but the displacement ellipsoid of the Gd3 atom becomes unphysical and therefore, the refinement was completed using the isotropic approximation.[10] The observed and calculated powder diffraction patterns are shown in Fig. 19.9.

The model of the Gd_5Ge_4 crystal structure, after all parameters have been refined, is shown in Fig. 19.10. This structure is built from distinct slabs, formed by Gd and Ge atoms; the slabs are infinite along both the *X* and *Z* directions, but they are limited

[9] The input data file, *Ch19Ex01a.inp*, and the file with diffraction data, *Ch19Ex01_MoKa.dat*, are located online.

[10] Final values of all parameters can be found in the data file *Ch19Ex01b.inp* online.

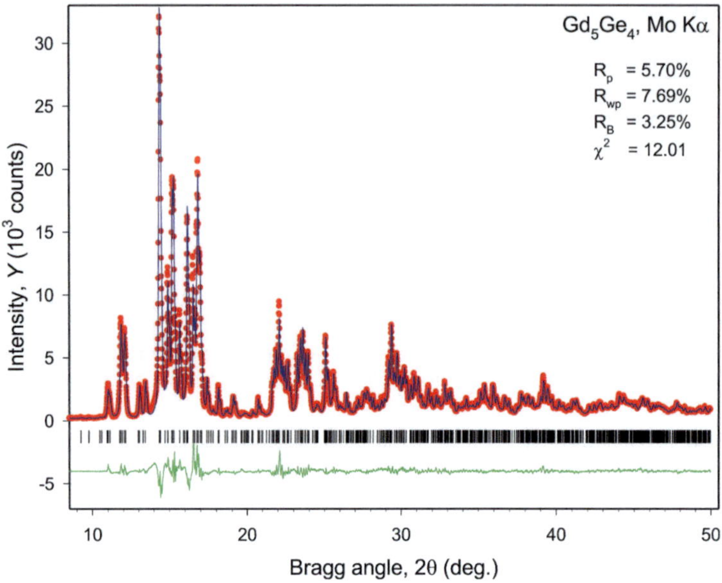

Fig. 19.9 The observed and calculated powder diffraction patterns of Gd5Ge4 after the completion of Rietveld refinement. A total of 809 independent Bragg reflections are possible in the examined range of Bragg angles

along the *Y* direction. In a way, this crystal structure is built from thin layers of tightly bound atoms stacked along the *Y*-direction, and the thickness of each layer (slab) is approximately 7 Å.

All interatomic distances are within normal limits and the Fourier map calculated after the completion of Rietveld refinement confirms that no additional atoms are present in the unit cell of this material. The refined coordinates of all atoms are nearly identical to those determined from a single-crystal diffraction experiment for the Sm_5Ge_4 compound.[11] Thus, we conclude that the crystal structure of Gd_5Ge_4 is solved correctly and it belongs to the Sm_5Ge_4-type of crystal structure in which Gd atoms occupy Sm-positions and Ge atoms are distributed in the corresponding Ge sites of the prototype.

[11] G.S. Smith, Q. Johnson, and A.G. Tharp, Crystal structure of Sm_5Ge_4, Acta Cryst. **22**, 269 (1967). https://doi.org/10.1107/S0365110X67000453

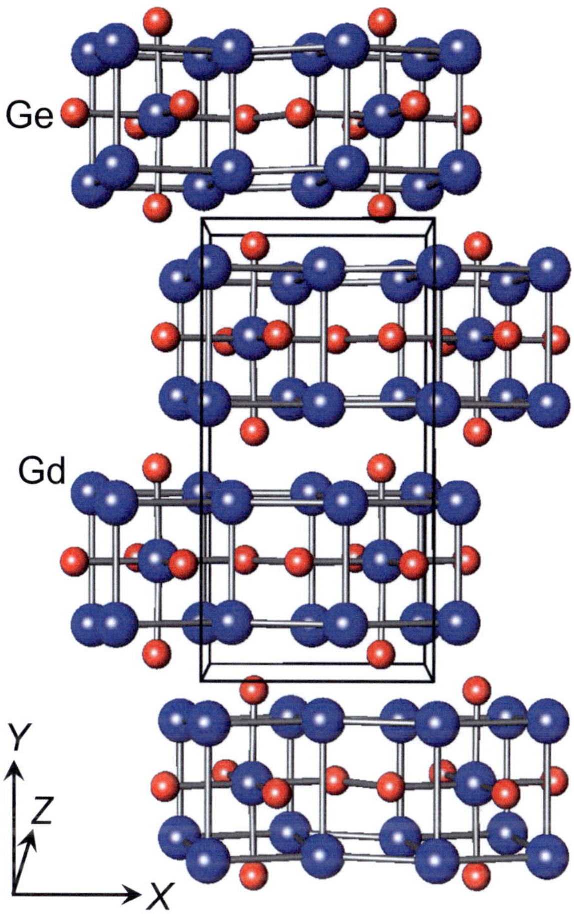

Fig. 19.10 The model of the crystal structure of Gd_5Ge_4 as determined by Rietveld refinement. The slabs, which are infinite along *X* and *Z* and are limited to ~7 Å thickness along *Y*, are shown as bonded Gd and Ge atoms

19.4.2 Gd_5Si_4

The initial coordinates of atoms were taken from Table 19.1 along with the unit cell dimensions and all profile parameters determined from Le Bail's refinement.[12] As already established in Sect. 19.2 (see Figs. 19.4 and 19.5), this model of the crystal structure is feasible but far from complete, which was evidenced by a poor agreement between the observed and calculated intensities. Thus, the Rietveld refinement strategy is slightly different from the earlier example as illustrated in Table 19.5.

Refinement of the scale factor results in poor residuals, as shown in Row 2 of Table 19.5. Considering Fig. 19.5, which shows no systematic deviations between

[12]The input data file, *Ch19Ex02a.inp*, and the file with diffraction data, *Ch19Ex02_MoKa.dat*, are located online.

Table 19.5 The progress of Rietveld refinement of the crystal structure of Gd_5Si_4 using X-ray powder diffraction data. Wavelengths used: $\lambda K\alpha_1 = 0.70932$ Å, $\lambda K\alpha_2 = 0.71361$ Å

Refined parameters	R_p	R_{wp}	R_B	χ^2
Initial (profile parameters from Le Bail, model from Table 19.1, overall $B = 0.6$ Å^2)	2×10^4	2×10^4	2×10^4	1×10^8
Scale factor	34.52	42.89	27.01	460.1
Scale and coordinates of all Gd atoms	10.54	14.43	7.12	52.2
Scale and coordinates of all Gd and Si atoms	8.08	10.71	4.82	28.8
Scale, *U*, *V*, *W*, η, asymmetry, *a*, *b*, *c*, zero shift, background, coordinates of all atoms, overall displacement parameter	6.77	9.21	3.67	21.4
All, plus individual isotropic displacement parameters and preferred orientation along [001]	6.57	9.07	3.54	20.7
Pearson-VII: All profile, then unit cell, coordinate and overall displacement parameters, preferred orientation along [001] (parameters were released sequentially)	6.16	8.28	3.16	17.3
Pearson-VII: All, plus individual isotropic displacement parameters and preferred orientation along [001]	6.07	8.19	3.02	16.9
Pearson-VII: All, plus individual isotropic displacement parameters of Gd; Si constrained to be identical; preferred orientation along [001]	6.13	8.22	3.13	17.1

the observed and calculated intensities as a function of Bragg angle, it is easy to conclude that the coordinates of atoms in the unit cell of Gd_5Si_4 deviate significantly from those assumed as in the model of the Sm_5Ge_4-type structure. In the unit cell of Gd_5Si_4, the Gd atoms are the strongly scattering species, and the Si atoms are the weakly scattering kind. Further, all profile and unit cell parameters have been determined quite precisely during the full-pattern decomposition. Therefore, we first proceed with refining the coordinates of heavy atoms, and if the result is satisfactory, we then include the coordinates of light atoms.

The refinement of the coordinates of gadolinium atoms together with the scale factor results in a considerable improvement of all residuals (Row 3 in Table 19.5) and the calculated pattern is now quite close to the measured profile (Fig. 19.11). When the coordinates of the Si atoms were included in the fit, further reduction of residuals is observed, thus indicating that the model of the crystal structure is nearly complete.

At this point, it is useful to calculate a Fourier map to verify that no other weakly scattering atoms (i.e., Si) have been missed. The electron density distribution has been calculated and it confirms that there are no additional atoms in the unit cell of Gd_5Si_4. Proceeding with the refinement of all relevant parameters in the order indicated in Table 19.5, the full-profile least-squares fit converges easily.[13]

The residuals are comparable to those obtained for Gd_5Ge_4. However, the displacement parameter of Si1 becomes negative, while the displacement parameter of Si3 is about twice than that of other atoms, as shown in Table 19.6. It is unfeasible

[13] Data file *Ch19Ex02b.inp* is available online.

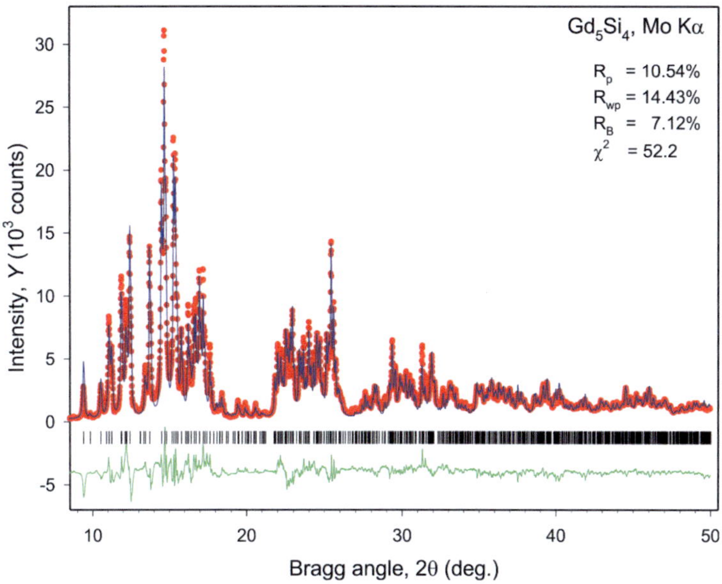

Fig. 19.11 The observed and calculated powder diffraction patterns of Gd_5Si_4 after only the coordinates of Gd atoms have been refined together with the phase scale factor. Compare this figure with Fig. 19.5

Table 19.6 Coordinates of atoms and individual isotropic displacement parameters in the crystal structure of Gd_5Si_4 fully refined by Rietveld technique using the pseudo-Voigt peak-shape function. The space group is Pnma. The unit cell dimensions are: $a = 7.4896(4)$, $b = 14.7544(8)$, $c = 7.7519$ (4) Å. All sites are fully occupied

Atom	Site	*x*	*y*	*z*	*B* (Å²)
Gd1	4(c)	0.3539(4)	¼	0.0099(3)	0.89(5)
Gd2	8(d)	0.0294(2)	0.0974(1)	0.1787(2)	1.18(3)
Gd3	8(d)	0.3159(2)	0.8780(1)	0.1837(2)	0.98(3)
Si1	4(c)	0.244(1)	¼	0.367(1)	−0.6(2)
Si2	4(c)	0.990(2)	¼	0.919(2)	1.0(3)
Si3	8(d)	0.159(1)	0.9510(6)	0.490(1)	2.2(2)

that some of the Gd atoms are mixed statistically with Si1 atoms because of the large difference in their atomic volumes. Refinement of the occupancy of the Si3 site does not result in any defects. As already explained before, this experimental artifact may be the result of the low scattering ability of Si when compared to that of Gd, coupled with small but unaccounted experimental errors that could be present in the data (see Sect. 18.3 describing the refinement of a related crystal structure of Nd_5Si_4).

Hence, we continue the refinement and employ a different peak-shape function. The use of the Pearson-VII function to represent peak shapes results in lower residuals (Rows 7 and 8 in Table 19.5). Nonetheless, individual isotropic parameters of Si1 atoms remain unphysical and we may conclude that this is due to the low

Table 19.7 Coordinates of atoms and individual isotropic displacement parameters in the crystal structure of Gd_5Si_4 fully refined by Rietveld technique using the Pearson-VII peak-shape function. The space group is Pnma. The unit cell dimensions are: $a = 7.4877(2)$, $b = 14.7496(5)$, $c = 7.7497(2)$ Å. All sites are fully occupied

Atom	Site	*x*	*y*	*z*	*B* (Å^2)
Gd1	4(c)	0.3544(2)	¼	0.0102(3)	0.88(4)
Gd2	8(d)	0.0289(2)	0.0987(1)	0.1790(2)	1.14(3)
Gd3	8(d)	0.3159(2)	0.8778(1)	0.1839(2)	0.94(3)
Si1	4(c)	0.246(1)	¼	0.369(1)	0.9(1)
Si2	4(c)	0.984(2)	¼	0.912(2)	0.9(1)
Si3	8(d)	0.158(1)	0.9518(5)	0.490(1)	0.9(1)

scattering power of Si and other errors present in the measured powder diffraction pattern. The errors were likely introduced during sample preparation, as it is easy to overlook inhomogeneities in the coverage of a flat sample holder with a powder when the specimen has been prepared by dusting. An additional argument, which supports the potential for problems with the specimen employed to collect powder diffraction data, follows from considering the physical properties of the silicide. According to Holtzberg et al. (see Footnote 6 on page 660), Gd_5Si_4 is ferromagnetic at room temperature (its Curie temperature is ~335 K). Thus, it is possible that an unusual preferred orientation exists in the powder sample, even though the material is magnetically soft.

All things considered, the simplest solution is to refine the displacement parameters of Si atoms in a pseudo-overall isotropic approximation by constraining them to be identical, as is often done with light elements, such as C, N, and O (see Chaps. 16–24). The resulting residuals (the last row in Table 19.5) are only slightly higher when compared to the refinement of the individual displacement parameters of all atoms.

The final free variables in the crystal structure of Gd_5Si_4 are listed in Table 19.7. A small difference in the unit cell dimensions obtained using different peak-shape functions is normal because peak asymmetry has been treated differently. The observed and calculated powder diffraction patterns after all parameters in the crystal structure of Gd_5Si_4 have been refined are shown in Fig. 19.12.

When the coordinates of atoms listed in Table 19.7[14] are compared to those from Table 19.4, the differences are obvious, especially in the values of the *x*-coordinates. It appears that all atoms are shifted along the *X*-direction when the Gd_5Ge_4 structure is compared with the Gd_5Si_4. The latter is shown in Fig. 19.13, from which it is clear that Gd_5Si_4 is built from the slabs that are essentially the same as those found in Gd_5Ge_4 (Fig. 19.10), except that Ge atoms are replaced with Si. The shifts along the *X*-direction result in short Si–Si distances (~2.76 Å), which are shown as bonded Si_2 pairs connecting the slabs in Fig. 19.13. No Ge–Ge bonds can be found between the slabs in the Gd_5Ge_4 structure, where the corresponding Ge–Ge distance exceeds

[14] Data file *Ch19Ex02c.inp* is available online.

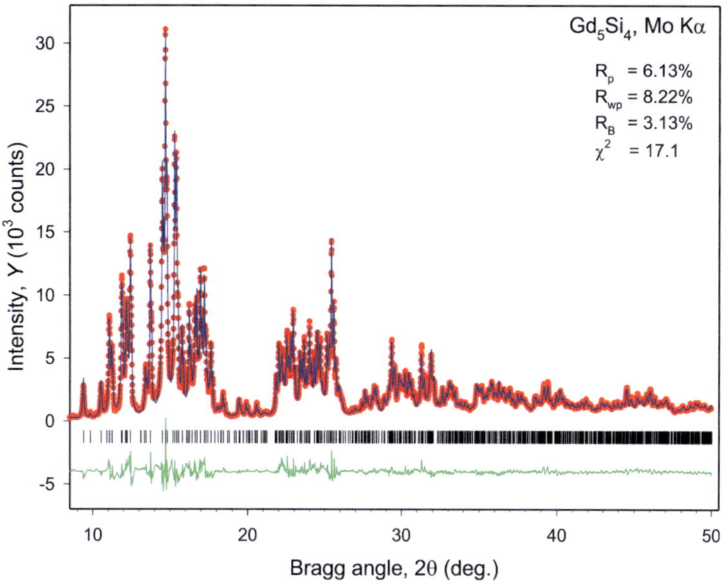

Fig. 19.12 The observed and calculated powder diffraction patterns of Gd_5Si_4 after the completion of the refinement using Pearson-VII peak-shape function. A total of 808 independent Bragg reflections are possible in the examined range of Bragg angles

3.6 Å. This variation in the interatomic distances is much larger than the difference in the atomic radii of Si (1.17 Å) and Ge (1.23 Å). Thus, the two crystal structures are closely related, but different.[15]

19.4.3 $Gd_5Si_2Ge_2$

As a starting model, we use the coordinates of atoms taken from Table 19.2 together with the unit cell dimensions and all profile parameters determined from Le Bail's full-pattern decomposition.[16] As already established in Sect. 19.3, this model of the crystal structure is also feasible but far from complete because the observed and calculated intensities do not match well. Thus, the refinement strategy is similar to the preceding example. The least-squares fit here may become complicated by the

[15] The distinct difference between the crystal structures of two materials has tremendous effect on their magnetic properties. The silicide is ferromagnetic below 335 K, while the germanide is antiferromagnetic below ~130 K [e.g. see V.K. Pecharsky and K.A. Gschneidner, Jr., $Gd_5(Si_xGe_{1-x})_4$: An extremum material, Adv. Mater. **13**, 683 (2001), https://doi.org/10.1002/1521-4095(200105)13:9<683::AID-ADMA683>3.0.CO;2-O]

[16] They are available online in the input file *Ch19Ex03a.inp* for LHPM-Rietica. The powder diffraction data are located in the file *Ch19Ex03_MoKa.dat*.

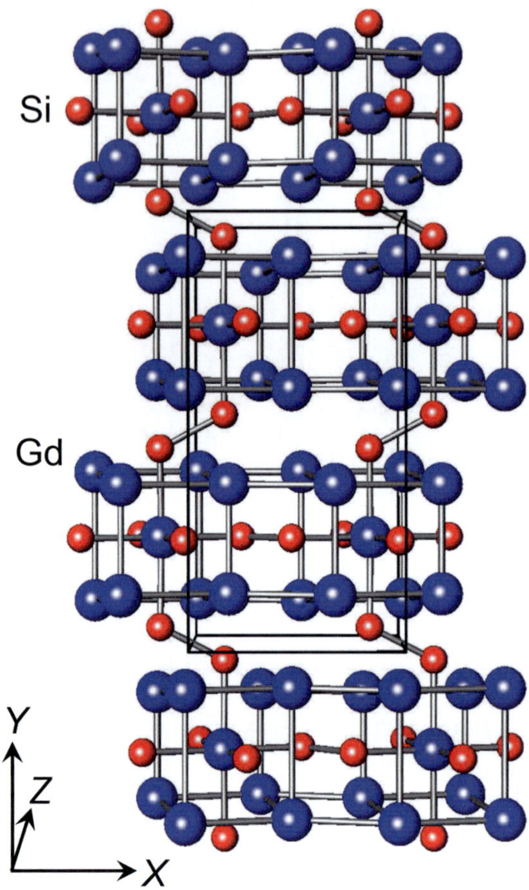

Fig. 19.13 The model of the crystal structure of Gd_5Si_4 as determined by Rietveld refinement. The slabs, which are infinite along *X* and *Z* and limited along *Y* to ~7 Å, are shown as bonded Gd and Si atoms. The major difference between this structure and Gd_5Ge_4 (Fig. 19.10) is the presence of short Si–Si interslab bonds (which are connected by gray lines) here and the absence of short Ge–Ge bonds between the slabs in Gd_5Ge_4

pseudosymmetry introduced when we derived the coordinates of atoms assuming a monoclinic distortion of the orthorhombic Gd_5Ge_4-type structure. The progress of Rietveld refinement is illustrated in Table 19.8.

The residuals steadily decrease when we include coordinates of Gd and then Si/Ge atoms into the least-squares fit, followed by profile, lattice, and displacement parameters as seen in Rows 2 – 6 in Table 19.8. A Fourier map, calculated at this point, reveals no additional atoms in the unit cell. When compared to the two earlier considered examples, an additional degree of freedom in the $Gd_5Si_2Ge_2$ structure is the distribution of the Si and Ge atoms among the corresponding lattice sites. Thus, as the next step we refine site occupancies constrained to full occupancy by Si/Ge (i.e., $g_{Ge} = 1 - g_{Si}$). Final refinement of all parameters converges to low residuals, indicating that the model of the crystal structure is correct and complete. The

Table 19.8 The progress of Rietveld refinement of the crystal structure of $Gd_5Si_2Ge_2$ using X-ray powder diffraction data. Wavelengths used: $\lambda K\alpha_1 = 0.70932$ Å, $\lambda K\alpha_2 = 0.71361$ Å

Refined parameters	R_p	R_{wp}	R_B	χ^2
Initial (profile parameters from Le Bail, model from Table 19.2, overall $B = 0.6$ Å^2)[a]	3×10^4	3×10^4	3×10^4	2×10^8
Scale factor [a]	19.5	25.2	15.0	131.4
Scale and coordinates of all Gd atoms [a]	9.19	11.9	6.85	29.4
Scale and coordinates of all Gd and Si/Ge atoms[a]	6.51	8.53	4.46	15.2
Scale, U, V, W, η, asymmetry, a, b, c, γ, sample displacement, background, coordinates of all atoms, overall displacement parameter, preferred orientation along [001][a]	5.14	6.82	2.72	9.78
All, plus individual isotropic displacement parameters of Gd atoms; Si/Ge atoms displacements in overall approximation; preferred orientation along [001][a]	5.10	6.76	2.67	9.61
All as above plus occupancies of Si/Ge sites[b]	5.04	6.69	2.61	9.43
All parameters together[b]	5.03	6.69	2.59	9.41

[a] Si and Ge atoms are distributed randomly and equally among four available crystallographic sites
[b] Constrained to full occupancy, i.e., $g_{Si} + g_{Ge} = 100\%$

Table 19.9 Atomic parameters in the crystal structure of $Gd_5Si_2Ge_2$ fully refined by Rietveld technique using pseudo-Voigt peak-shape function. The space group is $P112_1/a$. The unit cell dimensions are: $a = 7.5814(3)$, $b = 14.8039(6)$, $c = 7.7801(3)$ Å, $\gamma = 93.203(2)°$

Atom	Site	x	y	z	B (Å^2)	g (%)
Gd1	4(e)	0.3258(4)	0.2485(2)	0.0045(3)	0.91(5)	100
Gd2	4(e)	−0.0048(3)	0.0989(2)	0.1778(4)	0.88(6)	100
Gd3	4(e)	0.0180(3)	0.4023(2)	0.1799(4)	1.20(6)	100
Gd4	4(e)	0.3583(3)	0.8812(2)	0.1665(3)	1.05(5)	100
Gd5	4(e)	0.3286(4)	0.6226(2)	0.1768(3)	0.64(5)	100
Si1[a]	4(e)	0.211(1)	0.2496(5)	0.3706(8)	0.8(2)	42(1)
Si2[a]	4(e)	0.958(1)	0.2511(6)	0.896(1)	0.3(2)	65(1)
Si3[a]	4(e)	0.206(1)	0.9580(5)	0.470(1)	1.2(2)	44(1)
Si4[a]	4(e)	0.156(1)	0.5413(5)	0.474(1)	1.5(2)	50(1)

[a] Ge atoms occupy the same site with identical coordinates and displacement parameter. The occupancy is $g_{Ge} = 100 - g_{Si}\%$. Population parameters were refined with the common displacement parameter of all Si and Ge atoms to avoid potential correlations

parameters of individual atoms in the crystal structure of $Gd_5Si_2Ge_2$ are listed in Table 19.9.[17]

It is worth noting that the chemical composition of the material was not restricted in any other way except to maintain the full occupancy of all Si/Ge sites.[18] As follows from Table 19.9, the chemical composition of the powder is $Gd_5Si_{2.01}$

[17] They are also found in the file *Ch19Ex03b.inp* online.

[18] In some algorithms (e.g. GSAS), it is permissible to impose restrictions on chemical composition, if the latter is known, by setting and fixing the total number of atoms of a specific element in the unit cell during the refinement of population parameters.

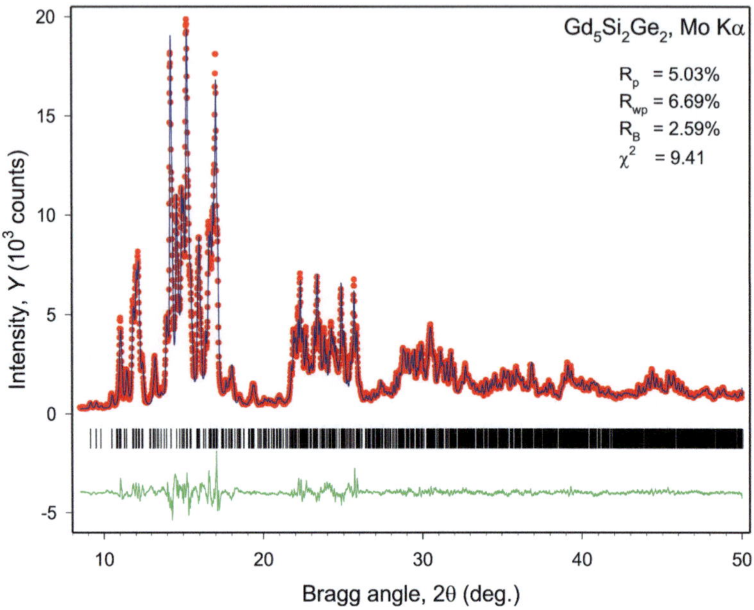

Fig. 19.14 The observed and calculated powder diffraction patterns of $Gd_5Si_2Ge_2$ after completion of Rietveld refinement. Compare this figure with Fig. 19.8. A total of 1532 independent Bragg reflections are possible in the examined range

$_{(4)}Ge_{1.99(4)}$, which is a nearly ideal match to the nominal composition of the as-prepared sample. The observed and calculated powder diffraction patterns are shown in Fig. 19.14.

The refined model of the crystal structure of $Gd_5Si_2Ge_2$ is shown in Fig. 19.15, from which it is clear that similar to both Gd_5Ge_4 and Gd_5Si_4, it is built from the same slabs. The $Gd_5Si_2Ge_2$ structure is an intermediate between the two parent compounds because here pairs of the slabs (B and C) are connected by short (Si,Ge)–(Si,Ge) bonds, while no similar bonds exist between the next neighboring slabs (A to B and D to C). It is interesting to note that these small, but distinct differences between the three closely related structures were established first from powder diffraction,[19] and only later were confirmed by a single-crystal diffraction experiment.[20] Indeed, the single-crystal diffraction investigation was complicated by the

[19] V.K. Pecharsky and K.A. Gschneidner, Jr., Phase relationships and crystallography in the pseudobinary system Gd_5Si_4-Gd_5Ge_4, J. Alloys Comp. **260**, 98 (1997). https://doi.org/10.1016/S0925-8388(97)00143-6

[20] W. Choe, V.K. Pecharsky, A.O. Pecharsky, K.A. Gschneidner, Jr., V.G. Young, Jr., and G.J. Miller, Making and breaking covalent bonds across the magnetic transition in the giant magnetocaloric material $Gd_5(Si_2Ge_2)$, Phys. Rev. Lett. **84**, 4617 (2000). https://doi.org/10.1103/PhysRevLett.84.4617

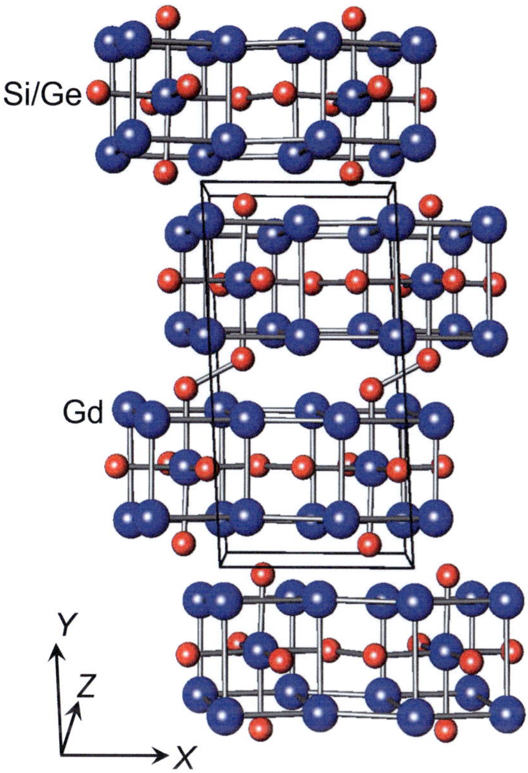

Fig. 19.15 The model of the crystal structure of $Gd_5Si_2Ge_2$ as determined by Rietveld refinement. The slabs, which are infinite along *X* and *Z* and limited along *Z* to ~7 Å are shown as bonded Gd and Si atoms. This structure is intermediate between Gd_5Ge_4 (Fig. 19.10) and Gd_5Si_4 (Fig. 19.13): only pairs of slabs (A – B) are connected with short (Si, Ge)–(Si,Ge) bonds, while no such bonds exist between the pairs (A – B and C – D)

inherent twinning of the monoclinic phase, which has no effect on the powder diffraction data.

19.5 Structure–Property Relationships

The series of compounds existing in the pseudobinary $Gd_5(Si_xGe_{1-x})_4$ system is exceptionally interesting because of the distinct role the crystal structure plays in defining the magnetic properties of the alloys. As discussed by Pecharsky and Gschneidner,[21] all Gd_5Si_4-type phases undergo a second-order paramagnetic to ferromagnetic transformation on cooling, which occurs without changing the crystal structure. However, both $Gd_5Si_2Ge_2$- and Sm_5Ge_4-type phases undergo ferromagnetic ordering simultaneously with a crystallographic phase transition, during which

[21] See V.K. Pecharsky and K.A. Gschneidner, Jr., $Gd_5(Si_xGe_{1-x})_4$: An extremum material, Adv. Mater. **13**, 683 (2001), https://doi.org/10.1002/1521-4095(200105)13:9<683::AID-ADMA683>3.0.CO;2-O, and references therein.

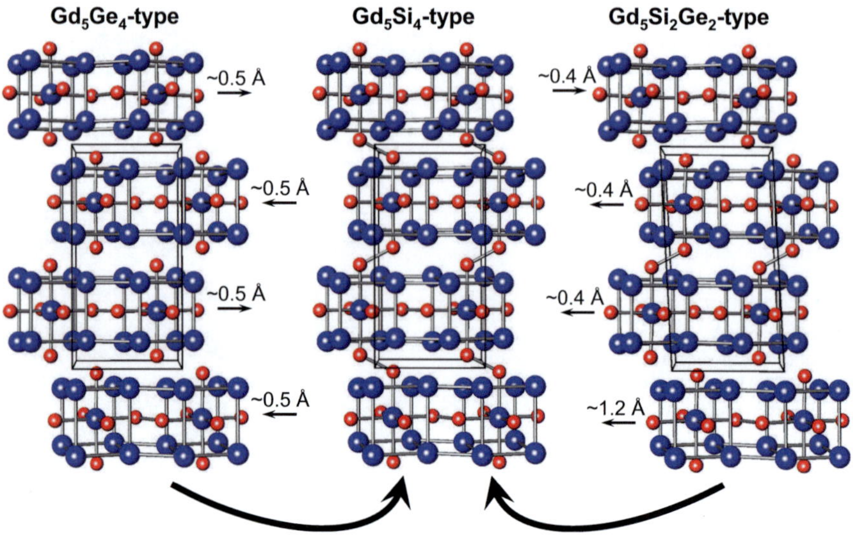

Fig. 19.16 The nature of the crystallographic phase changes when the Gd_5Ge_4-type (*left*) or the $Gd_5Si_2Ge_2$-type structures (*right*) transform into the Gd_5Si_4-type structure. Short black arrows indicate the directions and the numbers above the arrows the extent of the interslab shear

the slabs shift relavite to one another in a shear-like fashion, and all interslab bonds re-form, as shown in Fig. 19.16.

The coupled magnetic-crystallographic ordering, therefore, becomes a first-order transformation. In the ferromagnetic state, only the Gd_5Si_4-type crystal structure is stable. It is also important to mention that the crystallographic phase change as a function of temperature in this system was discovered first using powder diffraction data,[22] and 2 years later, it was confirmed from a single-crystal diffraction experiment by Choe et al. (see reference No. 20 in the footnote on page 678).

[22] L. Morellon, P.A. Algarabel, M.R. Ibarra, J. Blasco, B. Garcia-Landa, Z. Arnold, and F. Albertini, Magnetic-field-induced structural phase transition in $Gd_5(Si_{1.8}Ge_{2.2})$, Phys. Rev. B **58**, R14721 (1998). https://doi.org/10.1103/PhysRevB.58.R14721

Chapter 20
Crystal Structure of $NiMnO_2(OH)$: Combined X-Ray and Neutron Refinement

This is an example of a relatively simple inorganic compound—nickel manganese oxide hydroxide, $NiMnO_2(OH)$[1]—which has, though not perfect, crystallinity. Due to the peculiar shape of the crystallites, which grow as well-defined, elongated faceted needles (see inset in Fig. 20.1), the powder exhibits a tendency toward complex preferred orientation. This is a complicating factor, particularly when solving the structure from first principles.[2] This example also illustrates how to detect and refine hydrogen atoms from powder data, which may be a daunting task when using only X-rays.

Powder diffraction data were collected from a specimen prepared by thoroughly grinding the powder and screening it through a 38 μm sieve. Initially, a fast experiment was conducted in the range $2° \leq 2\theta \leq 70°$ using a continuous scan with a sampling step of 0.03°and a scan rate of 1°/min. Positions of 23 individual Bragg reflections at $2\theta \leq 60°$ were determined using a semimanual profile fitting. Indexing was performed using TREOR, resulting in an orthorhombic base-centered lattice with 27 possible reflections, $F_{20} = 135$ (0.003, 27).

[1] R. Chen, P.Y. Zavalij, M.S. Whittingham, J.E. Greedan, N.P. Raju, and M. Bieringer, The hydrothermal synthesis of the new manganese and vanadium oxides, $NiMnO_3H$, MAV_3O_7 and $MA_{0.75}V_4O_{10}{\cdot}0.67H_2O$ (MA = CH_3NH_3), J. Mater. Chem. **9**, 93 (1999), https://doi.org/10.1039/A805312J. The polycrystalline material was prepared by hydrothermally treating a mixture of Li_2CO_3, $N(CH_3)_4MnO_4$, and $Ni(CH_3COO)_2$ in a molar ratio 1:1.5:1 at 200 °C for 2 days.

[2] When preferred orientation effects are strong, the intensities of Bragg reflections become biased by systematic errors. Correcting for this bias is nearly impossible before solving the crystal structure, as the preferred orientation must be refined using an acceptable model. These errors are in addition to those introduced by the deconvolution of overlapped Bragg peaks.

P. Zavalij, V. Pecharsky, *Fundamentals of Powder Diffraction and Structural Characterization of Materials*, https://doi.org/10.1007/978-3-031-91504-8_20

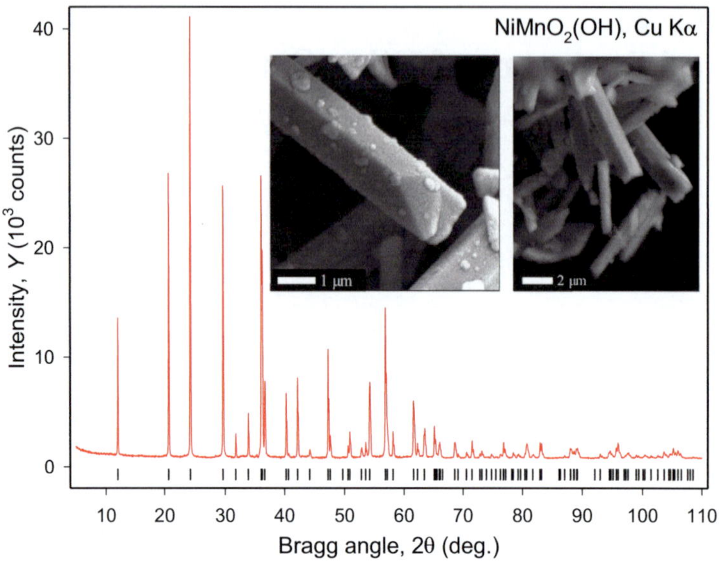

Fig. 20.1 Powder diffraction pattern collected from $NiMnO_2(OH)$ powder using Cu Kα radiation on a Scintag XDS2000 diffractometer. The experiment was conducted in a step-scan mode with a step size of 0.02° and a counting time of 30 s per step. The *vertical bars* indicate the calculated positions of the $K\alpha_1$ components of all possible Bragg reflections. The inset shows a scanning electron microscopy (SEM) image of the particle morphology in the as-received state

20.1 Observed Structure Factors from Experimental Data

To proceed with the structure solution, a high-quality powder diffraction pattern was collected in a weekend-long experiment, extending to $2\theta_{max} = 110°$ (Fig. 20.1).[3] The unit cell dimensions were refined using all 75 observed Bragg peaks resulting in the following values: $a = 2.8609(1)$ Å, $b = 14.6482(5)$ Å, $c = 5.2703(2)$ Å, $V = 220.86(2)$ Å^3, with a sample displacement of −0.123(6) mm for a 250 mm goniometer radius.

The analysis of systematic absences indicated three possible space groups: Cmcm, or one of its noncentrosymmetric subgroups, $Cmc2_1$ or C2cm (Ama2 in a standard setting). The pattern decomposition was performed using profile fitting on manually selected small ranges of Bragg angles.[4] For each group of peaks within the

[3] Data files *Ch20Ex01_CuKa.xy* and *Ch20Ex01_CuKa.raw* are available online.

[4] Semi-manual profile fitting was chosen over the full-pattern decomposition to allow better control over the integrated intensities extracted from groups of overlapped Bragg peaks, which were significantly broadened due to anisotropy (see Fig. 20.2). Small fitting ranges were selected visually, ensuring that they contained one or more distinct Bragg reflections clearly separated by the background. Full-pattern decomposition can be performed using Le Bail or Pawley techniques, both of which should converge to $R_p \approx 4.2\%$, $R_{wp} \approx 5.5\%$, and $\chi^2 \approx 4.1$. We encourage the reader to attempt this and use the extracted intensities to solve the crystal structure as an exercise.

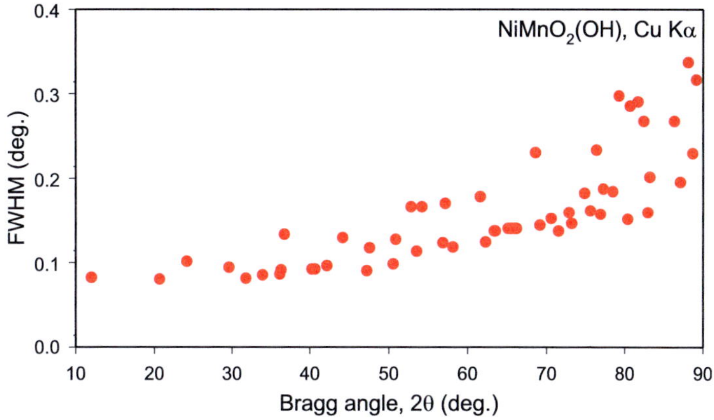

Fig. 20.2 Full width at half maximum as a function of Bragg angle, observed in the powder diffraction pattern of $NiMnO_2(OH)$

processed range, a least-squares profile fitting was conducted, refining both the positions and full widths at half maximum (FWHM) of potentially resolvable Bragg reflections. A parabolic background, a mixing parameter of the pseudo-Voigt function, and an asymmetry parameter were kept identical for all peaks within each selected range. The average R_p was approximately 2.5%.[5] To maintain stability of the nonlinear least-squares fitting, the FWHM was constrained to be identical for all peaks only in a few ranges at high Bragg angles. As shown in Fig. 20.2, the distribution of FWHMs is rather broad, which is attributed to anisotropic broadening caused by the peculiar shapes of the particles.

In illustrating the solution of this crystal structure, we use the extracted intensities of 30 observed peaks at $2\theta \leq 70°$, corresponding to a total of 36 possible Bragg reflections. Only two pairs of reflections overlap nearly completely and cannot be conclusively resolved.[6] Four unobserved reflections were assigned small intensity values, each amounting to about 1% of the intensity of the nearest observed peak. This was done because including all Bragg reflections, even those with near-zero intensity, increases the likelihood of solving the structure using direct phase determination methods. The inclusion of zero-intensity reflections is particularly important when working with small datasets, as is the case here with only 36 possible reciprocal lattice points.

[5] This value is considerably lower than R_p achievable during full-pattern decomposition, as extended background-only ranges are typically excluded from the semimanual profile fitting. Additionally, the independent treatment of positions and full widths at half maximum (FWHM) of Bragg peaks within the processed range allows for a better fit between the observed and calculated intensities.

[6] A key advantage of semimanual profile fitting is the ability to distinguish which Bragg peaks are resolvable with acceptable accuracy and which are not.

The indexed list of $|\mathbf{F}_{obs}|$ and the corresponding crystal data are stored in the SHELX format[7] for use in further structure determination. The intensities of all observed Bragg reflections[8] can also be utilized in solving the crystal structure with even greater success. We encourage the reader to use this data as a practical self-exercise.

20.2 Solving the Crystal Structure

Chemical composition of the crystals was established from microprobe and thermogravimetric (TGA) analyses. According to microprobe data, the ratio of Mn to Ni is 1:1. From TGA results, the amount of oxygen was determined to be 1.5 times that of both metals, thus suggesting a 1:1:3 stoichiometry yielding $NiMnO_3$ composition. The amount of hydrogen cannot be accurately established due to the complexity of the TGA trace, which has several different weight losses in both oxygen and nitrogen atmospheres.

The gravimetric density of the crystals was not measured, but the content of the unit cell may be established by using Eq. (15.5) and the expectation that the reasonable value of ρ should be between 4 and 5 g/cm^3. The estimated density assuming $NiMnO_3$ composition has a reasonable value of 4.86 g/cm^3 when $Z = 4$. The two closest numbers of formula units ($Z = 3$ or 5) are impossible due to the restrictions imposed by symmetry: in a base-centered lattice, sites with odd multiplicities are impossible. The next two closest numbers ($Z = 2$ or 6) result in unrealistically low and high densities, respectively. Thus, we assume that there are 4 Mn, 4 Ni, and 12 O atoms in the unit cell.

Considering space group Cmcm first, the multiplicity of the general site here is 16. The group also includes several eight- and fourfold sites, suggesting that at least one special position with the multiplicity 4 is occupied by oxygen. Both Mn and Ni atoms should occupy fourfold sites. The remaining eight O atoms can occupy one position with the multiplicity eight or two fourfold sites. The very short unit cell dimension, $a \cong 2.86$ Å, and the presence of mirror planes perpendicular to it further limits possible locations of atoms in this unit cell: all atoms must be located in the mirror planes. Thus, only the following sites 4(a): 0,0,0; 4(b): 0,½,0; 4(c): 0,*y*,¼; and 8(f): 0,*y*,*z* may be occupied. If *x* accepts a nonzero value, the distance between symmetrically related atoms becomes ~1.43 Å or shorter,[9] which is unrealistic. Another possibility is to use the noncentrosymmetric groups. The space group C2cm does not look promising because the absence of the mirror plane

[7] Online files Ch20Ex01.hkl and Ch20Ex01.ins.

[8] Online file Ch20Ex01_full.hkl.

[9] Mirror planes are spaced at ½*a*. Thus, every atom with the coordinates *x*,*y*,*z* has nearest symmetrically equivalent atoms at -*x*,*y*,*z* and 1-*x*,*y*,*z*. The pairs of atoms are separated by 2*xa* and (1-2*x*)*a*, respectively. The distances are at maximum when $x = ¼$, i.e., the spacings are $a/2 \cong 1.43$ Å.

perpendicular to *X*, which is a very short unit cell edge, is highly unlikely as discussed here. The noncentrosymmetric space group $Cmc2_1$ looks more promising as it has a fourfold special position in the mirror plane perpendicular to *X*: 4(a) with coordinates 0*yz*, where all atoms can be located.

The structure was solved[10] using SHELXS SHELX and partial least-squares refinement using SHELXL programs.[11] The centrosymmetric space-group symmetry Cmcm was tested first; however, several attempts with varying parameters produced no acceptable model. It may be difficult to recognize the incorrect selection of space-group symmetry based on a few failures to find the model, especially when relatively low-quality or truncated structure factor data are employed (e.g., those extracted from the powder pattern). If a solution at certain conditions in the selected space-group symmetry was not found, this does not necessarily mean that it does not exist. Often, it may be tricky to identify a true solution.

Taking into account that the number of formula units per cell (Z = 4) gives a preference to the space group $Cmc2_1$, it was chosen for the next attempt. At first, the direct phase-angle determination using SHELXS was attempted with all default parameters. The program automatically assigns heavy atoms to the peaks from the E-map. In this case, the first three peaks were initially treated as Mn. However, an analysis of interatomic distances indicated that the second strongest peak cannot be a metal, and therefore, this solution was dismissed.[12]

The next step, typically recommended when working with powder data and when default parameters do not yield an acceptable solution, is to decrease the minimum normalized structure amplitude (E_{min}) used in phase generation. While reducing E_{min} generally decreases the probability of phase relationships (see Eqs. 10.17, 10.20, and 10.21), it increases the number of reflections included in the process. In our case, decreasing E_{min} from the default 1.2 to $E_{min} = 1.1$ increases the number of reflections from 11 to 14. The best solution, shown in Table 20.1, includes the first two peaks, which are suitable as metals, and the next three peaks, which are appropriate as oxygen atoms.[13] Peaks starting from Q4 and below are unacceptable because they are too close to the already assigned peaks.

---

[10] The ready-to-use reflection file *Ch20Ex01.hkl* and crystal data file *Ch20Ex01.ins*, which can be used as input files for SHELXS, are available online.

[11] G. M. Sheldrick, Phase annealing in SHELXS-90: direct methods for larger structures, Acta Cryst. **A46**, 467 (1990), https://doi.org/10.1107/S0108767390000277; G.M. Sheldrick, Crystal structure refinement with SHELXL, Acta Cryst. **C71**, 3 (2015), https://doi.org/10.1107/S2053229614024218, which are distributed free for academic use (see home page at https://shelx.uni-goettingen.de/ for details).

[12] Generally, such a situation does not necessarily indicate that the model of the crystal structure cannot be completed using this solution. While the process may take longer and involve more difficulty in deciding which peaks to include and how to assign atom types to them, completion remains possible.

[13] The suitability of peaks as atoms has been assessed based on their relative heights on the E-map and the shortest interatomic distances. Although the distance between Mn2 and Q2 (1.59 Å) is quite short, it may still be acceptable at this stage of the structure solution, given that only a small number of reflections were included in the computation of both the phases and the E-map.

Table 20.1 Maxima localized from the E-map, refined for the best solution obtained by SHELXS, using 14 reflections with $E_{min} = 1.1$ and intensity data at $2\theta \le 70°$

Peak[a]	*x*	*y*	*z*	Height	Bond distances, Å		Comment
					Mn1[b]	Mn2[b]	
Mn1	0.0000	0.3018	0.4909	711.5			
Mn2	−0.5000	0.4934	0.7409	508.9			
Mn3	−1.0000	0.4550	1.0060	343.3		2.04, 2.08	O1
Q1	0.0000	0.4109	0.7040	162.3	1.96	1.89	O2
Q2	−0.5000	0.3482	0.4986	154.5	1.59	2.48	O3
Q3	−0.5000	0.4254	1.0030	103.6			
Q4	−0.5000	0.2987	0.1932	86.4			
Q5	−0.5000	0.2276	0.3901	60.1			

[a] Mn is indistinguishable from Ni at this stage because Mn has 25 electrons and Ni has 28, resulting in similar scattering factors for both atoms

[b] This column lists the distances to the indicated atom in Å

Table 20.2 Results of structure refinement (SHELXL) and the difference Fourier peaks computed using all reflection data below $2\theta = 70°$ ($R_F = 25\%$)

Atom/ peak	*x*	*y*	*z*	Height	Bond distances, Å				Comment
					Mn1[a]	Mn2[a]	O2[a]	O3[a]	
Mn1	0.0000	0.3021	0.4777						
Mn2	0.5000	0.4904	0.7232						
O1	0.0000	0.4384	1.0096			2.10,2.21			
O2	1.0000	0.3822	0.4185		1.22	2.67			b
O3	0.5000	0.3255	0.5629		1.54	2.56			b
Q1	0.5000	0.3961	0.4660	4.5	1.99	1.94	1.47	1.16	O2
Q2	0.5000	0.2192	0.4574	4.0	1.88			1.66	O3
Q3	0.0000	0.3931	0.7511	2.9					
Q4	1.0000	0.2146	0.7047	2.8					

[a] This column lists the distances to the indicated atom in Å

[b] Deleted

The obtained model was refined by using SHELXL to $R_F = 25\%$, which resulted in the coordinates listed in Table 20.2. Two oxygen atoms, O2 and O3, converged into locations that are too close to Mn1 and therefore were eliminated from the model. A difference Fourier map was calculated using phase angles determined by Mn1, Mn2, and O1 and the first two strongest maxima are at reasonable distances to both metal atoms.

This is a good point to address the near-blind inclusion of O2 and O3 in the previous step, which was done without a proper analysis of the model's geometry. It is essential to always evaluate bond angles and coordination polyhedra, in addition

Table 20.3 Results of structure refinement (SHELXL) and the difference Fourier peaks computed using all reflection data below 2θ = 70° (R_F = 9.5%)

Atom/ peak	*x*	*y*	*z*	U_{iso}/ height	Bond distances, Å Mn1[a]	Mn2[a]	O2[a]	O3[a]	Comment
Mn1	0.0000	0.3047	0.4759	0.022					
Mn2	0.5000	0.5004	0.7235	0.032					
O1	0.0000	0.4496	0.9497	0.000		2.01,2.16			
O2	0.5000	0.3920	0.4445	0.029	1.93	1.96,2.17			
O3	0.5000	0.2177	0.5605	0.007	1.97,2.22				
Q1	–0.500	0.4302	0.9673	1.9		1.65	1.46		b
Q2	0.5000	0.4249	0.6171	1.5		1.24		1.03	b

[a] This column lists the distances to the indicated atom in Å
[b] Distances are unreasonably short

to bond distances.[14] The insertion of "incorrect" atoms could be more detrimental than the simple removal of these atoms at a later point. Indeed, this may disable solving a structure as a whole. Therefore, only those atoms, which are rational from chemical and physical points of view should be included/added to the model. If there is a suspicion about a partially built model, a conventional Fourier map should be computed and used to improve the coordinates of already known atoms and verify their positioning in the unit cell (e.g., see Sect. 17.2.1, where the improper positioning of the Rh atom was easily detected from a Fourier map calculated in the space group I4/mmm).

The addition of two new oxygen atoms improves the geometry of the model, and the subsequent refinement reduces R_F to 9.5% (Table 20.3). Overall, this result provides a great confidence in the correctness of the model. All peaks on the following difference Fourier map (with only two are listed in Table 20.3) are too close to the already present atoms.

Thorough readers who try to use the full array of diffraction data in order to verify the solution of this simple crystal structure will easily find out how important is the completeness of the data: the whole model may be obtained from an E-map employing phase angles, generated by using E_{min} = 1.1.[15] The refinement of atomic parameters, however, converges to R_F = 27%, which is much higher than R_F = 9.5%

[14] Analysis of bond angles is usually unnecessary in the case of intermetallic structures, where bonding is predominantly metallic. However, knowing bond angles is critical when solving structures with principally covalent bonding.

[15] The full array of the individual structure factors can be found online in the file Ch20Ex01_full.hkl. Identical Bragg reflections have slightly different values of $|F_{hkl}^{obs}|$ when compared to Ch20Ex01.hkl because the two files were created during independent profile-fitting attempts. Readers can use this file, or they can generate their own list of the observed structure factors by using Le Bail or Pawley full-pattern decomposition approaches. In this case, anisotropic peak broadening should be refined to achieve good convergence between the observed and calculated powder diffraction profiles.

obtained for the data array truncated at $2\theta = 70°$. This increase in R_F is associated with the weaker and broader peaks and with a substantial overlap at high Bragg angles, which disable accurate determination of the individual integrated intensities. A similar situation was observed in the intermetallic Nd_5Si_4, see Chap. 18.

It is important to recognize that low R_F alone does not provide sufficient evidence for the validity of a structural model.[16] Far more crucial indicators of a model's accuracy are its geometry—such as bond distances, angles, and coordination polyhedra—and the crystal's electroneutrality. In this example, the structure consists of square pyramids for Mn1 and octahedra for Mn2, which share oxygen atoms and form a three-dimensional framework as shown later in Fig. 20.8. Furthermore, this figure clearly demonstrates that there are no mirror planes perpendicular to *Z* and confirms the correctness of the space group $Cmc2_1$. The resulting model appears reasonable and complete, except Mn and Ni atoms must be distinguished, and locations of hydrogen atoms should be established. The latter may require a different type of experimental data, while the former should be possible during Rietveld refinement of the structural parameters using the existing X-ray diffraction data.

20.3 A Few Notes About Using GSAS

Computations described here and in the following few chapters have been performed using GSAS (General Structure Analysis System), one of the most advanced implementations of the Rietveld refinement approach combined with a variety of computational crystallographic routines, which has been developed by A.C. Larson and R.B. Von Dreele with graphic user interface EXPGUI created by B.H. Toby.[17] However, the latter does not cover all commands, and many must still be executed only through a DOS-based text command interface. Since 2001, GSAS has not been officially supported. Note that, right before the release of this book, it was discovered that the original GSAS and EXPGUI versions are not longer available on the official sites. Therefore, in order to work within the GSAS environment, the only option is to use the newer GSAS-II, which is being developed and maintained by B.H. Toby and R.B. Von Dreele.[18] GSAS-II not only retains all the features of the original GSAS but also introduces numerous new capabilities and remains under continuous development. Note that although the examples provided in Chaps. 20–25 are based on

[16] R_F and profile residuals (Eqs. 15.18–15.22) can be easily reduced by increasing the number of adjustable parameters in an entirely unreasonable structural model. Therefore, these residuals should be used to assess the fit between the observed and experimental data, rather than as the sole measure of the validity of the underlying crystal structure.

[17] A.C. Larson and R.B. Von Dreele, General Structure Analysis System (GSAS), Los Alamos National Laboratory Report, LAUR 86-748 (2004); B. H. Toby, EXPGUI, a graphical user interface for GSAS, J. Appl. Cryst. **34**, 210 (2001). https://doi.org/10.1107/S0021889801002242

[18] B.H. Toby and R.B. Von Dreele, GSAS-II: the genesis of a modern open-source all purpose crystallography software package, **46**, 544 (2013). https://doi.org/10.1107/S0021889813003531

using the original version of GSAS, they can be replicated with GSAS-II using the same input files.[19]

Both the GSAS and GSAS-II are freely available and can be downloaded along with the installation instructions, manuals, and several examples.[20]

One of the peculiarities of the GSAS is the use of the instrumental parameters file. The latter contains default values of peak-shape parameters along with other instrumental and sample factors, including the wavelength, $K\alpha_2/K\alpha_1$ intensity ratio, default zero-shift or sample-displacement corrections, etc. The instrumental parameters file can be created for a specific instrumental setup, for example, a combination of divergence, Soller, and receiving slits in a given data collection geometry, and used as the default or starting values of profile parameters in the Rietveld refinement.[21]

It is important to note that not all profile parameters should or can always be routinely refined, as this depends on the quality of the specific pattern and/or sample crystallinity. For example, U, V, and W parameters, which define the instrumental part of the FWHM as a function of Bragg angle (see Eq. 8.26 and Fig. 8.14), can be kept fixed assuming that their values in the instrumental parameters file were thoroughly determined using a standard with a high degree of crystallinity, for example, LaB_6 (NIST SRM-660 or 660a). Another example is when Gaussian and Lorentzian components of peak broadening, which correspond to grain-size and strain contributions (see Eqs. 8.29–8.33), respectively, correlate severely and cannot be refined simultaneously. The latter may be due to an insufficient quality of diffraction data, low resolution, or both. In such cases, only one parameter should be refined because both produce similar broadening effects.

Overall, when initial parameters are significantly far from the correct values, and especially with noticeable correlations between certain free variables, Rietveld refinement may struggle to converge. When this happens, the calculated parameter shifts are much greater than needed and the minimization diverges, or in other words, the refinement process moves the free variables away from a global minimum (e.g., see Fig. 15.4, right). There are several ways of stabilizing the refinement and reducing the risk of the undesired divergence. One of them is numerical damping

[19] • GSAS files *.raw* and *.exp* can be imported through the "Powder Data" and "Phase" menus, respectively.

- Powder data from a *.dat* file can be imported using the FullProf file format.
- Powder data from an *.xy* file could be imported using the Topas file format.
- File. inp contains phase info in Rietica format and cannot be imported into GSAS or GSAS-II.

[20] GSAS and GSAS-II along with manuals and examples are available at https://subversion.xray.aps.anl.gov/trac/EXPGUI/wiki/GSASIntro and https://advancedphotonsource.github.io/GSAS-II-tutorials/, respectively.

[21] Other computer programs handle default settings similarly. For instance, in LHPM-Rietica, defaults can be specified for various diffractometers or experimental setups and then applied as the initial parameters of each dataset included in the processing.

of the nonlinear least squares by using the Levenberg–Marquardt technique, see Eq. (15.16) on page 560 and relevant discussion.

Another way of avoiding the "out-of-control" least squares is by applying a damping multiplier, $0 < d \leq 1$, to the calculated shifts for all or some of the refined variables, not necessarily to those that correlate. The damping multiplier in the GSAS is specified by using a numerical constant (D), which varies from 0 (default) to 9. The shifts (δx_i) are computed as $\delta x_i = \Delta x_i d = \Delta x_i(10 - D)/10$, where Δx_i is the shift of the ith free variable as determined from Eq. (15.10), and δx_i is the shift of the ith free variable, which is added to the current value of x_i. Thus, d is 1 for $D = 0$ and 0.5 for $D = 5$. The latter constant was employed in examples found in the following few chapters and all free variables except for the scale factor and background parameters were dampened.[22]

Both damping techniques stabilize Rietveld refinement but obviously require more least-squares cycles for its completion. It is worth noting that according to the GSAS default (which can be changed), the refinement process is considered converged to a global minimum when the maximum shift (Δx_i) observed among all free variables is less than 1/100th of the corresponding standard deviation (σx_i), i.e., when $|\Delta x_i|/\sigma x_i < 0.01$, $i = 1, 2, \ldots p$ and p is the number of free variables.[23]

The structure and properties of both the metal oxides and their intercalates, which are used as examples in the following few chapters, result in high anisotropy in crystal shapes. Distinct plate-like (see Fig. 21.1) or needle-like (Fig. 20.1) particle shapes lead to a highly random distribution of particle orientations, even after thorough grinding and screening. The state of the specimen often cannot be adequately described using a simple single-parameter preferred orientation model, so two preferred orientation axes or a spherical harmonics expansion should be employed. Additionally, the anisotropy of the particle shapes almost necessarily leads to anisotropic broadening of diffraction peaks, which must also be accounted for to achieve a good fit.

Unless otherwise noted, the initial values of peak-shape parameters were obtained from a refinement using the experimental powder diffraction pattern of LaB_6, measured on a Scintag XDS2000 diffractometer with Cu Kα radiation and a liquid nitrogen-cooled Ge(Li) solid-state detector under typical experimental settings.[24] It is important to note that all profile parameters in GSAS are in centidegrees (i.e., the corresponding values in degrees have been multiplied by 100). The following settings and initial parameters were used:

---

[22] Examples considered in Chaps. 16–18 did not require damping, yet LHPM-Rietica and other commonly available Rietveld refinement computer codes foresee either or both damping approaches described here. Applying damping cannot be deleterious even when it is unnecessary. It leads to more least-squares cycles required to achieve convergence (global minimum) but simultaneously prevents accidental "out-of-control" least squares, thus enabling to use starting parameters, which are relatively far from correct values.

[23] The IUCr imposes the requirement $|\Delta x_i|/\sigma x_i < 0.05$ to ensure the completeness of refinements based on single crystal data. In Rietveld fits, a value of $|\Delta x_i|/\sigma x_i < 0.1$ is considered satisfactory.

[24] These are provided in the file *Scintag.prm*, which is available online.

- The peak-shape function was a Thompson modified pseudo-Voigt,[25] referred as No. 2 in GSAS (also see Sect. 8.5.1 and relevant equations).
- Bragg peaks were extended over the range where their calculated intensity was greater than or equal to 0.5% of the intensity at the peak maximum.
- The instrumental part of FWHM: $U = 0.7104$, $V = -0.9565$, $W = 1.7318$; $P = 0$.
- Isotropic peak broadening due to grain size and strain with $X = 2.2952$, $Y = 3.9551$, and $X_a = 0$ and $Y_a = 0$.
- Peak asymmetry $\alpha = 2.5471$.
- Porosity and absorption effects were initially accounted for using the Suortti approach (see Eq. 8.55) with parameters $a_1 = 0.4$ and $a_2 = 0.4$. These two parameters tend to correlate strongly and were refined only when the quality of the pattern was sufficiently high.
- The sample-displacement parameter S_s was calculated for each case from the displacement s (in mm), obtained during unit cell refinement as: $S_s = -36000s/(\pi R)$ or $S_s = -144s/\pi$ for the goniometer radius $R = 250$ mm.
- Cu Kα radiation wavelengths used were 1.540562 Å for the Kα_1 component and 1.544390 for the Kα_2 component.
- The initial value of the phase scale factor was always set to 1.[26]
- The initial background was set to a constant value of 100 counts or, in some difficult cases, fitted manually and kept fixed during initial refinement steps.
- Unless otherwise noted, a six-parameter shifted-Chebyshev function was used to fit the background at later stages of refinement.
- The initial atomic displacement parameters were always set to $U_{iso} = 0.015$ Å^2, which is equivalent to $B_{iso} = (=8\pi^2 U_{iso})$ or approximately 1.2 Å^2.
- Fractional population factors in GSAS are treated as g-values (see Eq. 16.1), with site multiplicities automatically accounted for and not adjustable. Site populations, however, can be refined when needed. For example, a population factor $g = 0.75$ for an A atom in a site with multiplicity 4 means that 75% of the site is occupied and that there are 3 A atoms in the unit cell. Naturally, the fractional population factor cannot exceed unity or be less than zero. If the refined value falls outside the range $0 \leq g \leq 1$, this usually indicates incorrect atom type assignments or improperly located atoms.

---

[25] P. Thompson, D.E. Cox, and J.B. Hastings, Rietveld refinement of Debye–Scherrer synchrotron X-ray data from Al_2O_3, J. Appl. Cryst. **20**, 79 (1987). https://doi.org/10.1107/S0021889887087090

[26] Phase scales are treated differently in various implementations of the Rietveld algorithm, such as in GSAS and in LHPM-Rietica. In LHPM-Rietica, the calculated absolute intensity is scaled (normalized) to match the observed relative intensity, meaning the scale factor is applied exactly as shown in Eqs. 15.29 and 15.30. In GSAS, the observed relative intensity is scaled to match the calculated absolute scattered intensity. In other words, the scale factors in GSAS and LHPM-Rietica are related as $K_{GSAS} = 1/K_{Rietica}$.

20.4 Completion of the Model and Rietveld Refinement

This example demonstrates how to refine complex preferred orientation and distinguish chemical elements with similar scattering factors, such as Ni (28 electrons) and Mn (25 electrons), using X-ray powder data. Additionally, we will see how much easier this distinction becomes when neutron diffraction data are available, as well as how neutron data help locate hydrogen atoms in the unit cell. Finally, we will consider some important geometrical aspects of the interpreting of the structural data.

20.4.1 Initial Refinement Steps

The initial model of the crystal structure is presented in Table 20.3. Both independent metal atoms are treated as manganese (Mn1 and Mn2) and the hydrogen atom is absent from this model. Therefore, our goal is to distinguish between the Mn and Ni atoms, locate the H atom using Fourier maps, and obtain accurate positional, atomic displacement, and profile parameters.

The initial stages of Rietveld refinement are often both crucial and challenging, as the starting values for structural and profile parameters may deviate significantly from the correct values. As a result, nonlinear least-squares fitting tends be less stable during these early stages compared to the end of the refinement, when most parameters are closer to their accurate values. As mentioned earlier, variables should be refined in a proper sequence, typically starting with the most critical parameters and gradually adding other relevant variables. Throughout the process, it is important to monitor how the initially refined parameters evolve. Any parameters that correlate strongly or begin to diverge should be excluded from the refinement or possibly constrained.

The following initial parameters were used to begin this refinement:[27,28]

- Default profile parameters from the instrumental parameter file[29] as described in Sect. 20.3.

[27] The starting model of the crystal structure with all the necessary parameters is found online in files *Ch20Ex01a.exp* (which is the main GSAS data file, which contains all structural, instrumental, and other parameters needed for refinement) and in *Ch20Ex01a.cif*. The latter is a Crystallographic Information File (CIF file), which records all information in a standard format acceptable by the majority of crystallographic programs and is required by the majority of technical journals for publication of the structure determination results.

[28] The experimental pattern is located in the file *Ch20Ex01_CuKa.raw*—the profile (histogram) file—which contains experimental powder pattern in a standard GSAS format. Note that this format is also suitable for LHPM-Rietica.

[29] The file *Scintag.prm* is available online.

Table 20.4 The progress of Rietveld refinement of the crystal structure of $NiMnO_2(OH)$ first using X-ray and later both X-ray and neutron powder diffraction data

Refined parameters	R_p	R_{wp}	R_B	χ^2
Initial[a]	36.9	50.8	99.7	350
Scale factor only (Fig. 20.3)	18.2	25.9	36.0	90.9
Scale, background, unit cell dimensions, grain size (X)	14.9	22.4	34.0	68.1
All of the above plus preferred orientation (PO) for [010] axis and then adding another PO for [100] axis	9.6	13.5	12.2	24.8
All of the above plus strain (Y) instead of X, PO1/PO2 ratio, asymmetry (α), coordinates of all atoms, U_{over} (Fig. 20.4)[b]	7.4	10.6	8.9	15.3
All of the above plus Mn2 was changed to Ni1 (5 cycles), then individual U_{iso} for Mn1 and Ni1	7.3	10.5	9.6	15.0
All of the above plus S_s, profile parameters, grain size, strain together with their anisotropy (X_a and Y_a)	5.9	8.0	6.8	8.75
Only scale, background, unit cell dimensions, and absorption, a_1 and a_2, in the Suortti approximation	6.0	8.1	6.3	8.79
All of the above plus coordinates, U_{iso} for Mn1 and Ni1, $U_{overall}$ for O, PO[010], PO[100], X, X_a α	6.0	8.0	6.7	8.77
All of the above plus U, V, W, Y, Y_a. Final (X-ray only), see Fig. 20.5	5.1	6.6	6.7	5.99
Combined final: X-ray	5.1	6.7	6.7	n/a
Combined final: Neutrons	4.0	5.0	24.4	n/a
Combined final: Total (Fig. 20.7a, b, Table 20.5)[c]	5.0	6.5	n/a	5.85

[a] Files online: *Ch20Ex01_CuKa.raw*, *Ch20Ex01a.exp*, and *Ch20Ex01a.cif*
[b] Files online: *Ch20Ex01b.exp* and *Ch20Ex01b.cif*
[c] Files online: *Ch20Ex01c.exp* and *Ch20Ex01c.cif*

- Sample shift, $S_s = 5.64$ to represent sample displacement $s = -0.123$ mm, which was obtained together with the unit cell dimensions during the lattice parameter refinement.
- Space group $Cmc2_1$ and unit cell dimensions $a = 2.8609$ Å, $b = 14.6482$ Å, $c - 5.2703$ Å, as determined earlier.
- Structure model from Table 20.3 with overall isotropic displacement parameter $U_{iso} = 0.015°A^2$.

Initially, only the scale factor has been refined, resulting in the residuals shown in Row 2 in Table 20.4. The calculated pattern matches the observed data quite well, considering the unrefined model, as can be seen from Fig. 20.3. The strong calculated Bragg reflections correspond to the strong observed peaks and the weak reflections to the low observed intensity peaks, which is a reasonable confirmation of the initial model. The quite good initial approximation of both the unit cell dimensions and sample displacement is clear from the inset in Fig. 20.3.

The subsequent refinement step included six background coefficients for a shifted-Chebyshev polynomial, unit cell dimensions, and grain-size parameter (X). Only a little improvement in the fit results. Next, the preferred orientation was refined using the March–Dollase approach (see Sect. 8.6.6). At first, the preferred orientation axis was chosen along the [010] direction. This direction is perpendicular

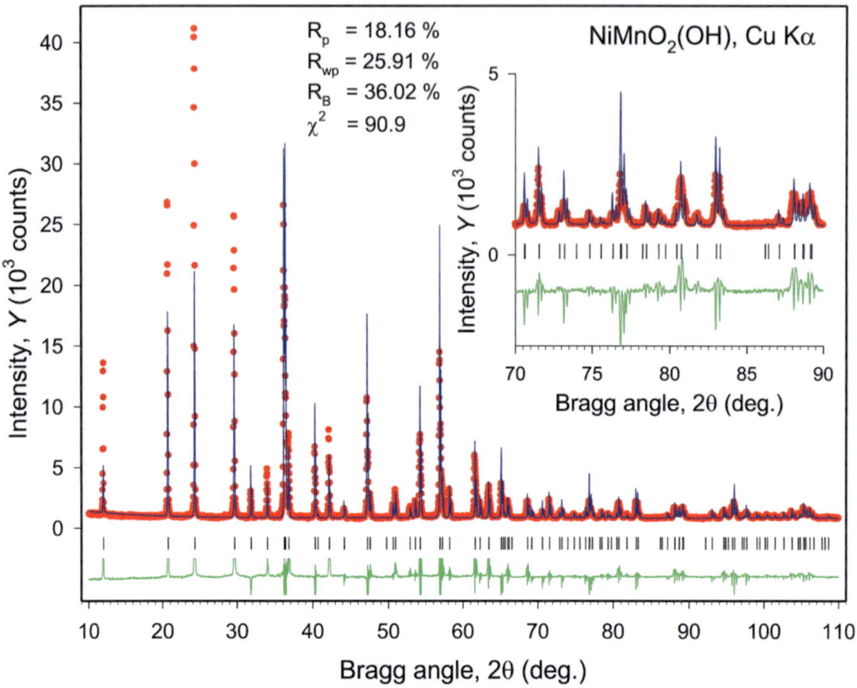

Fig. 20.3 The observed and calculated powder diffraction patterns of $NiMnO_2(OH)$ after the initial Rietveld refinement with only the scale factor determined. The inset highlights the range between 70° and 90° 2θ. The difference ($Y_i^{obs} - Y_i^{calc}$) is shown using the same scale as both the observed and calculated data, but the plot is truncated to fit within the range [−1500, 1500] for clarity

to the metal-oxide layers (Fig. 20.8) and coincides with the longest unit cell edge, *b*. The preferred orientation parameter, τ_{010}, was refined to $\tau_{010} = 0.80$, which corresponds to the preferred orientation magnitude of 2.73.[30] A second preferred orientation axis, [100], was added after the first, and this choice was based on the presence of metal-oxide chains along the shortest unit cell edge, *a*. Both preferred orientation parameters were refined together with other variables already included in the least squares, resulting in a substantial lowering of all residuals (Row 4 in Table 20.4). The preferred orientation axes could be easily predicted in this structure by comparing the unit cell dimensions and simple geometrical analysis of the model. In many instances, the longest unit cell dimension (in this material the *b*-axis), which is perpendicular to the layers formed in the crystal structure, is also parallel to the shortest dimension of the plate-like crystallites.

[30]The magnitude of the preferred orientation is the ratio between the maximum and the minimum correction factors, which in this case is the ratio between the correction factors for reflections whose reciprocal lattice vectors are parallel to $\mathbf{d}^*_{010}$ and those, which are perpendicular to $\mathbf{d}^*_{010}$, i.e., $T_{\parallel}/T_{\perp}$.

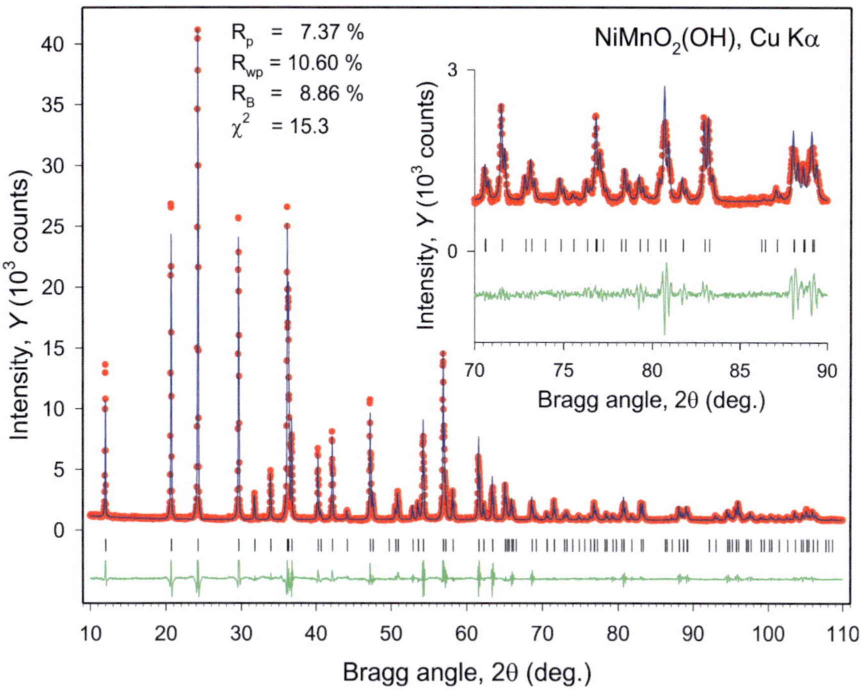

Fig. 20.4 The observed and calculated powder diffraction patterns of $NiMnO_2(OH)$ after preferred orientation, coordinates of all atoms, and the overall displacement parameter were refined in addition to the scale factor, unit cell dimensions, background, grain size and strain effects, and peak asymmetry. The insert highlights the range between 70° and 90° 2θ

Similarly, the shortest unit cell dimension is usually parallel to the chain-like formations in the structure (if any) and, simultaneously, to the longest dimension of the needle-like crystallites. In $NiMnO_2(OH)$, the a axis is much shorter than the two others: the needle-like crystallites are elongated along the [100] direction, with the additional preferred orientation axis [010] perpendicular to the flat sides of the needles (see the inset in Fig. 20.1).

When the two preferred orientation axes are assumed, the ratio between them should be refined as well. This was done subsequently, together with the refinement of the coordinates of individual atoms, overall isotropic displacement parameter, $U_{overall}$, and peak asymmetry, α. The resulting fit is substantially improved, as shown in Fig. 20.4. It is clear, however, that there are still some differences between the observed and calculated intensities, as well as in the peak shapes (e.g., see the inset in Fig. 20.4, where some calculated peaks appear too narrow when compared with the observed peak shapes).

20.4.2 Where Is Mn and Where Is Ni?

The next structure determination step is to distinguish the Mn and Ni atoms in two positions, presently treated as manganese. There are several possible ways of accomplishing this task:

- Refining individual isotropic atomic displacement parameters, where a larger value indicates the lighter atom, and a smaller or even negative displacement parameter suggests the heavier atom. This method works well in refinements using single-crystal diffraction data or highly precise powder data, typically for chemical elements with substantial differences in the scattering ability (number of electrons), see Sect. 16.3.3 for an example.
- Refining site population factors: similar to the earlier approach, this method is a more accurate way to test an atom's scattering power, which is directly proportional to the number of electrons, and, therefore, the population factor.
- Conducting the refinement using all possible combinations of elements and selecting the best model based on the resultant figures of merit. While this approach can be time-consuming if many permutations are possible. However, in the case of $NiMnO_2(OH)$, only two possibilities exist: Mn1 is Mn and Mn2 is Ni, or Mn1 is Ni and Mn2 is Mn.
- Analyzing the geometry of the model: this approach is almost always used in molecular compounds, even when single-crystal data are available. The analysis involves prior knowledge of possible bond distances, angles, coordination polyhedra, etc.
- Using additional information beyond Cu Kα X-ray diffraction data, such as anomalous scattering near the K-absorption edge of one of the metals, neutron diffraction data (see later), and/or spectroscopic results.

Despite a variety of available methods, some may not always yield clear results. In this case, the first three approaches based on X-ray diffraction data failed to clearly differentiate between Mn and Ni atoms, as the differences in atomic displacement parameters, site populations, and standard figures of merit were not statistically significant. However, the different environments of the two independent metal atoms suggest that a geometrical analysis may be helpful.

Thus, the bond valence sum method,[31] which is primarily used to differentiate between the oxidation states of chemical elements rather than the elements themselves, was employed. The calculated bond valence sum should be as close as possible to the oxidation state it represents. In our model, the bond valence sum technique yielded 2.93 for the first atom, treated as Mn^{3+}, and 1.96 for the second

[31] Bond valence sum is calculated using interatomic distances and empirical bond valence parameters tabulated for each type of the bond. The analysis was conducted using VaList software [A.S. Wills and I.D. Brown, VaList, CEA, France (1999)], https://www.iucr.org/resources/other-directories/software/valist; I.D. Brown, VALENCE: a program for calculating bond valences, J. Appl. Cryst. 29, 479 (1996). https://doi.org/10.1107/S002188989600163X

atom, assumed to be Ni^{2+}. The next closest result was 2.78 for Ni^{3+} in the first position and 2.61 for Mn^{3+} in the second metal site (see Table 20.3). Therefore, bond valence sum clearly reveals that the first metal (Mn1) is actually Mn, while the second atom (Mn^2) is Ni, with oxidation states of 3+ for Mn and 2+ for Ni. The final chemical composition is, therefore, $Ni^{2+}Mn^{3+}O_2(OH)$, confirming the presence of a hydroxyl group in the compound. In $NiMnO_3$, both Mn and Ni atoms must have oxidation states of 3+ to maintain charge balance. Thus, all subsequent refinement steps included Mn and Ni in their correct sites, with atomic displacement parameters refined independently.

20.4.3 Finalizing the Refinement of the Model Without Hydrogen

Including both grain-size and strain contributions to the full width at half maximum (X and Y), along with their anisotropic components (X_a and Y_a), noticeably improved the fit (Table 20.4). Setting the porosity and absorption effects as free variables using the Suortti approach, while keeping most of other parameters fixed to avoid correlations, adjusted the values from 0.40 and 0.40 to 0.32 and 0.51, respectively, without improving the figures of merit. Finally, the U, V, and W parameters, representing the instrumental contribution of the FWHM, were refined until the full convergence was achieved. The noticeable improvement of the profile figures of merit, after refining U, V, and W suggests that their preset values were incorrect. It is also worth noting that X and X_a were kept fixed during the final few least-squares cycles due to their strong correlation with Y and Y_a. It was unnecessary to refine both pairs, as we had no intent to analyze grain-size distribution or microstrain effects. The resulting observed and calculated diffraction patterns are shown in Fig. 20.5.

20.4.4 Locating Hydrogen Atoms

The results of the last refinement can be considered conclusive if the location of a single independent hydrogen atom in the unit is not a concern.[32] As expected, it was impossible to unambiguously locate hydrogen using X-ray data. Therefore, neutron powder diffraction data, collected on a powder diffractometer at the McMaster

[32] Due to the low X-ray scattering ability of hydrogen atoms, their effect on the intensity of powder diffraction patterns measured using conventional X-ray sources is usually negligible, especially in the presence of relatively strongly scattering atoms, such as Mn and Ni. As a result, localizing hydrogen atoms from X-ray powder diffraction data often presents a serious and sometimes inexplicable challenge. However, hydrogen atom positions are important in crystallography because they reveal the nature of hydrogen bonds, which are critical for understanding the stability of both inorganic and organic crystals.

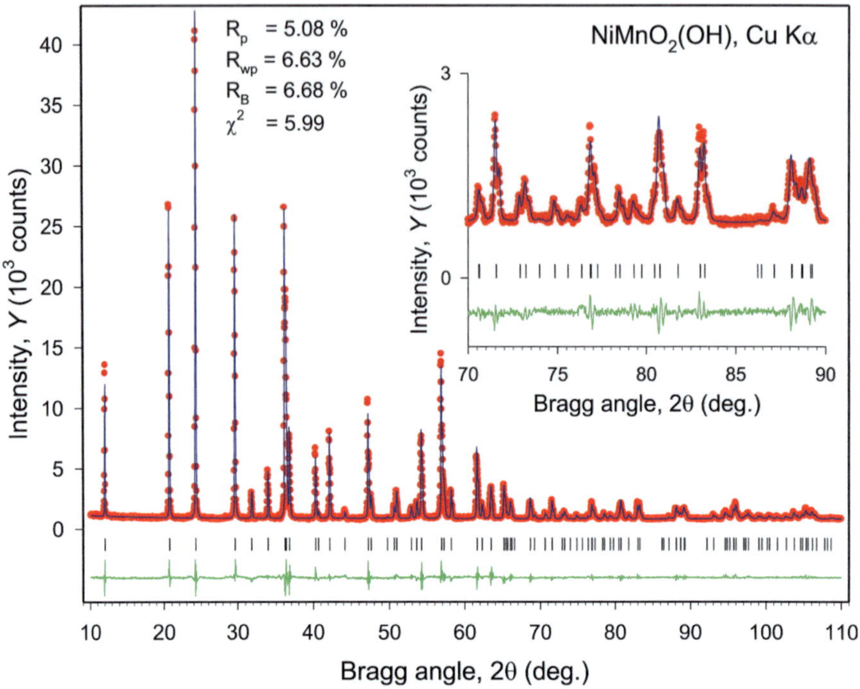

Fig. 20.5 The observed and calculated powder diffraction patterns of NiMnO$_2$(OH) after the completion of Rietveld refinement using only X-ray powder diffraction data (the hydrogen atom is still missing from the model). The inset highlights the range between 70 and 90° 2θ

University's nuclear reactor using thermal neutrons with $\lambda = 1.3920$ Å, were employed.

It is important to note that a normal sample containing hydrogen, rather than its deuterium-substituted analogue, was employed in this experiment. The presence of ^1H instead of ^2H causes a substantial diffuse scattering and significantly increases the background. However, this ensures that both the neutron and X-ray diffraction data were collected from exactly the same compound.

The availability of neutron diffraction data allows for the combined X-ray and neutron Rietveld refinement.[33] The following neutron scattering lengths (b) were used: $b_{Mn} = -3.73$, $b_{Ni} = 10.3$, $b_O = 5.803$, and $b_H = -3.739$ (all are in fm).[34] The negative values of the scattering lengths of Mn and H make it easier to distinguish them from other elements easily. After several cycles of refinement, a

[33] In order to carry out the combined refinement in GSAS, the neutron powder diffraction pattern (online file *Ch20Ex01_Neut.raw*) should be added as the second histogram using the neutron instrumental parameter file (data file *Neutron.prm* is also located online).

[34] GSAS employs neutron scattering lengths divided by 10, i.e., the units are 10^{-12} cm. Also note that the incoherent scattering length of ^1H is 25 fm, which is quite large.

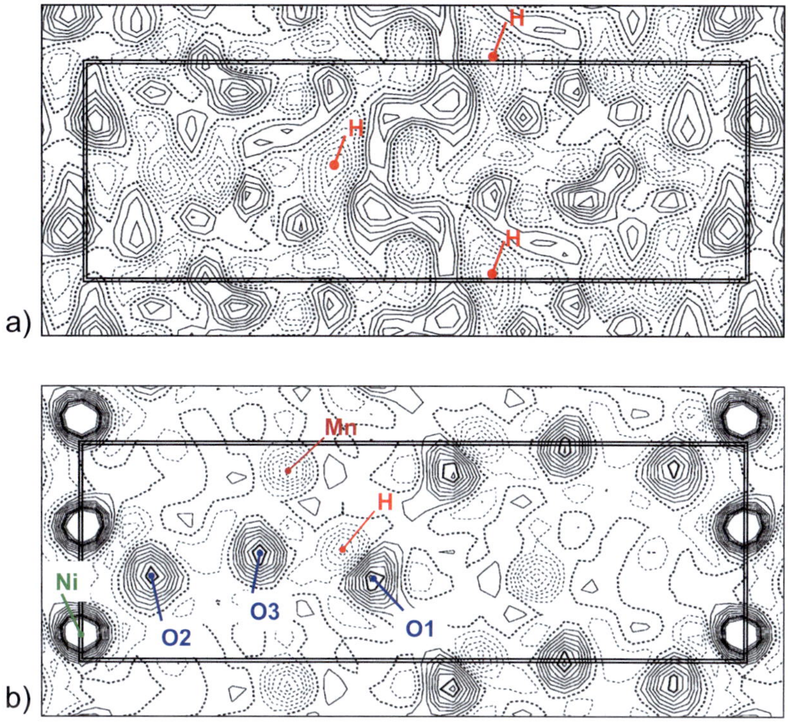

Fig. 20.6 Distributions of the nuclear density in the unit cell of $NiMnO_2(OH)$ at $x = 0$: (**a**) the difference Fourier map calculated using $|\Delta F| = |F_{obs}\text{-}F_{calc}|$ and phase angles calculated without the H atom ($\rho_{min} = -2.4$ fm/Å^3, $\rho_{max} = 2.4$ fm/Å^3, $\Delta\rho = 0.4$ fm/Å^3, $\rho(H) = -2.4$ fm/Å^3); (**b**) the conventional Fourier map computed using $|F_{obs}|$ and phase angles calculated with the H atom included ($\rho_{min} = -12$ fm/Å^3, $\rho_{max} = 18$ fm/Å^3 (high ρ on Ni atoms not shown), $\Delta\rho = 2$ fm/Å^3, $\rho(H) = -10$ fm/Å^3, $\rho(Mn) = -12$ fm/Å^3). *Solid lines* show positive values, *thin dotted lines* negative, and *thick dotted lines* indicate the zero level

strong agreement between the observed and calculated patterns was achieved. This result definitively proves that the Mn and Ni positions were identified correctly: their scattering factors have opposite signs and, when switched, the computed intensities are significantly different.

The difference Fourier map was calculated using the neutron data and is shown in Fig. 20.6a. The position of the H atom appears as the deepest minimum on the map. However, this minimum is only slightly deeper than other extremes, many of which could be considered noise due to the relatively low accuracy of the experimental data. Confirmation of the H position was made using chemical intuition and the crystal structure model, particularly because the geometry of the hydrogen bond $O1 - H\cdots O3$ is nearly ideal. The positions of all atoms in the model were confirmed on the conventional Fourier map, as shown in Fig. 20.6b.

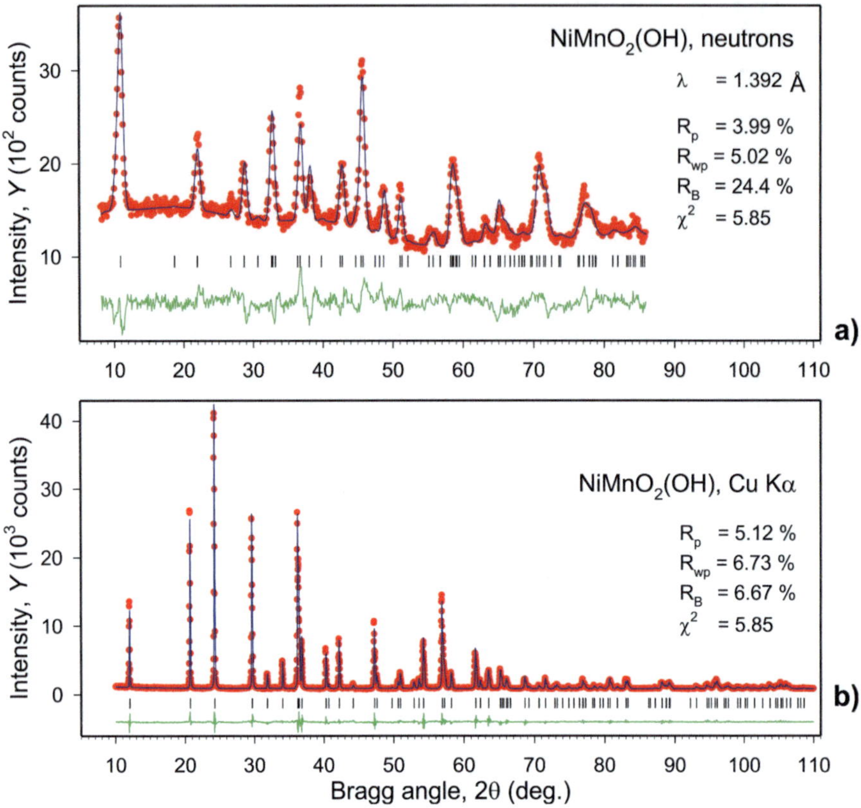

Fig. 20.7 The observed and calculated powder diffraction patterns of $NiMnO_2(OH)$ after the completion of combined Rietveld refinement using neutron (a) and X-ray (b) powder diffraction data

20.4.5 Combined Rietveld Refinement

The final Rietveld refinement, combining both neutron and X-ray powder diffraction data, was performed until complete convergence was achieved. The resulting observed and calculated powder diffraction patterns are illustrated in Fig. 20.7, and the refined parameters of the individual atoms are listed in Table 20.5.[35] The agreement between the observed and calculated X-ray data (Fig. 20.7b) is only slightly inferior to that obtained when only X-ray data were used in the refinement (Fig. 20.5). This is likely due to the contribution of the high background in the neutron pattern (Fig. 20.7a), caused by the incoherent scattering of hydrogen. The relatively large Bragg residual calculated using neutron data is also linked to the

[35]They can also be found online in the files *Ch20Ex01c.exp* and *Ch20Ex01c.cif.*

Table 20.5 Atomic parameters and interatomic distances (in Å) after the completion of the combined Rietveld refinement based on both the X-ray and neutron powder diffraction data collected from $NiMnO_2(OH)$ powder. The refined chemical composition is $NiMnO_{3\text{-}\delta}\,(OH)_\delta$, where $\delta = 0.62(5)$. The unit cell parameters are: $a = 2.86112(4)$, $b = 14.6516(1)$, $c = 5.27097(5)$ Å, $V = 220.959(7)$ Å^3, the space group is $Cmc2_1$

						Bond distances, Å		
Atom	Site	x	y	z	U_{iso}[a]	Mn[b]	Ni[b]	H[b]
Mn	4 (a)	0	0.30556 (6)	0.3800[c]	0.0164 (5)			
Ni	4 (a)	½	0.4989 (3)	0.618 (2)	0.0159 (4)			
O1	4 (a)	0	0.4480 (2)	0.876 (4)	0.0124 (8)		$2.07(1)_2$, $2.11(1)_2$	1.10 (4)
O2	4 (a)	½	0.3909 (2)	0.382 (3)	0.0124 (8)	$1.900(2)_2$	2.01(1), 2.13 (1)	
O3	4 (a)	½	0.2308 (3)	0.490 (2)	0.0124 (8)	$1.892(3)_2$, 2.126 (7)		1.73 (4)
H[c]	4 (a)	0	0.387(3)	0.998 (8)	0.0124 (8)			

[a] Displacement parameters of oxygen and hydrogen atoms were refined under the overall isotropic approximation
[b] The subscript after the distance indicates how many times this bond occurs for this particular central atom
[c] The refined population parameter of the hydrogen atom is $g = 0.62(5)$. The hydrogen bond characteristic angle O1-H···O3 is 143(4)°

fairly low quality of the neutron diffraction data, where the strongest peak-to-background ratio is less than 2.5 (see relevant discussion in Sect. 15.7.1).

The population parameter of hydrogen was refined to a value of 62(5) %, indicating the chemical composition of the material as $NiMnO_3H_\delta$, or $NiMnO_{3\text{-}\delta}(OH)_\delta$, where $\delta = 0.62(5)$. Therefore, a fraction of the Mn atoms must be in the 4+ oxidation state, a result confirmed by the magnetic susceptibility measurements. The preferred orientation parameters, refined for both axes [010] and [100], yielded values of 0.74 and 1.40, respectively, resulting in texture factors ranging between 0.52 and 2.10, corresponding to the preferred orientation magnitude of about 4.

It appears that all pieces of this crystallographic puzzle are now in place, and they agree with each other as well as with all available information. These include the chemical composition and the oxidation states of the metal atoms; the overall crystal structure, including the distribution of Mn and Ni atoms, and the quantity and positions of hydrogen atoms; the geometry, which covers bond lengths, coordination polyhedra, and hydrogen bonding; and basic magnetic properties, which confirm a mixture of Mn^{3+} and Mn^{4+} in the material. Thus, the fully determined and refined crystal structure of $NiMnO_{3\text{-}\delta}(OH)_\delta$ is chemically and physically consistent, and its model is illustrated in Fig. 20.8.

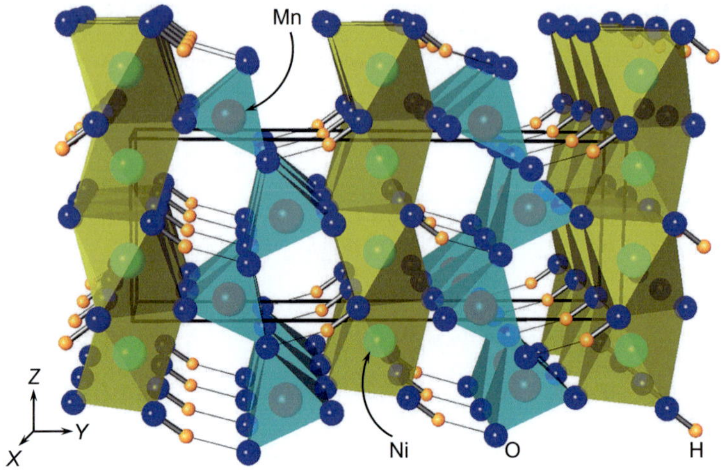

Fig. 20.8 The model of the crystal structure of $NiMnO_{3-\delta}(OH)_{\delta}$. The covalent O1-H bonds are shown as *cylinders*, and the H···O3 hydrogen bonds are depicted using *thin lines*

Chapter 21
Crystal Structure of *tma*V_3O_7: Direct Methods in the Presence of Strong PO

This example illustrates an unconventional case of solving a crystal structure that shows relatively broad Bragg peaks that are substantially overlapped at high angles, in addition to a significant preferred orientation. The latter occurs due to the distinct platelet-like shape of the crystals. The powder diffraction pattern (Fig. 21.1) was collected from a tetramethylammonium (*tma*) trivanadate powder—a black, graphite-like crystalline material that was thoroughly ground and screened. The specimen was prepared by filling a 1 mm deep cavity of a sample holder without applying any pressure (see Sect. 12.1) to minimize preferred orientation effects.

Despite all precautions in the sample preparation, the pattern (Fig. 21.1) contains two distinct Bragg peaks that are substantially stronger than the others. The first peak, at ~9.4° ($d = 9.308$ Å), is shown at one-fourth of its height, while the second, at ~19.1° ($d = 4.640$ Å), has intensity approximately four times lower than the first but is about three times stronger than any other Bragg reflection. The intensities of the remaining Bragg reflections are below 10% of the strongest. The *d*-spacing ratio of the two strongest peaks is 2.006, clearly indicating that they belong to the same zone—for example, 001 and 002 in the 00 *l* zone (or, more generally, they are related as *hkl* and 2*h*2*k*2*l*, respectively). Combined with the markedly planar shape of the crystals (inset in Fig. 21.1), these features strongly suggest the presence of significant preferred orientation, which may complicate solving the structure and refining structural and profile parameters. Conversely, the fact that the two strongest reflections belong to the same zone can be used to correct the observed peak positions for

P.Y. Zavalij, T. Chirayil, and M.S. Whittingham, Layered tetramethylammonium vanadium oxide $[N(CH_3)_4]V_3O_7$ by X-ray Rietveld refinement, Acta Cryst. **C53**, 879 (1997), https://doi.org/10.1107/S0108270197000474; *tma* – tetramethylammonium $[N(CH_3)_4]^+$. The material, in a form of a black crystalline powder, was prepared by hydrothermal treatment at 185°C of a mixture of V_2O_5, *tma*OH, and LiOH in a 1:2:1 molar ratio and acidified with CH_3COOH to pH = 6.5.

P. Zavalij, V. Pecharsky, *Fundamentals of Powder Diffraction and Structural Characterization of Materials*, https://doi.org/10.1007/978-3-031-91504-8_21

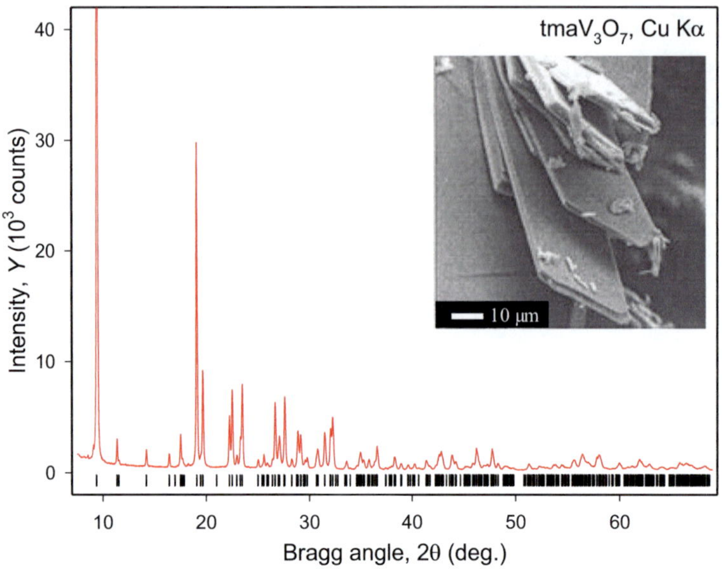

Fig. 21.1 Powder diffraction pattern collected from the $tma\mathrm{V_3O_7}$ powder using Cu Kα radiation on a Scintag XDS2000 diffractometer. The experiment was carried out in a step-scan mode with a step 0.02° and counting time 60 s per step. The *vertical bars* indicate calculated positions of the $K\alpha_1$ components of all possible Bragg reflections. The inset shows the SEM image of particle morphology in the as-received state

the sample-displacement or zero-shift errors during the ab initio indexing, as was done in the original work.[1]

The first 41 peaks below 39° 2θ were indexed using the ITO program in a monoclinic unit cell ($M_{20} = 37$), which provided the best solution and was the only one with all peaks indexed. Unit cell refinement resulted in $a = 18.453$ Å, $b = 6.560$ Å, $c = 8.437$ Å, $\beta = 91.12°$, and $V = 1021.1$ Å^3. Analysis of the systematic absences reveals the following conditions: $h + l = 2n$ for $h0l$ reflections and $l = 2n$ for 00 l reflections, which unambiguously points to $P2_1/n$ (or $P2_1/c$ in standard setting) as the only possible space-group symmetry that satisfies the reflection conditions. This makes indexing result highly probable, especially considering that $P2_1/c$ is one of the most common space groups observed in both natural and synthetic materials (see Sect. 3.4.4).

[1]This example illustrates an interesting approach that may be helpful in the indexing from first principles. Assume that two patterns were collected from the same powder: one from a specimen with minimal or no preferred orientation and the other with artificially induced strong texture. If the texture axis coincides with one of the principal crystallographic directions (e.g., [001], [010], or [001]), comparing two patterns may provide critical information about the indices of certain peaks whose intensity is most affected (increased or reduced). Once their indices are determined by analyzing the ratios between the corresponding *d*-spacings, the problem of finding the remaining lattice parameters is simplified by eliminating one unknown.

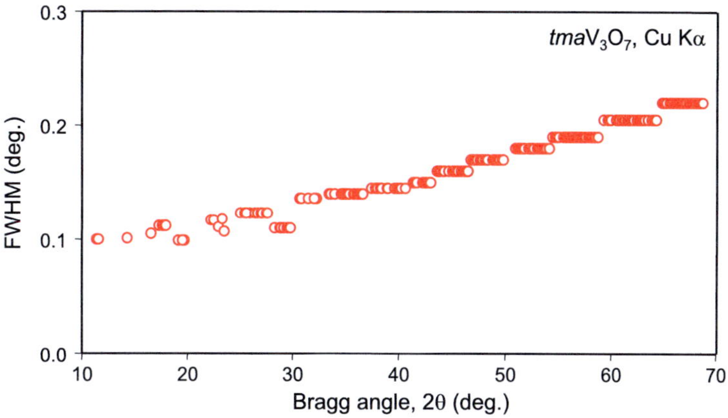

Fig. 21.2 Full width at half maximum as a function of Bragg angle observed (below $2\theta = 35°$) and linearly interpolated (above $2\theta = 35°$) in the powder diffraction pattern of *tma*V_3O_7 shown in Fig. 21.1

21.1 Observed Structure Factors

Due to the complexity of the pattern, multiple overlaps (e.g., about 90 Bragg reflections are possible in the range of the first 40 observed peaks below 40° 2θ), and the relatively broad peaks, the pattern decomposition was carried out separately in about 20 consecutive ranges using semimanual profile fitting. For each group of Bragg reflections, located within the manually selected ranges of the powder diffraction pattern as described in Footnote 4 on page 682, the least-squares profile fitting was performed by refining the individual 2θ position and integrated intensity of each peak. Common to each range were parabolic background, full width at half maximum, and the mixing and asymmetry parameters of the pseudo-Voigt function. The average R_p was around 2.5%. When 2θ exceeds 35°, peak widths increase to over 0.15°. Combined with the increasing density of peaks, this makes it quite difficult to refine both profile and positional parameters independently. Therefore, the values of full width at half maximum were linearly extrapolated from previous ranges (Fig. 21.2), and only the individual intensities and positions of Bragg peaks were included as free least-squares parameters.

This type of profile fitting is, in fact, a modification of the Pawley decomposition technique, performed without restricting peak positions by unit cell dimensions and to small ranges rather than the full pattern. The observed intensities of 236 individual peaks up to $2\theta = 69°$ were determined in this way with a total of 425 Bragg reflections possible in this range of Bragg angles. Unobserved reflections were assigned small intensity values (equal to about 1% of the intensity of the nearest observed peaks) to improve the chances of solving the structure using direct methods.[2]

21.2 Solving the Crystal Structure

The chemical composition of the powder was established from the following data: thermogravimetric analysis in an oxygen atmosphere shows a ~4% weight gain around 200 °C, followed by a sharp weight loss of 19.2% at ~320 °C. The orange product obtained after the TGA was identified as vanadium pentoxide (V_2O_5) from its powder diffraction pattern, using the Powder Diffraction File. The decomposition temperature, indicated by the weight loss, corresponds to the loss of tetramethylammonium. Thus, the chemical composition of the residue, along with the temperature and weight changes, was used to deduct a possible formula of the material, the simplest of which is *tma*V_3O_7, which gives a calculated weight loss of 19.5 wt. %. This is in good agreement with the experimentally observed 19.2%. The oxidation states of vanadium in $[V_3O_7]^-$ are 4+ for two vanadium atoms and 5+ for the third. A weight gain of 4% in an O_2 atmosphere can be explained by the oxidation of all vanadium to V^{5+} and by the formation of *tma*V_3O_8. The theoretical weight gain of 4.7% is slightly higher than that observed experimentally (4%), most likely because the oxidation reaction was not completed before the decomposition began.

The density was not measured, but assuming Z = 4, which is reasonable for the $P2_1/n$ space-group symmetry where the multiplicity of the general site is 4, the calculated density is 2.20 g/cm^3. Other Z values either do not match the site multiplicities (e.g., Z = 3 or 5) or they result in the unrealistically high densities (e.g., Z = 8). Thus, the final chemical composition, *tma*V_3O_7, appears to be the only feasible choice and was employed in the structure determination using direct phase recovery methods.

As mentioned earlier, pattern decomposition resulted in 236 distinguishable observed peaks. Therefore, a substantial number of unresolved Bragg reflections remain, partially due to significant peak broadening.[3] Additionally, there are 12 light atoms (C, N, and O, not counting 12 hydrogen atoms) per 3 vanadium atoms in the formula unit. The ratio of light-to-heavy atoms is 4:1. A fraction of electrons in the heavy atoms is 41%, which may be insufficient to recover the entire structure from the phase angles calculated using only the V atom positions, especially considering the relatively poor resolution of the diffraction data. Therefore, it is highly unlikely

[2]Experimental data are available online in the following files: GSAS and tabular (XY) formats, files *Ch21Ex01_CuKa.raw* and *Ch21Ex01_CuKa.xy*, respectively; the indexed list of the individual structure factors $|F_{obs}|$ is in the file *Ch21Ex01.hkl*, and relevant crystal data are in the file *Ch21Ex01.ins*. The latter two files are in SHELX format.

[3]Structure amplitudes of unresolved Bragg reflections were determined by dividing the total intensity among all overlapped reflections equally.

Table 21.1 Maxima localized from the E-map refined for the best solution obtained by SHELXS using 425 reflections data below 70° 2θ; $E_{min} = 1.1$

Atom/peak	*x*	*y*	*z*	U_{iso}/height	Bond distances, Å V1[b]	V2[b]	V3[b]	Comments[a]
V1	0.794	0.176	0.019	429.9				
V2	0.722	0.683	0.069	342.8				
V3	0.785	−0.112	−0.252	281.2				
Q1	0.797	0.189	−0.211	178.5	1.94		2.01	O1a
Q2	0.736	0.482	−0.031	151.7	2.31	1.59	1.96	O2a
Q3	0.892	0.192	0.068	137.1	1.85			O3a
Q4	0.664	0.019	−0.193	133.3			2.46	Deleted
Q5	1.040	0.270	−0.308	132.3				N5a
Q6	0.862	−0.295	−0.284	124.2			1.89	O6a
Q7	0.815	−0.129	0.037	121.1	2.05	2.15		O7a
Q8	0.760	0.263	0.242	116.8	2.07	1.71		O8a
Q9	0.685	0.118	0.040	114.7	2.06			O9a

[a] Here and in the subsequent tables, the names assigned to atoms consist of the chemical symbol of the element, E-map or Fourier map peak number, and a letter (if any), which indicates the sequential number of the calculated Fourier map: a—corresponds to the first iteration, b—second, and so on
[b] This column lists the distances to the indicated atom in Å

that the Patterson method is suitable in this case, even though it is less sensitive to data quality. Thus, this structure solution is far from trivial. Considering all factors, direct phase determination methods should be employed to identify a more substantial portion of the model than just the three independent vanadium atoms.

The structure was solved using SHELXS, based on the extracted structure amplitudes of 425 possible reflections below 2θ = 70°. The direct phase determination with $E_{min} = 1.1$ resulted in the E-map containing an acceptable model with three "heavy" peaks (Table 21.1) that were automatically assigned to vanadium. The distances from the vanadium atoms to all but one of nine strongest peaks correspond well to V–O bonds. The only exception is Q4, which is too far from all V atoms and too close to Q6 (1.32 Å). The Q4–Q6 distance is not listed in Table 21.1.

Since the Q6 distance to V3 is nearly ideal (1.89 Å), the Q6, but not Q4, was included in the trial model. One of the peaks, Q5, is not in contact with vanadium or oxygen atoms and was treated as nitrogen. Peaks following Q9 have conflicting distances with the first 12 strongest maxima. The analysis of bond angles and coordination polyhedra confirms the validity of this model. Its composition, V_3O_7 plus one isolated light atom (presumably nitrogen from the tetramethylammonium molecule), aligns well with the TGA data, symmetry restrictions, and estimated density.

The least-squares refinement of this model using SHELXL yields R_F value of 42%, as shown in Table 21.2. All oxygen atoms, as well as the isolated nitrogen atom, appear reasonable since their distances to the vanadium atoms fall within

Table 21.2 The atomic coordinates after structure refinement using SHELXL, along with peaks from the difference Fourier map that may indicate possible locations of carbon atoms. All data below 70° 2θ were used, yielding R_F value of 42%

Atom/				U_{iso}/	Bond distances, Å				
peak	*x*	*y*	*z*	height	V1[a]	V2[a]	V3[a]	N5[a]	Comments
V1	0.791	0.178	1.015	0.016					
V2	0.722	0.669	1.088	0.046					
V3	0.782	−0.106	0.759	0.039					
O1a	0.793	0.164	0.797	0.003	1.84		1.81,2.10		
O2a	0.755	0.450	0.993	0.000	1.91	1.75	2.24		
O3a	0.884	0.224	1.052	0.005	1.76				
O6a	0.846	−0.192	0.757	0.008			1.31		
O7a	0.781	−0.159	0.968	0.089	2.25	1.87			
O8a	0.759	0.154	1.230	0.012	1.93	1.58			
O9a	0.698	0.156	1.000	0.108	1.72				
N5a	0.014	0.259	0.793	0.005					
Q3	0.060	0.301	0.903	2.6				1.27	C3b
Q8	0.064	0.247	0.630	2.4				1.66	C8b
Q9	−0.036	0.440	0.761	2.1				1.53	C9b
Q11	−0.001	0.072	0.736	2.1				1.34	C11b

[a] This column lists the distances to the specified atom in Å

acceptable limits. What is not shown in Table 21.2 is that the coordination polyhedra of the vanadium atoms are chemically and physically sensible: vanadium oxide forms a layer, consisting of VO_5 pyramids and tetrahedral VO_4 units, as illustrated in Fig. 21.6. This layer is similar to those observed in other vanadates. Thus, the model presented in Table 21.2 can be considered a good approximation for the subsequent Rietveld refinement, which is discussed later in the text. After calculating the difference Fourier map, analysis of the observed maxima suggests that four peaks—though not all the strongest—can be interpreted as carbon atoms. These peaks exhibit reasonable bond distances and form a distorted tetrahedral arrangement around the previously identified nitrogen atom, as expected for the tetramethylammonium $[N(CH_3)_4]^+$ cation.

The result of structure refinement after including carbon atoms is shown in Table 21.3. Neither the residuals nor the distances improve, and one carbon atom, C8b, moves too far from the nitrogen. The inconsistency in C8b's location can be identified even without the refinement through careful analysis of the nitrogen environment and examining the C–N–C bond angles. This example highlights, once again, the importance of considering chemical, physical, and geometrical criteria in solving and refining crystal structures, especially when quality, resolution, or completeness of the data is limited by the technique or the specimen.

The high value of R_F and poor convergence of the model—for instance, with the C–N and O6a–V3 bond distances worsening instead of improving—are attributed to

Table 21.3 Results of structure refinement using SHELXL, with an R_F value of 42%. The coordinates of all atoms were shifted by the vector (−½, 0, −½) compared to those in Table 21.2

Atom/ peak	x	y	z	U_{iso}/ height	Bond distances, Å V1[a]	V2[a]	V3[a]	N5[a]	Comments
V1	0.292	0.182	0.513	0.011					
V2	0.224	0.665	0.585	0.042					
V3	0.277	−0.108	0.257	0.050					
O1a	0.291	0.193	0.298	0.000	1.81		1.86, 2.02		
O2a	0.257	0.462	0.485	0.000	1.96	1.70	2.17		
O3a	0.389	0.230	0.546	0.018	1.82				
O6a	0.339	−0.181	0.249	0.051			1.24		
O7a	0.272	−0.137	0.473	0.068	2.15	1.84	1.84		
O8a	0.262	0.152	0.727	0.000	1.91	1.60			
O9a	0.191	0.189	0.503	0.030	1.87				
N5a	−0.493	0.241	0.310	0.000					
C3b	−0.449	0.308	0.373	0.000				1.05	
C8b	−0.425	0.223	0.142	0.070				1.90	Incorrect
C9b	−0.507	0.353	0.206	0.010				1.17	
C11b	−0.508	0.078	0.226	0.000				1.31	

[a] This column lists the distances to the specified atom in Å

both the low resolution and suspected strong preferred orientation, which was discussed earlier in this chapter but not yet been confirmed. In the original paper (see Note on page 703), an arbitrary scale factor of two-third was applied to the intensities of all $h00$ reflections to reduce preferred orientation effects. No corrections have been applied so far in this example, yet the crystal structure model remains reasonable. At this stage of the structure determination process, geometrical, chemical, and physical considerations play a far more important role than the residuals.

21.3 Completion of the Model and Rietveld Refinement

This example demonstrates the completion of the model and Rietveld refinement of a complex structure containing inorganic vanadium oxide layers (with ten independent V and O atoms) intercalated with tetramethylammonium (*tma*) ions. The latter consists of 17 independent atoms: four carbon, one nitrogen, and 12 hydrogen. The refinement is based on conventional powder diffraction data, which are of high quality but still affected by strong and unavoidable preferred orientation (see plate-like shapes in the inset of Fig. 21.1). Additionally, we provide basic information on restraints[4] that can be applied to known bond lengths and valence angles to improve both the stability of the least squares and the reasonableness of the model.

The following initial parameters were used for model completion and refinement:

- The initial crystal structure model was taken from Table 21.3.
- Default profile parameters from the instrumental parameter file[5] (see Sect. 20.3).
- The sample shift parameter $S_s = 8.94$ corresponds to a sample displacement of $s = -0.195$ mm, which was determined earlier during the least squares.
- The space group is $P2_1/n$ with the unit cell dimensions: $a = 18.482$, $b = 6.5526$, $c = 8.4297$ Å, $\beta = 91.103°$, as determined earlier.
- The overall isotropic displacement parameter is $U_{iso} = 0.015$ Å^2.

21.3.1 Unrestrained Rietveld Refinement

Initially,[6] only the scale factor and six coefficients for a shifted-Chebyshev polynomial approximation of the background were refined. The resulting residuals, shown in the second row of Table 21.4, are higher than expected for a nearly complete model, where all nonhydrogen atoms were thought to be located. As shown in the inset of Fig. 21.3, both the calculated intensities and peak shapes deviate significantly from the observed values. The intensity mismatch may be partially attributed to significant preferred orientation, which is expected due to the highly anisotropic shapes of the crystals (see the inset in Fig. 21.1) and the likely easy cleavage of particles along planes parallel to the vanadate layers, expanded by the *tma* molecules.

Therefore, the subsequent refinement included the grain-size broadening parameter (X) and the preferred orientation along the [100] axis, which resulted in some improvement of the fit. Rietveld minimization continued by incorporating porosity effects, initially refining them with most other parameters fixed, and later releasing them. This led to further improvement, reducing the R_{wp} to approximately 15%. However, refining the grain-size and strain broadening effects, their anisotropy, and the coordinates of atoms in the organic molecule proved challenging due to

[4] A *constraint* is an exact mathematical relationship existing between two or more parameters that eliminates one or more variables from the least-squares refinement. For example, $y = x$, and $B_{33} = B_{22} = B_{11}$, eliminate y, and B_{33} and B_{22}, respectively (also see Eq. 16.2). A *restraint*, however, is additional information subject to a probability distribution. For example, chemically (but not symmetrically) identical bond lengths in tetramethylammonium: $\delta_{N\text{-}C1} = \delta_{N\text{-}C2} = \delta_{N\text{-}C3} = \delta_{N\text{-}C4} = 1.55 \pm 0.05$ Å (see Sect. 15.7.3 for details).

[5] The file *Scintag.prm* is available online.

[6] The starting model can be found online in the files *Ch21Ex01a.exp* and *Ch21Ex01a.cif*, while the experimental powder pattern is available in the file *Ch21Ex01_CuKa.raw*.

Table 21.4 The progress of Rietveld refinement for the crystal structure of *tma*V_3O_7

Refined parameters	R_p	R_{wp}	R_B	χ^2
Initial[1]	60.7	68.6	99.4	535
Scale, shifted-Chebyshev polynomial background (Fig. 21.3)	35.7	44.7	45.7	228
Plus grain broadening (X) and preferred orientation, PO, with [100] texture	33.0	41.7	41.6	198
Plus unit cell dimensions and asymmetry, α	22.9	30.2	35.0	104
Added O8; coordinates and U_{indiv} included, Table 21.5	15.2	19.7	20.3	45.1
Deleted O7 and C2	15.5	20.2	19.1	47.2
Added C2 and replaced C3 from a difference Fourier map (Fig. 21.4)[a]	13.3	17.7	15.6	36.5
Strain broadening (Y), asymmetry (α), and sample shift	13.2	17.1	13.9	33.4
Scale, background, and unit cell dimensions: first with porosity (absorption) a_1 and a_2 and then with PO [100], X, sample shift, and overall U_{iso}	11.8	15.2	13.4	26.4
Excluded 8 to 12° 2θ; PO [010], then Y, α, X_a, Y_a	10.0	13.4	16.8	16.1
Plus coordinates of V and O	8.6	11.6	15.4	12.1
Plus coordinates of N and C	6.1	8.2	13.2	6.12
Plus PO ratio, individual U_{iso}(V), overall U_{iso}(O) and U_{iso} (N, C) (Fig. 21.5)[b]	4.7	6.3	11.9	3.59
Plus restraints on C–N distances and C–N–C angles (Table 21.6)[c]	4.9	6.6	12.6	4.01

[a] Files online: *Ch21Ex01b.exp* and *Ch21Ex01b.cif*
[b] Files online: *Ch21Ex01c.exp* and *Ch21Ex01c.cif*
[c] Files online: *Ch21Ex01d.exp* and *Ch21Ex01d.cif*

significant correlations: the resulting shifts of the free least-squares parameters drove the solution away from a global minimum.

One of the reasons for the instability in the least-squares refinement is the presence of an extremely intense low Bragg angle peak at 2θ ≈ 9.5°, which is heavily influenced by various systematic errors and is significantly larger than all other reflections. The high intensity of this peak translates in its disproportionate contribution to the least-squares calculation (see Eq. 15.28), thus strongly affecting the refinement results. From this point onward, the low Bragg angle range below 12° 2θ was excluded from the least-squares minimization. This decision noticeably stabilized the least-squares refinement, though it has little impact on the figures of merit—a result that was somewhat unexpected.[7]

Next, a second preferred orientation axis [010] was included into the minimization. This decision was based on the crystal structure model, which shows the orientation of chains in the vanadate layers, and on the observed crystallite shapes, where the plates are elongated in one direction. The refinement was completed by

[7] When a single and tremendously intense peak is present in the data, all residuals may become quite low when the fit of the strongest peak is well-fitted. Even though the remaining peaks may not be fitted well, their intensities have little effect on all figures of merit because the denominators in Eqs. (15.18)–(15.21) become defined by a few extremely large numbers.

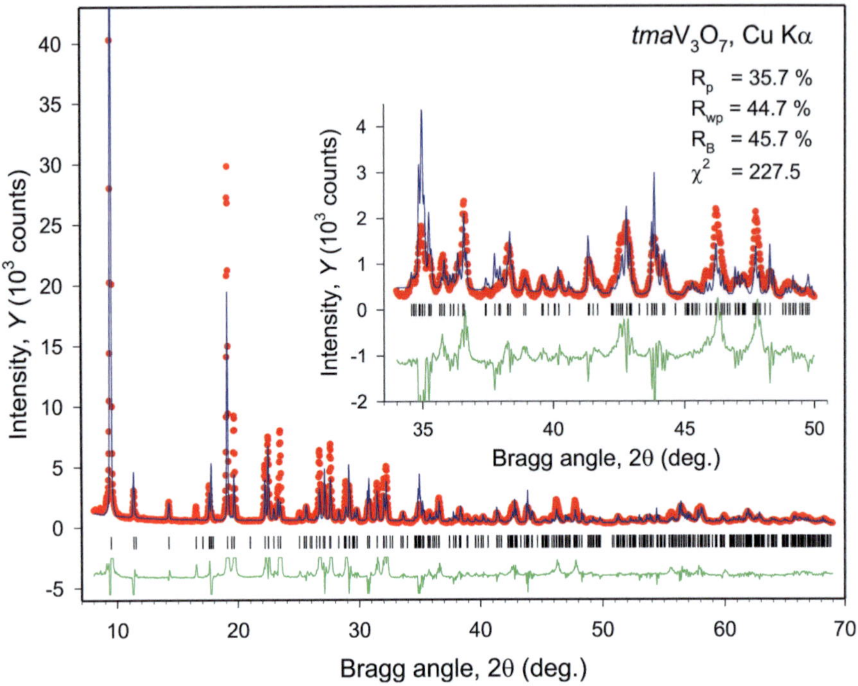

Fig. 21.3 The observed and calculated powder diffraction patterns of *tma*V_3O_7 after the initial Rietveld least-squares refinement, where only the scale factor and shifted-Chebyshev polynomial background were refined. The difference ($Y_i^{obs} - Y_i^{calc}$) is displayed on the same scale as both the observed and calculated data, but the plot is truncated to fit within the range [−1500, 1500] for clarity. The ordinate is reduced to approximately one-third of the maximum intensity to better illustrate low-intensity Bragg peaks. The inset highlights the range between 34 and 50° 2θ

gradually releasing the coordinates of atoms in the inorganic framework (V and O atoms) and then in the organic molecule (N and C atoms), the individual U_{iso} for the metal atoms (V1 – V3), and overall U_{iso} for the oxygen atoms and the organic molecule (two different overall parameters). These were refined together with all profile parameters except for *U*, *V*, *W* and *P*. The resulting fit is quite good, as shown in Fig. 21.5.

21.3.2 Rietveld Refinement with Restraints

Despite the reasonable fit between the observed and calculated intensities, an analysis of the interatomic distances and bond angles reveals that the geometry of the tetramethylammonium ion, though acceptable considering data quality,

Table 21.5 The Rietveld refinement results after adding O8 and refining the atomic coordinates and individual isotropic displacement parameters of all atoms in the tmaV_3O_7 crystal structure model. The R_{wp} value is 19.7%. Values highlighted in bold indicate issues with the model

						Bond distances, Å			
Atom	Site	*x*	*y*	*z*	U_{iso} [a]	V1 [b]	V2 [b]	V3 [b]	N1 [b]
V1	4(e)	0.2954	0.1840	0.5117	0.039				
V2	4(e)	0.2278	0.6740	0.5781	0.017				
V3	4(e)	0.2837	−0.1077	0.2538	0.041				
O1	4(e)	0.2792	0.171	0.279	−0.038	1.98		1.84,1.88	
O2	4(e)	0.2358	0.480	0.493	0.007	2.23	1.47	2.18	
O3	4(e)	0.3987	0.229	0.556	0.043	1.96			
O4	4(e)	0.3664	−0.203	0.254	0.078			1.65	
O5	4(e)	0.2686	−0.095	0.483	0.010	1.91	1.88	1.96	
O6	4(e)	0.2585	0.153	0.733	0.049	2.01	1.61		
O7	4(e)	0.2540	0.203	0.524	**0.600**	**0.78**			
O8	4(e)	0.3542	0.216	0.931	0.040		1.54		
N1	4(e)	0.5150	0.243	0.280	−0.011				
C1	4(e)	0.5554	0.305	0.407	−0.060				1.36
C2	4(e)	0.5643	0.286	0.128	−0.068				**1.61**
C3	4(e)	0.4928	0.415	0.217	0.015				1.31
C4	4(e)	0.5034	0.064	0.234	−0.057				1.25

[a] The individual isotropic displacement parameters vary considerably due to the model being incompletely refined. A value that is an order of magnitude larger than the others suggests an incorrectly positioned atom

[b] This column lists the distances to the specified atom in Å

quantity,[8] and complexity, is slightly outside the expected range. The geometry was improved by applying restraints in GSAS.[9] The C–N bond lengths were restrained to 1.500 ± 0.015 Å, and the C–N–C angles were restrained to 109.5 ± 1.5°. The weights for bonds and angles were manually set to 4 and 6, respectively, after testing their influence on the least-squares refinement. While applying these restraints slightly increased all residuals, as seen by comparing the last two rows in Table 21.4, it substantially improved the geometry of the organic molecule

[8] [1]Due to the large number of lightly scattering atoms, the diffracted intensity of this powder specimen becomes extremely low above 2θ = 70° ($\sin\theta/\lambda < 0.37\text{A}^{-1}$). In a typical powder diffraction experiment, it is necessary to collect the data to $\sin\theta/\lambda \cong 0.5\text{A}^{-1}$, preferably to even higher values.

[9] Restraints set limits on the bond distances and/or angles, which are known and are desired to reach or keep during the least-squares minimization (see Sect. 15.7.3). The influence of restraints, compared to a straightforward profile fitting, is controlled by adjusting the weight, which can be increased to bring the geometry closer to the desired values. However, if the restraints are incorrect, the fit may significantly worsen.

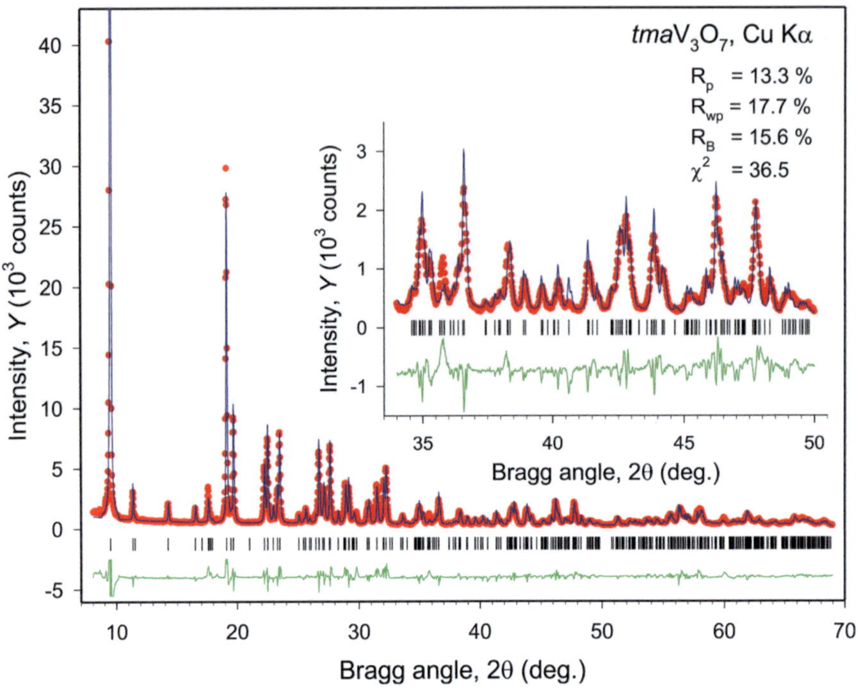

Fig. 21.4 The observed and calculated powder diffraction patterns of tmaV$_3$O$_7$ after all nonhydrogen atoms have been placed. The difference ($Y_i^{obs} - Y_i^{calc}$) is displayed on the same scale as the observed and calculated data, but the plot is truncated to fit within the range [−1500, 1500] for clarity. The ordinate is reduced to approximately one-third of the maximum intensity to better illustrate low-intensity Bragg peaks. The inset highlights the range between 34° and 50° 2θ

(Table 21.6). Before the restrained refinement, the C–N bond lengths were in the 1.48–1.67 Å range with the average bond length of 1.56 Å. The C–N–C angles varied from 100°−127°, showing a considerable deviation from the ideal tetrahedral angle. After the refinement with restraints, the C–N bond lengths fall within the 1.50–1.55 Å range, and the bond angles are between 106° and 114°, both of which are acceptable. It is worth noting that, instead of restraints, a rigid-body refinement of the *tma* molecule—whose geometry is well-known—could also be applied. The rigid-body approach is another method of imposing geometrical restrictions and is available in GSAS.

A notable feature of this refinement is the strong preferred orientation along two axes, [100] and [010], with parameters $\tau_{100} = 0.76$ and $\tau_{010} = 1.38$ in a ratio of 2:1. The preferred orientation multipliers range from 0.58 to 2.08, resulting in a total magnitude of approximately 4, similar to the case described in Sect. 21.3.1.

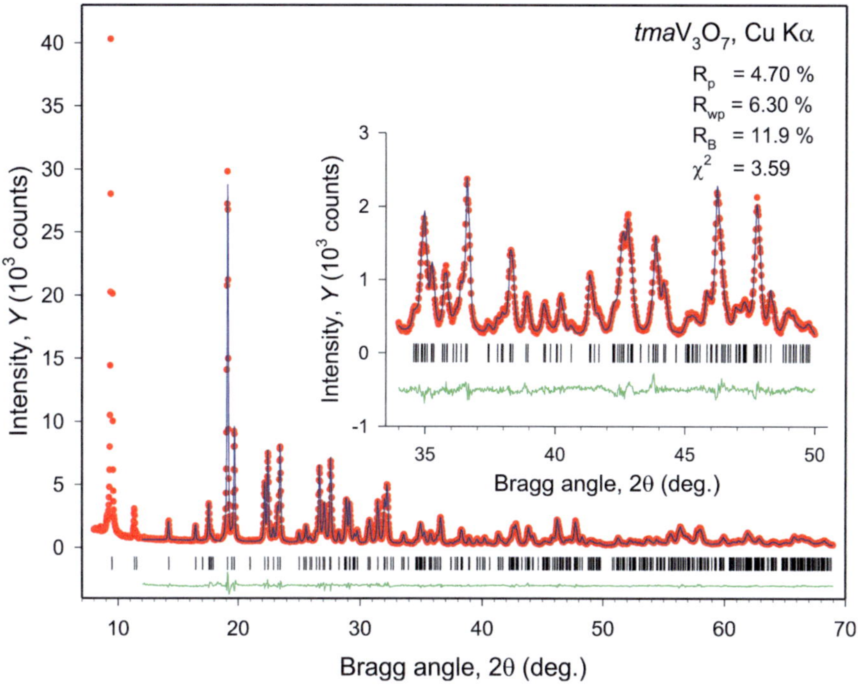

Fig. 21.5 The final Rietveld plot of tmaV$_3$O$_7$ data. The low Bragg angle range ($2\theta < 12°$) was excluded from the refinement because it contains the strongest peak, which is an order of magnitude more intense than the others and is heavily affected by experimental errors, potentially distorting the refinement

The final Rietveld refinement produces a reasonable structural model (Fig. 21.6), which fits well as a new member in the V_3O_7-based structure series, where vanadium oxide layers differ only in the orientations of the square pyramids and tetrahedra.[10] Later, this structure was solved using single-crystal data,[11] which further validates the structure determination from the powder data discussed here. It is worth mentioning that the structure from obtained single-crystal data was not of higher quality, and perhaps even somewhat lower, due to

[10] P.Y. Zavalij, F. Zhang, and M.S. Whittingham, Crystal structure of layered *bis*(ethylenediamine)-nickel hexavanadate as a new representative of the V_6O_{14} series. Acta Cryst. **B55**, 953 (1999). https://doi.org/10.1107/S0108768199005467

[11] P.Y. Zavalij, S.T. Lutta, and M.S. Whittingham, CCDC 2388052: Experimental Crystal Structure Determination, 2024. https://doi.org/10.5517/ccdc.csd.cc2l4yyn

Table 21.6 The atomic parameters and interatomic distances (in Å) in the $tma$$V_3O_7$ model, fully refined using the Rietveld technique in the GSAS environment. The unit cell parameters are a = 18.4878(5) Å, b = 6.5552(2) Å, c = 8.4318(2) Å, β = 91.131(2)°, V = 1021.66(7) Å^3, and the space group is $P2_1/n$[a]

					Bond distances, Å			
Atom	x	y	z	U_{iso}	V1	V2	V3	N1
V1	0.2956(2)	0.1816(8)	0.5149(5)	0.027(2)				
V2	0.2228(2)	0.6793(8)	0.5781(4)	0.018(2)				
V3	0.2864(2)	−0.1121(6)	0.2487(5)	0.028(2)				
O1	0.2843(6)	0.178(2)	0.288(2)	0.014(1)	1.93(1)		1.93(1), 1.92(1)	
O2	0.2465(5)	0.436(2)	0.476(2)	0.014(1)	1.92(2)	1.87(1)	2.00(1)	
O3	0.3814(5)	0.213(2)	0.560(1)	0.014(1)	1.636(8)			
O4	0.3677(4)	−0.203(2)	0.234(1)	0.014(1)			1.624(9)	
O5	0.2702(5)	−0.106(2)	0.475(2)	0.014(1)	1.97(1)	1.88(1)	1.94(1)	
O6	0.2555(5)	0.164(2)	0.731(2)	0.014(1)	1.99(1)	1.65(1)		
O7	0.3627(6)	0.202(2)	0.928(1)	0.014(1)		1.59(1)		
N1	0.5287(4)	0.267(1)	0.254(1)	0.034(4)				
C1	0.5664(6)	0.287(2)	0.412(1)	0.034(4)				1.498(7)
C2	0.5849(6)	0.280(2)	0.125(2)	0.034(4)				1.519(7)
C3	0.4676(5)	0.427(2)	0.236(2)	0.034(4)				1.547(7)
C4	0.4927(6)	0.063(2)	0.245(2)	0.034(4)				1.495(7)

[a] All crystallographic data can be found online in the files *Ch21Ex01d.exp* and *Ch21Ex01d.cif*

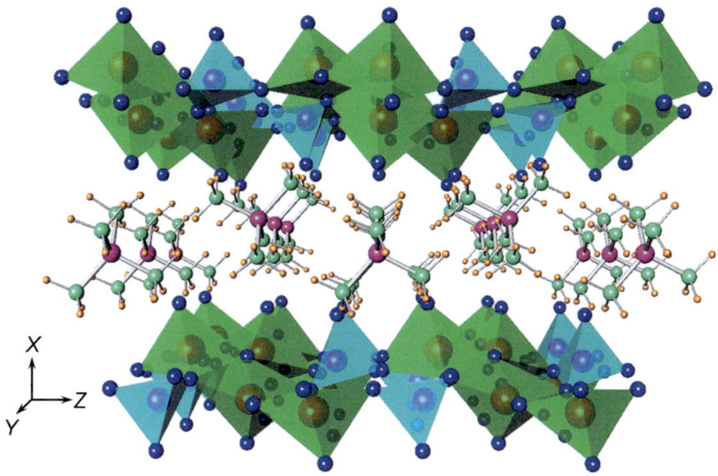

Fig. 21.6 The final model of the *tma*V_3O_7 crystal structure highlighting the layered vanadium oxide framework and the intercalated *tma* ions. The V1 and V3 atoms (*red spheres*) are located within the *green* [VO_5] square pyramids, the V2 atoms (*purple*) are positioned inside the *blue* [VO_4] tetrahedra, the O atoms are shown in *blue*, N atoms in *magenta*, C atoms in *light green*, and H atoms as *small orange spheres. The* H atoms positions were computed assuming the sp^3 hybridization of carbon and trans-configurations of methyl groups

the crystals being very thin (<5 microns) plates exhibiting pseudo-merohedral twinning. As a result, all atomic displacement parameters had to be restrained, not only to ensure reasonable anisotropy but also to correspond to the rigid-body motions of molecules and polyhedra. Final R_F value was 7.88% which is, as expected, somewhat higher than typical.

Chapter 22
Crystal Structure of $ma_2Mo_7O_{22}$: Solved Using Structural Analogy

A white crystalline powder, prepared by hydrothermal treatment at 200 °C of a mixture of molybdic acid (H_2MoO_4) and methylammonium (*ma*) chloride (CH_3NH_3Cl) in a 1:2 molar ratio, acidified with hydrochloric acid (HCl) to pH = 3.5, resulted in a complex powder diffraction pattern shown in Fig. 22.1. The pattern was indexed in the monoclinic crystal system as was discussed in Sect. 14.11.2. The space group C2/c (or its acentric subgroup Cc) was established from the analysis of the systematic absences, and the unit cell dimensions were refined using 120 resolved reflections with 2θ below 60°: $a = 23.0648(6)$ Å, $b = 5.5134(2)$ Å, $c = 19.5609(5)$ Å, $\beta = 122.931(1)°$, with the sample displacement $\delta = -0.098$ (3) mm for a 250 mm goniometer radius. The unit cell volume is 2087.8 Å^3.

The Powder Diffraction File search was unsuccessful, prompting further analysis and a structure solution. Thermogravimetric analysis (TGA) in an oxygen atmosphere revealed sharp 7.3 wt.% weight loss at 300 °C. The powder diffraction pattern collected from the solid residue after the TGA confirmed the formation of molybdenum oxide (MoO_3). Assuming the following decomposition reaction:

$$(CH_3NH_3)_mMo_nO_{3n+k} \rightarrow nMoO_3 + mCO_2 + m/2N_2 + 3mH_2O \qquad (22.1)$$

It can be shown that the observed weight loss nearly precisely corresponds to m:n ratio 2:7 and k = m/2 = 1. This ratio is further supported by the white color of the material, which indicates that Mo is in the 6+ oxidation state. Therefore, the chemical composition of the substance is $(CH_3NH_3)_2Mo_7O_{22}$.

P.Y. Zavalij and M.S. Whittingham, The layered methylammonium molybdate $(CH_3NH_3)_2Mo_7O_{22}$ from X-ray powder data, Acta Cryst. C53, 1374 (1997). https://doi.org/10.1107/S010827019700601X

P. Zavalij, V. Pecharsky, *Fundamentals of Powder Diffraction and Structural Characterization of Materials*, https://doi.org/10.1007/978-3-031-91504-8_22

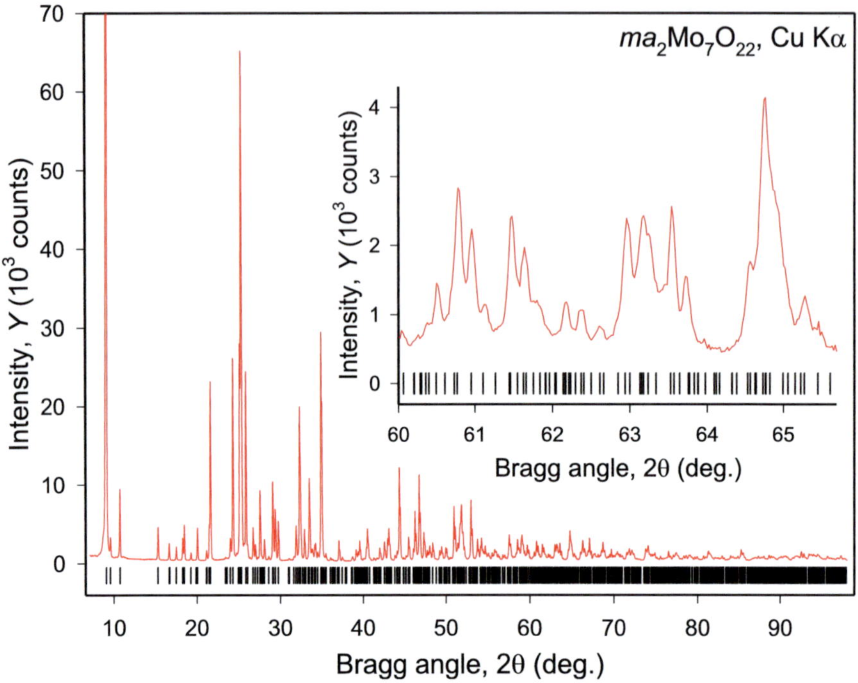

Fig. 22.1 Powder diffraction pattern collected from a ground $ma_2Mo_7O_{22}$ powder using Cu Kα radiation in step-scan mode with a step size of 0.02°. The counting time was 30 s/step in the range $7° \leq 2\theta \leq 67°$ and 60 s/step for $67 \leq 2\theta \leq 98°$. The counting time was increased at higher Bragg angles to improve the counting statistics for a large number of weak Bragg peaks possible in this range. To ensure consistency of intensity measurements, the high Bragg angle portion of the diffraction pattern was scaled to 30 s/step counting time for further processing. The *vertical bars* indicate calculated positions of the $K\alpha_1$ components of all possible Bragg reflections. The inset shows an expanded view from 60° to 65.7°, which contains 76 possible reflections

22.1 Possible Model of the Crystal Structure

The relatively large volume of the monoclinic unit cell results in considerable complexity of the diffraction pattern, even a base-centered lattice, as shown in Fig. 22.1. There are approximately 10 reflections per degree at $2\theta \cong 50°$, and about 20 reflections per degree at $2\theta \approx 100°$.

The total number of possible Bragg reflections below $2\theta = 98°$ is 1032. Therefore, before attempting an ab initio structure solution, a more thorough search of the relevant databases was conducted. A search based on the unit cell dimensions and volume yielded no results. However, after searching the ICSD database for a matching stoichiometry of the molybdate anion,[1] two compounds, both with the space-group symmetry C2/c were found. These compounds, along with the title compound, are listed in Table 22.1.

Considering the unit cell dimensions listed in Table 22.1, the title compound likely has a structure closely related to both cesium and thallium heptamolybdates. It is possible that they are isostructural, with methylammonium substituting for metal cations in the crystal lattice. A thallium compound can also be found in the Powder Diffraction File, record No. 30-1349. It is clear why all powder pattern and unit cell dimensions searches failed: the unit cell volumes are quite different due to the variations in the *a*-axis length. The latter is not surprising considering that methylammonium ions are expected to occupy larger cavities, both due to their size and the weaker interactions of a hydrophobic methyl group with oxygen atoms. Given the expected similarity among the three crystal structures listed in Table 22.1, the atomic coordinates of the Mo_7O_{22} layer from the Tl compound were used as the initial model of the metal-oxygen framework of the methylammonium compound (Table 22.2 and Fig. 22.2).

If our hypothesis about the relationship between the crystal structures of $Tl_2Mo_7O_{22}$ and $ma_2Mo_7O_{22}$ is correct, refining the relevant atomic parameters and completing the model—specifically finding the coordinates of missing methylammonium groups—should be easier done after the Rietveld refinement, which is described in Sect. 21.2.

22.2 Rietveld Refinement and Completion of the Model

Structure determination should be completed by locating the methylammonium ($CH_3NH_3^+$) ion from a difference Fourier map and refining all relevant parameters. Once located in the unit cell, the N atom is clearly distinguished from carbon by much shorter distances it forms with oxygen, due to N–H···O hydrogen bonding.

The following parameters were used at the beginning of this refinement:

- The initial model of the structure (the molybdenum oxide layer) was taken from Table 22.2.
- The default profile parameters were taken from the instrumental parameter file[2] as described in Sect. 20.3.
- The sample shift parameter $S_s = 4.49$ for the sample displacement $s = -0.098$ mm was obtained along with the unit cell dimensions during lattice parameters refinement.

---

[1] First, 1174 compounds containing Mo and O were found. Second, the list was narrowed to 151 structures that have monoclinic base-centered or body-centered lattices. Third, "Mo_7O_{22}" text in the chemical formula was searched. Searching all ICSD records for specific text would take a considerable amount of time.

[2] The file *Scintag.prm* is available online.

Table 22.1 Unit cell dimensions for Mo_7O_{22}-containing compounds found in the ICSD database

Compound	*a* (Å)	*b* (Å)	*c* (Å)	β (°)	*V* (Å^3)	ICSD	Reference
$Cs_2Mo_7O_{22}$	21.54(1)	5.537(3)	18.91(1)	122.71(3)	1897.7	1887	[a]
$Tl_2Mo_7O_{22}$	20.512(6)	5.526(2)	19.460(6)	125.20(3)	1802.4	343	[b]
$ma_2Mo_7O_{22}$	23.0648(6)	5.5134(2)	19.5609(5)	122.931(1)	2087.8		

[a] B.M. Gatehouse and B.K. Miskin, The crystal structures of cesium pentamolybdate, $Cs_2Mo_5O_{16}$, and cesium heptamolybdate, $Cs_2Mo_7O_{22}$, Acta Cryst. **31**, 1293 (1975). https://doi.org/10.1107/S0567740875005080

[b] P. Tolédano, M. Touboul, and P. Herpin, Structure crystalline de l'heptamolybdate de thallium(I), $Tl_2Mo_7O_{22}$, Acta Cryst. **B32**, 1859 (1976). https://doi.org/10.1107/S0567740876006535

Table 22.2 Coordinates of Mo and O atoms in the unit cell of $ma_2Mo_7O_{22}$ based on the model of the crystal structure of $Tl_2Mo_7O_{22}$. The space-group symmetry is C2/c

Atom	Site	*x*	*y*	*z*
Mo1	4(e)	½	0.1375	¼
Mo2	8(f)	0.4081	0.0553	0.4751
Mo3	8(f)	0.3732	0.5000	0.0935
Mo4	8(f)	0.0533	0.1061	0.1378
O1	8(f)	0.3140	0.0880	0.3882
O2	8(f)	0.1496	0.1560	0.2199
O3	8(f)	0.0779	0.1750	0.4549
O4	8(f)	0.3995	0.1770	0.0266
O5	8(f)	0.3906	0.2330	0.1641
O6	8(f)	0.0393	0.3870	0.0677
O7	8(f)	0.0234	0.4480	0.1977
O8	8(f)	0.4960	0.4500	0.3196
O9	8(f)	0.4361	0.3800	0.4157
O10	8(f)	0.1395	0.2290	0.3577
O11	8(f)	0.2820	0.4630	0.0033

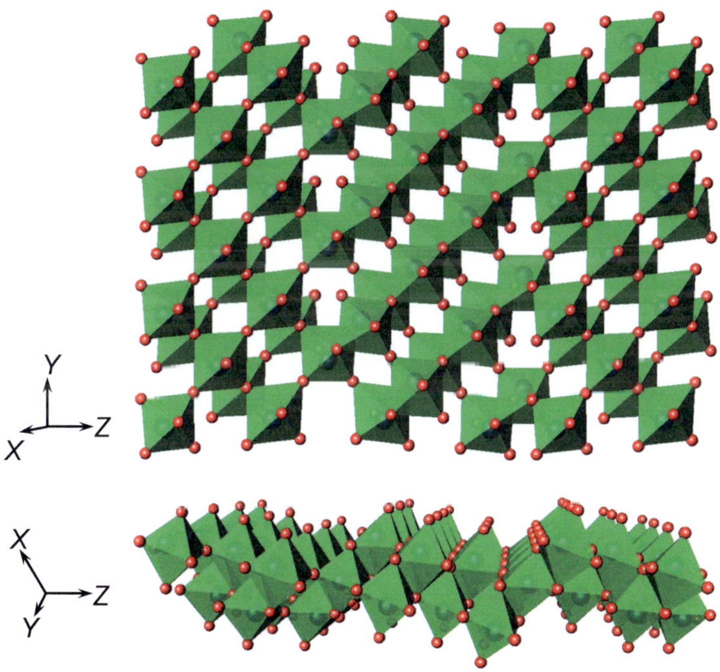

Fig. 22.2 Metal-oxide layer in the crystal structure of $maMo_7O_{22}$, shown in two different orientations. The Mo atoms are represented as *large dark spheres* inside the octahedra, and the O atoms are shown as *small red spheres*

Table 22.3 The progress of Rietveld refinement of the crystal structure of $ma_2Mo_7O_{22}$

Refined parameters	R_p	R_{wp}	R_B	χ^2
Initial model from $Tl_2Mo_7O_{22}$ (Mo and O only)[a]	93.3	93.5	98.8	1579
Scale and background (Fig. 22.3)	37.6	46.8	38.2	396
Plus unit cell, *X*, and preferred orientation (PO) along [100]	34.2	43.1	36.9	335
Plus coordinates, U_{iso}, and PO along [010]	19.4	25.6	15.2	120
Scale, background, PO, *X*, asymmetry (α), and porosity *a*1, *a*2	13.7	18.8	12.0	64.3
Scale, background, PO, *X*, α, coordinates, U_{iso} (Fig. 22.4)[b]	12.9	17.9	10.6	58.5
Same as above with the region 2θ < 15° excluded	10.5	14.5	11.8	31.8
Plus PO ratio and X_a	7.2	9.4	7.1	13.5
X, *Y*, α, sample shift, N1 and C1 located and added	5.4	7.4	5.9	8.3
All profile parameters *X*, *Y*, α, sample shift, X_a, Y_a, with fixed PO ratio (Fig. 22.5 and Table 22.4)[c]	5.3	7.2	5.9	7.8

[a] The starting model with all necessary parameters can be found online in the files *Ch22Ex01a.exp* and *Ch22Ex01a.cif*. Experimental data are available online in the data files *Ch22Ex01_CuKa.xy* and *Ch22Ex01_CuKa.raw*

[b] Files online: *Ch22Ex01b.exp* and *Ch22Ex01b.cif*

[c] Files online: *Ch22Ex01c.exp* and *Ch22Ex01c.cif*

- The space group is C2/c, and the unit cell dimensions are $a = 23.065$ Å, $b = 5.5134$ Å, $c = 19.561$ Å, $\beta = 122.93°$ as determined earlier.
- The overall isotropic atomic displacement parameter was set to $U_{iso} = 0.015$ Å^2.

Initially,[3] only the phase scale factor and six coefficients of the shifted-Chebyshev polynomial used to approximate the background were refined, resulting in relatively high residuals, as listed in Table 22.3. The poor fit is mainly due to a significant mismatch between the observed and calculated intensities as shown in Fig. 22.3.

It is highly likely that the intensity mismatches are caused by the relatively crude initial structural model as the atomic coordinates were taken directly from the Tl-based structure without adjusting for the lattice distortion along the **a**-axis (compare the unit cell dimensions of both materials in Table 22.1). Additionally, considerable preferred orientation could be expected due to the distinctly layered structure (see Fig. 22.2). Therefore, the subsequent refinement included unit cell dimensions, grain-size contribution to peak broadening (*X*), and preferred orientation (PO) along the [100] direction, which is perpendicular to the Mo_7O_{22} layers. Some improvement of the fit has been observed as a result. Unlike in the two examples considered in Chaps. 20 and 21, the preferred orientation is not as pronounced here, possibly due to less severe particle cleaving during grinding.

A noticeable improvement, especially in the Bragg residual, was observed when the coordinates of all atoms were refined, along with the isotropic displacement

[3] The starting model with all necessary parameters can be found online in the files *Ch22Ex01a.exp* and *Ch22Ex01a.cif*. Experimental data are available online in the data files *Ch22Ex01_CuKa.xy* and *Ch22Ex01_CuKa.raw*.

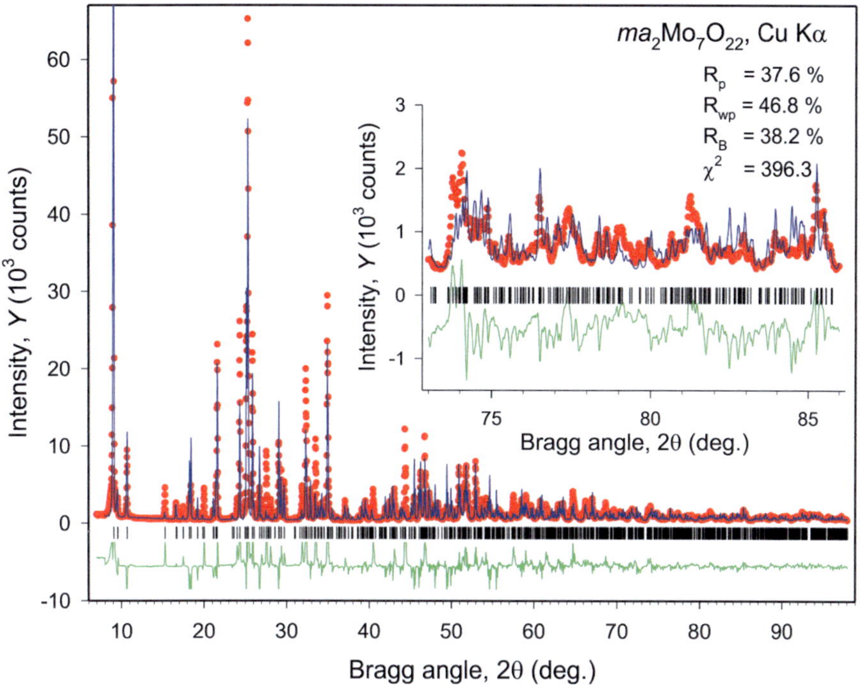

Fig. 22.3 The observed and calculated powder diffraction patterns of $ma_2Mo_7O_{22}$ after the initial Rietveld least-squares minimization with only the scale factor and the background refined. The difference ($Y_i^{obs} - Y_i^{calc}$) is shown using the same scale as both the observed and calculated data, but the plot is truncated to fit within the range [−3000, 3000] for clarity. The ordinate is reduced to ~¼ of the maximum intensity to better illustrate low-intensity Bragg peaks. The inset highlights the range between 73 and 86° 2θ

parameters: individual for the Mo atoms and a single parameter, common for all oxygen atoms. A second preferred orientation parameter along the [010] axis was also added, but its effect on the fit improvement was quite small. A slightly negative overall isotropic displacement parameter, $U_{iso} = -0.002(4)$ Å^2, for the O atoms likely indicates a contribution from specimen porosity, which, to a certain extent, also incorporates other unaccounted systematic errors, such as absorption and the beam size exceeding the sample dimensions at low Bragg angles due to an improper selection of the divergence slit aperture. Therefore, the porosity effect was optimized in a subsequent refinement using two parameters (a_1 and a_2) in the Suortti approximation. The refinement was carried out after isotropic atomic displacement parameters were set to 0.015 Å^2 for all molybdenum and 0.020 Å^2 for all oxygen atoms, with all atomic parameters kept fixed. The two porosity coefficients were refined to $a_1 = 0.31$ and $a_2 = 0.16$ starting from initial values of 0.40 and 0.40, respectively.

Afterwards, they were kept fixed through the reminder of the Rietveld refinement. Next, the individual atomic parameters were released and re-refined. This substantially improved the fit as shown in Fig. 22.4.

Regardless of the considerable improvement, some mismatches between the observed and computed intensities remain. At this point, we are still missing one nitrogen and one carbon from the methylammonium ion, not counting six hydrogen atoms, which are nearly impossible to locate from X-ray powder data (see Chap. 20). Pinpointing the locations of carbon and nitrogen in the unit cell and incorporating them into the model should indeed improve the fit. However, another potentially deleterious effect on the overall fit is the presence of a large intensity peak at low Bragg angle ($2\theta \cong 9°$), in front of which there is a much smaller and broader impurity peak. Hence, similar to the earlier example, the range below $2\theta = 15°$ was excluded from further refinement. It is important to note that this exclusion eliminates ~400

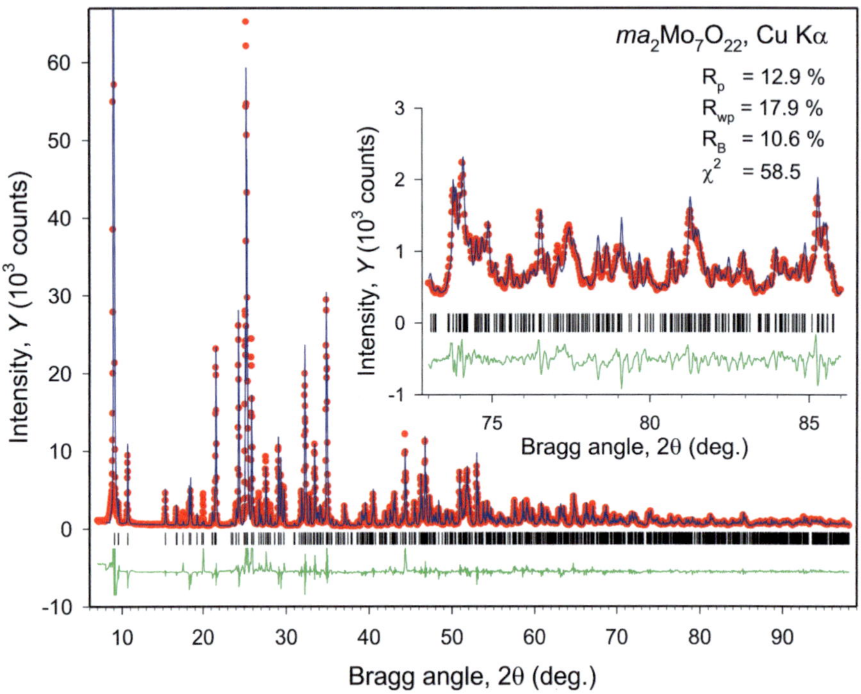

Fig. 22.4 The observed and calculated powder diffraction patterns of $ma_2Mo_7O_{22}$ after accounting for preferred orientation. The individual atomic parameters of the Mo and O atoms were refined along with several profile parameters and correction for porosity effect. The difference ($Y_i^{obs} - Y_i^{calc}$) is shown using the same scale as both the observed and calculated data; however, the plot is truncated to fit within the range [−3000, 3000] for clarity. The inset highlights the range between 73° and 86° 2θ

points (~9%) from the profile, which contains more than 4600 data points in total, but it removes only 3 (~0.3%) of about 1000 possible Bragg reflections. From the perspective of the structural model, such truncation of the experimental data is valid and is often employed in structure determination from powder diffraction data. With this modification, followed by several least-squares minimization steps, the profile residuals decrease, but R_B is slightly increased.

The subsequent Rietveld refinement of the preferred orientation ratio and anisotropic peak broadening further improved the fit, resulting in an R_{wp} value of 9.4%. At this stage, a difference Fourier map was generated, revealing two peaks that were notably stronger than the others. Their geometry closely matched the positions of the missing nitrogen (N1) and carbon (C1) atoms of the methylammonium ion. After incorporating these atoms into the model and completing the Rietveld refinement, the final fit (Fig. 22.5) was highly satisfactory.

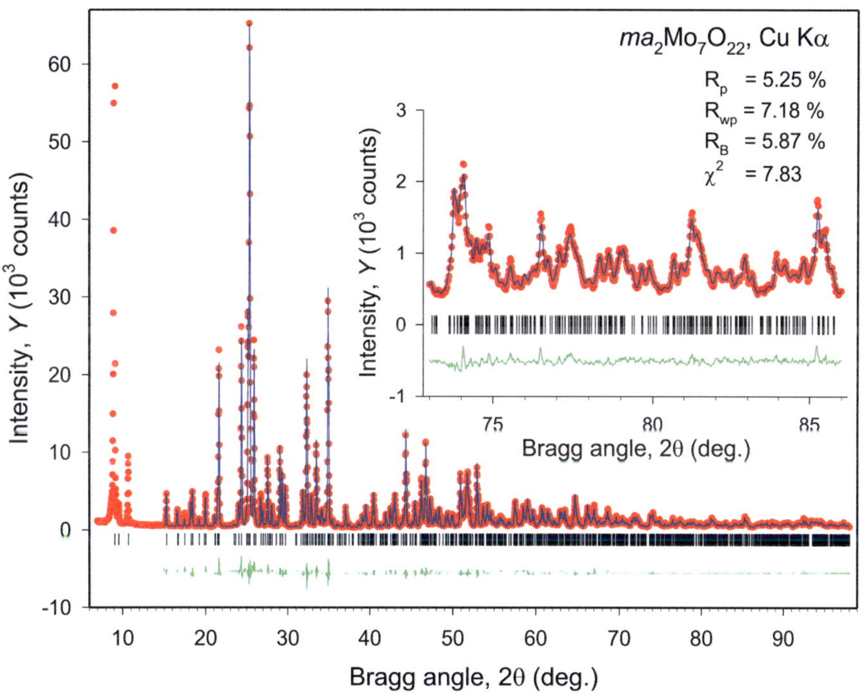

Fig. 22.5 The final Rietveld plot of $ma_2Mo_7O_{22}$ refinement. The range below 15° 2θ was excluded from the final refinement due to the presence of an extremely strong peak—nearly an order of magnitude stronger than all other peaks. This peak is highly susceptible to various experimental errors and could significantly distort the least-squares minimization

Attempts to adjust potentially refinable parameters did not result in a statistically significant improvement in the fit. In some cases, these adjustments caused the solution to deviate from a global minimum or pushed free variables outside the acceptable range. Additionally, certain combinations of simultaneously refined variables led to divergence in the least-squares minimization, despite an already good fit. For instance, incorporating preferred orientation using a spherical harmonics approximation resulted in only a negligible correction factor of 1.006, which did not improve the fit. When individual isotropic displacement parameters were refined for the light atoms (O, C, and N), the results were chaotic, ranging from unexpectedly high values to unphysical negative ones, without reducing the residuals. Similarly, refining the anisotropic displacement parameters for the heavy Mo atoms produced unphysical results, such as open ellipsoids or abnormally strong anisotropy, leading to highly elongated or flattened ellipsoids. Therefore, the fit shown in Fig. 22.5 can be considered final, and the resulting crystal structure model is provided in Table 22.4.[4]

The correctness of the crystal structure is nearly certain, as the Mo_7O_{22} layer is isotypical to both Tl- and Cs-based compounds (see Table 22.1). A nearly identical layer is also found in an intercalate containing a larger organic molecule, $RMo_7O_{22}{\cdot}H_2O$, where R = 4,4′-bipyridinium $(H\text{-}NC_5H_4\text{-}C_5H_4N\text{-}H)^{2+}$.[5] The orientation of the methylammonium ion (in other words, the recognition of the nitrogen and carbon atoms) is also convincing because the N atom is much closer to the O atoms than the C atom, which is due to the formation of strong N–H···O hydrogen bonds. The C–H···O interactions are much weaker, as follows from the comparison of bond lengths: 2.8–2.9 Å for N···O and 3.3 Å or more for the C···O distances. The C–N bond length, 1.46(3) Å, is within the expected range (compared to single-crystal data for maV_3O_7, where the C–N bond length is 1.487(8) Å). Stacking of the layers with the intercalated methylammonium ions forming N–H···O hydrogen bonds, which hold the layers together, is illustrated in Fig. 22.6.

The peculiarity of this refinement is the relatively large unit cell with 17 nonhydrogen atoms in the asymmetric unit, resulting in 1035 possible independent Bragg reflections and a total of 75 free least-squares variables, respectively. This example also illustrates that the location and the quality of the determination of light C and N atoms, which form a small organic molecule encapsulated between massive molybdenum oxide layers, are quite reliable, provided that sufficient quality powder diffraction data are available. The preferred orientation in this case is less

---

[4]The final model is available online in the files *Ch22Ex01c.exp* and *Ch22Ex01c.cif*.

[5]P.J. Zapf, R.C. Haushalter, and J. Zubieta. Crystal engineering of inorganic/organic composite solids: the structure-directing role of aromatic ammonium cations in the synthesis of the "step"--layered molybdenum oxide phase $[4,4'\text{-}H_2bpy][Mo_7O_{22}]{\cdot}H_2O$, Chem. Comm. 321 (1997). https://doi.org/10.1039/A607022A

Table 22.4 Final atomic parameters and interatomic distances (in Å) of $ma_2Mo_7O_{22}$ obtained from Rietveld refinement using GSAS. The refined unit cell parameters are: $a = 23.0707(3)$, $b = 5.51522(7)$, $c = 19.5669(2)$ Å, $\beta = 122.930(1)°$, V = 2089.68(5) Å^3, space group C2/c[a]

Atom	x	y	z	U_{iso}	Bond distances (Å)			
					Mo1	Mo2	Mo3	Mo4
Mo1	½	0.1231(7)	¼	0.015(1)				
Mo2	0.4211(1)	0.0676(5)	0.4783(2)	0.015(1)				
Mo3	0.3912(1)	0.4777(6)	0.0970(1)	0.019(1)				
Mo4	0.0480(1)	0.1020(5)	0.1380(2)	0.014(1)				
O1	0.3381(7)	0.094(3)	0.4034(9)	0.014(1)		1.67(1)		
O2	0.1309(8)	0.157(3)	0.216(1)	0.014(1)				1.71(2)
O3	0.0663(7)	0.146(3)	0.454(1)	0.014(1)		1.98(2)	1.97(2)	2.26(1)
O4	0.4146(9)	0.161(3)	0.037(1)	0.014(1)		1.76(2)	2.33(2)	
O5	0.4087(7)	0.219(3)	0.168(1)	0.014(1)	2*1.90(1)		1.88(2)	
O6	0.0358(7)	0.382(3)	0.066(1)	0.014(1)		1.98(2), 1.76(2)		2.00(2)
O7	0.0224(6)	0.439(3)	0.201(1)	0.014(1)	2*1.66(2)			2.48(2)
O8	0.4961(7)	0.432(3)	0.319(1)	0.014(1)	2*2.20(2)		2.21(1)	1.89(2)
O9	0.4472(8)	0.379(3)	0.422(1)	0.014(1)		2.29(2)		1.74(2)
O10	0.1191(7)	0.198(3)	0.360(1)	0.014(1)			1.56(2)	
O11	0.3124(6)	0.425(3)	0.010(1)	0.014(1)			1.71(1)	
N1[b]	0.2855(8)	0.991(4)	0.153(1)	0.014(4)				
C1[c]	0.265(1)	0.143(4)	0.196(2)	0.014(4)				

[a] All crystallographic data are available online in the files *Ch22Ex01c.exp* and *Ch22Ex01c.cif*
[b] Hydrogen bond distances for N1 are: 2.97(3) Å to O5, 2.88(3) Å to O2, 2.85(3) Å to O10, and 2.76(2) Å to O11
[c] C1 distances: 1.46(3) Å to N1, 3.31(4) Å to O2, 3.25(3) Å to O1, and 3.39(3) Å to O2

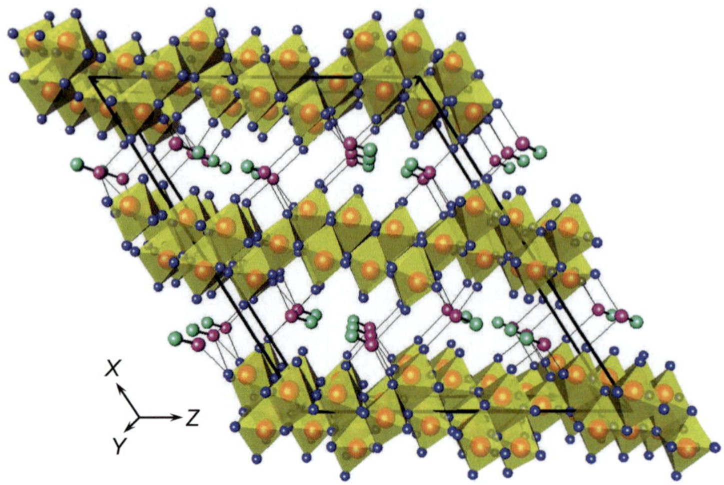

Fig. 22.6 The crystal structure of $ma_2Mo_7O_{22}$ shown along the Y-axis in a perspective view. The Mo atoms are depicted as *large orange spheres* inside translucent octahedra formed by oxygen atoms (*small blue spheres*). The N atoms are shown as *magenta spheres,* and the C atoms as *green spheres*. The N–[H]···O hydrogen bonds are shown using *thin lines*, without depicting the hydrogen atoms. Unlike in the crystal structure of the $tmaV_3O_7$ (see Chap. 21), the coordinates of hydrogen atoms could not be located from Fourier maps nor can be easily computed because the methylammonium ion can rotate around the C–N bond, although this rotation is restrained by the formation of N–H···O hydrogen bonds, which can be used to predict their location

severe compared to the two preceding examples: $\tau_{100} = 0.690(3)$ and $\tau_{010} = 1.091(4)$ with an approximate 1:2 ratio, which corresponds to the correction range between 0.7 and 1.8, or to the preferred orientation magnitude of about 2.6. The latter value is still significant, and it may be partially responsible for the unphysical anisotropic displacement parameters of the Mo atoms.

Chapter 23
Crystal Structure of $Mn_7(OH)_3(VO_4)_4$: Making Sense of Complex Disorder

In this example, we demonstrate the solution of a moderately complex inorganic structure, featuring ten independent atoms occupying various sites in a noncentrosymmetric hexagonal space group. Several sites are partially occupied, providing an opportunity to learn how to sensibly refine the chemical composition (a simpler example has been considered earlier in Chap. 16). During the least-squares minimization, partially vacant sites are identified, and their populations are first refined independently, then with reasonable restrictions on chemical composition, until complete convergence is achieved. This example also illustrates how to apply certain constraints available in GSAS.

A diffraction pattern (Fig. 23.1) collected using a powder prepared from brown rod-like hollow crystals produced hydrothermally[1] (Fig. 23.1, inset) was indexed employing TREOR and using 16 peaks below 40° 2θ in the hexagonal crystal system with $a = 13.255$ Å, $c = 5.265$ Å, and $V = 801.1$ Å^3. The F_N figure of merit, $F_{16} = 335$ (0.0021, 23), is exceptionally high. ITO indexing produced the same result but in a C-centered orthorhombic lattice with $a_{ortho} = a_{hex}$, $c_{ortho} = c_{hex}$, and $b_{ortho} = a_{hex}\sqrt{3}$. Unit cell refinement using 150 reflections observed below 130° 2θ resulted in highly accurate unit cell parameters (see Table 23.1) and a sample displacement of −0.112 (3) mm for a 250 mm goniometer radius.[2] The analysis of systematic absences suggests the following possible space groups $P6_3/mmc$, $P6_3mc$, $P\bar{6}2c$, $P31c$, or $P\bar{3}1c$.

This is idealized composition, the refined composition is $Mn_{7-2/3x}(OH)_3(VO_4)_{4-x}(V_2O_7)_{x/2}$, x = 0.20; see: F. Zhang, P.Y. Zavalij, and M.S. Whittingham, Synthesis and characterization of a pipe-structure manganese vanadium oxide by hydrothermal reaction, J. Mater. Chem., 9, 3137 (1999). https://doi.org/10.1039/A907465A.

$Mn_6^{2+}Mn_{1-y}^{3+}(OH)_3(VO_4)_{4-x}(V_2O_7)_{1/2x}$; $x = 0.181(4)$, $y = 2/3x$

---

[1] The material was prepared by hydrothermal treatment of V_2O_5, $Mn(CH_3COO)_2$, and $N(CH_3)_4Cl$ taken in 1:1:4 molar ratio at 165 °C for 3 days.

[2] The maximum absolute difference between the observed and calculated 2θ was 0.005°, which is an exceptionally low value.

P. Zavalij, V. Pecharsky, *Fundamentals of Powder Diffraction and Structural Characterization of Materials*, https://doi.org/10.1007/978-3-031-91504-8_23

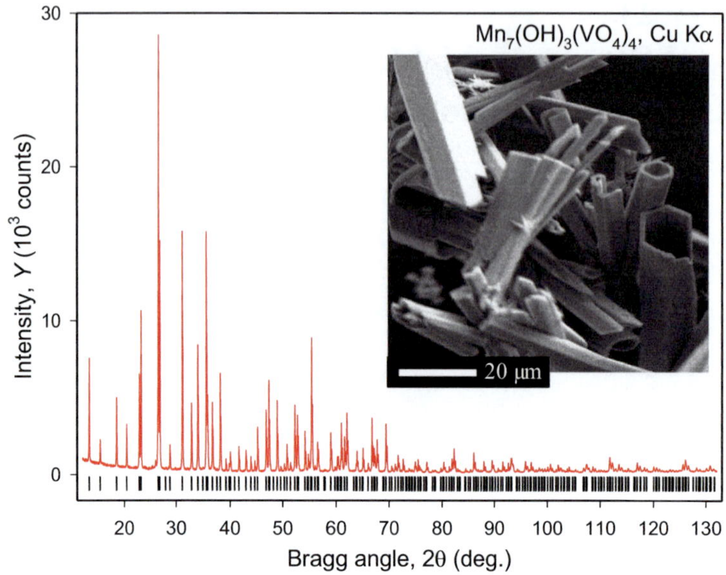

Fig. 23.1 Powder diffraction pattern collected from ground $Mn_7(OH)_3(VO_4)_4$ powder (screened through a 38 μm sieve) using Cu Kα radiation on a Scintag XDS2000 diffractometer. Data were collected in a step-scan mode with a step size 0.01° and counting time 15 s/step in the range of $7° \le 2\theta \le 70°$ and 30 s/step for $70° < 2\theta \le 132°$ to improve counting statistics of weak reflections at high Bragg angles. The high Bragg angle range was scaled to a constant counting time of 15 s/step for subsequent data use. The *vertical bars* indicate calculated positions of the $K\alpha_1$ components of all possible Bragg reflections. *The inset* shows a scanning electron microscopy image of peculiar empty hexagonal-pipe particle morphology in its as-received state (Powder diffraction data are available in the files *Ch23Ex01_CuKa.xy* and *Ch23Ex01_CuKa.raw* online)

Table 23.1 Comparing the unit cell dimensions of potentially closely related compounds identified through searches of the ICSD and PDF databases, with the title compound listed in the third row

Compound	*a* (Å)	*c* (Å)	*V* (Å³)	ICSD	PDF	Reference
$Zn_7(OH)_3(VO_4)_3SO_4$	12.8130(6)	5.1425(2)	731.15	402,888	88–1317	[a]
$Mn_{6-x}(OH)_3(HPO_3)_4$	13.1957(6)	5.1770(3)	780.68	75,269	47–0868	[b]
$Mn_7(OH)_3(VO_4)_4$ (Unit cell dimensions as refined from full profile fitting)	13.2294(1)	5.25529(7)	796.54	88,468	50–1796	

[a] K. Kato, Y. Kanke, Y. Oka, and T. Yao, Crystal structure of zinc hydroxide vanadate (V) $Zn_7(OH)_3(SO_4)(VO_4)_3$, Z. Kristallogr. **213**, 26 (1998). https://doi.org/10.1524/ncrs.1998.213.14.26

[b] M.P. Attfield, R.E. Morris, and A.K. Cheetham, Synthesis and structures of two isostructural phosphites, $Fe_{11}(HPO_3)_8(OH)_6$ and $Mn_{11}(HPO_3)_8(OH)_6$, Acta Cryst. **C50**, 981 (1994). https://doi.org/10.1107/S0108270194001368

23.1 Solving the Crystal Structure

Thermogravimetric analysis produced complex traces in both the oxygen and nitrogen atmospheres, with gradual weight losses approximately 1% and 4%, respectively. The powder diffraction pattern of the thermal decomposition product indicates as a mixture of $Mn_2V_2O_7$ and Mn_2O_3. The available data only allow a qualitative assumption regarding the absence of organic or water molecules, while also suggesting the presence of a small amount of hydroxyl groups due to of the continuous weight loss.[3] In general, it can be assumed that this compound contains Mn cations, OH^- groups, and individual or corner-sharing $[VO_4]^{3-}$ tetrahedra, similar to those found in V_2O_7, V_4O_{12}, or $(VO_3)_n$. The conclusion is based on the color, since all other oxidation states or coordination environments of V would result in black, dark-green, or dark-blue crystals. This reasoning demonstrates how various chemical and physical information may be used when considering composition, predicting, proposing, or solving the structure.

An identification attempt using the Powder Diffraction File was unsuccessful, as no acceptable matches were found. However, the high quality of the powder diffraction data should be sufficient to solve the structure from first principles using either Patterson or direct methods. Since a fully automated structure solution was not available, the ICSD database was searched in the following sequence:

- Searching all compounds containing oxygen and one or both of the metals, Mn and V, yielded 3413 entries.
- Filtering for hexagonal and primitive trigonal systems was considered, thus reducing the number of entries to 204.
- Restricting the search to unit cell volumes between 700 and 900 Å^3, within ~100 Å^3 of the target compound, narrowed the list to 16 compounds.
- Twelve of these belong to a different diffraction class, and two have *c*/*a* ratios, where *c* is much greater than *a*.
- Two remaining entries belong to the $P6_3mc$ space-group symmetry and have similar unit cell dimensions as shown in Table 23.1.

Note that $Mn_{6-x}(OH)_3(HPO_3)_4$ has a unit cell dimension and volume similar to those of the title material—much closer than the Zn-containing compound—and is also included in the PDF file. It should have been identified using a less restrictive search. In fact, a DIFRAC.EVA[4] full-pattern search-match with default settings and no composition restrictions matches the Mn compound but not the Zn one.

The Zn-containing structure can be easily modified to represent the crystal structure of Mn–OH–VO_4 (the composition derived earlier) by substituting Zn

---

[3]Compounds containing organic molecules or water of crystallization usually demonstrate rapid weight loss, while hydroxyl groups are lost slowly over a broad temperature range.

[4]Bruker DIFFRAC.EVA (2020). https://www.bruker.com/en/products-and-solutions/diffractometers-and-x-ray-microscopes/x-ray-diffractometers/diffrac-suite-software/diffrac-eva.html

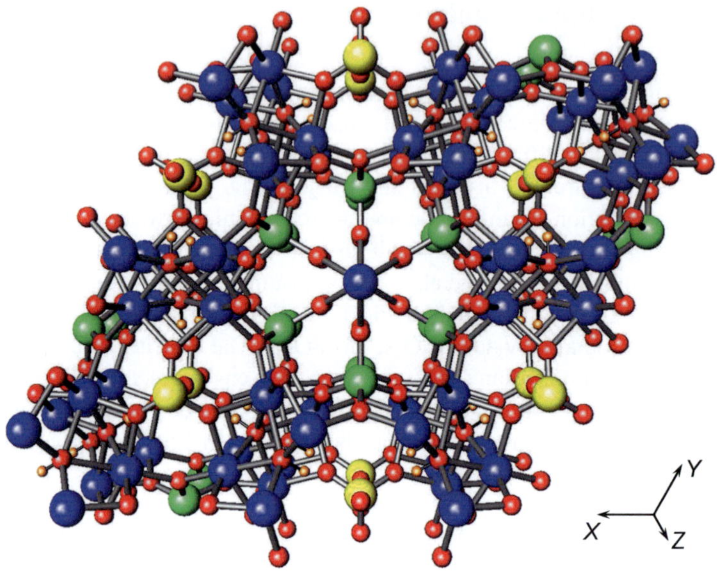

Fig. 23.2 The model of the crystal structure of $Mn_7(OH)_3(VO_4)_4$, derived from $Zn_7(OH)_3(VO_4)_3SO_4$, assuming that Mn atoms substitute for Zn (*large blue spheres*), and V atoms occupy positions of both V (*large green spheres*) and S (*large yellow spheres*). Oxygen atoms are shown as medium-sized *red spheres*, and hydrogen atoms are depicted using *small orange spheres*

with Mn and S with V. Therefore, it was selected as the initial model in the Rietveld refinement (see Fig. 23.2 and Table 23.2). The second structure was not considered promising, as it consists of Mn cations, hydroxyl and HPO_3 groups; the latter have a different geometry. It was not tested, as the first model resulted in a successful solution, as discussed in the following section.

23.2 Rietveld Refinement

The following parameters were used at the start of the refinement:

- The initial structural model was taken from Table 23.2, and the default profile parameters were taken from the instrumental parameter data file[5] (Sect. 20.3).
- The sample shift parameter, $S_s = 5.13$, corresponds to a sample displacement of $s = -0.112$ mm, which was obtained together with the unit cell dimensions at an earlier stage of the structure solution.

[5]The file *Scintag.prm* is available online.

Table 23.2 Coordinates of Mn, V, and O atoms in the unit cell of $Mn_7(OH)_3(VO_4)_4$ as derived from the model of the crystal structure of $Zn_7(OH)_3(VO_4)_3SO_4$. The space-group symmetry is $P6_3mc$

Atom[a]	Site	x	y	z
Mn1 (Zn1)	12(d)	0.4266	0.0802	0
Mn2 (Zn2)	2(a)	0	0	0.8217
V1 (V1)	6(c)	0.1513	−0.1513	0.0257
V2 (S1)	2(b)	⅓	⅔	0.7479
O1	12(d)	0.0676	0.3460	0.8469
O2	6(c)	0.8090	−0.8090	0.815
O3	6(c)	0.5280	−0.5280	0.715
O4	6(c)	0.3967	−0.3967	0.642
O5	6(c)	0.9243	−0.9243	0.571
O6	2(b)	⅓	⅔	0.024

[a] Symbols in parentheses indicate the corresponding atoms in the parent $Zn_7(OH)_3(VO_4)_3SO_4$ structure

- The space group is $P6_3mc$, with unit cell dimensions of $a = 13.229$ Å and $c = 5.2553$ Å.
- The overall isotropic displacement parameter, $U_{iso} = 0.015$ Å^2.

Initially,[6] only the phase scale factor and six coefficients of a shifted-Chebyshev polynomial were refined to approximate the background, resulting in a reasonable fit, as shown in Fig. 23.3. The residuals, listed in Table 23.3, were quite low, especially considering that the crystal structure model was adapted from a different compound, where the geometry is expected to differ (e.g., the SO_4 group is smaller than the VO_4 group). Combined with a potential for a preferred orientation in the specimen,[7] this leads to some obvious, but not severe, mismatches between the observed and calculated intensities (see the inset in Fig. 23.3).

The subsequent refinement included unit cell dimensions, grain-size peak broadening (X), peak asymmetry (α), and preferred orientation along [001]. The [001] axis is a typical preferred orientation axis in both hexagonal and tetragonal crystal systems. Refining all parameters resulted in a noticeable improvement in the fit, reducing the weighted profile residual by nearly 10%—from 27% to 16.6%. This significant change is primarily due to adjustments in three parameters: X increased from 0.023 to 0.044, asymmetry α decreased from 0.025 to 0.010, and preferred orientation along the [001] axis shifted from 1 to 1.25 (with correction factors ranging from 0.5 to 1.4, corresponding to a preferred orientation magnitude of 2.8). The unit cell dimensions remained nearly unchanged.

[6] The initial model with all other relevant parameters is available online in the files *Ch23Ex01a.exp* and *Ch23Ex01a.cif*.

[7] This effect is not expected to be significant due to the three-dimensional metal oxide framework (see Fig. 23.2), unlike the layered structures in the three previous examples. However, the obvious anisotropy in the crystallite shapes (see the inset in Fig. 23.1) suggests that the possibility of a complex preferred orientation cannot be entirely ruled out.

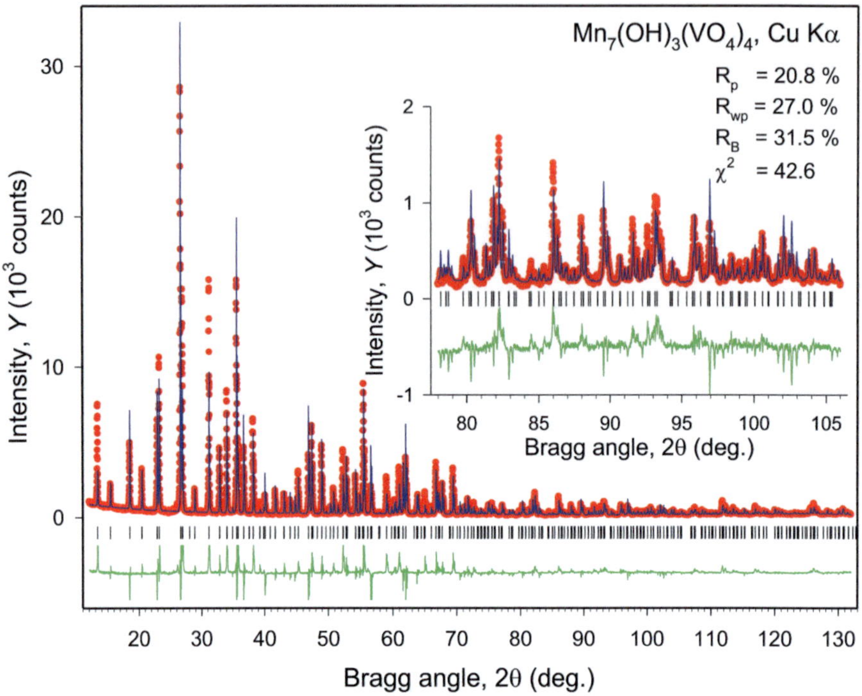

Fig. 23.3 The observed and calculated powder diffraction patterns of $Mn_7(OH)_3(VO_4)_4$ after the initial Rietveld minimization, where only the scale factor and the background were refined. The difference ($Y_i^{obs} - Y_i^{calc}$) is displayed on the same scale as the observed and calculated data, but the plot is truncated to fit within the range [−2000, 2000] for clarity. The inset highlights the range between 78° and 106° 2θ

Refining coordinates of all atoms except Mn2 and H[8] and individual isotropic atomic displacement parameters resulted in only slight improvement, achieving an additional 3% reduction in the fit. While all individual isotropic displacement parameters were positively defined, the U_{iso} values for Mn2, V2, and O6 were

[8]The only potentially free variable in the coordinate triplet of Mn2 (z) was constrained to $z = 0.8217$, a somewhat unusual value taken from the original paper on $Zn_7(OH)_3(SO_4)(VO_4)_3$, to maintain a fixed origin along the Z-axis. The origin of coordinates is not fixed in the space group $P6_3mc$ due to the absence of symmetry elements (planes or axes), perpendicular to Z, or centers of inversion. As a result, the z-coordinates, if none is fixed, become highly correlated, allowing the entire structure to shift back and forth by any translation along the Z-axis. A similar situation was discussed in the case of $CeRhGe_3$ (see Chap. 17), where the space group *I4mm* lacks symmetry elements that fix the origin of coordinates along the Z-axis. The coordinates of the H1 atom were constrained to be identical to those determined in the original Zn-based structure, as refining them is unfeasible with the current X-ray powder diffraction data due to the low scattering ability of a single-electron hydrogen atom (also see Chap. 20).

Table 23.3 The progress of Rietveld refinement of $Mn_7(OH)_3(VO_4)_4$

Refined parameters	R_p	R_{wp}	R_B	χ^2
Initial	82.3	83.7	99.4	409
Phase scale and background (Fig. 23.3)	20.8	27.0	31.5	42.6
Plus unit cell, grain-size broadening (*X*), peak asymmetry (α), and preferred orientation (PO) along [001]	13.3	16.6	18.7	16.1
Plus coordinates and individual U_{iso} of all atoms	11.1	13.7	14.4	11.0
Overall U_{iso} (Mn and V), overall U_{iso} (O), population parameter (*g*) for Mn2, V2, and O6 (Fig. 23.4)[a]	11.0	13.8	14.3	11.1
Plus strain broadening, *Y* and Y_a, and sample displacement	10.7	13.4	13.4	10.5
Porosity, a_1 and a_2, then coordinates of all atoms, overall U_{iso} (Mn, V) and overall U_{iso} (O); population parameters restricted as *g*-(V2a) = *g*(O6) = 1 − *g*(V2b)	10.7	13.3	13.7	10.4
Plus *W*, *X*, *Y*, asymmetry, sample displacement, X_a, Y_a, spherical harmonics of eighth order (six free variables), coordinates of all atoms	9.7	12.3	11.9	8.85
Plus 12 background coefficients and transmission (Fig. 23.6, Table 23.4)[b]	9.0	11.7	10.7	7.98

[a] Files online: *Ch23Ex01b.exp* and *Ch23Ex01b.cif*
[b] Files online: *Ch23Ex01c.exp* and *Ch23Ex01c.cif*

noticeably higher and continued to increase, indicating potential vacancies at these positions.[9]

The potential deficiency makes sense structurally, as the suspected defect sites are located along the tunnels around the sixfold and threefold axes, leaving the main framework intact. At this stage, the overall U_{iso} were assigned separately to all metal and oxygen atoms, and the occupancies of the three suspicious sites were refined (see Fig. 23.4).

23.3 Determining Chemical Composition

A difference Fourier map, calculated at this stage, reveals an additional small electron density maximum in the tetrahedral cavity near the partially occupied V2 site.[10] Therefore, it is reasonable to assume that the V2 site splits into two

[9] As sinθ/λ increases, both the atomic scattering functions and temperature factors decrease exponentially (see Sects. 9.1.3 and 9.1.2, respectively). Therefore, the unreasonably high isotropic displacement parameters of certain atoms suggest that the scattering ability of these sites is reduced. Unlike in $LaNi_{4.85}Sn_{0.15}$, where some sites may be occupied by Ni, Sn, or their statistical mixtures (see Chap. 16), in $Mn_7(OH)_3(VO_4)_4$, the most reasonable explanation is that the suspected sites are only partially occupied.

[10] This peak exhibits a reasonable tetrahedral configuration formed by the oxygen atoms, but it is located too close to the existing V2 site. Given the deficiency at the V2 site, it is plausible that the structure contains two types of $[VO_4]$ tetrahedra, distributed randomly throughout the structure.

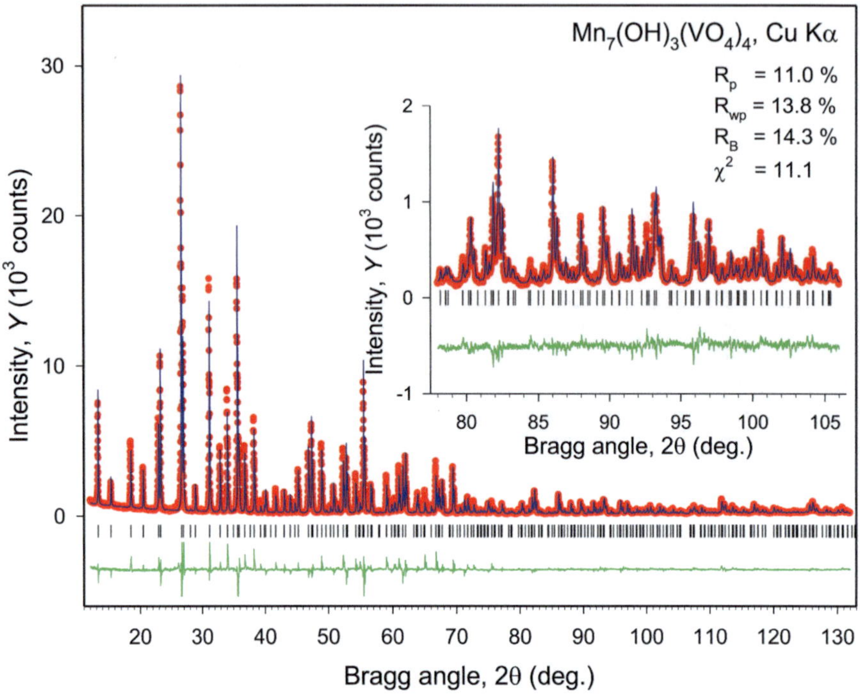

Fig. 23.4 The observed and calculated powder diffraction patterns of $Mn_7(OH)_3(VO_4)_4$ after refining preferred orientation, individual atomic parameters, and occupancies of Mn2, V2, and O6. The difference ($Y_i^{obs} - Y_i^{calc}$) is truncated to fit within the range [−2000, 2000] for clarity. The inset highlights the range between 73° and 106° 2θ

independent, partially occupied positions, with coordinates that distribute V atoms randomly between two adjacent tetrahedral positions, rather than simply indicating vanadium deficiency. These two sites are labeled V2a (corresponding to the former V2) and V2b (corresponding to the Fourier peak). Refinement of this model led to a slight improvement in the fit. Subsequently, additional profile parameters (Y, Y_a, and sample displacement) were included in the refinement. This was followed by the typical procedure of refining porosity using the Suortti approximation, with atomic coordinates and U_{iso} fixed. The porosity parameters were then fixed for the remainder of the refinement.

The prohibitively short distance between V2a and V2b clearly requires the following relationship to hold :

$$g_{\text{V2a}} + g_{\text{V2b}} = 1, \text{and in general}, g_{\text{V2a}} + g_{\text{V2b}} \leq 1 \tag{23.1}$$

Furthermore, an analysis of the refined occupation factors for V2a, V2b, and O6 suggests the following relationship:

$$g_{V2a} = g_{O6} \tag{23.2}$$

These two relationships can be easily programmed in GSAS and most Rietveld software by using the constraint apparatus, which was briefly discussed earlier (see Sect. 16.3.3 and Eq. 16.2). Since constraints affect only the parameter shifts determined during every least-squares refinement cycle, and not the actual parameter values, the latter must be manually synchronized before imposing constraints. For example, when the computed shift for g_{V2a} is 0.02, the new values of the constrained parameters Eqs. (16.2, 23.1, and 23.2) will be calculated as follows:

$$g_{V2a} = g_{V2a} + 0.02, g_{V2b} = g_{V2b} - 0.02, \text{and } g_{O6} = g_{O6} + 0.02 \tag{23.3}$$

If, before the start of the constrained refinement, g_{V2a} and g_{O6} are not equal (e.g., they are 0.6 and 0.8, respectively), then after adding the shifts according to Eq. (23.3), their values become 0.62 and 0.82. Therefore, if needed, parameters constrained in this manner should be manually matched: in this case, both g_{V2a} and g_{O6} should be set to identical values, and g_{V2a} and g_{V2b} should sum to one.[11]

The relationships between the occupancies in this crystal structure have both chemical and physical significance. The V2 atom and the surrounding four oxygen atoms (three O4 and one O6) in a fully ordered structure form a chain, as shown in Fig. 23.5a, where the occupied and empty tetrahedra alternate. Vanadium atoms can also occupy pairs of corner-sharing tetrahedra, forming the well-known V_2O_7 group (hatched in Fig. 23.5b). When a "mistake" occurs and a vanadium atom "jumps" to the next empty site, the corner-sharing pair of empty tetrahedra appears, in which the shared corner (*pink square*) must be vacated because the corresponding oxygen atom is no longer bound to any vanadium atom. Therefore, vacancies on the O6 sites, constrained as shown in Eq. (23.2), should exist, as was confirmed by the independent refinement of g_{O6}.

The subsequent refinement included profile parameters *X*, *Y*, peak asymmetry X_a, Y_a, sample displacement, and transparency shift. Preferred orientation was switched from the March–Dollase model to the eighth-order spherical harmonics expansion (six variables total), and 12 coefficients of the shifted-Chebyshev polynomial background approximation were employed. A reasonably good fit, shown in Fig. 23.6, was achieved as a result.

The preferred orientation correction was accounted for in two ways during the refinement. First, the March–Dollase approach with one texture axis [001] resulted in $\tau = 1.247(2)$ and correction coefficients ranging from 0.52 to 1.39, which gives a preferred orientation magnitude of 2.70. Second, the eighth-order spherical harmonics expansion, which corresponds in this crystal system to six adjustable

[11] Another example of synchronizing constrained parameters can be given by considering the coordinates of V1 and O2—O5 atoms in this structure. All of them are located in the 6(c) sites, with the coordinates of the independent atom $x\bar{x}z$. Hence, when entering the respective coordinate parameters, it is necessary to ensure that $y = -x$.

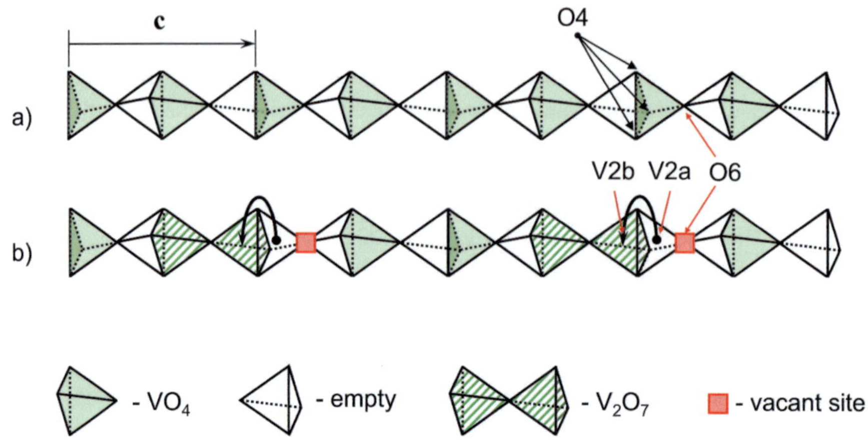

Fig. 23.5 Chains of tetrahedra in the $Mn_7(OH)_3(VO_4)_4$ structure. The fully ordered chain (**a**) with the composition VO_4, where vanadium occupies every other tetrahedron (*green*), with the remaining being vacant (*white*). The disordered chain (**b**) shows some vanadium atoms occupy pairs of neighboring tetrahedra, forming V_2O_7 groups (*hatched*), thus creating vacancies on the O6 site (*pink square*) and resulting in the chemical composition $(VO_4)_{1-x}(V_2O_7)_{x/2}$

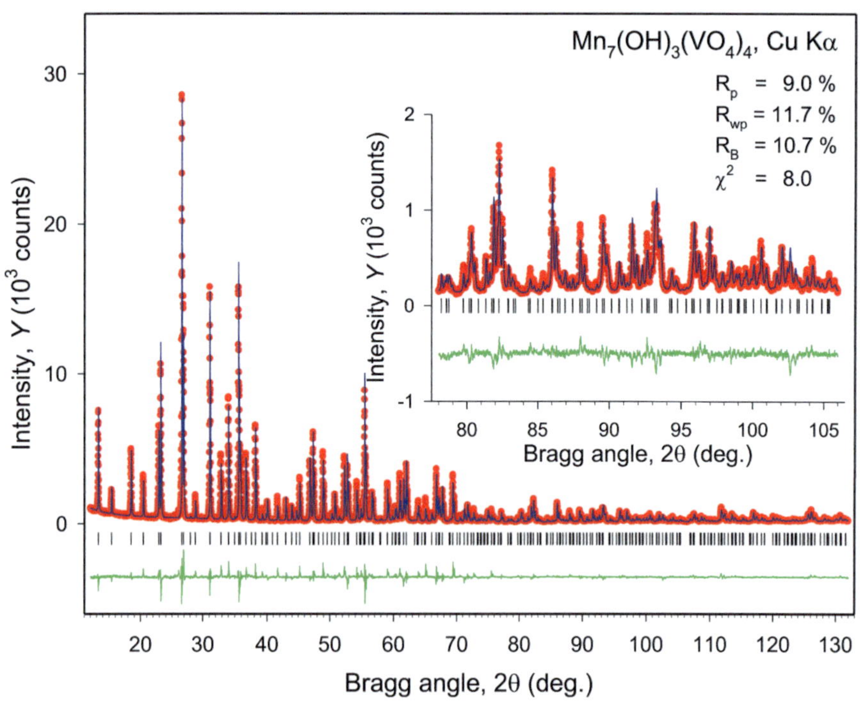

Fig. 23.6 The final Rietveld plot of $Mn_7(OH)_3(VO_4)_4$ data. The inset highlights the details between 78° and 106° 2θ

parameters (200, 400, 600, 606, 800, and 806) was attempted with the March–Dollase preferred orientation correction (1) left as is but fixed (i.e., the spherical harmonics were in addition to the March–Dollase model) or (2) eliminated. Both approaches resulted in practically an identical outcomes, except for the magnitudes of the coefficients. In the second case, the correction coefficients ranged from 0.61 to 1.54, corresponding to a preferred orientation magnitude of 2.52.

The fully refined model of the crystal (coordinates, population, and individual isotropic displacement parameters of atoms) and interatomic distances in the crystal structure of $Mn_7(OH)_3(VO_4)_4$ can be found in Table 23.4. It is shown in Fig. 23.7 as the arrangement of the coordination polyhedra of the Mn and V atoms.

In addition to the previously discussed vacancy relationships among V2a, V2b, and O6 atoms, the refined deficiency at the Mn2 site is nearly equal to two-third of

Table 23.4 The final atomic parameters[a,b] and interatomic distances (in Å) in the crystal structure of $Mn_7(OH)_3(VO_4)_4$ obtained from Rietveld refinement using GSAS. The fully refined unit cell parameters are: $a = 13.2292(1)$, $c = 5.25467(6)$ Å, $V = 796.42(2)$ Å^3, the space group is $P6_3mc$[c]

					Bond distances (Å)			
Atom	Site	*x*	*y*	*z*	Mn1	Mn2	V1	V2a
Mn1	12(d)	0.42649(8)	0.07921(8)	0.0061(8)				
Mn2	2(a)	0	0	0.8217[d]				
V1	6(c)	0.15041(6)	−0.15041(6)	0.031(1)				
V2a	2(b)	⅓	⅔	0.775(1)				
V2b	2(b)	⅓	⅔	0.497(5)				
O1	12(d)	0.0731(3)	0.3380(4)	0.863(1)	2.211(4), 2.191(5)		$1.602(5)_2$	
O2	6(c)	0.8108(3)	−0.8108(3)	0.811(2)	2.122(4)		1.720(7)	
O3	6(c)	0.5271(3)	−0.5271(3)	0.710(2)	2.178(6), 2.392(6)			
O4	6(c)	0.3987(2)	−0.3987(2)	0.638(2)	2.172(4)			$1.661(6)_3$
O5	6(c)	0.9199(3)	−0.9199(3)	0.586(2)		$2.214(7)_3$, $2.302(8)_3$	1.637(6)	
O6	2(b)	⅓	⅔	0.118(3)				1.81(2)
H1[e]	6(c)	0.433	−0.433	0.156				

[a] U_{iso} is 0.0145(2) Å^2 for metal atoms and 0.0140(8) Å^2 for O atoms

[b] The population +6: $g_{V2a} = g_{O6} = 0.819(4)$, $g_{V2b} = 0.181(4) = 1 - g_{V2a}$, $g_{Mn2} = 0.873(4) = (1 - 2/3g_{V2b})$. All other sites are fully occupied

[c] Full set of crystallographic data is available online in the files *Ch23Ex01c.exp* and *Ch23Ex01c.cif*

[d] This parameter was kept constant at all times to maintain a fixed origin of coordinates along the Z-axis. Although GSAS enables automatic fixation of the origin of coordinates, the refinement may become unstable, especially when the fit is far from the best

[e] Taken from the Zn-based compound and kept constant during the refinement

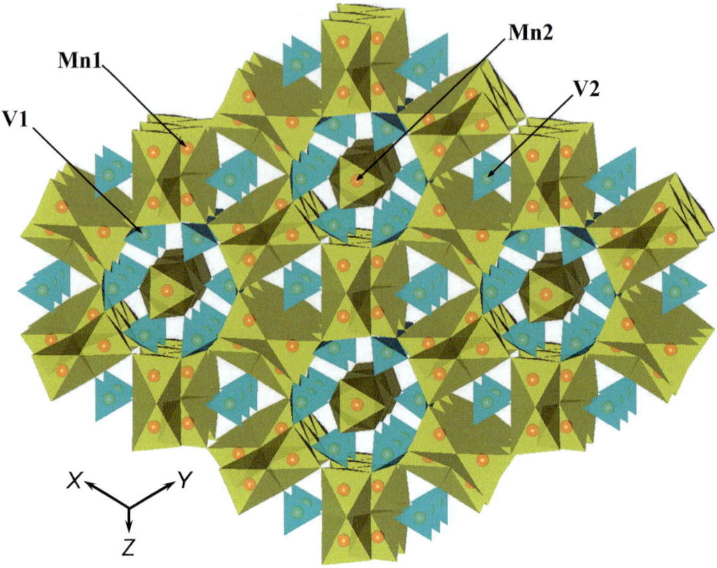

Fig. 23.7 The model of the crystal structure of $Mn_{7-y}(OH)_3(VO_4)_{4-x}(V_2O_7)_{x/2}$ shown along the Z-axis. The Mn1 sites (Mn^{2+}) and partially occupied Mn2 sites (Mn^{3+}) are shown as *orange spheres* inside the *golden* [MnO_6] octahedra, while the V1 and V2 sites are shown as *green spheres* inside the *blue* [VO_4] tetrahedra. Hydrogen atoms from the hydroxyl groups are not shown in this figure

the vacancies observed on the V2b site. Although the experimentally established relationship is approximate, when followed exactly (i.e., when $g_{Mn2} = (1 - 2/3g_{V2b})$), the oxidation state of manganese at the Mn1 site is 2+, while it is (but may not always be) 3+ at the Mn2 site.[12] All things considered, the following expression describes the chemical composition of $Mn_7(OH)_3(VO_4)_4$:

$$Mn_6^{2+}Mn_{1-y}^{3+}(OH)_3(VO_4)_{4-x}(V_2O_7)_{1/2x}; \quad x = 0.181(4), \quad y = 2/3x$$

This crystal structure and composition have been confirmed by single-crystal diffraction data, yielding $R_F = 2.2\%$ and x = 0.204(3).[13] This result not only validates crystal the crystal structure determination from powder data but also illustrates that the accuracy achieved using powder data might be comparable to that of single-crystal data, at least in some cases. A small difference is acceptable, as a tiny single crystal may not fully represent a larger polycrystalline sample, particularly in the case of occupational disorder.

[12] Note that the oxidation state of Mn2 may not always be 3+, and therefore, the occupation of Mn2 site may not correlate with the occupation of V2b site.

[13] P.Y. Zavalij, S.T. Lutta, and M.S. Whittingham, CSD 2387690: Experimental Crystal Structure Determination (2024). https://doi.org/10.25505/fiz.icsd.cc2l4l8m

Chapter 24
Crystal Structure of $FePO_4$: Solved Using Structural Analogy and Optimization

This example illustrates how a model of a crystal structure can be derived based on a suspected analogy with related compounds, followed by geometry optimization to enhance and improve the deduced structural model. Such a complex approach in this case has been adopted due to poor crystallinity of the material, which results in a low resolution of its powder diffraction pattern (see Fig. 24.1), where the full widths at half maximum range from 0.25° to 0.55°. Furthermore, the pattern is relatively complex, with as many as 255 Bragg reflections possible below 37.5° 2θ when Mo Kα radiation is employed.

As we discover in this chapter, there are only six atoms in the asymmetric unit. However, the Rietveld refinement of the model is complicated by the inadequate quality of the powder diffraction data. This limitation even precludes an unrestrained refinement of the optimized model, and similarly to the example considered in Chap. 21, restraints are imposed on the geometry of the PO_4 groups. We, therefore, take this opportunity to illustrate the role of restraint weighting in Rietveld refinement.

The title compound was prepared by thermal decomposition of the monoclinic dihydrate $FePO_4 \cdot 2H_2O$. The solid-state preparation reaction is likely responsible for the poor crystallinity, and consequently, peak broadening.[1] The inadequate crystallinity of the material results in the insufficient accuracy of both peak positions and intensities. A serious lack of resolution in this powder diffraction pattern, due to the physical state of the powder, causes considerable problems in both the indexing and

Y. Song, P.Y. Zavalij, M. Suzuki, and M.S. Whittingham, New iron(III) phosphate phases: Crystal structure, electrochemical and magnetic properties, Inorg. Chem. 41, 5778 (2002). https://doi.org/10.1021/ic025688q

[1]This example is also an excellent illustration of a case where the physical state of a material precludes single crystal diffraction analysis, making a powder diffraction experiment the only option for a solution of its crystal structure.

P. Zavalij, V. Pecharsky, *Fundamentals of Powder Diffraction and Structural Characterization of Materials*, https://doi.org/10.1007/978-3-031-91504-8_24

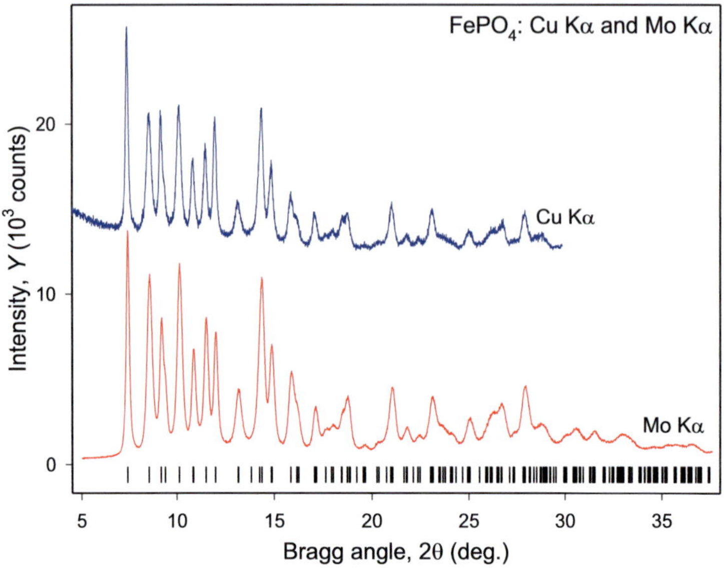

Fig. 24.1 Powder diffraction patterns collected from monoclinic $FePO_4$ using Cu Kα radiation on a Scintag XDS2000 diffractometer (*top*, step scan, 0.02° step) and Mo Kα radiation on a rotating anode Rigaku TTRAX diffractometer (*bottom*, step scan, 0.01° step). Bragg angles in the pattern collected using Cu Kα radiation have been converted to match Mo Kα radiation. The two patterns are shown with a substantial displacement along the intensity axis for clarity. The *vertical bars* indicate calculated positions of Bragg reflections for the locations of $K\alpha_1$ components (Experimental data, collected using Mo Kα radiation, are online: *Ch24Ex01_MoKa.xy* and *Ch24Ex01_MoKa.raw*)

structure determination. When experimental data collected using Cu Kα radiation were employed in the ab initio indexing, ITO and TREOR runs did not result in a reasonable solution.

A single-crystal diffraction experiment conducted on a small low-quality single crystal yielded only about 20 detectable Bragg reflections, all at low angles. This truncated array of data was insufficient even for a reliable automatic indexing, but after a visual inspection, a monoclinic or orthorhombic lattice with the unit cell dimensions $a \approx 5.5$, $b \approx 7.5$, and $c \approx 8.0$ Å was clearly noticeable. A solution with similar unit cell dimensions and a monoclinic angle around 95° was found by employing ITO using Mo Kα data with 19 of 20 low Bragg angle peaks indexed. The unit cell dimensions are $a = 5.489(1)$, $b = 7.493(1)$, $c = 8.055(1)$ Å, $\beta = 95.81$ (1)°. The monoclinic symmetry of the lattice and the systematic absences (likely corresponding to the space group $P2_1/n$) are the same as in the parent hydrate, $FePO_4{\cdot}2H_2O$.

24.1 Building and Optimizing the Model of the Crystal Structure

The identical symmetry and similar unit cell dimensions of the hydrated and anhydrous iron phosphates are observed in the two orthorhombic modifications (see Table 24.1), which have closely related crystal structures—featuring the same bonding in the $FePO_4$ frameworks, except for the water of crystallization in the hydrated compound (for more details, see the reference listed in Note on page 743). This fact can be used to solve the crystal structure of the anhydrous monoclinic compound, if $FePO_4$ connectivity remains intact in the two monoclinic modifications as well. As illustrated in Fig. 24.2, the coordinates of Fe, P, and four independent O atoms from the hydrated monoclinic compound were incorporated as the initial model of the crystal structure of the anhydrous phosphate (Model A, listed in Table 24.2). Water molecules, present in monoclinic $FePO_4{\cdot}2H_2O$, were ignored in Model A.

Unfortunately, a straightforward Rietveld refinement of Model A fails because it is far from reality, in addition to the low resolution of powder diffraction data.[2] Therefore, the initial Model A must be improved before attempting the Rietveld refinement. The improvement was achieved using the following two approaches to geometry optimization:

1. The first optimization attempts were conducted using routines included in the Materials Studio[3] suite of crystallographic programs: DMol3,[4] a molecular optimization technique based on a density functional theory (DFT) quantum

Table 24.1 The comparison of unit cell dimensions of the orthorhombic and monoclinic modifications of hydrated and anhydrous iron phosphates

Compound	Space group	*a* (Å)	*b* (Å)	*c* (Å)	β (deg.)	*V* (Å^3)
$FePO_4{\cdot}2H_2O$	Pbca	9.867	10.097	8.705	–	867.3
$FePO_4$	Pbca	9.171	9.456	8.675	–	752.4
$FePO_4{\cdot}2H_2O$	$P2_1/n$	5.307	9.755	8.675	90.16	449.1
$FePO_4$	$P2_1/n$	5.489	7.493	8.055	95.81	329.7

[2]Note the considerable contraction along the *b*-axis, Table 24.1, and the expected distortion of the $FePO_4$ framework upon dehydration, Fig. 24.2a, b.

[3]Materials Studio is a software for modeling materials developed and distributed by BIOVIA (formerly Accelrys and currently part of 3DS). https://www.3ds.com/products/biovia/materials-studio

[4]B. Delley, J. Chem. Phys. **92**, 508 (1990), https://doi.org/10.1063/1.458452; B. Delley, J. Chem. Phys. **94**, 7245 (1991), https://doi.org/10.1063/1.460208; B. Delley, J. Phys. Chem. **100**, 6107 (1996), https://doi.org/10.1021/jp952713n; and B. Delley, J. Chem. Phys. **113**, 7756 (2000), https://doi.org/10.1063/1.1316015

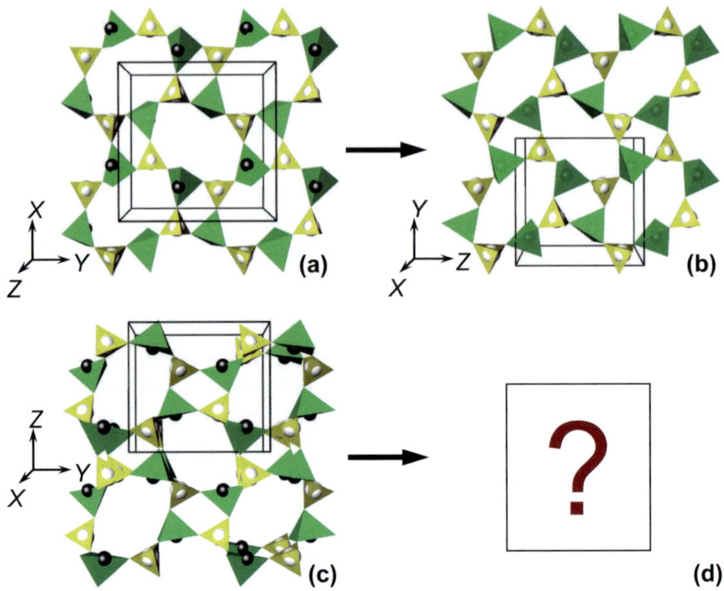

Fig. 24.2 Analogy between the crystal structures of orthorhombic hydrate $FePO_4{\cdot}2H_2O$ (**a**) and anhydrous orthorhombic $FePO_4$ (**b**) motivates the use of the monoclinic hydrate $FePO_4{\cdot}2H_2O$ structure (**c**) as the initial model for monoclinic anhydrous $FePO_4$ (**d**). The crystal structures are shown as packing of the corresponding [XO_4] polyhedra of iron (*black spheres inside green tetrahedra*) and phosphorus (*white spheres inside yellow tetrahedra*) atoms without the oxygen atoms from water molecules that complete the coordination polyhedra of iron in the hydrates. Oxygen atoms are in the corners of the polyhedra

Table 24.2 Coordinates of Fe, P, and O atoms in the unit cell of $FePO_4$, Model A, as assumed from the model of the crystal structure of monoclinic $FePO_4{\cdot}2H_2O$

Atom	Site	*x*	*y*	*z*
Fe	4(e)	0.0914	0.6739	0.6916
P	4(e)	−0.0870	0.3509	0.6839
O1	4(e)	−0.1164	0.5068	0.6700
O2	4(e)	0.1659	0.3221	0.7638
O3	4(e)	−0.0944	0.2814	0.5268
O4	4(e)	−0.3002	0.2937	0.7831

mechanical approach, and CASTEP[5] (Cambridge Sequential Total Energy Package), which is the ab initio quantum mechanical DFT method that enables the modeling of properties of solids. However, when the $FePO_4$ Model A was

[5] M. C. Payne, M. P. Teter, D. C. Allan, T. A. Arias, and J. D. Joannopoulos, Rev. Mod. Phys. **64**, 1045 (1992), https://doi.org/10.1103/RevModPhys.64.1045, V. Milman, B. Winkler, J. A. White, C. J. Pickard, M. C. Payne, E. V. Akhmatskaya, and R. H. Nobes, Int. J. Quant. Chem. **77**, 895 (2000), https://doi.org/10.1002/(SICI)1097-461X(2000)77:5<895::AID-QUA10>3.0.CO;2-C; and M. D. Segall, P. L. D. Lindan, M. J. Probert, C. J. Pickard, P. J. Hasnip, S. J. Clark, M. C. Payne, J. Phys.: Cond. Matt. **14**, 2717 (2002), https://doi.org/10.1088/0953-8984/14/11/301

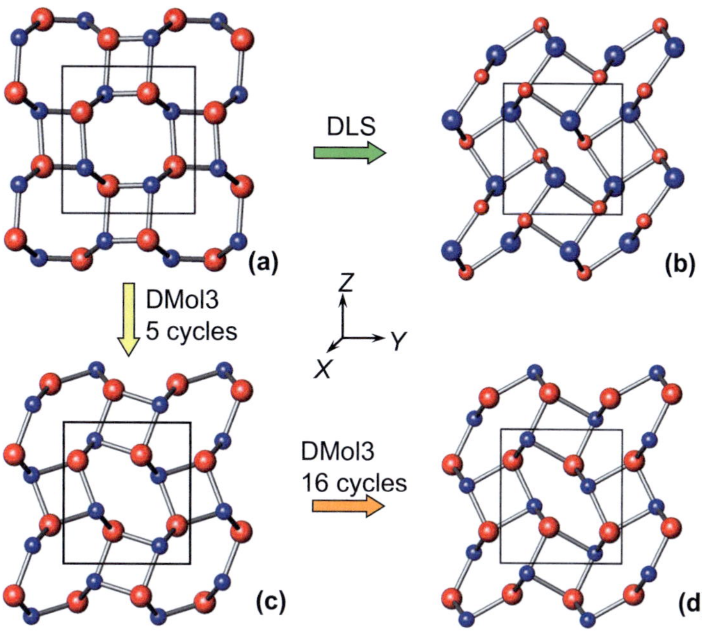

Fig. 24.3 Backbone models of the crystal structure of the anhydrous monoclinic $FePO_4$ projected along the *X*-axis: the initial Model A derived from the monoclinic hydrate $FePO_4{\cdot}2H_2O$ (**a**); Model B1 after five optimization cycles as $AlPO_4$ using DMol3 (**c**); Model B after final optimization as $AlPO_4$ (16 cycles) using DMol3 (**d**); Model C optimized using DLS-76 (**b**)

employed, these optimizations were unstable and did not converge in a reasonable number of cycles. Therefore, an attempt was made to optimize Model A, in which Fe^{3+} is substituted with Al^{3+}. Both the latter and the former often form similar or isostructural compounds due to the same oxidation states and nearly identical radii. Yet Al, as a *p*-element, has a simpler electronic structure. The optimization performed by DMol3 successfully converged in 16 cycles, resulting in Model B.[6] The initial Model A, intermediate Model B1 (obtained after five optimization cycles), and final Model B are compared in Fig. 24.3. It is easy to see the contraction of the octagonal ring where the water molecule was located in the hydrate, which makes excellent chemical and physical sense. Changes in the geometry of the coordination polyhedra also make perfect sense, as the strongly distorted tetrahedra in Model A (which are actually parts of the octahedra in the hydrate) are optimized into almost ideal tetrahedral configurations of oxygen around iron atoms.[7]

---

[6]The optimization run takes approximately 24 h on a PC equipped with a single 2 GHz processor.

[7]It is worth noting that energy minimization was carried out without applying any geometrical restrictions. Nonetheless, quite reasonable geometry resulted.

Table 24.3 Coordinates of Fe, P, and O atoms in the unit cell of anhydrous $FePO_4$, Model B, obtained from DMol3 optimization[a]

Atom	Site	*x*	*y*	*z*
Fe	4(e)	0.4385	0.7795	0.0764
P	4(e)	0.5544	0.4012	0.2674
O1	4(e)	0.5001	0.5792	0.1797
O2	4(e)	0.7980	0.4120	0.3802
O3	4(e)	0.5793	0.2536	0.1386
O4	4(e)	0.3450	0.3548	0.3720

[a] In order to directly compare models A and B, the atomic coordinates of the former should be transformed as: $x'_A = x_A + ½$, $y'_A = y_A$, $z'_A = z_A - ½$ because atomic coordinates undergo several transformations during the optimization process

2. In parallel to the quantum mechanical optimization, in which multiple attempts took many days of computing and analyzing the results, a purely geometrical optimization was also conducted using the DLS-76 (distance least squares) program.[8] This program minimizes the differences between the existing and desired distances, with the Fe–O and P–O set to 1.88 and 1.53 Å, respectively. Additionally, the O–O distances were set to 3.07 and 2.50 Å for $[FeO_4]$ and $[PO_4]$ tetrahedra, respectively. The process converges very quickly, resulting in Model C,[9] which is quite similar to Model B obtained from DMol3. However, we note that the latter was achieved without any restrictions imposed on the geometry of the crystal structure. Therefore, if the geometrical restrictions in the DLS attempt are incorrect or not accurate enough, the algorithm may not (and likely will not) converge to a reasonable model. The final Model C is compared with the initial Model A and Models B1 and B from the DMol3 optimization in Fig. 24.3. This model was not tested further because Model B was successful.

Model B, obtained from DMol3 optimization, is shown in Table 24.3 and was used as the initial approximation in Rietveld refinement discussed below.

24.2 Rietveld Refinement

The experimental powder diffraction pattern was collected on a rotating anode Rigaku TTRAX powder diffractometer using monochromatized Mo Kα radiation from 5° to 50° 2θ in a step-scan mode with a 0.01° step and a counting time of 10 s/step. The following parameters were employed at the beginning of this refinement:

[8] Ch. Baerlocher, A. Hepp, and W.M. Meier, DLS-76, a program for the simulation of crystal structures by geometric refinement. Institute of Crystallography and Petrography, ETH: Zurich, Switzerland (1997).

[9] The optimization took approximately 2⁄100th of a second on a PC with a single 2 GHz processor.

- The initial structure model derived and optimized above (Table 24.3).
- The default profile parameters from the instrumental parameter file[10] obtained from the refinement of the LaB_6 standard, as described in Sect. 20.3.
- The space group $P2_1/n$ and the unit cell dimensions $a = 5.489$, $b = 7.493$, $c = 8.055$ Å, and $\beta = 95.81°$ (see Table 24.1).
- The sample shift parameter $S_s = 3.99$ for the sample displacement $s = -0.087$ mm obtained together with the unit cell dimensions at the stage of lattice parameters refinement.
- The overall isotropic displacement parameter $U_{iso} = 0.015$ Å^2.

Only the range from 5° to 37.6° 2θ was used in all calculations.[11,12] The background was manually approximated using a straight line, with two coefficients of a shifted-Chebyshev polynomial employed in the process. Automatic background determination was not feasible at the beginning of the refinement due to significant overlap of the Bragg peaks. Therefore, only two data points—the first and the last within the range—were used for background estimation. The initial refinement of the phase scale factor resulted in R_{wp} of 62.7% (Table 24.4) and a poor fit, indicating that all calculated Bragg peaks were too narrow. Consequently, both peak-broadening parameters, *X* and *Y*, were manually increased by a factor of 10, which reduced the weighted profile residual to 35.6%, yielding an acceptable and tolerable fit, as shown in Fig. 24.4.

Both the complexity and low resolution of the experimental data, along with the possibility of far-from-ideal coordinates for some or all atoms in the optimized crystal structure model,[13] present a dilemma when selecting the next set of parameters for a subsequent Rietveld refinement. While it is clear that profile parameters need further improvement, the inadequacy of the initial fit, combined with the low data resolution, may hinder their unambiguous refinement. Furthermore, atomic coordinates likely deviate significantly from their true values, as evident in Fig. 24.4, which shows substantial discrepancies between the observed and calculated intensities for many Bragg reflections.

[10]The file *Rigaku.prm* is available online.

[11]This range of Bragg angles corresponds to a $2\theta_{max}$ of approximately 89° when using Cu Kα radiation. In this case, the use of the instrument with Mo Kα radiation was justified by the large goniometer radius (285 mm), which offers potentially high resolution, and by the significant amount of Fe in the material, which creates a substantial fluorescent background when Cu Kα radiation is used.

[12]The initial model, including all necessary crystallographic parameters, can be found online in the files *Ch24Ex01a.exp* and *Ch24Ex01a.cif*, with the experimental data available in the file *Ch24Ex01_MoKa.raw*.

[13]It is worth noting that the quantum chemical optimization was performed after Fe was substituted by Al.

Table 24.4 The progress of Rietveld refinement for the crystal structure of anhydrous $FePO_4$ using low-resolution X-ray powder diffraction data

Refined parameters	R_p	R_{wp}	R_B	χ^2
Initial	75.2	77.2	98.6	1414
Scale factor, linear background adjusted manually	57.4	62.7	67.4	933
Peak broadening *X* and *Y* multiplied by 10 then phase scale (Fig. 24.4)	29.4	35.6	32.6	301
Plus two background coefficients, coordinates, PO_4 group soft-restrained with weight 4	13.3	16.8	22.6	67.6
Same but soft restraints with weight increased to 10	11.1	13.8	13.8	46.2
Plus grain-size broadening, *X*	8.5	11.1	10.7	30.1
Plus unit cell dimensions, PO [010], overall U_{iso} (Fig. 24.6)[a]	8.1	10.8	9.9	28.1
Plus strain broadening, *Y*, then asymmetry and sample displacement	7.3	9.7	6.2	22.7
Four coefficients of the background	6.5	8.6	4.1	17.9
X, *Y*, peak asymmetry, X_a, Y_a (Fig. 24.7, Table 24.5)[b]	5.2	7.1	3.6	12.3

[a] Files online: *Ch24Ex01b.exp* and *Ch24Ex01b.cif*
[b] Files online: *Ch24Ex01c.exp* and *Ch24Ex01c.cif*

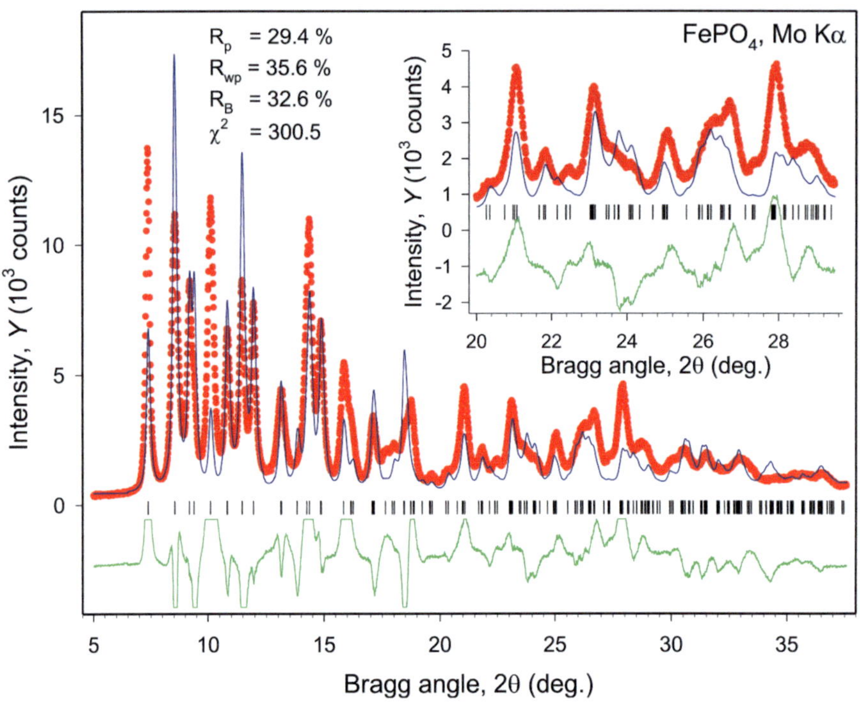

Fig. 24.4 The observed and calculated powder diffraction patterns of anhydrous $FePO_4$ after the initial Rietveld least-squares minimization with only the scale factor and linear background refined. The difference ($Y_i^{obs} - Y_i^{calc}$) is shown using the same scale as the observed and calculated data, but the plot is truncated to fit within the range [−2000, 2000] for clarity. The inset highlights the range between 20° and approximately 29° 2θ

Table 24.5 Atomic parameters[a] and interatomic distances (in Å) in the crystal structure model of anhydrous monoclinic $FePO_4$, obtained from Rietveld refinement using GSAS. The unit cell dimensions are: a = 5.4856(6), b = 7.4882(8), c = 8.0626(9) Å, β = 95.694(8)°, V = 329.56 (5) Å^3, space group P2$_1$/n[b]

					Bond distances (Å)	
Atom	Site	*x*	*y*	*z*	Fe	P
Fe	4(e)	0.3891(5)	0.8060(4)	0.0585(3)		
P	4(e)	0.5878(9)	0.4539(5)	0.2680(6)		
O1	4(e)	0.471(1)	0.638(1)	0.225(1)	1.864(7)	1.544(6)
O2	4(e)	0.835(2)	0.464(1)	0.384(1)	1.944(8), 2.224(8)	1.569(7)
O3	4(e)	0.639(2)	0.362(1)	0.112(1)	1.857(8)	1.488(7)
O4	4(e)	0.402(2)	0.344(1)	0.362(1)	1.801(8)	1.565(7)

[a] Overall U_{iso} = 0.018(1) Å^2

[b] Full crystallographic data can be found online in the files *Ch24Ex01c.exp* and *Ch24Ex01c.cif*

Hence, the next refinement step included a linear background and the coordinates of all atoms.[14] Anticipating significant challenges with the least-squares minimization and a high likelihood of deviating from a global minimum, restraints were applied to preserve the well-known geometry of the phosphate group (PO_4).[15] The initial geometry, obtained from quantum chemical optimization, was nearly ideal: P–O bond lengths ranged from 1.52 to 1.54 Å, and O–P–O angles ranged from 107.8° to 110.2°. The following restraints were imposed: a P–O distance of 1.53 ± 0.01 Å and O–P–O angles of 109.5 ± 2.0°; the weight was set to 4. The first five cycles of the refinement significantly improved the fit, lowering R_{wp} by more than 20%, down to 16.8%. However, this reduction came at the cost of deteriorating the PO_4 geometry: P–O distances now range from 1.43 to 1.61 Å, and the O–P–O angles vary from 103° to 117°. The Fe–O distances remain acceptable, ranging from 1.83 to 1.95 Å, although one additional elongated Fe–O bond of 2.27 Å has emerged.

To improve the geometry, the restraint weight factor was increased to 10, and several subsequent least-squares cycles were performed. The weighted residual further decreases, and most importantly, the geometry of the PO_4 group was restored. This adjustment's effectiveness is demonstrated in Fig. 24.5, which shows the relative shifts of all atoms as a function of least-squares cycle number. It is clear that setting the weight to 4 did little to stabilize the convergence, whereas

[14] The positions (coordinates) of atoms in the unit cell are the strongest contributors to the computed integrated intensities of Bragg reflections, assuming that preferred orientation effects are weak. For this powder, preferred orientation was expected to be minor (and was later confirmed) due to the small particle sizes and predominantly isotropic particle shapes.

[15] A thorough reader should be able to verify the correctness of this statement by attempting Rietveld refinement without imposing restraints.

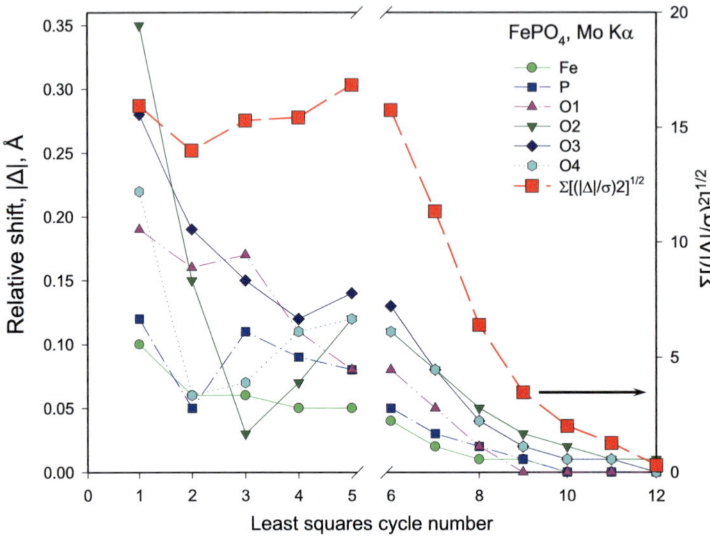

Fig. 24.5 Relative shifts of individual atoms (left-hand scale) and the average shift-to-standard deviation (Δ/σ) ratio during the first 12 cycles of least-squares refinement of atomic coordinates in the crystal structure model of anhydrous monoclinic $FePO_4$. Both the P–O distances and O–P–O angles were restrained with a weight of four during the initial five cycles. The weight was increased to 10 starting from cycle 6. The initial five cycles show erratic shifts for the P and O atoms. From the sixth cycle onward, the shifts of all atoms steadily decrease and approach zero after cycle 12

increasing the weight to 10 led to a rapid reduction of the magnitudes of atomic displacements over just a few refinement cycles.

The subsequent refinement included unit cell dimensions, the grain-size peak-broadening parameter X, preferred orientation in the March–Dollase approximation with the [010] texture axis,[16] and the overall isotropic displacement parameter U_{iso}, in addition to the linear background, phase scale, and coordinates of individual atoms, which reminded restrained with the weight set to 10.[17] In this case, the preferred orientation correction was insignificant. The corresponding parameter ($\tau_{010} = 0.986$) results in correction factors ranging from 0.98 to 1.04, making it negligible. This refinement leads to a clear improvement of the geometry of the crystal structure. The P–O bond lengths now range from 1.48 to 1.58 Å, and the O–P–O angles vary from 102° to 115°. The profile fit (Fig. 24.6) is also improved, with the weighted residual reduced to 10.8%.

To further improve the fit, the following parameters were sequentially included into the least-squares minimization: the strain peak-broadening parameter Y, peak

[16] The [010] axis was chosen after trying the three major crystallographic directions as potential preferred orientation axes.

[17] The same weight was maintained throughout the remainder of this Rietveld refinement.

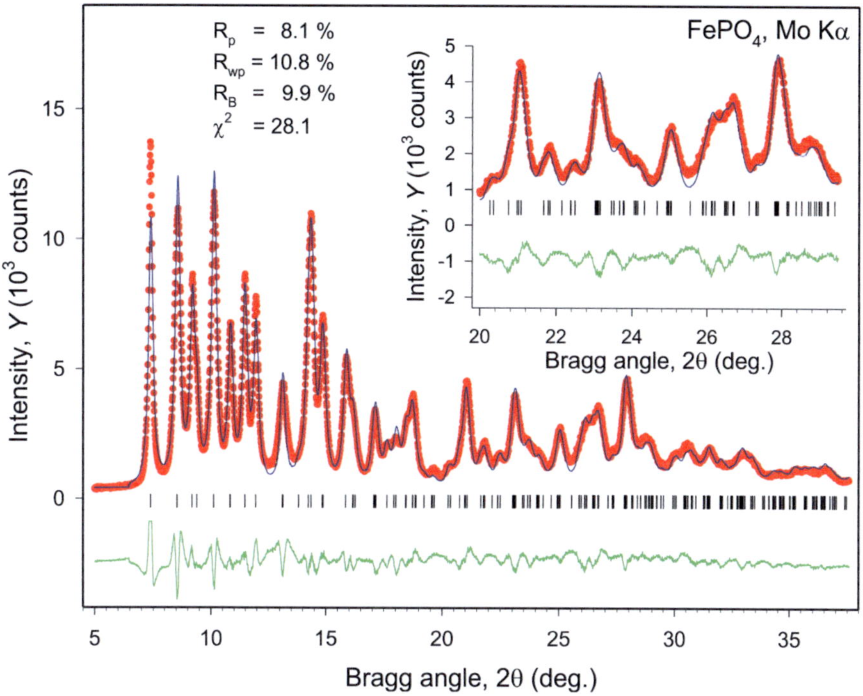

Fig. 24.6 The observed and calculated powder diffraction patterns of anhydrous $FePO_4$ after the refinement of atomic coordinates, restrained to match the ideal geometry of the PO_4 group with a weight of 10, along with linear background, preferred orientation, grain-size peak-broadening parameter, and unit cell dimensions. The inset highlights the range between 20° and ~29° 2θ

asymmetry, α, and sample displacement. At this point, the linear background was replaced by a fourth-order shifted-Chebyshev polynomial and refined with all other profile parameters fixed. Finally, all relevant parameters were refined together. Convergence was achieved, and the final fit, illustrated in Fig. 24.7, is quite satisfactory given the poor resolution of the powder diffraction pattern.

The resulting structural parameters are listed in Table 24.5.[18] The improvement in the PO_4 geometry is significant: the final P–O bond lengths range from 1.49 to 1.57 Å with the average 1.54 Å. The O–P–O angles also show good agreement, ranging from 107.4° to 113.9°. The coordination of the Fe atom remains distorted, and its polyhedron has transformed from a tetrahedron (the result of

[18]The complete geometrical characteristics of this crystal structure are available online in the file *Ch24Ex01c.cif*.

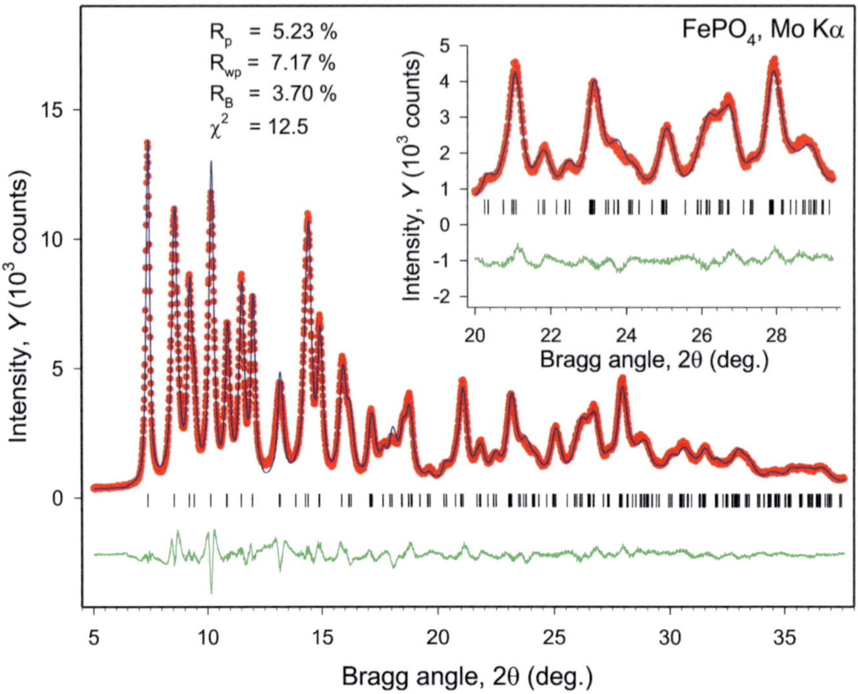

Fig. 24.7 The final Rietveld plot of $FePO_4$ data. The inset highlights the details between 20° and ~29° 2θ. The true background is quite difficult to determine due to heavily overlapped Bragg peaks, especially at 2θ > 20°

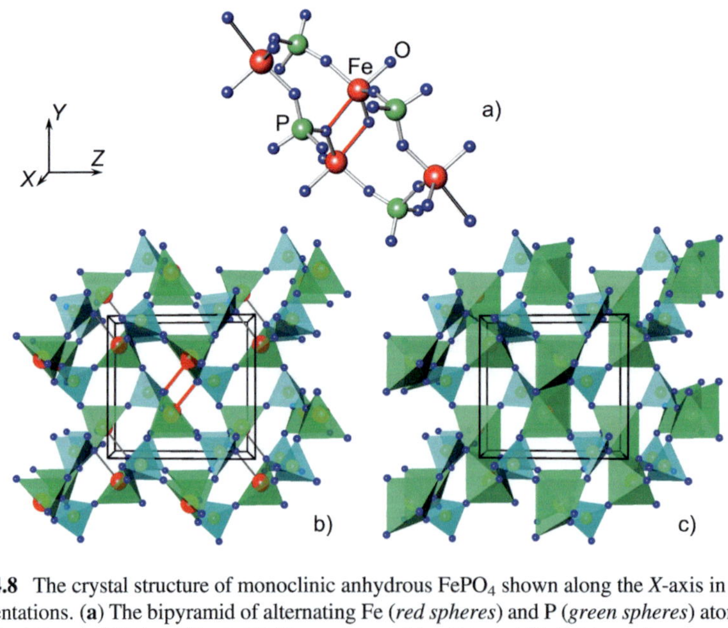

Fig. 24.8 The crystal structure of monoclinic anhydrous $FePO_4$ shown along the *X*-axis in various representations. (**a**) The bipyramid of alternating Fe (*red spheres*) and P (*green spheres*) atoms with weak Fe–O bonds (*thick red lines*). (**b**) The packing of the distorted [FeO_4] (*green*) and nearly ideal [PO_4] (*blue-gray*) tetrahedra. (**c**) The packing of the stretched [FeO_5] trigonal bipyramids (*green*) and [PO_4] (*blue*) tetrahedra

geometry optimization) into a trigonal bipyramid, as shown in Fig. 24.8a. The latter is commonly observed in Fe(III) compounds. One of the Fe–O bonds remains elongated, depicted by dark lines extending across the octagonal tunnel in Fig. 24.8b, which illustrates the distorted oxygen tetrahedra surrounding the Fe atoms. Figure 24.8c highlights the presence of [FeO_5] trigonal bipyramids.

Chapter 25
Crystal Structure of Acetaminophen: Solved Using Direct-Space Methods

In the final section, we demonstrate the determination of the crystal structure of a simple organic compound using powder diffraction data collected on a standard laboratory X-ray powder diffractometer. The compound is *N*-(4-hydroxyphenyl) ethanamide, *p*-HO–C_6H_4–NH–$COCH_3$, a well-known active ingredient in pain relievers. It is commonly referred to as acetaminophen, *p*-hydroxyacetanilide, Panadol, paracetamol, or Tylenol®. The material is readily available, and its molecular structure, as shown in Fig. 25.1, is relatively simple. We hope that this example is useful to readers interested in the structures of pharmaceuticals, particularly since imperfect crystallinity, as observed in this specimen, is common in such compounds.

The powder was prepared by crushing and gently grinding two 500 mg Tylenol® pills purchased at a local drug store.[1] The powder diffraction pattern, shown in Fig. 25.2, was collected using a D8 Advance powder diffractometer equipped with SolX point detector using Cu Kα radiation. Moderate peak broadening is evident: peak widths at half maximum range from 0.13° at low Bragg angles to 0.25° when 2θ approaches 40°. At higher angles, the broadening becomes more significant, but it cannot be quantified as the peak intensities deteriorate as rapidly with the increasing density of Bragg peaks. Such a rapid reduction in scattered intensity is typical for organic compounds, primarily due to the quickly decaying atomic scattering factors of C, N, O, and H (see Fig. 9.4 showing the atomic scattering functions of O and H), as well as the characteristically large thermal displacements of light atoms (see Fig. 9.2). As a result, only about 40 distinguishable Bragg peaks can be reliably identified in the pattern, despite nearly 130 possible reflections below 47° 2θ. Together, these complications make solving and refining this crystal structure particularly challenging, essentially requiring the use of direct-space techniques.

[1] The inactive ingredients listed on the package, such as carnauba wax, castor oil, cellulose, corn starch, magnesium stearate, and titanium dioxide, are not detected by X-ray powder diffraction, evidently due to amorphous state and/or small quantities.

P. Zavalij, V. Pecharsky, *Fundamentals of Powder Diffraction and Structural Characterization of Materials*, https://doi.org/10.1007/978-3-031-91504-8_25

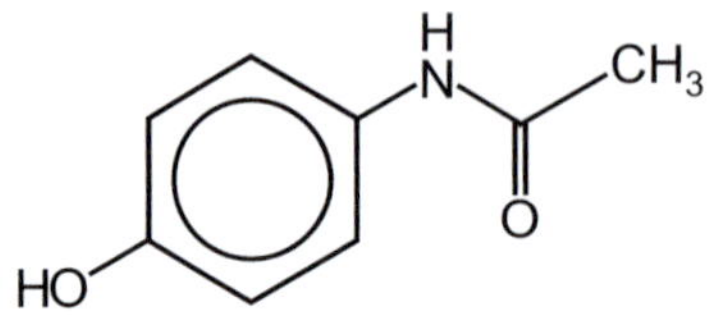

Fig. 25.1 Illustration of the molecular structure of acetaminophen, $C_8H_9NO_2$. The carbon and hydrogen atoms in the phenyl ring are not labeled for clarity

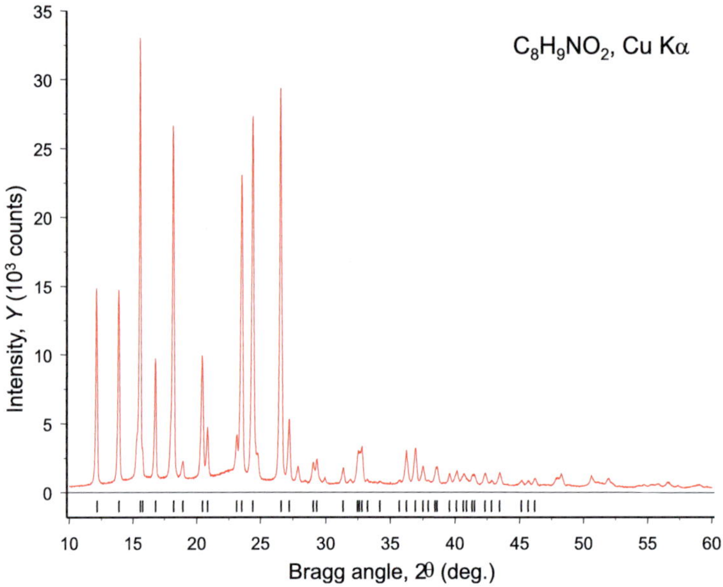

Fig. 25.2 Powder diffraction pattern collected from two ground Tylenol® pills using Cu Kα radiation on a Bruker D8 Advance diffractometer equipped with SolX detector. The pattern was recorded from a spinning flat sample using a step scan over the range of Bragg angles from 10° to 60° 2θ, with a 2θ step of 0.01° and counting time of 13 s/step (The total experiment time was 18 h using SolX point detector. The same counting statistics can be achieved in less than an hour with a position-sensitive detector, such as LynxEye. While the point detector results in a much lower background, particularly at low Bragg angles, it also reduces instrumental broadening of the Bragg peaks by about 10–15%, which is crucial in this case due to a significant peak overlap and the presence of an amorphous component.). The vertical bars indicate positions of the $K\alpha_1$ components of the Bragg peaks observed below 47° 2θ (The diffraction data are available online in the files *Ch25Ex01_CuKa.xy* and *Ch25Ex01_CuKa.raw*)

Here, we employ the simulated annealing method incorporated in FOX, which is also used for automatic peak search and ab initio indexing. Finally, the Rietveld refinement using GSAS demonstrates extensive use of restraints.

FOX ("Free Objects for Crystallography") is a free, open-source application for ab initio structure determination developed by Favre-Nicolin and Černý.[2] We selected this application to further illustrate direct-space and general optimization methods because FOX is readily available,[3] relatively simple to use, and suitable for all types of structures (molecular and extended frameworks) and chemical compounds (intermetallic, inorganic, organic, and metal–organic). It works with both powder and single-crystal diffraction data. Simulated annealing and parallel tempering methods implemented in FOX are briefly described in Sect. 15.1.3, and this section provides a practical example of applying them to solve the crystal structure of acetaminophen. As throughout this book, only a few basic features relevant to this example are discussed here. This chapter is not intended to substitute for a tutorial or a full application manual, which the reader should consult before attempting to follow our steps or independently solve this or any other crystal structure. FOX is continuously being revised and developed.[4] For example, not so long ago, it was expanded with several important features, including automatic peak search, ab initio indexing, and Le Bail full-profile fitting, all of which are discussed in Sect. 25.1.

25.1 Ab Initio Indexing and Le Bail Fitting

Once positions of the Bragg peaks are determined, ab initio[5] indexing of the pattern shown in Fig. 25.2 can be performed using any of the classic indexing applications (TREOR, ITO, and DICVOL). However, FOX is convenient because it includes a built-in automatic peak-search utility, followed by two ab initio indexing methods: a

[2] V. Favre-Nicolin, R. Černý, FOX, "Free objects for crystallography": a modular approach to ab initio structure determination from powder diffraction, J. Appl. Cryst. **35**, 734 (2002), https://doi.org/10.1107/S0021889802015236; R. Černý, V. Favre-Nicolin, FOX: A friendly tool to solve nonmolecular structures from powder diffraction, Powder Diffraction, **20**, 359 (2005), https://doi.org/10.1154/1.2135314; V. Favre-Nicolin, R. Černý, A better FOX: using flexible modeling and maximum likelihood to improve direct-space ab initio structure determination from powder diffraction, Z. Kristallogr. **219**, 847 (2004). https://doi.org/10.1524/zkri.219.12.847.55869

[3] The FOX program, manual, tutorials, and other information are available at: https://objcryst-fox.readthedocs.io/en/latest/index.html

[4] FOX software can be downloaded from: https://objcryst-fox.readthedocs.io/en/latest/index.html. Note that using different versions of any software can yield different results. However, in case of FOX, this is primarily because it employs simulated annealing method based on a Monte Carlo algorithm that generates random numbers. Conversely, obtaining different numerical results does not imply that the structure has not been solved.

[5] Even though the crystal structure of acetaminophen is well-known, we assume, for the purpose of illustrating the capabilities of direct space techniques, that neither the unit cell nor the crystal structure is known. We limit prior knowledge to the connectivity of atoms in the molecule, which is well-known and can be established through non-crystallographic methods.

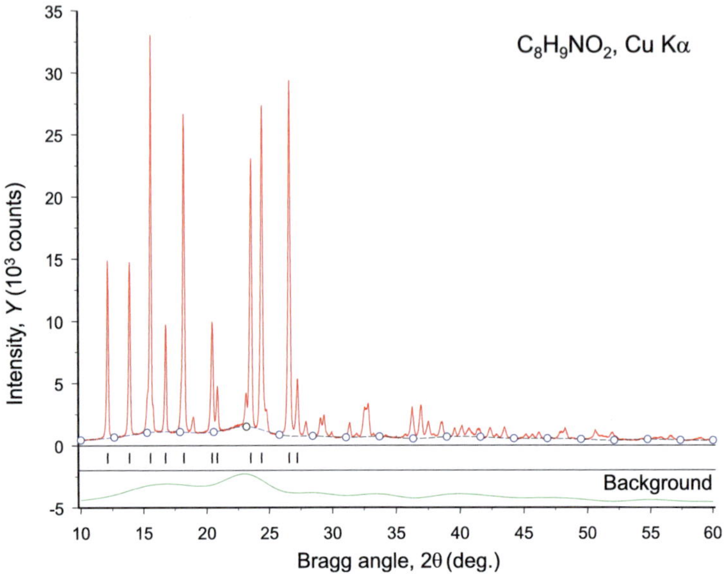

Fig. 25.3 Experimental powder diffraction pattern of acetaminophen (*red solid line*) and the background (*blue dashed line*) determined automatically using the 20-point Bayesian spline interpolation. The *green solid line* at the *bottom* shows the background enlarged by a factor of 2 along the intensity axis revealing the presence of amorphous phases. The *vertical bars* indicate positions of the $K\alpha_1$ components of the Bragg peaks found by the automatic peak-search routine

differential evolution algorithm[6] and another is a new implementation of DICVOL (see Sect. 14.10.2)—the successive dichotomy of volume algorithm, specifically written for FOX. In this section, the latter is used along with Le Bail full-pattern fitting, which confirms the correctness of the indexing solution and establishes space-group symmetry. The profile parameters from the automated and semi-automated Le Bail fitting are also necessary for structure determination, as they are not optimized during the solution process.

Indexing and structure solution in FOX begin by creating a new crystalline phase, importing experimental data to generate a new pattern, and setting required parameters, such as radiation (Cu $K\alpha_{1,2}$) and the polarization factor (0.5). The next step is determining the background, which can be done either by an automatic Bayesian spline interpolation or through user-defined points. In this case, the default setting of 20 points produces a satisfactory background approximation, as shown in Fig. 25.3, and reveals the presence of an amorphous component, likely resulting from the inactive ingredients (see Footnote 1 on page 757).

[6]For more details see: R. Černý, V. Favre-Nicolin, J. Rohlíček, M. Hušák, Z. Matěj, R. Kužel, Expanding FOX: Auto-indexing, grid computing, profile fitting, Commission on Powder Diffraction, Newsletter No. 35, 16 (2007). https://www.iucr.org/__data/assets/pdf_file/0013/21622/CPD_35_total.pdf

Bragg peaks positions in FOX can be determined either by using the automatic peak-search routine incorporated in the FOX[7] or by manually marking peaks on the plot, or both. In this case, the automatic peak search yields 11 strongest, well-resolved peaks below 30° 2θ, marked by vertical lines in Fig. 25.3.[8] It is worth noting that adding weak or overlapped Bragg peaks typically decreases the chances of successful indexing. If manual additions are made, they should be followed by profile fitting to obtain more reliable peak positions.[9]

The ab initio indexing performed using FOX's default settings[10] and the 11 automatically identified peaks (Fig. 25.3) yields a single solution with a very high figure of merit[11] (415.9) in a primitive monoclinic unit cell with $a = 7.102$, $b = 9.390$, $c = 11.708$ Å, $\beta = 97.41°$, and $V = 774.3$ Å^3. This solution is reasonable because, on the one hand, any indexing figure of merit higher than 50 is typically worth further examination, and, on the other hand, the gravimetric density calculated using $Z = 4$ (the multiplicity of a general site in a primitive monoclinic lattice), $\rho = 1.296$ g/cm^3, is perfectly "normal" for an organic compound. Further confirmation of the indexing solution is provided by the Le Bail full-pattern fitting using the obtained monoclinic cell. In FOX, pattern fitting is performed over 20 cycles (20 + 10 when executed from the Cell Explorer window) by sequential refinement of the following groups of parameters:[12]

- Zero shift
- Constant full width at half maximum
- Variable full width at half maximum
- Gaussian–Lorentzian mixing parameter
- Asymmetry
- Displacement and transparency shifts
- Background
- Unit cell

Le Bail profile fitting, using all the parameters listed, results in an R_{wp} of 29.2% after the first 20 cycles. The residual drops rapidly to $R_{wp} = 7.27\%$ over the next 10 cycles, indicating a good fit between the observed and calculated profiles. Notably, profile

[7] Manipulation of Bragg peaks (searching, adding, removing, importing, etc.) and indexing can be accessed from the menu, which is invoked by right-clicking in the FOX window that displays the powder pattern (see the FOX application manual for details).

[8] These peaks can also be imported (loaded) into FOX from the data file Ch25Ex01.txt, available online.

[9] The current version of FOX does not have an option of fitting individual Bragg peaks. This can be done using other applications and then the results can be imported into FOX.

[10] Quick tab in FOX's Cell Explorer window.

[11] The figure of merit used in FOX is called "score" and is derived from M_N (see Sect. 14.4.2) according to the following equation: score $= M_N{}^* d_{last}$, where d_{last} is the d-spacing of the last peak used in the indexing (V. Favre-Nicolin, Private communication, 2008).

[12] By default, all parameters are free variables, but refinement of any or all groups can be suppressed.

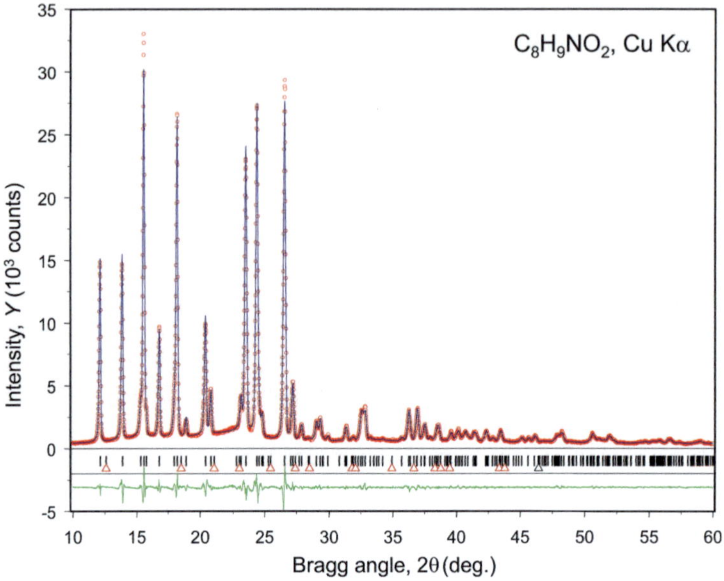

Fig. 25.4 Le Bail fit showing the observed (*red circles*) and calculated (*blue solid line*) diffraction patterns of acetaminophen. The *vertical bars* indicate calculated positions of the $K\alpha_1$ components of Bragg peaks. The *open red triangles* indicate locations of reflections that are forbidden in the space group $P2_1/n$. The *green solid line* at the *bottom* represents the difference between the observed and calculated profiles

fitting executed from the Cell Explorer window (as done in this case) is performed in two stages, with 20 and 10 refinement cycles, respectively. Additionally, FOX automatically determines the 2θ range used for the Le Bail fit, which in this case was about 45° (max $\sin\theta/\lambda = 0.25$).

The resulting R_{wp} is low enough to explore possible space-group symmetry by analyzing indices of Bragg reflections observed in the pattern. This analysis suggests the space group $P2_1/n$, as all observed (nonzero intensity) $0k0$ reflections have k even (i.e., $k = 2n$), and all observed $h0l$ reflections have $h + l = 2n$. Bragg reflections that do not meet these conditions, such as $0k0$ with $k = 2n + 1$ and $h0l$ with $h + l = 2n + 1$ (marked with the triangles in Fig. 25.4), have zero intensity, as they are forbidden by symmetry and, therefore, systematically absent. The subsequent 20 refinement cycles in space group $P2_1/n$ further reduce the R_{wp} to 5.73%. This reduction further supports the correct choice of the space group (noting that this group, or $P2_1/c$ in its standard setting, is the most common group found in organic structures; see Sect. 3.4.4). When other space groups, $P2/m$ or $P2_1/m$ (or their noncentrosymmetric subgroups $P2$, $P2_1$, or Pm) with fewer forbidden reflections are tested, only a slightly lower residual ($R_{wp} = 5.66\%$) is obtained. This small reduction occurs because adding forbidden reflections into the fit introduces additional free intensity variables, naturally reducing discrepancies between the observed and calculated profiles. In contrast, when space groups with incorrect forbidden reflections are tested using the same unit cell, for example, $P2_1/a$ or $P2_1/c$, this leads to much higher residuals, $R_{wp} > 40\%$.

By repeating the full-profile fit several times and using the entire 2θ range (with a maximum sinθ/λ set to 0.325), the following figures of merit are achieved: $R_{wp} = 5.27\%$ and $R_p = 4.79\%$. This corresponds to a very good fit, as shown in Fig. 25.4. There is still room for improvement, such as increasing the number of background points to better describe the amorphous component. However, the goal of this full-profile fitting is not to obtain the most precise integrated intensities, but rather to establish a reasonable set of profile parameters that will not be optimized during the structure solution process. At this point, these parameters appear acceptable.

25.2 Solving the Crystal Structure

After the unit cell and its symmetry are established—achieved with great confidence in this case due to the high indexing figure of merit and an excellent Le Bail fit between the observed and calculated profiles—the structure solution can be attempted. This is done using direct-space methods in two steps. First, an initial structural model (or structure description) is created based on the chemical and physical knowledge of the compound. Second, the model is optimized to minimize a cost function, which, in the case of powder diffraction, is typically the difference between the observed and calculated profiles. It is worth noting that other cost functions may also be used, as discussed later in this section.

25.2.1 *Creating a Model*

The initial model can be created in several ways, depending on the available information about a compound or its class:

- The simplest model consists of a set of atoms of the correct types and quantities, determined from the known or estimated chemical composition and/or gravimetric density of a material. This type of model is typically used for intermetallic or relatively simple inorganic compounds and may include minimum interatomic distance constraints, known as anti-bump restrictions.
- The model can be further improved by specifying known types of polyhedra or other building blocks, such as groups of polyhedra (e.g., a pair of tetrahedra in P_2O_7), solvent molecules, etc. This approach requires more knowledge and certain assumptions about the structure, but specifying the building blocks can significantly reduce the number of parameters to optimize. It is particularly useful for modeling complex intermetallic compounds, inorganic frameworks, and metal–organic compounds.

Fig. 25.5 The numerical scheme identifying all atoms in the molecule of acetaminophen. The *dashed ellipses* encircle two flat rigid groups and the solid rectangle encloses a common bond around which the rigid groups may rotate relatively to one another

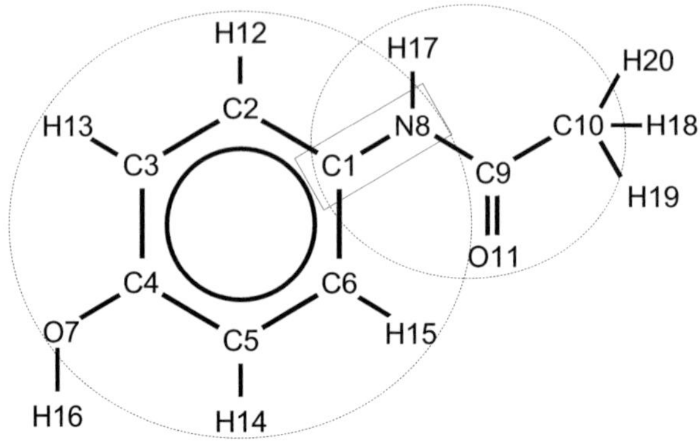

- Lastly, if the connectivity of whole molecules is known, it can be incorporated into the initial model. The molecule may be rigid (e.g., benzene), consist of several rigid groups (e.g., a sugar molecule), or have no rigid groups (e.g., hexane). These types of models are commonly used for organic or metal–organic compounds.

Similar to constrained Rietveld refinement (see Sect. 15.7.3), using a rigid-body approach significantly reduces the number of parameters to optimize, which, in turn, speeds up the process of finding the correct structure. This is especially useful for solving complex structures in a reasonable amount of time. For example, a molecule consisting of ten atoms treated independently would require optimization of 30 coordinate parameters (3 per atom). However, treating the same molecule as a rigid body reduces this to just six parameters: three for the location of the center and three for the orientation of the molecule in the unit cell—a fivefold reduction. When dealing with flexible molecules or groups, the rigid-body approach cannot be used, but restraints can be applied instead, which increases the number of observations. While restraints add to the computational time, they also help find the minimum in the global optimization process more quickly, requiring fewer trial models.

The acetaminophen molecule can be defined by two rigid groups: one consisting of atoms O7, C1 through C6, H12 through H15 and N8, and the second consisting of atoms N8, C9, C10, O11, and H17 as shown in Fig. 25.5. The two groups, encircled by the dashed ellipses, do not include the hydrogen atoms from the hydroxyl (H16) and the methyl group (H18, H19, H20), as they can freely rotate around the C4–O7 and C9–C10 bonds, respectively. Both rigid groups can also rotate freely around the common C1–N8 bond, which is enclosed within a solid rectangle. By ignoring the hydrogen atoms in the freely rotating functional groups, there are seven parameters to optimize: six for one of the rigid groups and one additional parameter that defines the conformation of the molecule, expressed as the rotation angle (φ) between the two rigid groups. Therefore, instead of optimizing or refining all atoms independently—which would require a total of 33 coordinate parameters for 11 nonhydrogen atoms—the same result can be achieved using only seven independent variables. When molecules are relatively simple, like acetaminophen, a restrained model can also be optimized quickly, as demonstrated in this example.

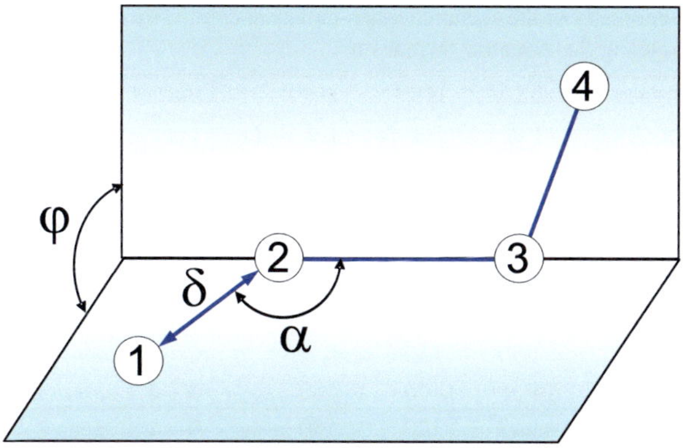

Fig. 25.6 Illustration of how to derive the location of a new atom (atom 1) from the triplet of known atoms (atoms 2, 3, and 4) by specifying the distance between atoms 1 and 2 (δ_{12}), the angle α_{123}, and the torsion angle φ_{1234}. The torsion or dihedral angle describes the conformation of four atoms 1, 2, 3, and 4, and it is defined as the angle between the two planes formed by atoms 1, 2, 3 and 2, 3, and 4, respectively

There are several ways to define the geometry (known or expected) of a molecule, group, polyhedron, or any other building unit. The simplest method is to specify coordinates of each atom using crystallographic, Cartesian, spherical, or other coordinate systems. This is easily done when the unit is incorporated from a known structure. However, when the knowledge of the unit's geometry is based on analogy with similar compounds and consists only of expected bond lengths and angles, this approach becomes much more complex and can be time-consuming. In such cases, the Z-matrix formalism can be applied to describe the structural unit using only bond lengths and angles. This relies on the fact that, once the locations of any three atoms are known, the position of a fourth atom can be uniquely defined by its distance, bond angle, and torsion angle relative to the known atoms, as illustrated in Fig. 25.6.

The locations of the first three atoms (as shown in the first three rows of Table 25.1) are defined as follows:

- The first atom (C1) is placed randomly, for example, at the origin (0,0,0).
- The second atom (C2) is positioned at a specified distance (δ_{C2-C1}) from the first, typically along the *X*-axis.
- The third atom (C3) is placed at a certain distance (δ_{C3-C2}) from the second atom or alternatively from the first, forming a specified angle ($\alpha_{C3-C2-C1}$), for instance, in the *XY* plane.

The fourth and following atoms are positioned using distances, angles, and torsion angles, with any of the three already defined atoms as a reference. As shown in Table 25.1, the geometry of the acetaminophen molecule can be specified as follows: the first column lists the label of a new atom, based on the labeling scheme from Fig. 25.5, the second column shows labels of the triplet used to define the

Table 25.1 The geometry of the acetaminophen molecule defined by distances (δ), angles (α), and torsion angles (φ) using the Z-matrix formalism

Atom	Reference	δ (Å)	α (°)	φ (°)	Entries in the Z-matrix file[a]
I (new)	Atoms J,K,L	I–J	I–J–K	I–J–K–L	Type J δ K α L φ
C1	–	–	–	–	`C 1`
C2	C1	1.39	–	–	`C 1 1.39`
C3	C2 C1	1.39	120	–	`C 2 1.39 1 120.`
C4	C3 C2 C1	1.39	120	0	`C 3 1.39 2 120. 1 0.`
C5	C4 C3 C2	1.39	120	0	`C 4 1.39 3 120. 2 0.`
C6	C5 C4 C3	1.39	120	0	`C 5 1.39 4 120. 3 0.`
O7	C4 C3 C2	1.38	120	180	`O 4 1.38 3 120. 2 180.`
N8	C1 C2 C3	1.43	120	180	`N 1 1.43 2 120. 3 180.`
C9	N8 C1 C2	1.34	130	60	`C 8 1.34 1 130. 2 60.`
C10	C9 N8 C1	1.51	115	180	`C 9 1.51 8 115. 1 180.`
O11	C9 N8 C10	1.22	123	180	`O 9 1.22 8 123. 10 180.`
H12	C2 C1 C6	0.93	120	180	`H 2 0.93 1 120. 6 180.`
H13	C3 C2 C1	0.93	120	180	`H 3 0.93 2 120. 1 180.`
H14	C5 C4 C3	0.93	120	180	`H 5 0.93 4 120. 3 180.`
H15	C6 C5 C4	0.93	120	180	`H 6 0.93 5 120. 4 180.`
H16	O7 C4 C3	0.82	110	180	`H 7 0.82 4 110. 3 180.`
H17	N8 C9 C10	0.82	115	0	`H 8 0.82 9 115. 10 0.`
H18	C10 C9 N8	0.93	109	60	`H 10 0.93 9 109. 8 60.`
H19	C10 C9 N8	0.93	109	180	`H 10 0.93 9 109. 8 180.`
H20	C10 C9 N8	0.93	109	300	`H 10 0.93 9 109. 8 300.`

[a] First and second rows of the file (not shown in the table) contain the name (acetaminophen) and the number of atoms (20), respectively. Sub-columns labeled "J," "K," and "L" show sequential number of an atom (row) in this list and correspond to atoms J, K, and L in the second column "Reference Atoms J,K,L"

corresponding atom in the first column, and the following three columns describe the distances, angles, and torsion angles, respectively. The last column illustrates the corresponding Z-matrix entries in the Z-matrix file used in this example to import the acetaminophen geometry into FOX for structure determination.

The Z-matrix file can be easily created manually using Table 25.1 as a guide when dealing with simple molecules or building blocks. However, when structural units are complex and their geometry is adopted from other known structures, the Z-matrix can be generated automatically from the atomic coordinates of a different structure using tools, such as FOX. To do this, a known structure containing the desired molecule or group in CIF format can then be opened in FOX (or other suitable application) and the desired molecule or group saved as a Z-matrix file in one of the two available formats.[13]

[13] The format used here is the Fenske-Hall Z-matrix. The alternative format uses atom names instead of sequential numbers of atom and includes an additional column describing atom types (chemical elements). The standard file extensions for these two formats are .fhz and .zmat, respectively.

The geometry of acetaminophen used here is idealized and is based on parameters from the orthorhombic polymorph of acetaminophen.[14] The torsion angle C9–N8–C1–C2 between two rigid groups was set arbitrarily to 60°. The geometry of acetaminophen, with nearly the same accuracy, can also be derived from bond lengths and angles found in similar functional groups, which are very well-known in organic chemistry. Alternatively, it can be obtained from sources such as the Cambridge Structure Database or various handbooks.

25.2.2 Optimizing the Model (Solving the Structure)

Solving the crystal structure using this method involves searching for the location, orientation, and conformation of the structural units that form the molecular model, which best matches the observed data. This is achieved by minimizing a specific cost function. In FOX, the cost function used in conjunction with powder diffraction is the difference between observed and calculated profiles, represented as the weighted profile residual (R_{wp}), defined in Eq. (15.19). The corresponding Bragg residual (R_B, Eq. 15.20), based on the integrated intensities obtained from either a single-crystal or powder diffraction experiment (i.e., "observed" intensities obtained from Le Bail full-pattern decomposition), is also well-suited as a cost function. In addition to, or even instead of, experimental diffraction data, other cost functions can be used. These include restraints on distances and angles, anti-bump distances keeping proper intermolecular contacts, and bond valence sums[15] defining polyhedra. In principle, the optimization of the crystal structure can be performed without experimental diffraction data. For example, restraints, anti-bump distances, and bond valence sum cost functions—serving as observed data in this case—can be used to optimize a hypothetical structure geometrically, such as when a different metal atom replaces the one in the original structure. This process also assumes optimization of the unit cell parameters.

The generated Z-matrix can be imported into FOX following indexing and the Le Bail fit.[16] Atoms imported from the Z-matrix are automatically converted into Cartesian coordinate system. When the structural model is exported, for example, into a CIF file, the Cartesian atomic coordinates are converted into crystallographic fractional coordinates. FOX automatically generates restraints on bond lengths and angles (but not torsion angles) upon import, simplifying the start of the optimization process for the restrained model. If a rigid-body approach is needed, rigid groups

[14] M. Haisa, S. Kashino, H. Maeda, The orthorhombic form of *p*-hydroxyacetanilide. Acta Cryst. **B30**, 2510 (1974). https://doi.org/10.1107/S0567740874007473

[15] See Footnote 31 on page 696, Chap. 20.

[16] The Z-matrix file in Fenske-Hall format (Ch25Ex02.fhz), shown in the rightmost column of Table 25.1, is available online. If the reader has not performed the indexing and profile fitting, the corresponding file, Ch25Ex02.xml, is also accessible online and can be opened in FOX before importing the model from the Z-matrix file.

must be imported or created, and appropriate restraints manually setup to replace the automatically generated ones. This is particularly useful for complex structures with large rigid groups, as it minimizes the number of optimized parameters and speeds up the structure solution process or ensures faster convergence. However, since the acetaminophen molecule is relatively simple and the diffraction data are of high quality, setting up rigid-body optimization may take more time and effort compared to using the automatically generated restraints.

As an exercise, the reader is encouraged to import both rigid groups separately[17] and perform structure solution without restraining the configuration of the entire molecule, such as the C1–N8 bond length and the corresponding angles. During optimization, both the optimized structure and the powder data (experimental and calculated) can be monitored in real time using Crystal and Powder Pattern windows, respectively. It is quite engaging to observe the animated changes in the crystal structure and the improvements in the profile fit as the optimization progresses. For example, when two independent rigid-body groups (C1–O7 and N8–O11) are optimized independently, the optimization may get trapped in a false minimum,[18] where the methyl group (C10), instead of the amino group (N8), is positioned next to the phenyl ring (C1). After some time, the optimization shifts into another minimum, with the carboxyl group (O11) next to the phenyl. Finally, after a few more seconds of bouncing between false minima, the global minimum is reached with the amino group correctly attached to the phenyl ring.

Despite the success of the solution using rigid bodies, we will now demonstrate the same process using a restrained model. Once again, the bond lengths and corresponding angles are generated automatically, minimizing the required input. However, the torsion angles are not automatically restrained and are allowed to vary freely during the optimization. If there is a strong rationale, restraints can be applied to the torsion angles. It is worth mentioning that, in this particular case, only torsion angles that define the relative orientation of the rigid groups, as well as the orientation of the hydroxyl (O7) and methyl (C10) groups are not restrained. All other torsion angles are restrained through the bond length and bond angles. It is important to note that optimizing the hydroxyl and methyl groups at this stage of the structure-solving process is unreliable due to their minimal impact on the cost function (R_{wp}). However, we allow the optimization to proceed, as no assumptions can be made about their orientations of these groups at this stage.

To proceed with the optimization process after importing the model's Z-matrix, the following parameters and options should be set:

[17] Files Ch25Ex02a.fhz and Ch25Ex02b.fhz containing acetaminophen rigid groups are available online.

[18] Since the optimization process and the initial model are randomized, the correct orientation of both groups may be achieved on the first attempt—with some luck. However, multiple attempts and further optimization cycles (at least 10^6) may be required to observe the false "alternative" solutions described in this paragraph.

1. In the Crystal tab:

   (a) Disable dynamic occupancy corrections. This option should only be disabled if there is certainty that the optimized structural units cannot share common atoms or be located on the symmetry elements.[19]
   (b) Randomizing the model is recommended (though not mandatory), as the initial model derived from the Z-matrix is, by default, placed in the origin of the unit cell, interacting with the center of symmetry.
   (c) Optionally, open the Crystal window to observe the animated structure during optimization.

2. In the Powder Diffraction tab:

   (a) Set an appropriate global isotropic displacement parameter (B_{iso}) that applies all atoms but do not optimize it at this stage. For organic compounds, B_{iso} typically ranges from 3 to 6 Å^2. In this refinement, a global B_{iso} of 4 Å^2 was used.
   (b) Optionally, open the Powder Pattern window to monitor the convergence of the experimental and calculated profiles.

3. Using the Objects menu, create a "New Monte Carlo Object" and a "Global Optimization" window will appear, in which:

   (a) Select Optimized Objects, first Crystal and then Powder Pattern.
   (b) Leave the Parallel Tempering optimization algorithm as is. The alternative is Simulated Annealing.[20]
   (c) Set the number of trials per run to 500,000 instead of the default 10,000,000, which is excessive for such a simple structure.
   (d) Leave other parameters and options at their default values, or refer to the FOX's manual for their meaning.
   (e) Save the project to allow restarting the solution process from this point without going through the setup process again.
   (f) Start a single optimization run and monitor the minimization of the cost function in the new window that appears, as well as structural changes and fitting of the observed and calculated patterns in the corresponding windows. R_{wp} and R_p residuals can be monitored in the Powder Diffraction tab.

---

[19] Dynamic occupancy correction is an advanced feature in FOX that allows adjustment of the occupancies of superimposed atoms, such as when the optimized structural units are located on finite symmetry elements or share atoms, **for instance**, when using polyhedra to model extended frameworks. When two or more atoms overlap (i.e., are too close to each other), their occupancies are set to 1/N, where N is the number of overlapped atoms. It is reasonable to disable this correction for the acetaminophen molecule since it has no own symmetry elements and clearly cannot overlap with itself. Disabling this option can lead to an improved cost function, since uncorrected overlapped molecules will have a stronger impact on the residuals, although this is not mandatory.

[20] See details of the algorithms in Sect. 15.1.3, FOX application manual, and references therein.

Upon the completion of the optimization (which can be stopped at any moment), the R_{wp} and R_p residuals are 10.81% and 8.68%, respectively. These low residuals, along with a very good match between the observed and calculated profiles and the absence of unrealistically short intermolecular contacts, provide solid confirmation of the solution's correctness. The resulting structure can be saved in CIF format and further analyzed using different applications. The resulting tilt between the two acetaminophen rigid groups, or the C9–N8–C1–C2 torsion angle, is 17° (or −17°) if a second molecule that is related by the center of inversion was saved. This agrees well with the same torsion angle in the orthorhombic modification (18°) and lies within the range of torsion angles observed in other known structures of acetaminophen solvates that can be found in the Cambridge Structure Database.

Small discrepancies between the observed and calculated profiles occur mainly in the middle of several of the strongest peaks, which may be due to inadequate peak-shape parameters, a complex amorphous background, and/or preferred orientation. Therefore, further optimization was undertaken starting from the already optimized structure, but this time including the global B_{iso} and the preferred orientation along the [101] axis in the optimization. The latter corresponds to the indices of the worst-fitted strong peak. It should be noted that the peak-shape parameters, including the background spline, are not adjusted during the structure optimization and must be refined separately using the Le Bail profile fitting procedure and Bayesian optimization, respectively. Repeating the optimization process improves the fit, resulting in $R_{wp} = 8.34\%$ and $R_p = 6.58\%$. The global B_{iso} increases to 6.14 Å^2, yet this value remains within acceptable limits. The torsion angle C9–N8–C1–C2 changes slightly (to 15°), and there is no significant preferred orientation along the [101] axis. If there are any doubts about the correctness of the structure or about assumptions regarding the initial model, the Fourier map can be calculated and displayed as three-dimensional contour plots using the Crystal window.

Further improvement of the fit is possible; however, it makes little sense to continue, as the fit is already quite good, as shown in Fig. 25.7.We can now proceed to the next step—the Rietveld refinement—which should allow us to adjust the structural and profile parameters together. The optimized structure of acetaminophen, saved in a CIF file, is used as the initial model in the Rietveld refinement.

As noted earlier, this relatively simple example is intended as an introduction to the application of direct-space methods for solving crystal structures, and in particular, to using the freely available FOX software. The latter has been used extensively over the last couple decades to solve the structures of a wide variety of chemical compounds. A long, though far-from-complete, list of references for the crystal structures solved using FOX can be found online.[21] Selected recent, nontrivial examples of powder structure determination using a variety of methods are provided in the Additional Reading section at the end of this chapter.

[21] https://objcryst-fox.readthedocs.io/en/latest/biblio-structures.html

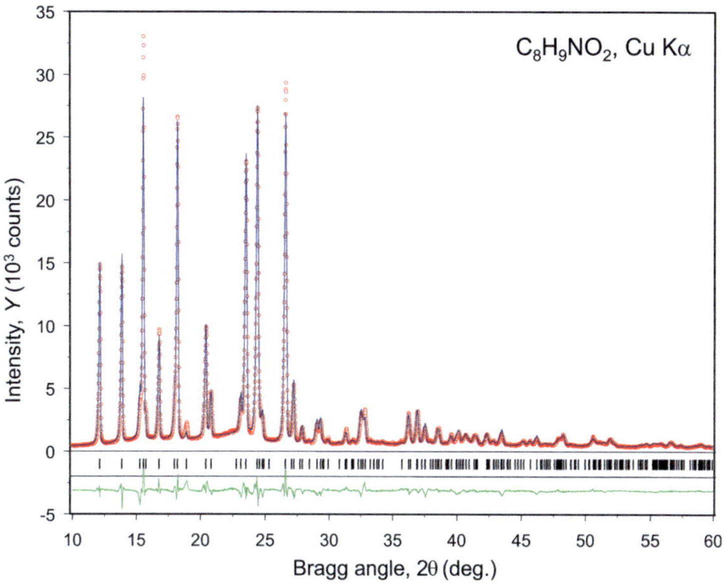

Fig. 25.7 Result of solving the acetaminophen structure using FOX parallel tempering optimization, showing the observed (*red circles*) and calculated (*blue solid line*) diffraction patterns. The *vertical bars* indicate the positions of the $K\alpha_1$ components of calculated reflections. The oscillating curve at the *bottom* shows the difference between the observed and calculated profiles

25.3 Restrained Rietveld Refinement

The Rietveld refinement of acetaminophen is performed using GSAS and EXPGUI applications to apply restraints on distances, angles, and planar groups. It begins with the import of the crystal structure solved using FOX and the experimental diffraction pattern.[22] The refinement is then carried out in three steps:

1. Full-profile fitting without a structural model (also known as the full-pattern decomposition), which can be done using either the Le Bail method or the method employing equally weighted structure factors available in GSAS. This step is needed to obtain accurate starting values of profile parameters because of the complexity of the diffraction pattern arising from the amorphous background and a substantial peak broadening in the presence of multiple peak overlap. Profile parameters obtained from the Le Bail fit while solving the structure using FOX cannot be easily imported into GSAS (at least not all of them) since peak-shape functions used in both applications are somewhat different.

[22]The corresponding files in CIF and GSAS formats are available online as *Ch25Ex03a.cif* and *Ch25Ex01_CuKa.raw*.

2. Once the profile parameters, including the background function, are satisfactory, the next step is to switch to Rietveld refinement using a structural model. The latter was established quite accurately during the structure solution step and is therefore not refined at this stage, except for the overall isotropic displacement parameter. Profile and background functions are further optimized. Additionally, other nonstructural features may be explored or tested before refining the crystal structure itself. These are, for example, preferred orientation, absorption, and porosity effects and the type of the peak-shape function.
3. The final step is the Rietveld refinement of the crystal structure along with other nonstructural parameters. In this step, the refinement starts with setting up and adjusting the geometric restraints and concludes with the crystallographic and chemical analyses of the resulting structure.

The Le Bail profile fitting began with defining the background due to its complexity. The background curve was initially extrapolated from about 20 manually selected points using a shifted-Chebyshev polynomial with 12 coefficients. The initial values of the profile parameters were taken from the instrumental parameters file,[23] determined using the LaB_6 standard and a modified pseudo-Voigt peak-shape function.[24] Initially, the coefficient X of the Lorentzian grain-size broadening was adjusted to approximately match the peak widths. During the first step, only this parameter was refined. Then, other profile parameters were gradually released. First was the transparency shift (acetaminophen is a low-absorbing material and this correction is greater than sample displacement), followed by the unit cell parameters, background coefficients, the coefficient Y of strain broadening, and asymmetry parameters. The latter did not converge and were fixed again for the time being. The least-squares refinement was extremely unstable; therefore, automatic adjustment of the Marquardt damping (see Sects. 15.5 and 20.3) was enabled. In this process, the damping parameter is estimated from the correlation coefficients to improve convergence, but at the cost of a longer (often very long) refinement. Significant mismatch between the observed and calculated intensities and peak shapes at this point resulted in a relatively high R_{wp} of about 14%. Yet, the fit was good enough, so that the structural model was added into the refinement.

In the next step, the Rietveld refinement of the acetaminophen crystal structure was conducted. However, due to a poor profile fit, new parameters had to be introduced gradually and carefully, while monitoring correlation coefficients and the parameters themselves to ensure meaningful values. The crystal structure was not refined in this step, except for the isotropic displacement parameters, which were

[23] The file is available online as *D8advance.prm*.

[24] This function (number 3 in GSAS) has 19 profile coefficients as parameterized in: P. Thompson, D. E. Cox, and J.B. Hastings, Rietveld refinement of Debye–Scherrer synchrotron X-ray data from Al_2O_3, J. Appl. Cryst. **20**, 79 (1987), https://doi.org/10.1107/S0021889887087090, and with asymmetry correction by L.W. Finger, D.E. Cox, and A.P. Jephcoat, A correction for powder diffraction peak asymmetry due to axial divergence, J. Appl. Cryst. **27**, 892 (1994), https://doi.org/10.1107/S0021889894004218, see Sects. 8.5.1 and 8.5.2 for details.

constrained to be identical for all atoms. Thus, the following parameters were added to the refinement: asymmetry, specimen displacement, empirical extension of microstrain anisotropy (see Eq. 8.34 on page 199, Chap. 8), and preferred orientation in the spherical harmonics approximation. Despite these additions, the fit improved slowly, still showing discrepancies in the peak shapes, as reflected by a relatively high profile residual $R_{wp} = 12.85\%$, compared to the structural $R_B = 7.52\%$. At this point, a different peak-shape function (number 4 in GSAS), which employs anisotropic microstrain broadening suggested by P. Stephens (see Eq. 8.35 on page 199, Chap. 8), was applied. Additionally, the background function was expanded to 36 coefficients, the maximum possible number. After numerous refinement cycles, these changes resulted in a much better agreement between the calculated and observed profiles, with $R_{wp} = 8.99\%$, $R_p = 6.91\%$, and $R_B = 6.68\%$.

Finally, the refinement of the crystal structure was undertaken by releasing the atomic coordinate parameters and setting restraints for bond lengths and angles for all 20 bonds and 31 bond angles.[25] The bond lengths were restrained to be within 0.005 and 0.002 Å from the expected values for nonhydrogen and for hydrogen atoms, respectively. The angle restraints were initially set to 1°. The expected values for bond lengths and angles were those used in the structure solution.[26] Two flat, rigid groups of acetaminophen were also restrained by applying planar restraints to keep atoms within 0.02 Å of the plane.

From this point onward, the refinement converged much faster, and the structural parameters were analyzed. It appeared that the hydroxyl group (O7–H16) does not form a hydrogen bond[27] even though it could potentially form one with the carboxyl oxygen atom O11 from the neighboring molecule. This was corrected by restraining the H17···O11 distance to be within 0.001 Å from the estimated $\delta_{H\cdots O}$ value of 1.85 Å. The latter was set to such a small value in order to force the hydroxyl group to adopt the desired orientation. However, in further refinements, this restraint was relaxed to 0.05 Å, yet the hydrogen atom remained perfectly still and in the correct position. This shows that light atoms are often trapped in false minima, and additional effort and chemical knowledge are required to prevent or correct this.

Initially, the bond angles around the C1 atom, C2–C1–N8 and C6–C1–N8, were set to the ideal 120° instead of the 124° and 116° observed in other acetaminophen structures. Therefore, the restraints on these angles, as well as on some other angles, were relaxed to 4°. Further refinement brought these angles to within 0.3° of the values observed in other structures. This serves as additional confirmation of the reliability of the Rietveld refinement, assuming everything is done correctly.

In the final Rietveld refinement, the following parameters were refined:

[25] Due to the complex and time-consuming nature of the preliminary refinement, readers wishing to focus on the application of restraints may use the up-to-this point GSAS EXP file, available online as *Ch25Ex03.exp*.

[26] These values were adopted from the orthorhombic modification of acetaminophen but are also well known for organic compounds. They can be viewed in the Restraint Window in FOX.

[27] As a rule of thumb, all hydroxyl and amino groups tend to form hydrogen bonds in the presence of electron donors.

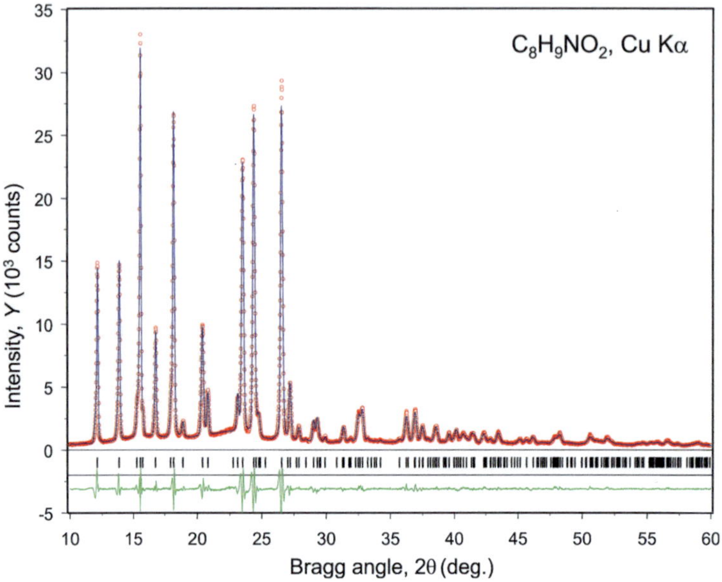

Fig. 25.8 The results of the Rietveld refinement of the acetaminophen structure using GSAS, showing the observed (*red circles*) and calculated (*blue solid line*) diffraction patterns. The *vertical bars* indicate the positions of the $K\alpha_1$ components of the calculated Bragg peaks. The line at the *bottom* shows the difference between the observed and calculated profiles

- Scale factor
- Thirty-six background coefficients
- Seventeen of 21 profile parameters; four were fixed as they approached zero
- Atomic coordinates for all atoms retrained by 21 distance, 31 angular, and two planar restraints
- Isotropic displacement parameters: one for each of the rigid groups and one for the OH group

The final refinement converged to $R_{wp} = 7.71\%$, $R_p = 5.80\%$, and $R_B = 3.48\%$. The observed and calculated profiles along with their difference, as shown in Fig. 25.8, exhibit a very good fit in overall. However, minor discrepancies in the peak shapes are still noticeable on the difference curve in the vicinities of several strong peaks. Therefore, the Rietveld refinement of this structure was also conducted using a fundamental parameters approach in TOPAS.[28] This resulted in better residuals $R_{wp} = 6.68\%$, $R_p = 5.14\%$, and $R_{Bragg} = 2.66\%$, mainly due to a more realistic peak-shape function, as the fundamental parameter approach was used to model the shape of the peaks. The crystal structure (atomic parameters) was not refined.

[28] Bruker AXS: TOPAS V3: General profile and structure analysis software for powder diffraction data. User's Manual, Bruker AXS, Karlsruhe, Germany (2005).

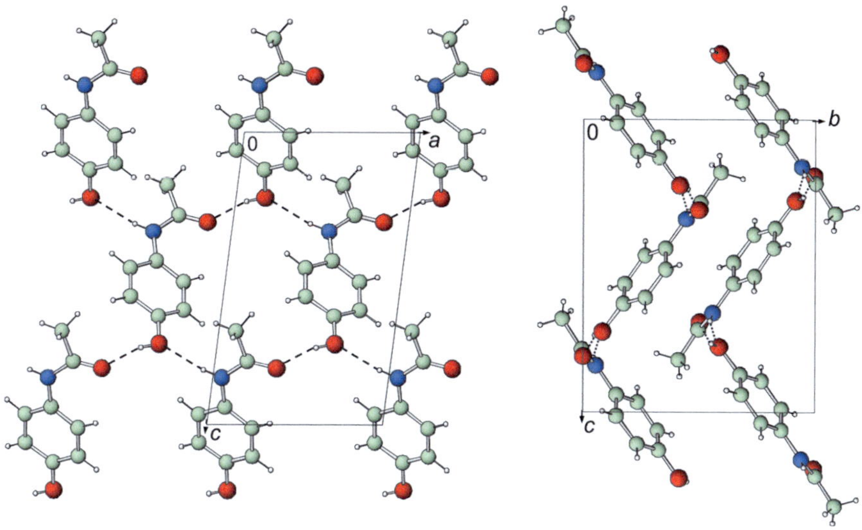

Fig. 25.9 The layer of H-bonded acetaminophen molecules (*left*) and the packing of zigzag layers (*right*). Oxygen atoms are *red*, nitrogen *blue*, and carbon *green*. Hydrogen atoms are depicted as *small white circles* and hydrogen bonds as *dotted lines*

Despite the clearly better fit, we do not discuss the refinement settings and results in detail, as it is unlikely that TOPAS is available to all our readers.

The fully refined structure[29] is shown in Fig. 25.9, illustrating the arrangement of the acetaminophen molecules linked in two-dimensional layers by hydrogen bonding and the stacking of these zigzag layers perpendicular to the *b*-axis. Finally, it should be noted that the structure shown in Fig. 25.9 is in good agreement with multiple published single-crystal data describing the crystal structure of the monoclinic polymorph of acetaminophen.[30]

25.4 Chapters 15–25: Additional Reading

1. W. I. F. David and D. S. Sivia, Extracting integrated intensities from powder diffraction data, in: Structure determination from powder diffraction data. IUCr monographs on crystallography 13. W. I. F. David, K. Shankland, L.B. McCusker, and Ch. Baerlocher, Eds., Oxford University Press, Oxford (2002)

[29] CIF file with the final structure is available online as *Ch25Ex03b.cif*.

[30] M. Haisa, S. Kashino, R. Kawai, and H. Maeda, The Monoclinic Form of p-Hydroxyacetanilide, Acta Cryst. **B32**, 1283 (1976), https://doi.org/10.1107/S0567740876012223; R. Anitha, M. Gunasekaran, S. Suresh Kumar, S. Athimoolam, and B. Sridhar, Single crystal XRD, vibrational and quantum chemical calculation of pharmaceutical drug paracetamol: A new synthesis form, Spectrochim. Acta, Part A, **150**, 488 (2015), https://doi.org/10.1016/j.saa.2015.05.091

2. T. Wessels, Ch. Baerlocher, L. B. McCusker, and W.I.F. David, Experimental methods for estimating the relative intensities of overlapping reflections, in: Structure determination from powder diffraction data. IUCr monographs on crystallography 13. W. I. F. David, K. Shankland, L.B. McCusker, and Ch. Baerlocher, Eds., Oxford University Press, Oxford (2002)
3. R. Peschar, A. Etz, J. Jansen and H. Schenk, Direct methods in powder diffraction—basic concepts, in: Structure determination from powder diffraction data. IUCr monographs on crystallography 13. W. I. F. David, K. Shankland, L.B. McCusker, and Ch. Baerlocher, Eds., Oxford University Press, Oxford (2002)
4. C. Giacovazzo, A. Altomare, M. C. Burla, B. Carrozzini, G. L. Cascarano, A. Guagliardi, A. G. G. Moliterni, G. Polidori and R. Rizzi, Direct methods in powder diffraction—applications, in: Structure determination from powder diffraction data. IUCr monographs on crystallography 13. W. I. F. David, K. Shankland, L.B. McCusker, and Ch. Baerlocher, Eds., Oxford University Press, Oxford (2002)
5. M. A. Estermann and W. I. F. David, Patterson methods in powder diffraction: maximum entropy and symmetry minimum function techniques, in: Structure determination from powder diffraction data. IUCr monographs on crystallography 13. W. I. F. David, K. Shankland, L.B. McCusker, and Ch. Baerlocher, Eds., Oxford University Press, Oxford (2002)
6. C. J. Gilmore, K. Shankland and W. Dong, A maximum entropy approach to structure solution, in: Structure determination from powder diffraction data. IUCr monographs on crystallography 13. W. I. F. David, K. Shankland, L.B. McCusker, and Ch. Baerlocher, Eds., Oxford University Press, Oxford (2002)
7. J. Ruis, Solution of Patterson-type syntheses with the direct methods sum function, in: Structure determination from powder diffraction data. IUCr monographs on crystallography 13. W. I. F. David, K. Shankland, L.B. McCusker, and Ch. Baerlocher, Eds., Oxford University Press, Oxford (2002)
8. K. Shankland and W. I. F. David, Global optimization strategies, in: Structure determination from powder diffraction data. IUCr monographs on crystallography 13. W. I. F. David, K. Shankland, L.B. McCusker, and Ch. Baerlocher, Eds., Oxford University Press, Oxford (2002)
9. P. G. Bruce and Y. G. Andreev, Solution of flexible molecular structures by simulated annealing, in: Structure determination from powder diffraction data. IUCr monographs on crystallography 13. W. I. F. David, K. Shankland, L.B. McCusker, and Ch. Baerlocher, Eds., Oxford University Press, Oxford (2002)
10. L. B. McCusker and Ch. Baerlocher, Chemical information and intuition in solving crystal structures. IUCr monographs on crystallography 13. W. I. F. David, K. Shankland, L.B. McCusker, and Ch. Baerlocher, Eds., Oxford University Press, Oxford (2002)
11. D. Louër, Advances in powder diffraction analysis, Acta Cryst. **A54**, 922 (1998). https://doi.org/10.1107/S0108767398007363

12. D.M. Poojary and A. Clearfield, Application of X-ray powder diffraction techniques to the solution of unknown crystal structures, Acc. Chem. Res. **30**, 414 (1997). https://doi.org/10.1021/ar960143j
13. K.D.M. Harris and M. Tremayne, Crystal structure determination from powder diffraction data, Chem. Mater. **8**, 2554 (1996). https://doi.org/10.1021/cm960218d
14. C. Giacovazzo, Direct methods and powder data: State of the art and perspectives, Acta Cryst. **A52**, 331 (1996). https://doi.org/10.1107/S0108767395013651
15. J.A. Kaduk, Use of the Inorganic Crystal Structure Database as a problem solving tool, Acta Cryst. **B58**, 370 (2002). https://doi.org/10.1107/S0108768102003476
16. V. Favre-Nicolin and R. Černý, FOX, "Free objects for crystallography": a modular approach to ab initio structure determination from powder diffraction, J. Appl. Cryst. **35**, 734 (2002). https://doi.org/10.1107/S0021889802015236
17. R. Černý, V. Favre-Nicolin, FOX: A friendly tool to solve nonmolecular structures from powder diffraction, Powder Diffraction. 20, 359 (2005). https://doi.org/10.1154/1.2135314
18. V. Favre-Nicolin, R. Černý, A better FOX: using flexible modeling and maximum likelihood to improve direct-space ab initio structure determination from powder diffraction, Z. Kristallogr. **219**, 847 (2004). https://doi.org/10.1524/zkri.219.12.847.55869
19. The Rietveld method. IUCr monographs on Crystallography 5, R.A. Young, Ed., International Union of Crystallography, Oxford University Press, Oxford (1993)
20. H.M. Rietveld, Line profiles of neutron powder-diffraction peaks for structure refinement, Acta Cryst. **22**, 151 (1967). https://doi.org/10.1107/S0365110X67000234
21. H.M. Rietveld, A profile refinement method for nuclear and magnetic structures, J. Appl. Cryst. **2**, 65 (1969). https://doi.org/10.1107/S0021889869006558
22. A. K. Cheetham, Structure determination from powder diffraction data: an overview, in: Structure determination from powder diffraction data. IUCr monographs on Crystallography 13, W.I.F. David, K. Shankland, L.B. McCusker, and Ch. Baerlocher, Eds., Oxford University Press, Oxford (2002)
23. Real-Space and Hybrid Methods for Structure Solution from Powders. Commission on Powder Diffraction, Newsletter No. 35, (2007). This and other newsletters available at: https://www.iucr.org/resources/commissions/powder-diffraction/newsletter
24. K. Goubitz, P. Čapková, K. Melánová, W. Mollemana, H. Schenka, Structure determination of two intercalated compounds $VOPO_4{\cdot}(CH_2)_4O$ and $VOPO_4{\cdot}OH{-}(CH_2)_2{-}O{-}(CH_2)_2{-}OH$; synchrotron powder diffraction and molecular modeling, Acta Cryst. **B57**, 178 (2001). https://doi.org/10.1107/S0108768100015603
25. A. J. Mora, A. N. Fitch, B. M. Ramirez, G. E. Delgado, M. Brunelli, J. Wright, Structure of lithium benzilate hemihydrate solved by simulated annealing and

difference Fourier synthesis from powder data, Acta Cryst. B**59**, 378 (2003). https://doi.org/10.1107/S0108768103004634
26. P. Derollez, N. T. Correia, F. Danède, F. Capet, F. Affouard, J. Lefebvre, M. Descamps. Ab initio structure determination of the high-temperature phase of anhydrous caffeine by X-ray powder diffraction, Acta Cryst. **B61**, 329 (2005). https://doi.org/10.1107/S010876810500546X
27. J. Pérez, J. L. Serrano, J. M. Galiana, F. L. Cumbrera, A. L. Ortiz, G. Sánchez, J. García, Structure determination of di-μ-hydroxo-bis[(2-(2-pyridyl)phenyl-κ^2N,C^1)palladium(II)] by X-ray powder diffractometry, Acta Cryst. **B63**, 75 (2007). https://doi.org/10.1107/S0108768106044843
28. M. Rukiah, M. Al-Ktaifani, Poly[(μ^2-2,2-dimethylpropane-1,3-diyl diisocyanide)-μ^2-nitrato-silver(I)]: a powder study, Acta Cryst. **C64**, m170 (2008). https://doi.org/10.1107/S0108270108005143
29. J.S.O. Evans, S. Bénard, P. Yu, R. Clément, Ferroelectric alignment of NLO chromophores in layered inorganic lattices: structure of a stilbazolium metal-oxalate from powder diffraction data, Chem. Mater. **13**, 3813 (2001). https://doi.org/10.1021/cm011155o
30. C. Lupu, J.-G. Mao, J.W. Rabalais, A.M. Guloy, J.W. Richardson, Jr. X-ray and neutron diffraction studies on "$Li_{4.4}Sn$", Inorg. Chem. **42**, 3765 (2003). https://doi.org/10.1021/ic026235o
31. M. Rukiah, J. Lefebvre, O. Hernandez, W. van Beek, M. Serpelloni, Ab initio structure determination of the Γ form of D-sorbitol (D-glucitol) by powder synchrotron X-ray diffraction, J. Appl. Cryst. **37**, 766 (2004). https://doi.org/10.1107/S0021889804016206
32. D. Grebille, S. Lambert, F. Bourée, V. Petrícek, Contribution of powder diffraction for structure refinements of aperiodic misfit cobalt oxides, J. Appl. Cryst. **37**, 823 (2004). https://doi.org/10.1107/S0021889804018096
33. I.Yu. Zavaliy, R. Černý, I.V. Koval'chuck, A.B. Riabov, R.V. Denys, Synthesis and crystal structure of κ-$Zr_9V_4SH_{\sim 23}$, J. Alloys Comp. **404–406**, 118 (2005). https://doi.org/10.1016/j.jallcom.2004.12.168
34. P. Bhojane, A. Le Bail, and P.M. Shirage, A quarter of a century after its synthesis and with >200 papers based on its use, '$Co(CO_3)_{0.5}(OH)\cdot 0.11H_2O$' proves to be $Co_6(CO_3)_2(OH)_8\cdot H_2O$ from synchrotron powder diffraction data, Acta Cryst. **C75**, 61 (2019). https://doi.org/10.1107/S2053229618017734
35. S.J. Hong, J. Lia, and M.A. Subramanian, Crystal structure refinement of magnesium zinc divanadate, $MgZnV_2O_7$, from powder X-ray diffraction data, Acta Cryst. **E77**, 588 (2021). https://doi.org/10.1107/S2056989021004503
36. J. Schoiber, D. Söllinger, V. Baran, T. Diemant, G.J. Redhammer, R. J. Behmc, and S. Pokranta, Resolving the structure of $V_3O_7\cdot H_2O$ and Mo-substituted $V_3O_7\cdot H_2O$, Acta Cryst. **B78**, 637 (2022). https://doi.org/10.1107/S2052520622006473
37. Z.R. Butler, J.A. Kaduk, A.M. Gindhart, and T.N. Blanton, Crystal Structure of Fosfomycin Tromethamine, $(C_4H_{12}NO_3)(C_3H_6O_4P)$, from Synchrotron Powder Diffraction Data and Density Functional Theory, Crystals, **9**(8), 384 (2019), https://doi.org/10.3390/cryst9080384

25.5 Chapters 15–25: Problems[31]

1. The compound $Mn_5Si_3O_{12}$ crystallizes in the space group $Ia\bar{3}d$ with lattice parameter of a = 11.85 Å. The measured gravimetric density ρ is 4.4 g/cm^3. Calculate the number of formula units in the unit cell and the number of atoms of each kind. Additionally, suggest which sites in the unit cell could be occupied by the different types of atoms.
2. The compound $Co_2Mn_3O_8$ crystallizes in the space group $Pmn2_1$ with lattice parameters a = 5.743, b = 4.915, and c = 9.361 Å. Assuming a reasonable density for a 3d metal oxide (3–6 g/cm^3), determine the number of formula units in the unit cell and calculate the X-ray density of the material.
3. Cobalt oxide, CoO, crystallizes in the cubic crystal system, space group $Fm\bar{3}m$, and lattice parameter a = 4.26 Å. The measured gravimetric density of the oxide is ρ = 6.438 g/cm^3. Using only these data, solve its crystal structure by determining positions of atoms that make chemical and physical sense, while ensuring reasonable interatomic distances.
4. The compound $TaMn_2O_3$ crystallizes in the hexagonal crystal system and belongs to the space group P6/mmm with lattice parameters a = 5.321 Å and c = 3.578 Å. The measured gravimetric density of the material is ρ = 6.30 g/cm^3. Using only these data, solve the crystal structure of the material by determining positions of atoms that make chemical and physical sense, while ensuring reasonable interatomic distances.
5. Hexamethylenetetramine (*hmta*) molecule, $C_6H_{12}N_4$, has a tetrahedral configuration, with nitrogen atoms occupying the corners. These nitrogen atoms are linked to each other via six methylene (CH_2) groups, which are located above the midpoints of the edges of the tetrahedron, as shown in Fig. 25.10. The compound crystallizes in the cubic crystal system, space group $I\bar{4}3m$, and a lattice parameter of a = 7.05 Å. The measured gravimetric density is ρ = 1.33 g/cm^3. Assuming that the tetrahedron is ideal and that the C–N distances are 1.49 Å, solve this crystal structure (i.e., determine the coordinates of nonhydrogen atoms) using the data provided, including the symmetry of the lattice and *hmta* molecule. To simplify calculations, consider the following: the distance from the center (X) of the tetrahedron to the C atoms is $\delta_{X\text{–}C}$ = 1.72 Å and to the N atoms, $\delta_{X\text{–}N}$ = 1.49 Å.
6. Consider the powder diffraction pattern shown in Fig. 25.11, which was collected from an intermetallic compound with the approximate stoichiometry $SrSi_2$ using a Rigaku TTRAX rotating anode powder diffractometer using Mo Kα radiation. The density of the alloy was measured in a pycnometer and it is 3.3(1) g/cm^3. The pattern was indexed, and possible space groups were established during the solution of Problem 3 in Chap. 14. Experimental data

[31] Instrumental parameters for GSAS Rietveld refinement are available online in files *InstCuKa.prm* and *InstMoKa.prm*.

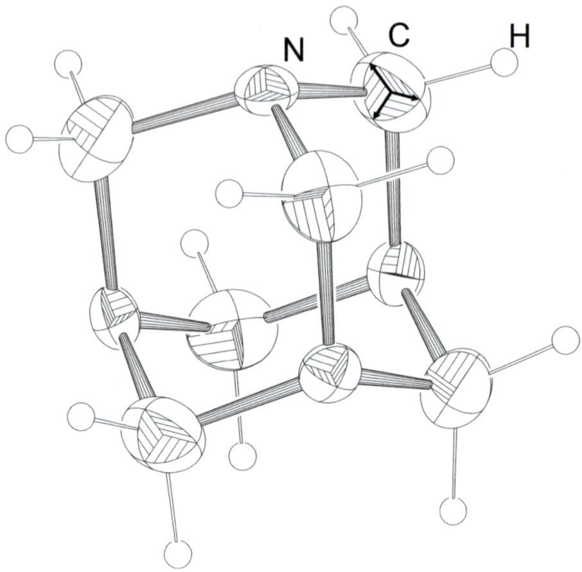

Fig. 25.10 The molecule of hexamethylenetetramine, shown using displacement ellipsoids of carbon and nitrogen atoms

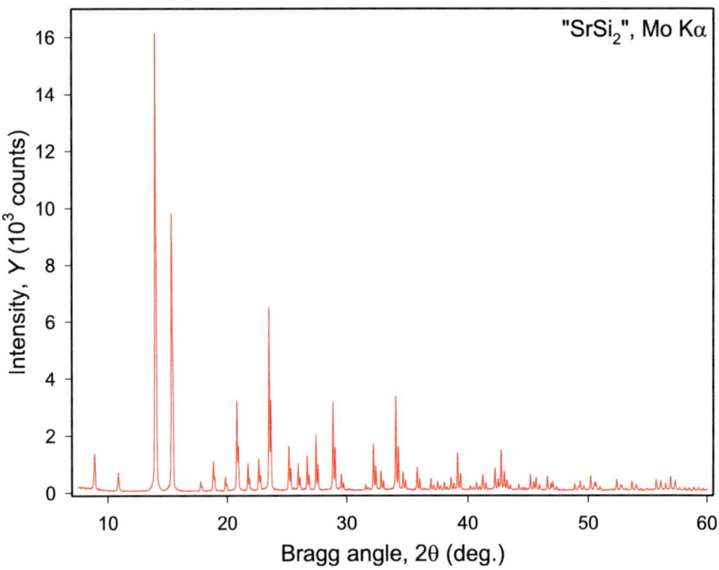

Fig. 25.11 Powder diffraction pattern collected from a ground powder with the approximate stoichiometry $SrSi_2$ on a rotating anode Rigaku TTRAX diffractometer. The data were collected with a step $\Delta 2\theta = 0.015°$

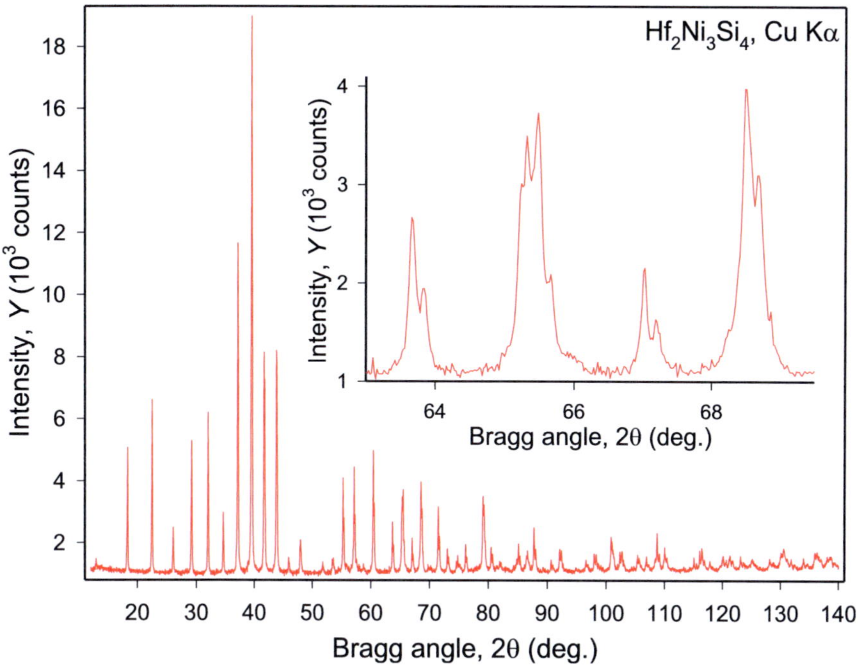

Fig. 25.12 Powder diffraction pattern collected from ground $Hf_2Ni_3Si_4$ powder using an HZG-4a diffractometer. The data were collected with a step size of $\Delta 2\theta = 0.02°$. The inset shows splitting of some Bragg peaks, indicating the need for a relatively large orthorhombic lattice to index this seemingly high-symmetry pattern (low Bragg angle peaks appear regularly spaced). (Data courtesy of Dr. L.G. Akselrud)

can be found in the data files *Ch25Pr06_MoKa.xy* and *Ch25Pr06_MoKa.raw* online. Solve this crystal structure from first principles.

7. Consider the pattern from Problem 1 in Chap. 14 (Fig. 14.25). Powder diffraction data were collected in the range of Bragg angles from 20° to 140° on an HZG-4a powder diffractometer using filtered Cu Kα radiation. The data can be found in the files *Ch25Pr07_CuKa.xy* and *Ch25Pr07_CuKa.raw* online. Solve the crystal structure of this material, given that its gravimetric density, measured pycnometrically, is $\rho = 7.7$ g/cm^3.
8. Consider the powder diffraction pattern collected from ground $Hf_2Ni_3Si_4$ powder, shown in Fig. 25.12. The pattern has been indexed in the orthorhombic crystal system, with the unit cell dimensions $a = 5.18$ Å, $b = 13.65$ Å, and $c = 6.85$ Å. An analysis of the systematic absences indicates that the following groups of reflections have nonzero intensity only at the reflection conditions listed here:

hkl, $h + k = 2n$
$hk0$, h and $k = 2n$

$h0l$, h and $l = 2n$
$0kl$, $k = 2n$
$h00$, $h = 2n$
$0k0$, $k = 2n$
$00l$, $l = 2n$

The pycnometric density is 8.8(5) g/cm^3. Solve the crystal structure of this material without using handbooks or databases. Powder diffraction data can be found in the files *Ch25Pr08_CuKa.xy* and *Ch25Pr08_CuKa.raw* online.

9. The compound $VO(CH_3COO)_2$ crystallizes in the orthorhombic crystal system with $a = 14.066$ Å, $b = 6.878$ Å, $c = 6.925$ Å. Its gravimetric density is less than 2 g/cm^3. The powder diffraction pattern (data files *Ch25Pr09_CuKa.xy* and *Ch25Pr09_CuKa.raw*) is shown in Fig. 25.13. The extracted structure amplitudes (data file *Ch25Pr09_CuKa.hkl*) are also available online. Solve the crystal structure of this material.
10. Complete the structure determination and perform Rietveld refinement of the model of $SrSi_2$, which you solved in Problem 6. The experimental powder diffraction pattern is available online in the file *Ch25Pr10_CuKa.raw*.
11. Complete the solution of the crystal structure and perform Rietveld refinement of the model of $LaNi_{11.4}Ge_{1.6}$ from Problem 7. The experimental powder diffraction pattern is available online in the file *Ch25Pr11_CuKa.raw*.

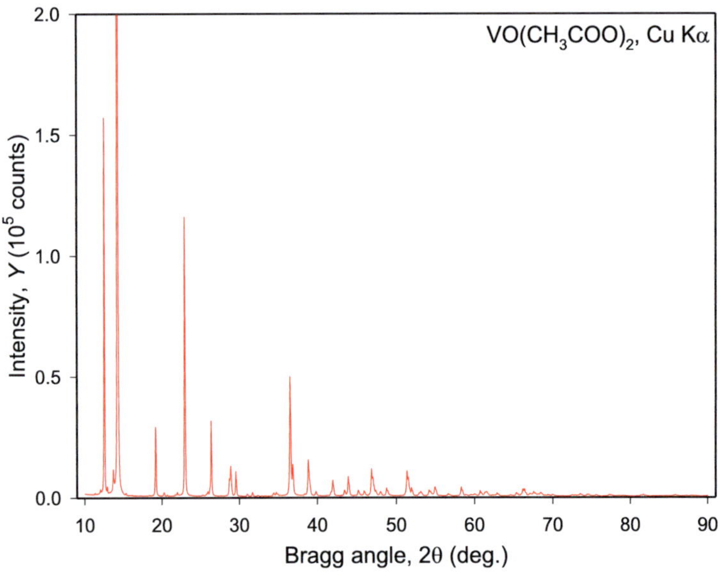

Fig. 25.13 Powder diffraction pattern collected from a $VO(CH_3COO)_2$ powder using a Scintag XDS2000 diffractometer. The data were collected with a step size of $\Delta 2\theta = 0.02°$. The strongest peak is shown at approximately one-quarter of its full intensity

12. Complete the solution of the crystal structure and perform Rietveld refinement of the model of $Hf_2Ni_3Si_4$ from Problem 8. The experimental powder diffraction pattern is available online in the file *Ch25Pr12_CuKa.raw*.
13. Perform Rietveld refinement of hexamethylenetetramine, $C_6H_{12}N_4$, using the model established in Problem 5. The experimental powder diffraction pattern is available online in the file *Ch25Pr13_CuKa.raw*.
14. Complete the solution of the crystal structure and perform Rietveld refinement of the model of vanadyl acetate, $VO(CH_3COO)_2$, using the model established in Problem 9. The experimental powder diffraction pattern is available in the file *Ch25Pr14_CuKa.raw*.
15. Complete the solution of the crystal structure and perform Rietveld refinement of the model of manganese oxide, MnO_2, which crystallizes in the space group $P4_2/mnm$ with lattice parameters $a = 4.41$ Å and $c = 2.88$ Å. The gravimetric density of the material is $\rho = 5.10$ g/cm^3. Assume that manganese atoms occupy the site 2(a): 0,0,0. The experimental powder diffraction pattern is available online in the file *Ch25Pr15_CuKa.raw*.
16. Solve the crystal structure and perform Rietveld refinement of the model of NaV_2O_5, which crystallizes in the space group Pmmn with lattice parameters $a = 11.317$ Å, $b = 3.611$ Å, and $c = 4.807$ Å. It is known that V_2O_5 belongs to the same space-group symmetry with the unit cell dimensions $a = 11.51$ Å, $b = 3.564$ Å, and $c = 4.368$ Å. The atomic coordinates in V_2O_5 are:

    V in 4(f): 0.10, ¼, 0.90
    O1 in 4(f): 0.10, ¼, 0.53
    O2 in 4(f): −0.07, ¼, ~0.00
    O3 in 2(a): ¼, ¼, ~0.00

    The experimental powder diffraction pattern is available online in the file *Ch25Pr16_CuKa.raw*.
17. Complete the solution of the crystal structure and perform Rietveld refinement of the model of tungsten oxide peroxide hydrate, $WO_2(O_2)(H_2O)$, which crystallizes in the space-group symmetry $P2_1/n$ with lattice parameters $a = 12.07$ Å, $b = 3.865$ Å, and $c = 7.36$ Å, $\beta = 102.9°$. The coordinates $x = 0.680$, $y = 0.066$, $z = 0.364$ of W have been determined from a Patterson map. Note that tungsten typically exhibits octahedral or square-pyramidal coordination, with the peroxide group, O–O, counted as a single ligand. The experimental powder diffraction pattern is available online in the file *Ch25Pr17_CuKa.raw*.
18. Locate the missing water molecule and perform Rietveld refinement of zinc vanadate ($Zn_3(OH)_2V_2O_7{\cdot}2H_2O$), which crystallizes in the space group $P\bar{3}m1$ with lattice parameters $a = 6.05$ Å and $c = 7.19$ Å, starting from the following model:

    Zn in 3(e): ½, 0, 0
    V in 2(c): 0, 0, z, $z = 0.25$
    O1 in 2(d): ⅔, ⅓, z, $z = 0.88$

O2 in 6(i): x, $2x$, z, $x = 0.15$, $z = 0.82$
O3 in 1(b): 0, 0, ½

The experimental powder diffraction pattern is available online in the file *Ch25Pr18_CuKa.raw*. Note that the data have been affected by significant sample-displacement error, approximately 0.2 mm for a 250 mm goniometer radius.

Chapter 26
Supplementary Examples and Exercises Using Published Data

The previous ten chapters (Chaps. 16–25) provide examples of various compounds, complete with detailed, step-by-step instructions and comprehensive explanations for conducting crystal structure determination from powder diffraction data. This process begins with ab initio indexing and unit cell refinement, continues with the determination of space-group symmetry and structure solution, and concludes with crystal structure refinement using the Rietveld method. Readers are not required to complete all ten exercises. Instead, it is suggested to focus on a selection of examples closely related to the compounds in their own research, whether by chemistry or complexity, to develop sufficient expertise.

The next step is to apply the skills learned in this book to real-life, unknown compounds. However, this may still be too advanced, particularly for beginners. In such cases, it is recommended first to perform a few exercises on crystal structure determination using published powder diffraction data. On the one hand, this approach is more challenging than simply following a detailed tutorial. On the other hand, it provides a known correct final answer—unit cell, symmetry, and structure—enabling users to verify whether their process and progress are on the right track. Yet, detailed descriptions of the intermediate steps are typically not included in the publication. To accomplish this, a digitized experimental diffraction pattern (full profile) is required in addition to the published crystallographic data. However, the majority journals—except for those focused on crystallography, such as *Acta Crystallographica* or *Powder Diffraction*—do not make experimental diffraction patterns available online[1] or deposit them into databases like the ICDD Powder Diffraction File. Furthermore, even when full patterns are deposited, access to the powder diffraction database is often available only at selected crystallographic laboratories.

For this reason, we compiled 20 datasets[2] used for crystal structure determination from powder diffraction data and the results were published in peer-reviewed

[1] Unlike single-crystal data, which are routinely deposited in the CCDC Crystal Structure Database by most journals, powder diffraction patterns are not as widely shared.

[2] Obtained and analyzed by the authors.

P. Zavalij, V. Pecharsky, *Fundamentals of Powder Diffraction and Structural Characterization of Materials*, https://doi.org/10.1007/978-3-031-91504-8_26

journals. The corresponding compounds are listed in Table 26.1, with references provided in the same order. The table is conditionally sorted into three groups—simple, medium, and complex—based on the complexity of the structure and refinement. This classification is designed to help the readers select exercises of

Table 26.1 List of published structures for supplementary exercises organized by complexity

N	Chemical formula	Space group	V (Å^3)	N_{atm}[a]	N_{dis}	N_H	Comments
	Simple						
1	$VO(CH_3COO)_2$	$Cmc2_1$	670	6			
2	$WO_2(O_2)(OH_2)$	$P2_1/n$	335	6			
3	ε-$VOPO_4$	$P2_1/n$	328	7			
4	$Al_2(OH)_3VO_4$	C2/m	547	8			
5	MnV_2O_5	Pnma	394	8			
6	$NaB(C_2O_4)_2$	Cmcm	657	8			
	Medium						
7	$WO_2(O_2)(OH_2)\cdot1.66H_2O$	C2/m	431	6			Partial occupation of water
8	$VO(OCH_2CH_2O)$	C2/c	860	10		4	Positioning of H atoms
9	$Au_{0.75}Ni_{0.25}OSn_4$	Ccca	473	3	1		Mixed site occupation
10	$K_{0.27}MnO_2\cdot0.54H_2O$	R-3m	151	4	1		Mixed site occupation
11	$LiB(C_2O_4)_2$	Pnma	637	11			Pseudo-hexagonal cell[b,c]
12	$Zn_3(OH)_2(V_2O_7)\cdot2H_2O$	P-3m1	228	7		1	Locating of H atom
13	HIO_3	$P2_12_12_1$	264	5		1	Charge flipping solution
	Complex						
14	$Li_{2x}V_{1-x}O_{2-x}\cdot0.5H_2O$, $x = 1/6$	I4/mmm	217	5	3		Heavy-atom method
15	$[N(CH_3)_4]Mo_4O_{12}$	C2/m	1435	13			Complex structure and PO[b]
16	$(CH_3NH_3)_{0.75}V_4O_{10}\cdot0.67H_2O$	C2/m	468	10	3		Significant disorder
17	$[V_{7.2}O_8(OH)_8](Cl)_{0.77}(H_3O)_{2.3}$	I4/m	329	7	4	1	Significant disorder
18	$[N(CH_3)_4]V_8O_{20}$	C2/m	523	11			Disordered intercalated ion
19	$Mo_{5.25}O_{17.25}\cdot(H_2O)_{1.4}$	$P6_3/m$	358	5	3		Pseudo-cubic cell[c]
20	$(NH_4)_{0.94}Mo_{5.29}O_{17.29}$	$P6_3/m$	362	6	4		Significant disorder

[a] N_{atm} number of atoms in the asymmetric unit cell, N_{dis} number of disordered sites, N_H number of located H atoms

[b] Preferred orientation

[c] Cell parameters ratio suggests higher symmetry than the actual symmetry of the structure

References to Table:

1. C. Weeks, Y. Song, M. Suzuki, N.A. Chernova, P.Y. Zavalij, M.S. Whittingham, The one dimensional chain structures of vanadyl glycolate and vanadyl acetate, J. Mater. Chem. **13**, 1420 (2003). https://doi.org/10.1039/b208100h

2. B. Pecquenard, S. Castro-Garcia, J. Livage, P.Y. Zavalij, M.S. Whittingham, R. Thouvenot, Structure of Hydrated Tungsten Peroxides $[WO_2(O_2)H_2O]\cdot nH_2O$, Chem. Mater. **10**, 1882 (1998). https://doi.org/10.1021/cm980045n

(continued)

3. Y. Song, P.Y. Zavalij, M.S. Whittingham, ε-$VOPO_4$: Electrochemical synthesis and enhanced cathode behavior, J. Electrochem. Soc. **152**, A721 (2005). https://doi.org/10.1149/1.1862265
4. B. Pecquenard, P.Y. Zavalij, M.S. Whittingham, Hydrothermal synthesis and characterization of a new aluminum vanadium oxide hydroxide $Al_2(OH)_3(VO_4)$, J. Mater. Chem. **8**, 1255 (1998). https://doi.org/10.1039/A709151F
5. F. Zhang, P.Y. Zavalij, M.S. Whittingham, Hydrothermal synthesis and electrochemistry of a manganese vanadium oxide, γ-MnV_2O_5, Electrochem. Commun. **1**, 564 (1999). https://doi.org/10.1016/S1388-2481(99)00113-7
6. P.Y. Zavalij, S. Yang, M.S. Whittingham, Structures of potassium, sodium and lithium bis (oxalato)borate salts from powder diffraction data, Acta Cryst. **B59**, 753 (2003). https://doi.org/10.1107/S0108768103022602
7. Same as 2.
8. Same as 1.
9. L. Zavalij, A. Zribi, R.R. Chromik, S. Pitely, P.Y. Zavalij, E.J. Cotts, Crystal structure of $Au_{1-x}Ni_xSn_4$ intermetallic alloys, J. Alloys Comp. **334**, 79 (2002). https://doi.org/10.1016/S0925-8388(01)01780-7
10. R. Chen, P. Zavalij, M.S. Whittingham, Hydrothermal synthesis and characterization of $K_xMnO_2{\cdot}yH_2O$, Chem. Mater. **8**, 1275 (1996). https://doi.org/10.1021/CM950550+.
11. Same as 6.
12. P.Y. Zavalij, F. Zhang, M.S. Whittingham, The crystal structure of new zinc pyrovanadate $Zn_3(OH)_2V_2O_7{\cdot}2H_2O$ from X-ray powder data, Acta Cryst. **C53**, 1738 (1997). https://doi.org/10.1107/S0108270197009700
13. T. Wu, P.Y. Zavalij, M. Zachariah, Crystal structure of a new polymorph of iodic acid, δ-HIO_3, from powder diffraction, Powder Diffraction, **32**(4), 261 (2017). https://doi.org/10.1017/S0885715617000859
14. T. Chirayil, P. Zavalij, M.S. Whittingham, Hydrothermal synthesis and characterization of "$Li_xV_{2-\delta}O_{4-\delta}{\cdot}H2O$", Solid State Ionics, **84**, 163 (1996). https://doi.org/10.1016/0167-2738(96)00004-5
15. J-D. Guo, P. Zavalij, M.S. Whittingham, Hydrothermal synthesis of a new molybdate with a layered structure, $(NMe_4)Mo_{4-\delta}O_{12}$, Chem. Mater. **6**, 357 (1994). https://doi.org/10.1021/cm00040a004
16. R. Chen, P.Y. Zavalij, M.S. Whittingham, J.E. Greedan, N.P. Raju, M. Bieringer, The hydrothermal synthesis of the new manganese and vanadium oxides, $NiMnO_3H$, MAV_3O_7 and $MA_{0.75}V_4O_{10}{\cdot}0.67H_2O$ ($MA = CH_3NH_3$), J. Mater. Chem. **9**, 93 (1999). https://doi.org/10.1039/A805312J
17. N.A. Chernova, J.K. Ngala, P.Y. Zavalij, M.S. Whittingham, Formation of magnetic clusters in the strongly frustrated hollandite vanadium oxide $V_{7.22}O_8(OH)_8Cl_{0.77}(H_3O)_{2.34}$, Phys. Rev. B, **75**, 14,402 (2007). https://doi.org/10.1103/PhysRevB.75.014402
18. T. Chirayil, P.Y. Zavalij, M.S. Whittingham, Synthesis and characterization of a new vanadium oxide, *TMA*V_8O_{20}, J. Mater. Chem. **7**, 2193 (1997). https://doi.org/10.1039/A704624C
19. J-D. Guo, P. Zavalij, M.S. Whittingham, Metastable Hexagonal Molybdates: Hydrothermal Preparation, Structure and Reactivity, J. Solid State Chem. **117**, 323 (1995). https://doi.org/10.1006/jssc.1995.1280
20. Same as 19.

desired complexity and is only loosely based on the number of atoms, the presence and severity of disorder, as well as data quality, challenges in indexing, or structure solution. Additional analytical data, such as TGA and FTIR, can be obtained from the original publications via corresponding links in the references list.

Finally, the crystallographic data and original experimental diffraction patterns are available for download[3] as *CH26Ex##.cif* and *CH26Ex##.xy* text files, where *##* corresponds to the example number in Table 26.1. The CIF files provide lattice parameters, symmetry, atomic coordinates with occupancy factors, and atomic displacement parameters, while the XY files present the experimental profile in a two-column format: 2θ (in degrees) and intensity.

[3] www.springer.com/FPDSCM3

Index

P. Zavalij, V. Pecharsky, *Fundamentals of Powder Diffraction and Structural Characterization of Materials*, https://doi.org/10.1007/978-3-031-91504-8

W

X

Z

MIX
Papier aus verantwortungsvollen Quellen
Paper from responsible sources
FSC® C105338

If you have any concerns about our products,
you can contact us on
ProductSafety@springernature.com

In case Publisher is established outside the EU,
the EU authorized representative is:
Springer Nature Customer Service Center GmbH
Europaplatz 3, 69115 Heidelberg, Germany

Printed by Libri Plureos GmbH
in Hamburg, Germany